Advanced Control Foundation:
Tools, Techniques and Applications

Advanced Control Foundation:

Tools, Techniques and Applications

Terrence Blevins
Willy K. Wojsznis
Mark Nixon

ISA™

Notice

The information presented in this publication is for the general education of the reader. Because neither the author(s) nor the publisher has any control over the use of the information by the reader, both the author(s) and the publisher disclaim any and all liability of any kind arising out of such use. The reader is expected to exercise sound professional judgment in using any of the information presented in a particular application.

Additionally, neither the author(s) nor the publisher has investigated or considered the effect of any patents on the ability of the reader to use any of the information in a particular application. The reader is responsible for reviewing any possible patents that may affect any particular use of the information presented.

Any references to commercial products in the work are cited as examples only. Neither the author(s) nor the publisher endorses any referenced commercial product. Any trademarks or tradenames referenced belong to the respective owner of the mark or name. Neither the author(s) nor the publisher makes any representation regarding the availability of any referenced commercial product at any time. The manufacturer's instructions on use of any commercial product must be followed at all times, even if in conflict with the information in this publication.

Printed in the United States of America.
10 9 8 7 6 5 4 3

ISBN: 978-1-937560-55-3

ISA
67 Alexander Drive
P.O. Box 12277
Research Triangle Park, NC 27709

Library of Congress Cataloging-in-Publication Data in process

Dedication

This book is dedicated to Karen Blevins, Susan Wojsznis, and Nancy Nixon, who have provided encouragement and support throughout our careers.

Acknowledgments

The authors wish to express their appreciation to Grant Wilson and Peter Zornio for supporting advanced control research and work on this book, and to Jim Nyquist, John Berra, Mike Sheldon, Duncan Schleiss, John Caldwell, Dawn Marruchella, Darrin Kuchle, Jay Colclazer, Nathan Pettus, Bruce Greenwald, Jim Hoffmaster, Ron Eddie and Gil Pareja from Emerson Process Management for their inspiration and support of advanced control initiatives. Also, we would like to express our deepest thanks to Bud Keyes, former Senior Vice President of Technology, Emerson Process Management for his initiative in establishing the DeltaV advanced control program. In our work, we have benefited from communications with Karl Åström from Lund University, Tom Edgar from the University of Texas, Joe Qin from the University of Southern California, and Dale Seborg from the University of California, Santa Barbara on both basic and advanced control topics. The authors wish to thank Joanne Salazar, Glenn McLaughin, and Deborah Franke for their help with the web site for this book and Brenda Forsythe and Jim Sipowicz for the creative book cover design. We want to thank Joan Forbes and Cary Laird for their thorough review of the first draft and many editing suggestions. We also want to thank technical editor Scott Bogue for his excellent review of the final draft. In addition, we appreciate the input provided by ISA, and thank Susan Colwell, Manager, Publications Development, ISA, for her support in the publication of this book.

Over the years we have benefited from working with many others in the design and implementation of advanced control tools. The authors are deeply indebted to the developers of the advanced control tools that were the basis for writing this book. This includes Vasiliki Tzovla, Ron Ottenbacher, Dirk Thiele, Ashish Mehta, Yan Zhang, Peter Wojsznis, Ian Nadas, Paul Muston, Bob Havekost, Paul Daly, Adam Qui, Chris Worek, Ken Beoughter, Dan Christensen, Ling Zhou, Ian Lloyd, and Vivi Hidayat. Also, we would like to recognize the valuable contribution of Tom Aneweer, Dennis Stevenson, Dick Seemann, Yang Zhang, Steve Morrison, Mike Ott, Chuck Johnston, Randy Reiss, Greg McMillan, Todd Maras, Michael Boudreau, David Rehbein, Pat Dixon, Quay Finefrock, Shelli Callender, and Sai Ganesamoorthi. We gratefully acknowledge the support of the many customers we have worked with in field testing advanced control products. In particular, the following individuals supported evaluation of these tools in their plants: John Traylor, Texas Eastman; Mark Sowell, Solutia; Romeo Ancheta, Husky Energy; Derrick Vanderkraats, Canfor; Bruce Johnson and Efren Hernandez, Lubrizol; Bruce Eldridge, Frank Seibert, Ricardo Dunia, Eric Chen, Robert Montgomery, and Stephen W. Briggs, University of Texas, Austin, Pickle Research Center; Scott Broadley, President, Broadley-James; and Dan Coyne and Paul Oram, BP. Our special thanks goes to the Lubrizol team led by Bob Wojewodka for their feedback on the development of data analytics.

The authors greatly appreciate the effort of Emerson Process Management Systems and Solutions control engineers progressing advanced control applications in petrochemical, oil and gas, chemical, life science, pulp and paper, power, and other industries. Our thanks go to James Beall, Lou Heavner, Pete Sharpe, Doug White, Jim Dunbar, Elizabeth Alagar, and many others.

It has been gratifying to work with Terry Chmelyk, Don Umbach, John Peterson, and Mike Begin and the rest of Spartan Control's team in field testing new advanced control technologies. The lime kiln example in this book is an example of their work in this area. Also, we greatly appreciate the work done by the Spartan Control and Emerson Process Management teams in applying real-time optimization and MPC in a wide variety of applications. The application examples derived from the valuable field experience of Chris Hawkins, George Buchanan, Andrew Riley from Emerson Process Management, UK, Terrance Chmelyk, Saul Mtakula, Manny Sidhu, and Carl Sheehan from Spartan Controls, Barry Hirtz from Canfor, and by Stewart McLeod from Catalyst Paper greatly enriched this book

We are indebted and thankful as well to many of those not listed here who impacted the development of Emerson Process Management's advanced control products or contributed to the successful application of these products and inspired us to write this book.

Contents

About the Authors

Terrence L. "Terry" Blevins has been actively involved in the application and design of process control systems throughout his career. For more than 15 years, he worked as a systems engineer and group manager in the design and startup of advanced control solutions for the pulp and paper industry. Terry was instrumental in the establishment of Emerson Process Management's Advanced Control Program. From 1998-2005 Terry was the team lead for the development of DeltaV advanced control products. He is the Fieldbus Foundation team lead for the development and maintenance of the Function Block Specification and editor of the SIS Architecture and Model Specifications. In this capacity, Terry is involved in the adoption of Fieldbus Foundation function block work by international standards. Terry is the US expert to the IEC SC65E WG7 function block committee that is responsible for the IEC 61804 function block standards. He is a voting member and chairman of ISA SP104-EDDL (Electronic Device Description Language) committee and is the technical advisor to the United States Technical Advisory Group (USTAG) for the IEC65E subcommittee. He is also a member of the USNC TAG (IEC/SC65 and IEC/TC65). Terry authored "An Overview of the ISA/IEC Fieldbus," Section 11, Standards Overview, Fifth Edition of the *Process/Industrial Instruments and Controls Handbook* and coauthored four sections in the Fourth Edition of the *Instrumentation Engineer's Handbook, Process Control and Optimization*. He coauthored the ISA bestselling books *Advanced Control Unleashed* and *Control Loop Foundation*. He has over 45 patents and has written over 70 papers on process control system design and applications. Terry received a Bachelor of Science in Electrical Engineering from the University of Louisville in 1971 and a Master of Science in Electrical Engineering from Purdue University in 1973. In 2004, he was inducted into Control Magazine's Process Automation Hall of Fame.

Presently, Terry is a principal technologist in the future architecture team of DeltaV Product Engineering at Emerson Process Management.

Phone: (512) 418-4628 E-mail: terry.blevins@emerson.com

Willy Wojsznis has been involved in the development of advanced control products over the last twenty years focusing on model predictive control, data analytics, and auto tuning. Over the previous nearly 25-years of his career he developed computer control systems and applications in the cement, steel, mining, and paper industries. His professional work resulted in a number of successful and innovative advanced control products, over thirty patents, and over forty technical papers. He received a control engineering degree (EE) from Kiev Technical University in 1964, an M.S. in Applied Mathematics from Wroclaw University in 1972, and a Ph.D. from Warsaw University of Technology in 1973. He coauthored the ISA bestselling book *Advanced Control Unleashed.* In 2010, he was inducted into *Control Magazine*'s Process Automation Hall of Fame. Presently, Willy is part of the DeltaV future architecture team. He conducts applied research in the areas of optimization, adaptive control, data analytics, and model predictive control.

Phone: (512) 418-7475 E-mail: willy.wojsznis@emerson.com

Mark Nixon has been involved in the design and development of control systems throughout his career. Mark started his career as a systems engineer working on projects in oil and gas, refining, chemicals, and pulp and paper. In 1998, he moved from Canada to Austin, TX, where he has held a variety of positions in both research and development. From 1995 to 2005 Mark was lead architect for DeltaV. In 2006 he joined the wireless team, taking a very active role in the development of the WirelessHART specifications and the development of the IEC 62591 standardization. Mark's current research includes control using WirelessHART devices, data analytics for batch processes, wireless technology in the process industry, web based user interfaces, operator interfaces, and advanced graphics. He is currently active in the Center for Operator Performance (http://www.operatorperformance.org), WirelessHART, ISA88, Foundation Fieldbus standards (http://www.fieldbus.org/), and ISA101. He has written numerous papers and currently holds over 70 patents. He coauthored *WirelessHART: Real-Time Mesh Network for Industrial Automation* and has made contributions to the *Industrial Instruments and Controls Handbook* and *Modern Measurement and Final Element Essentials for the Process*

Industry. He coauthored the ISA bestselling book *Control Loop Foundation*. In 2012, he was inducted into *Control Magazine*'s Process Automation Hall of Fame. Mark received his Bachelor of Science in Electrical Engineering from the University of Waterloo in 1982.

Phone: (512) 418-7445 E-mail: mark.nixon@emerson.com

Foreword

Process control has become increasingly important in the process industries as a consequence of global competition, rapidly changing economic conditions, and more stringent environmental and safety regulations. High-performance measurement and control systems have become an essential part of industrial plants. To address this environment, the authors of *Advanced Control Foundation: Tools, Techniques and Applications* have assembled a set of approaches that are at the cutting edge of process control practice.

The authors all come from Emerson Process Management, and I have had the distinct pleasure of collaborating with them on a variety of process control projects during the past 12 years. They are well-respected engineers and members of the Process Automation Hall of Fame by virtue of their past contributions to computer process control, with over 100 years of collective experience in this field. Two of the authors previously helped write the successful ISA book *Advanced Control Unleashed* (2003), which presents techniques, such as model predictive control, fuzzy logic control, and artificial neural networks to improve performance over what may be obtained by basic feedback and feedforward control.

The new book covers timely topics, such as advanced multi-loop techniques, control system performance evaluation, adaptive and model-based tuning, fuzzy logic and neural networks, wireless control, continuous and batch data analytics, model predictive control, and real-time optimization. It also includes companion software that can be used in a workshop setting. The authors do not take a theoretical approach, but rely on the foundational underpinnings of process control in presenting the various topics.

Representative applications are presented for a variety of industries, including ammonia manufacturing, pulp and paper, refining, batch specialty manufacturing, power generation, and life sciences. The presented applications reflect the most recent progress in the area of advanced process control. Most of the presented approaches were developed in research collaboration of the University of Texas at Austin, University of California, Santa Barbara, and Emerson Process Management. The book is well-suited for independent study by industry professionals, and I congratulate the authors on producing this useful compendium of advanced control techniques.

Professor Thomas F. Edgar
University of Texas, Austin, TX

1 Introduction

The mathematical basis for many of the advanced control techniques in use today was known long before the technology was available to incorporate these techniques into process control systems. With the introduction of the first process control computers in the 1970s, it became possible to deploy advanced control as a layered application. Today, advanced control products are available as embedded applications within modern process control systems or as layered applications that may be easily added to older control systems. In this book, examples of commercially available products illustrate advanced control tools that are commonly used in the process control industry.

The authors' work in the design, development, and field testing of advanced control software products can in many cases be traced to the establishment in the early 1990s of an advanced control research and development program within Emerson Process Management. Many of the advanced control products embedded in the DeltaV control system came out of this program. The ISA book *Advanced Control Unleashed – Plant Performance Management for Optimum Benefit*, published in 2003, included examples based on advanced control software products embedded in the DeltaV control system. Since then we have seen significant improvements made in advanced control tool capabilities and in user interfaces, improvements that make it easier to design and commission advanced control solutions. Also since then, new advanced control applications have been introduced for batch and continuous processes. This book provides a fresh look at some of the latest advanced control technologies that are available to the process industry.

In this work, we benefited from the insight of many people we have worked with over the years in applying advanced control. Our goal in writing this book is to address concepts and terminology that are needed to understand and apply advanced control. The book will provide insight that will be helpful in applying advanced control techniques to improve plant performance. Whether the reader is working as a process control engineer in a manufacturing plant, in a controls group within an engineering department, or in an instrumentation department within a manufacturing plant, the information in this book sets a solid foundation for implementing advanced control solutions.

Today, the process control systems installed in the process industry can be quite large and complex in design. Therefore, a person who is familiar with traditional control techniques may be reluctant to take the time needed to explore and work with the advanced control tools that are part of the control system. This is unfortunate, because in many cases advanced control techniques can provide significant advantages over traditional control techniques. The application examples in each chapter are designed to address multiple aspects of the design, implementation, and commissioning of advanced control solutions and how advanced control techniques compare to traditional control techniques.

There are many aspects of advanced control, and this book is structured to allow engineers, managers, technicians, and others who are new to advanced control to quickly get up to speed on the technology and how it is applied. Experienced control engineers who are familiar with advanced control products will benefit from the application examples and solutions to workshops included in the book. The material is presented in a manner that is independent of the control system manufacturer. The workshops can be completed using a web browser to access the web site that has been created for this book.

In general, the world of process control is broken into the control of batch and continuous processes. In this book we focus primarily on continuous process advanced control techniques, which in many cases can also be applied to batch processes. Many of the examples and workshops illustrate the control of both continuous and batch processes.

The chapters in this book can be read in any order. If the reader has worked with advanced control tools in the past, it is possible to read only those chapters that address the topics of most interest. Even so, it is a good idea to at least review all chapters and associated workshops to gain a better appreciation of the latest features of advanced control tools. In most cases these added features make it easier and faster to design and commission advanced control systems. Many of the concepts on justifying advanced control introduced in the first portion of the book establish a basis for understanding that can be helpful in assessing the benefits provided by techniques presented in later sections of the book. Also, with this understanding it should become clear that an advanced control strategy can and should be designed and justified independent of the advanced control software product that is selected to implement the solution.

In each chapter on specific advanced control techniques, examples are used to show a typical application interface that is used to configure and commission advanced control solutions. When advanced control is embedded in the control system, the advanced control application interface can be fully integrated into the distributed control system. Also, in some cases, example displays are shown that can be used by the operator to access an advanced control solution. A significant part of the operator interface is directly tied to the control strategy. For example, the alarm system is often carefully designed to focus operator attention on the highest priority alarms first. Although the overall design of operator interface displays for advanced control implementation is beyond the scope of this book, many of the examples illustrate the working relationship between the control strategy and the operator interface displays

The authors assume that the reader may not have worked with advanced control systems and concepts but does have a solid understanding of conventional control techniques and software tools for control implementation within a distributed control system. Thus, each chapter covers concepts and terminology that are helpful in working with advanced control. For example, the chapters on maximizing return on control system investment and evaluating control system performance address the basic principles that must be understood to estimate the economic benefits that can be achieved using advanced control products. In looking at whether advanced control is appropriate for a manufacturing plant, it is important to keep in mind the plant's requirements as well as the economic savings that may be achieved by using advanced control techniques. For example, what are the production and quality objectives that must be met? When the control goals and economic savings are clearly understood, it is usually a straightforward process to determine if advanced control products can be economically justified and, if so, to choose the appropriate advanced control technique.

Background material on economic justification for an investment in advanced control is included in the first part of the book. Examples are used to illustrate techniques that the authors have found useful in justifying improvements that can be achieved within a variety of process operations.

Having established a background in basic concepts and terminology, the book then addresses specific techniques such as on-demand or adaptive tuning that can be used to improve the operation of an existing control system. Later chapters address advanced control techniques such as fuzzy logic control and model predictive control, which may be used not only to replace traditional control techniques but to significantly extend control system functionality and the benefits obtained. As we address these new areas, the advantages of advanced control techniques will become obvious.

At various points in the chapters on advanced control techniques, the reader has an opportunity to see an example of advanced control product by viewing the workshop solutions. The workshop solutions are provided on the web site http://www.advancedcontrolfoundation.com/ and can be accessed without special software. The only requirement is access to a high speed Internet connection and a web browser. In the Appendix we provide detailed directions on how to access the web site. The workshop is designed to reinforce learning of the material in the chapter. The workshops are structured to be fun as well as informative. After viewing the workshop solution, it may be helpful to go back through the chapter if the workshop solution was not clearly understood. A dynamic process simulation is built into each of the book's workshops to show a realistic representation of the type of process response that may be achieved with the advanced control techniques presented in this book. The authors have found that during the pre-commissioning checkout of an advanced control system it is often helpful to create a dynamic process simulation that interacts with the control system. When a reasonable process simulation is tied to the control system, the control operation and the process response will usefully simulate those encountered in actual plant operation. Chapter 14 is dedicated to techniques that can be used to create a process simulation using the same tools that are commonly available within most process control systems.

After the advanced control techniques that are most commonly used in industry are covered, techniques that can be used to integrate advanced control into an existing control system are addressed in Chapter 15. Some of the common issues that must be addressed in doing this are discussed in this chapter. Examples are used to illustrate how advanced control can be layered on top of an existing control system.

Throughout the book, as advanced control techniques are introduced, the manner in which each technique can be applied to address various process control requirements is illustrated using process examples and workshops. It should become clear by studying the examples how these control techniques can be applied to other processes with similar requirements. It is our hope that the understanding achieved by reading this book and viewing the workshop solution will set a foundation for addressing other control applications.

The reader should feel free to contact the authors at their email addresses with any questions about the book or about the use of the web-based workshops.

All royalties from this book will be given directly to universities and educational programs to promote and enhance the understanding of advanced process control. A beneficiary of each year's royalties will be chosen by the authors.

2

Maximizing Return on Control System Investment

To maximize the return on control investment, the design of a new control system or the improvement of an existing control system should take into account the desired system performance and the installation and long term maintenance costs. Having a clear idea of the control system's objectives (as defined below) is critical in achieving long term benefits from the control system. When a modern digital control system has been installed, then the flexibility of these systems allows changes to be quickly made in process control implementation. However, such changes often have a direct impact on control system documentation, operator training, display configuration, and, in some cases, the field devices that will be required to implement the changes. The ease of making changes in a modern digital system does not reduce the need to clarify the control system objectives early in a project and to evaluate the expected benefits and the installation and long term maintenance costs.

The objectives of process control system changes generally fall into one of three categories:

1. Response to an economic incentive (reduced cost or increased revenue)
2. Safety and environmental compliance
3. Equipment protection

One category of objectives of a new or upgraded control system is response to an economic incentive by an improvement in plant operating efficiency through better utilization of feedstock materials (i.e., higher yields), a reduction in off-specification product or a reduction in energy consumption. In addition, when there is a sufficient demand for a product produced by a plant, an increase in revenue can be achieved by increasing plant throughput. A second category is safety and environmental compliance. The safe operation of a plant often requires that critical operating parameters be maintained within their normal range. The control system is designed with functions that, in the event of a process upset, act to avoid the disruption that can be introduced by a shutdown initiated by a separate safety system. However, if safe operating conditions are not maintained by the control system, the safety system takes action. The control logic and field devices that make up a safety system operate independently of the process control system. In this way, the safety system acts as an independent arbiter of actions taken by the control system and thus is ultimately responsible for the safety of the plant. If the distributed control system has been recently installed, then the safety system may be incorporated into the distributed control system and share resources such as the operator interface and environment for control configuration.

One example of how the functions provided by the control system are essential for environmental compliance is in keeping emitted gases and waste streams within limits that are set by environmental standards. Newer distributed control systems incorporate interface capability such as OPC for exchange of process control data. This capability may be used to access analyzers measurements used to prove environmental compliance. Such measurements may be monitored and included in control strategies designed to maintain operating conditions required to avoid exceeding limits set by environmental standards.

The third category of control system objectives is protecting the process equipment. Certain plant operating conditions may result in process equipment damage but do not pose a safety or environmental risk. In these cases, the process control system is often responsible for taking corrective action to avoid equipment damage. For example, a pump that is used to transfer liquid from a tank may run dry if the tank is emptied. This can cause damage to the pump. Since this condition does not create an unsafe operating condition, it will not be acted on by the safety system. However, replacing a pump can be costly and production could be disrupted if the pump cannot perform its normal function. In such a case, the control system can automatically shut off the pump to avoid equipment damage. Modern distributed control systems often make it easier to implement multi-loop control strategies that can be more effective than single loop control in maintaining the plant operating conditions needed to protect process equipment.

Since the cost of implementing advanced control must be justified, in this chapter we will focus on the economic incentive for control system improvement and the areas where advanced control may be required to achieve control system economic objectives. The reduction in process variation that may be achieved using advanced control techniques is often used to increase plant throughput or operating efficiency. However, the economic incentive to reduce process variation and maintain the process at or below an operating limit can also be driven by the need to avoid fines associated with violating limits set by environmental standards.

2.1 Economic Incentive

If a control system can reduce variations in process operation, it is possible to shift the plant to a more efficient point of operation. The economics of plant operation can be impacted by process variation when production is limited by equipment capacity or when maximum production and operating efficiency are achieved at a specific operating condition. When production is *process-limited,* that is, limited by equipment capacity, maximum production is achieved when the control system minimizes process variations and adjusts operating targets to allow the equipment to be operated closer to its operating constraint. For example, the constraint in the throughput to an exothermic reactor may be the available capacity for heat removal. When maximum production is achieved by maintaining specific operating conditions, the process is referred to as exhibiting a *global production maximum*. In such cases, any variation from these specific operating conditions will reduce production.

The generation of steam in a boiler is an example of a process that exhibits a global production maximum at a given throughput. The amount of excess air in the flue gas leaving a boiler has a direct impact on boiler efficiency and the cost of steam generation. When the amount of air delivered to the boiler is greater than that required for complete combustion, the excess air leaving the boiler decreases the boiler efficiency because air is being heated in the combustion process. As the amount of air delivered to the boiler is reduced and the excess air level in the flue gas decreases, at some point only enough air will be provided to support complete combustion of the fuel. If the combustion air level is reduced further, unburned fuel will exit the boiler in the flue gas and boiler efficiency will diminish.

As illustrated in Figure 2-1, for a given firing rate the best operating efficiency is achieved at a specific excess air operating point. Excess air is the amount of air in a combustion process greater than the amount

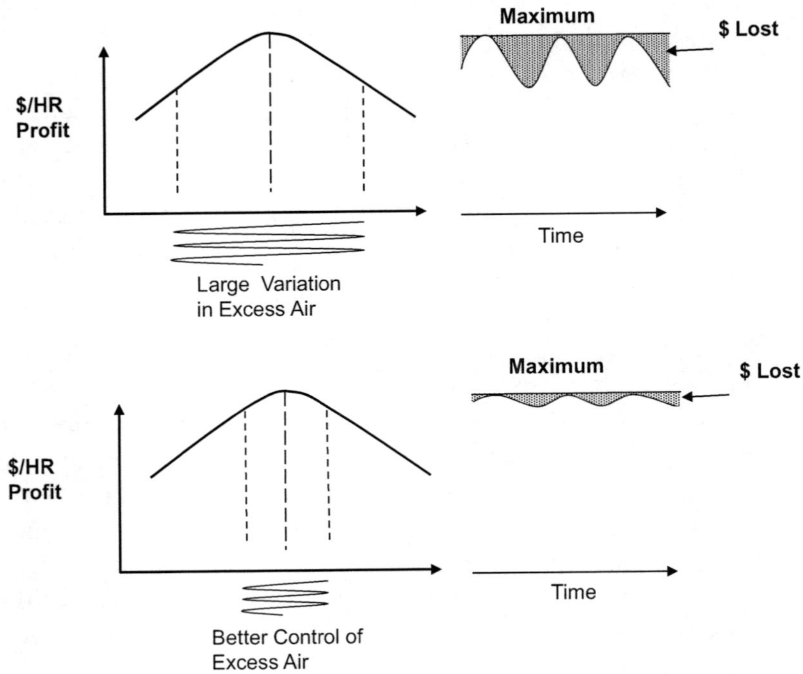

Figure 2-1. Operating at Global Production Maximum

theoretically required for complete oxidation. As shown in Figure 2-1, as the excess air varies above or below this point of best operating, the efficiency is reduced, which impacts the profit of plant operation.

Any variation from this operating point will result in decreased operating efficiency (and, it should be noted, possible environmental noncompliance). If the control system can reduce variation in the excess air level, the boiler operating efficiency can be increased. For a given firing rate, the impact of excess air level on boiler efficiency will exhibit a global production maximum.

In boiler control, a measurement of the O_2 concentration in the flue gas exiting the boiler can be used as an indirect indication of the boiler's excess air level. The boiler control system must minimize variation in this key process parameter. If O_2 control has not been implemented, variations in the excess air level will occur with changes in the process inputs, such as variations in fuel BTU/lb value or a change in the air density because of a change in the outside temperature. The O_2 setpoint in boiler combustion control is often characterized as a function of the load (demand) on the boiler because the most efficient operating point may change as a function of the firing rate, which changes with changes in load. Examples of how advanced control techniques have been applied in boiler control to reduce the cost of steam generation are included in Chapter 13.

We have seen that if a process (such as this steam boiler) is characterized by having a global production maximum, improved operation can be achieved through a control system that is capable of maintaining the process at the best operation point. Note, however, that, even with such a system, operating efficiency can actually be worse if the setpoint is incorrect for the current operating conditions which, in the case of a boiler, may vary with steam demand. In such cases, the setpoint should be set as a function of the operating conditions. For example, the setpoint value can be set as a function of the process throughput. A key point in control system installation and maintenance is that decreasing variability is desirable, but it is also important to specify the operating setpoints needed to achieve best process performance. If this is not done, the benefits of a new or upgraded control system may not be fully realized. The impact of operating at a wrong setpoint value is illustrated in Figure 2-2.

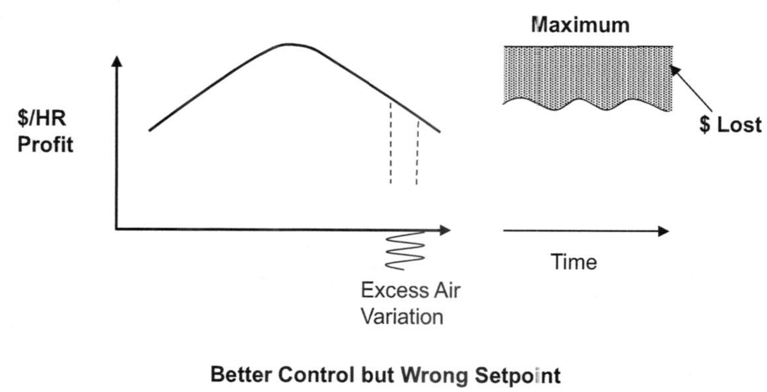

Figure 2-2. Operating Away from the Global Production Maximum

Shifting our attention now to the matter of equipment capacity, when a plant is designed the equipment is sized so that, at the design production rate, the equipment in each area of the plant is operating at near-maximum capacity. If plant production is not limited by the market, that is, if an increase in production can always be sold, revenue may often be maximized by running the plant at maximum throughput. Even so, as plant production is increased above the design production rate, one or more pieces of equipment may become a bottleneck. If production cannot be increased because of an operating constraint such as the physical limit in the equipment, it is often possible to increase plant production without exceeding the physical limit by reducing variation in the process, as illustrated Figure 2-3.

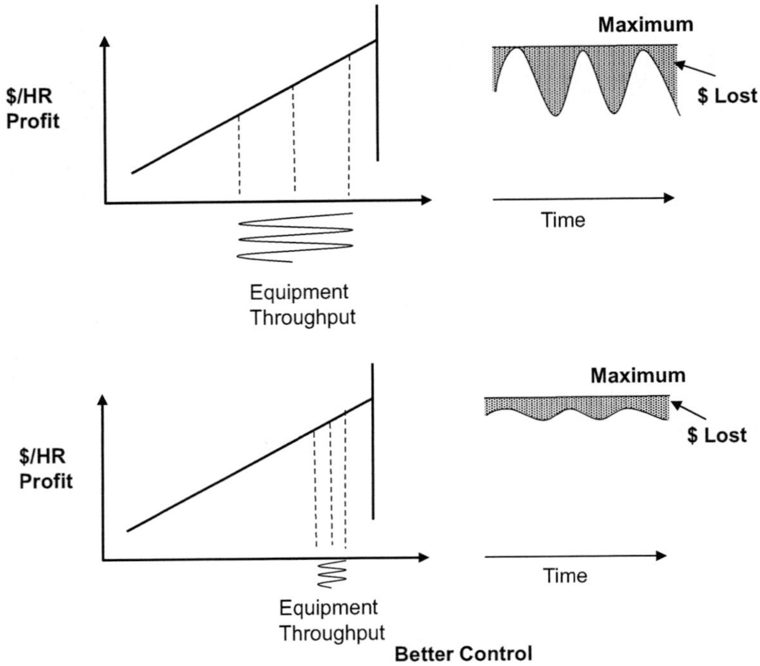

Figure 2-3. Production Maximum at Equipment Physical Limit

As mentioned earlier, the term *process-limited* is often used to describe the condition where the maximum production of a plant is determined by physical limitations in one or more processes. In such cases, it is important to explore the source of the production bottleneck. That is, which process area or piece of equipment limits plant production? One way to determine this is to discuss the reasons that plant production cannot be increased with the control room operators who are responsible for the minute to minute operation of the plant.

If the control system allows wide variations in the process, the target production rate must be set at a value that ensures that the physical limit of each piece of equipment or process area is never exceeded. Similarly, to avoid exceeding product specification limits, the target values for product quality may be set to avoid exceeding specification limits. For example, by improving the control system to reduce process variation, plant production setpoint may be increased without exceeding the physical limits associated with the process area or equipment that is the plant bottleneck, as illustrated in Figure 2-4. This simple concept of reducing process variation to allow a shift in production rate or quality parameter targets is often the basis used to justify the cost of upgrading or installing a better control system in a plant.

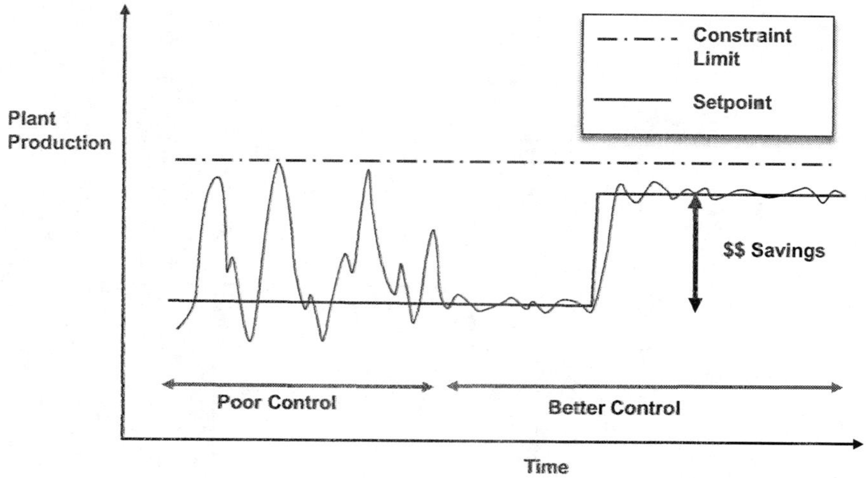

Figure 2-4. Production Maximum at Limit

2.1.1 Ammonia Plant Example

The impact of process variation on plant economics can be illustrated by examining the operation of a typical ammonia plant. The feedstock to an ammonia plant is natural gas, and plants are located in places with an abundance of natural gas. Even though the price of natural gas has recently dropped significantly, it is still a challenge for plants located in North America to compete in the world market, and there is a strong economic incentive to operate these plants in the most efficient manner.

The process flow through an ammonia plant is shown in Figure 2-5. In the primary reformer section, the natural gas feedstock (methane, CH_4) is combined with steam (H_2O) and heated to a very high temperature in the presence of a catalyst. As a result of the reaction with the catalyst, this feedstock is converted into hydrogen (H_2) and carbon monoxide (CO). In the secondary reformer, air (approximately 21% O_2, 78% N_2, and some inert gases such as argon) is added and in a combustion process, most of the carbon monoxide combines with the oxygen, leaving CO_2, H_2, and N_2. Any unreacted CO is converted into CO_2

in the shift reactor. In the absorber units (Figure 2-5), the CO_2 is removed, leaving nitrogen and hydrogen. Any unabsorbed CO_2 is then converted into methane (CH_4) in the methanator.

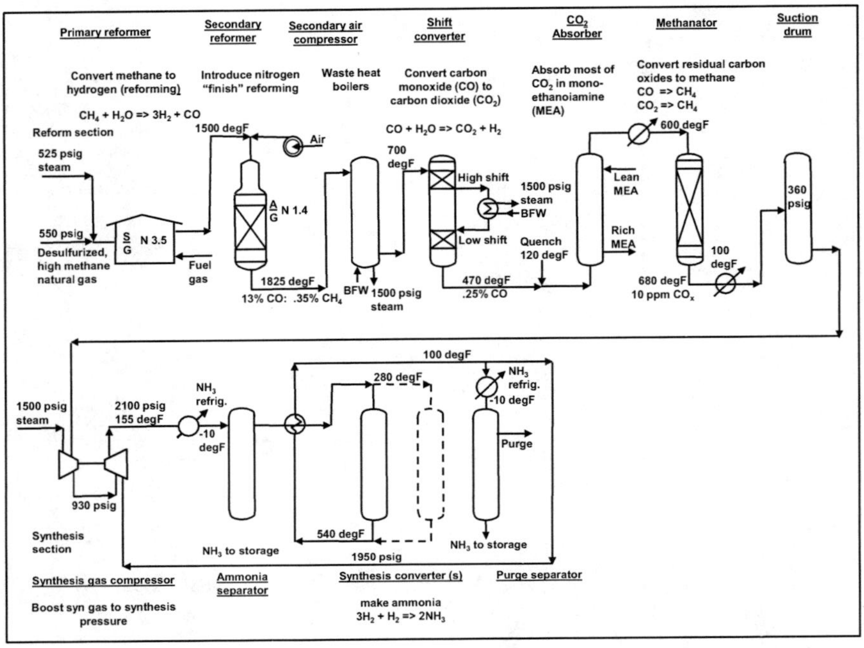

Figure 2-5. Ammonia Plant Example

In the downstream process, the feed stream of hydrogen, nitrogen, and traces of methane and inert gases is compressed. In the synthesis loop, this feed stream is passed over beds of catalyst in the synthesis converter, where the hydrogen and nitrogen combine to form ammonia (NH_3). A separator is used to remove ammonia from the synthesis loop. Any accumulation of methane and inert gases in the synthesis loop is removed using a purge stream.

The control system provided with a new ammonia plant is often designed to control normal day-to-day plant operation. This type of basic control system is quite simple and allows the plant to be quickly commissioned and brought up to designed operating conditions. However, once stable plant operation has been achieved, it is often necessary to invest in added control capability to maximize operating efficiency and production (which is important, as mentioned, in North American plants). Before the introduction of distributed control systems, it was necessary to add a computer to the control system to provide this additional control capability. At one time, some process control suppliers offered special ammonia plant control packages, which included services to engineer and install computer-based control systems for plant optimization. Today, however, if a plant has a modern distributed control system, the advanced control capability of the control system, such as adaptive loop tuning and model predictive control, can be sufficient to implement control designed to increase plant operating efficiency and production.

A significant improvement in ammonia plant operating efficiency and production can often be realized through the addition of model predictive control in four or five areas of plant operation. This added control directly impacts the operation of critical units within the plant. In most cases, it is possible to achieve an investment payback of only a few months; thus, the controls that can be added to an ammonia plant to improve plant efficiency and production serve as a good example of controls designed to maintain plant operation at a global production maximum and to maintain throughput at levels that maximize production within equipment physical limits.

The temperature within the synthesis converter has a large impact on the reaction rate for the conversion of hydrogen and nitrogen into ammonia, an example of a process that exhibits a *global production maximum*. Disturbances to the operating conditions within the synthesis loop can impact the synthesis bed temperature. Examples of such disturbances are changes in methane and inert gas concentration, the flow rate through the synthesis converter, and the relative concentrations of hydrogen and nitrogen. In the absence of a more capable control system, the manual valves on a quench stream to the converter as shown in Figure 2-6 are typically manipulated by an operator to make rough adjustments to the synthesis catalyst bed temperature, which can result in a wide variation of the temperature within the synthesis converter. The impact of the synthesis bed temperature on ammonia production for a constant plant feed rate is illustrated in Figure 2-6.

In the scenario shown in Figure 2-6, the economic incentive would be classified as a global maximum since greatest production is achieved at a specific bed temperature under normal operating conditions. At a given feed rate, ammonia plant production can be reduced by any reduction in the synthesis bed temperature from its optimal value. There is a strong economic incentive to install control capability that will automatically regulate the synthesis converter quench valve to maintain the bed temperature at setpoint. This automatic control will immediately compensate for any changes in the operating conditions. The capability of model predictive control to compensate for delay in

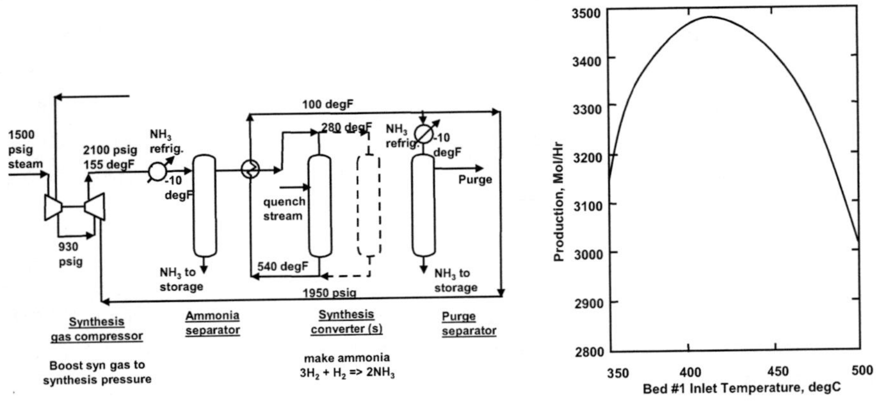

Figure 2-6. Impact of Synthesis Bed Temperature on Plant Production

process response and measured disturbances makes it practical to implement synthesis bed temperature control.

The improvement in efficiency that can be achieved through automatic control of the synthesis bed temperature can be calculated based on reducing the average deviation from the best operating point (optimal value). For example, based on the curve shown in Figure 2-6, an increase in the average operating temperature could increase plant production by approximately 0.4%/°C, as shown below:

$$\frac{\%production\ change}{change\ in\ bed\ \#1\ temperature} = \frac{\left[\dfrac{\dfrac{(3441-3413)mole\ NH_3}{hour}}{3413\dfrac{mole\ NH_3}{hour}}\right]}{(430-410°\text{C})} * 100$$

$$= \frac{0.4\%\ production\ change}{1\ degree\ change\ in\ bed\ temperature}$$

Since an ammonia plant can produce over 1000 tons of ammonia per day, even a small increase in production for a given feed rate can significantly increase the revenue generated by the plant.

An example of *process-limited* production is the synthesis loop. For a given feedstock rate, the amount of ammonia produced increases with the synthesis loop pressure as illustrated in Figure 2-7. The pressure in the synthesis loop is directly affected by operating conditions including the plant feed rate, the concentration of methane and inert gases in the feed stream, and the rate of conversion in the synthesis loop.

Operating efficiency is maximized by running at the highest synthesis loop pressure that falls within the pressure limits of the process area's vessels and pipes, limits that are determined by the physical strength of the materials used. In the absence of automatic control, the pressure in the synthesis loop is controlled by an operator adjusting a valve to change the purge flow from the synthesis loop. When the operator manually adjusts the purge value to prevent an over-pressure condition, wide variations can occur in the synthesis loop pressure.

Much tighter control of loop pressure can be achieved by using automatic control to adjust the purge valve to maintain a pressure setpoint. Thus, synthesis loop pressure control may be used to maximize production by running as close as possible to the maximum pressure limit.

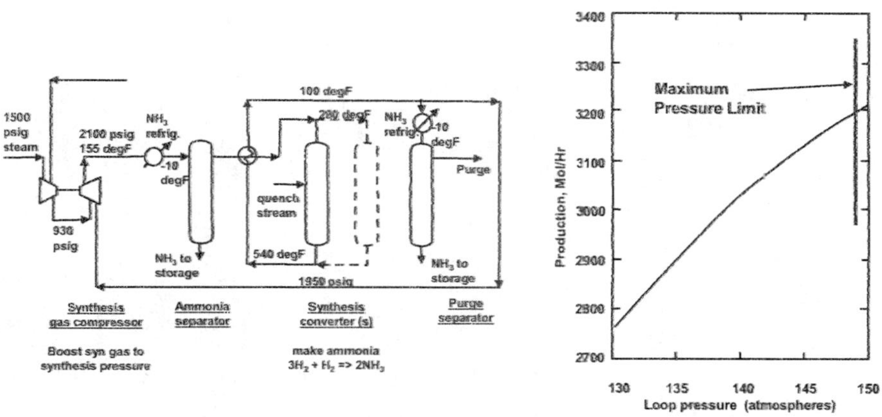

Figure 2-7. Synthesis Loop Pressure Control at a Limit

An increase in synthesis loop pressure provides a 0.05%/Atm increase in production, as shown below:

$$\frac{\%production\ change}{change\ in\ pressure(atm)} = \frac{\left[\dfrac{(3205.6-3142.26)mole\ \dfrac{NH_3}{Hr}}{3142.26\ mole\ \dfrac{NH_3}{Hr}}\right]}{(150-146)atm}$$

$$= \frac{0.05\%\ production\ change}{1\ atm\ change\ in\ pressure}$$

Another example of process-limited production is the primary reformer. In the front end of an ammonia plant, the temperature that is maintained in the primary reformer tubes impacts the amount of the natural gas feedstock that is converted into hydrogen and carbon dioxide. The conversion rate increases with higher temperatures. The proximity of the conversion rate to 100% depends on the tube temperature.

Unconverted natural gas is known as methane leakage. As the methane leakage increases, the purge flow in the synthesis loop must increase to maintain synthesis loop pressure at a safe level, reducing ammonia production. For a given feed rate, the impact of methane leakage on ammonia production is shown in Figure 2-8.

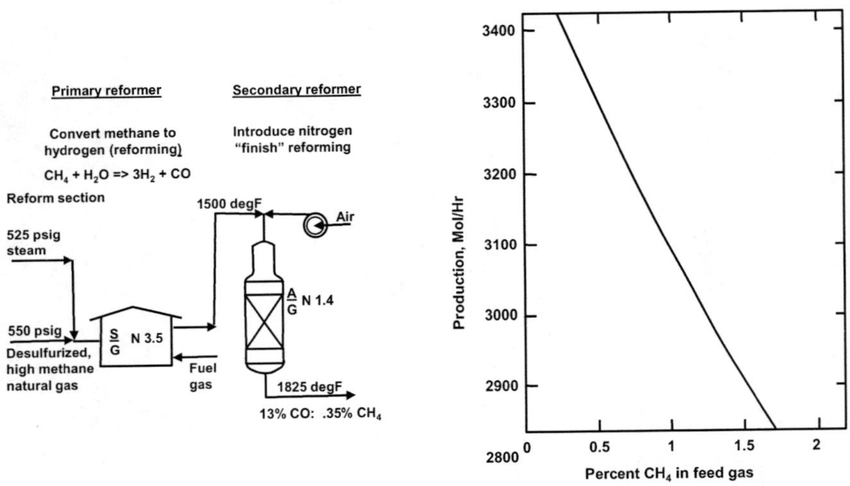

Figure 2-8. Primary Reformer Methane Leakage

The natural gas and steam input streams to the primary reformer flow through tube bundles that are filled with a catalyst. Banks of these tube bundles, often referred to as *harps*, are contained inside refractory-lined walls. Burners located at the bottom of the primary reformer heat the gases flowing through the tube bundles. In the absence of automatic control, fuel gas valves are manually adjusted by a plant operator to determine the energy input to the burners and thus the temperature of the gases flowing through the tube bundles.

Through the application of automatic control, it is possible to reduce the variation in the primary reformer tube temperature and thus reduce methane leakage by operating the primary reformer closer to its temperature limit. This type of control maximizes production within the temperature limitations for the primary reformer. For a given feed rate, even a small reduction in methane leakage can have a large impact on plant production.

At 1% CH_4 *leakage*

$$\frac{\%production\ change}{change\ in\ CH_4\%} = \frac{\left[\dfrac{(3303.9 - 2909.1)mole\ \dfrac{NH_3}{Hr}}{2909.1\ mole\ \dfrac{NH_3}{Hr}}\right]}{(0.5 - 1.5)\ \%CH_4} * 100$$

$$= \frac{12.7\%\ production\ change}{\%\ change\ in\ CH_4\ Leakage}$$

In summary, within an ammonia plant there are some operation areas where control can be classified as global production maximum and others where production is process limited, that is, limited by equipment capacity. The preceding examples help to demonstrate how, through advanced control, an ammonia plant's efficiency can be improved, and can serve as a guide for developing analogous cost saving and production increase solutions for other processes.

2.2 Reducing Process Variation – Achieving Control Objectives

As shown in the previous section, two things are critical in achieving economic objectives: defining the optimal operating point, and reducing process variation to allow operation closer to the optimal conditions. A variety of traditional and advanced control tools can be used to reduce process variation. Through the installation and maintenance of a control system and advanced control solutions focused on minimizing process variations it is possible to derive maximum economic benefits from process operation.

2.2.1 Single Loop Control

During the design and construction of a new plant there are strong economic incentives to minimize the number of transmitters, automatic regulating valves, and control loops that must be installed and commissioned. Thus, in most cases the majority of the automatic control installed in a new plant is based on single loop PID feedback control. Flow control is a very common application of single loop PID feedback control as illustrated in Figure 2-9.

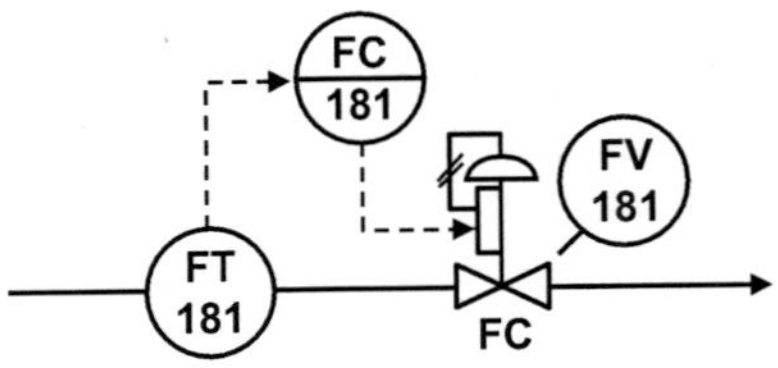

Figure 2-9. Single Loop Feedback Control

To help minimize the process variation that is experienced with single loop PID control, manufacturers have introduced advanced tools that allow control performance to be automatically evaluated (as will be discussed in Chapter 3, Evaluating Control System Performance). These tools can be used to identify measurement and actuator problems that can impact performance that may be achieved using single loop feedback control. Also, some tools automatically identify where process variation could be reduced by tuning (adjusting) parameters associated with the PID (that is, the gain, reset, and rate parameters) to achieve the best controller response to setpoint and disturbance input changes. The procedure used to set the PID tuning parameters can be challenging when the process response is slow or includes significant process delay. Thus, most modern process control systems provide some capability to automatically establish loop tuning (as addressed in Chapter 4, On-Demand Tuning). When the process gain or dynamics change with changes in operating conditions, advanced control tools for adaptive control can be applied as will be addressed in Chapter 5, Adaptive Tuning.

The impact single loop setup and tuning can have on plant operation is illustrated by considering liquid storage tank control. Liquid storage tanks are used in various ways to control inventory within a plant. Intermediate liquid storage tanks (surge tanks) are commonly installed between processing areas of the plant. Under normal operating conditions, these storage tanks allow each process area to be operated for a short period of time independent of the production rate that can be achieved in other process areas. Any off-spec or recycled material, or material that is a by-product of manufacturing, can also be transferred to storage for further processing. In addition, a process area may have a requirement to maintain a specific quantity of liquid inventory for correct operation.

The inventory in intermediate storage tanks between process areas can be maintained by an operator manually coordinating the throughput of the various process areas. However, improved coordination of process areas can often be achieved by automating the control of intermediate storage. Also, in most cases, automatic control is required to achieve specific liquid inventory levels within a process area.

Any imbalance in the area production rates within a plant will be reflected by a change in the level of the surge tanks between process areas. Thus, by maintaining the surge tank level within its upper and lower operating limits, it is possible to maintain a balance in the production rates. When a downstream process area is made up of a continuous process, its throughput can be automatically adjusted based on the surge tank level, as illustrated in Figure 2-10.

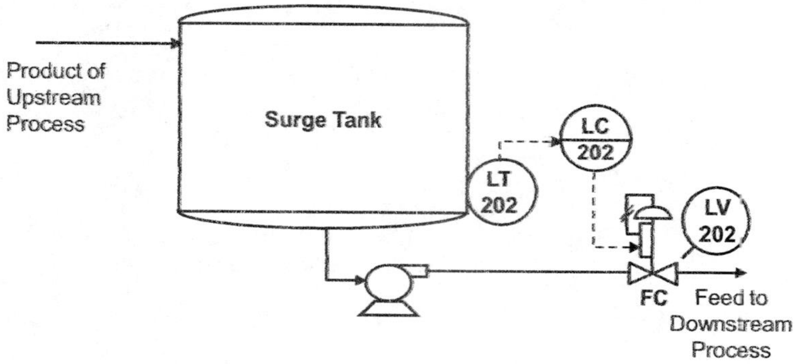

Figure 2-10. Surge Tank Level Control

To avoid abrupt changes in throughput in the downstream process caused by changes in surge tank level, the level controller (LC 202) can be configured for proportional-only control, with the bias setting based on the normal plant production rate. The range over which this floating-level control adjusts the downstream flow is determined by the controller gain and bias. Alternatively, PI control can be used where the reset determines the time required for any change in the upstream process throughput to be reflected in the downstream process throughput. The tuning of surge tank level control can have a direct impact on the variation in flow seen by the downstream process as illustrated in Figure 2-11.

2.2.2 Multi-Loop Techniques

When single loop control is not sufficient to minimize process variation or to maintain the process at an operating limit, multi-loop techniques such as feedforward, cascade, and override control should be applied. When the flow rate associated with multiple process inputs

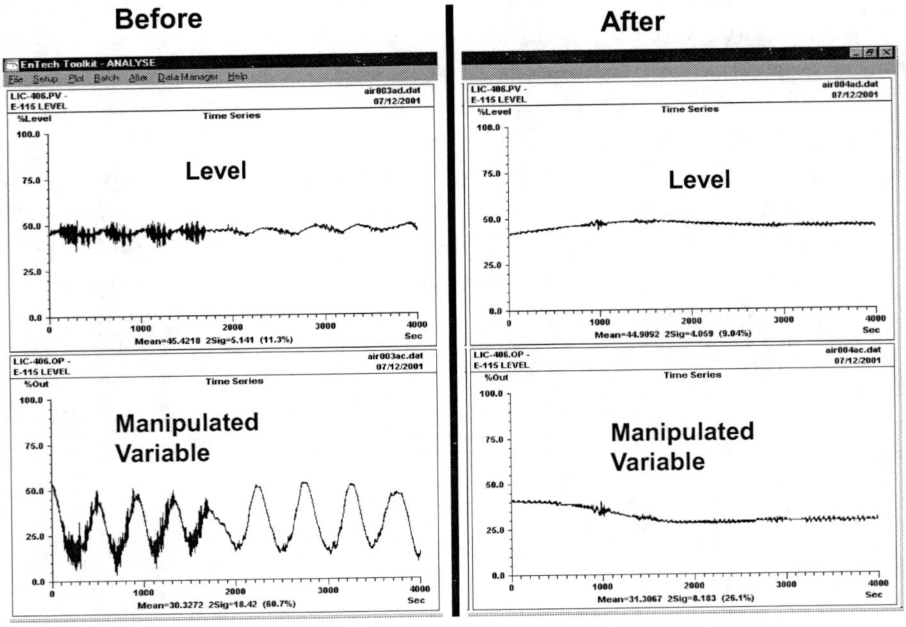

Figure 2-11. Impact of Tuning of Surge Tank Level Control on Downstream Process

must be coordinated to achieve correct processing conditions, split-range, ratio, and valve position control should be used. A concise overview of these traditional multi-loop techniques is provided in this section. As with single loop control, the setup and tuning of the PID feedback that is utilized in multi-loop control has a direct impact on how effective these techniques will be in reducing process variation. Thus, advanced control tools for performance evaluation, on-demand tuning and adaptive tuning can be applied to improve the performance of process control using multi-loop techniques.

Feedforward Control

One of the fundamental limitations of single-loop control is that control action is based on the error between the setpoint and the controlled parameter. If there is no error or if there is a change in error, the manipulated parameter will remain at its last value. To correct for process disturbance input using single-loop control, the disturbance input must impact the controlled parameter. Only *after* a change is seen in the controlled parameter will the manipulated parameter be adjusted to correct for the disturbance. However, if the control loop includes measurement of the process disturbance at its point of origin, then feedforward control may be used to anticipate and minimize the downstream impact of the disturbance. A measured disturbance can be fed forward to the control system so that action can automatically be taken to correct for the measured disturbance before it impacts the controlled parameter.

Feedforward control is always implemented in conjunction with feedback control. For example, heater temperature control can be improved through the use of the feed temperature TT185A as a feedforward input, as shown in Figure 2-12.

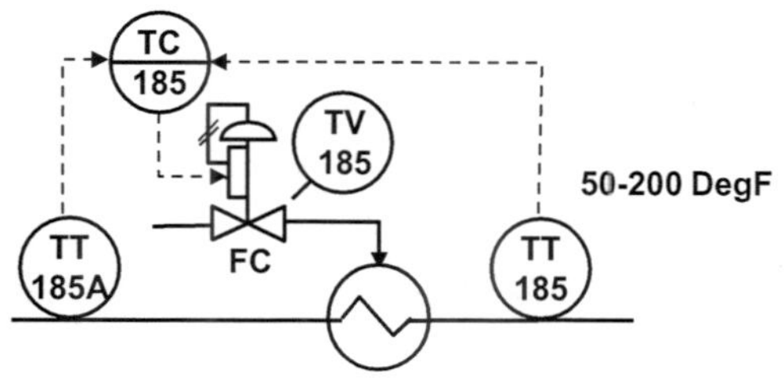

Figure 2-12. Feedforward Input in Heater Temperature Control

A key point when considering the use of feedforward control is that the process disturbance must be a measured parameter. If the disturbance input is unmeasured, a disturbance measurement must be added as a control system input to be able to apply feedforward control.

Cascade Control

Cascade control is applied when a process consists of two or more (sub) processes in series where a process is defined as a specific equipment configuration which acts upon inputs to produce output. To apply cascade control, the output of each of these processes must be measured. The input to the first process is directly manipulated by the secondary controller as shown in Figure 2-13. Since the output of each process in the series is the primary input of the next process in the series, any change in the manipulated input to the first process in the series will impact the output of the other processes.

The primary loop is configured to adjust the setpoint of the PID associated with the upstream process in the series. For the example boiler (Figure 2-14), the superheater outlet steam temperature is made up of two (sub) processes in series: attemperator and superheater tubes. The outlet temperature control implemented by the primary controller,

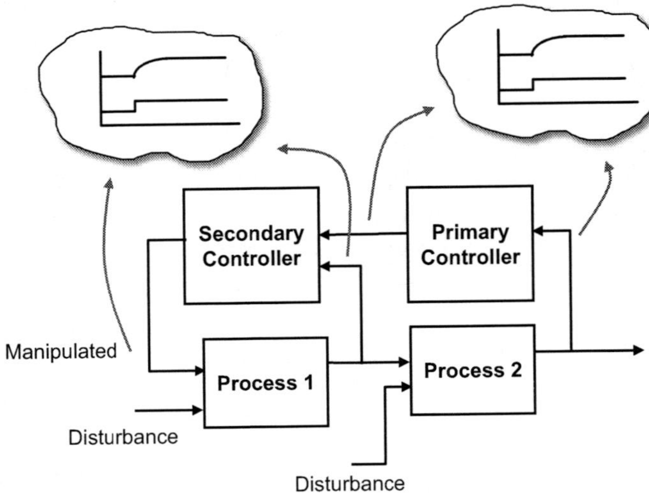

Figure 2-13. Cascade Control

TC187, adjusts the setpoint of the attemperator temperature control, TC186, as illustrated in Figure 2-14.

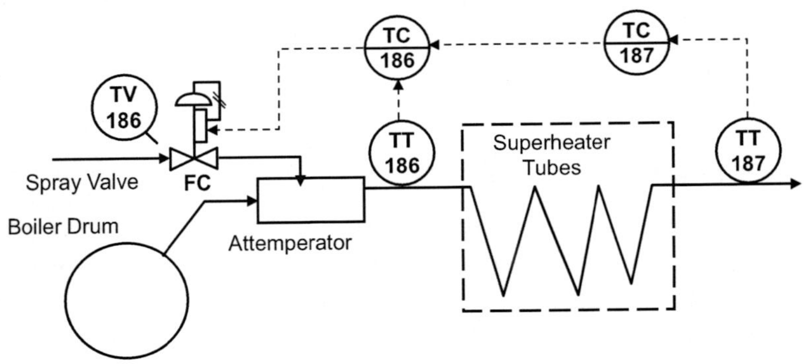

Figure 2-14. Cascade Control Example

One of the main reasons for implementing cascade control is that the secondary PID at each point in the cascade can react quickly to disturbance inputs to its associated process. If the PID responds quickly enough, changes introduced by disturbance inputs will have little or no impact on the downstream processes. The process associated with the secondary PID loop responds to changes in its manipulated input faster than the process associated with the primary loop, so the above condition is satisfied naturally. A feature of cascade control is that the secondary loop response time should be several times faster (most often 3–5 times) than the response time of the primary loop.

In some cases, cascade control can be implemented to compensate for the non-linear installed characteristic of a regulating valve. For example, the slave loop of a cascade might be associated with a flow process where the installed characteristic of the valve is non-linear. Because the flow process is capable of very fast changes in flow rate, the slave loop could be tuned to quickly adjust the valve to achieve the flow rate setpoint requested by the primary loop. The non-linear installed characteristic of the valve would have no impact on the tuning or response of the master loop. This is a real benefit if the process associated with the master loop is slow to change.

Override Control

To ensure proper process operating conditions, it may be necessary to maintain some measurable parameters such as vessel temperature, pressure, and level within certain operating limits (constraint limits). The implementation of override control is often the most effective way to maintain the process within its constraint limits. This control technique is quite important within the process industry. Tools that allow override control to be implemented are provided in most control systems.

Override control can be applied to a process that is characterized by having one manipulated parameter, one controlled parameter, and one or more constraint outputs. Figure 2-15 illustrates the implementation of override control in a compressor control scheme consisting of one constraint output, IT193, one controlled parameter, PT192, and a low selector, PY192, which manipulates the suction valve, PV192.

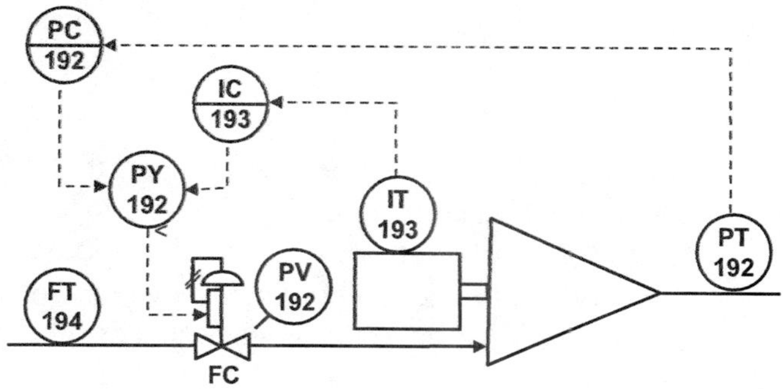

Figure 2-15. Override in Compressor Control

In this example, under normal operating conditions the pressure is maintained at setpoint by the selected PID, PC192. The override PID, IC193, takes an active role if the value of the constraint variable, motor current, approaches its setpoint. The setpoint of the override PID is set to the constraint limit, that is, a value that the constraint variable should never exceed. The point at which the override PID takes over control depends on the PID tuning and the rate of change of the constraint variable.

Ratio Control

For many processes, such as blending and boiler combustion, a key objective is to maintain the flow rates of two process streams in some proportion to one another. In such cases, ratio control can be applied. When ratio control is applied, one process input, the dependent input, is proportioned to the other process input, known as the independent input. The independent input is typically a process measurement. The proportion that is to be maintained between the inputs is known as the ratio. As the value of the independent input changes, through ratio control the other process input is changed to maintain the proportion of the inputs specified by the ratio setpoint. For example, a ratio of 1:1 would indicate that the two inputs are to be maintained in the same proportion.

In nearly all ratio control applications, the ratio controller establishes the setpoints of the flow controllers rather than the valve positions, as illustrated in Figure 2-16. Thus, any non-linearity in installed characteristics associated with valves is addressed by the flow controllers and has no impact on the ratio controller's ability to maintain the ratio setpoint.

When implementing ratio control, the control designer can choose the independent input. In many cases, the flow that is normally largest is selected as the independent input because this input also can be used to set process throughput.

Split-Range Control

One of the most common ways to address multiple process inputs is known as split-range control. Using this approach, a splitter block is used to map the controller output to multiple manipulated process inputs. The splitter block can be used to define a fixed relationship between the controller output and each manipulated process input as illustrated in Figure 2-17.

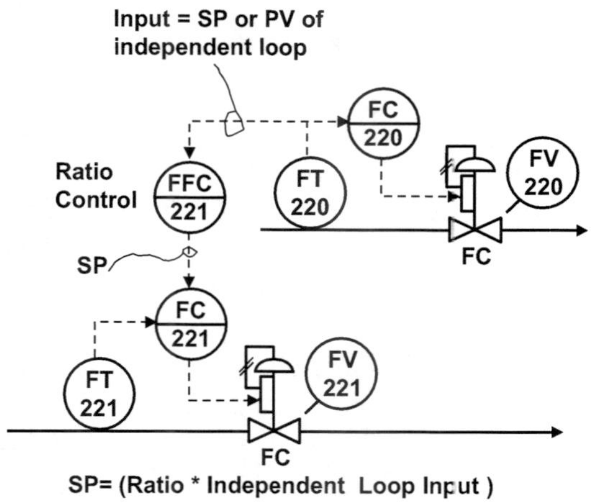

Figure 2-16. Ratio Control Example

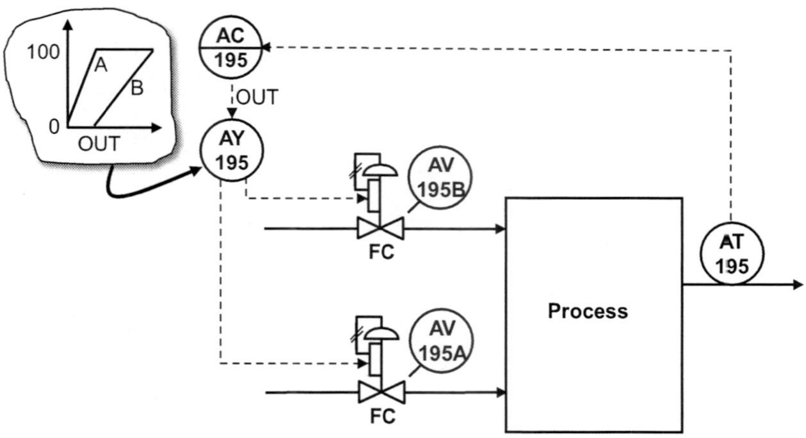

Figure 2-17. Split-Range Control Example

The splitter defines how each valve is sequenced as the controller output changes from 0 to 100%. From the controller's perspective, it appears as only one manipulated process input and the splitter is considered to be part of the process. Once the splitter has been configured, the PID block used with the splitter can be commissioned and operated in the same manner as a single-loop controller. Through the use of the splitter block, the two valves are sequenced in a fixed way so that they appear as one valve to the controller.

Valve Position Control

Precise metering requirements can be met with a small valve, but a small valve may not have the flow capacity needed to maintain the plant production rate. To achieve precise flow rate control over a large flow range, it is often necessary to use both a large and a small valve at the same time. As fine adjustments in flow rate are made with the small valve, the large valve is automatically adjusted to keep the small valve within its operation range of 0–100%. Valve position control can be used to achieve this type of small and large valve regulation.

Valve position control is a straightforward technique that can be easily implemented in most modern control systems. The basic components of a valve position control strategy are illustrated in Figure 2-18.

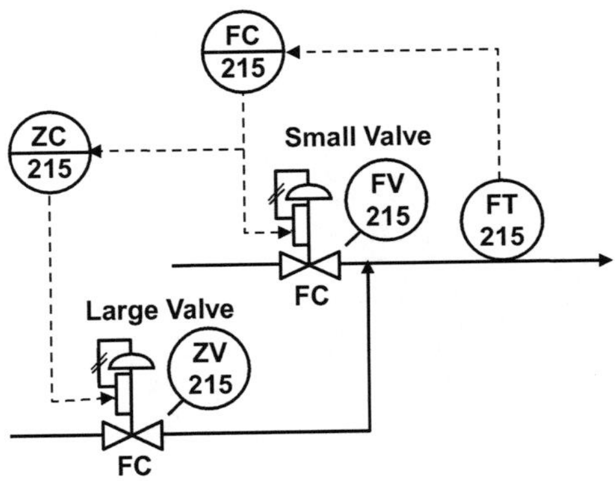

Figure 2-18. Valve Position Control Example

In this example, the feedback controller directly adjusts the small valve input to the process. The output of the feedback controller used to manipulate the smaller valve is also used as the control parameter of an integral-only (I-only) controller that manipulates the large valve. Let's assume for this example that the setpoint of the I-only is set to 50%. If the feedback controller moves the small valve away from 50% open for a sufficiently long period of time, the I-only controller will gradually adjust the large valve to drive the small valve back to its normal operating point of 50%. At an operating point of 50%, the feedback controller can make adjustments in either direction to achieve its setpoint. The tuning of the I-only controller is very slow; a position change in the large valve appears as a load disturbance to the feedback controller that adjusts the small valve.

2.3 Advanced Control

Single-loop PID feedback and multi-loop traditional control techniques such as feedforward, cascade, override, split range, and valve position control can be used to address a wide variety of control requirements. However, if a process is characterized by non-linear response, then adaptive control may be required to improve control performance as described in Chapter 5. When a faster response to setpoint and load disturbance cannot be achieved using PID control, then fuzzy logic control may be required as detailed in Chapter 6. An intelligent PID capability may be needed to improve recovery from process saturation or to use wireless transmitters in closed loop control as addressed in Chapter 8. When a quality parameter needed in control cannot be measured on-line, then on-line property estimation may be utilized as described in chapters 7, 9, and 10. In power house installations, best operating efficiency may be achieved through the use of online process optimization as described in Chapter 13. More complex processes are often characterized by extremely long process delays, a high degree of process interaction, and multiple operating constraints and measured process disturbances. The control of these more complex processes is effectively addressed by advanced control techniques such as model predictive control (MPC). (MPC for both small and large process is more fully addressed in Chapters 11 and 12). Pulp bleaching and reformer temperature control are examples of two applications that benefit from the use of model predictive control.

2.3.1 Pulp Bleaching

In the bleaching area of a paper pulp mill, as illustrated in Figure 2-19, bleaching chemicals are used to achieve a desired level of pulp brightness. Since the reaction after chemical addition is very slow, the bleach tower is designed for long residence times. Because of the process delay associated with movement of material through the bleach tower, a brightness measurement made after the bleach tower may not reflect a change in the bleaching chemical addition for 30–60 minutes, depending on the pulp feedstock flow rate.

To accommodate changes in the feedstock consistency, that is, changes in solids ("dry stock") concentration, and in feedstock flow rate, the bleaching chemical addition rate can be set by ratio control of chemical/dry stock. However, to effectively adjust this ratio based on deviations of brightness from setpoint, it is necessary to use process modeling and control techniques such as model predictive control that account for the very long process delay. Without such capability the brightness must be controlled by the operator manually setting the ratio control of chemical/dry stock.

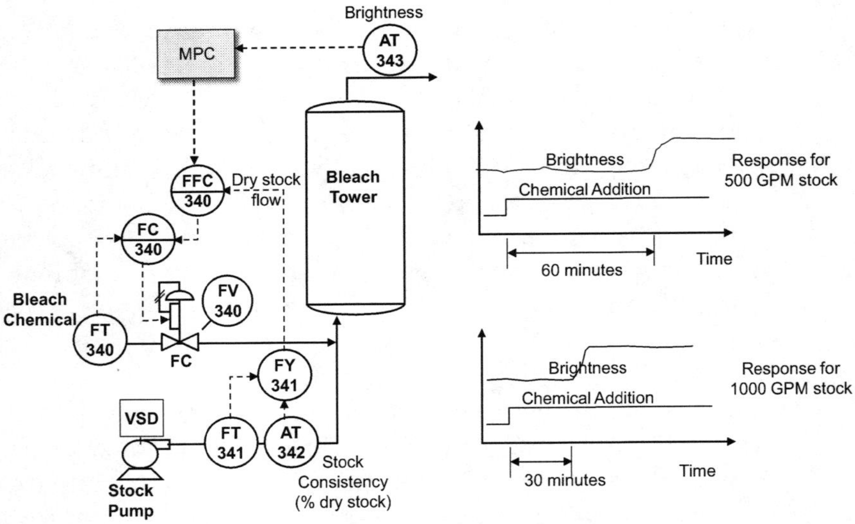

Figure 2-19. Pulp Bleaching

2.3.2 Primary Reformer Temperature Control

In an ammonia plant (as described in 2.1.1), the primary reformer is made up of multiple tube bundles, or "harps," which contain a catalyst. Natural gas-fueled burners are located between each harp at the bottom of the reformer as illustrated in Figure 2-20.

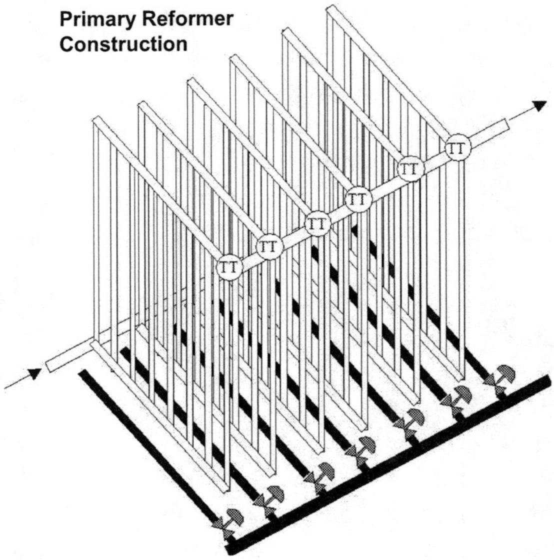

Figure 2-20. Primary Reformer Construction

The gas temperature within the harp tubes is approximately 1500°F (815°C). However, the temperature will change with the feedstock rate and also with changes in other operating conditions such as the outside air temperature. To efficiently drive the reforming process to completion while minimizing methane leakage, it is important to maintain a high, consistent process gas temperature within the harp.

In this example of process-limited production, there is a maximum temperature which, if exceeded, will cause the metal that makes up the tubes to distort and eventually rupture. When a tube fails in this manner, it is necessary to shut the plant down while the tube is replaced. Thus, for efficiency, it is important to operate at a high tube temperature to reduce methane leakage but without reaching temperatures that cause tube failure.

The outlet temperature of each harp is measured. However, any adjustment of one burner will impact the outlet temperature of adjacent harps. Such interactions and constraints imposed by the maximum tube temperature and measured process disturbances are best addressed using model predictive control, as illustrated in Figure 2-21.

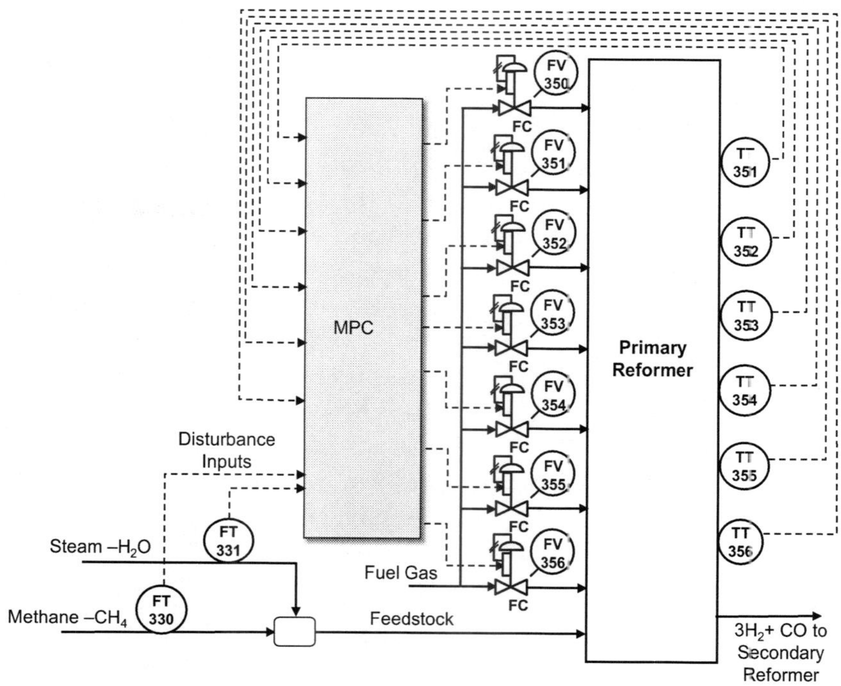

Figure 2-21. Primary Reformer Temperature

2.4 Balancing Complexity with Benefits

When a process control system is designed, the level of complexity required to meet the system objectives must be justified with measurable benefits to plant operation. This applies whether the objective is to provide basic control functionality for safety, environmental compliance, or equipment protection, or to improve the efficiency of plant operations (the economic incentive). In all cases, the company or plant must decide if the complexity and expense associated with a new control system or with adding new field devices or more advanced control capabilities is justified by the improvement in efficiency, output, safety, and/or environmental compliance, as illustrated in Figure 2-22.

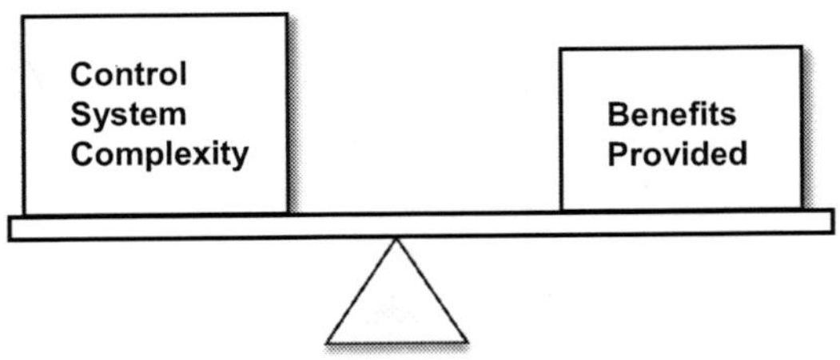

Figure 2-22. Control System Benefits Balance

A number of different advanced control tools and control techniques will be introduced in the following chapters of this book. Examples are used to illustrate how the application of advanced control provides benefits that justify the cost of the added complexity this introduces into the control system. Tools and techniques discussed in later chapters such as on-demand tuning, adaptive tuning, and inferential measurements may be used to improve the performance of traditional single loop and multi-loop control that are described in this chapter. PID modifications that expand its capability and advanced control techniques, such as model predictive control, on-line optimization, and data analytics, will be addressed in some detail. In general, advanced control techniques and tools should be considered when:

- The control objectives cannot be achieved through the improvement of traditional control techniques.
- Traditional control strategies are more difficult to maintain at optimal performance because of their complexity.
- Advanced control solutions provide greater economic benefits and improved product quality, safety, compliance, and equipment protection.

By documenting the economic and operations benefits that can be achieved through the application of advanced control, it may be possible to justify the cost of installing and maintaining advanced control tools.

BIBLIOGRAPHY

1. Edgar, T. F., Himmelblau, D.M., and Lasdon, L. *Optimization of Chemical Processes,* 2nd ed. New York City, NY: McGraw-Hill, 2001.
2. Lipták, B. *Instrument Engineers' Handbook – Process Control and Optimization, Boca Raton, FL,* CRC, 2006.
3. Seborg, D. E., Edgar, T. F., and Mellichamp, D. A. *Process Dynamics and Control,* Hoboken, NJ: John Wiley & Sons, Inc., 2004.

3 Evaluating Control System Performance

The operation of industrial process control systems often involves the control of multiple unit operations with hundreds of measurements and control loops, as illustrated in Figure 3-1. Although these systems are designed to automatically archive operational data in an historian, it is time consuming to examine historical data to determine if there has been a change in control system performance that could impact plant throughput or the quality of the final product. Thus, in many cases process and control engineers and instrument technicians at the plant level spend a great deal of time addressing the most critical control and measurement problems that plant operations personnel have identified as being potentially disruptive to production or quality. There is usually no time or resources to examine the control performance to proactively address problems that could later impact operation efficiency or limit plant production.

This manpower limitation is especially true in many North American plants, where over the past 20 years corporate engineering staff have been virtually eliminated. Thus, the operating plants can no longer call on the resources that corporate engineering traditionally provided to analyze and correct measurement and control problems that limit plant production and efficiency. In this environment, it is important that tools are available within the control system to automatically evaluate control performance and to quickly identify and correct the source of any measurement or control problem.

The field-based, scalable process control systems and fieldbus devices initially introduced in the late 1990s (Figure 3-1) are designed to support embedded tools for process analysis and measurement and control problem diagnostics. These tools can be used to analyze both fast

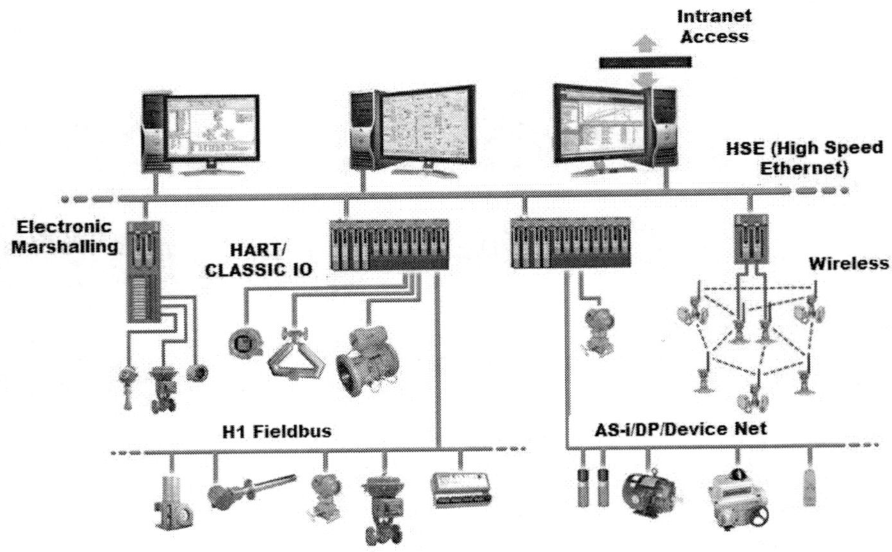

Figure 3-1. Example Distributed Control System

and slow processes. The core technology used in some of these systems supports continuous monitoring and detection of abnormal control and field device operation. For older distributed control systems, a similar capability can be provided by layering performance monitoring applications on top of the existing control system. In some cases, however, the communication speed of a layered application can limit the system's ability to monitor faster control loops. These limitations are inherent in any layered applications designed for older distributed control systems and programmable logic controller (PLC) systems.

In a modern distributed control system, embedded performance monitoring tools can be used to monitor control loops and automatically detect degradation in loop performance or an abnormal condition in a measurement, actuator, or control function block. By using automated system performance monitoring and built-in diagnostic tools, the typically limited resources of plant maintenance can be used to resolve measurement, actuator, and control function block problems. As a result, maintenance costs can be reduced or a higher overall level of system performance can be maintained and process variability reduced. Any increase in overall system performance or reduction in variability can lead directly to greater plant throughput, greater operating efficiency, and/or improved product quality.

To achieve the maximum benefit, the performance monitoring tools that are installed in control systems to evaluate plant control performance should contain features to support both **analysis** and **diagnostics**.

- **Process Analysis** is the examination of a process, its elements, and their relationships.
- **Process Diagnostics** is the investigation of the cause or nature of a process condition or problem.

The assessment of control performance is an ongoing function involving both process analysis and process diagnostics, as illustrated in Figure 3-2.

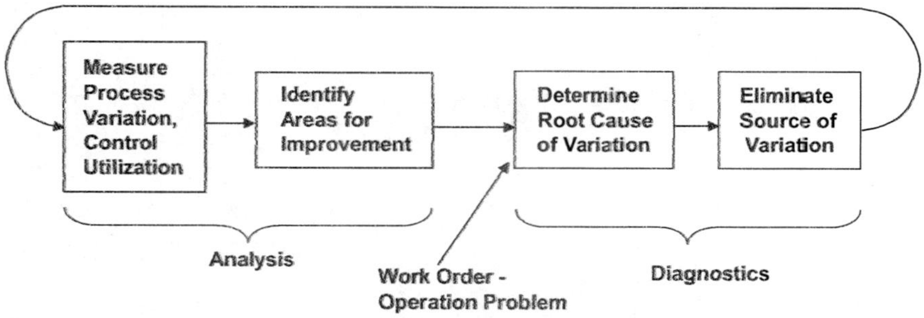

Figure 3-2. Assessment of Control Performance Is an Ongoing Function

Most performance monitoring tools provide both an instantaneous view and a historic view that summarizes control loop utilization by process area. Control loop utilization (often simply called control utilization) is the percentage of time a control loop operated in its designed ("normal") mode of operation over some specified period. If the utilization is lower than normal, the user can select a detail view of the process area that shows the utilization for each control loop. An example of a control performance monitoring tool's functionality is illustrated in Figure 3-3. The figure also shows some of the information that is automatically collected about control system performance.

Report generation is an important feature of performance monitoring tools. Reports can be used to gain the management support that is required to address low control utilization or to determine the source of excessive process variation that impacts production or product quality. Depending on the purpose of the report, there are usually several reporting options: an Overview Report, Area Performance Report,

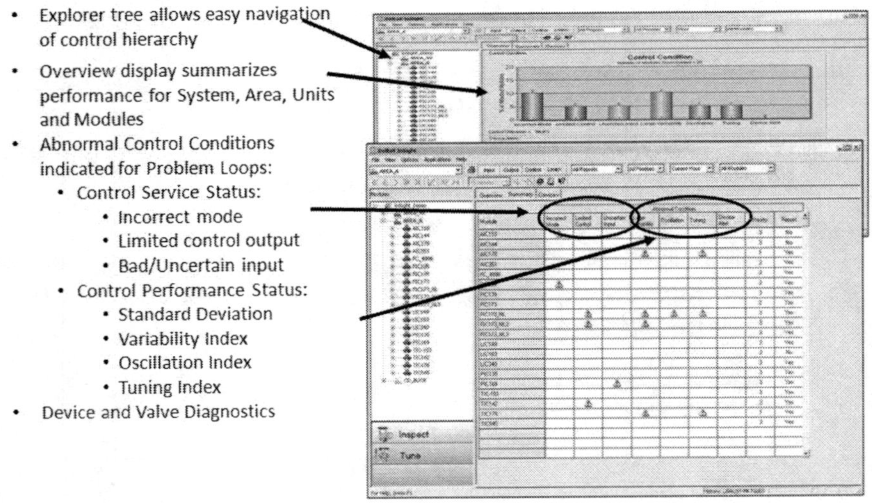

Figure 3-3. Tools for Examining Control Loop Utilization

Detail Control Loop Report, and Customized Report. An example of an Area Performance Overview Report is shown in Figure 3-4.

In summary, through the use of performance monitoring tools, it is possible to effectively analyze and diagnose control performance with minimum manpower. Applying such tools can have a significant positive impact on plant throughput and product quality.

3.1 Evaluating Control Performance

A plant operator is usually responsible for managing one or more process areas, with (potentially) hundreds of control loops and process measurements in each area. The operator's job is made easier if the control loops in these process areas are in automatic mode and can compensate automatically for disturbances and maintain the operating conditions the operator specifies through the control loop setpoints. Thus, the operator will place a control loop in a manual mode of operation only if there is a measurement or control problem that prevents the control loop from satisfying operational or performance requirements. Therefore, in evaluating measurement and control performance, a good starting point is to quantify current and average control loop utilization in each process area of the plant.

Most performance monitoring tools are designed to provide both an instantaneous value and an historic view that summarizes control loop utilization by process area. If the utilization is lower than some estab-

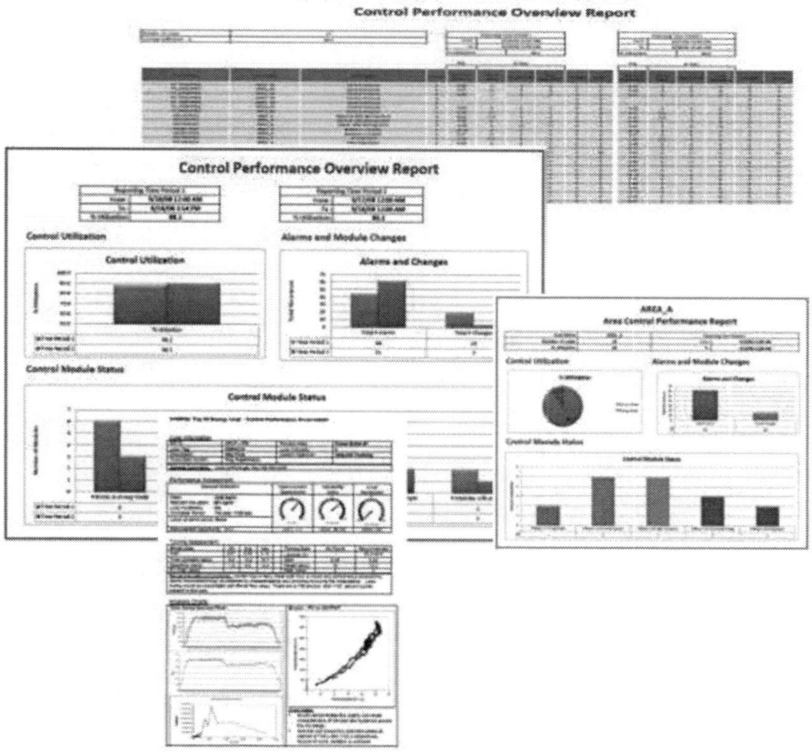

Figure 3-4. Example Area Overview Performance Report

lished limit, the user can select a detail view of the process area that shows the utilization for each control loop.

Control utilization can be determined by comparing the Actual and Normal mode attributes of the mode parameter. The mode parameter of the PID block determines the source of the block setpoint and the source of its output. This parameter is a critical part of the operator's interface to the PID block and consists of four mode attributes:

- Target – The mode of operation requested by the operator
- Actual – The mode of operation that can be achieved based on the target mode and the status of the PID block inputs
- Permitted – The mode(s) available to the operator for a given application
- Normal – The normal operating mode of the PID

The Actual mode attribute can take on a value other than that specified by the Target mode attribute. This occurs when an internal condition or input status indicates that the mode of operation requested by the operator through the Target mode attribute cannot be achieved. Under normal operating conditions, the Target and Actual mode attributes match. When the Actual mode attribute changes to Local Override (LO), the output tracking or an internal application such as auto-tuning is setting the block output. Similarly, if the path to the process is lost, the Actual mode attribute changes to Initialization Manual (IMan).

Actual Mode	What It Means
Override (LO)	Track or auto-tuning is active and in control of the output value
Initialization Manual (IMan)	The forward path to a physical output is broken and the output is tracking the downstream block

When the Actual mode attribute of a PID is IMan, the path to the process is incomplete. The path to the process can be incomplete if a downstream block is taken out of Cascade mode. The path to the process can also be incomplete because of a physical condition such as the hardware failure of a downstream block or an actuator failure that is manipulated by the PID.

It can be shocking to examine control utilization in a plant where performance tools have not been installed and/or not used. For example, in the mid-1980s a control survey was conducted at a major pulp and paper plant where the control system had recently been updated to the latest distributed control system. At that time performance tools were not available to automatically evaluate control system performance. However, it was possible to manually evaluate control utilization using a snapshot of the plant operation. The results of this survey are shown in Table 3-1.

Table 3-1. Pulp and Paper Mill – Control Utilization

Process Area	Control Loops	Normal Mode	Utilization
Bleach Plant	78	60	76%
Power House	185	130	70%
Pulp Mill	174	116	66%
Paper Mill	236	134	56%

After management saw the results of this survey, an instrumentation team was formed to investigate loops that were not running in their normal (design) mode. This team was responsible for making sure measurement, control valve, and process problems were addressed in a timely fashion. The resulting reduction in variability led to significant improvements in plant throughput and product quality. Two years later the plant set a new production record.

Even today, similar problems can be found in process areas equipped with the latest control systems and field instrumentation. For example, a survey of seven areas in a petrochemical complex was recently conducted using performance monitoring tools that were embedded in the control system. A summary report of the control utilization for this plant is shown in Table 3-2.

Table 3-2. Petrochemical Complex – Control Utilization Summary Report

System	Loops	Utilization
PX	471	67.3%
APS&VPS, CLE, Sulphur Recovery	469	59.7%
Refinery	478	60.9%
IGCC/Auxiliary Boiler	946	52.7%
Ethylene	1355	77.5%
FCCU	475	48%
C4	164	68.9%

Once the plant management at this large petrochemical complex became aware of the low control utilization in key processing areas, manpower and funding were provided to investigate and correct the measurement, control, and process problems that were preventing the operator from using the control as designed. This led to significant operational improvements.

These examples demonstrate that the path to improving control performance should start with an assessment of control utilization. The automatic collection of control utilization statistics by the control system is of major benefit in identifying problems in measurement, control, or process design and operation.

In plants that do not have a control system that provides embedded control performance monitoring, this capability can be layered onto the existing control system. When a control loop is identified as not being fully utilized, the reason for poor utilization should be found and resolved. Once the issues of control utilization have been addressed it is possible to start examining the variability in control parameters, expressed by its standard deviation. Where the variation is large enough to impact plant production or product quality, the source of the variation should be investigated and steps taken to minimize it. In the following section we explore steps for improving control utilization and for minimizing process variation when a control loop is working in its normal (designed) mode of operation.

3.2 Improving Control Utilization

When the utilization of a control loop is found to be low during normal plant operation, the measurement or control problems that prevent the control loop from functioning in its normal (designed) mode of operation should be further explored, as illustrated in Figure 3-5. To achieve full utilization, effective communication between maintenance and operations is important.

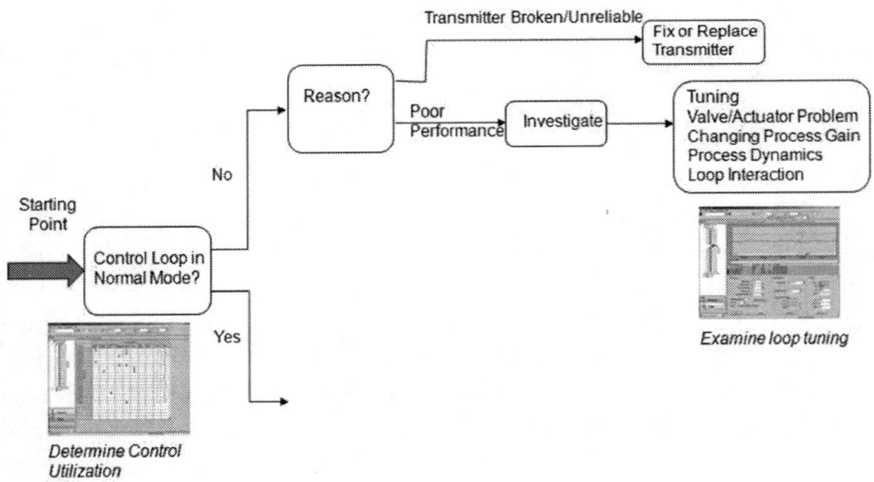

Figure 3-5. Resolving Problems that Impact Control Utilization

When a control loop is routinely not operated in normal mode (which is most often automatic mode), further investigation is needed to determine the cause of the poor control utilization. Some of the most common problems that can impact control loop operation in automatic mode are:

- Measurement health, i.e., quality of the measurement

- Incorrect tuning
- Valve/actuator malfunction
- Changing process gain
- Split range setup
- Process gain
- Process dynamics
- Loop interaction

Control performance tools provide information that guides the user in solving the performance problem. However, where there is no clear indication of the source of the problem, it may be necessary to test and observe the loop operation in both manual and automatic modes.

3.2.1 Transmitter Problems

One of the most obvious reasons for poor control utilization is a broken or unreliable transmitter or sensor. If the control system is designed to be consistent with the international function block standard, IEC 61804, the status attribute is always communicated with the measurement value as illustrated in Figure 3-6 and provides a direct indication of measurement condition. The percentage of time the measurement had a status of Bad, Uncertain, or Limited can be used to determine measurement health. The Target and Actual attributes (i.e., values or meaning) of the Mode parameter defined by the standard can be used to determine if a control loop is running in its designed mode of operation. Thus, the status and mode attributes of a function block are the basis for evaluating measurement health and control utilization.

The status attribute that accompanies a measurement value provides an indication of the quality of the measurement. For a measurement input provided by a traditional transmitter, the status attribute is determined by the analog input card. For a digital value provided by the transmitter, the status attribute can be provided by the transmitter. The status attribute value consists of eight bits. The most significant two bits are used to indicate the quality of the measurement classified as: Bad, Uncertain, Good – Control, and Good – Measurement. The last two bits of the status attribute, the lower-priority bits, are used to indicate high or low limit, no limit, or constant. The four middle bits of the status attribute are used to show why the quality is good, bad, or uncertain. The health of a measurement may be assessed based on the percent of time its status was Bad or Uncertain or a Good but an active limit condition.

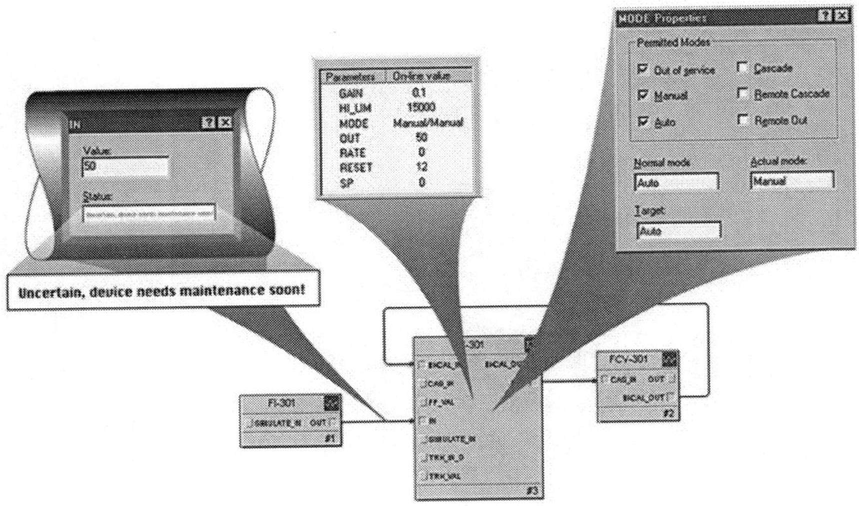

Figure 3-6. Standard PID Parameters

After a measurement value and status are generated by the analog input block, the other blocks that use this measurement input process the status and use it in some way, and propagate a status to the downstream blocks. If the input block status is Bad, the Actual mode of the block that uses a measurement provided by the input block is unable to achieve an Automatic mode of operation and thus the Normal and Actual mode attributes will indicate that control is not being utilized as designed.

Most problems associated with a measurement can be quickly addressed by calibration or replacement of the transmitter.

3.2.2 Incorrect Tuning

Loop tuning has a significant impact on control loop utilization and performance. Controller tuning parameters can be easily changed over a wide range to adjust loop operation for various process requirements. However, some expertise in tuning is required even when advanced control tools are used, and there is a chance that the behavior of a particular process will be unaccounted for in the tuning that was selected for the control loop. These are important reasons for considering loop tuning as a primary suspect in degraded performance.

During the design of a new process area or plant, initial loop tuning can be specified based on the type of measurement associated with the control loop, that is, whether pressure, level, temperature, flow, or some other process variable to be controlled. In most cases, these default settings may be close enough to support the initial startup of a new process area or plant. However, to minimize process variations from setpoint and to minimize the response time to setpoint and disturbance input changes while providing stable operation over a variety of operating conditions during or after startup, it is necessary to set the PID tuning based on the observed process gain and process dynamics for each control loop. Unfortunately, the instrument engineer involved in plant commissioning or plant operations may not have the opportunity to tune loops and, as a result, plant operation does not achieve maximum efficiency.

In an operating plant, various approaches can be taken by plant personnel to modify control loop tuning to improve control performance. The most direct and reliable way to manually or automatically establish PID block tuning is to calculate the tuning based on the identified process gain and dynamics as described in the following sections. With older control systems, the analog single loop controllers and programmable logic controllers (PLCs) used may not include tools that allow the PID tuning to be automatically established based on the observed process response to input changes. With modern control systems, some capability is normally provided to automatically establish loop tuning.

When tuning control loops, natural questions might include: "What is the best tuning for a particular loop? Should control loop performance be measured based on the observed process response to a setpoint change, or on the observed process response to a change in a load disturbance? What factors should be considered when judging loop control performance?"

The answers depend on the objective for the loop. If the setpoint is constantly being changed, the loop's response to setpoint changes would be a key factor in evaluating its performance. However, if the loop setpoint is always maintained at a fixed value, the response to disturbance inputs would be most important. If the response to both setpoint changes and disturbance inputs is important, tuning must be adjusted to achieve a balance in control response.

When selecting the PID gain, it is important to consider that the process gain can change with the operating conditions. To ensure stable operation over the entire operating range of a control loop, the normal practice is to establish the PID tuning when the process is operating in the region with the highest process gain. As a result, higher variation in the controlled parameter can be seen when the process transitions to the low gain region of operation.

3.2.3 Valve/Actuator Diagnostics

When a control loop is placed in automatic control, it is easy to detect if a valve or actuator is not responding correctly to the control system. This can be done by observing the response of the controlled parameter to control system changes in the PID output as illustrated in Figure 3-7.

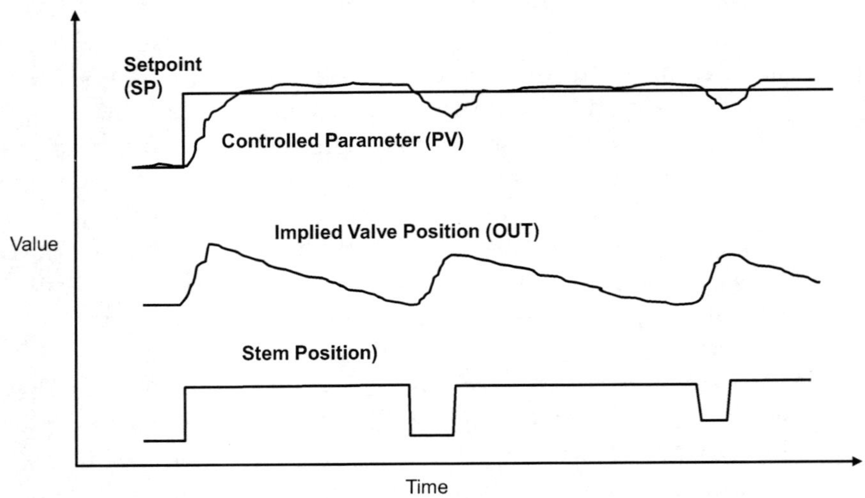

Figure 3-7. Impact of Sticky Valve on Control

The most common problems in commissioning a control system can often be traced to a valve without a positioner, or to a positioner that has not been properly installed or has malfunctioned. The rule of thumb is that to achieve best control performance, all regulating valves should be equipped with a properly functioning positioner. Without a positioner, the level of control performance that can be achieved is very limited when a valve is sticking—which is inherent in most valves.

The cyclic behavior caused by a sticky valve (with no positioner) or a malfunctioning positioner cannot be eliminated through tuning. Changes in tuning will only impact the period of the cycle. The only way to eliminate this type of behavior is to install a valve positioner and ensure that it is functioning properly.

3.2.4 Changing Process Gain

From a control perspective, it is highly desirable that the process gain is constant. If the process gain is constant, then the same proportional gain can be used over the entire operating range of the control loop. Using a control valve as an example, if the valve characteristic was not properly selected based on the process requirements, the valve installed characteristic could be non-linear.

As illustrated in Figure 3-8, which shows an example of a non-linear valve installed characteristic, the process gain varies from 0.5 to 4; that is, the process gain changes by a factor of eight.

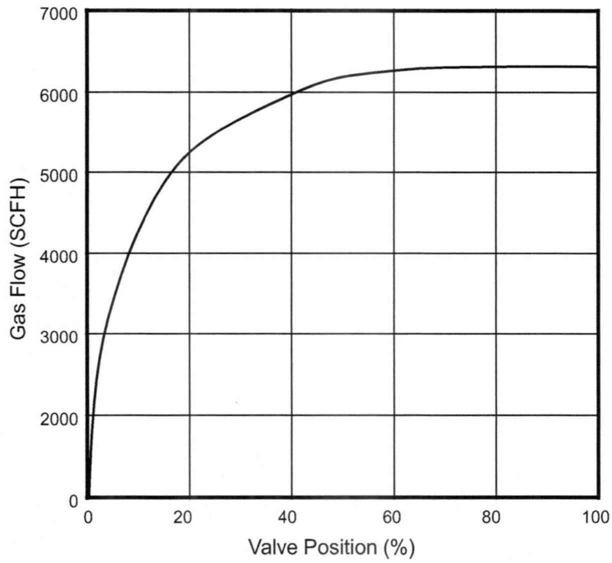

Figure 3-8. Non-Linear Valve Installed Characteristic

For such non-linear installed characteristics, the impact of the changing process gain with valve position can often be seen by trending the controlled parameter (PV), setpoint (SP) and controller output (OUT) to the value as shown in the following example, Figure 3-9.

As previously discussed, stable operation can be achieved by tuning the control loop in the high gain region, but doing this will result in slow response when the process is operating in the low gain region. To compensate for the changes in process gain, a characterizer block (SGCR) can be installed between the PID and Analog Output blocks as illustrated in Figure 3-10.

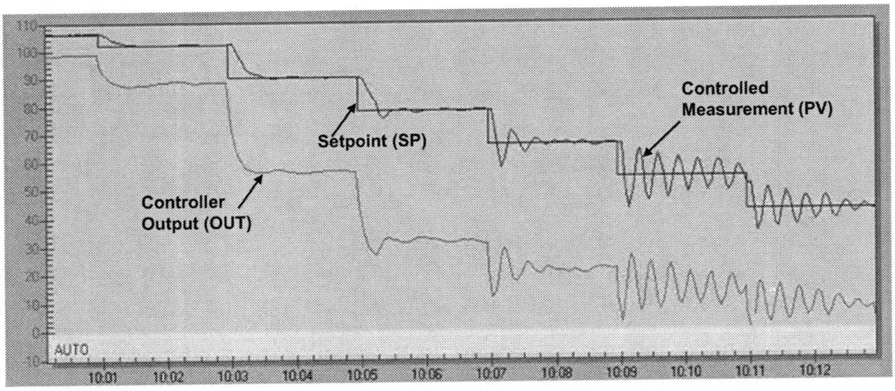

Figure 3-9. Example of the Impact of Non-Linear Process Gain on Control Loop Operation

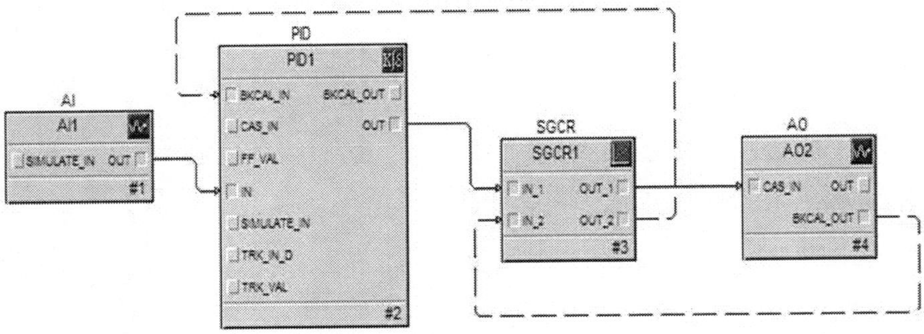

Figure 3-10. Characterizer Block Used to Compensate for a Non-Linear Installed Valve Characteristic

The relationship between the primary inputs and the output of the characterizer block can be defined over the valve's operating range. The curve defined by the characterizer's points on the graph appears as the inverse of the plot of the valve's installed characteristic. An example of the setup of the characterizer block is shown in Figure 3-11.

Because of the time and understanding that are necessary to install, commission, and maintain a control loop that includes a characterizer, this approach is not commonly found in the process industry. To improve control performance, the valve installation may be modified to provide linear installed characteristics. For example, the valve trim

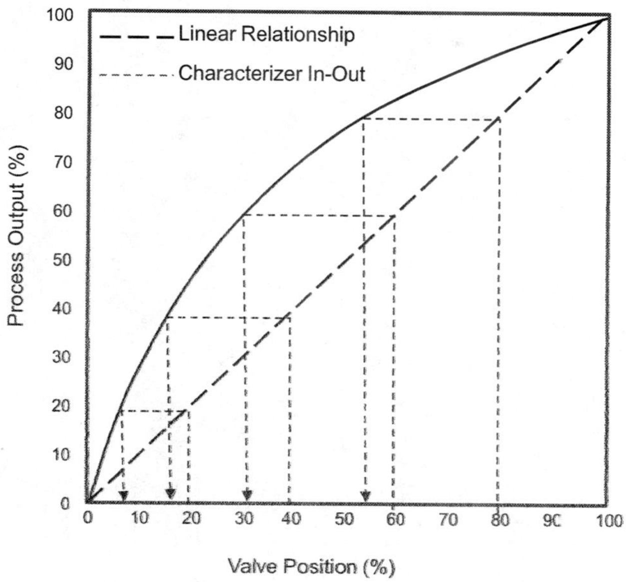

Characterizer

IN	OUT
0	0
5	1.5
10	3
15	5.5
20	8,5
25	10
30	11.5
35	13
40	17
45	20
50	22
55	26
60	31
65	36
70	42
75	48
80	54
85	63
90	74
95	86
100	100

Figure 3-11. Characterizer Block Setup

can be modified or a digital valve positioner installed that supports characterization.

3.2.5 Incorrect Split Range Setup

One of the most common ways of addressing multiple process inputs is split range control. A splitter block, which appears as one valve to the PID block, can be used in the control strategy to define a fixed relationship between the controller output and each manipulated process input. An example of split range control is illustrated in Figure 3-12.

To achieve consistent control behavior, the splitter block setup must account for the gain associated with each process input. Unfortunately, in many installations the split range setup was arbitrarily defined during commissioning as an equal split among the process inputs. Such a setting can result in sluggish or unstable operation when the controller output transitions from one input to another input.

The manner in which the splitter block setup accounts for the gain associated with each process input may be illustrated by considering a steam header pressure control application. One or more power boilers may be used to meet the process steam requirements of a plant. Also, this steam may be used in turbine generators to meet some or all of the

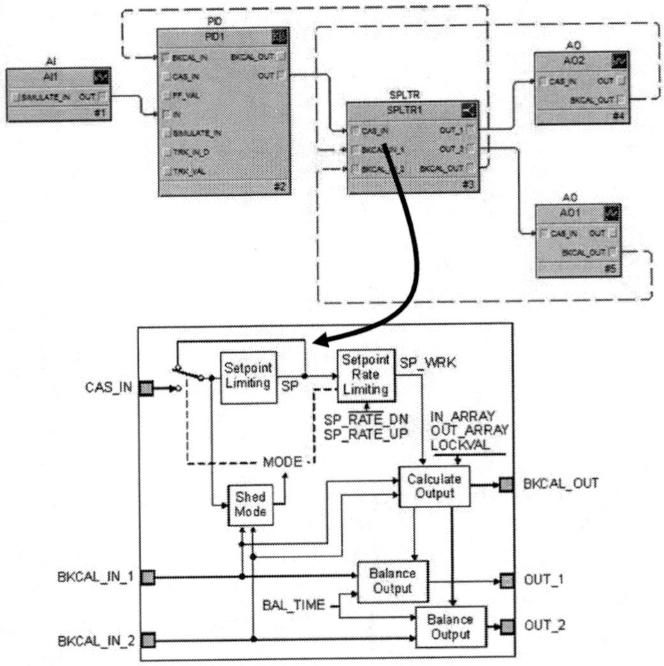

Figure 3-12. Split Range Control Block Example

plant requirements for steam. Steam turbines used to generate electricity are powered by high pressure steam. For example, in newer installations, the supply header may be at 1475 psi. The pressure of the steam is reduced as it flows through the turbine, allowing lower pressure steam to be extracted at various points. This lower pressure steam may be used to meet the plant's process steam demands. For example, at one point, the extraction may be regulated to meet the steam demands of a 400 psi header. Extraction valves associated with the turbine are automatically regulated to maintain the lower header pressure constant. To allow the plant to continue operation if a turbine or generator fails and must be shut down, pressure reducing valves (PRVs) between the high pressure header and the lower pressure header can be adjusted to meet the lower pressure header steam demand and to maintain the lower header pressure constant.

This can be accomplished by using a splitter block in conjunction with a PID block to adjust the pressure-reducing valves as illustrated in Figure 3-13.

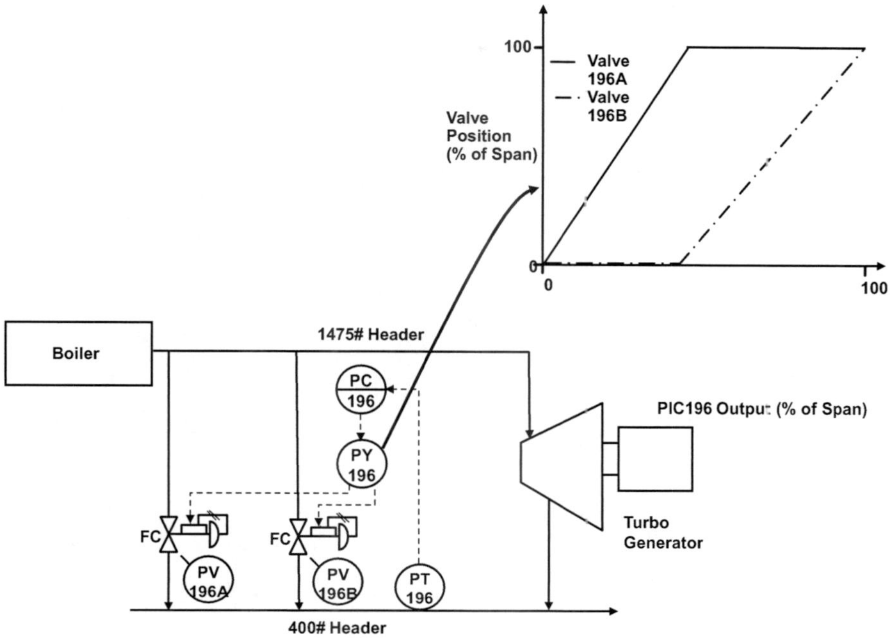

Figure 3-13. Steam Header Pressure Control Example

If the valve sizes or operating conditions differ from loop to loop it is necessary to characterize the splitter to compensate for these differences through the setup of the characterizer block.

When a loop that includes a split range output has low utilization, examine the assumptions made in the split range setup and, if necessary, modify the setup to provide a constant process gain as seen by the controller.

3.2.6 Loop Interaction

In some cases, the manipulated parameter of one control loop can impact the controlled parameter of another control loop as illustrated in Figure 3-14.

To allow both loops to operate in automatic mode, loop interaction is most often addressed by simply detuning one of the control loops by reducing the proportional gain. The valve (or another final control element) associated with the detuned loop will change position slowly. Thus, the two loops will tend not to interact but at the expense of higher variability in the de-tuned control loop.

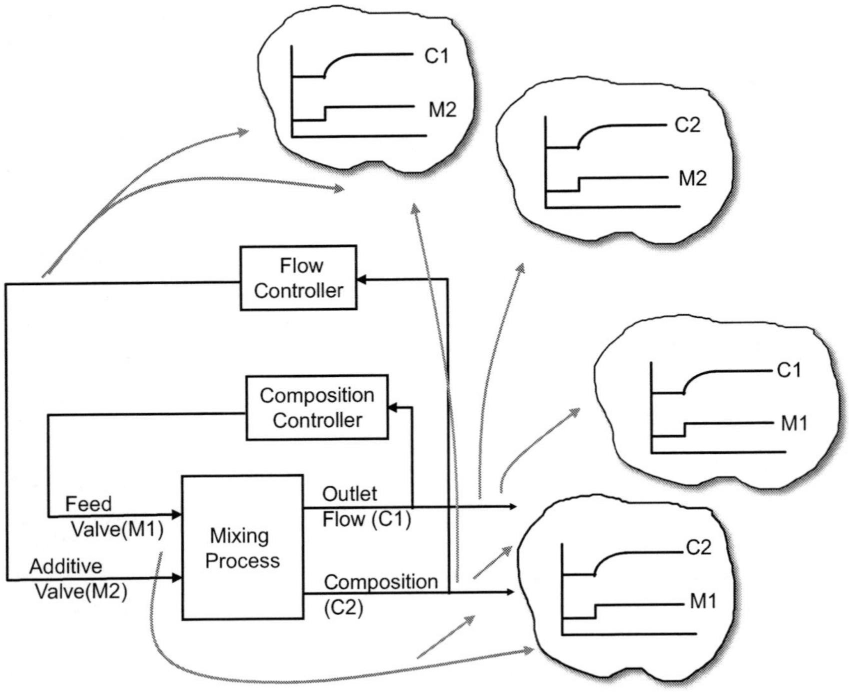

Figure 3-14. Interactive Loops Example

3.3 Addressing Process Variability

Even though a control loop has high utilization, traditional control techniques may not be sufficient to achieve the reduction in process variability necessary to increase production or to achieve product quality targets. As addressed in Chapter 2, product quality and manufacturing efficiency can be impacted by variation in key process parameters. Also, when production is process limited, throughput can be increased by reducing process variation and operating closer to the process or equipment limit. The degree to which a control loop must reduce process variability depends in many cases on the objective of the control loop. Thus, the limits set in a performance monitoring application for detecting excessive variability should be adjustable based on the control loop objective. For example, very wide limits can be appropriate when floating level control is implemented to use process storage capacity to absorb process variability.

Performance monitoring tools can be used to automatically quantify the process variation seen while on automatic control. If the control system does not include features to automatically assess and report control loop variability, it may be possible to use the statistical calculations included in a plant historian or to add loop performance monitoring packages that connect directly to the DCS or to the plant historian through OPC (Object Linking and Embedding [OLE] for Process Control). The standard for the real-time plant data communication between control devices from different manufacturers) connectivity. Both methods monitor control loop variability and provide insight into those areas in which control performance could be improved through the use of advanced control products.

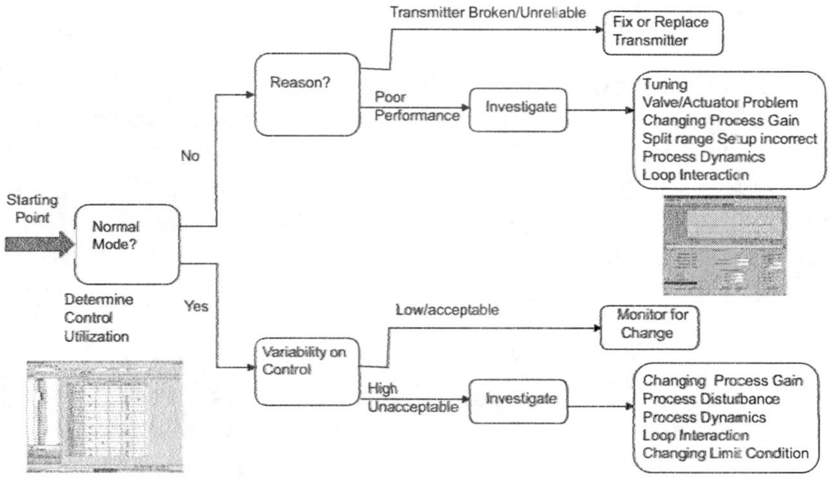

Figure 3-15. Detecting High Control Variability

When control utilization is found to be acceptable, that is, the control loop is in its designed mode of operation during normal process operation nearly 100% of the time, control performance can be examined based on the control variability (Figure 3-15). When high variation is found, further investigation is needed to determine the cause of poor control performance. Assuming that the problems associated with control utilization have been addressed, and the controller has been tuned correctly, some common problems that can impact control loop variation in automatic mode are:

- Changing process gain
- Unmeasured process disturbances
- Process dynamics
- Loop interaction
- Changing limit conditions

- Quality parameter is a lab measurement

3.3.1 Changing Process Gain and Dynamics

In some cases, the control performance tool can provide information that guides the user to the source of the control variation. Where there is no clear indication of the source of the problem, it may be necessary to observe loop operation in manual and automatic modes. As already discussed in this chapter, in many cases the advanced control tools detailed in the remaining chapters of this book may be required to reduce the process variation to levels needed to achieve best process operation.

If the variation in control is traced to a change in process gain, then as previously addressed in this chapter, a characterizer block can be installed between the control block and the control system output. Also, if the gain change is caused by incorrect split range setup, the splitter block's commissioning may need to be examined. However, if the control system supports adaptive control, then as discussed in Chapter 5 it is often much easier to use this capability to address changes in process gain. Adaptive control is particularly effective when the sources of changes in the process gain are not associated with the controller output. In some processes, for example, the process gain changes with plant throughput or through changes in an unmeasured parameter, such as the feed stream composition. Also, when process dynamics change along with gain, adaptive control may be the only way to correct for these changes and to provide the best tuning for all operating conditions.

3.3.2 Unmeasured Process Disturbances

Sometimes high process variability may be the result of unmeasured process disturbances. In these cases, the root source of the variation can often be found by cross correlation of the control parameter with measured upstream conditions that can potentially impact the control parameter. The root cause of poor control performance can often be traced to control design, tuning, measurement health, or actuator performance. For example, primary cascade loop performance can be degraded by the limited output of the secondary loop.

In some performance monitoring tools, the distribution of the variation in a measurement value or actuator position to be analyzed is plotted as a histogram. A bell-shaped distribution indicates that the source of variation is random in nature. A power spectrum is a frequency distribution of the components that make up a measurement or actuator signal over a selected period of time. Such information can be helpful in determining the magnitude and frequency of process noise or the frequency of disturbances caused by loop interaction or upstream processes. The influence of other control loops and upstream conditions on the variability of a control loop can be determined through cross-correlation. One example of how histograms, power spectrums, cross-correlations, and auto-correlations can be presented in performance monitoring systems is shown in Figure 3-16.

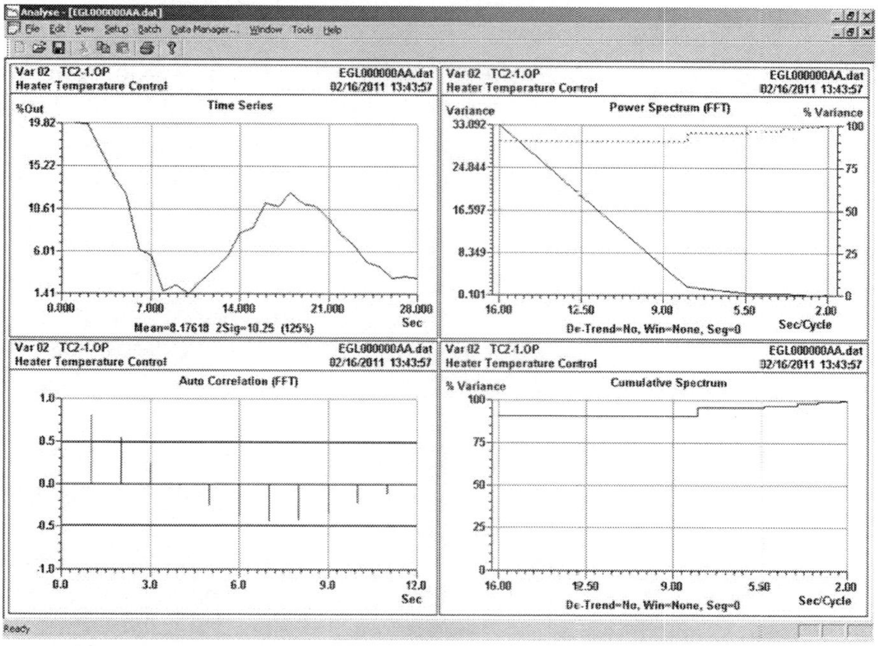

Figure 3-16. Display of Diagnostic Information

When the source of the process disturbance is indicated by a measured parameter then, as discussed in Chapter 2, this input can be added as feedforward input to the feedback control. By modifying the control by this means, the impact of the disturbance may be anticipated and accounted for in the control actions. When multiple measured parameters are known to impact the controlled parameters, or when there is a dynamic difference in the impact of a change in the measured disturbance and the manipulated parameter, Model Predictive Control (MPC) can be used to account for these multiple disturbances or the dynamic differences in their impact as discussed in Chapters 11 and 12.

3.3.3 Process Dynamics

In some cases, the dynamic response to setpoint changes and process disturbances may not meet the desired control objective when a process is characterized as being lag dominant, that is, when it has very little deadtime and a large time constant. If the objective is to reach setpoint in the minimum time with no overshoot, fuzzy logic control has been shown to provide an improvement over the response provided by PID. Improved performance is achieved through the non-linear action taken by the fuzzy logic control on large deviations of the control parameter from setpoint.

When the process is deadtime dominant, that is, the process deadtime is equal to or greater than the process time constant, even the best-tuned PID may be too slow to respond to setpoint changes and correct for process disturbances. As the ratio of deadtime to time constant increases, the proportional and reset gains must be decreased for stability. As a result, the control response to changes in setpoint and process disturbances is often slower than that required for best process operation. In such cases, control performance can be improved by replacing PID feedback control with Model Predictive Control as discussed in Chapters 11 and 12.

This improvement over PID feedback control performance is possible because the MPC algorithm is generated based on the process response to a step change in process inputs, known as the step response model, rather than on a pre-defined algorithm as is the case with the PID algorithm. When a change is made in an input of a process characterized by deadtime, unlike the PID algorithm, the MPC algorithm recognizes that this change will not be immediately reflected in the process. This recognition of the process response allows the MPC algorithm to better control a deadtime-dominant process.

3.3.4 Loop Interaction

When a process is characterized by multiple manipulated process inputs and multiple controlled process outputs, there is a potential for process interaction, that is, a change in one manipulated input can impact multiple controlled outputs. Detuning the PID control is the most common approach when the resulting interaction between control loops is significant. When MPC is used, the interaction of the manipulated inputs and controlled outputs may be addressed automatically while providing the best performance for all control loops.

As addressed in Chapters 11 and 12, the impact of each manipulated input parameter on each controlled output parameter is identified by the step response model used in MPC block generation. Thus, any interactions are automatically compensated for when using model predictive control.

3.3.5 Changing Limit Conditions

When using MPC for process control, and a process output is a constant parameter, the constraint parameter measurement is simply added as an input to the MPC block. The design and implementation of an override control strategy using an MPC block is thus much simpler than implementing override control using two PID blocks and a control selector block.

As addressed in Chapters 11 and 12, the step response of the constraint parameter to changes in the process input(s) is automatically identified when using the MPC application to test the process and to generate the step response model. Because the process step response is used to generate the MPC controller, the response of constraint parameters and controlled parameters to changes in the process inputs is automatically taken into account to minimize variation from setpoint while observing the constraint limits.

3.3.6 Quality Parameter – Lab Measurement

A quality-related parameter may be a lab measurement if an on-line analyzer for the measurement of a physical property or the composition of a liquid or gas stream is not commercially available, or if other conditions prevent the installation of an on-line analyzer. In these cases, the measurements must be obtained by lab analysis of a process grab sample. The time required to collect the grab sample and process it in the lab introduces a significant delay in obtaining the results of the analysis. The plant operator or control system often needs such measurements to maintain key quality-related parameters by making adjustments to plant operating conditions, thereby keeping product properties within specifications. Any delay introduced in processing a sample limits the usefulness of the measurement.

To obtain an immediate indication of the results of process changes that impact a quality-related parameter, it is often possible to use upstream measurements such as flow rate, temperature, and feedstock composition to calculate an estimated value of the quality parameter. For example, a dynamic linear estimator can be created using the calculation tools available in most distributed control systems, as discussed in Chapters 9 and 10. Also, some control systems support non-linear estimator blocks based on neural network technology, as discussed in Chapter 7. When a linear or non-linear estimator is available in the on-line control system, a collection of lab analysis results is used in conjunction with a history collection of upstream measurement conditions to create the estimator. Once the estimator has been created, lab analysis is used to verify the accuracy of the estimated value and to make corrections to compensate for unmeasured changes in the process that impact the property being estimated.

3.4 Application Example

Control system manufacturers and third party companies have introduced tools that allow the monitoring of process control and instrumentation. In addition to on-line information on control utilization, a variety of other information is normally provided to assist the user in determining the cause of, and solution to, poor control utilization. In this section we examine the process monitoring capability that is offered as a standard embedded capability in a commercial distributed control system. This performance-monitoring capability may be accessed from any operator or engineer station within the control system.

Similar to many other control applications, the user interface allows process measurement and control loops to be easily found in a tree structure that is organized as follows:

- Overview and Summary views that flag abnormal conditions detected in the modules associated with the plant, or with a selected process area, cell, or unit.
- Module view that shows the block(s) that are found to have one or more active abnormal conditions.
- Detail block view that provides more information on the conditions that influenced measurement, actuator, and control performance.

An overview of the plant is displayed when the performance monitoring application is first opened. From the explorer view at the left of the interface, the user can select an individual process area, unit, or cell in the plant for examination. The overview of a selected process area is shown in Figure 3-17.

As illustrated in this example, the number of times a module violates a key performance index is shown using a combination of graphic display and numeric data.

Figure 3-17. Example of a Process Area Overview

The following types of abnormal conditions are automatically detected in a module and reported in the overview display:

- The actual block mode is not in the Normal (designed) mode of operation.
- Control action is limited by a downstream condition.
- Input status indicates that the value is Limited or has a Bad or Uncertain status.
- Large variation is detected in a measurement or in the action taken by a control block.
- The output of a control block is oscillating.
- Control performance could be improved by re-tuning a control block in the module.
- A field device has detected a hardware or software failure.

By selecting the Summary tab, it is possible to see the specific modules in the selected area, unit, or cell. As illustrated in Figure 3-18, an icon is used in the Summary display to indicate that an abnormal condition has been detected within the module.

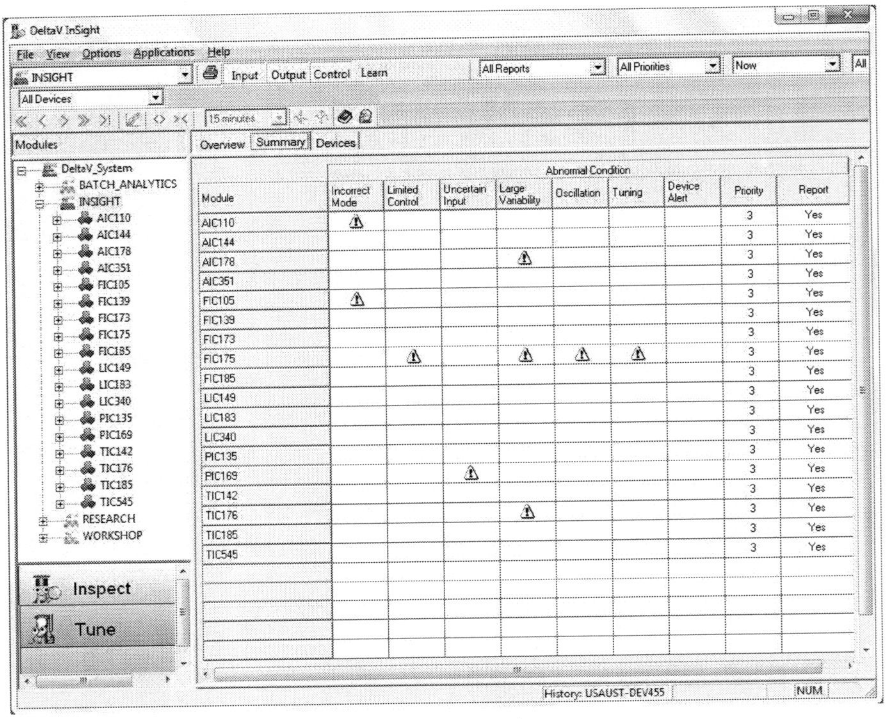

Figure 3-18. Summary View of a Process

In the Summary view, the individual modules that have an abnormal condition are listed in the right pane of the interface. When a module is selected in the Summary list, the individual I/O and control function blocks in that module are listed in the right pane of the module view as illustrated in Figure 3-19.

When an I/O or control function block is selected from the module view, the percent time that an abnormal condition existed within the block is shown in a block detail view as illustrated in Figure 3-20. If the percent time exceeds the configured percent time limit for this condition, the condition is flagged by a color change in the block detail view and an abnormal condition is flagged and is shown in the Overview and Summary views.

Through the Filter selections provided on the interface menu bar, the user can view information for the previous or current hour, shift, or

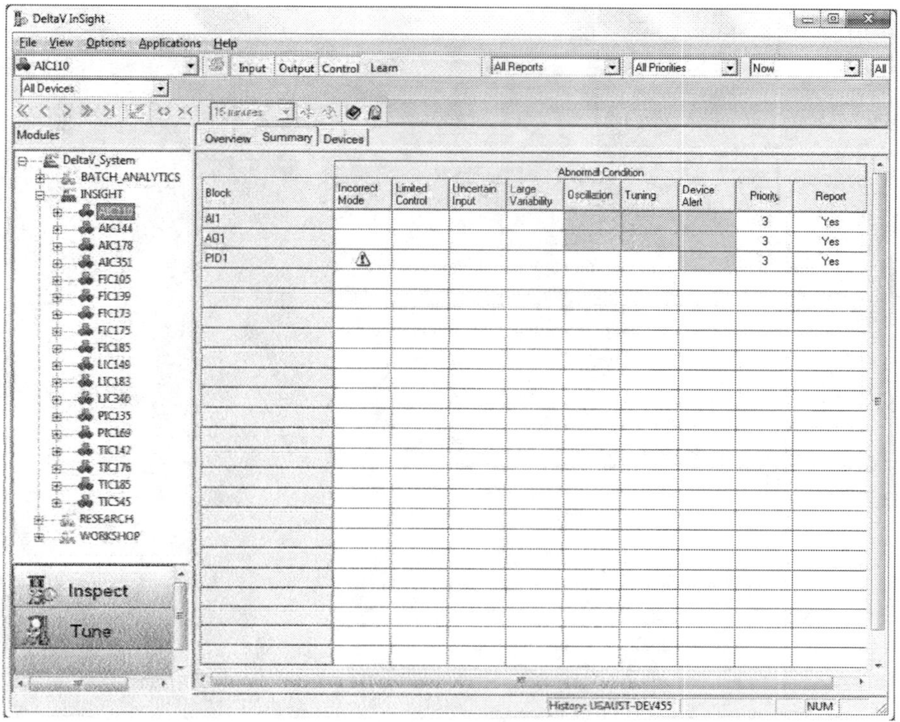

Figure 3-19. Module View

day. Also, the user can select the type of blocks to display and can set filters to show all modules or only those modules with abnormal conditions.

In many plants, it is common for a piece of equipment to be shut down at the end of a batch or be taken out of service for routine maintenance. During these periods of down time, the measurements on that equipment may indicate a Bad status. To prevent this information from being included in the performance monitoring view, the user can select an area and then indicate that the associated equipment is off-line. Such areas are clearly indicated in the equipment view and are removed from monitoring.

The user can find more information on an abnormal condition by using the detail tabs in the interface. Also, the user can schedule a variety of performance monitoring reports that summarize control utilization and instrumentation and control problems that were detected over the reporting period.

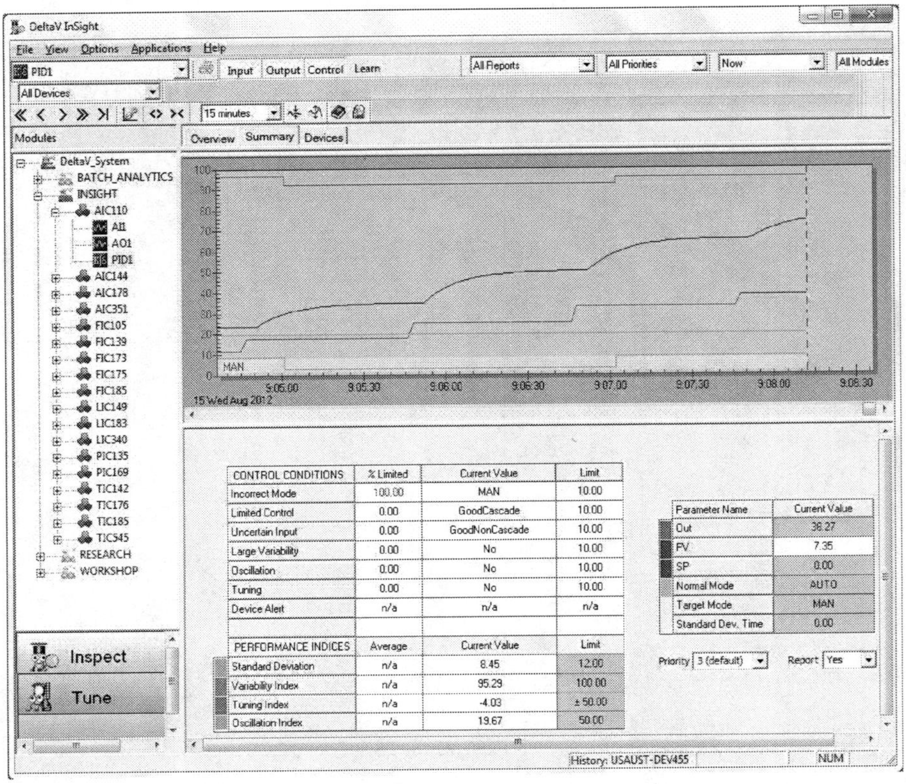

Figure 3-20. Detail View of Block Performance

3.5 Workshop Exercises – Introduction

Workshop exercises provided in this and later chapters are designed to allow the reader to view applications of advanced control tools. The dynamic process simulations used in these workshop demonstrations provide an environment that closely resembles that of an operating plant.

In a distributed control system, the measurement, calculation, and control functions performed by the system are commonly defined using a set of standard function blocks. During system configuration, an engineer selects an appropriate combination of blocks to perform the measurement, calculation, and control functions required in the plant. Information flow between these blocks is defined by configured connections between block inputs and outputs. For example, to create a simple control loop (Figure 3-21), an analog input block (AI), controller block (PID), and analog output block (AO) would be selected to access a transmitter measurement value, perform a PID control calculation, and then transfer the calculated output to the valve. Standard display components defined for these function blocks would be included in the operator interface to allow the operator to access this control loop.

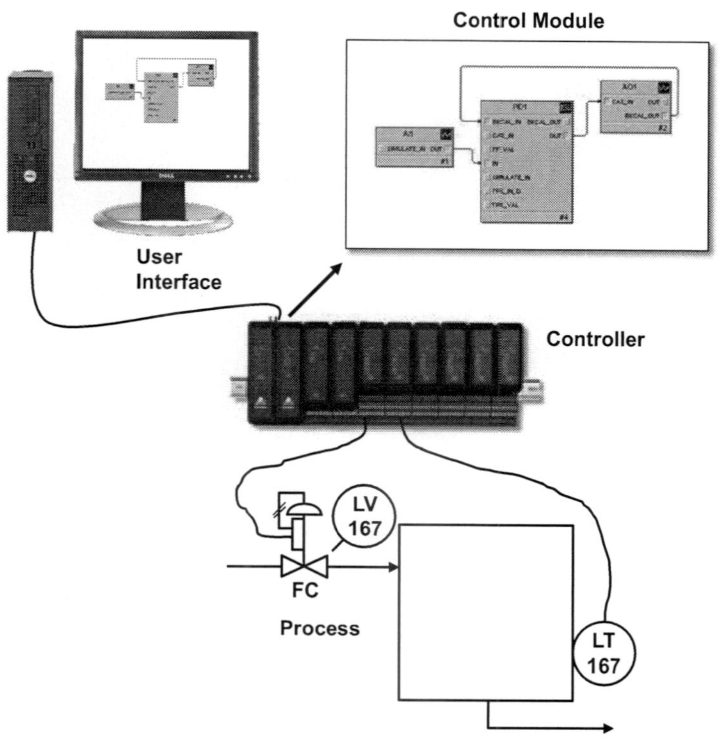

Figure 3-21. Simple Control Loop

To support the workshop exercises, dynamic process simulations have been created that are linked to the function blocks normally used to access field devices, as illustrated in Figure 3-22. Within the accuracy of the process simulation, the parameter values and interactions required when working with the function blocks and user interface will be the same as in an operating plant.

Using a web browser, it is possible to access the solution to the workshops defined in each chapter. Directions for accessing and using this web interface are provided in the Appendix.

For the following process performance monitoring workshop, the measurement and control modules that are assigned to one process area in a plant are simulated. The objective of this workshop is to show

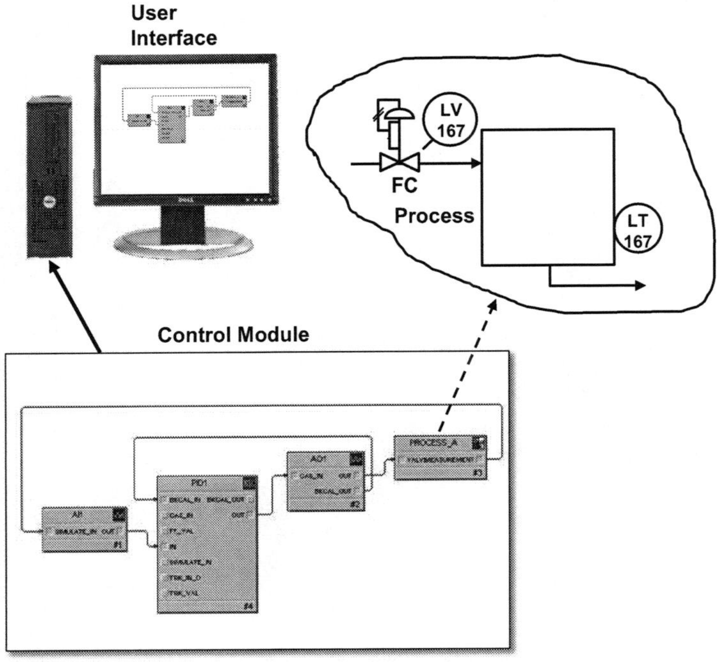

Figure 3-22. Simulated Process for Workshop Exercises

how performance monitoring tools may be used to identify operation problems that introduce variability in the process or that prevent the control system from operating in Automatic mode. The Process Performance Monitoring tool previously described in the application example will be used in this workshop to identify control or measurement problems.

3.6 Evaluating Control System Performance Workshop

This workshop is designed to illustrate some of the features that are typically available in commercial products for performance monitoring. The area addressed is shown in the Process tab. Access the book's web site http://www.advancedcontrolfoundation.com and select the Solution tab to view the workshop being performed.

> Step 1: From the Overview tab of the performance monitoring tools, determine how many of the modules in a selected area have at least one block that is not being utilized in its designed mode of operation (Figure 3-23).

Step 2: Access the Summary tab to find the modules that contain one or more blocks that are not being utilized in their designed mode of operation.

Step 3: Identify the modules that have one or more bad measurements based on the information provided in the area summary.

Step 4: Select a module with a bad measurement and then identify the block that has the bad measurement.

Step 5: Examine the block detail to determine the percent of time the measurement was bad.

Step 6: Using the area summary, find a module that has high variability that is not attributed to the PID tuning.

Step 7: Examine the PID detail to determine the standard deviation of the control parameter from setpoint.

Step 8: In the Area summary, find a module that has high variability caused by the PID tuning. Examine the tuning index in the PID block detail to determine if the current tuning is too aggressive or too sluggish.

3.7 Technical Basis

The preceding sections discussed the main aspects of control system performance and the practical indexes applied to inform operating personnel about performance. Many of those indexes are defined directly and can be understood easily. Others, such as variability index or tuning index, require more explanation and definitions that involve mathematics. This section provides information for those readers who want to expand their understanding of the subject beyond everyday operational needs and learn more detail about system operation. The main algorithms that create the control performance evaluation functionality discussed in sections 3.1 to 3.5 will be defined, with the added (but minimal) mathematical description required for a more rigorous understanding of the subject.

Subsequent chapters will be structured in a similar way, that is, after the discussion of operation and functionality and reference to the workshops, the algorithms and related information will be discussed at the end of the chapter. In this way, the book will separate optional

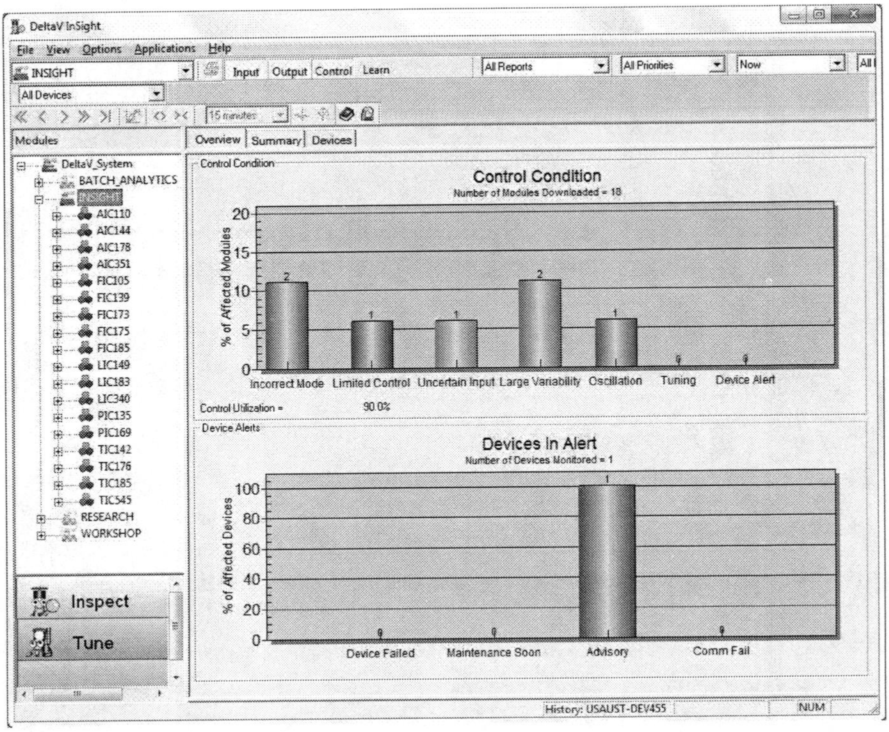

Figure 3-23. Example Process Area Overview

material (the Technical Basis) for advanced users from basic material for all users.

3.7.1 Control Performance Monitoring Infrastructure

An example of the information flow structure for control performance monitoring is shown in Figure 3-24. The structure consists of one server and multiple clients. The server is the central collection point for the information obtained from the controllers. The server calculates the percentage of time an abnormal condition is present for an hour/ shift/day, compares the results of these calculations with user-set percent time limits to detect an abnormal condition or an underperforming loop, and provides this information to clients for viewing upon request. The clients can access information only through the server.

The existence of an abnormal condition is determined within the controller based on function block parameters such as measurement status, back calculation input status, standard deviation, tuning index, and mode. These monitored conditions are automatically reported to the server on an exception basis.

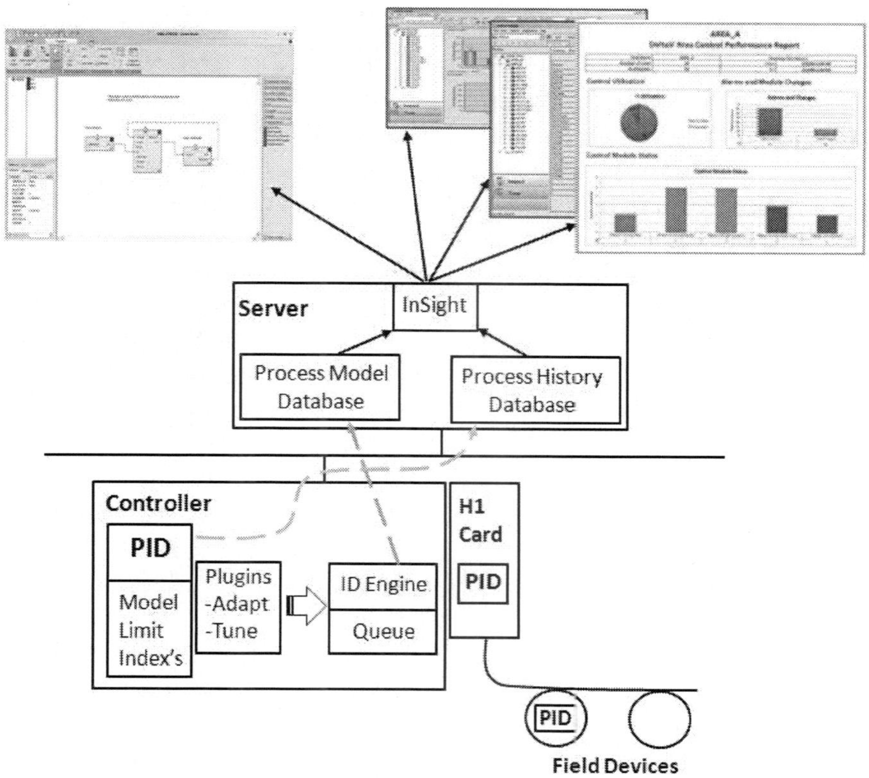

Figure 3-24. Information Flow Structure for Control Performance Monitoring (Note: More generically, InSight stands for Performance Monitoring/Adaptive Tuning Application)

The status and mode parameters are used by the function block to determine incorrect mode, Limited control, and Bad, Limited, or Uncertain measurement status. Other parameters are used within the block to calculate Standard Deviation, Tuning, and Variability Index, and to determine if these values exceed the limits set for the block. These conditions are communicated to the server in a single parameter only when a monitored condition changes state. Therefore, the communication load for reporting these parameters is usually very small. When the server is initially placed on-line, the current state of all monitored conditions is reported to the server by the controllers. From that point on, these parameters are reported by a controller only if one of the monitored conditions changes state.

3.7.2 Control Performance Evaluation Algorithms

Control performance evaluation has received significant attention from both academia and practitioners, particularly after the 1990s. Much of their work has focused on an assessment of control loop performance using a minimum variance controller as a reference. This approach was proposed by Harris (1989) and further developed by Desborough and Harris (1992, 1993), Eriksson and Isaksson (1994) and Isaksson (1996). Rhinehart (1992) explored simple ways of computing standard deviation from measurement-to-measurement deviations and filtering. Harris, Boudreau, and MacGregor (1996) and Huang, Shah, and Kwok (1997) provided guidelines for performance assessment of multivariable controllers. Shunta (1995) gave an excellent practical summary of performance monitoring.

This section provides details about the calculations used for evaluating control performance. Two control performance indexes are the most useful: *variability index* (*VI*) and *tuning index* (*TI*). While variability index is a measure of the quality of control, tuning index indicates the potential for improving control performance by updating controller tuning according to the identified process model and the selected or default tuning rule.

Through the variability and tuning index calculations, it is possible to better judge the plant control performance. The performance monitoring application shown in Figure 3-24 identifies control blocks whose variability index and total standard deviation exceed the configured limits. In most input/output function blocks, the total standard deviation and capability standard deviation are calculated as well. These two parameters are needed for variability index calculation. The conditions determined in the block are reported to the server. The server calculates performance, which is defined as the percent of time that control variability stays within limit, and utilization, which (as we have seen) is defined as the percentage of time that control blocks were operating in the normal (designed) mode.

Primary Performance Indexes

Standard deviation is the primary index for process variability assessment. In addition, standard deviation serves as a basis for calculating other, more sophisticated indexes. Standard deviation is calculated in every control and measurement block using a common formula (Equation 3-1), although in some implementations, the formula shown in Equation 3-3 is used instead to avoid a time-consuming square root calculation.

$$\sigma_{tot} = \sqrt{\frac{\sum_{i=1}^{n} (x(i) - \bar{x})^2}{n-1}} \tag{3-1}$$

where $\bar{x}$ is mean value of $x(i)$:

$$\bar{x} = \frac{1}{n}\sum_{i=1}^{n} x(i) \tag{3-2}$$

Standard deviation is computed continuously for a moving window of about 100 samples. In control blocks, either the working set point or the measurement value can be used as $\bar{x}$ depending on the block mode.

An alternate simplified formula for the standard deviation, which does not apply a square root calculation and is useful for an on-line application, is:

$$\sigma_{tot} = 1.25|\bar{\varepsilon}| \tag{3-3}$$

where $|\bar{\varepsilon}|$ is mean absolute error:

$$|\bar{\varepsilon}| = \frac{1}{n}\sum_{i=1}^{n} |x_i - \bar{x}| \tag{3-4}$$

Intermediate calculations are performed for each execution of the function block. The parameter values are then updated per every *n* executions of the function block. For a typical implementation, the update occurs after the function block executes 100 times. The measurement value is used in I/O blocks to calculate $\bar{x}$. In control blocks, either the working set point or the measurement value is used depending on the block mode.

The relation (Equation 3-3) between standard deviation and mean absolute error can be verified by computing a mean absolute value for the normalized Gaussian distribution, *p(x)*:

$$p(x) = \frac{1}{\sqrt{2\pi}}\int_{-\infty}^{\infty} e^{\frac{x^2}{2}} \tag{3-5}$$

$$|\bar{x}| = \frac{1}{\sqrt{2\pi}}\int_{-\infty}^{\infty} |x|e^{-\frac{x^2}{2}} = \frac{1}{\sqrt{2\pi}}\left(-\int_{-\infty}^{0} xe^{-\frac{x^2}{2}} + \int_{0}^{\infty} xe^{-\frac{x^2}{2}}\right) = \frac{1}{\sqrt{2\pi}}\left(e^{-\frac{x^2}{2}}\Big|_{-\infty}^{0} - e^{-\frac{x^2}{2}}\Big|_{0}^{\infty}\right) = \sqrt{\frac{2}{\pi}}$$

$$= .797\dot{8}845$$

Since for the normalized Gaussian distribution, standard deviation $\sigma = 1$

$$|\bar{x}| = .7978845\sigma \quad \text{or} \quad \sigma = 1.2533141|\bar{x}| \tag{3-6}$$

The capability standard deviation σ_{cap} is calculated as follows:

$$\sigma_{cap} = \frac{\bar{x}_{\Delta}}{1.128} \tag{3-7}$$

where $\bar{x}_{\Delta}$ is the average moving range for process parameter x with an estimated deadtime equal d scans:

$$\bar{x}_{\Delta} = \frac{1}{n-1-d}\sum_{i=2+d}^{n} |x(i) - x(i-1-d)| \tag{3-8}$$

Only the summing component associated with $\bar{x}$ and $\bar{x}_{\Delta}$ is performed for each execution. The division of the sum by n or n-1-d is done as a part of the σ_{tot} and σ_{cap} calculations only once every n executions (default $n = 120$). Capability standard deviation calculation requires that the sampling rate be fast enough. The requirements for the sampling rate are similar to those for the scan rate of a control loop. A practical estimate for selecting scan rate for control loops is sampling five or more times per process time constant.

To provide stable readings, new parameter values for all calculated parameters are filtered with coefficient f, $0 < f \leq 1$ as shown in Equation 3-9 for the mean value calculation:

$$\bar{x}' = \bar{x} + f\left(\frac{1}{n}\sum_{i=1}^{n} x_i - \bar{x}\right) \tag{3-9}$$

Variability Index

After σ_{cap} and σ_{tot} are calculated, the variability index (*VI*) is computed using formulas 3-10 and 3-11:

$$VI = 1 - \frac{\sigma_{lq} + s}{\sigma_{tot} + s} \tag{3-10}$$

where *s* is the sensitivity factor that stabilizes calculations, in particular for small values of σ_{tot}; default *s* = 0.1% of the variable scale. It has no effect on the result for higher values of σ_{tot}.

σ_{lq} is the minimum standard deviation that can be achieved with feedback control, and is defined as:

$$\sigma_{lq} = \sigma_{cap}\sqrt{2 - \left[\frac{\sigma_{cap}}{\sigma_{tot}}\right]^2} \tag{3-11}$$

The *variability index* reflects how close control performance comes to minimum variance control. In the implementation, variability index is expressed in percent.

There are several modified approaches for calculating variability index. The model-based approach assumes control loop and noise models and calculates for minimum variance control from the models' variability (Qin, 1998). Another approach estimates minimum achievable variability for PID, rather than for a minimum variance controller (Ko, Edgar, 1998). The estimated minimum control variability achievable by PID is used for developing the variability index.

Tuning Index

A PID controller with Internal Model Control (IMC) tuning applied to a first order-plus-deadtime process is equivalent to an IMC controller (Qin, 1998), (Ko, Edgar, 1998), therefore the minimum variance that the PID can achieve can be expressed as:

$$\sigma_{PID}^2 = \sigma_{lq}^2 + \sigma_{\Delta PID}^2 \tag{3-12}$$

Minimum PID variance is larger than minimum variance control σ_{lq}^2 on the term $\sigma_{\Delta PID}^2$.

$$\sigma^2_{\Delta PID} = \frac{(\tilde{c}_1 - c_1)^2}{1 - \tilde{c}_1^2}\sigma^2_w \qquad (3\text{-}13)$$

where

$\tilde{c}_1 = e^{-\frac{s}{\lambda}}$ = variance tuning term

s = sampling period

λ = required closed-loop time constant

c_1 = parameter of the assumed noise model $G_w = \dfrac{1 - c_1 z^{-1}}{1 - z^{-1}}$

$c_1 = e^{-\frac{s}{\lambda_{noise}}}$ where λ_{noise} is a white noise filter time constant $\lambda_{noise} < \lambda$

As a rough estimate it may be assumed $\lambda_{noise} \approx \min\{.2\lambda,\ .1\tau, s\}$.

σ^2_w = noise standard deviation

τ = process time constant

The capability standard deviation σ_{cap} calculated as in (3.7), (3.8) with assumed deadtime equal zero can be used as a practical estimate of σ_w.

Equation 3-12 is convenient for calculating a practical achievable minimum variance. The formula comprises two separate terms. The first term is a theoretical minimum variance σ^2_{lq}. This term can be calculated from the capability standard deviation as in Equation 3-7 using Equation 3-11. The second term is an increase in the variance due to the assumed realistic PID controller with IMC tuning. This term's calculation requires an estimate of three parameters: $\tilde{c}_1$, c_1, λ_{noise}.

The total variance σ^2_{tot} includes variance from stochastic and deterministic sources. The stochastic (random) variance comes from the measurement noise. The deterministic variance comes from the loop oscillatory operation, caused by process disturbances, process non-linearities or poor loop tuning – see section 3.3. The residual variance σ^2_{res} can be estimated as a difference between the current total variance and the minimum (non-compensating) PID variance for the assumed tuning:

$$\sigma_{tot}^2 = \sigma_{res}^2 + \sigma_{PID}^2 \tag{3-14}$$

The calculation of σ_{PID}^2 variance assumes a certain closed loop time constant λ. If the controller parameters in the actual PID are as assumed for σ_{PID}^2 variance calculation, σ_{res}^2 should be close to zero.

If $\sigma_{res}^2 > 0$, that is, $\sigma_{tot} > \sigma_{PID}$, one of the following conditions is possible:

1. The controller is poorly tuned. The actual λ used for the controller tuning is probably larger than that used in the formula (the tuning is too slow) or the controller tends to oscillate (inadequate Gain, Reset, or Rate).
2. The noise estimate made in the σ_{PID} calculation is low. It is possible that the noise content, which cannot be compensated for by the PID, is larger than the estimate.
3. The process model may not be exact and the σ_{PID} estimate does not correspond to a real process.
4. The assumed first order-plus-deadtime model may not be a perfect estimate for higher order real process models, process non-linearity, valve conditions etc.

In reality, therefore, $0 < \sigma_{res}^2 < \sigma_{resMAX}^2$ is accepted in most cases.

The non-compensated variance σ_{PID}^2 is smaller when tuning is better (in the variability sense). A smaller σ_{PID}^2 makes the total variance σ_{tot}^2 smaller.

The tuning index indicates the potential reduction of variability from stochastic disturbances due to the improved tuning. Because only non-compensated PID variability sources are included in the estimate, we may assume that $\Delta\sigma_{res}^2$ as in Equation 3-15 is a lower estimate of a possible variability change due to the different tuning of *PID1* and *PID2*.

$$\Delta\sigma_{res}^2 = \sigma_{PID1}^2 - \sigma_{PID2}^2 = \sigma_{MVC}^2 + \frac{(\tilde{c}_{PID1} - c_1)^2}{1 - \tilde{c}_{PID1}^2}\sigma_w^2 - \sigma_{MVC}^2 - \frac{(\tilde{c}_{PID2} - c_1)^2}{1 - \tilde{c}_{PID2}^2}\sigma_w^2$$

$$= \left(\frac{(\tilde{c}_{PID1} - c_1)^2}{1 - \tilde{c}_{PID1}^2} - \frac{(\tilde{c}_{PID2} - c_1)^2}{1 - \tilde{c}_{PID2}^2}\right)\sigma_w^2 \tag{3-15}$$

The ratio of the potential non-compensated PID variability reduction to the actual PID non-compensated variability is defined as Tuning Index *TI*, that is:

$$TI = \frac{\Delta\sigma_{res}^2}{\sigma_{PID1}^2} = \frac{\left(\frac{(\tilde{c}_{PID1} - c_1)^2}{1 - \tilde{c}_{PID1}^2} - \frac{(\tilde{c}_{PID2} - c_1)^2}{1 - \tilde{c}_{PID2}^2}\right)\sigma_w^2}{\frac{(\tilde{c}_{PID1} - c_1)^2}{1 - \tilde{c}_{PID1}^2}\sigma_w^2} = 1 - \frac{\frac{(\tilde{c}_{PID2} - c_1)^2}{1 - \tilde{c}_{PID2}^2}}{\frac{(\tilde{c}_{PID1} - c_1)^2}{1 - \tilde{c}_{PID1}^2}} \tag{3-16}$$

The applied max function normalizes *TI* in the range $-1 \le TI \le 1$.

$$TI = \max\left\{\left(1 - \frac{\frac{(\tilde{c}_{PID2} - c_1)^2}{1 - \tilde{c}_{PID2}^2}}{\frac{(\tilde{c}_{PID1} - c_1)^2}{1 - \tilde{c}_{PID1}^2}}\right), -1\right\} \tag{3-17}$$

In the implementation of the formula, the percent scale is usually used. To be consistent with the variability index scale, it is necessary to use the standard deviations ratio instead of the variances ratio:

$$TI = \max\left\{\left(1 - \frac{(\tilde{c}_{PID2} - c_1)\sqrt{1 - \tilde{c}_{PID1}^2}}{(\tilde{c}_{PID1} - c_1)\sqrt{1 - \tilde{c}_{PID2}^2}}\right), -1\right\}100\% \tag{3-18}$$

The tuning index estimates how much the non-compensating PID variability can be reduced by applying tuning *PID2* (it is tuning used in calculating variance σ_{PID2}^2). The negative *TI* indicates that the current *PID1* tuning provides smaller variability. In such cases the user might consider a tuning change that takes into account objectives other than variability objectives—for example loop stability and robustness or a required closed loop time constant.

The following chapters provide detailed information on a wide range of advanced control tools and procedures for identifying sources of process variability and improving control performance.

BIBLIOGRAPHY

1. Desborough, L. and Harris, T. J. "Performance Assessment Measures for Univariate Feedback Control," *Can. J. Chem. Eng.*, 70:1186, 1992.
2. Desborough, L. and Harris, T. J. "Performance Assessment Measures for Univariate Feedforward/Feedback Control," *Can. J. Chem. Eng.*, 71:605, 1993.
3. Eriksson, P. G. and Isaksson, A. J. "Some Aspects of Control Loop Performance Monitoring," Conference on Decision and Control (CDC) IEEE, Lake Buena Vista, FL, 1994.
4. Harris, T. "Assessment of Control Loop Performance," *Can. J. Chem. Eng.*, 67(10):856–861, 1989.
5. Harris, T., Boudreau, F., and MacGregor, J. F. "Performance Assessment of Multivariable Feedback Controllers," *Automatica*, 32(11): 1505–1518, 1996.
6. Huang, B., Shah, S. L., and Kwok, K.Y. "Good, Bad or Optimal? Performance Assessment of MIMO Processes," *Automatica*, 33(6): 1175–1183, 1997.
7. Isaksson, A. J. "PID Controller Performance Assessment," Control Systems Conference, Halifax, Canada, 1996, Conference Proceedings, pp. 163–169.
8. Ko, B. S. and Edgar, T. F. "Assessment of Achievable PI Control Performance for Linear Processes with Dead Time," In proceedings of American Control Conference (ACC), Philadelphia, PA, June, 1998, pp. 1548–52.
9. Qin, S. J. "Control Performance Monitoring – A Review and Assessment," *Computers and Chemical Engineering*, 23: 173–186, 1998.
10. Rhinehart, R. R. "A Cusum Type On-line Filter," *Process Control and Quality*, 2:169–179, 1992.
11. Shunta, J. *Achieving World Class Manufacturing Through Process Control,* Prentice Hall PTR, Upper Saddle River, NJ, (ISBN 0-13-309030-2), 1995.

4 On-Demand Tuning

During the commissioning of feedback control based on the PID algorithm, the PID tuning parameters (that is, the proportional, integral, and derivative gain) must be set to specific values to achieve the best controller response to setpoint and disturbance input changes. To minimize process variations and the response time to setpoint and disturbance input changes while providing stable operation, the PID tuning should be based on the observed process gain and process dynamics for each control loop. The instrumentation engineer involved in plant commissioning may not have correctly set the loop tuning, or the process operating conditions may have changed, and as a result plant operation does not achieve maximum efficiency. Fortunately, most distributed control systems (DCS) sold today include on-demand tuning support that can be used to automatically establish the correct loop tuning.

4.1 Process Identification

In most cases the on-demand tuning capability of a DCS, also known as auto-tuning capability, is based on the identification of the process model that matches the observed process response for a step change in a manipulated input. To identify the process response, it is necessary to test the process by initiating a change in the manipulated process input. The manner in which this is done varies with the on-demand identification technique used by the DCS manufacturer. The size of the change introduced in the manipulated process input must be large enough to easily distinguish the process response from any process noise that may be present. For a self-regulating process, the process response is most often approximated as first order-plus-deadtime, that is, the response is characterized by gain, time constant, and deadtime as illustrated in Figure 4-1.

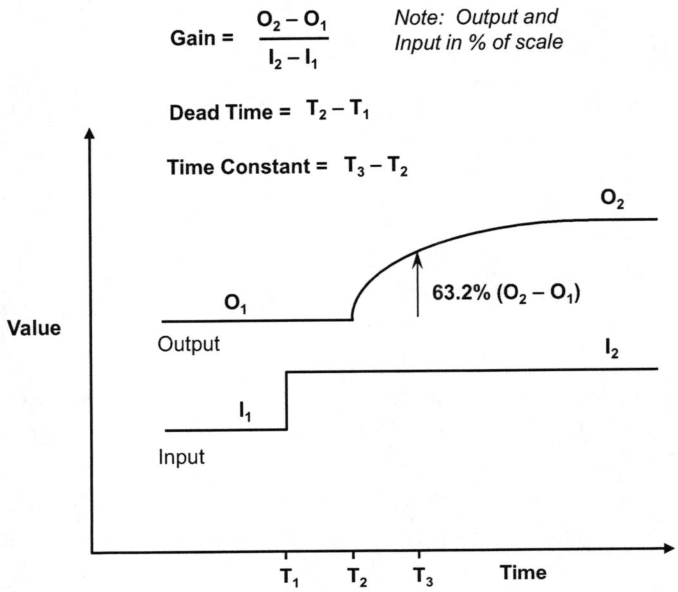

Figure 4-1. Characterization of a Self-Regulating Process

The response of an integrating process can be characterized by the process integrating gain and deadtime as illustrated in Figure 4-2.

Based on the identified response, tuning rules are applied to determine the recommended PID tuning.

When the on-demand tuning application resides in a DCS workstation, communications between the controller and the workstation can introduce variable delay and jitter in the observed process response. To allow the process to be accurately identified, the on-demand tuning functionality can be distributed between the workstation and the controller/field device as depicted in Figure 4-3.

Capturing process dynamics in the field (controller or device) instead of using test data transferred to the workstation eliminates bus communication for auto-tuning (see Section 4.1) and allows better process identification, particularly for the fastest loops. When auto-tuning is implemented in this manner, the tuning rules calculation may reside in the workstation. Process identification in the field device then requires only simple mathematical operations. Since the identification

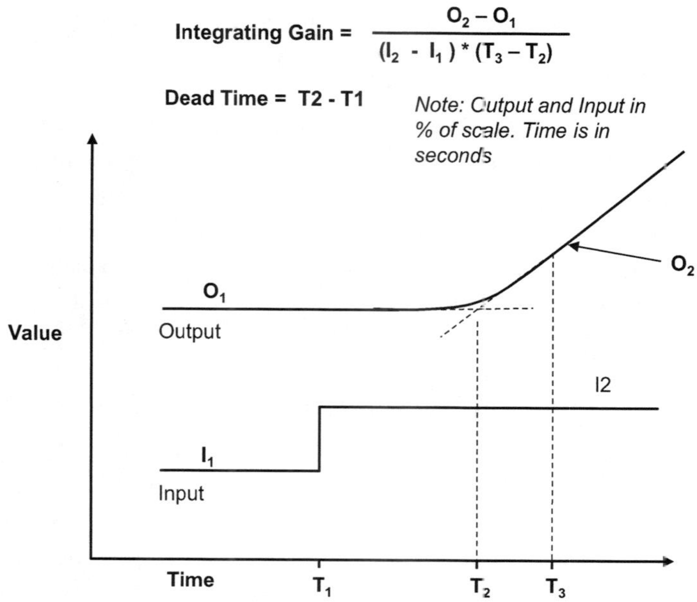

Figure 4-2. Characterization of an Integrating Process

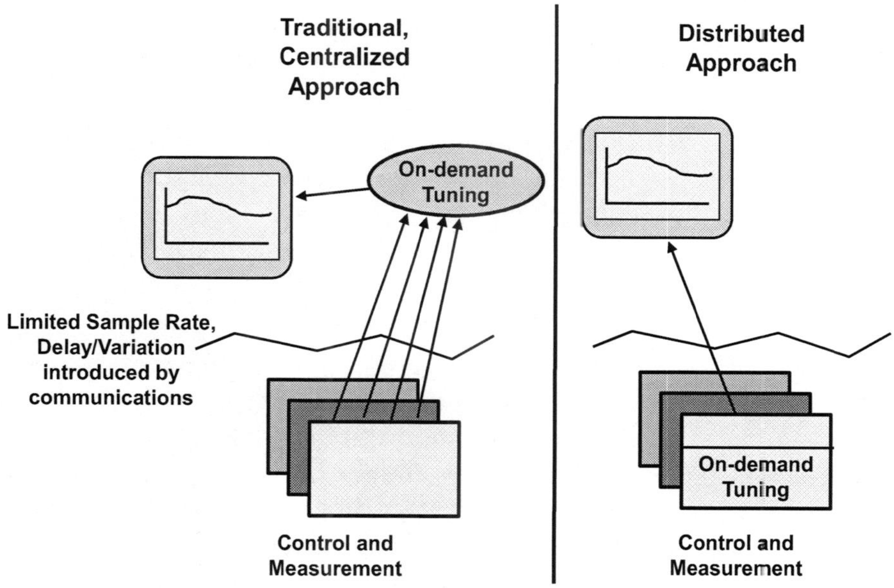

Figure 4-3. On-Demand Tuning Components

algorithm is no more CPU resource-consuming than the control algorithm, switching over to the identification algorithm does not result in an overall load increase in the field device.

Most on-demand tuning applications are designed to provide feedback indicating when process testing is in progress. Also, safety features are common and provide quick response to process conditions, such as too large a deviation of the controlled parameter from setpoint, which can indicate abnormal operating conditions, limit conditions, or significant process disturbance. These conditions will cause process identification to abort and normal control to resume.

Selecting Control Response

After the process characteristics are identified, the information is then used to calculate the recommended loop tuning. The tuning rules used to derive controller parameters from process characteristics can be chosen as appropriate for the specific process and the desired speed of response. The controller response to a setpoint or load disturbance change is determined by the controller tuning and can be classified as follows:

- **Overdamped response** – Controlled parameter is gradually restored without overshooting the setpoint value. In most cases an overdamped response is the best response for varying operating conditions that impact process gain and response time.
- **Critically damped response** – Controlled parameter is restored in minimum time without overshooting the setpoint value. While this response minimizes the time to respond to a change in setpoint or process disturbance, unstable control can be observed with changing operating conditions that impact process gain or response time.
- **Underdamped response** – Controlled parameter overshoots the setpoint value but eventually settles at the setpoint value. This response may minimize time to get back to setpoint but at the expense of stability and the controlled parameter overshooting to setpoint when responding to a change in setpoint or change in load disturbances.

Once the reset gain of the PID is set based on the identified process deadtime and time constant, it is possible to go from an overdamped response to a critically damped response or to an underdamped response by adjusting the PID proportional gain. Some on-demand tuning applications have a selection that changes the proportional gain to achieve a specified speed of response, for example, a fast, normal, or slow response.

As discussed in Chapter 3, when control loops are tuned, the following questions often arise: "What is the best tuning for a particular loop? Should control loop performance be measured based on the observed process response for setpoint change, or to a change in a load disturbance? What factors should be considered when judging control loop performance?"

The answers to these questions depend upon the loop objectives. If the setpoint is constantly being changed, the loop's response to setpoint changes would be a key factor in evaluating its performance. However, if the loop setpoint is always maintained at a fixed value, the response to disturbance inputs would be more important. If the response to both setpoint and disturbance inputs is important, tuning must be set to achieve a balance in control response. In many processes it is desirable to have an overdamped response. To achieve that response, the proportional gain used in the PID is typically much less than that required for an underdamped or critically damped response.

When evaluating the tuning automatically provided by an on-demand tuning application, it is important to consider that the process gain may change with the operating conditions. For example, the installed characteristic of a valve may provide a non-linear response to valve position in the valve operating range. The PID gain should be based on the operating conditions that provide maximum process gain. A change in process gain has the same effect on the control loop response as a change in the PID gain.

A mistake that is often made in tuning control loops is to examine loop control performance at only one operating point. For example, when a valve is operating at approximately 50% open, tuning can be adjusted to achieve an underdamped response. When the setpoint is changed to cause the valve to operate at 30% open, an entirely different response may be observed because the process gain has changed. As has been emphasized, it is wise to select a conservative PID gain setting, which will provide stable control even if the process gain changes with the setpoint. As a consequence, the "best" PID gain often gives a slower response to changes in setpoint and load disturbances. This conservative tuning will help ensure stable operation even if the process gain changes with changes in the operating setpoint. Most on-demand tools establish tuning based on the process characteristics and the user selection of speed of response.

To ensure stable operation over the entire operating range of a control loop, the normal practice is to establish the PID tuning when the process is operating in the region with the highest process gain. The worst thing that happens if the loop operating point changes to a low gain region is that the response to a setpoint change is sluggish, but stable. If a slower response does not meet process requirements, it may be necessary to use a characterizer to provide the response that would be provided by a final control element with linear installed characteristic.

When the process gain changes with operating conditions, the best control performance over the entire operating range would be achieved if the proportional gain were automatically changed based on the process gain at each point of operation. Some control systems provide a proportional gain scheduling feature that can be used to change the PID gain based on valve position or some other measured or calculated parameter that is related to the process gain. Setting up proportional gain scheduling is time consuming because the process gain must be established over the entire control operating range. However, as addressed in Chapter 5, adaptive control allows the tuning to be automatically changed to compensate for changes in process gain.

Application Example

As mentioned above, many of the newer control systems used in the process industry include on-demand tuning capability to automatically establish PID block tuning. On-demand tuning is based on automatically analyzing the controlled parameter's response to changes in the manipulated process input. Control system manufacturers have adopted different technologies to implement on-demand tuning. Because these tools are designed to work in a noisy process measurement environment, the tuning results are usually better than those achieved through manual tuning techniques. As an example, the on-demand tuning tool by one manufacturer will be examined in this chapter. Other tools may differ in technique used to introduce a change in the manipulated process input and the methods used to identify the process gain and process dynamics and to provide the recommended tuning.

Auto-tuning is designed to automatically determine the proportional, integral, and derivative gains and thus can save time in commissioning control loops. Since this capability will be used in many process applications, it must be designed to work with a variety of processes: self-regulating, integrating, fast or slow response. Protection is typically provided to compensate for noise in the measurements and the influence of unmeasured load disturbances. Also, for auto-tuning to be useful in initial plant startup, it is important to minimize the time required to tune a loop. The auto-tuning user interface will be used by a variety of people in the plant and must be designed to be easy to use.

The manufacturer can provide multiple ways to launch an auto-tuning tool. For example, one manufacturer's tool can be launched from the Start button, as an option in the function block view, or from the PID faceplate provided in the operator display. Also, when a PID block is selected in the performance monitoring application, selecting the Tune button in the lower corner of the interface accesses auto-tuning. When tuning is launched from the PID block's context menu, the auto-tuning interface is launched as shown in Figure 4-4.

The on-line values of the controlled parameter PV (top line in plot varying about the SP Value), the setpoint SP (top straight plot line), and the control output of the PID block (lower line in plots) are shown in the right portion of the screen in Figure 4-4 using specific colors. Also, a trend of these parameters is automatically displayed using the same colors. The current tuning of the PID is also shown in the right portion of the screen: proportional gain, reset (integral gain), and rate (derivative gain). The Test button is selected to initiate the PID tuning process. Once the tuning begins, the output of the PID is changed automatically from its initial value by the default step size (that is, the size of the step change) shown immediately above the Test button on the interface.

The default step size of 3% is appropriate for most processes. However, if the controlled parameter measurement contains significant process noise, a larger step size may be required. By increasing the step size, the resulting change in the controlled parameter is easier to detect and measure accurately. If the process gain is quite large, a smaller step change may be warranted to avoid the process upset often associated with a large change in a controlled parameter.

After the Test button has been selected and the initial step size for changes in the manipulated parameter has been specified, no further user input is required. Once a response is detected by the auto-tuning application and its magnitude achieves a certain predefined value, the manipulated parameter is changed in the opposite direction from its starting direction. After this second change in the manipulated parameter, no further action is taken until the controlled parameter responds to the manipulated parameter change. Each time the controlled parameter transitions through the value recorded at the start of the test, the manipulated parameter is automatically changed by the step size in the opposite direction from the last change made in the manipulated parameter, as illustrated in Figure 4-4.

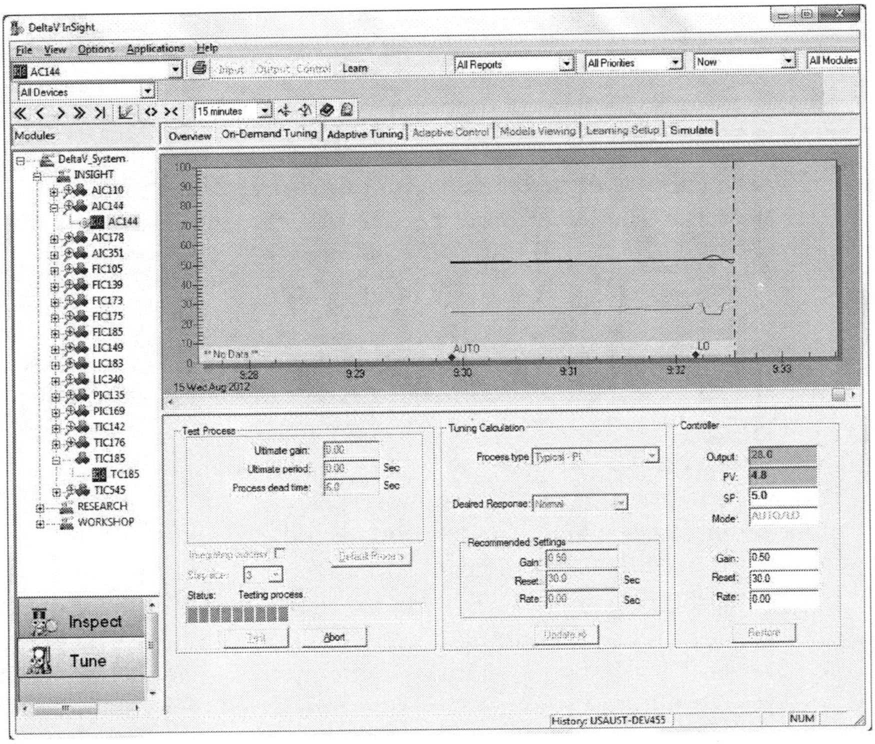

Figure 4-4. Automated Tuning of the Process

This technique for determining the parameters used for loop tuning is known as the relay oscillation technique (Åström and Hägglund, 1995 and 2006). The period of oscillation established by the control output switching defines the ultimate period.

- **Ultimate period** – Period of oscillation achieved for sustained oscillation when the proportional gain is set to the ultimate gain (with integral and derivative gain set to zero).

Based on the step size and the amplitude of the changes in the controlled parameter, it is possible for the auto-tuning application to calculate the ultimate gain.

- **Ultimate gain** – Proportional PID controller gain required to achieve sustained oscillation of the controlled and manipulated parameters (with integral and derivative gain set to zero).

Based on the ultimate gain, the ultimate period and the time required to see an initial response to a step change in the manipulated input, the auto-tuning application calculates the process gain, time constant, and deadtime (Wojsznis and Blevins, 1993). These parameters, as well as the ultimate gain and ultimate period, are shown in the left pane of the auto-tuner interface.

The ultimate gain and ultimate period can be determined manually by setting the integral and derivative gain to zero, then, with the PID in Automatic, gradually increasing the proportional gain in steps, changing the setpoint, and observing the response to the setpoint change. This procedure is repeated until sustained oscillation is observed (Ziegler and Nichols, 1962). However, with a slow-responding process this manual technique is very time consuming. The relay oscillation technique allows the process gain and dynamics to be determined quickly through automated testing. Also, the changes in the process input are known and thus can be selected as desired for a specific process to minimize the impact of testing on process variation.

Based on the process gain and dynamics identified through the automated test, the auto-tuning application can calculate a recommended setting for the PID proportional, integral, and derivative gains. To use the tuning determined by the application, the user simply selects the Update button, and the tuning values are automatically written to the PID block. Thus, with two clicks of the mouse, a control loop can be tuned.

To provide some flexibility in specifying the control loop response, the user can select the tuning method used to calculate the recommended tuning with the identified process response. This option is provided by selecting the Process Type and Desired Response in the Tuning Calculation area of the interface. Based on this selection and the identified process response, a recommended tuning setting is provided as illustrated in Figure 4-5.

Using the Tuning Method selection, the user can choose Typical PI or PID or select Pressure Loop, Temperature Loop, or Flow Loop. The recommended tuning is calculated based on the loop type selection and the identified process gain and dynamics. The tuning of a pressure loop might, for example, be different from the tuning of a temperature loop, even though each may have the same identified process dynamics and gain. In this way, the tuning may be matched to the process type. If the process type is left at the default (Typical PI), the tuning is calculated to provide a typical response that would be good for most processes. Similarly, the Desired Response selection allows the user to choose a slower or faster response to a change in setpoint and load disturbance.

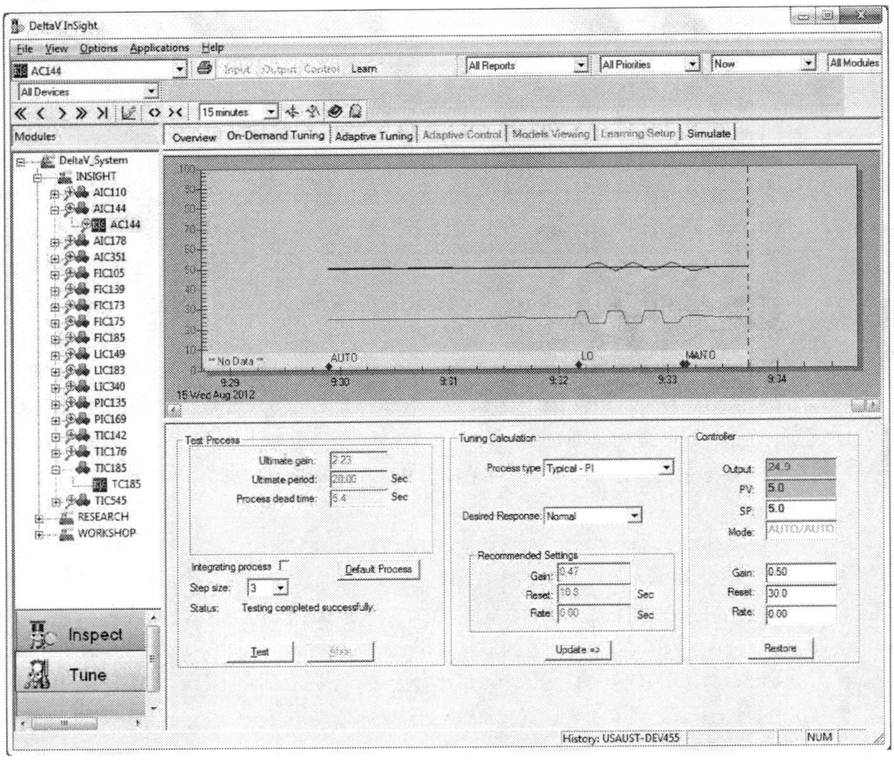

Figure 4-5. Tuning Recommendation Based on Tuning Method

4.1.1 Simulation of Loop Response

One of the most useful features provided by the auto-tuning application described above is the ability to examine the expected closed loop response before transferring the recommended tuning to the control loop. To view the expected closed loop response, the Simulate tab should be chosen after testing is complete. In response, a screen is displayed that includes a plot in the lower portion that shows the simulated closed loop response for a setpoint change, as illustrated in Figure 4-6.

By default, the closed loop response for a step change in setpoint is displayed. As illustrated in Figure 4-6, the change in the manipulated parameter and its impact on the controlled parameter are shown for the recommended tuning. When Disturbance is selected in the plot options, the closed loop response for a step change in an unmeasured process disturbance is displayed. If the process has been correctly identified during testing, the calculated (simulated) closed loop response should accurately reflect the actual closed loop response when the recommended tuning is transferred to the control loop.

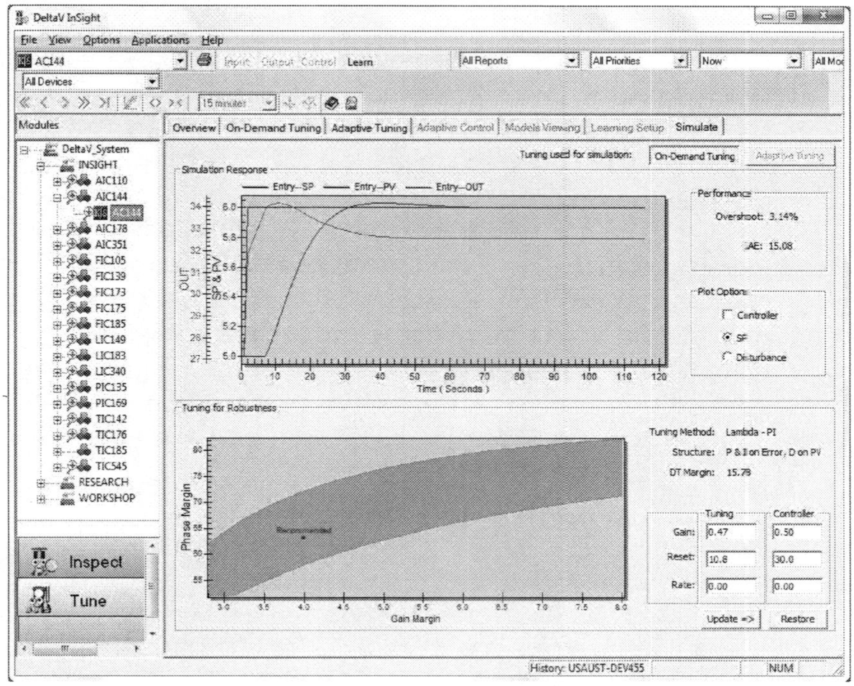

Figure 4-6. Simulation of Loop Response

The lower portion of the Simulate page contains a plot that shows the robustness of the closed loop control for recommended tuning.

- **Control Robustness** – A measure of how well control will cope with differences between the true process gain and dynamics and what was assumed in setting PID tuning.

The robustness plot (addressed in more detail in 4.3.3) shows the gain margin and phase margin for the recommended tuning. Also, the acceptable region of operation is shown by the dark band in the lower plot.

- **Gain Margin** – Change in process gain that would be required for an unstable closed loop response.
- **Phase Margin** – Added phase shift through the process that would be required for an unstable closed loop response.

(Gain margin and phase margin will also be addressed in more detail in 4.3.3.)

The robustness plot can be used to gain further insight into how the control loop will perform with the recommended tuning when the process gain or process dynamics change. The gain margin gives a direct indication of how much the process gain can change before the control loop becomes unstable. For example, if the recommended tuning is shown to have a gain margin of 3, when this tuning is applied, stable control will be observed even though the process gain changes by a factor of 3. If a gain margin of 5 is requested, that is, the process gain could change by a factor of 5, and the loop would still have stable control; a control engineer could click in the dark band on the robustness plot for a gain margin of 5 and see the tuning that provides this gain margin as well as a plot of the response. The impact of gain margin selection on the control response is shown in Figure 4-7.

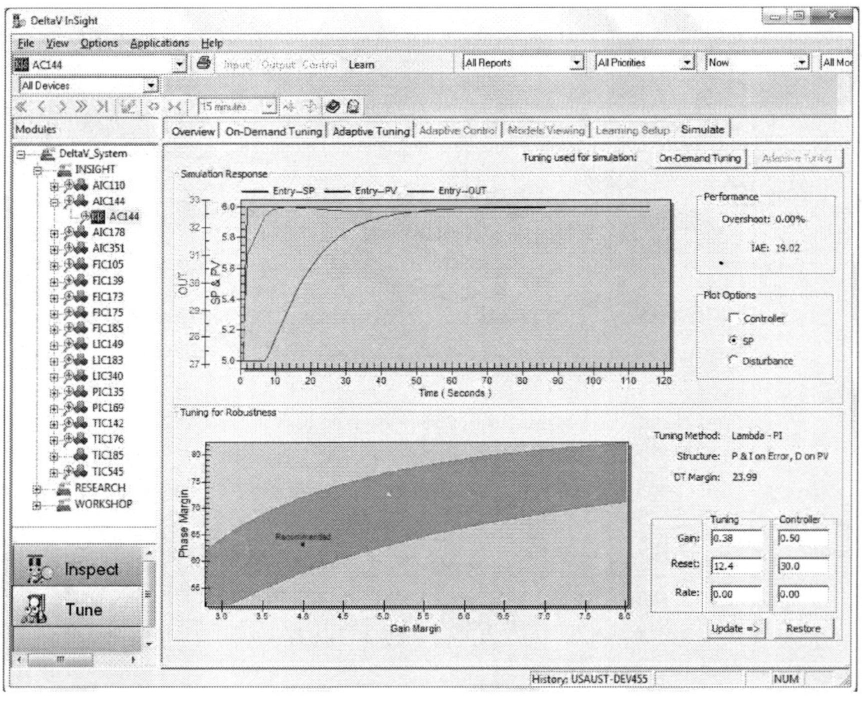

Figure 4-7. Impact of Gain Margin Selection on Control Response

Selecting an operation point in the left portion of the robustness plot provides higher proportional gain since the gain margin is smaller. Selecting an operation point in the lower portion of the robustness plot provides tuning with a lower phase margin. This means that if the process increases gain by more than the gain margin, an unstable response might be observed. Selecting a point associated with greater gain margin or phase margin makes control loop performance less susceptible to changes in process gain or dynamics.

The impact of a selection in the robustness plot is reflected in the plot of the response to a setpoint change, so by making different selections in the robustness plot it is possible to choose tuning that gives a desired response with a known gain and phase margin. Tuning determined by this method can be transferred to the recommended tuning through the selections provided in the upper portion of the screen. The robustness plot can be an effective tool not only for evaluating the recommended tuning but also for refining the recommended tuning to meet specific requirements.

4.2 On-Demand Tuning Workshop

This workshop illustrates some of the features that may be found in an on-demand tuning product. An example heater process is used in this workshop. Access the book's web site http://www.advancedcontrol-foundation.com and select the Solution tab to view the workshop being performed.

> Step 1: From the performance monitoring application, select the on-demand tuning interface. Change the PID mode to Manual and increase the PID output by 10%. Observe the process response shown in the trend of the PID control parameter, setpoint, and output.
>
> Step 2: Once the process has settled to a new steady state value, select Test and observe the change in the PID mode and the changes introduced in the PID output. As the test progresses, observe the process ultimate gain and period and the process gain, time constant, and deadtime that are identified based on the response.
>
> Step 3: Change the process type to PI and observe the recommended tuning.
>
> Step 4: Select the Simulate tab in the top portion of the interface and observe the predicted setpoint response provided by the recommended tuning. Select the Disturbance plot option to view the response to a change in a process disturbance.
>
> Step 5: Move the cursor over the green area of the robustness plot and left click in that area to select a spot

on the robustness plot. Observe the impact on the process response. Once a desired setpoint response is shown, select the Entry tuning as the recommended tuning.

Step 6: Select the Tuning tab and transfer the recommended tuning to the PID using the Update selection.

Step 7: Change the PID mode to Auto and change the setpoint by 10 °F. Does the observed process response match the predicted response seen in Step 5?

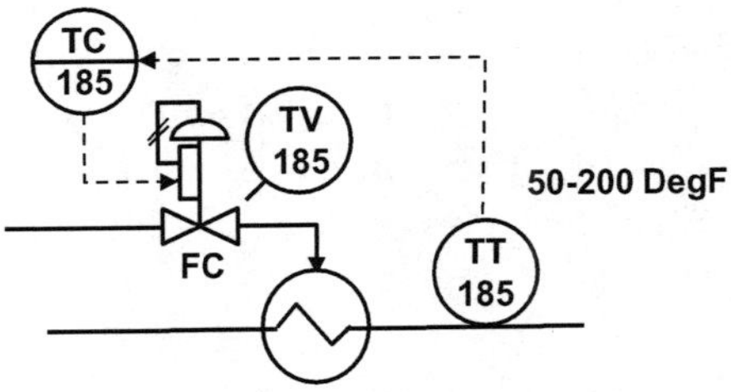

Figure 4-8. Steam Heater Process for PID Tuning Workshop

4.3 Technical Basis

Auto-tuners can provide a consistent automated procedure for finding the best tuning settings for a process control system. The procedure consists of three steps:

1. Test the process
2. Evaluate the process characteristics from test data
3. Calculate the controller parameters

The procedure emulates a typical manual tuning procedure. Let's take a look at an example of a manual procedure using the Ziegler-Nichols (ZN) method, the first industrial tuning procedure, introduced in 1942 (Ziegler, Nichols, 1962 and 1993). With this procedure a user has to gradually increase proportional controller gain to get limited-amplitude sustained oscillations, measure the period and the amplitude of the oscillations, and, finally, using ZN calculations, find the controller parameters. While control literature still cites this procedure verbatim, a significant practical modification that keeps the loop far from its stability limit is actually used in industry. In practice, the start of damped oscillations rather than sustained oscillations is sought. Even a single half wave suffices (McMillan, 1994, 2000). The period for these attenuated oscillations is longer than the loop actual ultimate period but is in the direction of a more conservative estimate of the integral time.

Another popular tuning technique, Lambda tuning, requires putting the controller in Manual and making a step change on the controller output, waiting for the full response, and calculating controller tuning parameters using Lambda tuning rules (Bialkowski, 1999).

While evaluating various auto-tuner techniques, both model-based and model-free, we should consider the following:

- Type of process where auto-tuning is to be applied (slow/fast, self- regulating/integrating/inverse response)
- Resistance to noise and disturbances
- Process disturbance magnitude required for process testing
- Process operation safety during tuning
- Tuning time
- Achieved controller performance

Both of the manual tuning techniques cited above can be applied to automatic tuning in principle; however, the requirements for process disturbance magnitude and safety are difficult to satisfy in practice. Increasing the controller gain to get just the start of an oscillation for the modified Ziegler-Nichols closed loop method requires close attention, and the process operation point may move during the time needed to reach steady state during an open loop test for Lambda tuning. In addition, a process disturbance will confuse the person making the test and measuring the results.

Åström and Hägglund (1984) and (1988) proposed using relay action to develop a process oscillation. This simple idea has many significant advantages and has been implemented in a number of industrial products (McMillan, Wojsznis, Meyer, 1993), (www.easydeltav.com). Applying relay action for tuning keeps the process under on-off control, which is inherently stable. The technique is robust and requires practically no settings. The amplitude of the oscillations is adjusted by relay step size and relay hysteresis.

Further advancement of the technique has been achieved by improving Ziegler-Nichols tuning formulas, by calculating a simplified first order-plus-deadtime process model from tuning data, and, finally, by identifying an exact process model (Yu, 1999).

4.3.1 Basics of Relay Oscillation Tuning

Process Testing

A two-state relay replaces the PID/PI controller for the period of process testing as shown in Figure 4-9. During testing the relay keeps the loop under two-state (on-off) control. The two-state controlled loop experiences oscillations with limited amplitude. The frequency of loop oscillations is the critical frequency, thus the period of these oscillations is the ultimate period. The amplitude of the oscillations and the relay step size are used for the calculation of the ultimate gain. The ultimate gain is defined as the proportional controller gain that brings the system to its stability limit. After these parameters are defined they are used to calculate controller parameters.

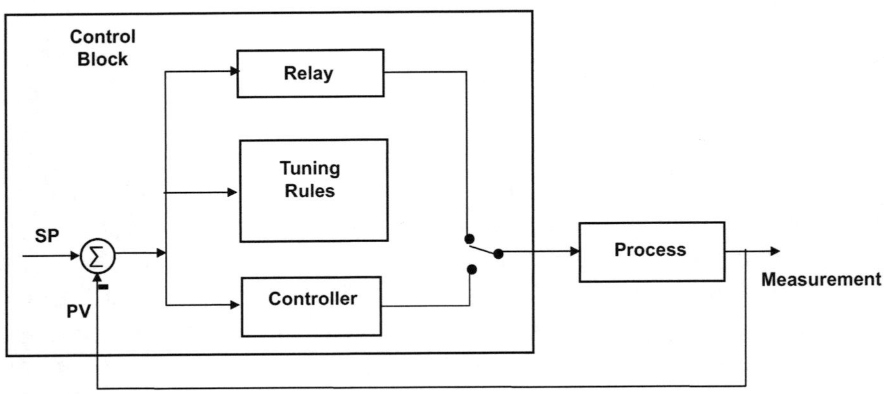

Figure 4-9. Relay Oscillation Method Block Diagram

A properly implemented relay oscillation auto-tuner adheres to the following rules:

1. Starts tuning when a loop is stable.
2. Collects data and computes various parameters during tuning and updates the calculation as tuning proceeds.
3. Supervises loop safety; interrupts tuning when loop safety is violated.
4. Uses averaging rather than instantaneous values in computing oscillation amplitude.

5. Applies noise protection techniques by using relay hysteresis, filters, and algorithmic noise protection.
6. Makes partial results available if tuning is interrupted.
7. Facilitates selection of various techniques for computing controller parameters.

Figure 4-10 illustrates a typical time plot of the relay output and the process variable (PV) during tuning. Note that after initialization the relay is triggered at the point when the PV passes through the set point. The relay amplitude ***d*** is typically 3 to 10% of the controller output range. The PV change ***a*** is largest during initialization (the first half-period of the sine wave). Typically, the PV change ranges from 1 to 3% of the PV range.

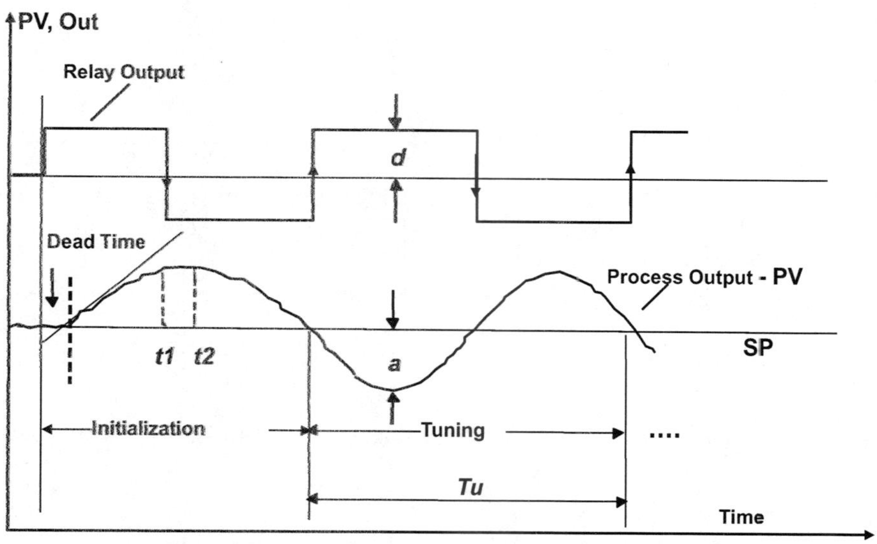

Figure 4-10. Relay Oscillation Tuning Plots

The oscillations must continue for at least one cycle after initialization. If more cycles are used for tuning, the average amplitude and period of the oscillations are used to determine the ultimate gain and ultimate period. Two cycles are usually adequate for defining the amplitude and period of the oscillations as the average values.

Evaluation of Process Characteristics

The frequency domain is used for loop analysis during tuning. Relay action is expressed by a describing function, $N(a, \varepsilon, d)$. The describing function approximates a non-linear element with a linear transfer function for the first harmonic frequency of the Fourier transform and is expressed at the critical frequency (Equation 4-1):

$$N(a, \varepsilon, d) = \frac{4d}{\pi\sqrt{a^2 - \varepsilon^2} + i\pi\varepsilon} \qquad (4\text{-}1)$$

where:

$N(a, \varepsilon, d)$ = relay-describing function at the critical frequency
ε = relay hysteresis
d = relay amplitude
a = amplitude of the oscillation of the process variable (PV)
i = imaginary unit vector

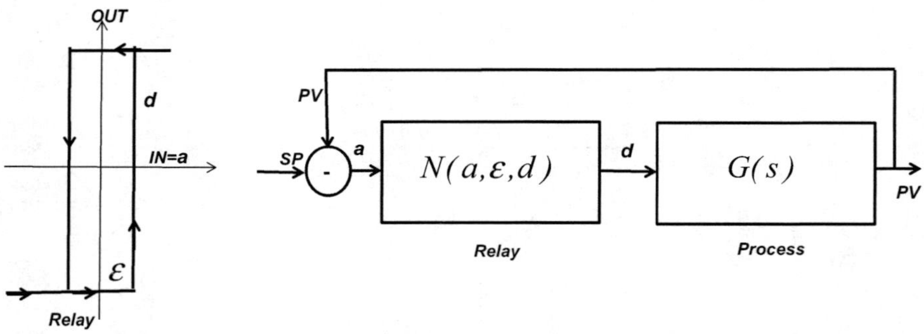

Figure 4-11. Relay with Hysteresis (left) and Loop Tuning Diagrams

Conditions for the oscillatory loop shown in Figure 4-11 are:

$$G(w_c)N(a, \varepsilon, d) = -1 \quad \text{or} \quad G(w_c) = \frac{-1}{N(a, \varepsilon, d)} \qquad (4\text{-}2)$$

$G(w_c)$ = process gain at the critical frequency

The Nyquist plot in Figure 4-12 shows the process transfer function $G(w)$ and negative inverse of the relay describing function with $\varepsilon = 0$ and with $\varepsilon > 0$. The cross-section of these two functions defines the loop oscillation parameters as shown in Equation 4-2.

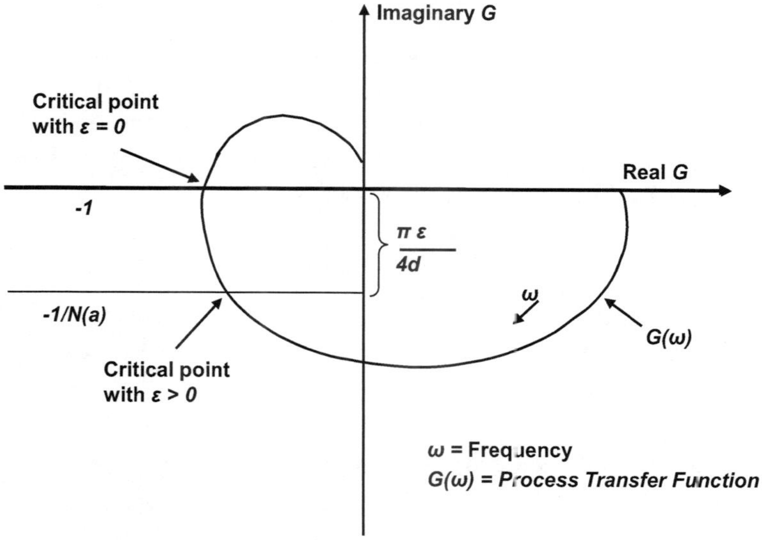

Figure 4-12. Relay Oscillation Tuning on Nyquist Plot

The $N(a, \varepsilon, d)$ amplitude as in Equation 4-1 is considered an approximation for the ultimate gain. Note that with $\varepsilon = 0$, $N(a, \varepsilon, d)$ is exactly the ultimate gain, as shown in Equation 4-3:

$$K_u = \frac{4d}{\pi a} \tag{4-3}$$

However, with hysteresis $\varepsilon > 0$, the gain and frequency determined by the oscillation are shifted from the original ultimate gain and ultimate period. As illustrated in Figure 4-12, the hysteresis used for noise protection introduces an error in defining the critical point on the Nyquist curve. To prevent the introduction of excessive errors, complementary noise protection techniques should be applied. A PV filter time constant equal to a properly selected loop scan period does not degrade loop performance and is beneficial to process test results. The application of an algorithmic noise protection technique disables relay switching at the start of each half period for about half of the initially defined deadtime (see Figure 4-10). The algorithmic noise protection technique has proven to be an effective means of noise protection without adversely affecting determination of the ultimate gain and ultimate period.

A significant advancement in relay oscillation tuning has been achieved by an estimate of the process deadtime from a linear regression to find the tangent line to the process step response curve (see the start of tuning in Figure 4-10). The tangent line is extrapolated to intercept the setpoint or process output value line at the tuning start. The time between the initial relay step and this intercept is the apparent deadtime. A complementary technique estimates deadtime as the delay of PV reaction after the relay output changes direction (see time interval t1–t2 in Figure 4-10). Finally, an apparent deadtime is calculated as the average of the two results.

Calculation of PID Controller Parameters

When the ultimate gain K_u and the ultimate period T_u have been identified, PID controller settings are determined by Ziegler-Nichols rules as in Table 4-1.

Table 4-1. Ziegler-Nichols Calculations of P, PI, and PID Controller Parameters

Controller\Parameter	K	T_i	T_d
P	$0.5K_u$		
PI	$0.4K_u$	$0.8T_u$	
PID	$0.6K_u$	$0.5T_u$	$0.1T_u$

Åström, Hägglund (1986) proposed more flexible rules with a phase and gain margin specification, as follows (Equations 4-4):

$$T_i = \frac{T_u}{4\pi\alpha}(\tan\phi + \sqrt{4\alpha + \tan^2\phi}) \tag{4-4}$$

$$T_d = \alpha T_i$$

$$K = K_u G_m \cos\phi$$

where

ϕ = desired phase margin

α = design selection of the ratio $T_d{:}T_i$

G_m = the desired inverse of gain margin (the default value .5)

K, T_i, T_d = controller gain, integral time and derivative time

Equations 4-4 give better tuning results than the original Ziegler-Nichols rules for processes with small deadtime-to-time constant ratios. If τ_d is defined during the process test deadtime, improved results are achieved over a wide range of process dynamics by using non-linear tuning rule estimators developed by Wojsznis, Blevins, and Thiele (1999). Non-linear estimators are intended to correct the following major deficiencies of Ziegler-Nichols tuning:

- Too short an integral time for processes with small deadtime-to-time constant ratios
- Excessive integral time for processes with large deadtime-to-time constant ratios
- Excessive controller gain for processes with small deadtime-to-time constant ratios

Non-linear tuning estimators have a general form as follows:

$$T_i = f_i(T_u, \tau_d)T_u; \qquad K = f_2(T_u, \tau_d)K_u; \qquad T_d = \alpha T_i; \tag{4-5}$$

Sigmoid functions are used for creating $f_1(T_u, \tau_d)$ and $f_2(T_u, \tau_d)$, a similar sigmoid function that is used for modeling a neural network (see Chapter 7).

The following formulas (Equations 4-6, 4-7) for the non-linear estimators were developed and tested by Wojsznis, Blevins, and Thiele (1999):

$$f_1(T_u, \tau_d) = a1 + \frac{b1}{1 + \exp-\left(\frac{T_u}{\tau_d} - c1\right)} \tag{4-6}$$

$$f_2(T_u, \tau_d) = a2 + \frac{(b2 - f1(T_u, \tau_d))}{c2} \tag{4-7}$$

a1, a2, b1, b2, c1, c2 are heuristic coefficients:

$$0.25 \le a1 \le 0.4 \quad 0.25 \le a2 \le 0.4 \quad b1 \sim 0.6, \quad b2 \sim 1.0, \quad c1 \sim 7.0, \quad c2 \sim 6.0$$

Equation 4-6 gives the value of the coefficient used for T_i computation, which smoothly varies from a minimal value of *a1* to a maximum value of *a1* + *b1* (upper plot in Figure 4-13). Equation 4-7 provides the coefficient for *K* computation in the range *[a2+(b2 – a1)/c, a2+(b2 – a1 - b1)/c]* (lower plot in Figure 4-13).

Let's take a closer look at the plot to better see the tuning dependencies. When ratio T_u/τ_d is close to 2, the process is deadtime-dominant.

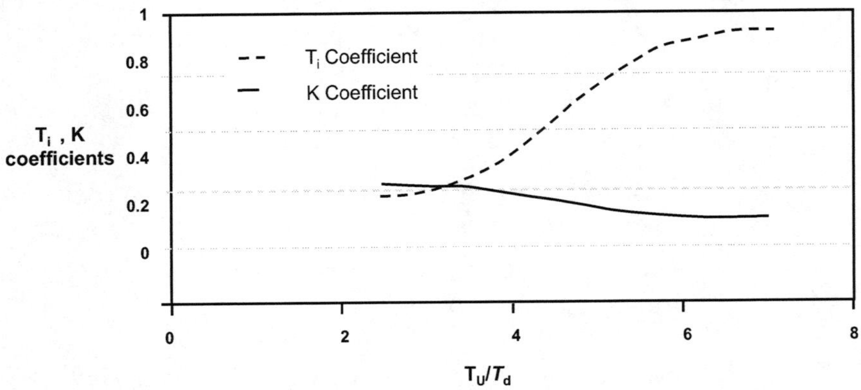

Figure 4-13. Non-Linear Functions for Computing Controller Integral Time and Gain

An increased T_u/τ_d ratio signifies a decreased τ_d/τ ratio, that is, a lag-dominant process.

The upper plot shows that for the deadtime-dominant process, $T_i \sim 0.35\ T_u$ while for a process with a small deadtime, $T_i \sim T_u$. The controller gain for a deadtime-dominant process is $K \sim 0.4\ K_u$, while for a process with a small deadtime it is $K \sim 0.27\ K_u$ (lower plot).

Implementation of non-linear tuning estimators results in significantly smoother, more robust loop responses that are close to the IMC or Lambda tuning (rather than to the Z-N quarter amplitude decay). Some typical step responses of loop tuning and loop step response for a second order process with $K_p = 1$, $\tau_1 = 10$ sec., $\tau_2 = 3$ sec., and $\tau_d = 2$ sec. are shown in Figure 4-14.

The auto-tuner gave the following controller settings for the loop in Figure 4-14: $K = 1.25$, $T_i = 12.36$ sec., and $T_d = 1.97$ sec., comparable to the IMC calculations with $\tau_{filter} = 13$ sec: $K \sim 0.87$, $T_i = 13$ sec. and $T_d = 2.3$ sec.

Finally, it should be noted that coefficients of the non-linear estimator can be modified to achieve a desired tuning. In this way, it is possible to achieve a tuning that provides an optimal setpoint response or the best disturbance rejection, or to a certain degree satisfies both objectives.

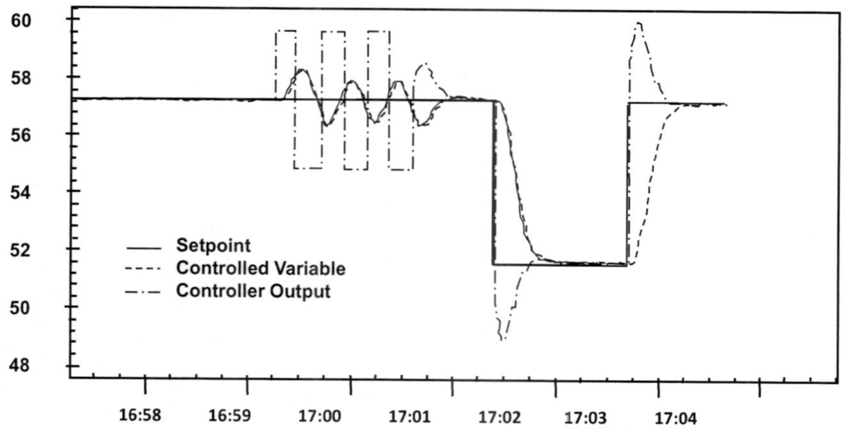

Figure 4-14. Tuning Plots and Loop Step Responses for a Process with $\tau_d/\tau \sim 0.2$, Loop Scan = 0.1 Sec.

Developing the Process Model

Having defined the process deadtime τ_d, ultimate gain K_u, and ultimate period T_u, it is possible to calculate the process time constant τ and process static gain K_p needed for model-based tuning.

One of the advantages of relay oscillation tuning is its applicability to both self-regulating and integrating processes. When the apparent deadtime is determined, the gain of an integrating process model is identified as the maximum slope of the process output during initialization – Figure 4-10. The calculation of a first order-plus-deadtime process model is detailed in Wojsznis and Blevins (1995), (Ott, Wojsznis, 1995) and (Kiong, Quing-Guo, Chieh, and Hägglund, 1999). Equation 4-8 is derived from the first order process model transformed into the frequency domain:

$$\tau = \frac{T_u}{2\pi}\tan\alpha \qquad \alpha = \pi - \frac{2\pi\tau_d}{T_u} < \frac{\pi}{2} \tag{4-8}$$

$$K_p = \frac{1}{K_u}\sqrt{1 + \frac{4\pi^2\tau^2}{T_u^2}}$$

where:

τ = process time constant

T_u = ultimate period

τ_d = process apparent deadtime
K_p = process static gain
K_u = ultimate gain

The process time constant in Equation 4-8 is expressed by a tangent function, which gives a good approximation when the deadtime is relatively large in relation to the time constant (that is, the argument of the tangent is less than $\pi/6$). For processes with insignificant deadtime, a large error results in the time constant computation, even for a small error in identified deadtime. In such cases the argument of the tangent is getting close to $\pi/2$ and a small error in argument changes the tangent value significantly.

The accuracy of model identification for a self-regulating process can be improved by identifying the process static gain through a closed loop step test. Following the relay oscillation test, a small setpoint change is made with the controller in automatic mode. The static gain is calculated as the ratio of percent change in setpoint to percent change in controller output between steady states.

$$K_s = \frac{\Delta SP}{\Delta u} \qquad (4\text{-}9)$$

where:

ΔSP = setpoint change (%)
Δu = controller output change from one steady state to a new steady state (%)

To detect valve hysteresis, the test should be repeated with a setpoint change in the opposite direction. Knowing the process static gain, ultimate gain, and ultimate period, the process deadtime and time constant are calculated from Equations 4-10:

$$\begin{aligned} \tau &= \frac{T_u}{2\pi}\sqrt{(K_p K_u)^2 - 1} \\ \tau_d &= \frac{T_u}{2\pi}\left(\pi - \tan^{-1}\frac{2\pi\tau}{T_u}\right) \end{aligned} \qquad (4\text{-}10)$$

Table 4-2 demonstrates the results of using these equations. The ultimate gain and ultimate period were determined for several first order-plus-deadtime process models using relay oscillation. The time constant and deadtime were calculated using Equations 4-10 and the actual static gain. The difference between actual and calculated time constant and deadtime averaged less than 5%.

Table 4-2. Comparison of Actual and Calculated Process Model from Static Gain, Ultimate Gain, and Ultimate Period

K_p actual	τ actual	τ_d actual	K_u	T_u	τ calc	τ_d calc
1.0	100.0	12.0	13.7	68.0	106.2	12.5
1.0	25.0	3.0	13.7	12.0	26.0	3.1
1.0	50.0	12.0	7.5	66.0	52.1	11.0
1.0	12.5	3.0	7.4	11.2	13.0	3.0
1.0	66.7	12.0	9.5	66.0	69.2	12.3
1.0	16.7	3.0	9.8	11.2	17.3	3.0
1.0	200.0	12.0	26.0	50.0	207.0	12.8
1.0	50.0	3.0	27.1	12.0	51.7	3.1
1.0	15.0	2.0	12.3	8.0	15.7	2.1
2.0	15.0	2.0	6.2	8.0	15.6	2.1
0.5	15.0	2.0	26.7	8.0	15.7	2.1

Better model accuracy is achieved by applying optimization techniques or least-square identifiers to the process test data. Using these techniques with the relay oscillation technique is limited by the requirement of process testing data collection faster than the original loop scan rate used in industry. The techniques are discussed in Chapter 5, 11, and 12.

4.3.2 Model-Based Tuning

Significant progress has been made with model-based tuning, particularly with Lambda and Internal Model Control (IMC) tuning (Chien and Fruehauf, 1990), (Bialkowski, 1999), (Bialkowski and Haggman, 1992). Both approaches result in a first order closed loop response to setpoint changes. A tuning parameter relating to the speed of response is used to vary the tradeoff between performance and robustness. Both approaches adjust the PID controller reset (or reset and rate) to cancel the process pole(s) and adjust the controller gain to achieve the desired closed loop response. IMC and Lambda tuning have become popular because oscillation and overshoot are avoided and control performance is specified in an intuitive way through the closed loop time constant. When there is mild to moderate interaction between loops, choosing the appropriate closed loop time constants with Lambda tuning can minimize the impact of the interactions.

IMC tuning rules for self-regulating processes use the three model parameters for a first order-plus-deadtime model and a user-defined filter time constant τ_{filter} to determine PID controller settings. The filter time constant should be set to the desired closed loop time constant. A larger value of the filter time constant gives a more damped tuning. Lambda tuning rules use a parameter λ, which is also used to set the desired closed loop time constant for a self-regulating process. The closed loop time constant is usually set to be longer than the open loop time constant, perhaps as much as two or three times longer. This is to ensure robustness in the event of inaccuracy in model identification and changing process conditions. Process analysis tools have been used to assist in choosing the closed loop time constant to achieve minimum variance control (Bialkowski, 1991). A proper closed loop time constant will attenuate slow disturbances without amplifying noise in the process measurement. In the following sections are formulas for IMC and Lambda calculations.

Self-Regulating Process

IMC Tuning Rules

PI Control:

$$K = \frac{\tau}{K_p(\tau_f + \tau_d)} \qquad T_i = \tau \tag{4-11}$$

where τ_f is a filter time constant equal to the required closed loop time constant.

PID Control (ISA standard form):

$$K = \frac{\tau + \tau_d/2}{K_p(\tau_f + \tau_d/2)} \qquad T_i = \tau + \tau_d/2 \qquad T_d = \frac{\tau\tau_d}{2\tau + \tau_d} \tag{4-12}$$

Lambda Tuning Rules

PI Control:

$$T_i = \tau \qquad K = \frac{T_i}{K_p(\lambda + \tau_d)} \tag{4-13}$$

where λ defines the desired closed loop time constant.

PID Control:

There is no direct form of Lambda tuning for the PID controller using a first order-plus-deadtime process model.

Integrating Process

Model-based control of integrating processes makes it possible to achieve averaging control. Rather than attempting to maintain tight control by aggressively moving the manipulated variable, averaging level control uses a surge tank, in the case of level control, to absorb infeed rate disturbances. By allowing the level to swing within limits, variability in the tank outlet flow is reduced, which minimizes disturbances to downstream processes. The IMC rule given below is intended for integrating processes with significant deadtime. The IMC and Lambda results are equivalent when there is no deadtime. For integrating processes, the tuning parameters τ_f and λ are the desired time for the disturbance effect to be arrested, that is, the time the process output begins to recover following a disturbance. Increasing the value of τ_f and λ makes control slower.

<u>IMC Tuning Rules</u>

PI control:

$$K = \frac{2\tau_f + \tau_d}{K_i(\tau_f + \tau_d/2)^2} \qquad T_i = 2\tau_f + \tau_d \tag{4-14}$$

where

$$K_i = \frac{\Delta y}{\Delta u \Delta t} = \text{ integrating process gain}$$

Δy	=	% change in the process output
Δu	=	% of step change in the process input
Δt	=	time interval over which Δy is calculated

<u>Lambda Tuning Rules</u>

PI control:

$$T_i = 2\lambda + \tau_d \qquad K = \frac{T_i}{K_i(\lambda + \tau_d)^2} \tag{4-15}$$

Model based tuning using the IMC or Lambda rules has some advantages over other tuning rules. Controllers are less sensitive to noise, valve life is prolonged, and process variability is minimized. However, the pole-zero cancellation approach results in poor load disturbance performance when the process time constant is long. A suggested means of overcoming this is to apply the integrating process model in place of the self-regulating model. Recall that the relay oscillation auto-tuning technique is capable of identifying both models for a self-regulating process – (Equation 4-8). Using the IMC rule for an integrating process can improve the performance of a self-regulating process with a long time constant.

Model-based tuning use is extended to the deadtime dominant processes (deadtime to time constant ratio is larger than 0.3), by applying Smith Predictor deadtime-compensating PID controller. Smith Predictor was first introduced by Otto Smith in 1957. The PI controller settings are determined using IMC or Lambda rules, applying the gain and time constant from the process model, with little or no deadtime.

The relay oscillation tuning method can also be used to auto-tune a fuzzy logic PID controller. The process parameters needed to tune the scaling factors of a fuzzy logic PID are the ultimate period, ultimate gain, apparent deadtime, and apparent time constant. Details are discussed in Chapter 6.

The discussed formulas 4-5 and 4-11 to 4-15 provide tuning for the ISA standard form controller. If a series PID controller is in use, the parameters should be appropriately converted:

$$K^S = \frac{K^n}{2}\left[1 + \sqrt{1 - \frac{4T_d^n}{T_i^n}}\right]; \quad T_i^S = \frac{T_i^n}{2}\left[1 + \sqrt{1 - \frac{4T_d^n}{T_i^n}}\right]; \quad T_d^S = \frac{T_i^n}{2}\left[1 - \sqrt{1 - \frac{4T_d^n}{T_i^n}}\right] \tag{4-16}$$

Note: Superscript *s* is used with series controller parameters, and superscript *n* is used with standard ISA controller parameters.

If *Td = 0* (PI controller), parameters for both controllers are identical.

The formulas to convert from series form to parallel are:

$$K^n = K^S\frac{T_i^S + T_d^S}{T_i^S}; \qquad T_i^n = T_i^S + T_d^S; \qquad T_d^n = \frac{T_i^S T_d^S}{T_i^S + T_d^S} \tag{4-17}$$

4.3.3 Robustness-Based Tuning

Before tuning calculations are accepted, it is important to know how much the process gain and deadtime can change before a loop becomes unstable. The answer is obtained from a robustness plot as shown in Figure 4-15. The robustness plot uses a logarithmic scale (Shinskey, 1990). The plot shows the stable area for a loop and a point with the loop's nominal deadtime and gain ratios. Around the point is an area of assumed changes of process deadtime and gain. The bounds are set for an increase in deadtime and gain, which move control toward the stability limit, and for a decrease in deadtime and gain, which move control toward being too sluggish. Two times the increase or decrease in the gain and deadtime is usually assumed. For a properly tuned loop (Loop 2) the area of assumed process-parameter change should be within the loop's stable area; otherwise (see Loop 1) the controller tuning parameters should be adjusted to expand the stable area.

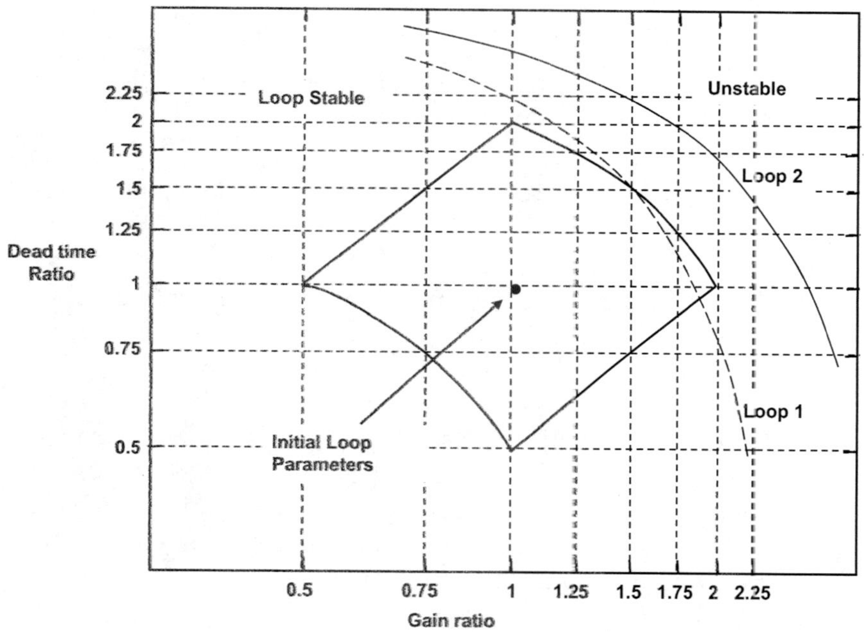

Figure 4-15. Robustness Plot – Loop Stable Area and Controller Operation Area

Gudaz and Zhang (2001) developed a practical way of graphical interactive tuning using a robustness map. The detailed explanation of the graphical interactive tuning follows. The robustness map uses loop gain margin and phase margin coordinates instead of process gain ratio and deadtime ratio coordinates as in the robustness plot.

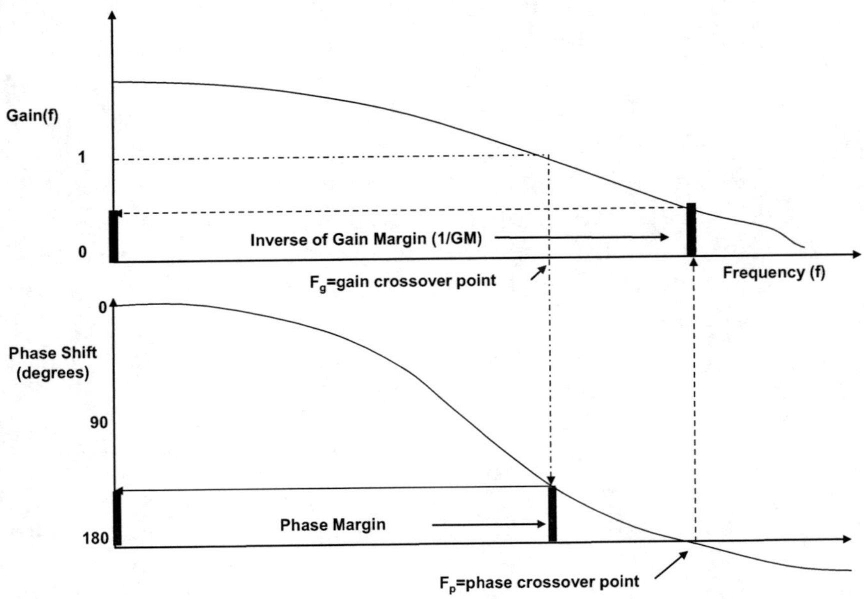

Figure 4-16. Gain Margin and Phase Margin on Bode Plot

It would be useful at this point to more fully define gain margin and phase margin and illustrate the definitions on the Bode plot in Figure 4-16.

Gain margin is the inverse of loop gain *K* at the phase crossover frequency, that is, when the loop phase shift is 180 degrees. It simply means how many times loop gain is increased to reach a stability limit at the crossover frequency. Since stability limit gain at the crossover frequency is equal to 1, we can write:

$$K^*GM = 1; \qquad GM = 1/K \tag{4-18}$$

Phase margin is the phase shift that is added to the loop phase shift at the gain crossover frequency to make a total phase shift of 180 degrees. The gain crossover frequency is the frequency when the loop gain is equal to 1.

A sample robustness map is shown in Figure 4-17. A robustness map presents a range of reasonable tuning parameters from which the user can select. Tuning objectives bound the tuning parameter range. The bounds for Figure 4-17 are defined by:

1. Tuning for near zero overshoot on a step change in setpoint (IMC tuning)
2. Near minimum integral absolute error (IAE) for a step disturbance

The robustness coordinates are in terms of gain margin and phase margin. A variety of available tuning parameters are in the delineated tuning area. Users can simply click a point to see a set of tuning parameters and view the responses. A setpoint performance range at a gain margin of 3 and 6 is shown in Figure 4-7 (the vertical number sequence on the robustness map correlates phase margin to response curve). Significant variation in the setpoint overshoot in Figure 4-18 for different tuning points on the robustness map in Figure 4-17 illustrates that not only are there performance and robustness tradeoffs to be made, there are performance tradeoffs between the speed of the approach to setpoint and overshoot as well.

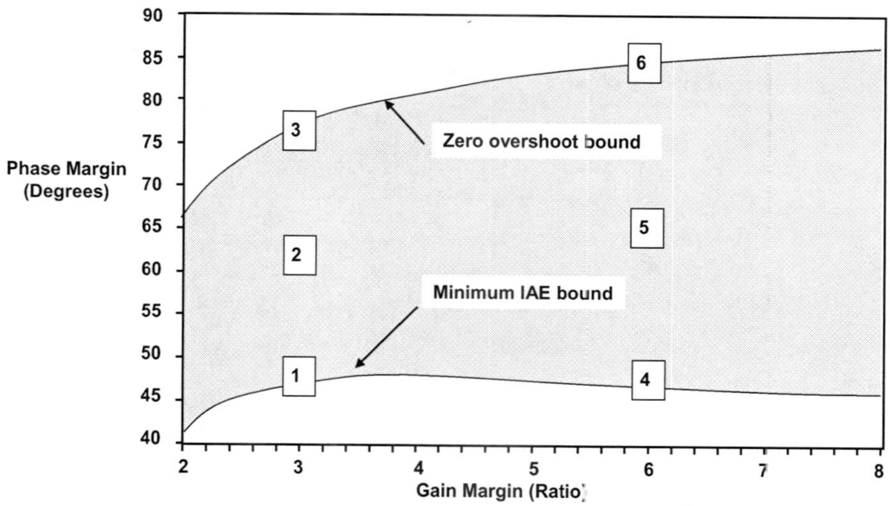

Figure 4-17. Robustness Map at Gain Margin = 3 and 6

There are no particular rules for choosing the response objectives. Using well-known tuning rules (such as Lambda and IMC) to establish a setpoint response boundary has proven to be a sound approach and presents rules for which there are a documented rationale and performance. Depending upon the presentation preference, more than two defining objectives could be presented and those that appear on the map could be user-selectable. Robustness map shape depends upon the process type and user-specified tuning preference and controller structure (PD, I, PID, etc.), regardless of the specifically defined tuning objectives. Some combinations (Proportional only, Integral only, for example) result in a robustness map that is simply a line.

While the robustness map is meant to define a range of tuning parameters, it is not necessarily meant to create hard constraints at the boundaries of that range. As a practical matter, a robustness map is intended to provide a tuning range from which good tuning is extracted.

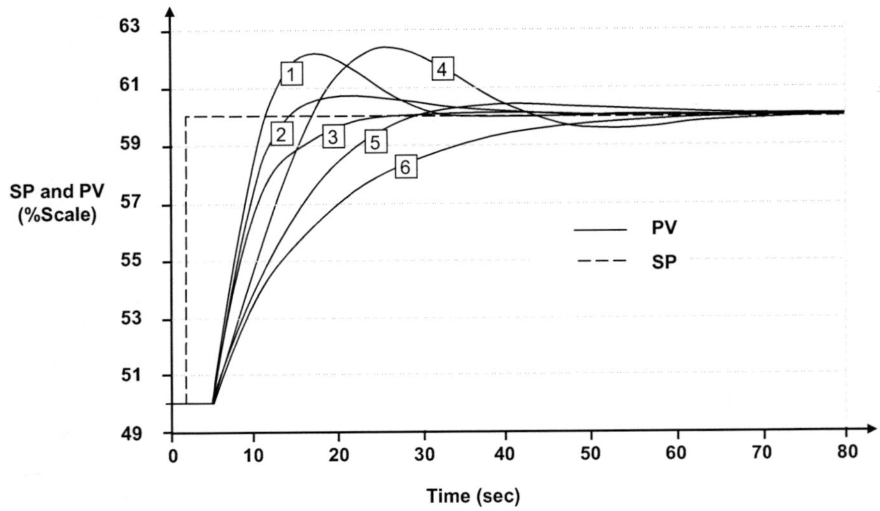

Figure 4-18. PV Responses to SP Step Change for Gain Margins from 1 to 6

For some process types, performance measurements other than IAE or overshoot are more relevant to the functioning of the loop. An example is an integrating process, where surge percentage and arrest time are essential performance indicators. In such cases the addition of these performance measurements will substantially enhance the robustness-based tuning utility.

Combining performance and robustness requires performance plots as well as the values of characteristic measurements. Response plots are less abstract than numerical measurements and provide a form of definition for the measurements. The simulation needs to show setpoint response and disturbance response. Disturbance responses should be in terms of defined types of process disturbances and could include noise. A robustness map provides the means for point-and-click tuning, but, without a plotted response, it is difficult to make a judgment as to whether or not the tuning is acceptable. Fast simulation and plotting allow the user to view a large number of responses in seconds. This makes the overall presentation more comfortable and interactive and enhances the robustness-based tuning utility use as a plant floor tuning tool.

4.3.4 Some Alternate Tuning Approaches

Most of the tuning techniques discussed in the preceding sections are commonly used in industrial practice. Model-based tuning rules in particular are easy to understand and interpret, making them the most popular despite their less than desirable performance on disturbance rejection. In recent years several improved versions of model-based tuning have been proposed. Below we present the basics of the two that seem to be the most promising approaches: direct synthesis tuning and simple analytic rules.

Direct synthesis tuning is based on the specification of the desired transfer function for the setpoint changes or disturbance rejection. The tuning rules for the setpoint changes developed by direct synthesis are identical with IMC tuning. Chen and Seborg (2002) applied direct synthesis and developed tuning rules for disturbance rejection which provide response for disturbances close to the Ziegler-Nichols tuning. The response to the setpoint is adjustable and may have some overshoot or no overshoot. The direct synthesis tuning rules for first order-plus-deadtime process and a PI controller are:

$$K = \frac{\tau^2 + \tau\tau_d - (\tau_f - \tau)^2}{K_p(\tau_f + \tau_d)^2} \qquad T_i = \frac{\tau^2 + \tau\tau_d - (\tau_f - \tau)^2}{\tau_f + \tau_d} \tag{4-19}$$

The reader can find more on direct synthesis tuning rules in Seborg, Edgar, and Mellichamp (2004).

It is apparent from Equation 4-19 that direct synthesis rules are more complex than IMC rules and not so intuitive. This might explain why these rules are not in wide practical use.

Simple analytic rules were derived by Skogestad (2003) using direct synthesis and then applying heuristic evaluation of how to modify the formulas to improve response on disturbance rejection. An objective was to develop simple and easy-to-memorize rules as well. The resulting formulas for the first order process and the integrating process and PI controller are only slighter more complex than IMC rules:

$$K = \frac{\tau}{K_p(\tau_f + \tau_d)} \qquad T_i = min\{\tau, 4(\tau_f + \tau_d)\} \tag{4-20}$$

$$K = \frac{1}{K_i(\tau_f + \tau_d)} \qquad T_i = 4(\tau_f + \tau_d) \tag{4-21}$$

PID series controllers used with a second order-plus-deadtime process should apply the derivative term $T_d^S = \tau_2$, τ_2 as the second lag of the process. The PID series controller gain and integral term are the same as those of the PI controller. More information on analytic tuning rules is in (Skogestad 2003).

The tuning applications for specific processes provide an opportunity and sometimes a need for rules adjustments. A good illustration of such an approach is improved control performance for slow, self-regulating processes by tuning these processes as integrators. An additional advantage of this tuning model is that only two model parameters are needed for the tuning, that is, only deadtime and the slope of the step response. This type of integrating model is easier to identify than a three-parameter model for a self-regulating process – see Equation 4-8, (Wojsznis, Blevins, 1995), and Figure 4-10.

BIBLIOGRAPHY

1. Åström, K. J. and Hägglund T. *PID Controllers: Theory, Design, and Tuning,* Research Triangle Park: ISA, 1995.
2. Åström, K. J. and Hägglund, T. *Automatic Tuning of PID Controllers,* Research Triangle Park: ISA, 1988.
3. Åström, K. J. and Hägglund, T. "Automatic Tuning of Simple Regulators with Specifications on Phase and Amplitude Margins," *Automatica*, 20, pp. 665–651,1986.
4. Åström, K. J. and Hägglund, T. *Advanced PID Control,* Research Triangle Park: ISA, 2006.
5. Åström, K. J. and Hägglund, T."A Frequency Domain Method for Automatic Tuning of Simple Feedback Loops," *Proceedings* of the 23rd IEEE Conference on Decision and Control, Las Vegas, Dec. 1984.
6. Bialkowski, W. L. and Haggman, B. "Quarter Amplitude Damping Method Is No Longer the Industry Standard," *American Papermaker*, Vol. 55, No. 3, pp. 56–58, March 1992.
7. Bialkowski, W. L. "Plant Analysis, Design, and Tuning for Uniform Manufacturing," *Process/Industrial Instruments and Control Handbook*, McMillan, G.K., editor, McGraw-Hill, New York, 1999.

8. Chen, D. and Seborg, D. "PI/PID Controller Design Based on Direct Synthesis and Disturbance Rejection," *Ind. Eng. Chem. Res.*, 2002, 41, pp. 4807–4822.
9. Chien, I. L. and Fruehauf, P. S. "Consider IMC Tuning to Improve Controller Performance," *Chemical Engineering Progress*, Vol. 86, No.10, pp. 33–41, October 1990.
10. *Emerson Process Management Auto Tuning* - www.easydeltav.com.
11. Gudaz, J. and Zhang, Y. "Robustness Based Loop Tuning," *Proceeding of* ISA Conference, Houston, September 2001.
12. Kiong, T. K., Quing-Guo, W., Chieh, H. C., and Hägglund, T. *Advances in PID Control*, Springer, London – New York, 1999.
13. McMillan, G. K. *Pocket Guide to Good Tuning*, Research Triangle Park: ISA, 2000.
14. McMillan, G. K. *Tuning and Control Loop Performance — A Practitioner's Guide*, 3rd Edition, Research Triangle Park: ISA, 1994.
15. McMillan, G. K., Wojsznis, W. K. and Meyer, K. "Easy Tuner for DCS," *Proceedings of* ISA Conference, Chicago, 1993.
16. Ott, M. and Wojsznis, W. "Auto-Tuning: From Ziegler-Nichols to Model Based Rules," *Proceedings Instrumentation and Control*: ISA/95 Conference, New Orleans, 1995.
17. Seborg, D., Edgar, T. F., and Mellichamp, D. A. *Process Dynamics and Control*, John Wiley & Sons, Hoboken, N.J., 2004.
18. Shinskey, F. G. "Putting Controllers to the Test," *Chemical Engineering*, Vol. 97, pp. 96–106, December, 1990.
19. Skogestad, S. "Simple Analytic Rules for Model Reduction and PID Controller Tuning," *Journal of Process Control*, 13, 2003, pp. 291-309.
20. Wojsznis, W. K. and Blevins, T. L. "System and Method for Automatically Tuning a Process Controller," U.S. Patent No. 5,453,925, 1995.
21. Wojsznis, W. K., Blevins, T. L., and Thiele, D. "Neural Network Assisted Control Loop Tuner," *Proceedings of IEEE Conference on Control Applications*, Hawaii, July, 1999.
22. Yu, C.-C. *Autotuning of PID Controllers*, Springer, London, New York, 1999.
23. Ziegler, J. G. and Nichols, N. B. "Optimum Settings for Automatic Controllers," *Transactions of the ASME*, Vol. 115, June 1993, pp. 220–222.
24. Ziegler, J. G. and Nichols, N. B. "Optimum Settings for Automatic Controllers," *Transactions of the ASME*, Vol. 66, Nov. 1962, pp. 759–768.

5 Adaptive Tuning

As addressed in Chapter 4, the introduction of on-demand tuning had a large impact on the manner in which control loops are commissioned. Using on-demand tuning, it is possible to quickly establish the tuning associated with processes that exhibit fast response, such as liquid flow and liquid pressure. However, the test time required to determine the tuning of a slow responding process can often be quite large. In addition, the speed of response of temperature, level, or composition within a vessel to process operating condition changes depends in many cases on the process flow rate and the volume of the vessel. Thus, if the process operating conditions change during the on-demand test, it might be necessary to repeat the test every time to ensure that the process gain and dynamics are accurately identified for the changed process conditions.

In some cases, the tuning established at one operating point may not provide the best control for the full operating range. For example, changes in flow rate can impact transport delay and the time required for heating and mixing operations in the process. Also, a non-linear installed valve characteristic will cause the process gain seen by the PID to change with the valve position. When multiple valves are adjusted by the PID using split range output, the process gain and dynamics associated with a change in one valve may be different from that seen when another valve is adjusted. To address these issues, adaptive tuning may be used to automatically identify the process gain and dynamics and establish loop tuning based on the operational loop setpoint or output changes.

Adaptive control can be used to improve loop tuning by:

- Establishing initial loop tuning without testing process, this is particularly useful when the process output responds slowly to controller output changes in the process input.
- Identifying changes in process gain and dynamics as the process operating conditions change.
- Automatically changing loop tuning to compensate for changes in plant operating conditions.

Process examples will be used to illustrate where adaptive tuning can be used to establish the tuning of a process that is characterized by slow response to input changes. Also, examples will be used to illustrate control applications where the process gain or dynamics associated with a control loop change with changes in plant operating conditions.

5.1 Adaptive Control – Examples

5.1.1 Continuous Reactor Control

The temperature within a continuous reactor responds slowly to a change in the coolant flow because of the volume of liquid within the reactor. For example, the reactor temperature loop TC305 (as illustrated in Figure 5-1) will exhibit a slow response to a change in the recirculation stream temperature maintained by TC306, by manipulating coolant flow.

The initial tuning of the temperature control loop TC305 can be established using auto-tuning. However, when process conditions change by a significant amount, the tuning established at one operating point may not be appropriate for best control. For example, if the reactor feed rate or the coolant temperature changes significantly during normal plant operation, adaptive control may be required to compensate for the changes in process gain and dynamics. A change in process gain and dynamics with a change in throughput is one of the reasons that control utilization decreases when there is a significant reduction in production rate.

5.1.2 Batch Reactor

Batch reactors are used in the process industry to produce a variety of products. The pressure in the reactor is often maintained by regulating a vent that is designed to release volatile materials that are generated by the reaction, as illustrated in Figure 5-2.

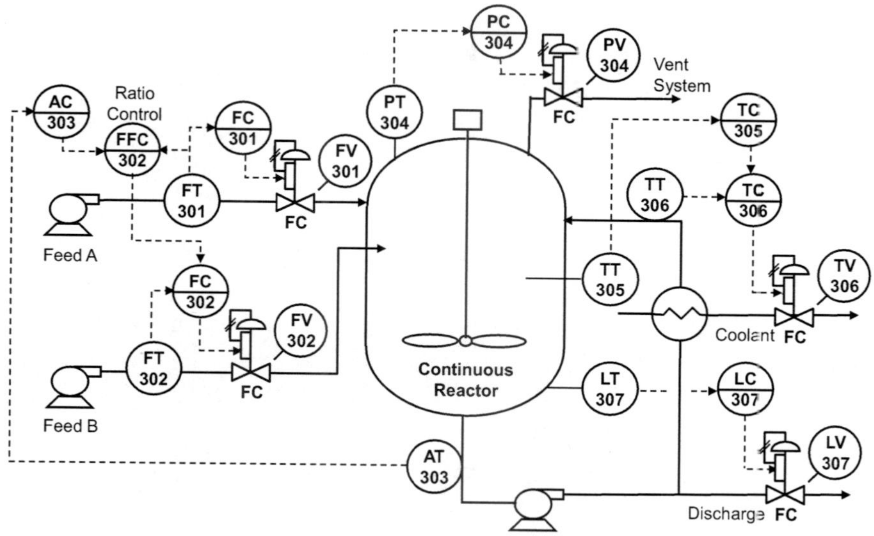

Figure 5-1. Continuous Reactor

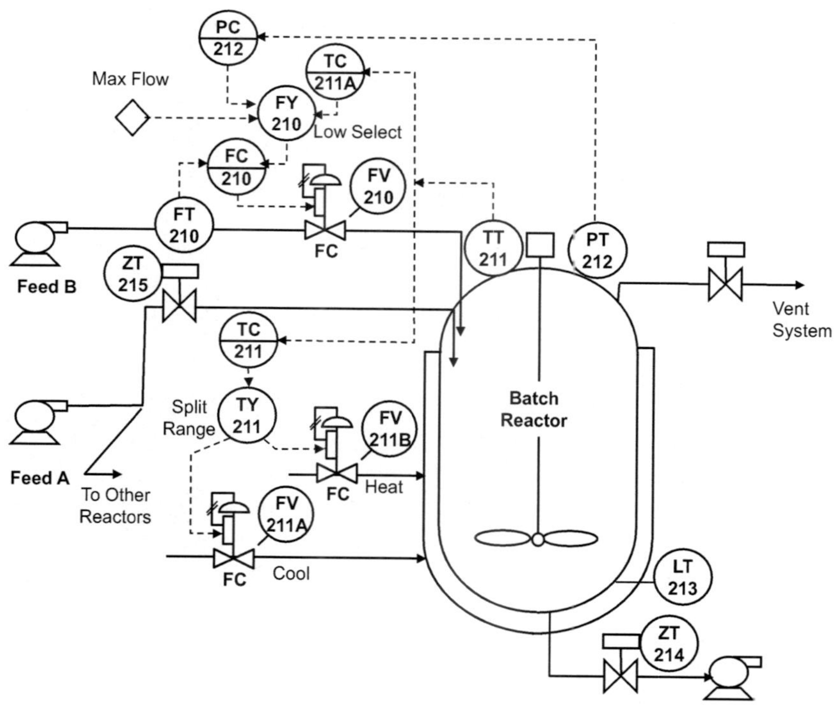

Figure 5-2. Batch Reactor

If feed material is added to the reactor throughout the processing of the batch, the head space is reduced for the volatile materials within the reactor. As a result, the process gain associated with reactor pressure increases throughout the processing of the batch for changes in the reactor level – Figure 5-3.

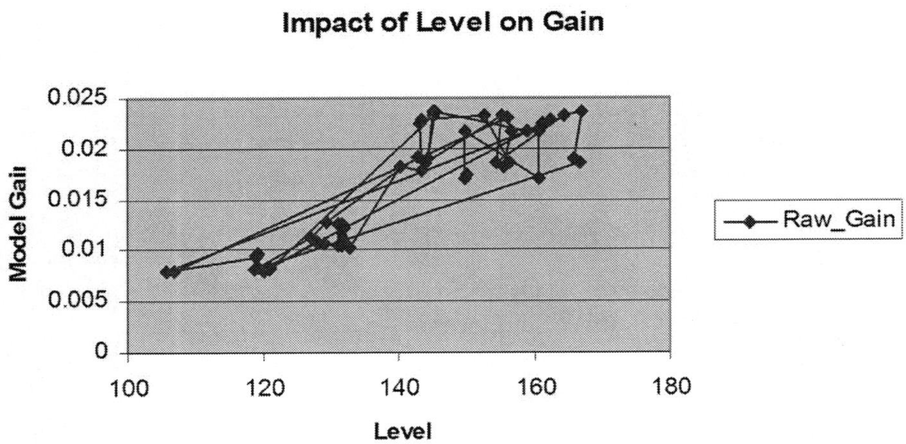

Figure 5-3. Impact of Level on Reactor Pressure Process Gain

Through the application of adaptive control it is possible to automatically modify the tuning of the pressure loop as a function of reactor level. When process gain is addressed by this means, responsive and stable pressure control can be achieved throughout the batch.

5.1.3 Hydrogen to Nitrogen (H/N) Ratio in Ammonia Production

The composition of a critical process stream is often the result of the reactions and mixing that take place in upstream units. Any changes made upstream to maintain the composition at setpoint may not be immediately reflected in an on-line analyzer because of significant transport delays and lags that may be introduced by the upstream processes. An example of this is the ammonia plant illustrated in Figure 5-4.

Any change in the air/gas ratio at the front end of the plant will not be immediately reflected in on-line analyzer H/N measurements used in the Makeup H/N and Synthesis H/N controls AC333 and AC334. Thus, it is challenging and can be time-consuming to test the process to establish the process gain and dynamics for these control loops. For

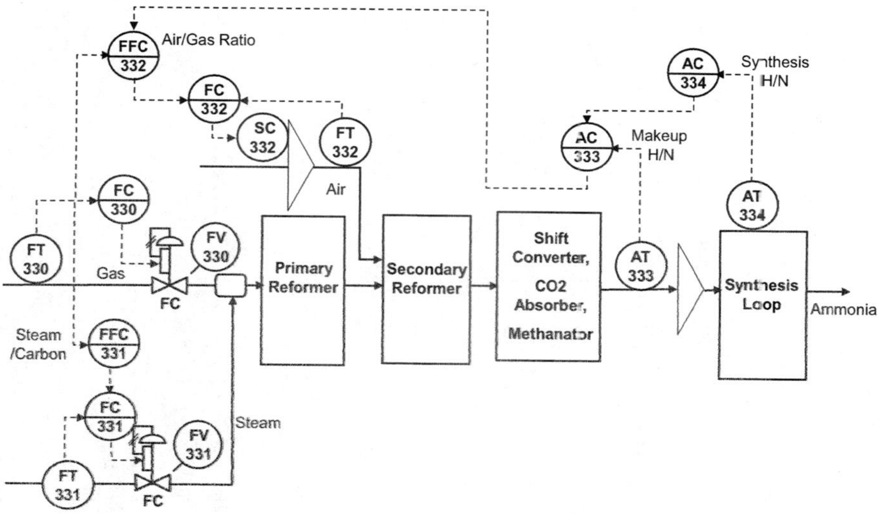

Figure 5-4. Ammonia Plant

example, the process time constant for a change in the air/gas ratio is often many hours. By using adaptive tuning, the process dynamics and controller tuning can be automatically established based on adjustments made by the operator during normal plant operation.

5.1.4 Neutralizer

In wastewater treatment and other neutralization applications in the process industry, it is quite common to control pH through the regulation of a reagent flow. An example of pH control implementation associated with a neutralizer is shown in Figure 5-5.

The challenge in implementing pH control is due to non-linear variances in pH measurement when the reagent is added. The changes in pH with the addition of acid or base reagent are described by a titration curve obtained through lab analysis or determined by observing the changes on-line – Figure 5-6 shows an example of the titration curve. Changes in the feed composition can impact the shape of the titration curve.

The relationship of pH and reagent addition can be automatically identified using adaptive control. Also, through the application of adaptive control it is possible to automatically change the PID tuning to compensate for the non-linear behavior.

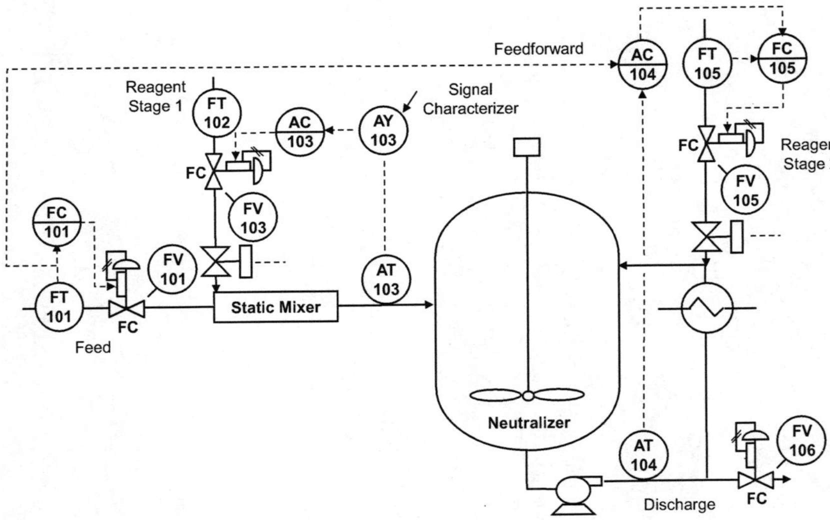

Figure 5-5. Neutralizer

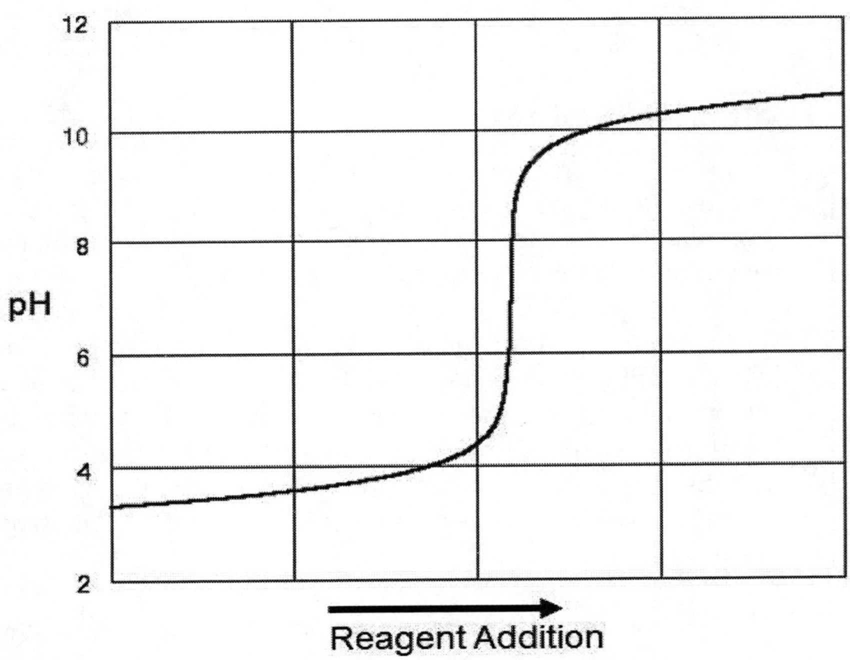

Figure 5-6. Example Titration Curve

5.1.5 Plant Master Control

The process gain associated with a loop can be impacted by the number of process units that are available to act on the PID output. An example of this is power house plant master control, where the number of swing boilers on-line impacts the process gain associated with the main header pressure control that is provided by the plant master controller as shown in Figure 5-7.

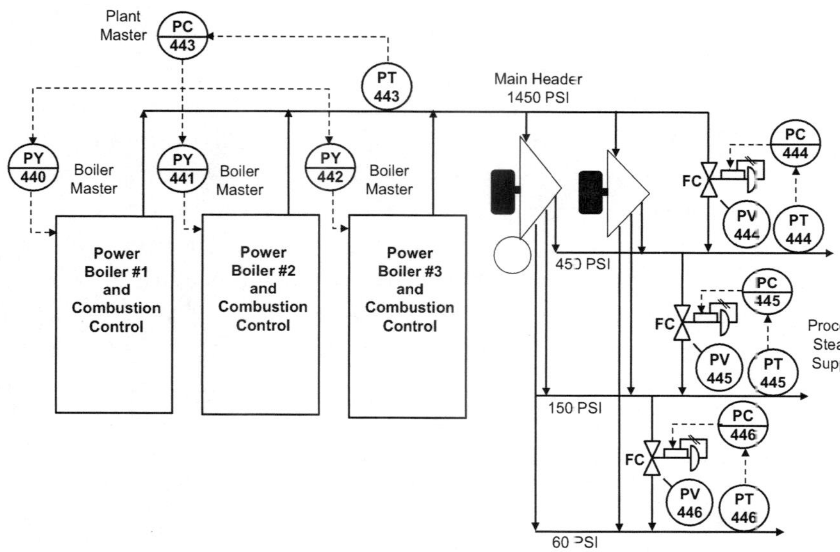

Figure 5-7. Power House Plant Master

By applying adaptive control, the plant master controller tuning can be modified to automatically compensate for the number of swing boilers on-line.

5.2 Application Example

Adaptive control has been the focus of control specialists and institutions for over 60 years. Many sophisticated techniques have been developed for adjusting controller parameters to compensate for process changes. Practical progress is, however, less impressive because the safety, reliability, and robustness requirements of practical applications are often not adequately addressed. Adaptation based on multi-model switching is a relatively new technology and is the basis for adaptive control in DeltaV DCS. The basic principle is to evaluate multiple models and select the model that best matches the process response to changing process inputs. This basic technique can be further enhanced through the use of parameter interpolation and sequential parameter adaptation. The following implementation of adaptive control serves as a good example of how adaptive control can be used to quickly identify a process model and how the results can be used to set loop tuning or continuously adjust loop tuning to compensate for changing process conditions.

5.2.1 Enabling Model Identification

The basis of adaptive control is the identification of the process step response based on operator changes in the loop operation. For example, when a control loop is in Automatic mode, changes made in the setpoint can trigger process model identification. If the loop mode has been changed to Manual, then changes by the operator in the loop output may trigger process model identification. When model identification is triggered, the value of the controlled parameter, setpoint, and control output are automatically captured and analyzed within the controller to determine the process characterization. Self-regulating processes are characterized by process gain, deadtime, and time constant. If the integrating process option has been selected in the interface, the response is identified by integrating gain and deadtime.

As new process models are identified, they appear in the on-line adaptive control interface. The consistency of the process model that has been identified is graphically indicated by an icon in the interface, where green is used to indicate a high degree of confidence in the model, and red is used to indicate that inconsistent models have been identified. Each time a new model is identified, a recommended tuning is provided based on the model and the user-selected tuning rule as shown in Figure 5-8.

When a high degree of confidence is indicated, the user can transfer the recommended tuning to the controller by simply selecting Update. Thus, when adaptive tuning is available in the control system, the process of tuning a control loop is greatly simplified.

In most process applications the setpoint of a control loop is changed rarely, and under normal operating conditions the mode of the control loop is always Automatic. To address such applications, an option is provided to allow a change in the PID output to be automatically injected on a periodic basis, developing process perturbations needed for the process model identification. The magnitude of the change in output should be small enough to not disrupt the process but large enough to allow the process to be identified in the presence of process noise. Thus, the frequency and magnitude of the output change can be

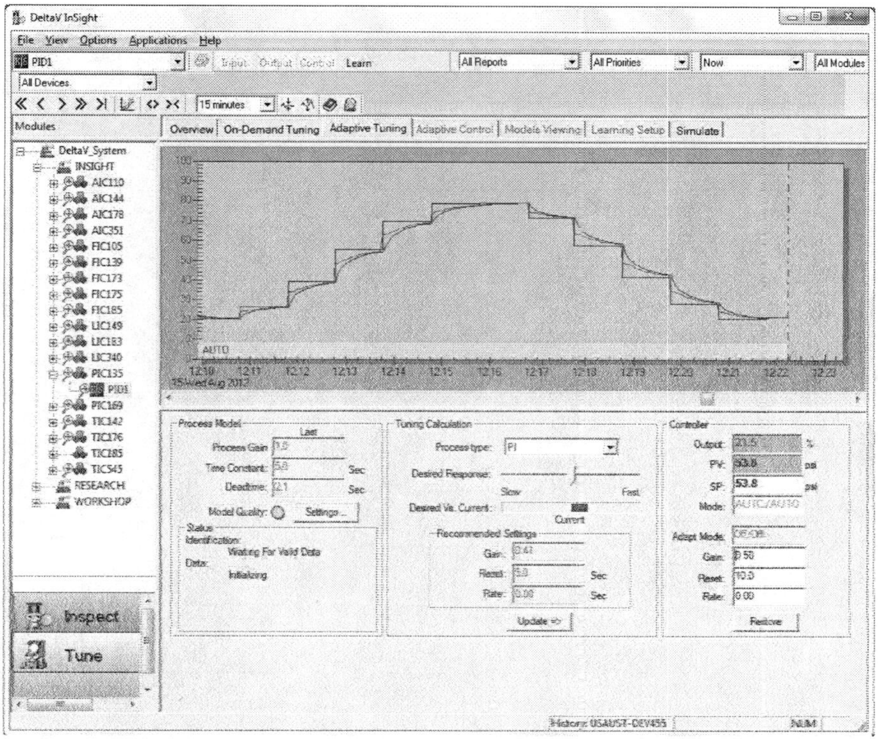

Figure 5-8. Tuning Based on Identified Process Model

set based on how quickly the process dynamics change and on the noise level in the process measurement. Also, the trigger point for changes in setpoint and output can be set through this option as shown in Figure 5-9.

The control loop response to disturbances or setpoint changes can vary because of changes in the operating conditions, such as production rate or non-linear valve installed characteristics that impact the process gain. Such changes can be observed using a real-time trend of the controlled parameter, manipulated parameter, and valve position. One example of the changes in performance that might be observed for non-linear valve installed characteristics is shown in Figure 5-10.

The process models identified by the adaptive control application can be examined to verify the observed variations in control performance. Since the last 200 process models that were identified are automatically saved by the adaptive control application, it is possible to see the impact of operating conditions on the process gain and dynamics. When the Model view is selected, the models that have been identified

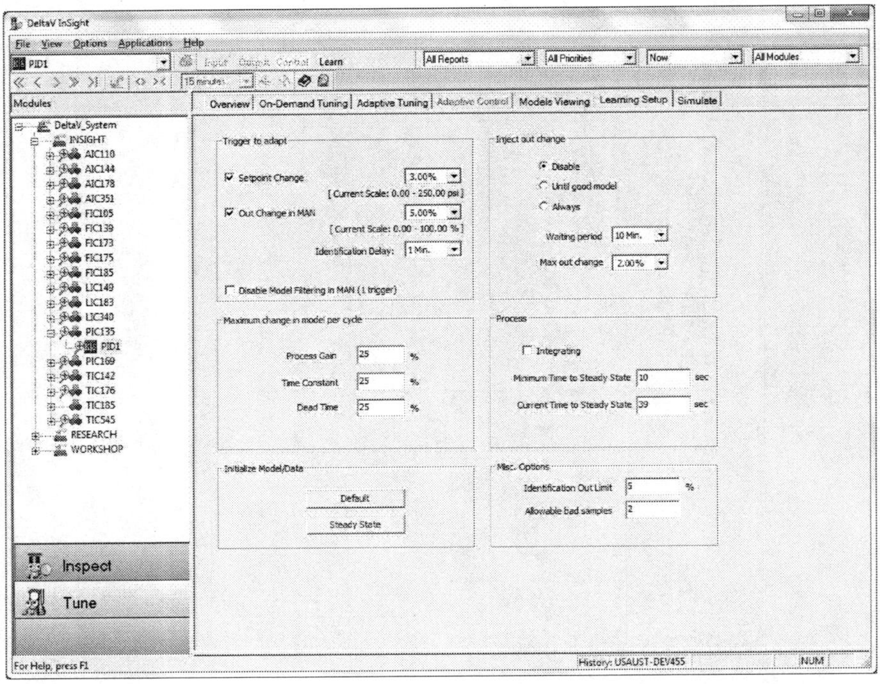

Figure 5-9. Adaptive Tuning Options

are listed along with the time of the setpoint or output change that triggered model identification. Also, the option is provided to plot some or all of the identified models against the selected source of process change, which is known as the state parameter. For example, in many cases the source of process change is associated with valve position and thus is one of the standard plot options, as illustrated in Figure 5-11.

When the option "Other" is chosen, any measured or calculated condition may be selected as the state parameter. For the batch reactor example previously discussed in this chapter, the level of the reactor would be selected as the state parameter when observing the models associated with reactor pressure control. For another example, when viewing the models identified for the plant master pressure control in a power house, the number of swing boilers that are in service is the main source of process change and thus would be identified as the state parameter.

To provide stable control for those control loops where the process gain is found to vary, the highest process model gain can be selected.

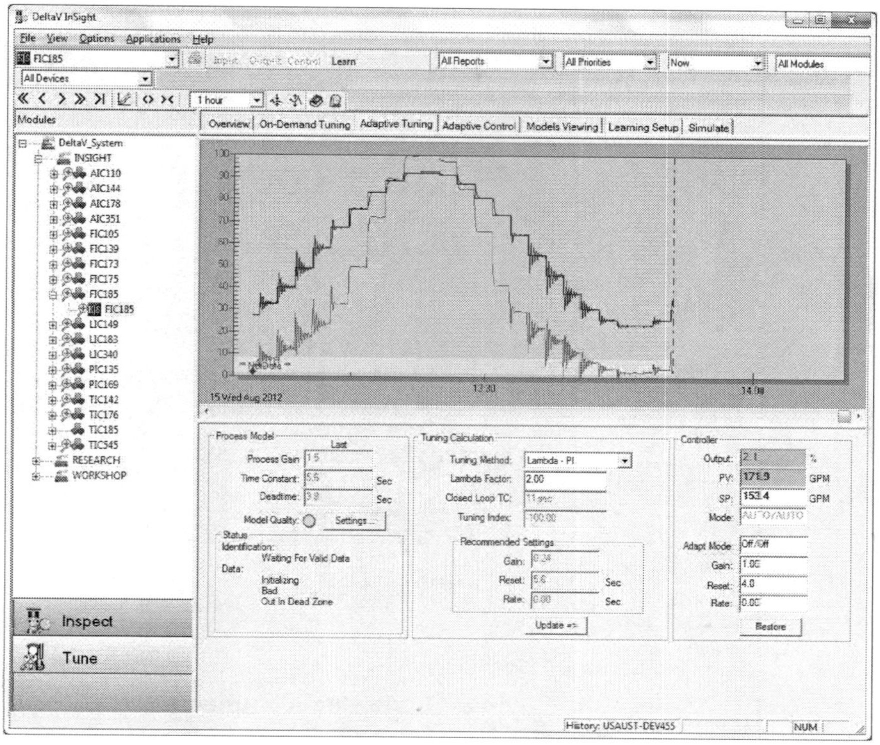

Figure 5-10. Example Loop Response for Non-Linear Valve Installed Characteristics

However, sluggish control performance will be observed when operating conditions cause lower process gain. To achieve best performance for all operating conditions, it is necessary to change the controller tuning as a function of the state parameter. When adaptive control is selected, the capability to apply models to loop tuning is automatically provided.

5.2.2 Applying Models to Loop Tuning

When the adaptive control option is enabled, the state parameter may be broken into as many as five regions. For each region the user can select the model or average of multiple models in that region to be the approved model for that region. Also, upper and lower limits can be specified in each region for each model parameter. The Adaptive Control view can be selected to view and approve a model and model limits for each region and the current value of the state parameter. Also, the manner in which adaptive control is used can be selected from this view. The options are:

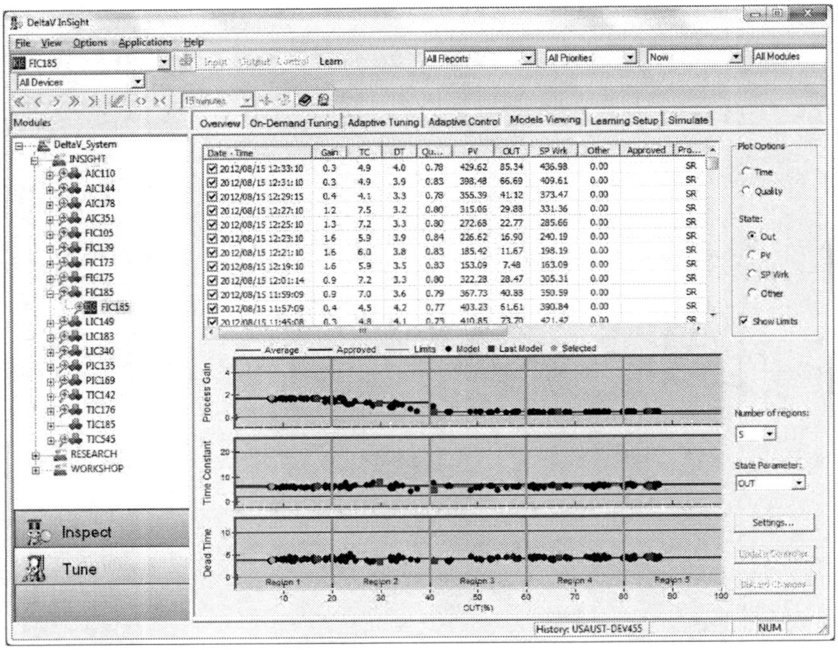

Figure 5-11. Viewing Identified Models

Off - the tuning set in the PID block is always used.

Partial - the tuning used in control is based on the approved model for the current value of the state parameter and the selected tuning rule.

Full - on transition of the state parameter from one region to another, the tuning is set based on the approved model for the new state and the tuning rule. While operating in this region, the model used to determine tuning will be adapted within the model parameter limits based on new models that are identified for that state.

Adaptation of the model used in control is automatically limited to the model limits specified for each operating region. Thus, the user can select the maximum change that can occur when full adaptive control is enabled. Through the use of partial or full adaptive control, good control performance may be achieved for all operating conditions as illustrated in Figure 5-12. The trends show an oscillatory response before the adaptation and significantly improved responses after model adaptation.

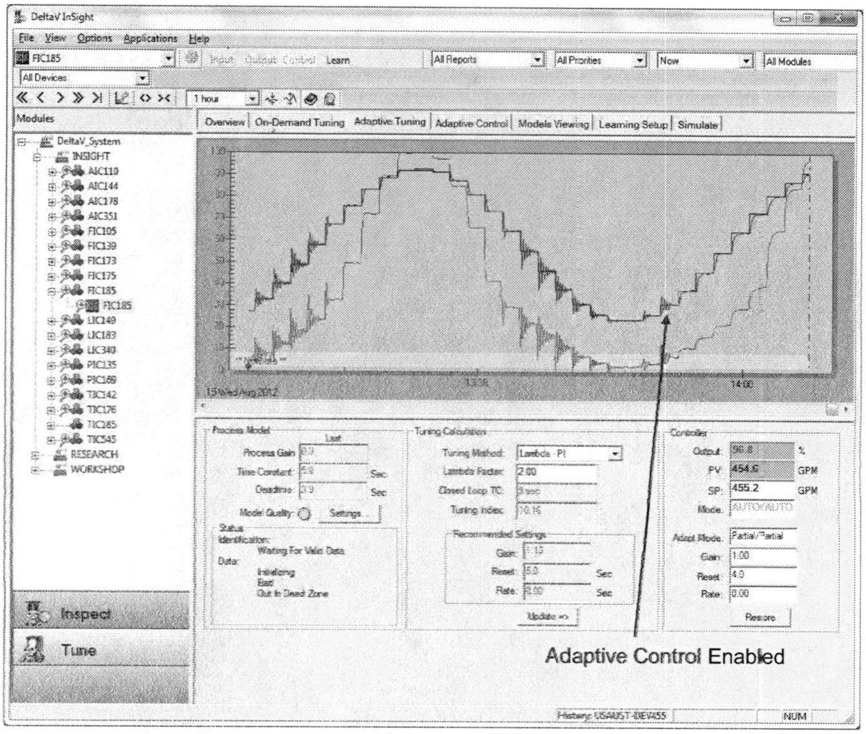

Figure 5-12. Impact of Adaptive Control on Control Performance

Control using the tuning specified in the PID is restored when adaptive control is set to Off.

5.3 Adaptive Tuning Workshop

This workshop provides several exercises that are used to further explore the use of adaptive control. A flow process with non-linear installed valve characteristics (Figure 5-13) is used in this workshop. Access the book's web site http://www.advancedcontrolfoundation.com and select the Solution tab to view the workshop being performed.

> Step1: Select the Tuning application and select the Adaptive Tuning tab to view the trend of the control operation for various flow setpoints. Notice that at the low end of the flow loop operating range the control response is very oscillatory, indicating a higher process gain in this region of operation.

Step 2: Select the Models Viewing tab to see how the process gain is changing as a function of the flow demand and associated region of valve operation. In the lower portion of the model view, note how the state parameter has been broken into five operating regions.

Step 3: Select the Adaptive Control tab and observe the control performance with adaptive control turned to Partial.

Step 4: Turn the adaptive control to Off and observe the impact this has on control performance. When adaptive control is enabled, a significant improvement should be noted in control performance.

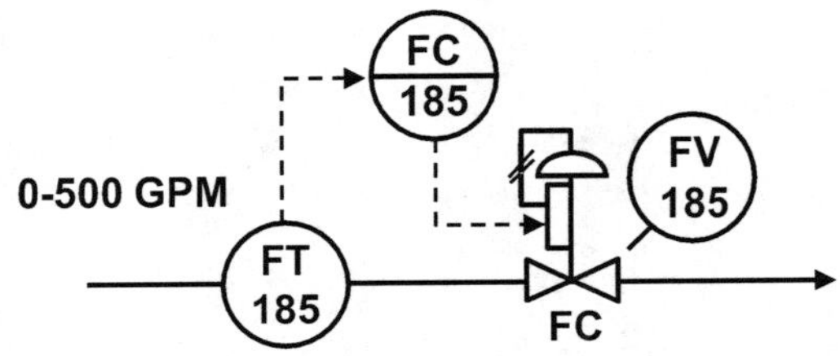

Figure 5-13. Flow Process for Adaptive Control Workshop

5.4 Technical Basis

A perfectly-tuned PID controller can degrade over time and perform poorly or become oscillatory. There are two main reasons for these changes:

1. The controlled process is non-linear, and the process operation has entered a region with significantly different process parameters than those used during the tuning.
2. The process parameters have changed since the auto-tuning was performed.

The use of a gain scheduler, which adjusts the PID controller gain as a function of the operating point, is often sufficient for solving the first problem. In more demanding applications, where not only process gain, but also lag and deadtime, depends on the operating point, the PID controller is auto-tuned in several ranges. Based on the point of operation at any given time, the associated process model is used to set the PID gain, reset, and rate. In between points where the process model was identified, the PID tuning is set based on weighted averages, taking into account the process model in adjacent ranges and the distance from the ranges boundaries. (Wojsznis, Borders, and McMillan, 1994).

To achieve the desired performance in the second case, the tuning should be repeated periodically or upon changes in controller performance, reflected in increased control variability. If this procedure is triggered and controlled automatically, the operation is called *adaptive tuning*. Adaptive tuning can also be applied in the first case.

We can classify simple and reliable adaptive tuners for PID controllers as shown in Figure 5-14.

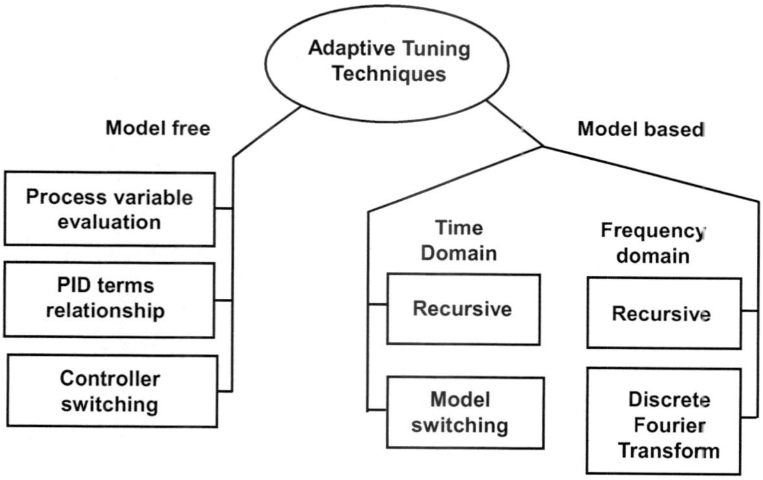

Figure 5-14. Adaptive Tuning Techniques for PID Controllers

There has been significant effort over the years to implement adaptive tuning. The next sections present a short overview of several promising adaptive tuning techniques. The main focus will be on *adaptation with model switching and parameter interpolation*, which is implemented as an industrial product and used in the workshop section for the presentation of the adaptive control. This technique therefore will be discussed in more detail than other model-based techniques or model-free tuning.

5.4.1 Model-Free Adaptive Tuning

Model-free adaptive tuning evaluates controller performance and adjusts the PID controller parameters to achieve the desired controller performance. A process model is not used in the calculations. Model-free tuning attracts the developers of PID self-tuners because this type of tuning gives the promise of fewer calculations and smaller process excitations than what is required for model-based tuning. These were major considerations leading to the development of the first PID self-tuning products, which were model-free adaptive tuners.

A classic example of model-free tuning is the PID adaptive controller Exact (Åström, and Hägglund, 1988), (Åström, and Hägglund, 1995) that adjusts the controller parameters based on process variable (PV) behavior. This controller adapts based on the loop response to a set-point change or a load disturbance, when the actual damping and overshoot of the error signal are recognized and the period of oscillation is measured. If there are no oscillations in the error signal, proportional control gain is increased, and integral and derivative times are decreased. If an oscillation occurs, damping and overshoot of the oscillation are measured and controller parameters are adjusted accordingly.

The adaptation principle is simple; however, it suffers from a few flaws:

1. Adaptation works only when control response is oscillatory.
2. When the control response is not oscillatory during quiet periods of operation, the changes in the controller parameters are too aggressive.
3. Adaptation requires more than one setpoint change to tune an over-damped controller.
4. The controller is tuned for an oscillatory response, with a small safety margin.

There are other known implementations of the PID adaptive controllers that exploit an oscillatory control error response and therefore have similar features (Åström, and Hägglund, 2006). An example is the UDC 6000.

A more sophisticated method of model-free adaptive control, where *the adaptation is based on controller parameter relationships*, avoids some of the pitfalls noted above. This method examines the relationships between the controller's proportional, integral, and derivative terms and adjusts the controller parameters to achieve the desired relationships (Maršik and Strejc, 1989). An adaptive system that exploits both PID parameter relationships and PV behavior (Wojsznis, Gudaz, and Blevins, 2001) is shown in Figure 5-15.

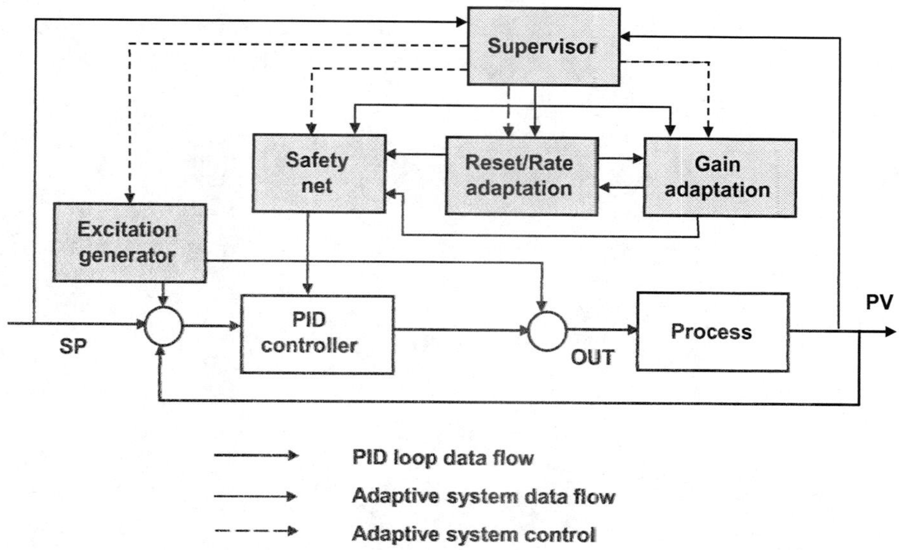

Figure 5-15. Model-Free Adaptive PID Control Loop

The adaptive tuner is composed of five functional components: Supervisor, Excitation Generator, Gain Adaptation, Reset/Rate Adaptation, and Safety Net.

The Supervisor coordinates the operation of all tuner components and tests the condition for starting the Gain Adaptation cycle:

$$|e(k)| > E_{min} \quad \text{or} \quad |\Delta e(k)| > \Delta E_{min} \qquad (5\text{-}1)$$

$$e(k) = SP(k) - PV(k)$$

$$\Delta e(k) = e(k) - e(k-1)$$

For Reset Adaptation, an additional condition should be satisfied:

$$e(k) * \Delta e(k) < 0 \qquad (5\text{-}2)$$

Both controller input (PV) and output (OUT) should be within their operating range and show no oscillations. If the Supervisor detects oscillations, it activates the Safety-Net component, which cuts controller gain K every time the PV crosses the SP line, that is:

$$\text{If } e(k) * e(k-1) < 0 \text{ then } K_c(k+1) = \eta K_c(k) \qquad \eta \approx .95 \qquad (5\text{-}3)$$

Adaptation works from setpoint changes made by the operator or injected by the Excitation Generator or load disturbances. When the conditions for gain adaptation are satisfied, the gain is adjusted according to the following formula:

$$\Delta K_c(k) = \chi K_c(k)(W(k) + W_{ref}) \qquad \chi = 0.02 \div 0.05 \tag{5-4}$$

χ is a heuristic coefficient defining the speed of gain adaptation

$$W(k) = \overline{sign\ e(k) sign[\Delta^2 e(k)]} \tag{5-5}$$

$$\Delta^2 e(k) = e(k) - 2e(k-1) + e(k-2)$$

$$sign\ e(k) = \begin{cases} -1 \text{ if } e(k) < 0 \\ 0 \text{ if } e(k) = 0 \\ 1 \text{ if } e(k) > 0 \end{cases} \tag{5-6}$$

W_{ref} is a value from the interval of –1 to +1. Practically, it was found by Maršik and Strejc (1989) that W_{ref} ~ -0.3 delivers an adaptive gain that is close to the model-based tuning.

W(k) is an oscillation index, which is negative for an oscillatory PV and positive for a normally damped and over-damped PV. *W(k)* is calculated as an average value of an expression $sign\ e(k)sign[\Delta^2 e(k)]$ over the adaptation update period, which is usually close to the loop settling time. As may be seen from Equation 5-4, $W(k) + W_{ref} > 0$ causes an increase in gain.

Reset Adaptation uses the PID controller property that a firm relationship between the increments of proportional and integral terms is specified:

$$\Delta P = \sum_k \Delta P_k; \quad \text{and} \quad \Delta I = \sum_k \Delta I_k; \qquad \beta = \frac{\Delta P}{\Delta I \alpha} \tag{5-7}$$

The summation is performed over an adaptation period or over the whole period when the adaptation conditions (5.1 and 5.2) are satisfied.

In Equation 5-7, a is set for a desired controller performance, and controller integral time is adapted to achieve this performance by adjusting $T_i(k)$ until $\beta = 1$, according formula:

$$\Delta T_i(k) = \gamma T_i(k)\left(\frac{1}{\beta} - 1\right); \tag{5-8}$$

$\gamma \sim 0.05$ defines the speed of reset adaptation.

Another simple concept in model-free adaptation is the *controller-switching adaptive technique,* which evaluates the performance of various controllers with a predefined set of parameters and selects the best performing controller (Morse, Pait, and Weller, 1994). The original idea assumes sequential evaluation of the controllers. Despite the simplicity of the concept, its significant deficiency is the time required for the evaluation of controllers in the sequence.

Safonov and Tsao (1997), Jun, and Safonov (1999), and Brozenec, Tsao, and Safonov (2001) introduced the concept of *a fictitious setpoint for PID controller* that assumes a simultaneous evaluation of the actual controller in the loop and all other controllers in the assumed controllers set. The set of candidate controllers is fitted to the process input and output's measured data over the period of adaptation. To make the fit of the same data set possible for various controllers, a setpoint value that is different from the true set point is calculated. The fictitious setpoint for an ideal PID controller is calculated by inverting the PID controller (Equation 5-9) as in Figure 5-16.

$$out = \left(k_p + \frac{k_i}{s}\right)(SP - PV) - \frac{sk_d}{s\tau_f + 1}PV \tag{5-9}$$

where τ_f is a filter time constant for the derivative term

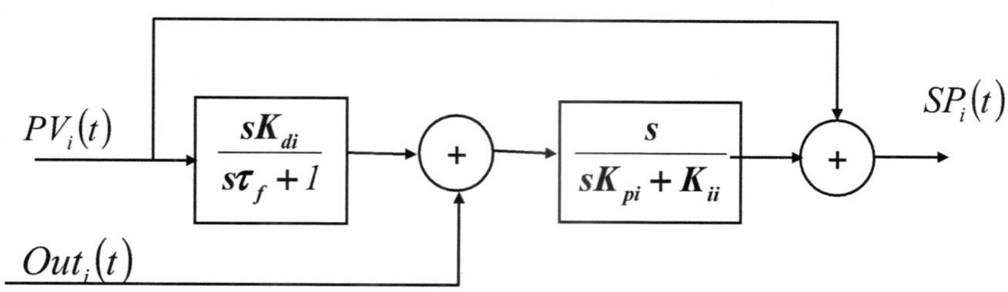

Figure 5-16. Fictitious Setpoint Calculation Diagram

In Figure 5-16, controller with actual parameters (Equation 5-9) is compared to the set of n potential optimal candidate controllers with proportional, integral, and derivative gains K_{pi}, K_{ii}, K_{di} $i = 1,\ldots n$.

After the fictitious setpoint is generated, the performance index is calculated as a squared control error relative to the fictitious set point. The performance index also accounts for the controller output move and the difference between the real and fictitious setpoints.

The controller with the minimum performance index is selected as the active controller for the next scan. The performance index is integrated over the adaptation period, which is defined as the time when reasonable changes in the process variable are detected.

5.4.2 Model-Based Recursive Adaptive Tuning

Model-based recursive adaptive tuning adjusts the process model to the changes in the process itself and then updates the controller parameters based on the model (Chien and Fruehauf, 1990). The process model is built and adapted in the time domain or the frequency domain. The most common classical approach for both the time and frequency domains is a recursive adaptation method of the type shown in Figure 5-17 that adjusts the model parameters every scan.

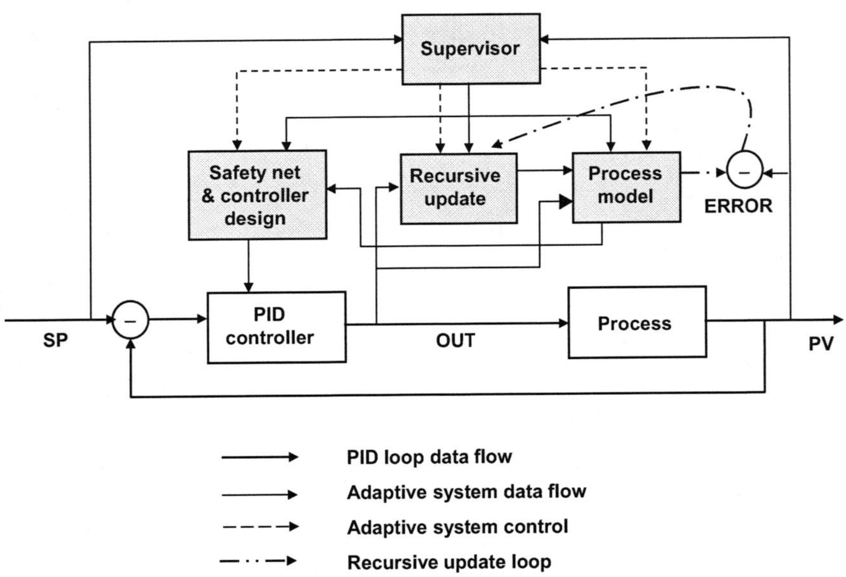

Figure 5-17. PID Loop with Recursive Adaptation

An adaptive PID loop contains Supervisor and Safety Net blocks, two typical blocks for any adaptive design. As soon as adaptation conditions – defined primarily by control error exist, the Supervisor activates a recursive model update mechanism. The recursive model update mechanism updates the corrective term first and then recursively updates the process model parameters:

$$\theta_k = \theta_{k-1} + P_k x_{k-1} e_k \qquad (5\text{-}10)$$

where:

θ_k	=	process model parameters vector
P_k	=	corrective term (matrix)
x_{k-1}	=	regression vector containing process output and input variables
e_k	=	error on the model output relative to true process output

The controller parameters are then recalculated based on the updated model. Typical problems of recursive model update include selection of initial parameters, insufficient excitation, filtering, parameters wind-up, and slow parameter tracking speed.

5.4.3 Discrete Fourier Transform Adaptation Technique

The Discrete Fourier Transform (DFT) is a basic operation used in many signal processing applications. The practical value of the DFT has been recognized in numerous areas of technology. There has been significant interest in the practical implementation of the DFT for process model identification, as presented by Lamaire, Valavani, Athans, and Stein (1991).

There are several self-tuning systems for industrial controllers that apply frequency domain tuning, though they don't use the DFT. Hägglund and Åström (1991) designed a controller that bases tuning on setpoint changes and natural process disturbances. The specific tuning frequency is selected by using band-pass filters on the process input and output. The frequency of the filters is set from an on-demand tuner, which defines the ultimate period using the relay-oscillation technique. The tuner defines the process gain for the tuning frequency, using a simplified recursive least squares estimator. The tuner can work from setpoint changes and from natural disturbances, and can include external excitation injected in the controller output or at the setpoint. A simplification of this design is achieved by using a discrete Fourier transform estimator instead of a recursive least squares estimator (Wojsznis, 1996). A DFT tuner design is shown in Figure 5-18.

DFT computations are applied on the input and output, and the subsequent transfer-function estimation for every filtered frequency is

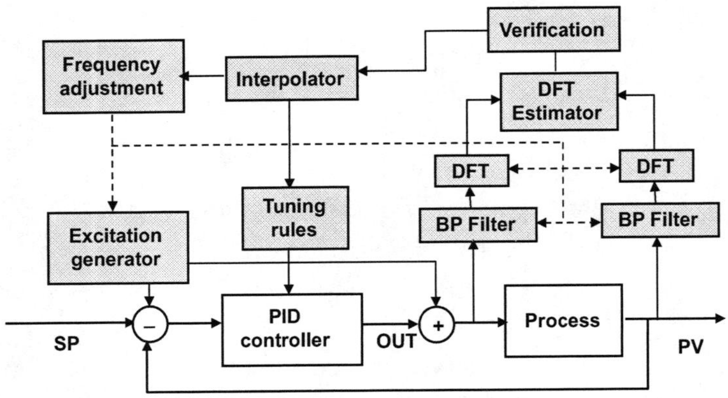

Figure 5-18. A DFT-Based Tuner with Multi-Frequency Estimator and Interpolator

defined. The Verification unit tests the monotonous process gain and phase decrease with frequency increase. The Interpolator defines the frequency and process gain for the point where the phase shift is about $-\pi$. From these two parameters, ultimate period and ultimate gain are calculated directly and used for PID controller tuning and for setting up a tuning frequency for the next cycle. The required excitation level depends on the process gain around the critical frequency and on valve performance. Loops with significant valve hysteresis and stickiness require an excitation level that is two to three times higher than the same loops with good valve performance. Though tuning from setpoint changes is possible, in most cases tuning from injected excitations gives consistent results and is considered a basic mode of tuner operation.

5.4.4 Adaptive Tuning with Model Switching and Parameter Interpolation

Logic-based model switching strategies have been proposed by many researchers as the way to implement adaptive control (Gendron, 1990), (Morse, Pait, and Weller, 1994). The approach uses an identifier-based parameterized controller consisting of two parameter-dependent subsystems, one an identifier whose primary function is to generate an output estimation error and the other an internal controller. The control signal fed back to the process is based on a current estimate of the process model. In general, such estimates are selected from a suitably defined admissible model set.

The overall strategy is based on the concept of "cyclic switching." Cyclic switching is employed with or without process excitation. Narendra and Balakrishnan (1997) considered an architecture with N identification models, operating in parallel. Corresponding to each model is a parameterized controller. At every instant, one of the models is selected by a switching rule, and the corresponding control input is used to control the process. Models are fixed or adaptive. The rationale for using fixed models is to ensure that there is at least one model with parameters sufficiently close to those of the unknown process. The approach yields the desired speed of adaptation but requires the use of a significant number of models. If, instead of switching models, the adaptation is performed through model-parameters interpolation, the number of models used is drastically reduced. Figure 5-19 shows a PID adaptive controller with model switching and parameter interpolation (Wojsznis, Blevins, and Wojsznis, 2003), (Wojsznis and Blevins, 2006).

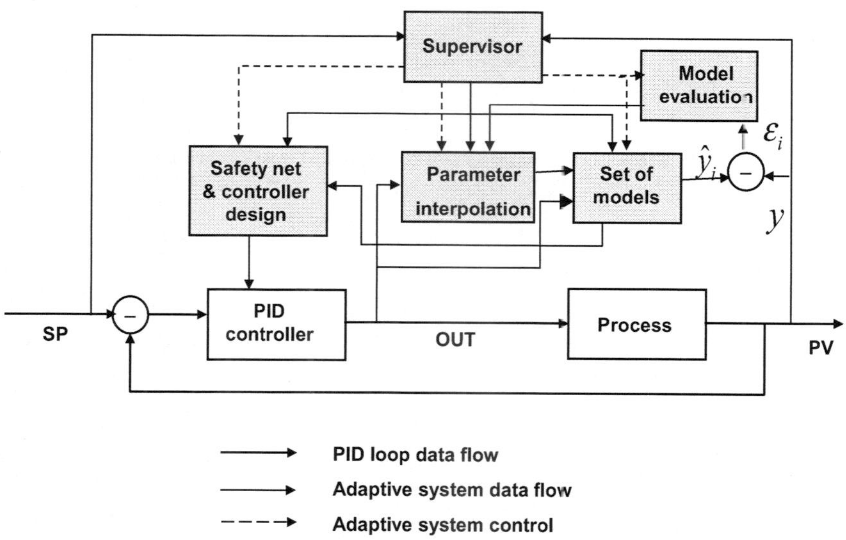

Figure 5-19. Adaptive PID Controller with Model Switching and Parameter Interpolation

The heart of the design is a set of models. The process identification is based on the evaluation of multiple models. Each model consists of three parameters ($p = 1,2,3$, i.e., gain, deadtime, lag). Assigning N values for every parameter, the model set has $M = N^3$ models. Let $v_p(n)$ signify *n-th* value of the parameter p.

$$p = 1, 2, \ldots, P \qquad n = 1, 2, \ldots, N$$

$P = 3, N = 3$ for the first order-plus-deadtime model and the assumed three values for every parameter.

The adaptive model identification operates in the following way:

The Supervisor detects changes in the process output (that is, the controlled variable $y = PV$), the setpoint SP or the manipulated process input ($u = OUT$).

If any changes exceed a minimum level, model evaluation starts. This consists of:

- Initialization of all models and adjustment of the models' output with the current process output
- Incremental update of all models based on the changes of the manipulated input to the process
- For every model i, computing squared error $e_i^2(k) = (y(k) - \hat{y}_i(k))^2$ for each scan k, where:

  $y(k)$ = the process output at time k

  $\hat{y}_i(k)$ = i-th model output at time k

The squared error is assigned for every parameter value of the model ***i*** if the parameter value is used in the evaluated model. A zero is assigned to any parameter value that is not used in the evaluated model. Next, the model $i + 1$ is evaluated. Again, squared error is assigned for every parameter value and accumulated with the previously assigned squared error for the corresponding parameter values. The model evaluation continues until all models are evaluated. As a result of the evaluation, every parameter value is assigned a sum of squared errors from all models in which this specific parameter value has been used. In the one scan k, therefore, every parameter type p with n-th value, that is, $v_p(n)$, has an assigned squared error $SE_p^n(k)$:

$$SE_p^n(k) = \sum_{i=1}^{M} \gamma_p^n e_i^2(k) \qquad (5\text{-}11)$$

where:

M = the number of models

$\gamma_p^n = 1$ if parameter p with value $v_p(n)$ is used in the model, otherwise $\gamma_p^n = 0$.

Model evaluation is repeated in the scan $k+1$ and the sum of the squared errors for every parameter value is added to the sum of the appropriate parameter values accumulated in the previous scans. The adaptation cycle continues through a declared number of scans (1 to K), or until there is enough excitation on the inputs. As a result of this procedure, every parameter type p with value $v_p(n)$ has an assigned sum of squared errors SSE_p^n over the period of evaluation:

$$SSE_p^n = \sum_{k=1}^{K} SE_p^n(k) \tag{5-12}$$

At the end of adaptation cycle, the inverses of SSE_p^n are calculated:

$$F_p^n = 1/SSE_p^n \qquad p == 1, 2, 3; \qquad n == 1, 2, 3; \tag{5-13}$$

An adaptive parameter value a_p for the parameter p is calculated as a weighted average of all values of this parameter $v_p(n)$ $n = 1, 2, 3$ used in evaluation.

$$a_p = v_p(1)f_p^1 + v_p(2)f_p^2 + v_p(3)f_p^3 \tag{5-14}$$

where f_p^n is the normalized F_p^n value in the range $0 - 1$:

$$f_p^n = F_p^n \Bigg/ \left(\sum_{1}^{N} F_p^n \right) \tag{5-15}$$

The coefficient f_p^n used for calculating the weighted average of the parameter p is inversely proportional to the sum of the squared errors assigned to the value n of that parameter. Therefore, if a parameter value is associated with large errors in modeling, the contribution of that parameter to the adaptive parameter value will be smaller. On the other hand, if a parameter value is associated with small errors in modeling, its contribution in the calculated adaptive parameter value will be larger. Calculated adaptive parameters define a new model set with center parameter values a_p and the range of parameter accepted changes for the adaptation cycle defined as $a_p \pm \Delta a_p$, $p == 1, 2, \ldots, P$.

Within that range, two parameters at a minimum should be defined for the next adaptation cycle. Normally, three parameters are used: $a_p - \Delta a_p, a_p, a_p + \Delta a_p$.

A first order-plus-deadtime process model in a discrete form is used:

$$\Delta y(k) = a\Delta y(k-1) + b\Delta u(k-1-d) \tag{5-16}$$

$$a = e^{-\frac{h}{\tau}}, \qquad b = k\left(1 - e^{-\frac{h}{\tau}}\right) = k(1-a) \tag{5-17}$$

where:

Δu	=	process input incremental change since adaptation started
Δy	=	model output incremental change
τ	=	model time constant
h	=	loop scan period
d	=	process delay in loop scan periods

With three values for every parameter of the model, there are $3^3 = 27$ models for evaluation.

Improved convergence and a reduction in the number of models can be achieved by the following amendments to the basic algorithm:

1. Performing parameter adaptation sequentially, one parameter at a time. In this way, the number of model combinations for the first order-plus-deadtime model has been reduced to 3x3 = 9 (Figure 5-20).
2. Using the original data set and performing adaptation iteratively by running the algorithm several times (Figure 5-21).

In an iterative adaptation procedure in which one parameter is updated over a calculation cycle, the updating is performed in a sequence of, first, process gain, then deadtime, and, finally, time constant over the model adaptation cycle (Figure 5-20).

Figure 5-21 explains parameter adaptation, showing how parameter adapted values are improved iteratively using the same set of test data. After each iteration, the parameter adaptation range is made smaller, improving the accuracy of the parameter identification.

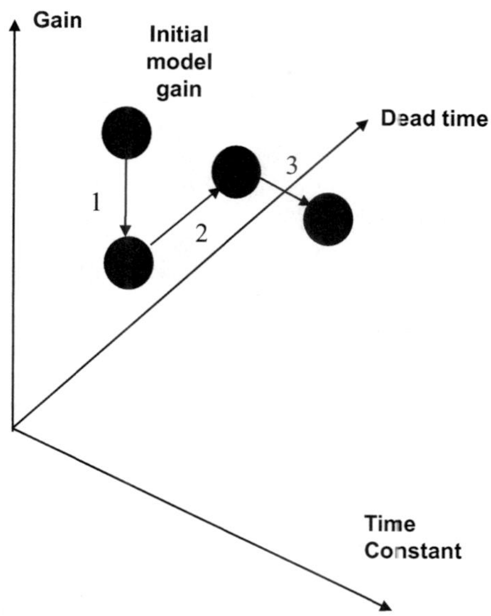

Figure 5-20. Sequence of Parameters Adaptation and Interpolation

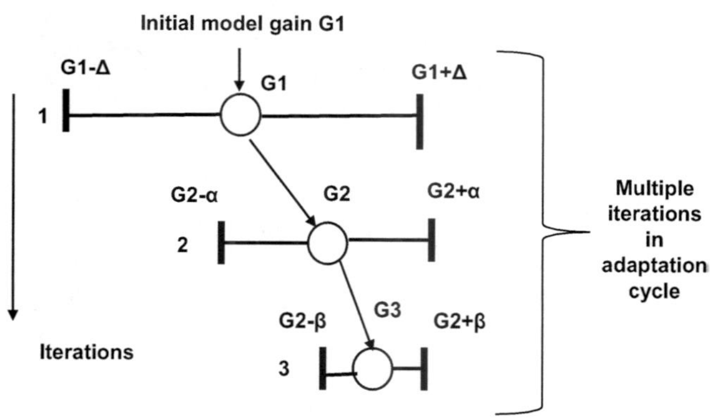

Figure 5-21. Iterative Parameter Adaptation

Figure 5-22 illustrates model adaptation, showing how process output is aligned with model outputs from three models.

After model adaptation completes, controller redesign begins using the first order-plus-deadtime process model. Any model-based tuning rules, typically Lambda or IMC, can be applied. If there are infrequent changes in the manipulated input, external excitations can be automat-

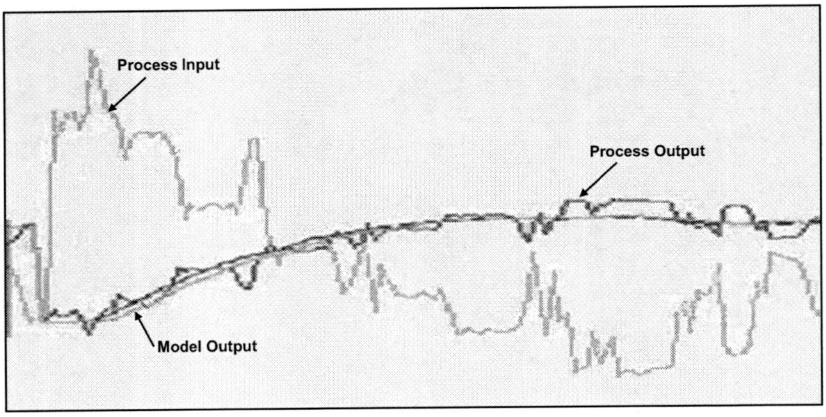

Figure 5-22. Adaptive Models Validation Plots

ically injected into the manipulated input in Automatic mode. A PID output pulse of amplitude 3–7% and duration of several scans is normally enough for a process model identification. This type of a short excitation does not cause a significant disturbance to the process output.

The same adaptation technique can be applied for the feedback and feedforward path, with the modeling structure as in Figure 5-23.

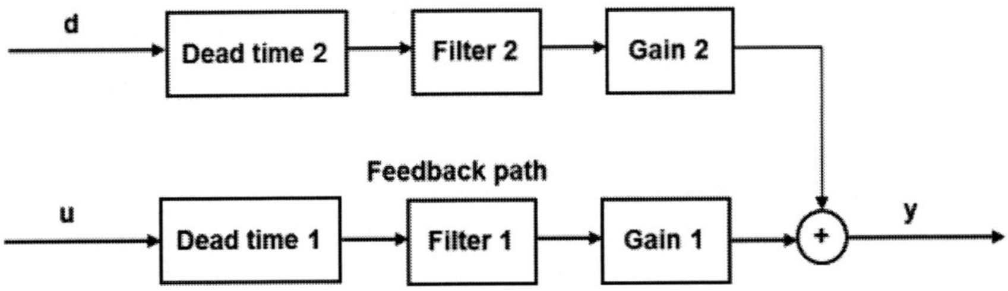

Figure 5-23. Model Structure Used in Adaptive Feedback/Feedforward PID Controller

Knowing the feedforward path model, the dynamic feedforward controller design formula (Equation 5-18) can be applied (Seborg, Edgar, and Mellichamp, 2004).

$$G_f = -\frac{K_f}{K}\frac{1 + s\tau}{1 + s\tau_f} \tag{5-18}$$

where:

G_f	=	feedforward controller transfer function
K_f, τ_f	=	static gain and time constant of the feedforward model
K, τ	=	static gain and time constant of the feedback model

The applied adaptive technique allows several levels of safety nets:

- Setting limits for the model parameters
- Setting ranges for model parameters adaptation in one adaptation cycle
- Validating models
- Detecting and preventing oscillation

High and Low limits for all model parameters are set automatically at initialization. Those limits can be changed by the operator.

During a particular adaptation cycle, the maximum range of parameter change is predefined and limited (typically +/- 25% of the parameter acceptable range.)

Model validation and preventing oscillations is a unique feature of the technique. The quality of the adaptive model can be calculated as the squared error between the calculated and actual process output. Residual errors are calculated for the three models – one for the model with a nominal parameter a_p the other two for the models with $a_p - \Delta a_p$, and $a_p + \Delta a_p$. This allows both evaluation of a particular model and comparative model validation.

Several criteria are used for the comparative model validation:

- Ratio of the maximum residual error to the minimum residual error. If the ratio is close to 1, a high noise level or a significant distance from the true model value is implied. If the ratio is high, a low noise level and/or fast convergence of the model parameters is indicated.
- Test to determine if the model with a middle value parameter value has the smallest error. Satisfaction of this criterion indicates that the true model value lies between the low and

high range of the parameter, and the interpolated parameter value should be close to the true value.

- Current model quality can be used for adjusting the filter, which applies only a portion of the currently adapted model for the filtered adaptive model update.

Finally, statistical model evaluation accounting for a certain number of adaptations can provide the next and more reliable level of the model validation. The factors taken into account in defining statistical model quality should be:

- Number of adaptations
- Average model quality
- The current and preceding adaptation's model quality
- The last adaptation's model parameter standard deviation

As it follows from the experience (http://www.easydeltav.com), an adaptive tuner product accepted in various industries requires a robust tuning technique, multilevel safety nets, user-friendly interface, and control performance reporting software (Chapter 3).

BIBLIOGRAPHY

1. Åström, K. and Hägglund, T. *Automatic Tuning of PID Controllers*, Research Triangle Park: ISA, 1988.
2. Åström, K. and Hägglund, T. *PID Controllers: Theory, Design, and Tuning*, Research Triangle Park: ISA, 1995.
3. Åström, K. J. and Hägglund, T. *Advanced PID Control*, Research Triangle Park: ISA, 2006.
4. Brozenec, T. F., Tsao, T. C., and Safonov, M. G. "Controller Validation," *International Journal of Adaptive Control and Signal Processing*, Vol. 15, 2001, pp. 431–444.
5. Chien, I-L. and Fruehauf, P. S. "Consider IMC Tuning to Improve Controller Performance," *Chemical Engineering Progress*, Vol. 86 (10), pp. 33–41, October 1990.
6. Gendron, S. "Improving the Robustness of Dead-time Compensators for Plants with Unknown or Varying Delay," *Preprints of the Control Systems 90 Conference*, Helsinki, 1990.
7. http://www.easydeltav.com/
8. Jun, M. and Safonov, M. G. "Automatic PID Tuning: An Application of Unfalsified Control," *Proceedings of the 1999 IEEE International Symposium on Computer Aided Control System Design*, Hawaii, USA, August 1999.

9. Lamaire, R. O., Valavani, L., Athans, M., and Stein, G. "A Frequency-domain Estimator for Use in Adaptive Control Systems," *Automatica*, Vol. 27, No. 1, 1991, pp. 23–38.

10. Maršik, J. and Strejc, V. "Application of Identification-free Algorithms for Adaptive Control," *Automatica*, Vol. 25, No. 2, 1989, pp. 273–277.

11. Morse A. S., Pait, F. M., and Weller, S. R. "Logic-Based Switching Strategies for Self-Adjusting Control," Presented at 33rd IEEE Conference on Decision and Control, Workshop Number 5. Lake Buena Vista, Florida, USA, December 1994.

12. Narendra, K. S. and Balakrishnan, J. "Adaptive Control Using Multiple Models," *IEEE Transactions on Automatic Control*, Vol. 42, No. 2, February 1997, pp. 177–187.

13. Safonov, M. G. and Tsao, T. C. "The Unfalsified Control Concept and Learning," *IEEE Transactions on Automatic Control*, Vol. 42, No. 6, Jun. 1997, pp. 843–847.

14. Seborg, D., Edgar, T. F., and Mellichamp, D. A., *Process Dynamics and Control*, John Wiley & Sons, Hoboken, N.J., 2004.

15. Wojsznis, W. K. and Blevins, T. L. "State Based Adaptive Feedback Feedforward PID Controller," U.S. Patent 7,113,834, 2006.

16. Wojsznis, W. K., Blevins, T. L., and Wojsznis, P. W. "Adaptive Feedback/ Feedforward PID Controller," Proceedings of ISA Technical Conference, Houston, 2003.

17. Wojsznis, W. K. "Discrete Fourier Transform Based Self-Tuning," *Advances in Instrumentation and Control*, Proceedings of ISA/96 Technical Conference, Chicago, October, 1996.

18. Wojsznis, W. K., Borders, G. T., Jr., and McMillan, G. K. "Flexible Gain Scheduler," *ISA Transactions* 33(1), May, 1994, pp. 35-41, Elsevier, New York.

19. Wojsznis, W. K., Gudaz, J., and Blevins, T. L. "Adaptive Model Free PID Controller," Paper presented at Fifth SIAM Conference on Control Applications, San Diego, CA, July 2001.

6 Fuzzy Logic Control

Fuzzy logic is a relatively new control technology. In the mid-1970s, Mamdani (1974) began to investigate fuzzy logic as an alternative to the PID controller (Kratowicz, 2006) and sparked the first practical applications of fuzzy logic in the process control industry (Shaw, 1998). Kratowicz (2006) designed a fuzzy logic controller that followed rules that were analogous to PI control rules. As a result, the rules generated an output by evaluating the error and the change in error. Membership functions that represent the degree of truth as an extension of valuation were created for error, change-in-error, and change-in-output. The shape, size, and overlap of membership functions were defined to be tunable parameters. Similar to the adjustment of PID proportional, integral, and derivative gain, more or less aggressive control for a given process can be achieved with fuzzy logic control through adjusting the scaling of membership functions. The fuzzy logic control proposed by Mamdani has since been simplified (Qin, 1994), and methods have been developed for basing the adjustment of membership function scaling on the identified process gain and dynamics. Today, based on these developments, many single loop digital controllers, and some distributed control systems, offer a version of fuzzy logic control with an auto-tuning capability. The control functionality that may be achieved using fuzzy rules and membership functions is similar to that available for PID control.

For some specific process applications, fuzzy logic control enables faster setpoint recovery with less overshoot than PID control for both setpoint and load changes (Figure 6-1). Fuzzy logic is best suited for controlling processes that are characterized by large time constants and little or no deadtime and where the control objective is to reach setpoint in the minimum time with no overshoot.

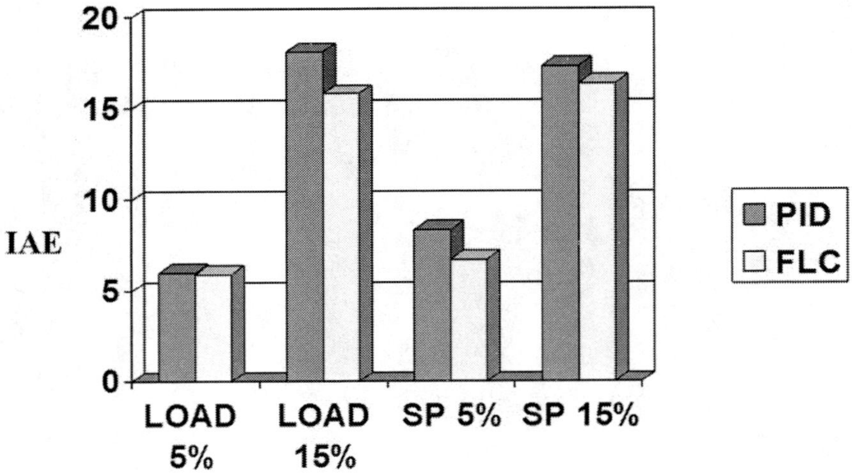

Figure 6-1. Control Performance, Fuzzy Logic vs. PID

Figure 6-1 compares the Integral of Absolute Error (IAE) on two cascade loops, one using traditional PID and the other using fuzzy logic control. Fuzzy exhibited less IAE for both load disturbance and setpoint changes. The difference in control response is illustrated in Figure 6-2.

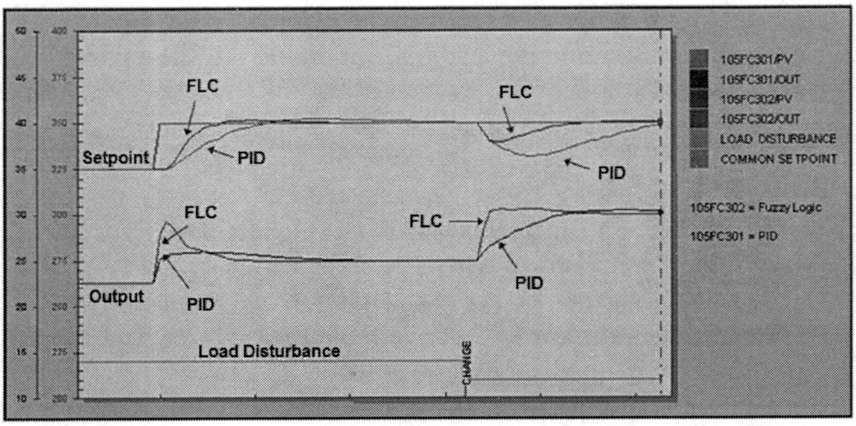

Figure 6-2. Fuzzy Logic Control Response vs. PID

To minimize the time needed to apply fuzzy logic control, most manufacturers offer a version of fuzzy logic control in which the rules and membership functions are pre-defined. Also, by restricting the fuzzy logic capability in this way, a manufacturer can provide support for auto-tuning. Thus, fuzzy logic control can be substituted for PID control. An example of a fuzzy logic function block that is designed as a replacement for PID control is shown in Figure 6-3.

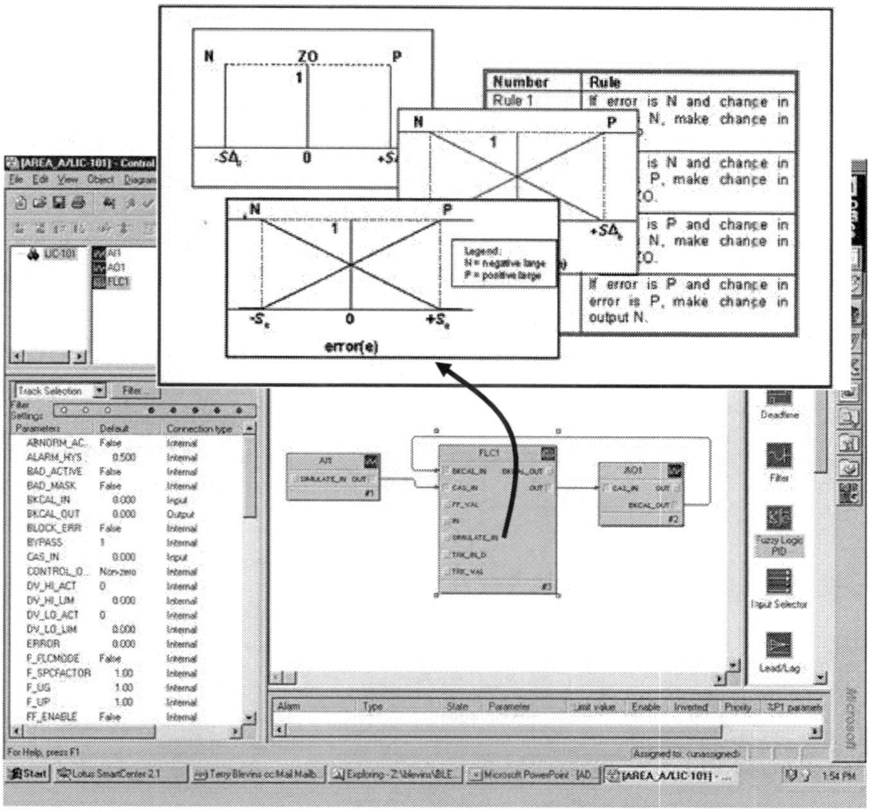

Figure 6-3. Predefined Fuzzy Logic Control Function Block

This type of single loop fuzzy logic control has been successfully applied to a variety of applications. Many of these applications involve temperature processes that are characterized by significant process lag with little or no deadtime.

Example 1 – Chiller Temperature Control

Chillers are commonly used for temperature control in many chemical processes. The ability for the temperature control to be able to correct for process disturbances and setpoint changes without overshooting setpoint is important to the process operation. One example of where fuzzy logic control has been used to provide this control response is the Kodak Park Refrigeration & Water Department, Rochester, New York. At that plant, a number of the refrigeration systems consist of several steam driven chiller units (Wildman, 1996). The objective for the largest system is to maintain water at a temperature of 0 °F (+/- 1 °F). Chiller units are frequently put on-line or taken off-line as necessary to meet the chilled water demand and to balance the steam load at the cogeneration plant. During chiller startup, the existing feedback PID controller typically allowed temperature overshoots and oscillation around the setpoint. These overshoots frequently led to automatic chiller shutdowns to prevent damage due to low evaporator pressure. Avoiding shutdowns and restarts is a major objective of the Refrigeration & Water Department.

It has been reported by the plant that after replacing the PID controller with fuzzy logic control, they observed a reduction in chiller water temperature variations and the elimination of temperature overshooting setpoint during chiller startup. Savings were realized by reducing the amount of time required to bring chiller units on-line automatically. Although operators are able to manually start units, even the best operators admit they cannot do the job as well manually as fuzzy logic does it automatically. Fuzzy logic control not only saves operators time, it produces precisely controlled water temperature, which results in cost savings in manufacturing operations. To quote one engineer, "We saw a kind of control we had never seen before. The water temperature came down at essentially the same rate as before, but it did not overshoot and did not oscillate around the setpoint. The temperature dropped right to the specification and just stuck there within plus or minus a tenth of a degree."

Example 2 – Cooler Temperature Control

Witco Corp. has reported that they struggled to maintain a constant 160 °F temperature for a newly made intermediate as it traveled from the production line to temporary storage tanks (Daly, 2006). An upstream distillation unit's level control constantly varied the flow rate of the surfactant traveling into the heat exchanger, while the heat exchanger's cooling water varied between 60 °F and 90 °F, depending on the season. For a little more than a year, Witco used a conventional PID loop with feedforward capability and adaptive tuning techniques to get the system's temperature variances down to ±5 to ±6 °F. However, Witco still required tighter temperature control for the line producing the intermediate, which is used in fabric softeners and other products. Overshooting the target temperature by 6 °F made the liquid too warm for transport and extended its stay in temporary storage tanks from less than a day to several days. Meanwhile, temperature undershoots also created problems. Given that the intermediate freezes at 150 °F, the liquid also begins to congeal when the temperature drops only a few degrees below the 160 °F target temperature. Once congealing starts, the heat exchanger loses efficiency until eventually it and the line become clogged with solid intermediate.

The results improved immediately after installing fuzzy logic for the temperature control. The line's temperature variance quickly fell within the desired ±3 °F band and stayed there. Fuzzy logic control maintains the temperature at plus or minus 2 °F (Keuer, 1997).

Example 3 – Dryer Temperature Control

Manufacturing carpets requires drying wet carpet from about 40% moisture to approximately 4% moisture (Kratowicz, 2006). Carpet is pulled through a seven-zone oven. Each zone has two gas-fired burners and two overhead blowers. Both overdrying and underdrying damage the product, leading to waste or reprocessing. The oven operates continuously with different batches of carpet sewn end-to-end to avoid shutdowns. Temperature and speed targets are unique for each grade of product.

One batch can have a significantly different weight and moisture load than the next batch. Transitions between grades present a difficult control problem. The operators know where to target oven temperatures to achieve correct drying. The difficulty is getting the temperature controls to respond fast enough without overshoot. Oven temperatures can vary from 80 to 340 °F, with setpoint changes as large as 100 °F. Sometimes waste carpet is sewn between carpet batches to act as a buffer while temperature control stabilizes. This extra work reduces throughput.

PID control did not meet the requirement to respond to setpoint changes in minimum time with no overshoot for the carpet manufacturer, and the PID controller was replaced with fuzzy logic. This reduced recovery time and eliminated transition overshoot. The fuzzy system was deployed only on the first zone, which has a tremendous impact on the entire oven because the burners in the first zone are larger and do most of the work. The auto-tuning function provided with the control automatically determined the scaling factors. The improved control in the first zone stabilized the process to such an extent that the PID loops in the other zones were left untouched.

The plant cited these benefits from fuzzy logic control:

- Elimination of product damage
- Flexibility to make quick changes
- Energy savings
- Ease of operation

The plant achieved documented gains of about 20% in capacity, or throughput, and rework was completely eliminated. Actual payback occurred within less than four months. Payback calculations did not include energy savings, which were also observed but not quantified. Expectations were met and exceeded.

6.1 Application Example

This example illustrates the use of a fuzzy logic control capability that is embedded in an industrial control system. The following topics are addressed:

- Fuzzy logic configuration
- Automated tuning
- Example of closed loop response

To allow consistent implementation of control applications, fuzzy logic control is implemented as a function block. The fuzzy rules and membership functions are predefined except for the scaling of the membership functions for error, change in error, and change in output. The fuzzy logic block replaces the PID block, as illustrated in Figure 6-4.

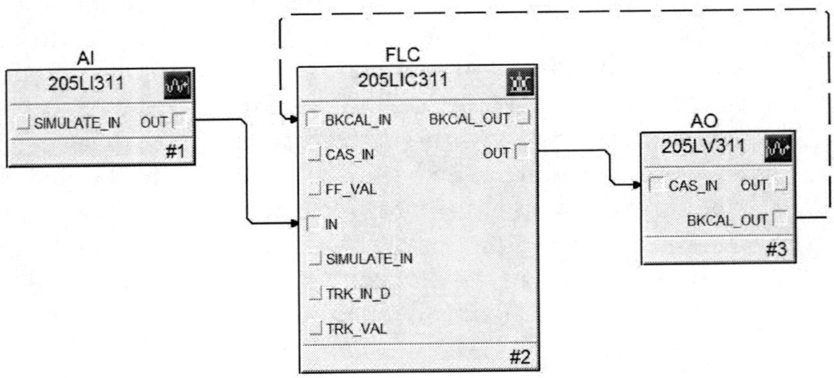

Figure 6-4. Fuzzy Logic Function Block for Control Applications

In this example, a tuning capability is provided to automatically set the fuzzy logic control block's scaling. The tuner interface is the same as that provided for PID control, except that the tuning parameter is for the membership functions' scaling, as shown in Figure 6-5.

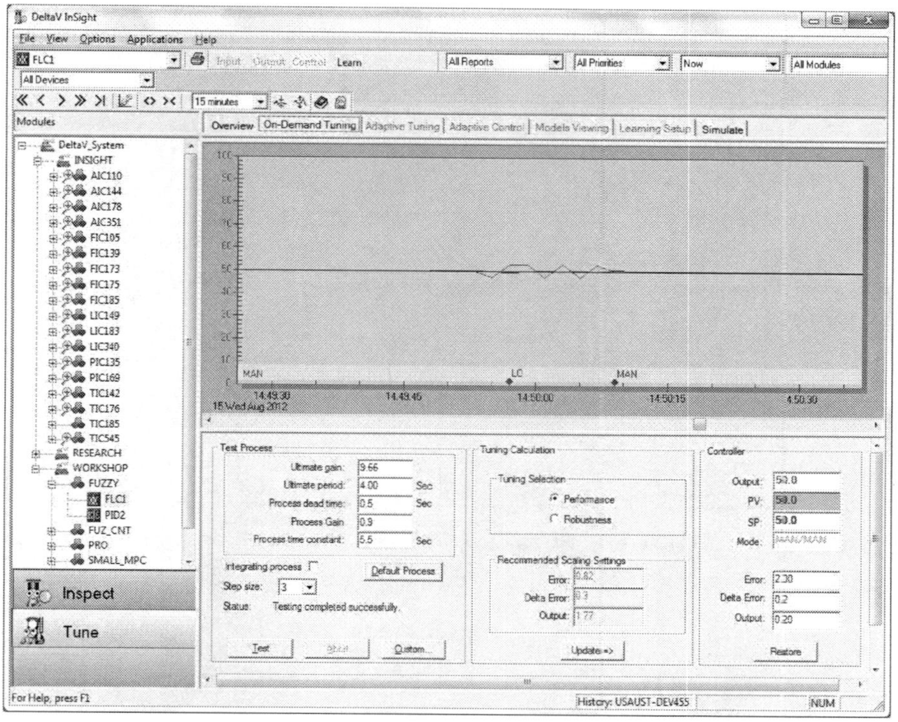

Figure 6-5. Tuner Interface for Fuzzy Logic Control

Similar to auto-tuning for PID, the process dynamics are automatically determined when the Test button is selected. The process is under two-state control during testing. The manipulated parameter is changed from its initial value by the specified step size. Based on the process dynamics, the recommended membership functions' scaling is calculated automatically.

Users can select the Simulate button to see the performance that this tuning will provide before using it in the process. The closed loop response to setpoint changes and load disturbances is shown. When the user changes the values of the recommended scaling factors, a new response is displayed as illustrated in Figure 6-5.

The recommended tuning is transferred to the fuzzy logic block by selecting Update.

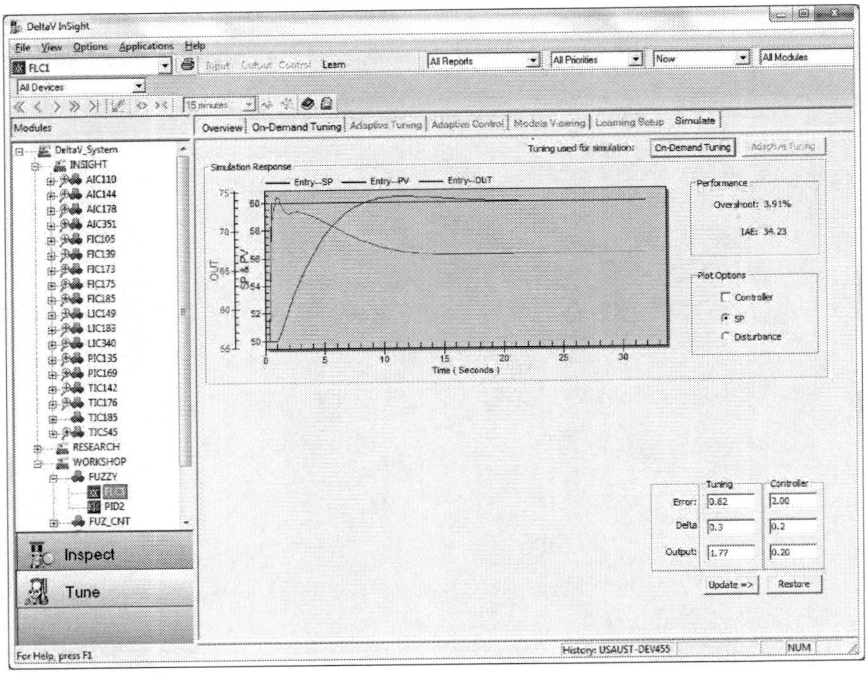

Figure 6-6. Fuzzy Logic Response for Recommended Tuning

Since fuzzy logic control is non-linear, the robustness of the control as measured by phase and gain margin cannot be applied as was shown for PID on-demand tuning in Chapter 4.

6.2 Fuzzy Logic Control Workshop

In this workshop, the control performance of fuzzy logic control will be compared to the performance of PID control when the control objective is to reach setpoint in minimum time with no overshoot. Two identical process heaters (Figure 6-7) are simulated with PID and fuzzy logic control. Access the book's web site http://www.advancedcontrolfoundation.com and select the Solution tab to view the workshop being performed.

Step 1: Open the workshop module and initialize the IAE calculation.

Step 2: Increase the temperature setpoint to the PID and Fuzzy blocks by changing the "SP" parameter value. Note the calculated IAE once both loops achieve setpoint and use the plot to observe which control reached the setpoint faster.

Step 3: Repeat Step 2, but this time decrease the "SP" value by 10. Note that the process and control response is the same as for a temperature increase.

Step 4: Reset the IAE calculation and then introduce an unmeasured process disturbance into both heaters by changing the "DISTURBANCE" parameter from 0 to 15. Note the IAE once both loops achieve setpoint and use the plot to observe which control was the first to reach setpoint.

Step 5: Repeat Step 3, but change the "DISTURBANCE" parameter value to 0. Note that the performance equivalent to that seen in Step 3.

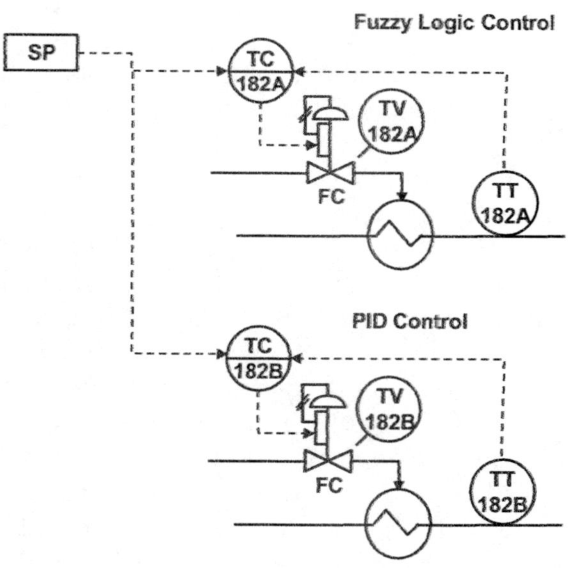

Figure 6-7. Processes for Fuzzy Logic Workshop

6.3 Technical Basis

6.3.1 Introduction to Fuzzy Logic Control

Fuzzy logic control (FLC) can perform the same functions as classical linear control: simple control feedback loops, interactive loops, or multivariable processes. In many cases fuzzy logic control can provide better performance, less process variability, or better robustness. The difference derives from calculations based on fuzzy logic. By applying fuzzy logic calculations, it is possible to design a non-linear controller without a detailed knowledge of operating point nonlinearity, as is required for a classical linear control design.

Fuzzy logic has its roots in classical two-valued "True" and "False" logic. Adding "fuzziness" makes it possible to better address real-life situations. To illustrate this connection and analogy, let's review a few basics of classical logic.

Classical logic operates on crisp sets. In most everyday kinds of thinking, and linguistic reflections of that, people use crisp sets to categorize things. Two simple examples of crisp sets (they are not actually used in industrial process control) illustrate the concept:

1. A set of temperature values above 33 °C belongs to the set "hot," and those values equal to or below 33 °C belong to the set "cold"
2. A person 6 feet tall belongs to the set "tall," but a person 5 feet 11.99 inches belongs to the set "short."

Is it reasonable to call a person of 5 feet 11.99 inches "short"? Is it reasonable to call a temperature of 32.99 °C "cold"? Applying common sense, clearly it is not. This type of question stimulated the search for more flexible expressions of such situations and led to the development of *fuzzy sets* and *multi-valued logic.*

In fuzzy sets, bounds are defined in a more flexible way. Set "hot" for example, is defined as including not only values above 33 °C, but also including any value 33 °C or lower that is "hot" to some degree, until it is fully "hot" at temperatures above 33 °C. The "degree of hotness" at 27 °C would of course be greater than at 26 °C, for example. To define the "degree of hotness," fuzzy logic applies a *membership function*. The membership function defines the degree of membership to a fuzzy set.

The above discussion is summarized in Table 6-1 and Figure 6-8.

Table 6-1. Classical Logic vs. Fuzzy Logic

	Sets	Membership	Logic Values	Operations
Classical Logic	Crisp sets	Unambiguous	"True," "False" or 1,0	Two-valued logic
Fuzzy Logic	Fuzzy sets	Various degrees of membership defined by the membership function	Any value in the range [0,1] defined by the membership function	Multi-valued logic

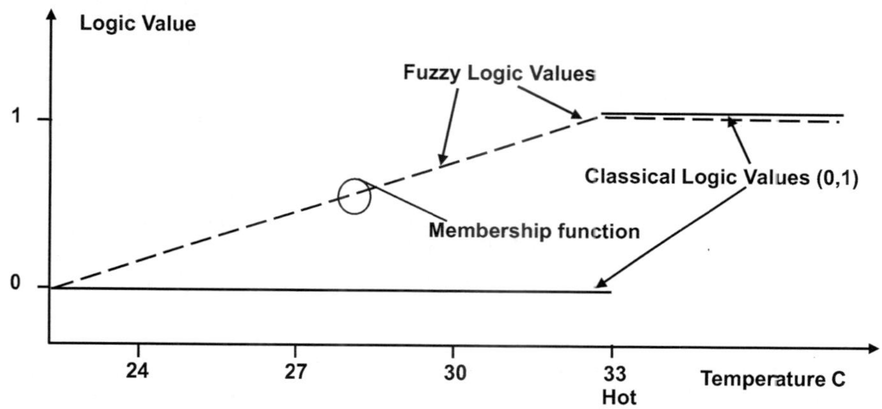

Figure 6-8. Logic Values in Classical Logic vs. Fuzzy Logic

In summary, we can say that in fuzzy logic the truth of any statement is a matter of degree. We can observe as well that classical logic is a special case of fuzzy logic with a step membership function. In other words, for a crisp set, elements of the set definitely do belong to the set, while in a fuzzy set elements of the set have a degree of membership in the set.

After fuzzy logic values are defined, the next step is to define fuzzy logic operations. Fuzzy logic operates on many values, and operations in fuzzy logic are based on multi-valued logic principles. Again using the comparison with classical logic, we introduce the basics of multi-valued logic operations in Table 6.2. Multi-valued logic uses the same operations as classical logic; the only difference is that the operations are performed on the real values in the range [0–1] instead of on 0 or 1 only.

Fuzzy logic operations can be illustrated with a simple example using already discussed membership functions for the height and temperature. Assume John is 6 feet 1 inch tall and today's temperature is 30 °C. We assign to 30 °C the membership function value 0.8. Applying classical logic to the question "Is it true that John is tall and today is hot?" we would answer "False." The fuzzy logic answer would be: "True with degree 0.8".

Table 6-2. Two-Valued Logic Operations vs. Multi-Valued Logic Operations

Operation	"A" AND "B"	"A" OR "B"	NOT "A"
Two-valued logic	min(A,B) min(0,0) = 0 min(0,1) = 0 min(1,0) = 0 min(1,1) = 1	max(A,B) max(0,0) = 0 max(0,1) = 1 max(1,0) = 1 max(1,1) = 1	Complement to 1 NOT(0) = 1-0 = 1 NOT(1) = 1-1 = 0
Multi-valued logic	min(A,B) Minimum of two values A, B in the range 0-1.	max(A,B) Maximum of two values A, B in the range 0-1.	Complement to 1 NOT(A) = 1-A $A \leq 1$

In the next sections we will learn more about fuzzy logic control by developing several fuzzy logic controllers. While summarizing the basic facts about fuzzy logic, it is worthwhile to honor Zadeh (1965), who developed fuzzy logic in the early 1960s, and Lukasiewicz (1963), who pioneered in developing multi-valued logic in the 1930s.

6.3.2 Building a Fuzzy Logic Controller

A typical fuzzy logic controller performs three basic operations (see Figure 6-9, more in Sugeno, 1985):

- Translating input signals into fuzzy logic values, or *fuzzification.*
- Applying multi-valued logic *inference rules (fuzzy rules)* to develop the control action.
- Retranslating the fuzzy logic values into continuous output signals, or *defuzzification.*

Fuzzification

As mentioned above, *fuzzy sets* and *membership functions* are used to translate controller input values into fuzzy values (Pedrycz, 1993), (Chen, Trung, 2001).

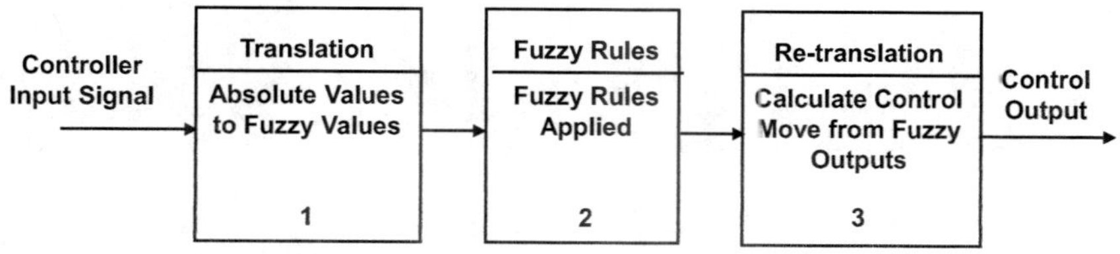

Figure 6-9. A Typical Fuzzy Logic Controller Diagram

At first we decide how we want to logically quantify control inputs by defining fuzzy sets. The typical fuzzy sets for input signals are Negative Large, Negative Medium, Negative Small, Zero, Positive Small, Positive Medium, and Positive Large. For simpler designs the fuzzy sets are Positive and Negative or Low and High.

The next step is to define membership functions for every fuzzy set, to establish boundaries for the set. Membership functions are defined analytically. Examples of simple membership functions for fuzzy sets Low and High are given by Equations 6-1 and 6-2 and presented graphically in Figure 6-10. Level is used as the control input. Prior to applying fuzzification, control input is normalized to the range 0–1.

$$M(High) = 1.0 * Level \qquad 0 \leq Level \leq 1 \tag{6-1}$$

$$M(Low) = 1.0 - 1.0 * Level \qquad 0 \leq Level \leq 1 \tag{6-2}$$

Equation 6-1 defines the membership function for fuzzy set High, and Equation 6-2 for fuzzy set Low. A level equal to 70% or .7 gives a degree of membership of .7 for the High membership function and .3 for the Low membership function. This can be expressed as M(Low) = .3; M(High) = .7 .

Rather than requiring full membership in one state or the other, fuzzy logic allows for a tank's level, for example, to be *somewhat high* (degree of membership = .7) and *somewhat low* (degree of membership = .3) at the same time. In other words, a statement is true only to a certain extent or degree.

Membership functions are of various types: linear, as in the above example; parabolic, exponential, or normal distribution. Linear membership functions are preferred in most applications due to their simplicity.

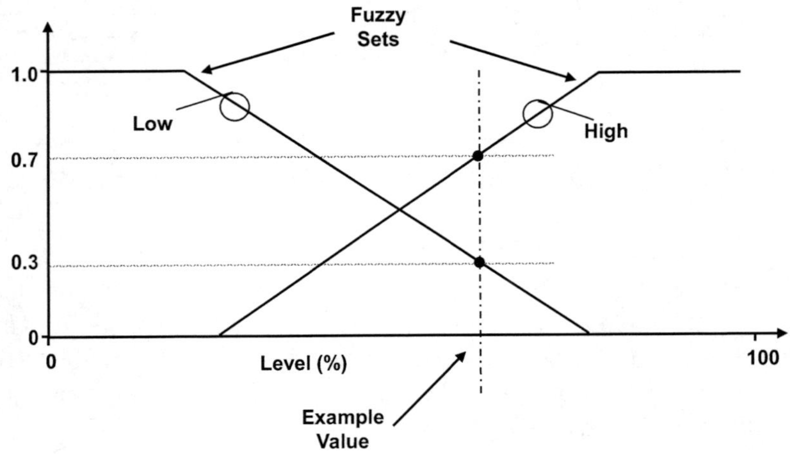

Figure 6-10. Membership Functions for Fuzzy Sets Low and High

Fuzzy Logic Inference Rules

An understanding of process behavior is of primary importance in developing fuzzy logic inference rules. Using process knowledge, fuzzifying inputs and outputs, and applying IF-THEN reasoning leads to fuzzy logic inference rules. Finally, applying fuzzy logic operations develops controller response.

Consider the example of developing a fuzzy logic controller to control the level in a tank through the outlet control valve, measuring inlet flow and tank level, as shown in Figure 6-11.

In this example, the membership functions describing the tank's level, the tank's inlet flow, and the outlet valve position might be defined as shown in Figure 6-12.

The inference rules for controlling the tank level are defined as in Table 6-3.

After applying the degree of membership for both level and inlet flow to the fuzzy rules for the tank outlet valve, as shown in Figure 6-13, we get fuzzified output values, that is, values for the outlet valve membership functions Closed, Normal, Open. Normal flow in this example is a flow that is not too low.

As can be seen from Figure 6-13, when applying the fuzzy logic AND function to input membership values, the output membership func-

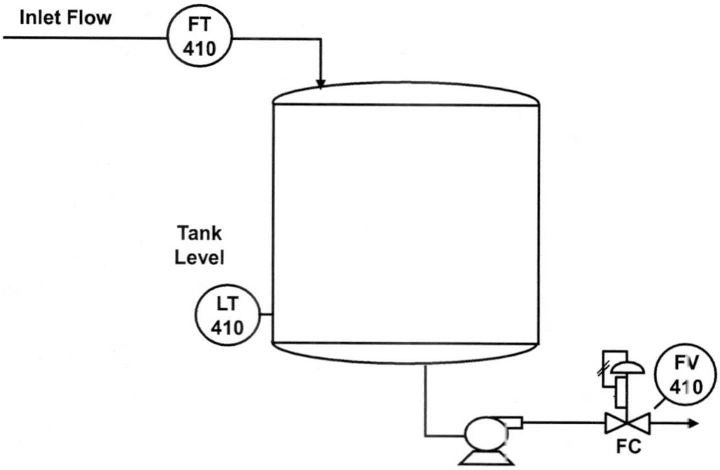

Figure 6-11. Tank Level Controlled by a Fuzzy Logic Controller

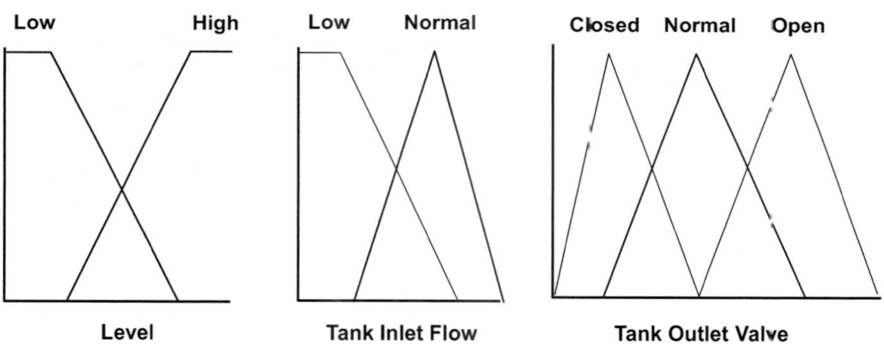

Figure 6-12. Tank Level Membership Functions

Table 6-3. Tank Level Control Inference Rules

Number	Rule
Rule 1	IF level is Low AND inlet flow is Normal THEN make outlet valve Closed.
Rule 2	IF level is High AND inlet flow is Normal THEN make outlet valve Open.
Rule 3	IF level is High AND inlet flow is Low THEN make outlet valve Normal.

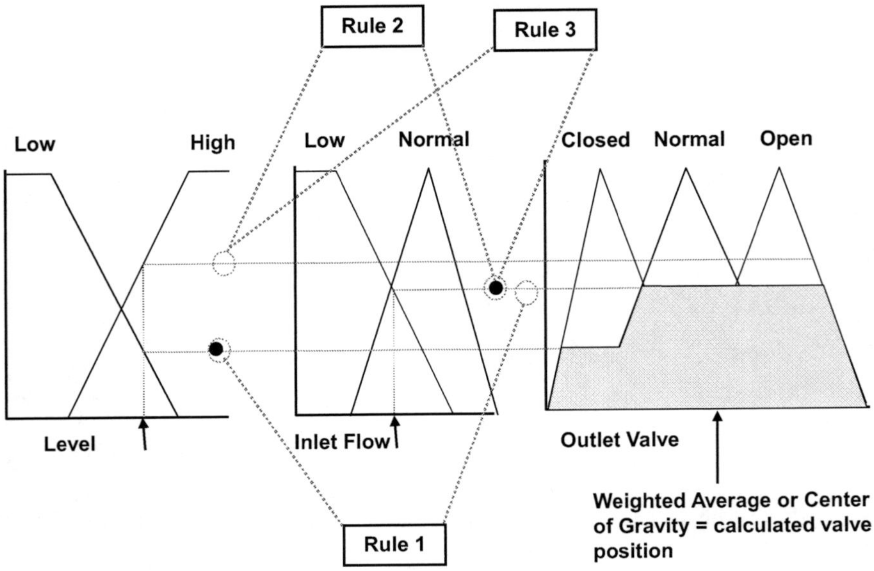

Figure 6-13. Fuzzy Logic Controller Operation

tion value is a minimum of two input membership values. The Valve Closed membership function value, for example, is equal to the Level Low membership function value, which is smaller than the value Flow Normal. This follows from Rule 1:

$$\begin{aligned} M(\text{ValveClosed}) &= \min[M(\text{LevelLow}), M(\text{FlowNormal})] \\ &= M(\text{LevelLow}) \end{aligned}$$

Defuzzification

Defuzzification translates fuzzy logic values into real control output values. The most common method of defuzzification is to calculate a weighted average of all the activated output membership functions. Referring to Figure 6-13, the weighted average of membership functions Closed, Normal, and Open determines the outlet valve position.

It is possible to build a significantly more complex controller, including a multivariable interactive controller, by applying the three-step development procedure for a fuzzy logic controller. The highest expected control performance, however, will require adjusting membership functions and scaling input/output. There are fuzzy logic tools available which simplify the effort, such as the Matlab® fuzzy logic toolbox.

For those with no fuzzy logic control experience, a pre-built fuzzy logic controller might be the most effective way to start the application of fuzzy logic. The next two sections discuss the design and use of a fuzzy logic controller that is functionally equivalent to a PID controller. This type of controller is available as a ready-to-use product in some DCS (www.easydeltav.com).

6.3.3 Fuzzy Logic PID Controller

Fuzzy logic PID control (FPID) provides an advanced alternative to standard PID control. FPID provides the control capability of PID with the added benefit of superior response to both setpoint changes and external load disturbances. By using fuzzy logic, FPID minimizes overshoot and provides good load-disturbance rejection. The scaling factors of FPID are automatically established using an auto-tuner as discussed by Qin (1994) and Qin, Ott, and Wojsznis (1998).

FPID operates by using predefined fuzzy rules, predefined membership functions, and adjustable parameters known as scaling factors. The FPID translates the loop's absolute values into fuzzy values by first calculating the scaled error (e) and the scaled change in error (Δe) and then the degree of membership in each of the predefined membership functions. It then applies the fuzzy rules and finally retranslates the values into a control move.

The FPID is available as a Fuzzy Logic Control (FLC) function block in a DCS (www. easydeltav.com). This function block supports mode control, signal scaling and limiting, feedforward control, override tracking, alarm limit detection and signal status propagation.

The nonlinearity built into the FLC function block reduces overshoot and settling time, achieving tighter control of the process loop. Specifically, the FLC function block treats small control errors differently from large control errors and penalizes large overshoots more severely. It also severely penalizes large changes in the error, helping to reduce oscillations.

FLC Membership Functions

The FLC function block uses three membership functions: the input signals are "error" and "change in error," and the output signal is the change in control action. The relationships among these three variables represent a non-linear controller. The nonlinearity results from a translation of process variables into a fuzzy set (fuzzification), rule inference, and retranslation of the fuzzy set into a continuous signal (defuzzification).

The two types of membership functions applied for error, change in error, and change in output are **Negative (N)** and **Positive (P)**. The membership functions' scaling (*Se* and *S*Δ*e*) and the error and change in error value, respectively, determine the degree of membership. First, error in percent is calculated relative to the setpoint and then scaled to the range [-1, 1]:

$$E_s = (PV - SP)/S_e \qquad -1 \le E_s \le 1 \tag{6-3}$$

Similar calculations are done for change in error:

$$\Delta E_s = [[PV(t) - SP(t)] - [PV(t-1) - SP(t-1)]]/S_{\Delta e} \qquad -1 \le \Delta E_s \le 1 \tag{6-4}$$

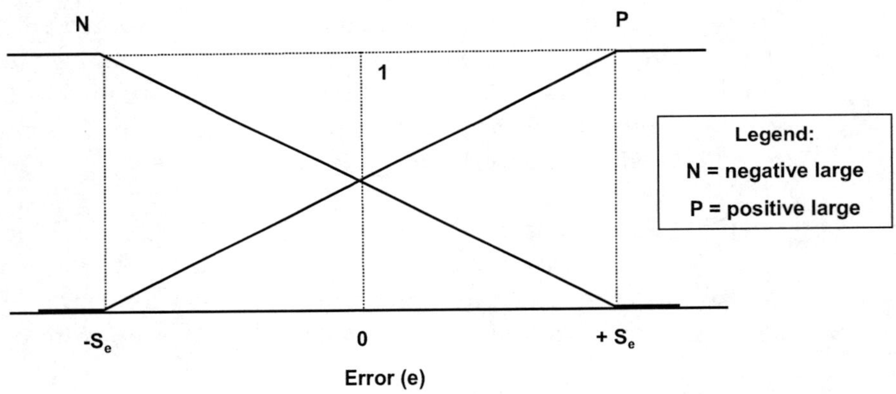

Figure 6-14. Error Membership Functions

Degrees of membership for positive error Me(P) and negative error Me(N) are calculated by using the following expressions and plotted in Figure 6-14:

$$Me(P) = 0.5 + 0.5E_s \qquad Me(N) = 0.5 - 0.5E_s \tag{6-5}$$

Membership functions for change in error are calculated in a similar way and plotted in Figure 6-15:

$$M\Delta e(P) = 0.5 + 0.5E_{\Delta s} \qquad M\Delta e(N) = 0.5 - 0.5E_{\Delta s} \tag{6-6}$$

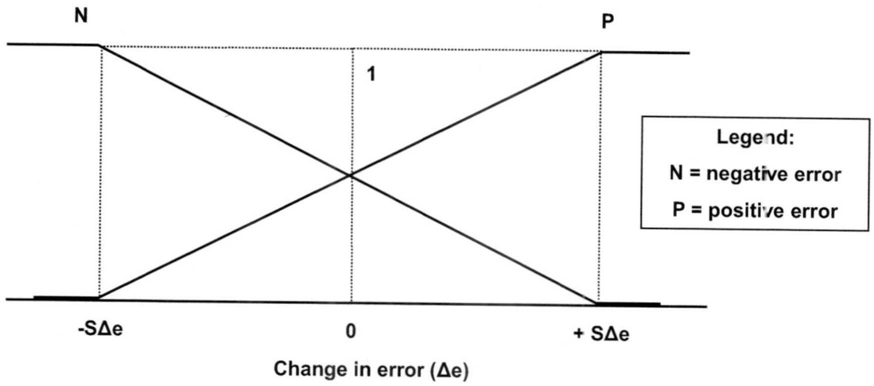

Figure 6-15. Change in Error Membership Functions

The changes in output membership functions are called *singletons.* They represent fuzzy sets with a single point and a membership function of 1. The membership scaling ($S\Delta u$) determines the magnitude of output change for an absolute value of scaled error and change in error equal to 1 with the same sign – Figure 6-16.

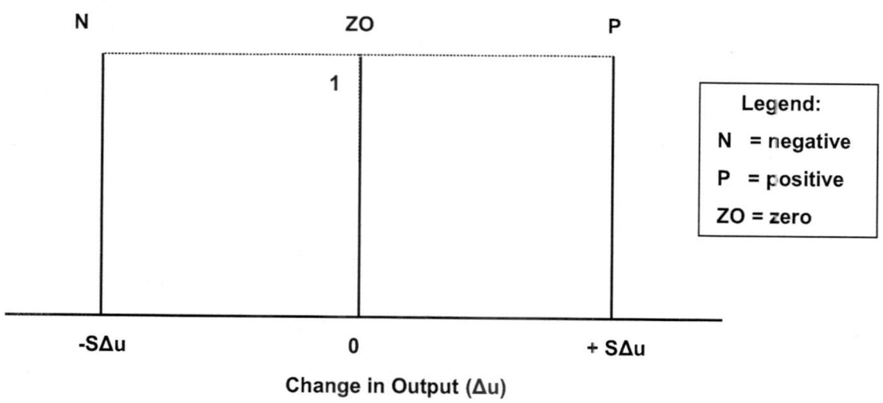

Figure 6-16. Change in Output Singleton Membership Function

Fuzzy Logic Inference Rules

There are four fuzzy logic IF-THEN inference rules that the FLC function block uses for a reverse acting fuzzy logic PI controller. The rules are listed in Table 6-4. The same rules are also written in a matrix form as in Table 6-5, where matrix cells represent fuzzy output membership functions.

Table 6-4. Fuzzy Logic Rules for Fuzzy PI Controller

Number	Rule
Rule 1	IF error is **N** AND change in error is **N** THEN make change in output **POSITIVE**
Rule 2	IF error is **N** AND change in error is **P** THEN make change in output **ZERO**
Rule 3	IF error is **P** AND change in error is **N** THEN make change in output **ZERO**
Rule 4	IF error is **P** AND change in error is **P** THEN make change in output **NEGATIVE**

Table 6-5. Fuzzy Logic Rules for Fuzzy PI Controller in a Matrix Form

Error\Change in Error	Negative	Positive
Negative	Positive	Zero
Positive	Zero	Negative

The membership functions for controller output are calculated by using Tables 6-4 and 6-5 and applying multi-valued logic operation rules:

$$\begin{aligned}
M\Delta u(P) &= \min(Me(N), M\Delta e(N)) \\
M\Delta u(N) &= \min(Me(P), M\Delta e(P)) \\
M\Delta u(Z) &= \max\{\min(Me(P), M\Delta e(N)), \min(Me(N), M\Delta e(P))\}
\end{aligned} \tag{6-7}$$

$M\Delta u(P), M\Delta u(N), M\Delta u(Z)$ are Negative, Positive and Zero membership functions for the change in output

Defuzzification

As mentioned, defuzzification uses the center-of-gravity method to determine the fuzzy scaled change in the output Δu_s:

$$\Delta u_s = \frac{M\Delta u(P) - M\Delta u(N)}{M\Delta u(P) + M\Delta u(Z) + M\Delta u(N)} \tag{6-8}$$

The change in output is calculated as:

$$\Delta u = \Delta u_s S_{\Delta u} \tag{6-9}$$

6.3.4 Fuzzy Logic Control Nonlinear PI Relationship

The two membership functions associated with each input variable and the single membership function for the output variable make the FLC function block non-linear in its response.

For regions where the absolute error is greater than the error scaling factor or the absolute change in error is greater than the change in error scaling factor, the values for error and change in error are clipped at the error scaling factor and change in error scaling factor, respectively. Figure 6-17 shows an example of an FLC gain curve that illustrates the change in controller gain for an FPID controller when $e = \Delta e$.

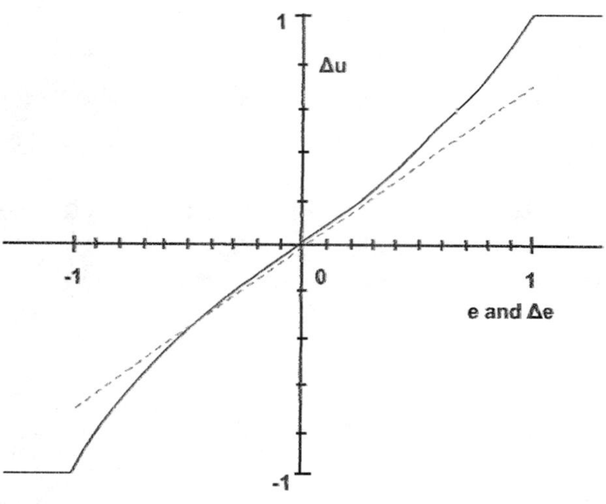

Figure 6-17. FLC Function Block's Gain Dependence on $e + \Delta e$ when $e = \Delta e$

The straight line through the origin shows the linear relationship of a conventional PI controller. As the error and change in error increase, the change in output of a conventional PI controller increases linearly. Note that the FLC function block's gain is similar to the PI controller's gain when the error and change in error are small. The FLC function block's gain increases gradually as the error and change in error increase.

The nonlinearity built into the FLC function block, with increased gain when the error and change in error increase and decreased gain when the error and change in error decrease, reduces overshoot and settling time, achieving tighter control of the process loop. To help anticipate a rapid change in the process, derivative action is provided in the feedback path of the loop, as shown in Figure 6-18, creating a fuzzy logic PID controller (FPID).

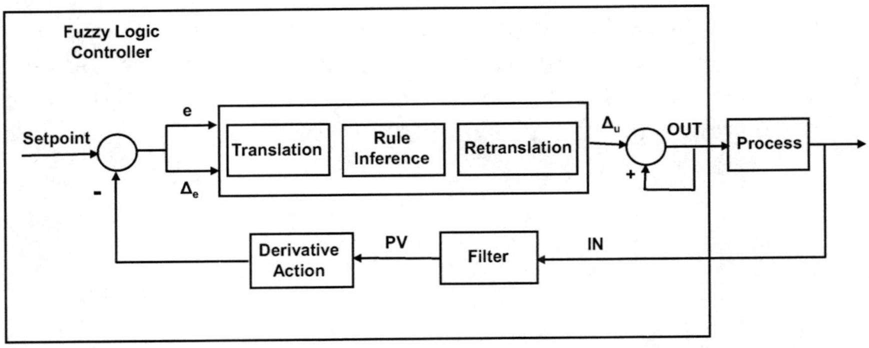

Figure 6-18. Fuzzy Logic Control with Derivative Action

As addressed in 6.3.3, the FLC function block treats small control errors differently from large control errors and penalizes large overshoots and large changes in the error, helping to reduce the oscillation.

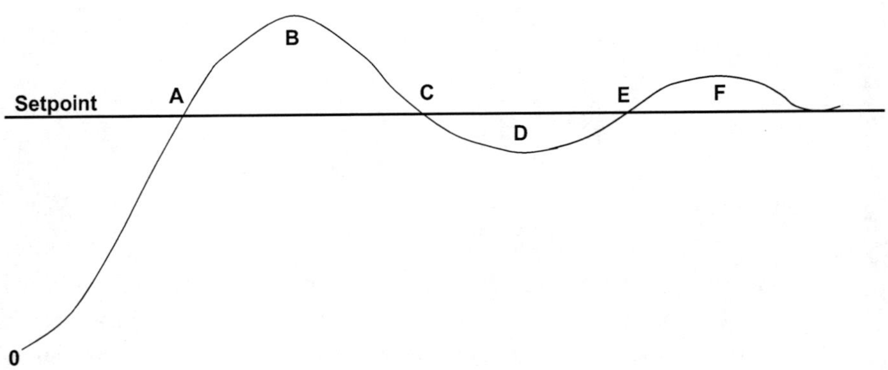

Figure 6-19. Process Variable Oscillation Example

Figure 6-19 shows how an FPID reacts to overshoot and oscillation. The FPID applies stronger control action prior to points B, D, and F where overshoot is significant and increasing, as compared to control action at points B, D, and F where the change in error is small, or after points B, D, and F where the change in error is negative. Similarly at points A, C, and E, control action is defined jointly by the minimum value of error and change in error.

This type of nonlinearity allows the FPID to provide better control performance than standard or nonlinear PID control. Following Qin (1994) we compare an FPID and a simple non-linear controller, an error-squared controller with the gain K_{esq} defined in Shinsky (1988):

$$K_{esq} = K_p(L + (1 - L)|e|) \tag{6-10}$$

where L is an adjusting parameter representing a weighting factor between the error-squared term and a linear term. The error-squared PID controller also tends to apply stronger action to larger control errors. However, a few differences between an FPID controller and an error-squared PID controller are immediately apparent:

- The error-squared PID controller applies stronger gain when the error is larger, while the FPID controller applies stronger gain when both the error and the change in the error are larger.
- The error-squared PID controller gain is extremely large given a very large controller error, while FPID controller gain has an upper limit.
- Some versions of the error-squared PID controller use a very small or zero value for L, which indicates that the controller gain around the origin is very small or zero. However, the FPID gain is designed (as in Equations 6-3 through 6-9) around the origin to give good performance.

A good way to represent FLC is as a control surface. Figure 6-20 shows a control surface for a fuzzy PI controller as a 3D plot of change in controller output depending on error and change of error.

This representation clearly shows the analogy between fuzzy logic controller and a non-linear controller, because control surface of a non-linear controller contains many irregular slopes like fuzzy logic controller surface shown in Figure 6-20. It is possible in principle to design any non-linear controller using a FLC control surface. Change in controller output would be defined by simply reading the values of the 3D plot, if error and change in error are known.

6.3.5 FPI and PI Relationships

Developing scaling factors for a fuzzy logic controller is a tedious task. Therefore an auto-tuner is used to establish the scaling factors S_e, SD_e and SD_u. The auto-tuning of FPI is possible because of the relation-

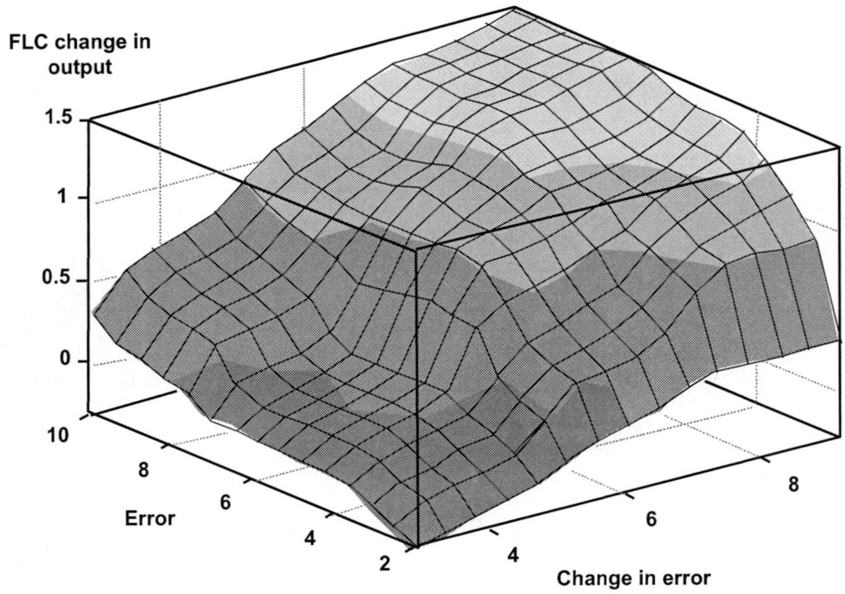

Figure 6-20. Control Surface of Fuzzy Logic Controller

ships between PI-controller tuning parameters (gain, reset) and FPI-scaling factors. For a small control error and a setpoint change of less than a nominal value D*SP*, the FPI-scaling factors are related to the proportional gain *K* and reset *Ti*, which would be used in a PI controller applied to control the same process as shown in Equations 6-11 through 6-13.

$$S\Delta e = \beta \Delta SP \tag{6-11}$$

$$S\Delta u = 2S\Delta eK \tag{6-12}$$

$$Se = \frac{T_i S\Delta e}{\Delta t} \tag{6-13}$$

where:

$S\Delta e$	=	change in error scaling factor
Se	=	error scaling factor
$S\Delta u$	=	change in controller output scaling factor
ΔSP	=	SP change relative to a nominal change (1%)
Δt	=	loop scan period

β = function of process deadtime (τ_d) and time constant (τ) with maximum value .5, as defined by Equation 6-14:

$$\beta = .2 + \frac{\tau_d}{\tau} \qquad .2 \le \beta \le .5 \tag{6-14}$$

6.3.6 Automatic Tuning of a Fuzzy PID Controller

The greatest advantage of fuzzy PID controllers as compared to the more complex fuzzy logic controllers is that the automatic tuning of a fuzzy PID controller is quite similar to that of a PID controller. Automatic tuning applies the relay oscillation technique (see Chapter 4) and exploits the relationships between PID and FPID addressed in section 6.3.5.

A fuzzy PID tuning technique has been developed by Qin (1994) and enhanced by Qin, Ott, and Wojsznis (1998).

An important feature of fuzzy PID tuning technique is that it takes error and change of error on the controller input into consideration and adjusts scaling factors appropriately to avoid saturation of the change in output as defined in (6.9). The formula $S\Delta e = \beta\Delta SP$ calculates the scaling factor for change in error, dependent on a setpoint change. The value of the factor is used until the PV comes to the SP, when the scaling factor for change in error is recalculated for $\Delta SP = 1$, and scaling factors for error and output are adjusted according to Equations 6-11 and 6-12.

In a slightly modified approach, ultimate gain and ultimate period obtained during relay oscillation tuning are used directly for calculating scaling factors without the intermediate calculation of PID controller parameters. The following formulas are applied:

$$Se = \frac{S\Delta e T_u}{2} \tag{6-15}$$

$$S\Delta u = 1.25 S\Delta e K_u \tag{6-16}$$

T_u and K_u are the ultimate period and ultimate gain of the loop

The formulas provide a good tuning over a range of .1 to .3 deadtime-to-time constant ratio. Tuning improvement over a wider range of deadtime-to-time constant ratio may be obtained by applying non-linear estimators as discussed in Chapter 4 – On-demand Tuning.

BIBLIOGRAPHY

1. Chen, G. and Trung T. *Introduction to Fuzzy Sets, Fuzzy Logic, and Fuzzy Control Systems,* Boca Raton: CRC Press, 2001.

2. Daly, R. "Fuzzy Logic: A Matter of Degrees," Published on the Internet: *Managing Automation,* Nov. 2006, web link: http://www.managingautomation.com/maonline/magazine/read/view/Fuzzy_Logic_A_Matter_of_Degrees_967.

3. Keuer, D. "Fuzzy Logic Smooths PID-Resistant Loop," *Control Engineering,* Vol. 44, No. 6, pp. 71–72, April 1997.

4. Kratowicz, R. "Fuzzy Logic: It's a Technology That Should be Crystal Clear to Control Engineers," Published on the Internet: *PlantServices.com,* March 2006, web link: http://www.plantservices.com/articles/2006/078.html?page=print.

5. Lukasiewicz, J. *Elements of Mathematical Logic,* London and New York: Pergamon Press/Warsaw, PWN-Polish Scientific Publishers, 1963.

6. Mamdani, E. H. "Application of Fuzzy Algorithms for Control of Simple Dynamic Plant," *Proceedings of the Institution of Electrical Engineers (London),* Vol. 121, No. 12, Dec. 1974, pp. 1585-1588.

7. Pedrycz, W. *Fuzzy Control and Fuzzy Systems,* Second edition, New York, N Y: John Wiley and Sons, 1993.

8. Qin, S. J., Ott, M., and Wojsznis, W. "Method of Adapting and Applying Control Parameters in Non-Linear Process Controllers," U.S. Patent No. 5,748,467, 1998.

9. Qin, S. J. "Auto-Tuned Fuzzy Logic Control," *Proceedings of American Control Conference,* Baltimore, 1994, Vol. 3, pp. 2465–9.

10. Shaw, I. S. *Fuzzy Control of Industrial Systems,* Boston: Klewer Academic Publishers, 1998.

11. Shinskey, F. G., *Proces Control Systems: Application, Design and Adjustment,* 3rd edition, New York: McGraw-Hill Books, 1988.

12. Sugeno, M. "An Introductory Survey of Fuzzy Control," *Information Sciences,* Vol. 36, 1985, pp. 59–83.

13. Wildman, K. "Fuzzy Logic Brings Kodak's Control Strategy into Focus," *Control,* Vol. 9, No. 11, Nov. 1996.

14. Zadeh, L. A. "Fuzzy Sets and Systems," *Information & Control,* Vol. 8, 1965, pp. 338–353.

7

Neural Networks for Property Estimation

One of the key objectives of a process control system is to maintain product quality parameters within specifications. An on-line analyzer may not be available to make a sampled or continuous measurement of a physical property or the composition of a liquid or gas stream, or other conditions may prevent their use. In such a case, the measurements must be obtained by lab analysis of a process grab sample. When a product quality measurement is available only from the lab, the delay associated with processing the sample and the lack of the measurement's continuous availability impacts the way this information can be used to control the process. In many cases, these lab measurements are used by the operator to manually make corrections in the process, such as changing feed flows or temperature that impact the parameter(s) reflected in the lab test. To obtain an immediate indication of the results of process changes that impact a quality-related parameter, it is often possible to use upstream measurements such as flow rate, temperature, and feedstock composition to calculate an estimated value of the quality parameter. Many of the newer control systems used in the process industry include the capability to estimate product quality parameters. Control system manufacturers have adopted different technologies to implement property estimation. To address the non-linear response of product quality parameters to changes in process inputs, the estimator can be based on a neural network model. This chapter addresses the development of property estimators that can be used with *non-linear* processes.

As discussed in Chapter 3, most process plants include one or more laboratories that are centrally located in the plant or distributed throughout the process areas. Labs are critical to the plant operation since lab analysis of raw materials (feedstock) and process product may be required to identify variations that could impact the process operation or product quality. In addition, automated process analyzers that provide continuous on-line measurements of the values associated with process product or feedstock properties that must be controlled may not be available. Such on-line analyzers may be too expensive, may have proven to be unreliable, or simply may not be practical because of the complexity of the sampling system or the complexity of the analysis. As a result, manual sampling and lab analysis are often a requirement.

When lab analysis is required to support plant operations, samples of the associated material are manually taken from the process and then transported to the lab for analysis. The collection of these samples, known as grab samples, is normally performed by the operator helper or a lab technician. The operator uses the results from the lab analysis of these samples to adjust the process to meet quality requirements. If the steps associated with an analysis are straightforward, then in some cases the operator helper can process the grab sample at a test stand physically close to the process. In such cases, the analysis is often referred to as "near line" analysis. The time required to collect the grab sample and process it in the lab introduces a significant delay in obtaining the results of the analysis. Often such measurements are needed by the plant operator or control system to maintain key quality-related parameters by making adjustments to plant operating conditions; thus, keeping product properties within specifications. Any delay introduced in processing a sample limits the usefulness of the measurement.

The PIDPlus described in Chapter 8, Intelligent PID, may be used with sampled, non-periodic measurement values obtained through lab analysis. Where the capability to automatically correct process operation based on lab results is supported, an interface in the control system is normally provided that allows the operator or lab technician to manually enter lab results into the control system. Alternatively, when a Lab Information Management System (LIMS) is in use, lab testing equipment can be interfaced directly with the control system to allow the results of the lab analysis to be automatically communicated to the control system as soon as the grab sample is processed.

Function blocks that utilize partial least square or multiple linear regression models to provide a linear estimator may be provided in the control system as detailed in Chapters 9 and 10 – Batch and Continuous Data Analytics. To address the nonlinearity associated with some properties, a non-linear estimator block based on neural network technology can be used to estimate the property as illustrated in

Figure 7-1 – see more details on function blocks in (Blevins, Wojsznis, Tzovla, and Thiele, 2002).

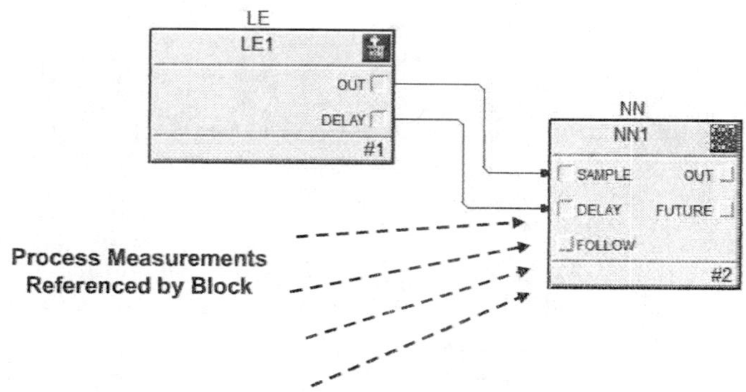

Figure 7-1. Non-Linear Property Estimation Based on Upstream Measurements

When a property estimator capability is available in a control system, a collection of lab analysis results is used in conjunction with a history collection of upstream measurement conditions to create the estimator. Once the estimator has been created, lab analysis can continue to be used to verify the accuracy of the estimated value.

Significant improvements in the tools used to develop neural networks for soft-sensor applications have been made over the last years. Soft sensors are inferential estimators, drawing conclusions from process observations when hardware sensors are unavailable or unsuitable. These tools are specifically designed to take into account the dynamic nature of the processes for which a soft sensor is to be developed. The difference in the traditional approach compared to what is available today is illustrated in Figure 7-2.

The estimator may also be designed to use lab data to make corrections that compensate for unmeasured changes in the process that impact the property being estimated. If this is not done, the prediction provided by a soft sensor will drift away from the true measurement value over time.

7.1 Example – Pulp and Paper Industry

Within the pulp and paper industry, a continuous digester is used to convert wood chips to pulp by chemically removing the lignin that

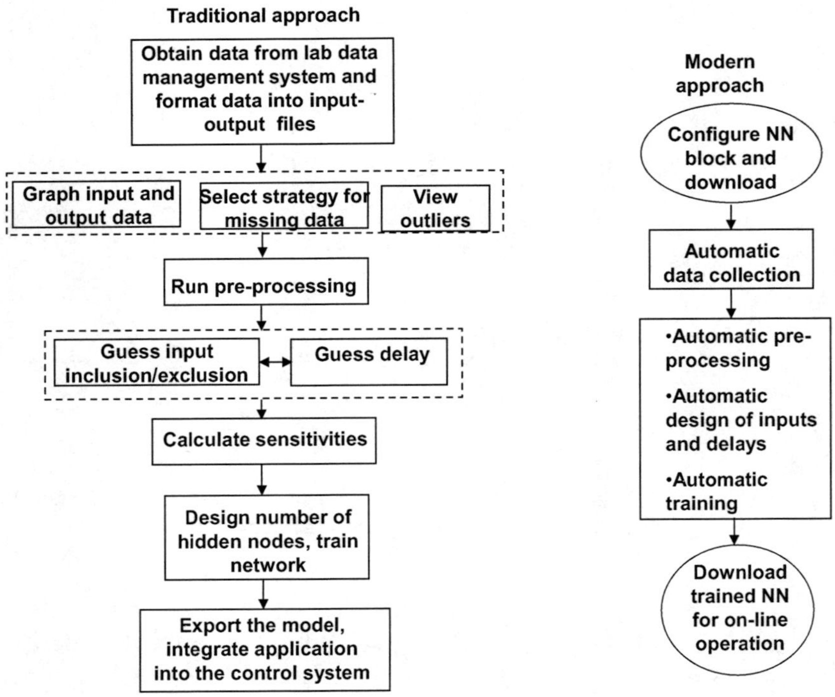

Figure 7-2. Steps in Neural Network Development

holds the cellulose fibers together. A typical continuous digester is shown in Figure 7-3. A neural network is used for this thermo-chemical process because of the need to monitor the degree of delignification as indicated by the Kappa number. The Kappa number is an indication of the residual lignin content or bleachability of wood pulp. Normally, the Kappa number is determined by a lab analysis of a grab sample of the pulp downstream of the digester because on-line measurements of Kappa are expensive and are often unreliable and inaccurate. If Kappa is determined by lab analysis of a grab sample, there is about an hour's delay before the results of the analysis are known. It should also be noted that the process's time delay is about four hours. Thus, a change in a process input that will not be reflected in the output for four hours. In such a process the ability to look into the future is a handy tool for making immediate corrections.

Figure 7-3 shows a simplified piping and instrumentation diagram of a continuous digester process. A digester can have more than 40 measured inputs that could potentially affect the Kappa analysis. By applying cross-correlation and sensitivity analysis, the number of inputs for the neural network may be reduced to five: cold blow flow,

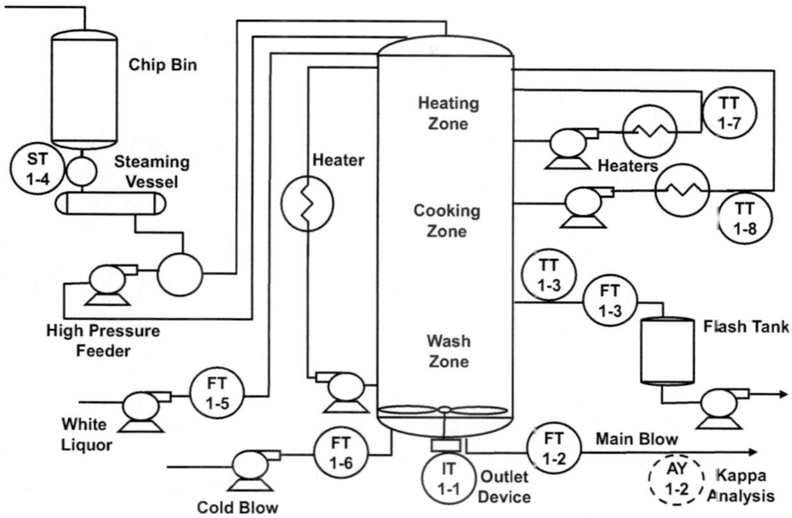

Figure 7-3. Continuous Digester Process

outlet device load, main blow flow, throughput, and lower extraction flow. The Kappa number estimated by the neural network (Fisher-Rosemount Intelligent Sensor Toolkit User Manual, 1997) agrees well with the lab analysis, as shown in Figure 7-4.

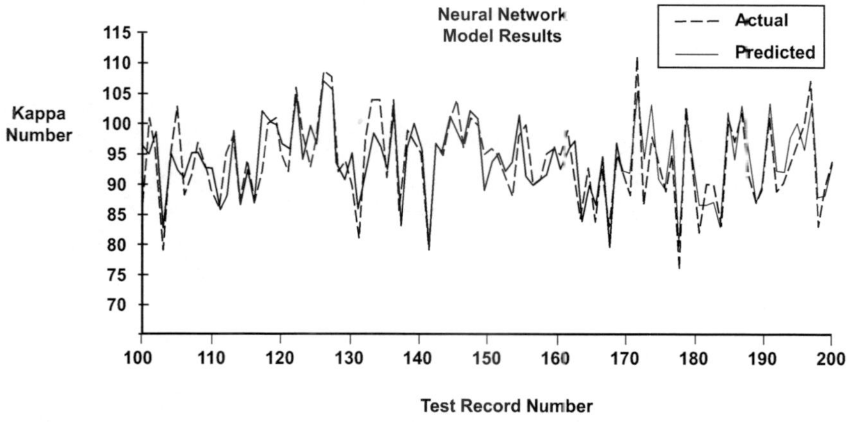

Figure 7-4. Kappa Lab Analysis and Virtual Sensor Estimate of Kappa

7.2 Property Estimator Application Example

As an example of the steps involved in working with a property estimator, the capability provided by one DCS manufacturer for property estimation based on a neural network model is examined in this chapter. Other manufacturers' tools may differ in the user interface provided for model development and deployment.

The development of a neural network model for property estimation becomes fast and easy when this capability is implemented as a function block in the control system. The following steps are required to commission property estimation:

- Configure neural network
- Select upstream conditions
- Screen data
- Determine input delay
- Examine input sensitivity
- Train neural network
- Verify model

As mentioned above, property estimation based on neural network technology is often preferred over linear estimators (see chapter 9) to address more complex dependences of the process involving multiple measurements that are related to the estimated value in a non-linear manner. In this example, an application for Kappa number estimation in a simulated process is created by adding a neural network (NN) function block to a module. When the neural network is used to estimate a measurement that is only available through lab analysis, a lab entry block is linked to the NN block. When the estimator is used as a backup for a continuous (on-line) analyzer or to fill in values from a sampled analyzer, the measurement input is provided to the sample input of the NN block, as shown in Figure 7-5.

To configure the neural network block, the upstream measurements that affect the estimated property must be identified. The user can do this by browsing the control system to find the measurements that are used, as illustrated in Figure 7-6.

When the module containing the neural network block is downloaded, communication connections to the inputs are automatically established within the control system. Also, all inputs to the block are automatically assigned to the historian for collection.

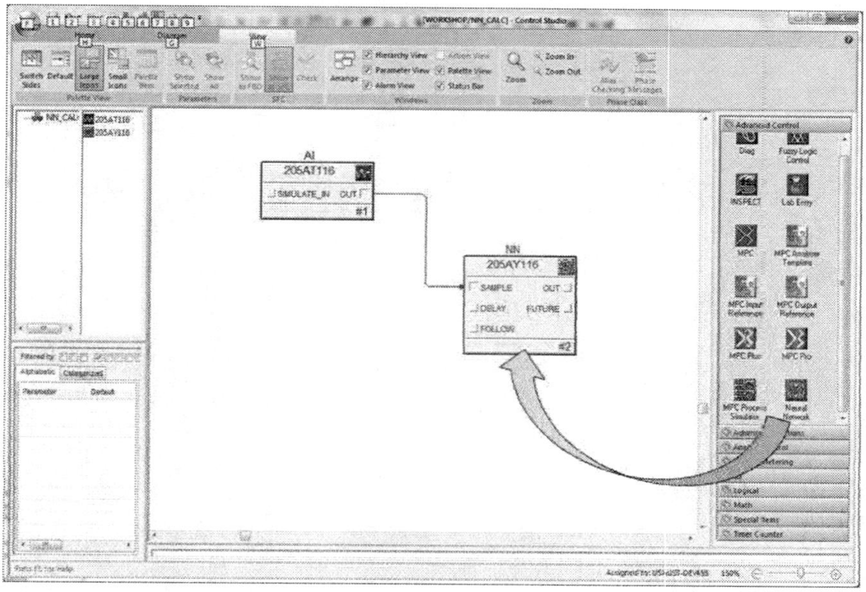

Figure 7-5. Neural Network Block for Estimation

To commission the neural network estimator, a special interface is provided to select and screen data that is required for training. Inputs to the neural network block are trended in this interface. From this trend view, the user can select the time frame over which the data will be used to train the neural network, as shown in Figure 7-7.

The user must provide an estimate of the time it takes for the process inputs to affect the estimated measurement. Once this information is provided and the data is selected, the user can automatically train the neural network by selecting the Autogenerate button. In response, a sensitivity analysis is automatically done for the configured inputs to identify the contribution of each input to the estimated measurement. Inputs that do not affect the measurement are automatically eliminated from the neural network. The results of the sensitivity analysis are presented to the user as illustrated in Figure 7-8.

As mentioned above, an important part of the sensitivity calculation and the training of the network is accurately accounting for the delay between an input change and its impact on the estimated measurement. This delay is automatically calculated by doing a cross-correlation between each input and the sample input values. The calculated cross-correlation and input delay are shown in the detail view for the sensitivity of each input, as shown in Figure 7-9.

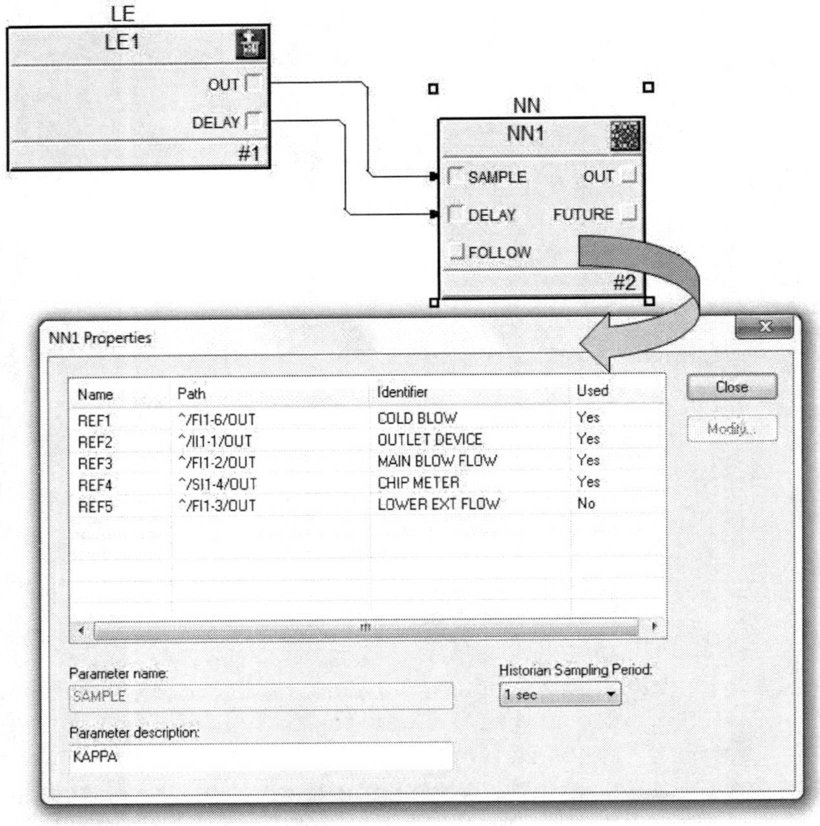

Figure 7-6. Identifying Upstream Measurements

As part of the training process, the number of hidden neurons that should be used for best results is automatically determined. Part of the data is used for training the neural network and part is used for testing. By comparing the results for the two sets of data, it is possible to automatically determine when to stop the training process, as shown in Figure 7-10.

Expert features allow the assumed expert user to select the inputs that are used to create the neural network. For example, the user may choose to calculate input sensitivity without training the neural network. In this case, the user can change the input delay or manually remove inputs that were automatically selected, as shown in Figure 7-11.

After generating and examining the input sensitivity, the expert user can train the neural network based on a selected input sensitivity.

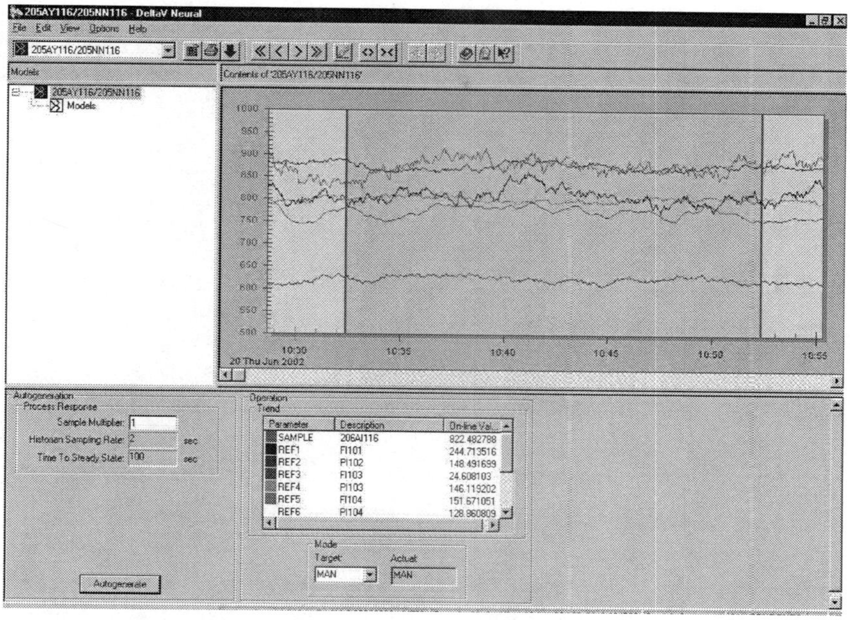

Figure 7-7. Selection of Training Data

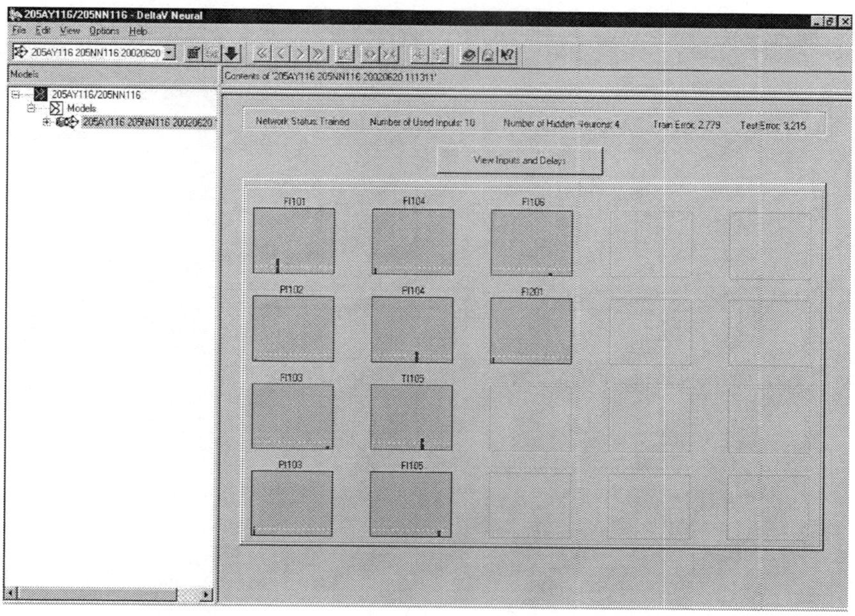

Figure 7-8. Sensitivity Analysis of Inputs

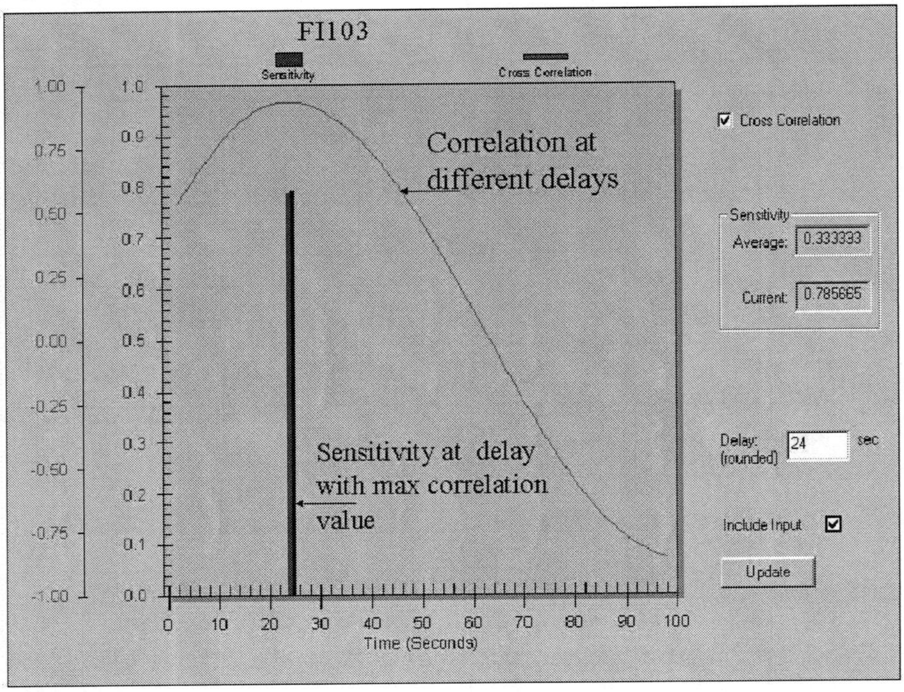

Figure 7-9. Determining Input Delay

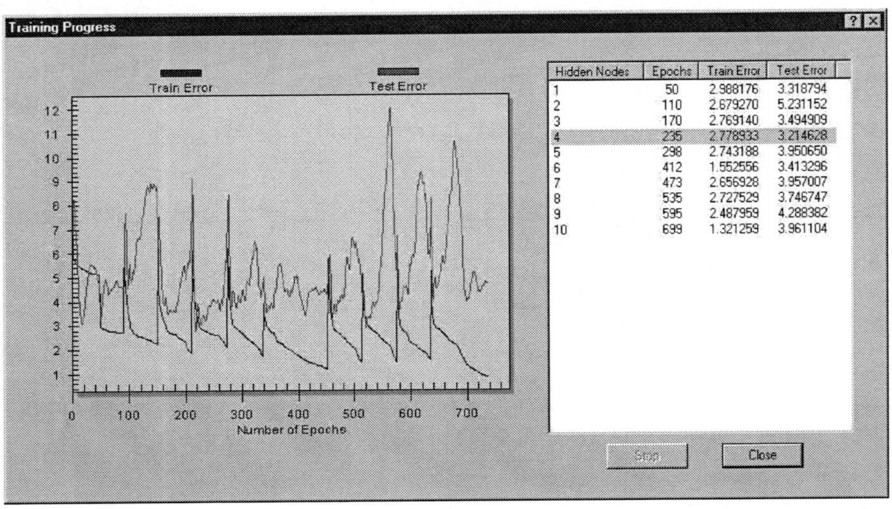

Figure 7-10. Automated Training Process

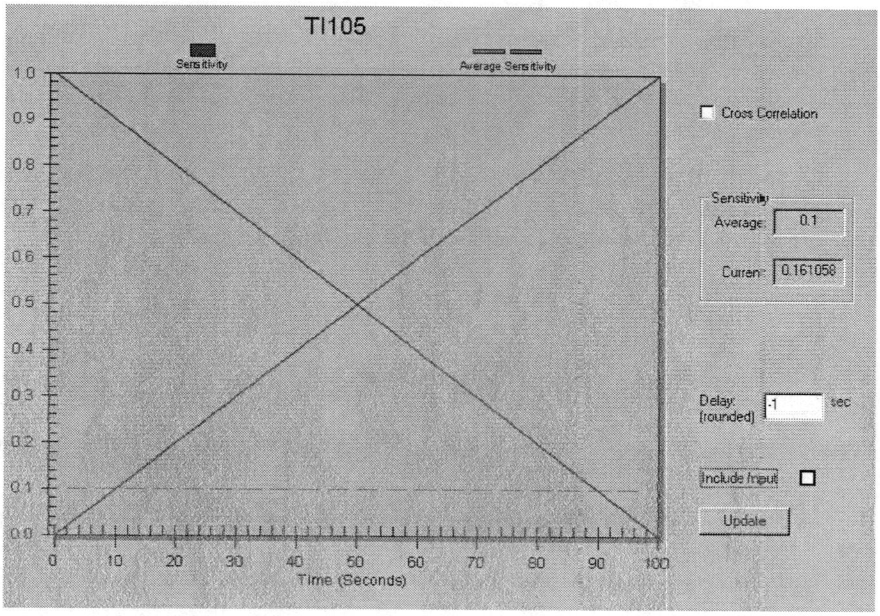

Figure 7-11. Expert Options for Input Screening

Prior to the network training, the expert user can modify some of the parameters used in training, as shown in Figure 7-12.

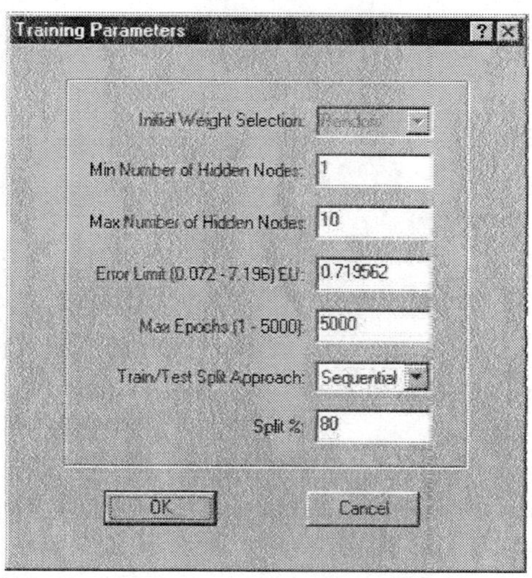

Figure 7-12. Training Options

The default training parameter value usually provides the best results. However, options such as the training and test data split are helpful in addressing special requirements associated with the test data.

Once a model has been developed, model accuracy can be tested by comparing the predicted (estimated) quality parameter to the actual values obtained from the lab or an on-line analyzer over a selected period of time as shown in Figure 7-13.

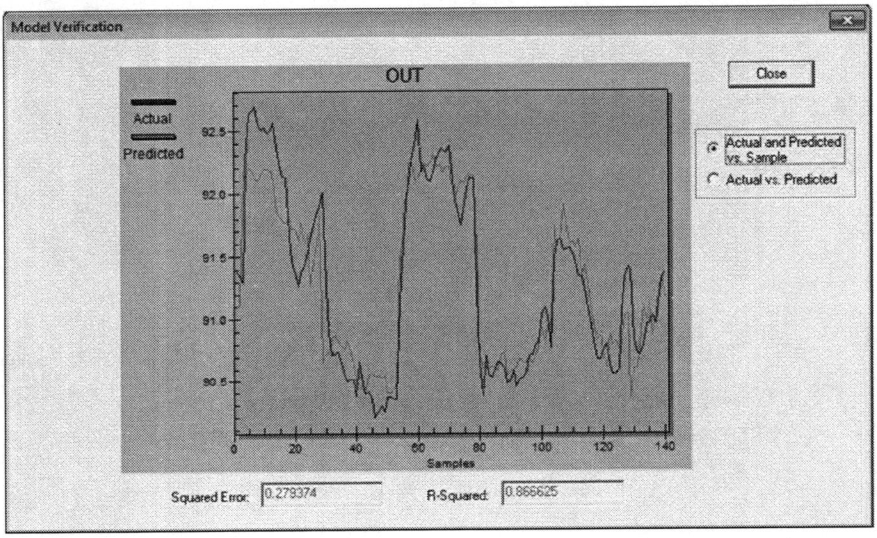

Figure 7-13. Model Verification

After the model for property estimation has been deployed, the capability of verifying the model can be used to determine whether the model needs to be updated to account for process changes.

7.3 Neural Networks for Property Estimation Workshop

This Neural Network workshop provides several exercises that are used to further explore neural network configuration, model development and verification. The prediction of the Kappa number for the Kamyr digester process (Figure 7-14) is illustrated in this workshop. Access the book's web site http://www.advancedcontrolfoundation.com and select the Solution tab to view the workshop being performed.

Step 1: Examine the module used in this workshop to learn how the neural network block was configured for the Kappa number prediction.

Step 2: Open the neural network application that is used for neural network modeling. Position the cursor over the trend window to choose the data that will be used in development of the neural network model.

Step 3: Select the Autogenerate button and then observe the model that is created from the selected data.

Step 4: Choose the Verify selection to determine how closely the predicted Kappa value matches that provided by the on-line analyzer.

Step 5: Download the module that contains the neural network block and observe the on-line current and future Kappa number prediction.

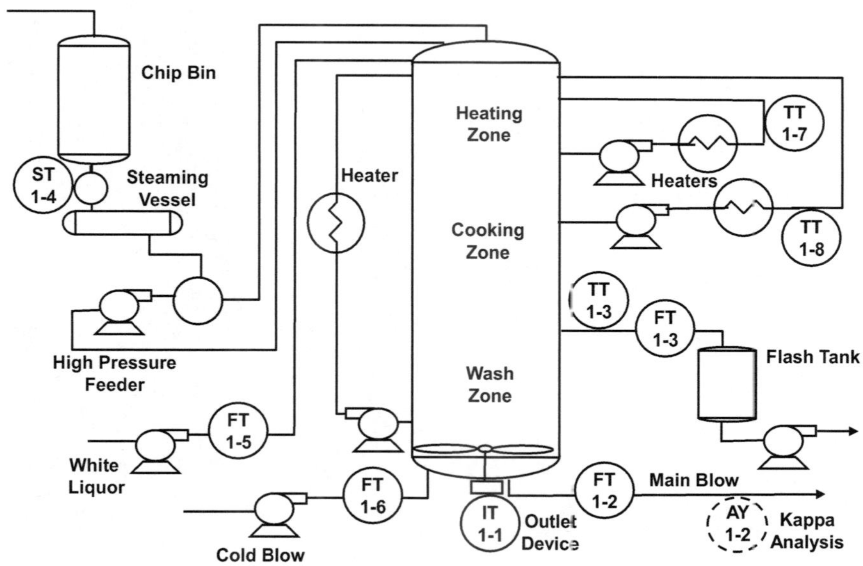

Figure 7-14. Kamyr Digester Process

7.4 Technical Basis

Neural networks (NN) are essentially non-linear function estimators that use process inputs to predict process output. Much of the initial neural network research project and development done some 50 years ago focused on emulating the network of neurons that make up the human brain. Significant progress took place after the development of feedforward NN with backward propagation training in 1983 (Rumelhart and McClelland, 1986), (Kosko, 1992). Since that time, much work has been done to adapt this technology for use in the process industry (Welstead, 1994), (Mehta, Ganesamoorthi, and Wojsznis, 2001), (Tzovla and Mehta, 2001), (Ganesamoorthi, Colclazier and Tzovla, 2000). The technical promise of neural network technology derives from the creation of universal approximators using a multi-layer network with a single hidden layer that can approximate any continuous function to any desired degree of accuracy (Qin, 1995), (Morrison and Qin, 1994).

Soft sensors that use neural networks must be adapted to the special requirements of the process industry. In particular, it is necessary to compensate for the delay in the process output for changes in upstream conditions. Thus, a neural network typically has one output (the predicted variable) and any number of upstream measurements as inputs with compensation for process delay. Figure 7-15 shows a three-layer feedforward NN.

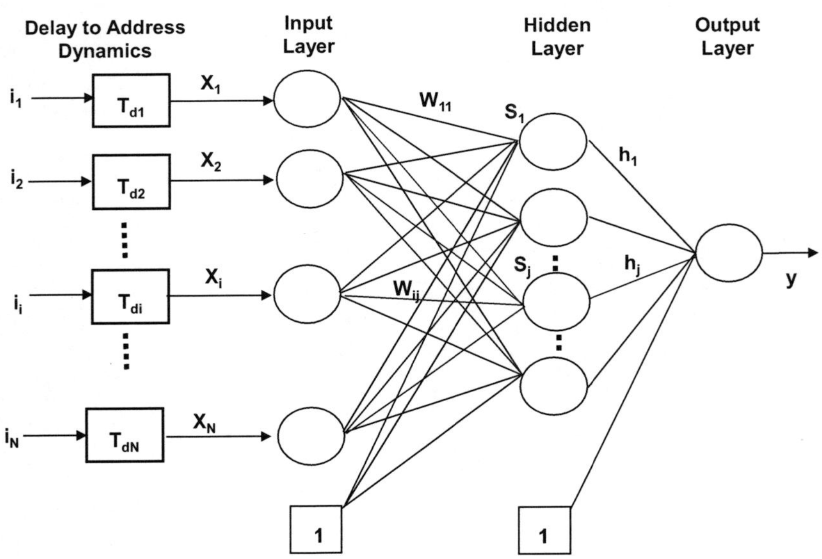

Figure 7-15. Three-Layer Feedforward Neural Network

In this example, the neural network supports n inputs, a single layer of hidden neurons, bias nodes at the input and hidden layers, and one output. Each input is delayed by a certain amount to allow the values used in the network to be time-coincident. The weight w_{ij} connects the ith node in the previous layer to the jth node in the next layer. Weighted values are summed at the node before being passed through an activation function.

For a sigmoidal hidden neuron, the summed node input S_j and the output h_j are given by Equation 7-1:

$$h_j = \frac{1 - e^{-S}}{1 + e^{-S_j}} \quad \text{where} \quad S_j = \sum_i w_{ij} x_i \qquad (7\text{-}1)$$

The output layer typically has a linear activation function, that is, a summation of the inputs to the output layer (Cybenko, 1989).

The structure of a neural network soft sensor is relatively simple and is implemented in real-time controllers with little processor loading. The real challenges in the design and implementation of tools used to create a working neural network are:

- Collecting historic data on the process inputs and the process output measurement to screen the potential inputs to the neural network. This data is also used to determine the neural network structure and the value of each parameter used in the neural network.
- Identifying the delay between an input and its impact on the process output predicted by the neural network.
- Determining which of the process inputs significantly affects the process output by calculating input sensitivity.
- Determining the weighting factors and the number of neurons included in the hidden layer or layers for best results.
- Validating the network.

7.4.1 Data Collection

Data collection is by far one of the most critical steps in the development of a neural network. When collecting data to be used to train the neural network, it is important that the inputs and outputs vary over their normal operating range. If the process output is available only as a lab analysis, this data must be merged with historical data on the input measurements to allow further analysis. Some simple rules that should be observed in data collection are:

- Only inputs that change during the time the data is collected impact the process and are identified in training. When possible, collect a uniform number of sample values over the operating range of the input.
- Make provisions to automatically exclude values that fall outside this normal range of operation. Statistical techniques often prove useful in defining the outlier boundaries. A good rule of thumb for valid data limits is *Mean +/- 3.5 * standard deviation*, which includes approximately 99.9% of the data in the given region.

Also, it is important that the user is able to flag regions of the historic data to indicate periods of abnormal operation so this data, for all inputs and outputs, will not be used in the development of the network.

7.4.2 Identification of Input Delay

A change in an upstream measurement may not be immediately reflected in a process output. Identifying this delay is the first step in creating a neural network model. The overall objective is to determine how much each process input needs to be delayed to allow best alignment with the output response. This can be done automatically by calculating the cross-correlation between each upstream measurement value selected as a neural network input and the process output value selected as the neural network output. The cross-correlation, Figure 7-16, is calculated for each process input and a process output.

The time shift, *K*, between the input and the output that produces the maximum cross-correlation coefficient is used as the process input delay that should be introduced into the input processing of the neural network, as shown in Figure 7-17.

The cross-correlation value indicates the magnitude (and sign) of the effect of the input on the output. For example, for a simple first order process, input at a delay that equals approximately (*deadtime + time constant/2*) has the most relevance, as the maximum correlation value occurs at that delay. Once the most significant delay is known, the input data is shifted by that delay.

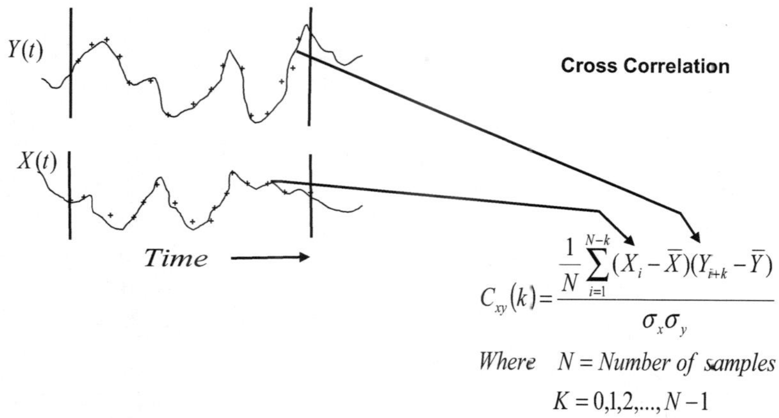

Figure 7-16. Cross-Correlation Calculation

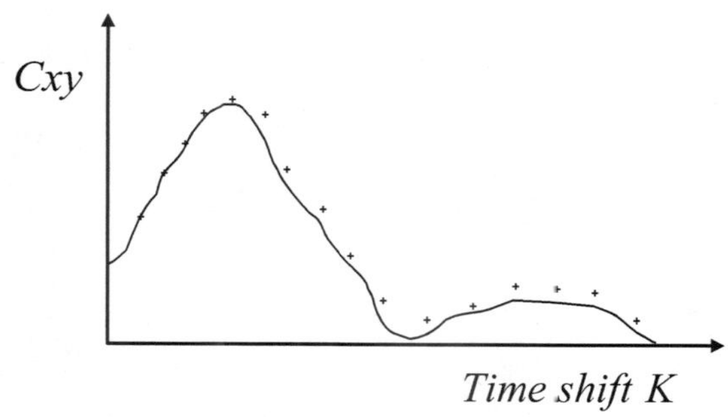

Figure 7-17. Determining Process Input Delay Using Cross-Correlation

7.4.3 Input Sensitivity

In the initial definition of the neural network, it may not be possible to know which of the upstream measurements influence the process output to be predicted by the soft sensor, and to what degree. Only those that have a significant impact should be included in the network. The input sensitivity is defined as the change in dependent variable (output) y for a unit change in independent variable (input) x, or mathematically:

$$S_x^y = \frac{\Delta y / y}{\Delta x / x} \qquad (7\text{-}2)$$

The initial sensitivity estimate is calculated from a simplified linear model, prior to developing the NN model. A linear model for computing the sensitivities can be obtained using the standard Projection to Latent Structures known as well as Partial Least Squares (PLS) algorithm, for example (Geladi and Kowalski, 1986), (Wold, Jellberg, Lundsted, Sjostrom, and Wold, 1987). The principle of PLS is outlined in the chapter 9; for more detail there are several good references (Masters, 1993), (Hornik, Stinchcombe, and Halbert, 1989). The delayed input values and process output are used in the development of this model. Using this model, the input sensitivity is calculated by changing the input by one unit while keeping all other inputs constant, and determining the output change.

In a multi-variable system, a higher value of sensitivity indicates that change in that input has higher influence on the output. The sum of all sensitivities is normalized to 1. Information on the sensitivity of the inputs to the output indicates their relative importance. The sensitivity value at the delay identified in the previous step is used to exclude inputs on which the output shows little or no dependence. One technique for doing this is to exclude inputs whose individual sensitivities are small compared to the average sensitivity.

7.4.4 Determining Input Weights

Having determined the inputs and delays to be used in the neural network, it is now possible to determine the weighting factors and the number of neurons included in the hidden layers to provide the best results. The weights in the network are initialized with random, small, non-zero values. Randomness ensures lack of bias, while small values give more freedom for modifying the weights to avoid saturation. For a given number of neurons in the hidden layer, the squared error between the calculated soft sensor output and the actual output measurement for one point in time is expressed as:

$$E_p = (y_p^{pred} - y_p)^2 \tag{7-3}$$

The cumulative error for a set of data is calculated as follows:

$$E = \sum_{p=1}^{data} E_p \tag{7-4}$$

By incrementing the weighing factor W_{ij} associated with input i and node j on the value ∂W_{ij}, and observing the change in cumulative error E (Figure 7-18), we can determine gradient,

$\frac{\partial E}{\partial W_{ij}}$, or approximately in final differences $\frac{\Delta E}{\Delta W_{ij}}$.

Gradient defines the change in the output for a unit change of the weight W_{ij}. The back propagation algorithm is used to calculate new weights that minimize the cumulative error in the direction of negative gradient (steepest descents) for each complete pass through the data set (an epoch):

$$W_{ij}^{new} = W_{ij}^{old} - \alpha \frac{\partial E}{\partial W_{ij}} \tag{7-5}$$

where α defines the step size of a change in the gradient direction, more popularly known as the learning rate. The speed of convergence is improved by modifying the back propagation algorithm to incorporate the conjugate gradient technique as described by Brent (1973). Rather than use a fixed step size, the new direction is based on a component of the previous direction. To avoid settling on one of local minima, the training algorithm is designed to automatically turn the direction remembrance on or off depending on whether the error is improving or not. A brand new direction starts each time this is done.

During the training of the neural network, the cumulative error for the training set of data will decrease monotonically, approaching a constant value in an asymptotic manner, as shown in Figure 7-18. However, if the cumulative error is calculated using a validation data set not used in training, at some point the error will begin to increase. Past the point where the training and validation errors begin to diverge, the neural network starts learning features specific to the training data set rather than to the general process. The goal of training is to learn to predict the output given real process inputs, not simply to memorize the training set. Predicting the output given real process inputs is known as generalization.

To detect overfitting, which happens when the model estimate for the test data is much worse than for the training data, a certain portion of the data set is kept aside for validation during the training phase. This is called the test set. At each epoch, as the weights are modified based on the error on the training set, the test set is used to detect when overfitting of the training set has occurred. At each epoch, the error on the test set (test error) for the new set of weights is calculated and compared with the best test error. In order to prevent training from stopping at a random choice of weights for which the test error is small, the algorithm runs for at least a fixed number of epochs before establishing any minima. Also, the training error is added to the test error to

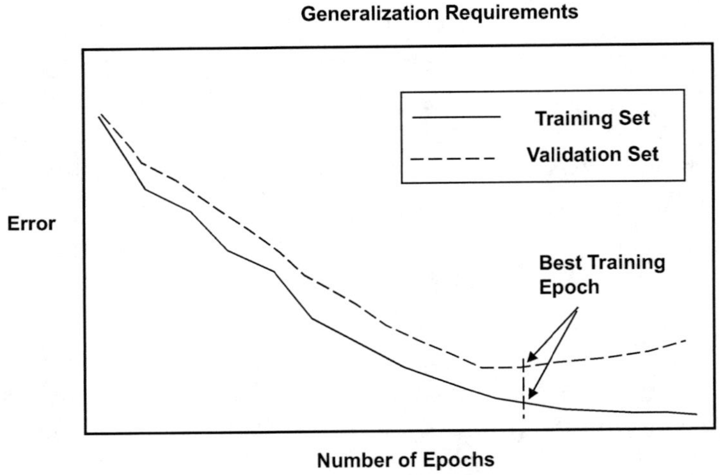

Figure 7-18. Neural Network Training and Test Errors vs. Number of Epochs

define a stringent minimum total error condition. This ensures that both the data sets have acceptable errors when the algorithm converges and the weights at the best epoch are picked at the end of training. More on iterative numerical algorithms is in (Press, Flannery, Teukolsky, and Vetterling, 2002) and (Polak, 1971.)

7.4.5 Nodes in the Hidden Layer

The number of nodes in the hidden layer has a large impact on the accuracy of the soft sensor. In general, a smaller number may lead to a poor fit, while a larger number may lead to overfitting the training set. This is indicated graphically in Figure 7-19, where overfitting occurs at the 5th hidden node, when test error increased despite decreased training error.

To determine the optimum number of hidden nodes, it is possible to train the network starting with one hidden node up to the maximum number of nodes, which is the number of inputs to the neural network. The minimum cumulative error is stored for each increment. If the difference between errors for different numbers of hidden nodes is within a tolerance level, the NN with a smaller number of hidden nodes is given preference. The algorithm then automatically picks the weights for the best combination of train/test error obtained. In the example considered in Figure 7-19, the NN with four hidden nodes is automatically chosen. By this means, network generalization is maintained while selecting automatically the best possible number of hidden nodes.

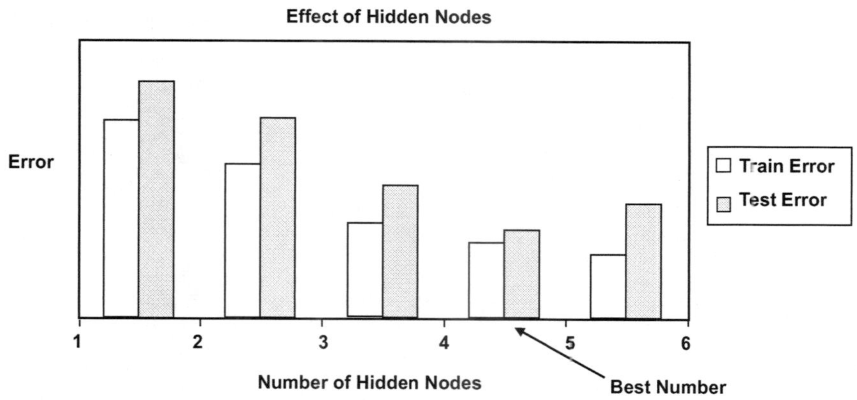

Figure 7-19. Impact of Number of Hidden Nodes on Error

7.4.6 Correction for Process Changes

A process output stream property that is predicted using a neural network and measured upstream conditions is automatically corrected for any error introduced by unmeasured disturbances and measurement drift. This correction factor is calculated based on a continuous measurement or sampled measurement of the stream provided by an on-line analyzer or a lab analysis of a grab sample. An adaptive correction of the neural network is created by incorporating this automatically-generated correction factor as an integral part of the neural network algorithm – see details in (Blevins, Tzovla, Wojsznis, Ganesamoorthi, and Mehta, 2004).

Two approaches are used to calculate the correction factor that must be applied to the NN prediction. Both are based on calculation of the prediction error using the time-coincident difference between the uncorrected predicted value and the corresponding measurement value. Depending on the source of the error, a bias or a gain change in the predicted value is appropriate. To avoid making corrections based on noise or short-term variations in the process, the calculated correction factor should be limited and heavily filtered; for example, equal to twice the response-time horizon for a change in a process input. During periods of time when a new process output measurement is not available, the last filtered correction factor is maintained.

A flag is set, if the correction factor is at the limit value. Also, a configurable filter on the corrected prediction value allows noise in the input measurements to be filtered. The basic implementation is shown in Figure 7-20.

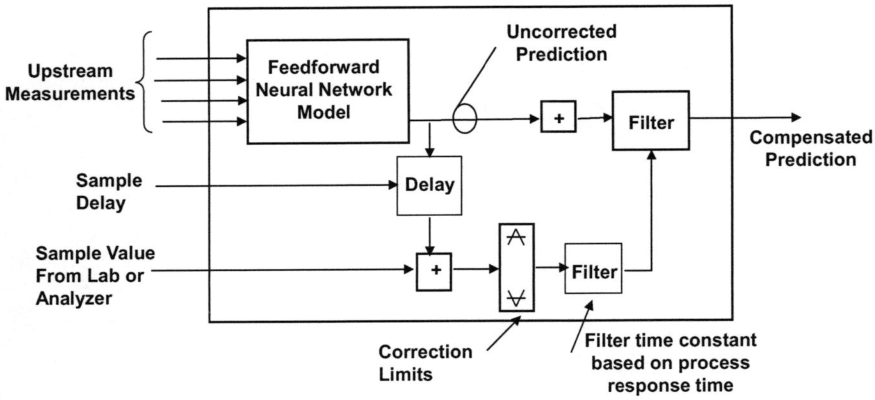

Figure 7-20. Adaptive Neural-Network Structure

Incorporating adaptive correction as part of the neural network simplifies implementation and dramatically improves neural network online performance.

BIBLIOGRAPHY

1. Blevins, T., Tzovla, V., Wojsznis, W., Ganesamoorthi, S., and Mehta, A. "Adaptive Predictive Model in Process Control Systems," U.S. Patent No. 6,760,716, 2004.
2. Blevins, T., Wojsznis, W., Tzovla, V., and Thiele, D. "Integrated Advanced Control Blocks in Process Control Systems," U.S. Patent No. 6,445,963, 2002.
3. Brent, R. *Algorithms for Minimization without Derivatives,* Englewood Cliffs: Prentice-Hall, 1973.
4. Cybenko, G. "Approximation by Superpositions of a Sigmoidal Function," *Mathematical Control, Signals, and Systems,* Vol. 2, 1989, pp. 303–314.
5. Fisher Rosemount Systems, "*Installing and Using the Intelligent Sensor Toolkit,*" User Manual for the ISTK on PROVOX Instrumentation, 1997.
6. Ganseamoorthi, S., Colclazier, J., and Tzovla, V. "Automatic Creation of Neural Nets for Use in Process Control Applications," in *Proceedings of ISA Expo/2000,* 21–24 August 2000, New Orleans, LA.
7. Geladi, P. and Kowalski, B. "Partial Least Squares Regression: A Tutorial," *Analytica Chimica Acta,* 185, 1986: pp. 1–17.
8. Hornik, K., Stinchcombe, M., and Halbert, W. "Multilayer Feedforward Networks Are Universal Approximators," *Neural Networks,* 2:5, 1989, pp. 359–366.
9. Kosko, B. *Neural Networks and Fuzzy Systems,* Englewood Cliffs: Prentice-Hall, 1992.

10. Masters, T. *Practical Neural Network Recipes in C++ London,* Academic Press, 1993.

11. Mehta, A., Ganesamoorthi, S., and Wojsznis, W. "Feedforward Neural Networks for Process Identification and Prediction," in *Proceedings of ISA Expo/2001,* September 2001, Houston, TX.

12. Morrison, S. and Qin, J. "Neural Networks for Process Prediction," *Proceedings of the 50th Annual ISA Conference,* New Orleans, 1994, pp. 443–450.

13. Polak, E. *Computational Methods in Optimization.* New York: Academic Press, 1971.

14. Press, W. H., Flannery, B., Teukolsky, B., and Vetterling, W. *Numerical Recipes in* C, New York: Cambridge University Press, 2002.

15. Qin, S. J. "Neural Networks for Intelligent Sensors and Control – Practical Issues and Some Solutions," *Progress in Neural Networks: Neural Networks for Control,* edited by D. Elliott, New York: Academic Press, 1995.

16. Rumelhart, D. and McClelland, J. *Parallel Distributed Processing,* Cambridge: MIT Press, 1986.

17. Tzovla, V. and Mehta, A. "Automated Approach to Development and Online Operation of Intelligent Sensors," in *Proceedings of ISA Expo/2001,* September 2001, Houston, TX.

18. Welstead, S. *Neural Network and Fuzzy Logic Applications in C/C++,* John Wiley and Sons, Inc., 1994.

19. Wold, S., Jellberg, S., Lundsted, T., Sjostrom, M., and Wold, H. "PLS Modeling with Latent Variables in Two or More Dimensions," *Proceedings, Frankfurt PLS Meeting,* September 1987.

8 Intelligent PID

PID remains the dominant technique used in the process industry for the implementation of feedback control. Most modern distributed control systems (DCS) are designed to support the distribution of control to a fieldbus environment, as specified in the Fieldbus Foundation specification. In a fieldbus environment, measurement, control, and calculations are represented as function blocks. Some major control systems have adopted these function blocks for use within the control system.

In some cases, however, traditional PID implementation and PID closed loop response to process disturbances and setpoint changes do not match plant operation requirements. Over the last 10 years, some significant advancements have been made in PID design that can improve plant performance. For example, it is quite common for operating conditions to cause the PID control to be limited during process start-up and normal plant operation. A modification of the PID algorithm now allows the PID to recover quickly from a limit condition. Also, with the introduction of wireless devices for field measurement, modifications have been made in the PID that allow the non-periodic, slow measurement values provided by a wireless device to be used in closed loop control. These improvements in the PID, along with already-discussed adaptive tuning, performance monitoring, and loop diagnostics, will be referred to as the intelligent PID. The chapter addresses the features of the intelligent PID extended algorithms and explains how these features can be used to improve plant operation.

Within the process industry, the standard or series forms of the PID have been adopted by the major distributed control system suppliers. Manufacturers of DCS have approached PID implementation in a variety of ways. Two basic approaches are taken in the implementation of the reset component. One approach is to realize the reset component using an integrator of the PID's error component. However, in many commercial products, the reset component is realized using a positive feedback network as detailed by Åström and Hägglund (2006), Blevins and Nixon (2010), and Rhinehart, et al. (2006). In a positive feedback network, the filter's time constant defines the reset time in seconds per repeat. These two reset implementations of the PID algorithm are illustrated in Figure 8-1.

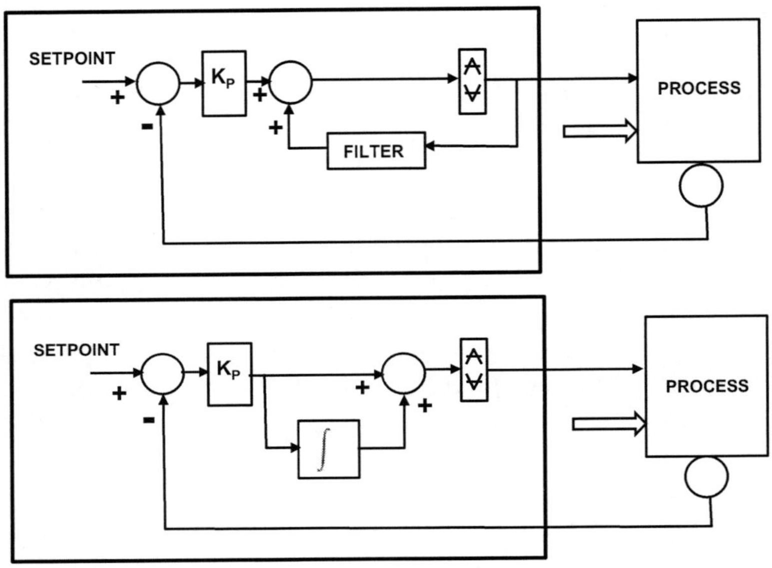

Figure 8-1. Conventional PI Controller Reset Implementation

For unconstrained operation, both approaches to PID implementation are mathematically equivalent. However, the use of a positive feedback network to create the reset component enables the implementation of external reset for use in cascade and override applications.

Another benefit of this approach is that the reset contribution is automatically prevented from winding up when the PID output is limited (i.e., operating at its upper or lower limit). However, the recovery from process saturation conditions is characterized by setpoint overshoot. In contrast, when the reset is implemented as an integrator, logic is used to avoid reset windup (i.e., the PID design must provide a way to avoid reset windup up when the PID output reaches an upper or lower limit). For example, logic can be included in the reset calculation to stop the integration of error that would drive the control output further into the limit. However, when the process is characterized by a noisy measurement or frequent disturbances, such an approach may prove to be ineffective.

Conventional PID design based on difference equations that approximate differential operators, or z-transform that are a discrete equivalent of the Laplace transform used in DCS controllers assumes that a new measurement value is available each execution and that control is executed on a periodic basis. When the measurement is not updated on a periodic basis, the calculated reset action may not be appropriate. If control is executed only when a new measurement is communicated, the result could be a delayed control response to setpoint changes and feedforward action on measured disturbances that occur between measurement updates. Also, as the PID execution period is increased, the basic assumptions made in the design of the PID reset and derivative calculation may no longer be valid.

The intelligent PID incorporates changes in its design that speed the recovery from process saturation and enable closed loop control to be done using the slow, non-periodic measurement sample values transmitted by a wireless device.

8.1 Recovery from Process Saturation

During plant start-up conditions, it is desirable to smoothly bring process temperatures, pressures, and flows to normal operating conditions in minimum time without process overshoot. During start-up, these adjustments in the process input must be made by the plant operator because of the limited range of process measurement. When process measurements are within their designed operating range, the plant operator depends on the PID to establish and maintain normal plant operating conditions.

At some point in plant operation, one or more conditions can limit PID operation for an extended period of time (that is a process saturation condition). When operating conditions change that allow the PID to resume normal operation, the PID recovery time (i.e., the time for the PID to reach setpoint) may not meet the plant processing requirement. For example, at high plant product rate it may not be possible for a heater (designed for a lower production rate) to maintain temperature setpoint even though the steam valve is fully open. When process saturation occurs due to equipment limitations or an abnormal operating condition, the associated PID control may operate at a limit condition for an extended period of time. Under this condition, the filter in the positive feedback network used to implement the reset component will settle at the limit value. When operating conditions that enable operation away from a limit condition change, PID control begins to take action only when the error signal changes sign (that is, the controlled parameter must transition through the setpoint before any control action is taken). As a result, when the process is recovering from a saturated process condition, the controlled parameter can overshoot the setpoint.

To allow control action to be taken before the controlled parameter transitions through setpoint, it is necessary to add rate action. However, if the control measurement is noisy, it may not be possible to use rate action. Thus, as described by Shinskey (2006), some manufacturers allow a user-specified "preload value" that is automatically substituted for the positive feedback term in the PID calculation when the output becomes limited, as shown in Figure 8-2.

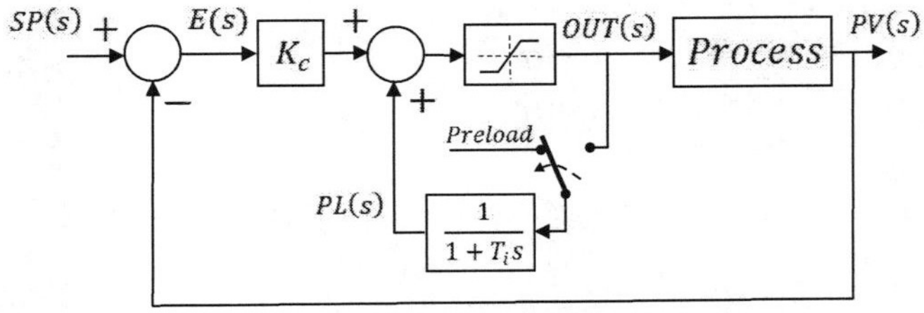

Figure 8-2. Addition of Preload when PID Output Is Limited

For such an implementation, the point at which the controller output begins to take action, when recovering from process saturation, is determined by the control error as well as the preload value. When the controller output begins to move away from the limit, the controller output is automatically used in the positive feedback network to allow normal reset action to resume.

This approach may be used to avoid overshooting setpoint when recovering from a process saturation condition. However, for slow rates of change in a process disturbance that caused the limit condition, the preload value may be too large and can cause premature control action. Thus, it is not always possible to choose a preload value that satisfies the different conditions that may exist in an operating plant.

Ideally, the magnitude and rate of change in the control error should determine when control action is taken for recovery from process saturation. The point at which control action is taken should be automatically determined in a manner that results in the controlled parameter coming to setpoint without overshoot. It is possible to modify the PID to provide such behavior (Blevins, 2012). When the PID controller output is limited for an extended period of time, a term known as variable pre-load is added to the OUT term used in the positive feedback network, as illustrated in Figure 8-3. This modification allows the PID to take action at the time needed to avoid overshoot. The rate at which the controlled parameter approaches setpoint on recovery from a saturation condition determines when and by how much the PID begins to take control action. Once the PID output is no longer limited, then normal PID function resumes.

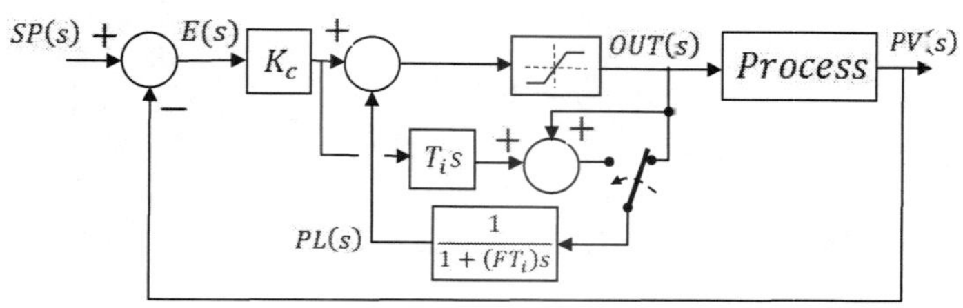

Figure 8-3. PI Control with Variable Preload Capability

One process application that demonstrates the benefit of variable preload is boiler outlet steam temperature control. If steam generation exceeds the attemperator capacity, the boiler outlet steam temperature will exceed the outlet setpoint with the spray valve fully open. When the boiler firing rate (and, with it, steam generation) is reduced, the spray valve opening should be cut back as the outlet temperature drops. The improved response for a 50% drop in steam generation when variable preload capability is provided in the PID is illustrated in Figure 8-4.

As illustrated by this example, a better response to major upsets can be achieved when this calculated value is used in place of a user-specified preload value for a process saturation condition.

8.2 Control Using Wireless Transmitter

Utilizing wireless communication to provide a measurement that will be used in closed loop control presents many technical challenges, as detailed in Han, et al. (2011). To reduce transmitter power consumption, it is desirable to minimize how often a measurement value is communicated. However, to avoid the restrictions of synchronizing the measurement value with the control, most multi-loop controllers

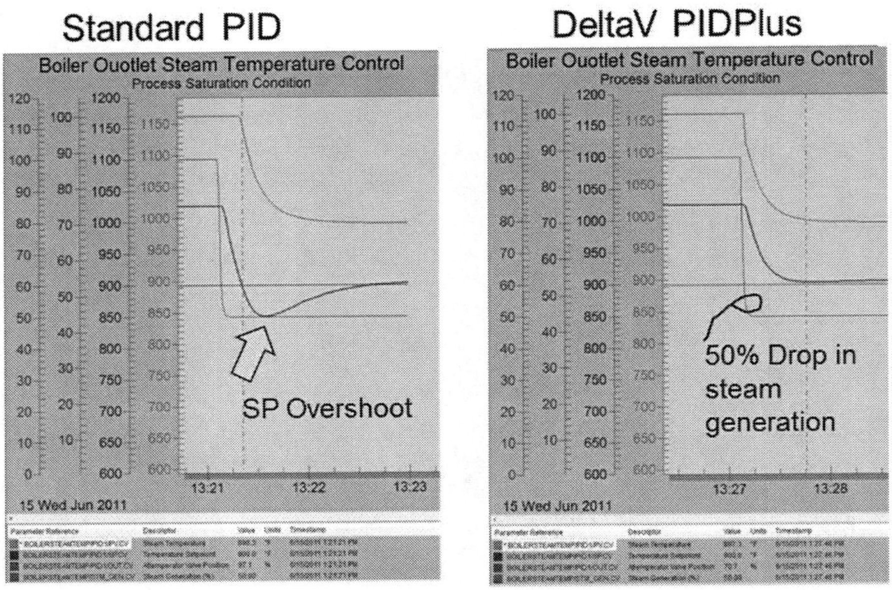

Figure 8-4. Response for 50% Drop in Steam Generation

are designed to over-sample the measurement by a factor of 2 to 10. Also, to minimize control variation, the rule of thumb is that feedback control should be executed 4 to 10 times faster than the process response time, where process response time is defined as the sum of the process time constant and process delay. To satisfy these requirements, the measurement value is often sampled at intervals much shorter than the process response time, as illustrated in Figure 8-5.

By synchronizing measurement and control execution, as is done in fieldbus devices based on Foundation Fieldbus, it is possible to eliminate the need to over-sample the measurement. However, if the traditional approach is taken in scheduling control 4 to 10 times faster than the process response, the power consumption associated with the transmission of the measurement value may be excessive for all but the slowest processes. Slowing down the control execution to reduce the power consumption associated with communication may increase control variability when the process is characterized by frequent unmeasured disturbances.

Ideally, the power consumption could be minimized by transmitting a measurement value only as often as required to allow control action to correct for unmeasured disturbances or changes in setpoint. The standard communication techniques that are supported by WirelessHART

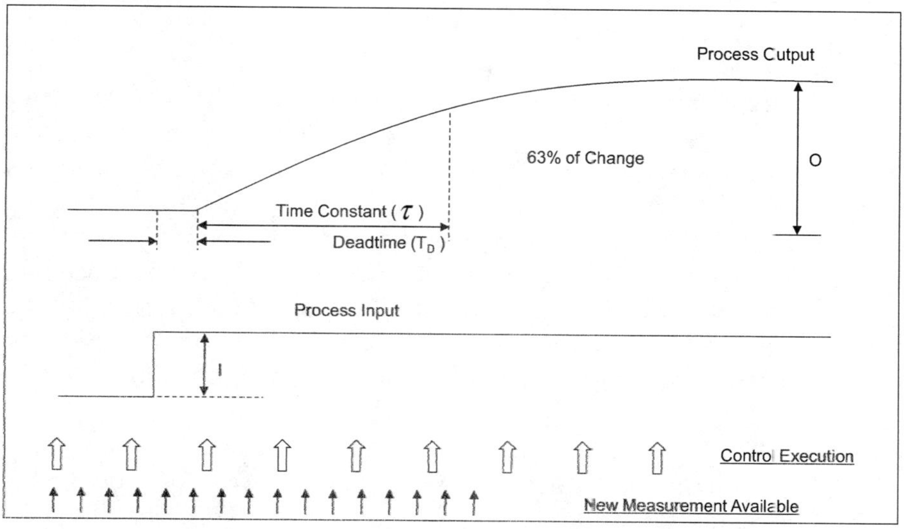

Figure 8-5. Measurement and Control Data Sampling Rate

devices are defined in the HART 7 specification that has been adopted as an international standard, IEC 62591 Ed. 1.0. The device may be configured to communicate new measurement values using one of five defined burst message-triggered modes. For control applications, the two communication modes that best fit control applications are:

- Continuous – The device wakes up at a configured update time, senses the measurement, and then communicates the value.
- Window – The device wakes up at a configured update time, senses the measurement, and then communicates its value if the specified trigger value is exceeded.

Window communication is the preferred mode for control applications since for the same update period, window communication always requires less power than continuous communication.

When window communication is selected, a new value will be communicated only if:

- The magnitude of the difference between the new measurement value and the last communicated measurement value is greater than a specified trigger value.

- The time since the last communication exceeds a maximum update period.

Thus, the measurement is communicated only as often as required to allow control action to correct for unmeasured disturbances or response to setpoint changes. For window communication mode, the user must specify an update period, a maximum update period, and a trigger value. The update period and the maximum update period are separate configured values. For best control performance in the event of process disturbances, the update period configured for window communications should be one-fourth of the process response time. If disturbance response is not critical, a longer update period can be used.

To provide best control, the PID algorithm can be modified to use measurement updates that are communicated much more slowly than four times over the process response time. Also, the PID algorithm must be able to work with the non-periodic measurement updates that are provided when window trigger mode is configured (Blevins, 2006), (McMillan, 2010). A product that implements this capability is the DeltaV system's PIDPlus, which is specifically designed to support control using wireless measurements. The PIDPlus implementation diagram is shown in Figure 8-6.

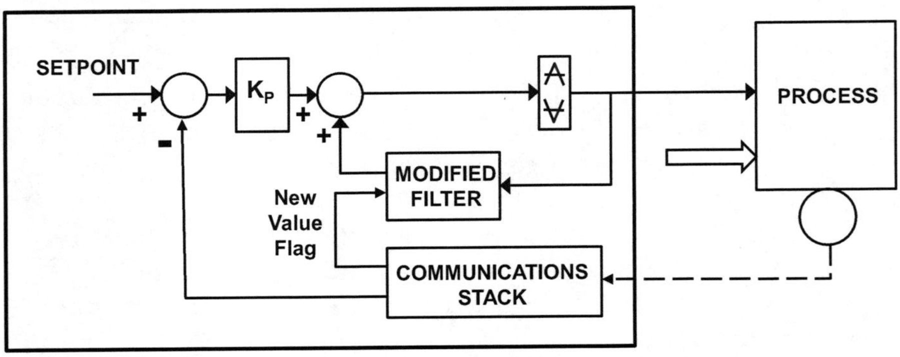

Figure 8-6. PIDPlus Implementation Diagram

The filter output used in the positive feedback network is calculated when a new measurement is received to directly reflect the process dynamic response. For example, when the Lambda PID tuning rules are used, the reset (filter time constant) is set equal to the process time constant plus the process deadtime.

Control execution is set much faster than measurement update. This permits immediate action on setpoint changes and updates in the DCS display. PIDPlus tuning is based on the process dynamics (for example, RESET = process time constant plus deadtime) and PIDPlus reset automatically compensates for variations in the measurement update rate. No change in PID tuning is required for varying update rates.

To further enhance the response for continuous changes in setpoint, the implementation of the PIDPlus algorithm can be modified, as shown in Figure 8-7.

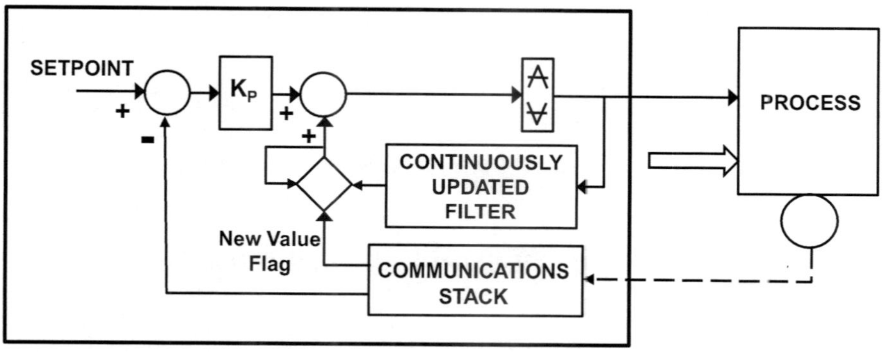

Figure 8-7. PIDPlus Implementation with Continuously Updated Filter

This PIDPlus implementation allows the reset calculation to automatically compensate for setpoint change and slow or variable communication of the new measurement value used in control. When PIDPlus is used, there is no need to modify tuning as the sampling rate changes; that is, reset is based strictly on the process dynamics.

The derivative component of the PIDPlus algorithm is also modified to account for the fact that a new measurement value is not available for each execution of the PID. The changes in the derivative calculation are shown in Figure 8-8.

The derivative contribution is only updated when a new measurement becomes available. Also, the calculation is based on the elapsed time since a new value was communicated.

The closed loop response of a PI controller modified for wireless communication is illustrated in Figure 8-9 for both setpoint changes and load disturbances. In this example, the wireless transmitter follows the rules for transmitting a new value. Also, the response is shown for a

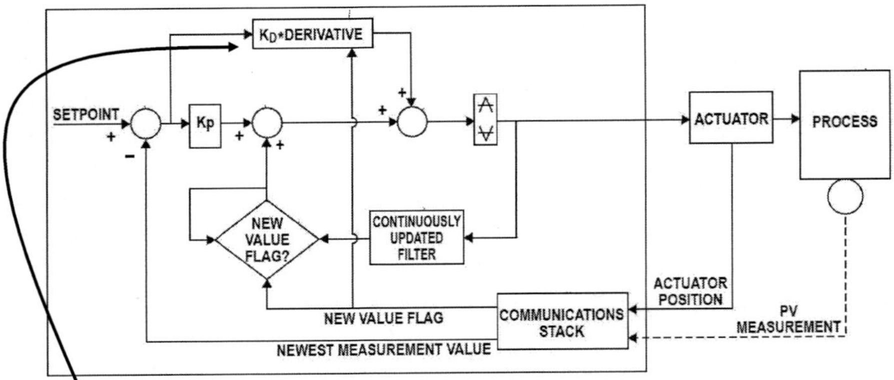

$$O_D = K_D \frac{e_N - e_{N-1}}{\Delta T}$$

where
e_N – current error
e_{N-1} – last error
ΔT – elapsed time since a new value was communicated
O_D – controller derivative term

Figure 8-8. Derivative Calculation in PIDPlus

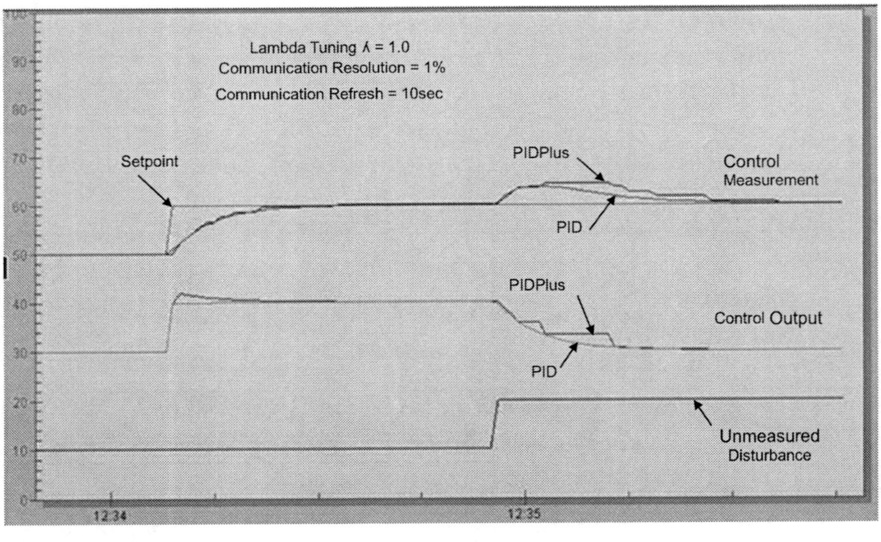

Figure 8-9. Response of PIDPlus Compared to Standard PID

standard PI controller where the measurement value is communicated as frequently as the PI control algorithm executes.

In this example, where the process time constant is much larger than process deadtime the number of communications during the duration of the test was reduced by over 96% when the rules for wireless communication as outlined above were followed. The impact of non-periodic measurement updates on control performance is minimized through the use of the modified PI algorithm for wireless communication is shown in Table 8-1 in terms of Integral of Absolute Error (IAE) for periodic vs. non-periodic measurement update.

Table 8-1. Comparison of PIDPlus and PID

Communications/Control	Number of Communications	IAE
Periodic Communications/ Standard PID Controller	692	123
Update Using Wireless Communications/PIDPlus	25	159

The power that must be supplied by the transmitter for data transmission can be significantly reduced when the Window communication and the PIDPlus algorithm are used with wireless transmitters. This reduction in power requirement increases the number of control applications that may be addressed using wireless transmitters (Chen, Nixon, Aneweer, Mok, Shepard, and Blevins, 2005), (Nixon, Chen, Blevins, and Mok, 2008).

The reliability of WirelessHART device communication has been well-established (Seibert, 2011). Even so, in the event of communication loss, the expected control behavior is of interest. A simulation environment may be created that allows the behavior of the PIDPlus for loss of communications to be compared to a PID with a wired transmitter where the measurement value is frozen for a period of time. The response observed when the measurement was lost during a setpoint change is shown in Figure 8-10.

The response observed when the measurement was lost after a process disturbance is shown in Figure 8-11.

As illustrated by these tests, the PIDPlus provides superior dynamic response under these loss measurement conditions. PID response is significantly worse and may not be acceptable in many process applications.

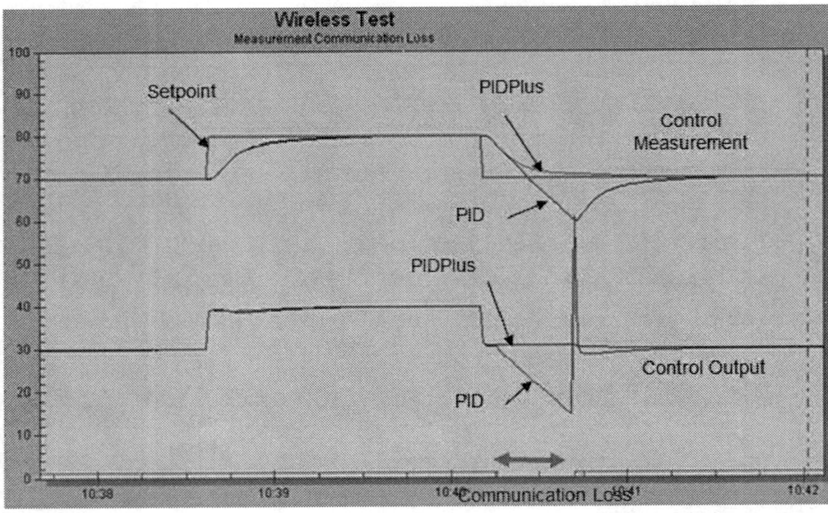

Figure 8-10. Response for Measurement Loss During a Setpoint Change

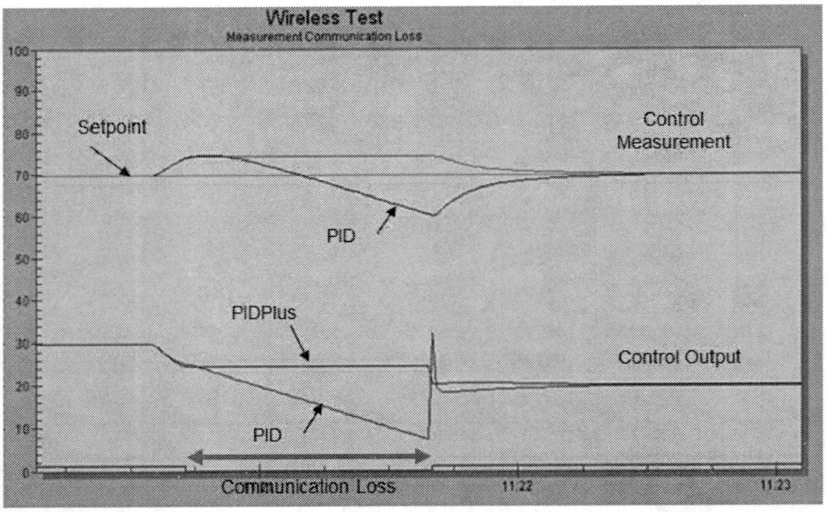

Figure 8-11. Response for Measurement Loss After a Process Disturbance

8.3 Application Examples

The added capability of intelligent PID to use slow and variable communications from wireless devices can be used to implement control in a wide variety of industrial applications. Similarly, the modifications that provide faster recovery from process saturation can be used to improve the operation of both batch and continuous processes.

These benefits are demonstrated by sample applications from the life science and specialty chemical industries.

- Bioreactor control using wireless devices
- Compressor surge control

The benefits of intelligent PID for these applications are typical of those that can be obtained using the advanced features of the PID to address control using wireless measurements and to provide faster recovery from process saturation.

8.3.1 Control of a Bioreactor Using Wireless Devices

A flexible manufacturing environment can be created through the use of portable skid-mounted process equipment that is interconnected and controlled with digital fieldbus technology. In some applications in the bioprocess (life science) industry, the introduction of a single-use vessel provides further flexibility in process design and operation. However, the fixed wiring practices of a traditional installation limit many of the potential benefits of these technologies. WirelessHART technology supports both monitoring and control applications in a flexible manufacturing environment, with one major benefit being the ability to move equipment with no impact on control system configuration.

Dedicated manufacturing sites that produce a single product are quite common within the life science industry. To achieve quality targets, the fixed equipment on such sites is often highly automated in order to maintain repeatable operating conditions and desired production rates. However, a significant portion of the time available for production is lost due to the need to sterilize in place (SIP) and clean in place (CIP) this equipment. The number of field devices associated with SIP and CIP often outnumbers those required for normal production, and the control system logic is more complex. Thus, there are significant economic benefits in moving to a skid-mounted, single-use process design that eliminates or minimizes the need for SIP and CIP. Such a design is in line with the trend towards a more flexible manufacturing environment that supports the production of multiple products. The advantage of skid-mounted, single use process design is that the cost of manufacturing can be spread over several products. Also, the production of smaller batches can be more easily justified in a flexible manufacturing environment.

Major manufacturers within the life science industry have installed skid-mounted, single use equipment in both upstream (e.g., cell culture) and downstream (e.g., Tangential Flow Filtration, TFF) processes. In most cases the deployment of skid-mounted, single use technology has been limited to product development or early phase clinical trial situations (Hartman 2009). However, as more experience is gained with larger scale units based on this design, the reduced initial costs and the flexibility of the design will drive the use of this technology into large scale manufacturing.

The stainless steel tanks that have traditionally been used for bioreactors can be replaced with a skid-mounted, Single Use Bioreactor (SUB) bag. The flexibility of a modern DCS allows the construction of modular, exchangeable components that can accommodate multiple differently-sized SUB bags and a variety of sensor types.

Rapid turnaround between bioreactor runs is one of the more appealing features of a SUB bag system. A new disposable bioreactor bag can be installed in an hour or two, depending on the size. As an alternative, a new bag holder—ready to go—can be wheeled over to the control system and hooked up to cables and coolant lines in even less time. Skid-mounted control systems have been designed to allow the user to connect to bag sizes from 50L to 500L size. Sterile sensors can be inserted into the bag via *Kleenpak* connectors embedded into the bag wall. In some bag designs, optical sensors are embedded into the bags themselves. Features like these and others enable the industry to utilize SUBs during product development (PD) and increasingly to migrate towards the use of SUBs in production.

The benefits of using WirelessHART transmitters with a single use bioreactor have been demonstrated by Broadley James, a major manufacturer of bioreactors for product development and production. A skid was instrumented with a 100L SUB with WirelessHART pH and temperature and pressure transmitters (Blevins, 2009). This SUB and its interfaced BioNet control system are shown in Figure 8-12.

The bioreactor pH and temperature were controlled over a series of batch runs using the WirelessHART measurements. The wireless pressure transmitter was used to monitor pressure within the bioreactor. The pH measurement was communicated on a 1-second exception reporting basis. The temperature was reported on a 2-second basis. A mammalian cell culture was used for each batch run. For purposes of comparison, wired pH and temperature measurements were also available during each batch run. The screen capture in Figure 8-13 shows the setpoint response of temperature control based on the WirelessHART input.

Figure 8-12. Single Use Bioreactor (SUB) with Wireless Instrumentation

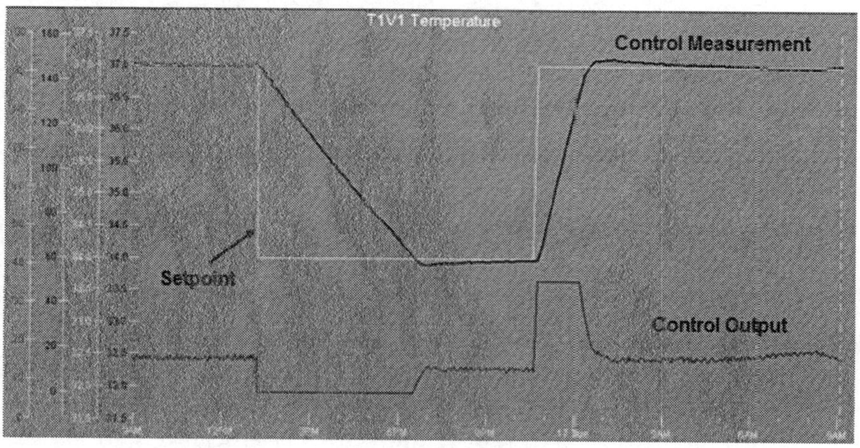

Figure 8-13. Wireless Temperature Control Response in SUB Unit

Similarly good performance was seen for pH control using the WirelessHART input. The response to 0.05 changes in pH setpoint is shown in Figure 8-14.

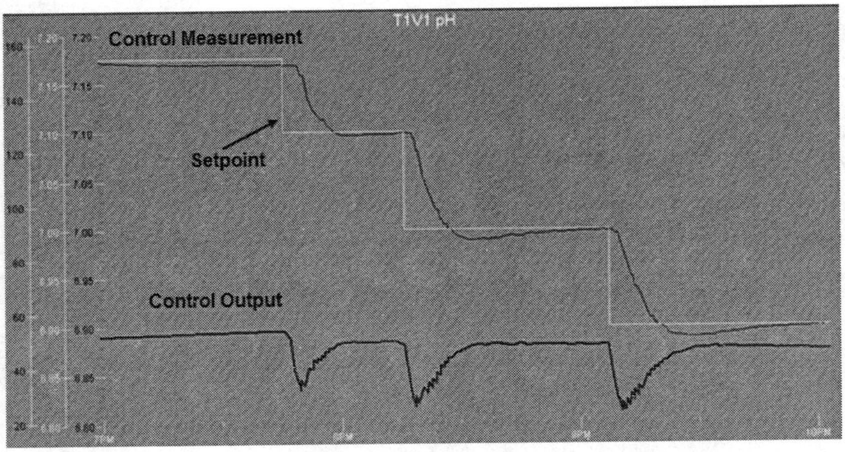

Figure 8-14. Wireless pH Control Response in SUB Unit

The performance of the observed pH and temperature control, using wireless measurements, matched that achieved using wired transmitters. In summary, while disposable bioreactor systems bring on new concerns and risks that need to be vetted and managed, such systems are in wide use and are changing the industry with notable cost-saving features and benefits.

8.3.2 Compressor Surge Control

The effectiveness of providing a variable preload to determine when control action is taken under saturated conditions has been tested and demonstrated in a surge control application. Such a capability is a standard feature of the DeltaV control system used in these tests. This capability was tested using a simulation of a "typical" compressor used in the chemical and refining industries.

To test the PID modifications designed to improve recovery from a limit condition, a dynamic high fidelity simulation was developed for a compressor and its associated downstream header. A series of tests was then conducted to evaluate the effectiveness of this new capability in preventing violation of surge limits for a 60% drop in process demand. In this example, the flow of gas through the compressor is regulated through adjustment of the compressor inlet guide vanes

(IGV). A dynamic simulation of the "typical" compressor and downstream header was implemented in one module within the control system. The process and the compressor anti-surge control addressed in these tests are shown in Figure 8-15.

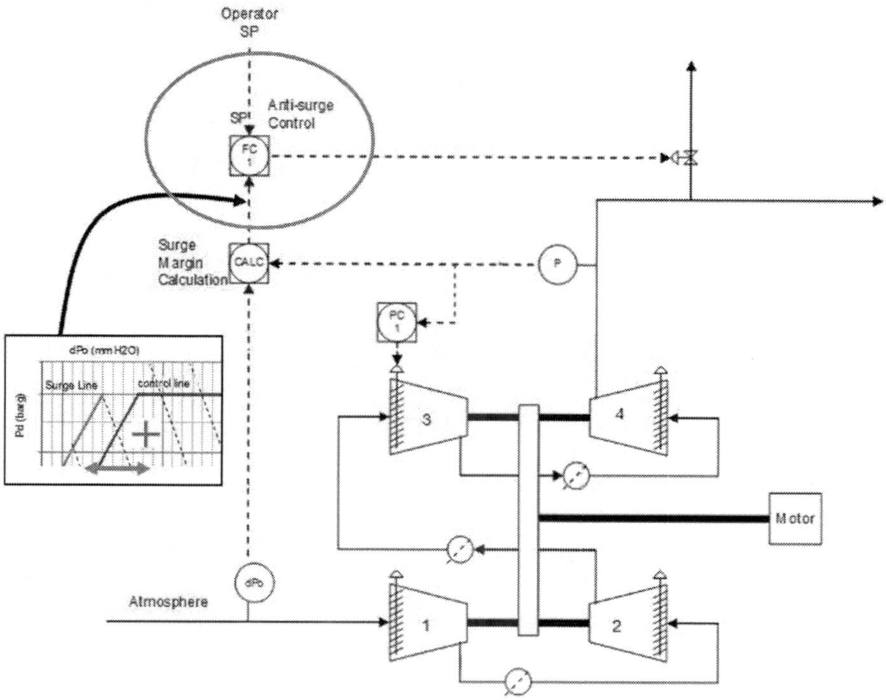

Figure 8-15. Compressor Surge Control

A process simulation module was created to allow the header demand to be adjusted and to thereby introduce a disturbance in the header pressure. Also, parameters were provided in the simulation that could be used to adjust the header volume and vent valve characteristics. The turbine simulation was based on an extrapolation of the compressor performance curves for 20, 30, 40, 50, 60, 70, 80, and 90% open IGV values as shown in Figure 8-16.

In the compressor simulation, IGV values that fell between these curves were determined by interpolation to provide values for all operating conditions.

Data on vent valve performance was used to determine the vent valve characteristics and flow capacity of the valve, CV. A best fit with the experimental data was achieved by applying quick opening valve

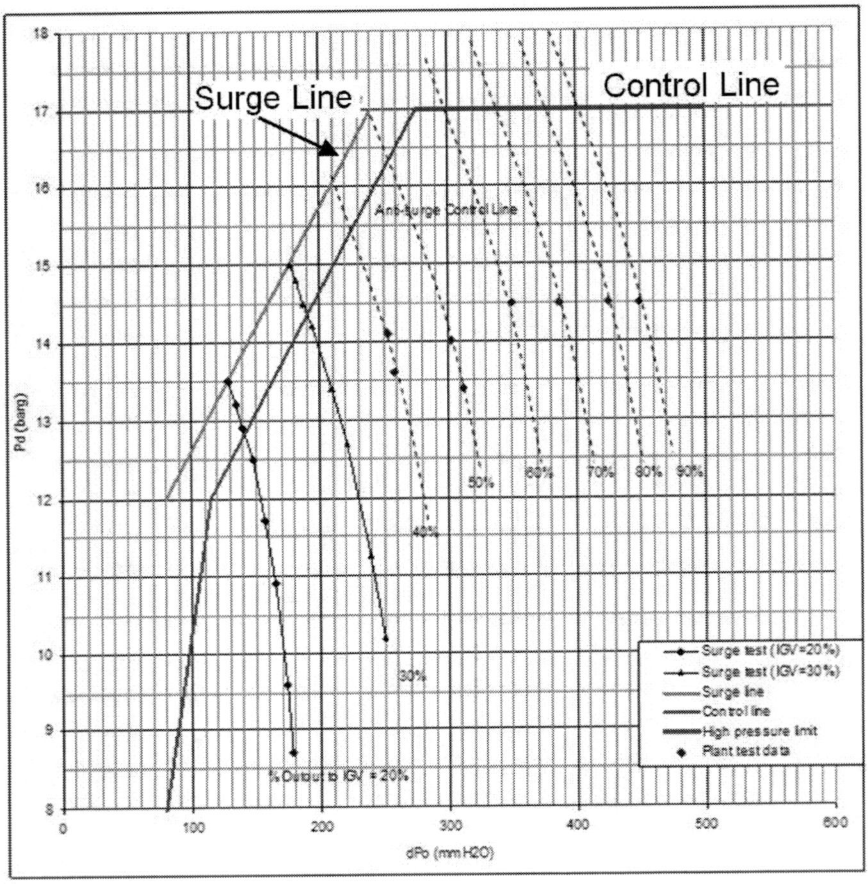

Figure 8-16. Typical Compressor Performance Curves

characteristics. Since flow data was provided as measured dP (mm H_2O) across the orifice (rather than actual flow value) the flow value used in the simulation is calculated in % of scale based on 100% flow at a dP of 600 mm H_2O.

Table 8-2. Vent Valve Performance – Load Test Data vs. Calculated Flow

Compressor Data for Constant IGV % and Vent Valve %					Calculated Vent Valve Flow (%)
IGV %	Vent %	Pressure Differential, Pd (barg)	Compressor Flow: $dPmmH_2O$	Compressor Flow: Flow (%)	
20	20	12.5	148	49.66	46.67
20	30	10.9	165	52.44	53.38
30	20	14.2	194	56.86	49.75
30	30	12.7	221	60.69	57.6
30	40	11.3	239	63.11	62.76

A copy of the compressor simulation module used in these tests is shown in Figure 8-17.

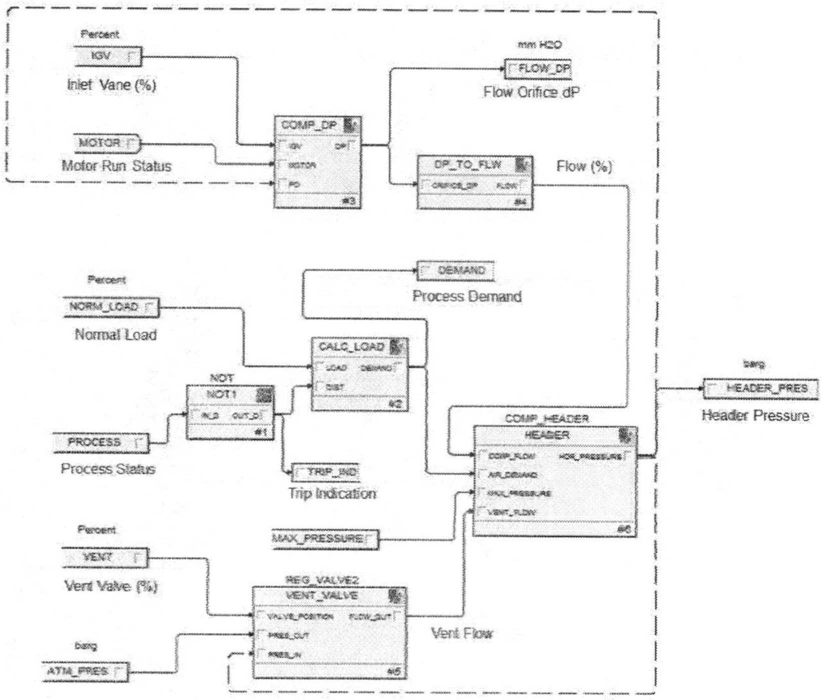

Figure 8-17. Compressor Simulation Module Used in Test

The compressor control used with the simulation is typical of that used for compressor surge control. The compressor controls were implemented in one control module scheduled to execute 10 times per second, 100 msec. execution rate. The control outputs in the compressor simulation modules were accessed using external references. Also, the process outputs calculated in the simulation modules were written to the SIMULATE_IN parameters of the associated analog and discrete (AI and DI) function blocks in the control module, using external references.

Three parameters were added to the control module to support testing of the control using the process simulation:

- MOTOR_STATUS – On/Off status of the compressor motor (1 = On Condition, 0 = Off Condition)

- PROCESS_STATUS – Normal operation (1), Shutdown of downstream process (0).
- NORM_LOAD – The normal demand of the downstream process in %; for example: 80%.

The Compressor control module is shown in Figure 8-18.

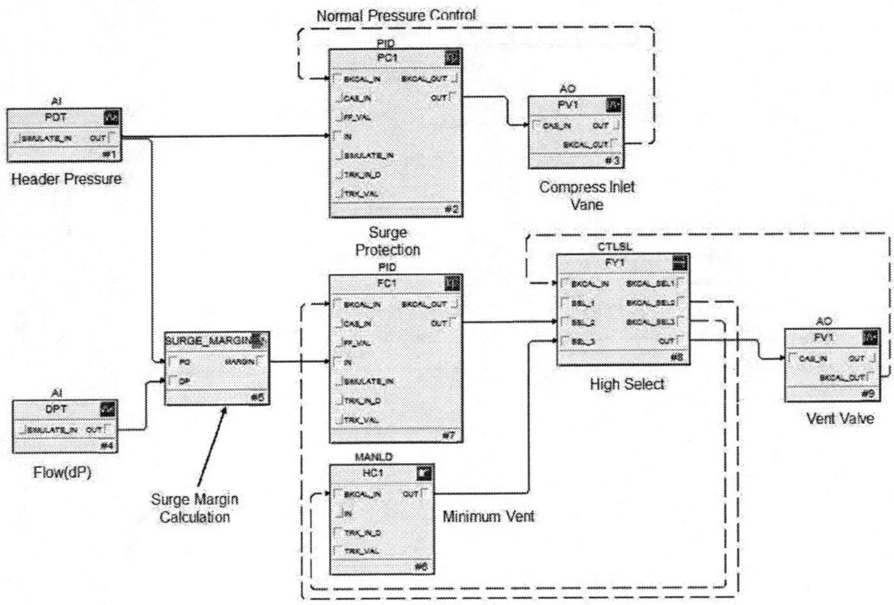

Figure 8-18. Compressor Control Module

The surge control response with no preload added to the PID during process saturation is shown in Figure 8-19. As seen in this trend, the compressor goes below the surge line, as shown in Figure 8-16, when preload is not used. Control action is taken only when the surge margin, the difference (expressed in dP) between the flow at the surge line and the current compress flow rate, drops below the target value specified by the surge control setpoint.

When the automatic addition of a variable preload under saturated conditions was enabled in the PID and the test repeated, an improved response was observed. As shown in the trend in Figure 8-20, the compressor surge margin setpoint of 35 mm H_2O was not violated.

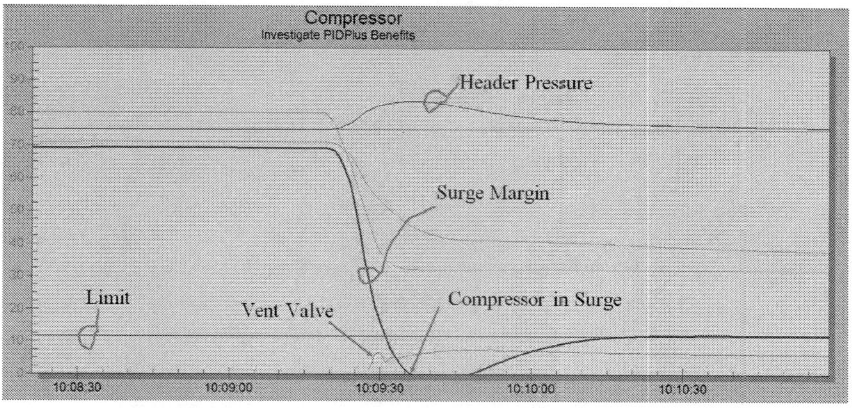

Figure 8-19. Control Response for 60% Reduction in Load with No Preload Applied

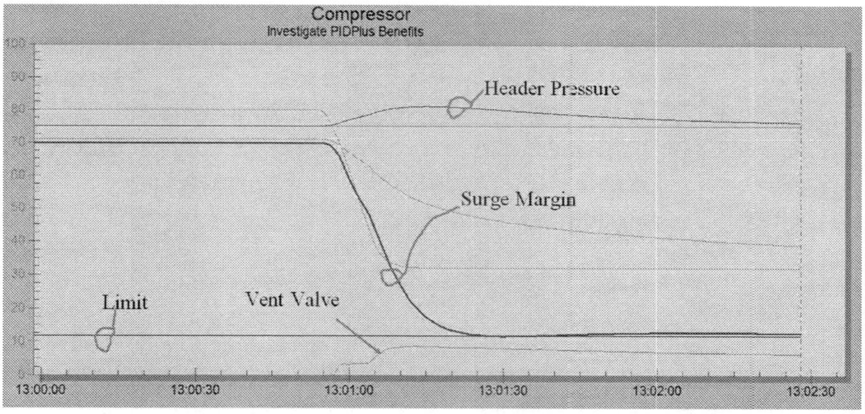

Figure 8-20. Control Response for a 60% Reduction in Load with Variable Preload

8.4 Intelligent PID Workshop

This workshop provides several exercises that can be used to further explore the control using a wireless measurement that is updated in a slow and non-periodic manner. In this workshop, a module will be used that includes two identical simulated processes (Figure 8-21) with a common source of disturbance. Access the book's web site http://www.advancedcontrolfoundation.com and select the Solution tab to view the workshop being performed.

Step 1: Open the module that will be used in this workshop and observe the control and simulated processes.

Step 2: Initialize the Performance Index (IAE) and then change the SP parameter of both control loops by 10%. Observe the control response using a plot of the setpoint, control measurements, and output.

Step 3: Note the IAE and the number of communications for the wireless and wired control. A significant difference should be seen in the number of communications for wired vs. wireless control that were required to respond to the setpoint change.

Step 4: Initialize the Performance Index and change the Disturbance input from zero to 10. Observe the response of the PID and PIDPlus to this unmeasured process disturbance.

Step 5: Note the IAE and the number of communications for the wireless and wired control. A significant difference should be observed in the number of communications for wired vs. wireless control that were required to respond to the unmeasured process disturbance.

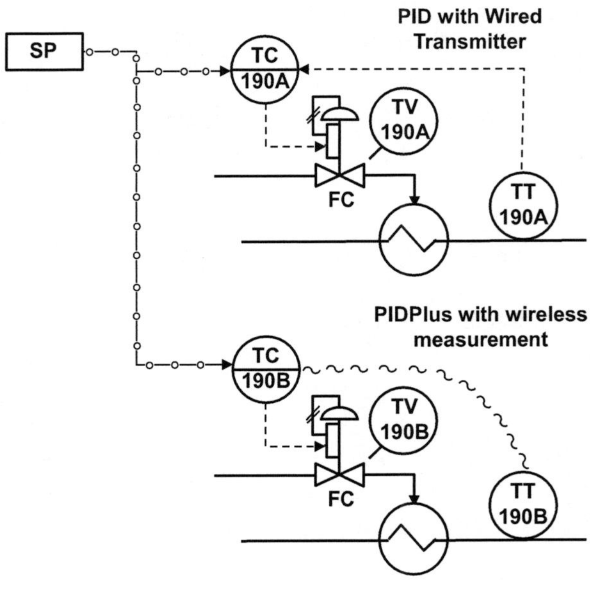

Figure 8-21. Intelligent PID Test Setup

8.5 Technical Basis

8.5.1 Wireless Control

Traditionally, a PI controller can be implemented using a positive feedback network to determine the reset contribution. Mathematically, it can be shown that the transfer function for this implementation is equivalent to the standard formulation for unconstrained control; that is, output is not limited.

$$\frac{O(s)}{E(s)} = K_P\left(1 + \frac{1}{sT_{Reset}}\right) \tag{8-1}$$

where

K_P	=	*Proportional Gain*
T_{Reset}	=	*Reset, seconds*

A significant advantage of the positive feedback network is that the reset contribution is automatically prevented from winding up when the controller output is high or low limited. When wireless transmitters are used in process control applications, the positive feedback network can be modified to accommodate non-periodic measurement updates. Specifically, the filter used in this network can be modified to maintain the last calculated filter output until a new measurement is communicated.

When a new measurement is received, a new filter output is calculated based on the last controller output and the elapsed time since the last measurement value was communicated (see Figure 8-6).

To account for the process response, the filter output can be calculated as follows when a new measurement is received (see Figure 8-7):

$$F_N = F_{N-1} + (O_{N-1} - F_{N-1}) * \left(1 - e^{\frac{-\Delta T}{T_{Reset}}}\right) \tag{8-2}$$

where

F_N	=	New filter output
F_{N-1}	=	Filter output last execution = filter output after last new measurement
O_{N-1}	=	Controller output last execution
ΔT	=	Elapsed time since a new value was communicated

The same results can be achieved by allowing the filter to execute continuously but to maintain the filtered value at the time of the last update until a new measurement is received. However, the advantage of running the filter continuously is that the impact of setpoint changes made between measurement updates can be accounted for in the filter value.

For PID wireless control, the same PI modifications for wireless communication are applied. In addition, the rate contribution to the PID output should be recomputed and updated only when a new measurement is received. The derivative calculation should use the elapsed time since the last new measurement.

8.5.2 Recovery from Process Saturation

For unconstrained PID operation, the P, D, and I terms work in a coordinated fashion to provide a positional PID algorithm. If we consider the case where a positive feedback network is used to implement the reset term, rate gain is zero, and proportional action is taken on error $E(s)$, the PID transfer function is as shown in Equation 8-1 or 8-4.

$$OUT(s) = E(s)K_c\left(1 + \frac{1}{T_i s}\right) \qquad (8\text{-}3)$$

$$\frac{OUT(s)}{E(s)} = K_c\left(1 + \frac{1}{T_i s}\right) \qquad (8\text{-}4)$$

Thus, for unconstrained conditions, standard PI operation is achieved. However, when the PID is at its output limit for an extended period of time, the reset filter output is unchanged, and the proportional contribution added to the limit value determines when the output moves away from the limit. For this limited condition, the output and error are dynamically related as shown in Equation 8-5.

$$OUT(s) = E(s) * K_c \qquad (8\text{-}5)$$

Thus, for the case where control is constrained, the controller output will not move from its limit until the error changes sign (that is, the process measurement goes past setpoint). However, to avoid overshooting the setpoint when recovering from process saturation, it is often necessary to take control action before the measurement reaches setpoint. When the PID output is not limited, the control measurement and controller output are related as shown for a first order process:

$$PV(s) = \frac{K_p OUT(s)}{1 + T_p s} \tag{8-6}$$

where:

K_p = Process gain
T_p = Process time constant

When the PID output is limited, the output that would have been calculated if the limit were not imposed may be calculated based on the control measurement:

$$OUT(s) = PV(s)\left(\frac{1 + T_p s}{K_p}\right) \tag{8-7}$$

If we assume that the controller gain is set for a Lambda factor of 1 and a pure lag process, that is, $K_c = \frac{1}{K_p}$ and $T_i = T_p$, then the controller output for unconstrained operation is:

$$OUT(s) = E(s) * K_c * (1 + T_i s) \tag{8-8}$$

However, when the PID output is limited, the proportional contribution and the preload value determine the PID calculated output:

$$OUT(s) = E(s) * K_c + PL(s) \tag{8-9}$$

By comparing Equations 8-8 and 8-9, it may be seen that a variable term $E(s)K_cT_is$ plus the limit value should be used rather than a fixed preload value to provide the same control response as provided when the output is not limited. This variable term can be substituted for the constant preload value shown in Figure 8-2 if the filter in the positive feedback network is not applied when the PID output has remained at its limit for an extended period of time.

From a practical standpoint, the filtering provided in the reset feedback path may be automatically reduced by a user-specified factor *F* when the PID output reaches its output limits for an extended period of time (Figure 8-3).

The PID output will begin to move from its limit at the point where the proportional contribution $E(s) * K_c$ is exactly cancelled by the added term $E(s) * K_c * T_is$. Ideally, the filtering in the positive feedback network should be set to 0 for an under-limit condition and to 1 for unconstrained operation. However, if the control measurement contains a significant level of noise, some filtering of the variable term $E(s) * K_c * T_i s$ may be required (that is, set to a value greater than 0 when the PID output is limited for an extended period of time) to avoid chatter at the limit.

When this variable preload value is used, and the PID output has been at a limit for an extended period of time, a rapid change in the control parameter can cause the PID output to start to move from the limit even though the control parameter has not reached setpoint. For a slow change in the control measurement, little if any control action will be taken before the control parameter reaches setpoint.

8.5.3 Extension to Include Rate

When a process is characterized as having two distinct dynamics (a second order process), the use of Rate in the PID may improve performance if the process measurement is relatively free of noise. For the case where Rate is applied with Proportional and Reset action, the PID output is:

$$OUT(s) = E(s)K_c\left(1 + \frac{1}{T_i s} + \frac{T_d s}{(\alpha T_d s + 1)}\right) \tag{8-10}$$

However, when the PID output is limited, and assuming α is small, the dynamic relationship of the PID output and control error may be approximated as follows when a variable preload value is used in the positive feedback network:

$$OUT(s) = E(s)K_c(1 + T_d s) + PL(s) \tag{8-11}$$

When the PID output is not limited, the controlled parameter $PV(s)$ and the PID output are related as follows, assuming a second order process:

$$PV(s) = \frac{K_p OUT(s)}{(1 + T_{p1})(1 + T_{p2})} \tag{8-12}$$

where:

K_p	=	process gain
T_p1	=	dominant process time constant
T_p2	=	secondary (faster) process time constant

Thus, when the PID output is limited, the output that would have been calculated if the limit were not imposed may be calculated based on the control measurement:

$$OUT(s) = PV(s)\frac{(1 + T_{p1})(1 + T_{p2})}{K_p} \tag{8-13}$$

If we assume that the controller gain is set for a Lambda factor of 1 ($K_c = \frac{1}{K_p}$), that the controller has been tuned to set the reset equal to the dominant time constant ($T_i = T_{p1}$), and that rate has been set equal to the secondary time constant ($T_d = T_{p2}$) as advised by Corripio and Smith (1970), the controller output and the control measurement are related as shown in Equations 8-14:

$$\begin{aligned} OUT(s) &= E(s) * K_c * (1 + T_i s)(1 + T_d s) \\ &= E(s) * K_c * (1 + (T_i + T_d)s + T_i T_d s^2) \\ &= E(s) * K_c * (1 + T_d s) + E(s)K * (T_i s + T_i T_d s^2) \end{aligned} \tag{8-14}$$

By comparing Equations 8-11 and 8-13, it may be seen that a variable term $E(s)K_c * (T_i s + T_i T_d s^2)$ plus the constant limit value should be used rather than a fixed preload value. This approach should provide the same control response as the response provided when the output is not limited. The control output will begin to move from its limit at the point where the proportional and rate terms are exactly cancelled by the variable preload plus the limit value. Ideally, the filtering in the positive feedback network should be removed when preload action is taken. However, if the control measurement contains a significant level of noise, some filtering of the variable term may be required to avoid chattering at the limit (that is, set to a value greater than zero when the PID output is limited for an extended period of time). This option can be especially useful for PID control if the measurement contains a significant level of noise.

BIBLIOGRAPHY

1. Åström, K. J. and Hägglund, T. *Advanced PID Control*, Research Triangle Park: ISA, 2006.
2. Blevins, T. and Nixon, M. *Control Loop Foundation – Batch and Continuous Processes*, Research Triangle Park: ISA, 2010.
3. Blevins, T. "Improving PID Control with Unreliable Communications," *Proceedings of ISA EXPO Technical Conference*, Houston, 2006.

4. Blevins, T. "PID Advances in Industrial Control," *Proceedings of IFAC Conference on Advances in PID Control PID'12,* Brescia, Italy, 28–30 March, 2012.

5. Caro, Dick. *Wireless Networks for Industrial Automation,* ISA, 2008.

6. Corripio, A. B. and Smith, C. L. "Mode Selection and Tuning of Analog Controllers," *Presented at 3rd ISA Biannual Instrument Maintenance Clinic*: Baton Rouge, Louisiana, 2 May 1970.

7. "DeltaV v.11 PID Enhancements for Wireless," *DeltaV Whitepaper,* Emerson, August 2010.

8. Han, S., Zhu, X., Mok, A. K., Chen. D, and Nixon, M., "Reliable and Real-time Communication in Industrial Wireless Mesh Networks," *In Proceedings, IEEE Real-Time Technology and Applications Symposium,* 2011: pp. 3–12.

9. Hartman, B. "Is Plastic the New Steel?" *Presented at the Emerson Exchange - Life Science Industry Forum,* Orlando, Florida, 2009.

10. Blevins, T., "Incorporating Wireless Devices into Single-Use Disposable Bioreactor Design," *Poster presented at 2009 Dhirubhai Ambani Life Sciences Symposium.*

11. Nixon, M., Chen, D., Blevins, T., and Mok, A. K. "Meeting Control Performance over a Wireless Mesh Network," *Proceedings of the 4th Annual IEEE Conference on Automation Science and Engineering (CASE 2008),* August 23–26, 2008, Washington, DC, U.S.

12. Nixon, M., "PIDPlus: An Enhanced PID Control Algorithm for Wireless Automation," *Presented at AS-74.3199 Wireless Automation,* Aalto University, Finland, 2012.

13. Rhinehart, R. R., Shinskey, F. G., and Wade, H. L. "Control Modes – PID Variations," *Instrument Engineers' Handbook, Process Control and Optimization,* Boca Raton, London, New York: CRC Press, 2006.

14. Shinskey, F. G. "The Power of External Reset Feedback," *Control,* pp. 53–63, May 2006.

15. Chen, D., Nixon, M., Aneweer, T., Mok, A. K., Shepard, R., and Blevins, T. "Similarity-based Traffic Reduction to Increase Battery Life in a Wireless Process Control Network," *Proceedings of ISA Technical Conference,* Houston, TX, 2005.

16. McMillan, G. "Wireless – Overcoming Challenges of PID Control & Analyzer Applications," *InTech,* Vol. 57, No. 4, pp. 16–21, Research Triangle Park: ISA, July/August, 2010.

17. Seibert, F. "WirelessHART Successfully Handles Control," *Chemical Processing* Vol. 74, No. 1, pp. 39–42, January, 2011.

9 Continuous Data Analytics

One of the key objectives of a process control system is to maintain product quality parameters within specification. In this chapter, we discuss how to fulfill this objective for continuous processes using data analytics. In many cases, the measurements required may extend beyond the basic physical properties (e.g., pressure, temperature, flow, and level.) Material properties, such as density or viscosity, may also be critical to process operation and the successful production of a product and, therefore, must be measured. As discussed in Chapter 3, such measurements are often needed by the plant operator or control system to meet quality objectives. On-line analyzers that automatically provide measurements of material properties to the control system may be too expensive, may have proven to be unreliable, or simply may not be available because of the complexity of the sampling system or the complexity of the analysis. As a result, most process plants include one or more labs that are centrally located in the plant or distributed throughout the process to provide an analysis of the material properties that must be controlled. The lab may also analyze raw materials that are used as feedstock to the plant to identify variations that could impact the process.

However, a gap often exists between the time a grab sample is taken for analysis and when the lab results are available to the plant operator. The delay associated with processing the sample in the lab impacts the speed at which corrections can be made for changes in feedstock (where feedstock is not tested prior to use) or in process operating conditions, which limits the usefulness of the measurement. Through the use of linear or non-linear estimators, it is possible to bridge this gap and thus improve plant operation as illustrated in Figure 9-1. Non-linear estimators were discussed in Chapter 7. Linear estimators are based on data analytic techniques, which may be applied as well for the process fault detection.

Figure 9-1. Bridging the Gap between Product Quality and Process Control

To obtain an immediate indication of the results of process or material changes that impact a quality-related parameter, it is often possible to apply data analytics to upstream measurements, such as flow rate, temperature, and feedstock composition, to calculate an estimated value of the quality parameter. Even when this is done, however, lab analysis is still required to verify the accuracy of the estimated value. It also allows corrections to be made in the calculation to compensate for unmeasured changes in the process that impact the property that is being estimated.

Through the use of on-line data analytics, on-line decision support for operations personnel can provide:

- Product quality predictions: Identify quality problems while there is time to make on-line corrections to prevent them.
- Early process fault detection: Detect abnormal process operation and/or equipment problems before they affect production. Provide root cause analytics to direct operations or maintenance personnel to quickly correct the cause of problems. (Mason and Young, 2002)

Various techniques may be used to provide a prediction of quality parameters associated with a continuous and batch process:

- Neural Network (NN) Model: For use with non-linear processes, as addressed in Chapter 7, Neural Networks. Application is restricted to the data range used in the development of the model.
- Multiple Linear Regression (MLR) Model: May be used with linear processes over a wide operating range. Can be used to extrapolate outside the data range used in model development.

- Partial Least Squares (PLS) Model: A linear technique that is preferred for use with collinear data. (Collinear means that multiple process input changes impact similar process output measurements.) Can be used to extrapolate outside the data range used in model development.

Process interactions that potentially impact quality parameter prediction and fault detection may be accounted for through the use of principal component analysis (PCA). This multivariate technique allows the fundamental underlying changes in the process to be identified. The benefit that this multivariate technique provides over the univariate statistical analysis for fault detection can be illustrated by considering two parameters of a collinear process at one point in time, as shown in Figure 9-2.

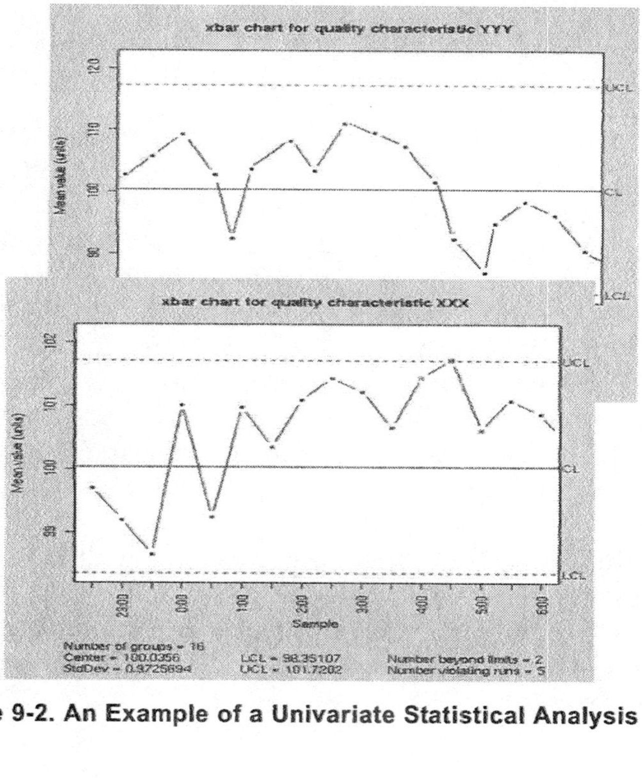

Figure 9-2. An Example of a Univariate Statistical Analysis Chart

Because the two variables are within their control limits, a univariate analysis in which each parameter value is compared to its control limits would indicate no fault. However, when the two parameters are plotted for time coincident points, it becomes clear that the points A and B in these plots do not follow the normal process variations as shown in Figure 9-3.

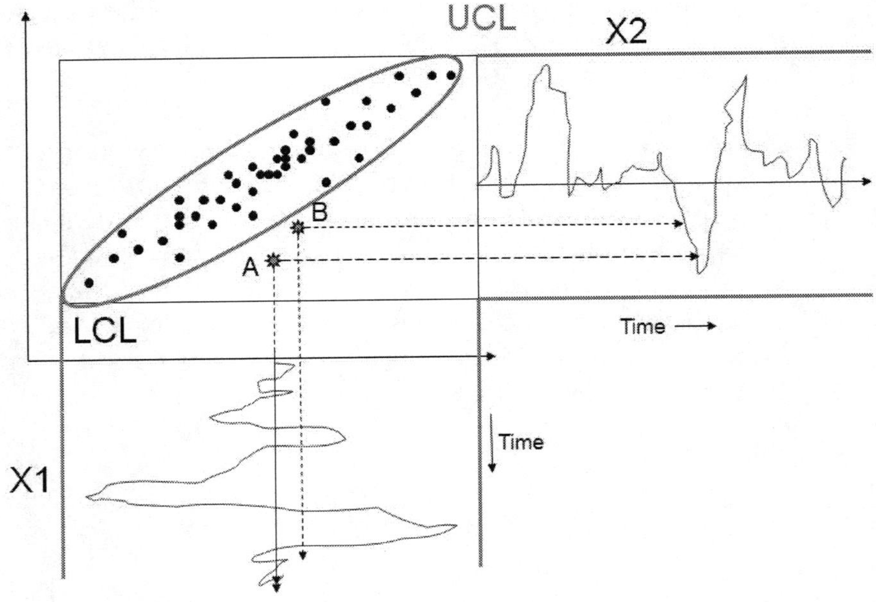

Figure 9-3. Detection of a Fault Condition in a Multivariate System

Because there are numerous process parameters in industrial processes, to be able to detect abnormal conditions a technique is needed that allows the interaction of all parameters to be characterized. Using principal component analysis, basic components that explain the normal process variation are determined by projecting process variation on the principal components. For this example, most of the variation in the process may be captured by two principal components, as illustrated in Figure 9-4.

In this example, the first principal component, labeled t1, explains the highest amount of variation of Pressure and Temperature in the process over time. The second component, labeled t2, explains the *next* highest amount of variation of Pressure and Temperature in the process over time.

Through the use of PCA, a model can be developed that allows the variables associated with a process to be represented on-line by a few principal components, thus reducing the dimensions of the data set and eliminating the effects of the interaction between variables. Through the application of Hotelling's T^2 statistic and the Q statistic to the reduced data set, it is possible to detect abnormal operation resulting from both measured and unmeasured/un-modeled disturbances (Chiang, Russell, and Braatz, 2001):

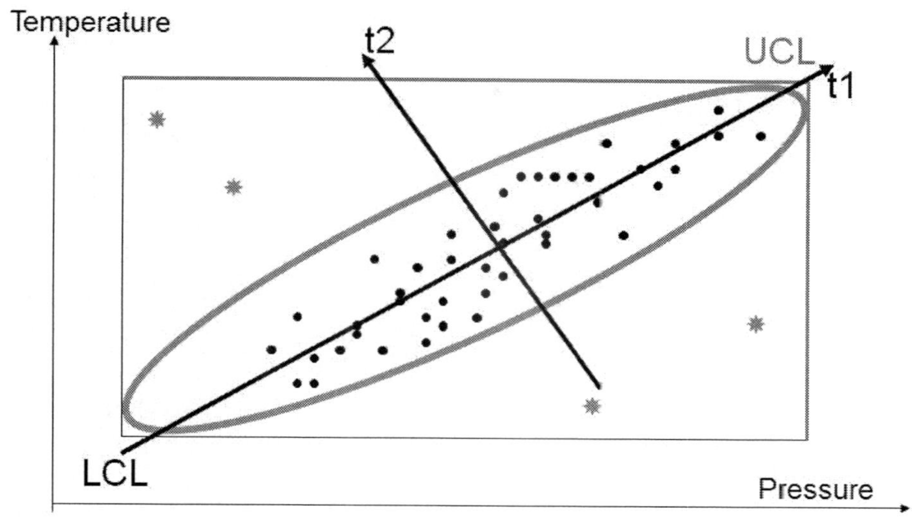

Figure 9-4. Principal Components Explaining Process Variation

- Measured and modeled disturbances: May be quantified through the application of Hotelling's T^2 statistic.
- Both measured and unmeasured unmodeled disturbances: The Q statistic, also known as the Squared Prediction Error (SPE), may be used.

Faults are determined by comparing these calculated statistics to an upper limit; an abnormal condition is indicated if the value exceeds the limit. For this example, a plot of the T^2 statistic clearly shows the fault condition in the data as illustrated in Figure 9-5. When a normalized limit is applied, any value above 1 indicates an abnormal condition.

Mathematically, any number of variables may be analyzed to develop a PCA model that allows process variation to be expressed in terms of a small number of principal components. For example, when considering three variables the same technique may be applied, allowing the variations to be expressed by two principal components as illustrated in Figure 9-6.

In three dimensions, similar to two dimensions shown in Figure 9-4, the first principal component explains the highest amount of the variation of Y_1, Y_2, and Y_3 over time. The second principal component explains the next highest – though significantly smaller than the first principal component – amount of the variation Y_1, Y_2, and Y_3 over time. The observations can therefore be projected onto a plane.

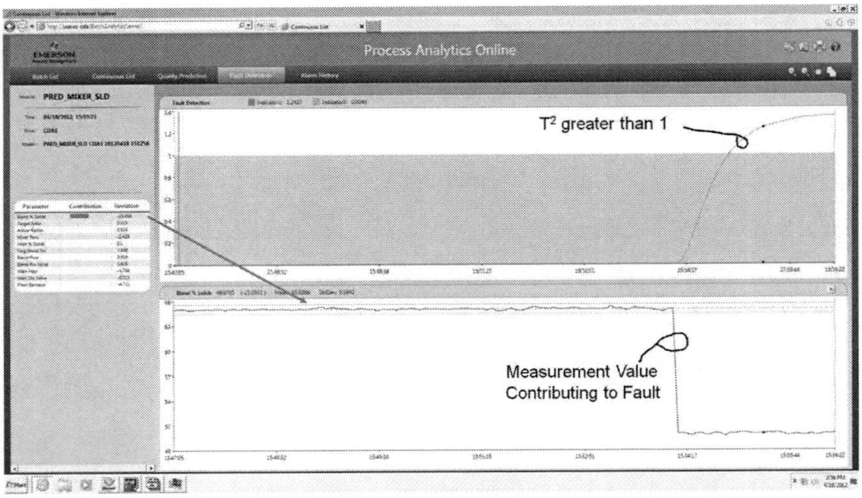

Figure 9-5. T^2 statistic plot (upper) for fault detection and related fault diagnosis plots: the contribution plot in the left pane and process parameter plot (bottom plot)

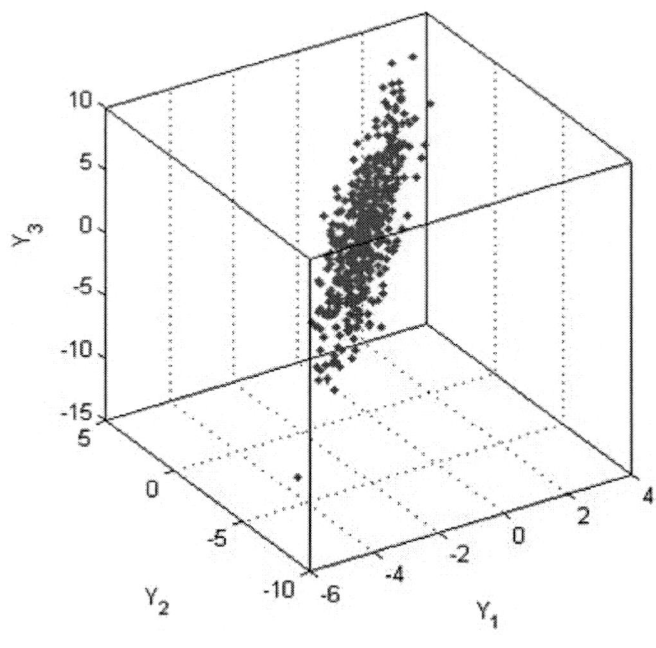

Figure 9-6. Analysis of Variation for Three Variables

If the number of principal components is increased, a larger percentage of the variation is explained (or modeled). However, the resulting model may not always work well with data that was not used in model development. Thus, the number of principal components used in model generation should be determined to give minimum error when the model is applied to test data (that is, data not used in model generation). For example, the PCA model that best characterizes a process with 30 measured variables may have five or six principal components.

Through the use of PCA, it is possible to develop a process model that provides the operator with the following information:

- An indication of a process fault condition that may impact the accuracy of the prediction
- A display of the input(s) that contributed most to the fault
- An indication of the expected value and normal range of variation vs. the current value of measurements used in the prediction

PLS applied for the quality prediction similarly as PCA reduces the dimensions of the data set and eliminates the effects of the interaction between variables. Once quality parameter prediction has been deployed, lab samples or on-line analyzer measurement (where available) of the quality parameter used for model development can continue to be used on-line with the model. This measurement of the quality parameter will be used to:

- Automatically correct prediction bias caused by unmeasured disturbances or changes in operating conditions
- Determine when the quality parameter prediction model needs to be updated

The PCA and PLS algorithms used for continuous processes are based on deviations from mean values of the parameters used in the data set.

In a few cases, a continuous process is operated at a constant throughput and only makes one product grade with the mean value of the process measurements will be nearly constant and traditional techniques for PLS, MLR, and PCA may be applied. In many cases, the throughput of a continuous process must be frequently changed to maintain an inventory level set by downstream processes or to meet market demand. Thus, a major constraint on the application of those linear analytic techniques is that the underlying technology is based on the deviation of process measurements from their mean value. Because a change in plant production rate or a change in product grade can cause mean values to shift, the continuous data analytics application used for model building and the on-line application must account for changes in the mean value of the measurements. It can be accomplished by the process state concept, which assigns specific states to a predefined process operating condition, that is plant production rate, product grade, range of selected parameter values, etc. Process state can be used to automatically modify mean values to accommodate changes in production rate or product grade. Additionally, the deviation values for the time required to transition from different throughputs and product grades are compensated.

A swing boiler in the power house area of a plant is one example of a process that must constantly respond to changes in throughput demand set by the plant master control. To satisfy these changes in demand, the air and fuel input and the ID (induced draft) fan speed must be changed in a coordinated manner to maintain O_2 (that is, excess air level) and draft in the boiler at constant levels. The variation in these parameters for step changes in steam demand is illustrated in Figure 9-7.

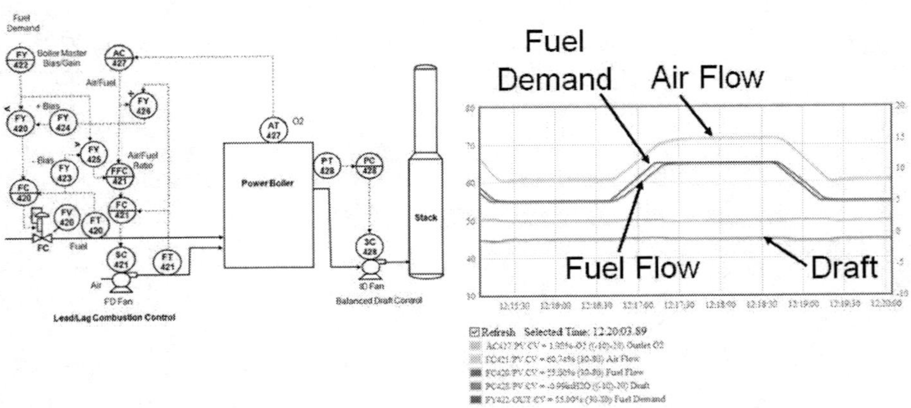

Figure 9-7. Swing Power Boiler – Operation to Meet Steam Demand

The process operating points may also change with changes to the product grade that is manufactured. One example of this is a continuous reactor in which the target for output composition is changed. To change the output composition, it is often necessary to change the operating point of one or more process inputs. In response to these changes in process inputs, other parameters such as cooling water flow, agitator power, and relief flow must also change to maintain batch temperature, constant agitation, and overhead pressure within specification limits. The variations in continuous reactor operation, when the outlet solids concentration target is changed from 40% to 35% and from 35% to 40%, are shown in Figure 9-8.

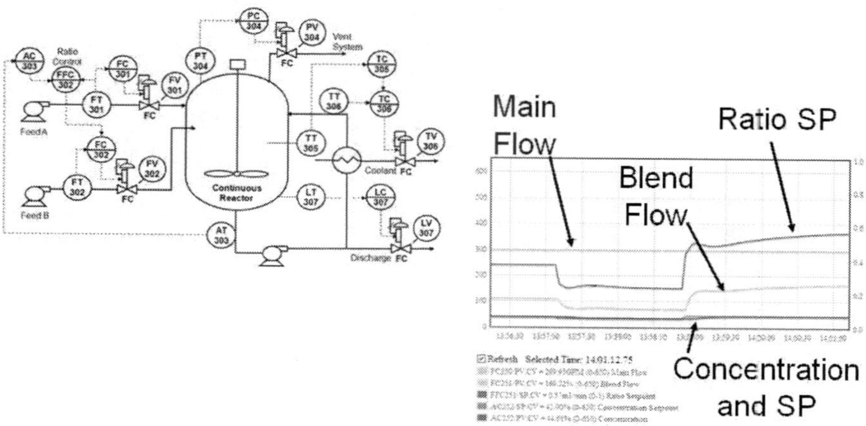

Figure 9-8. Continuous Reactor – Operation for Changes in Product Grade

To minimize deviations in quality parameter prediction, and to prevent false indications of a fault, the mean values used in data analytics should be changed to match those expected for a given throughput or grade. Also, as shown in Figures 9-7 and 9-8, the transition from one operating point to another may take some time after the throughput or grade is changed. To account for the variations in response time, some filtering should be applied. To account for variations in parameter mean values, some continuous data analytics products allow the user to configure a state parameter that reflects changes to the major disturbance to process operation. In some cases this will be a continuous parameter such as the production rate, but in other cases, it may be a discrete parameter, such as the product grade. In general, the user should select one measured or calculated parameter as the state parameter, indicate whether it is a continuous or discrete parameter, and indicate the number of process states that will be used in data analytics.

DeltaV (www.easydeltav) continuous analytic product, by default, uses a continuous parameter as a state parameter with five states assumed. When a set of data is selected for model generation, the range of variation in the state parameter is calculated. By default, this operating range is divided into equal segments that represent the states of the process considered in analysis. For a period – or periods – of time that the process operated within a state (as indicated by the state parameter value), the average value of the state parameter is calculated for each measurement and saved as a part of the model. For example, if the boiler steam demand shown in Figure 9-7 was configured as the state parameter and varied over a range of 25–75% in the data set selected for model building, the values collected for the five inputs included in the model might appear as illustrated in Table 9-1.

Table 9-1. Variation in Boiler Parameter Mean Values with State

State – Steam Demand	1	2	3	4	5
State Range	25-35	35-45	45-55	55-65	65-75
No. Sample in Range	210	340	150	85	30
Steam Demand – Ave	30	40	50	60	70
Fuel Flow – Ave	30	40	50	60	70
Air Flow – Ave	35	45	55	65	75
O_2 - Ave	2.5	2.5	2.5	2.5	2.5
Draft – Ave	-1	-1	-1	-1	-1
ID Fan Speed – Ave	20	28	38	50	65

During on-line operation, the instantaneous value of the state parameter is used with the average parameter values saved in the model to determine the deviations from mean used in fault detection and parameter quality prediction. For example, if the instantaneous value of the state parameter is equal to the average state parameter value for a range, then the average values for each parameter will be used as the mean value in calculating the parameter deviation from mean. In most cases the state parameter value will fall between the average state parameter values for the two states. In such cases the state parameter average values for these two states is used with the parameter average values to calculate the parameter mean values. The deviation from mean calculated for each parameter value at a point in time is filtered before it is used in model building or in on-line operation. A first order filter with a time constant based on the configured transition time (in seconds) is usually applied.

When the state parameter is a product grade (a discrete parameter with enumerated values, for example: 1–5), then in model development the parameter mean value for each state is calculated for the data samples that coincide with the state parameter being in that state. During on-line operation, the mean values for the current state saved in the model are used in calculating the deviation values used in prediction and fault detection. For example, the mean values that were established for a continuous reactor might appear as shown in Table 9-2.

Table 9-2. Variation in Continuous Reactor Mean Values with State

State	1	2	3	4	5
Grade Description	ADX201	ADX210	ADX215	ADX230	ADX240
Sample in State	210	340	150	85	30
Primary Flow – Mean	70	90	60	80	75
Secondary Flow – Mean	25	35	30	45	50
Prod. Conc. – Mean	35	40	30	45	42
Reactor Temp – Mean	210	215	205	211	200
Cooler Out Temp – Mean	180	185	174	200	190
Reactor Press – Mean	10	10	10	10	10

If the state parameter value is 3 (ADX215), then the mean value for primary flow is 60. If the state value changed to 1 (ADX201), then the mean value for primary flow is 70. Any changes in state will not be immediately reflected in the process parameters. To account for the time required for the process to respond to a state change, the deviation from mean is filtered based on the configured transition time.

The calculated deviation from mean values for a selected data set is used directly in the generation of the PCA model. In addition, the current deviation values are used on-line in fault detection using the PCA model. However, in the development of PLS models for quality parameter prediction, the deviation values must be shifted in time to account for the time required for a change in the input to impact the quality parameter. Similarly, these delays should be taken into account in the processing of the deviation values used for on-line quality parameter prediction.

The preferred means of determining the delay between a process input change and when its impact is reflected in the quality parameter is to find and apply a cross-correlation between the process input and the quality parameter. Identified delays are then used to create the time shifted data that are applied to the PLS model (without considering state).

An example of how the cross-correlation and delays may be presented to the user is illustrated by the following example, in which the Kappa number associated with the output of a Kamyr digester is predicted based on process inputs. The process and its associated inputs and outputs are shown in Figure 9-9.

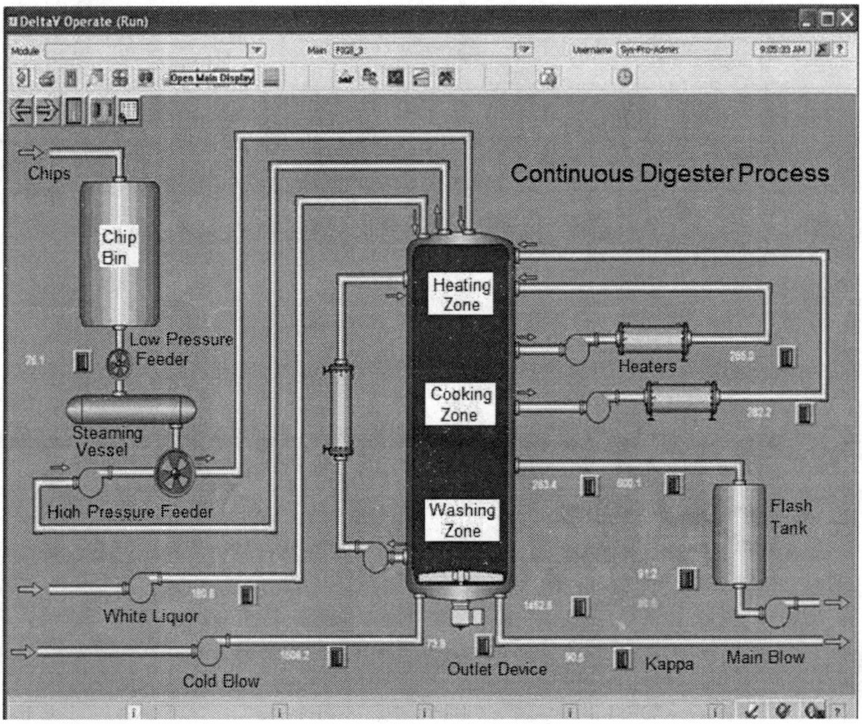

Figure 9-9. Kamyr Digester Process Example

In this example, the chip speed (Figures 9-9 and 9-10) is the state parameter because it sets the process throughput. The Kappa number of the product output may be measured using an on-line analyzer or by analyzing a grab sample in the lab. The delay associated with each input that is reflected in the Kappa number is automatically determined by doing a cross-correlation between the process inputs and the Kappa number contained in a selected data set. The results of this are displayed as shown in Figure 9-10.

In this example, by selecting a process input in this overview display, the user may see the cross-correlation between the process input and the Kappa number. Figure 9-11 illustrates the cross-correlation between the chip meter speed and the analysis of the Kappa number.

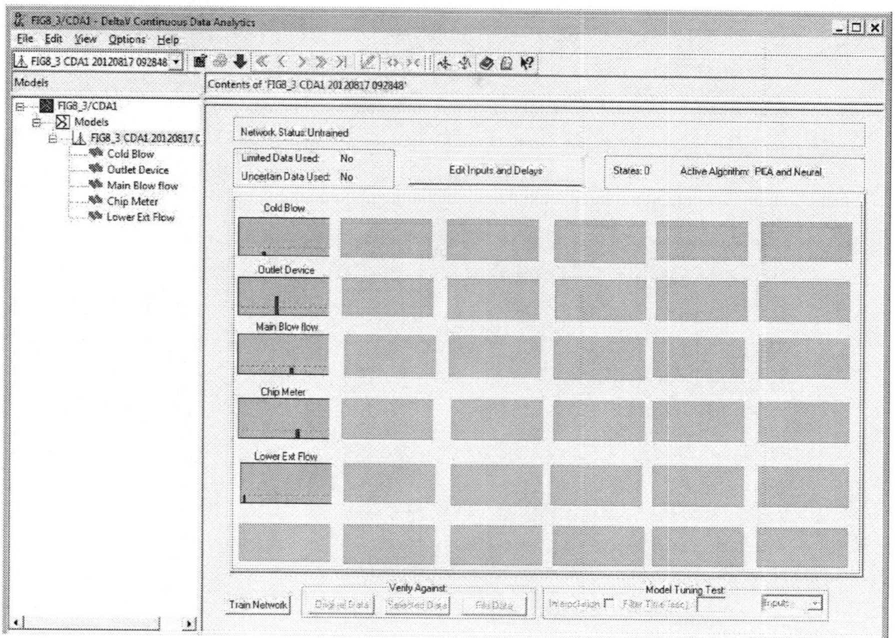

Figure 9-10. Delays Determined by Cross-correlation

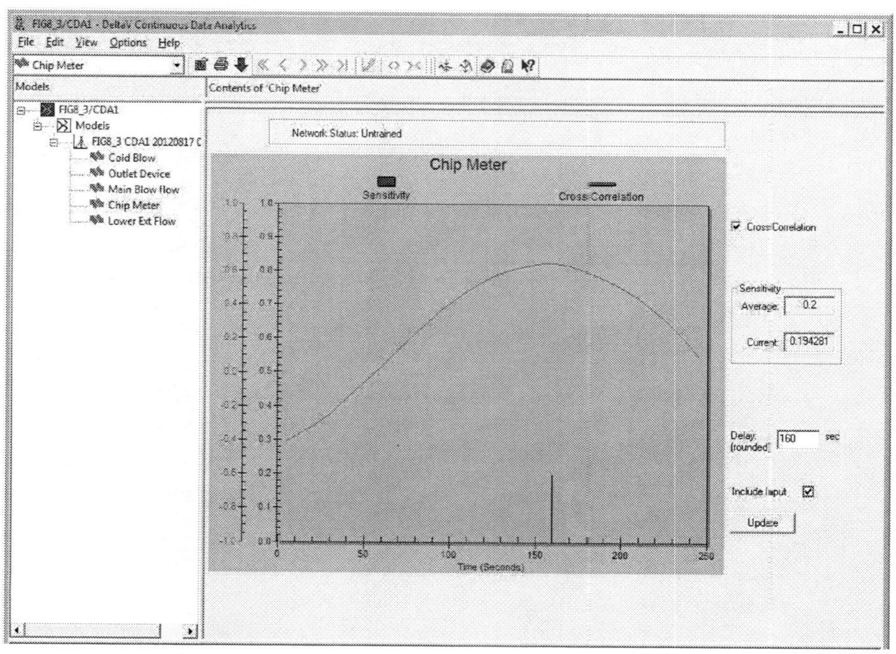

Figure 9-11. Cross-correlation between Chip Meter Speed and Kappa Number

As illustrated in Figure 9-11, the delay associated with a process input is established based on the time shift that provides maximum correlation. In this simulation of a Kamyr digester, the delay between the change in chip meter speed and the change being reflected in the Kappa number is 175 seconds. In an actual process the delay could be much longer.

Continuous data analytics may be applied to a wide variety of processes for fault detection and quality parameter prediction. One more example of where continuous data analytics has been applied is a two-stage flash skid process used for CO_2 recovery from the flue gas emitted by a power boiler shown in Figure 9-12.

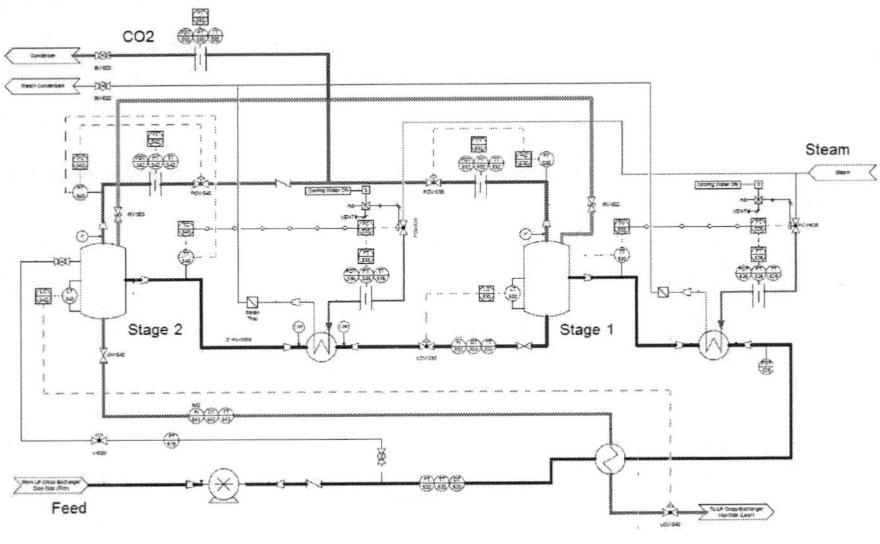

Figure 9-12. Two-Stage Flash Skid

For this process, the state parameter is the feed flow rate to the skid. As the feed rate changes, the pump pressure, gas vent flow from each flash vessel, steam flow to the heater required to maintain temperature, and the flow rates through the flash vessels and heat exchangers change accordingly. The primary quality parameters for the skid are CO_2 loading after the first flash vessel, and the measurement may be obtained by lab analysis of grab samples.

For model generation, the process time to steady state is estimated to be 600 seconds, and the transition time for changes in the state parameter was initially selected as 60 seconds. The presented results were achieved using a simulation of the skid process.

In this case, some inputs were automatically excluded from the model as a result of sensitivity analysis as shown in Figure 9-13.

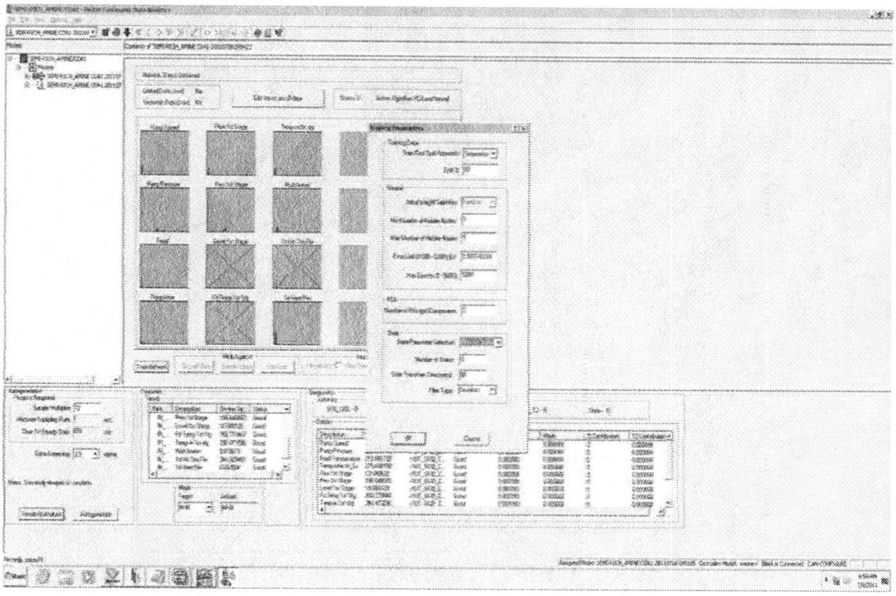

Figure 9-13. Analytic UI for Revising Parameters Selected for Model Generation

After model generation, model verification showed that the model accurately reflected the CO_2 loading as determined by lab testing, as shown in the trend Actual and Predicted values of the CO_2 loading vs. sample number (Figure 9-14).

9.1 Application Example

Several commercial products are available on the market for inferential quality parameter prediction, based on the model and process measurements. Some of the latest products incorporate PCA modeling for process fault detection, the ability to remotely monitor the process, support for on-line execution in the distributed control system controller, and features that allow this capability to be easily incorporated into an existing control system. A dynamic process simulation of a mixer process is used here to demonstrate the steps involved in devel-

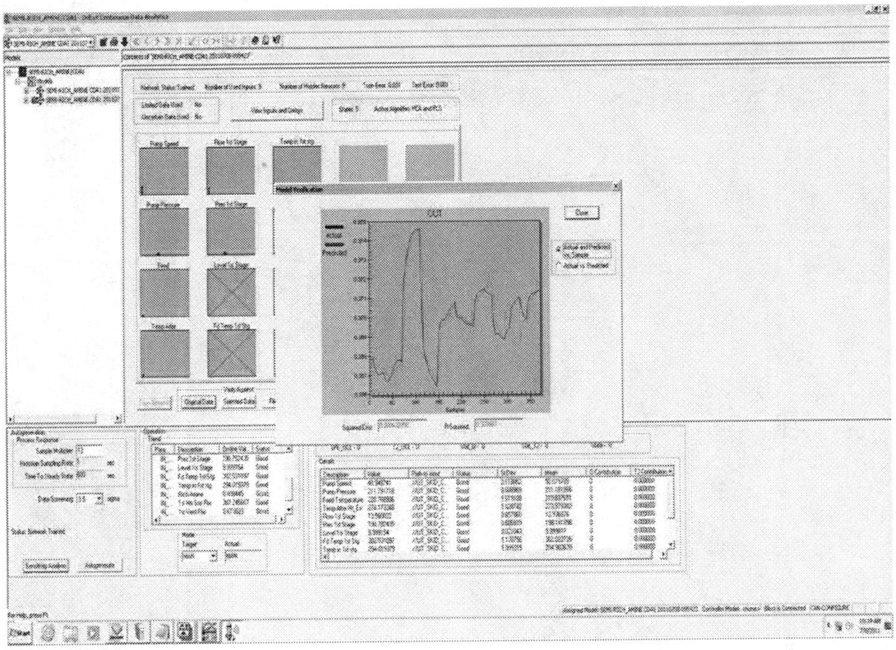

Figure 9-14. CO_2 Loading Model Verification

oping and deploying PCA and PLS models for quality parameter prediction and process fault detection. For the mixer process used for this example, the flow demand is the state parameter. A piping and instrumentation diagram of the mixer process is shown in Figure 9-15.

In a plant environment the flow demand is set by the operator. However, the simulation allows the flow demand and unmeasured disturbance to be automatically changed to allow operating data that includes normal process variation to be collected for model development.

In this example, the quality parameter is the % solids measurement after the mixer. To maintain the quality parameter at setpoint, the blend flow is ratioed to the main flow to the mixer. As the flow demand changes, the main flow and blend flow are changed to maintain the ratio. To achieve a target % solids concentration after the mixer this ratio target is manually adjusted to correct for changes in the solids concentration in the main and blend flows. The simulation is designed to allow various process faults to be introduced, in such a way that analytics fault detection can be easily tested and results verified.

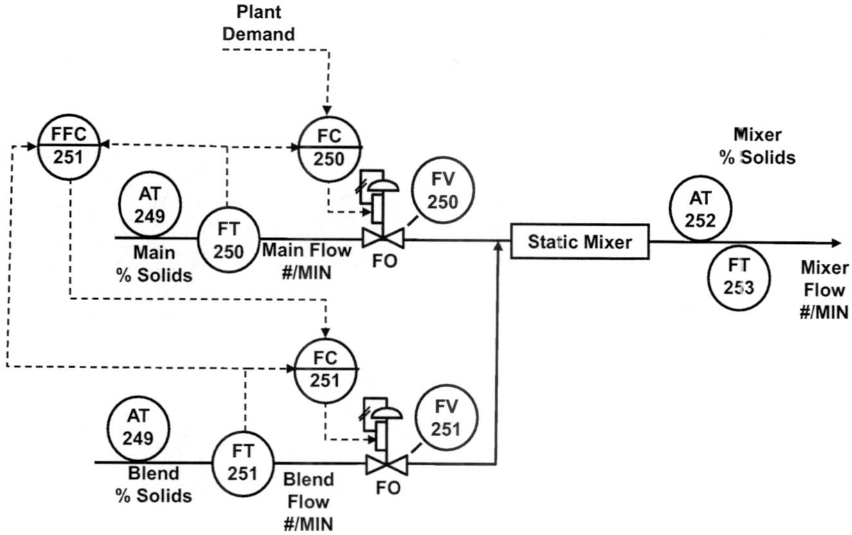

Figure 9-15. Mixer Process

When data are collected to train the data analytics models, normal process variations are automatically introduced into the process simulation. Once data analytics models have been developed, the process may be operated at normal operating conditions, or specific disturbances may be introduced to demonstrate how this fault impacts the quality parameter as shown in the operator interface.

9.1.1 Defining Model Inputs

The parameters associated with the mixer process that may be used to develop a PCA model for fault detection and a PLS model for quality parameter prediction are summarized in Table 9-3. It is assumed that the DEMAND.CV parameter is selected as the state parameter, and the property prediction will be used as a backup for the AT252 measurement, which is the on-line quality parameter.

Table 9-3. Mixer Parameters Used in Model Development

Parameter	Comment
DEMAND.CV	State Parameter
AT248	Main %, Solids %
FC250	Main Flow, #/Min
	Main Flow Valve, %
AT252	Mixer % Solids
	Target Ratio

Table 9-3. Mixer Parameters Used in Model Development

FFC251	Actual Ratio
	Target Blend Flow
FT253	Mixer Flow #/Min
AT249/OUT.CV	Blend % Solids
FC251/PV.CV	Blend Flow, #/Min
FC251/OUT.CV	Blend Flow Valve, %

To apply data analytics to the mixer application, a module must be created that includes a data analytics (DA) function block, illustrated in Figure 9-16.

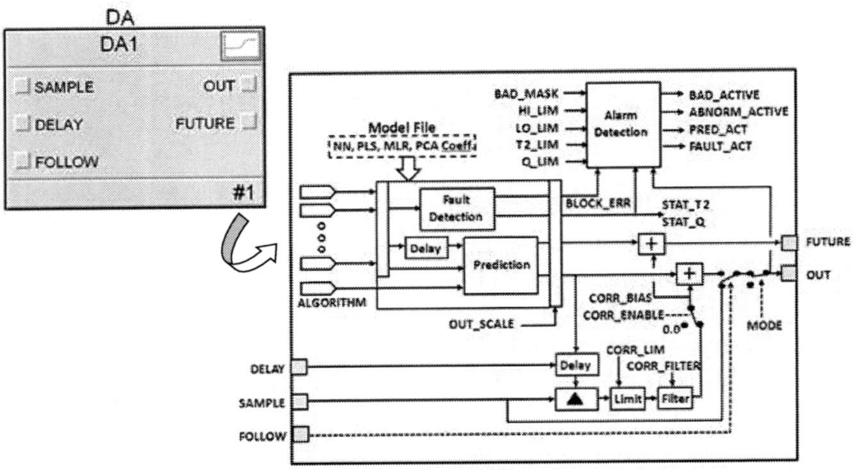

Figure 9-16. Data Analytics Function Block

The number of inputs used by the data analytics block can be configured by right clicking on the block and selecting Properties. A maximum of 36 inputs may be used by the data analytics block. In response, the dialog shown in Figure 9-17 is provided for configuring the measurement input.

As a part of the input configuration, user-friendly names can be defined for the measurement inputs and the quality parameter and used in the data analytics application. Once a module containing the data analytics block has been configured and saved, the inputs and some calculated parameters associated with the data analytics block are automatically assigned to the control system historian.

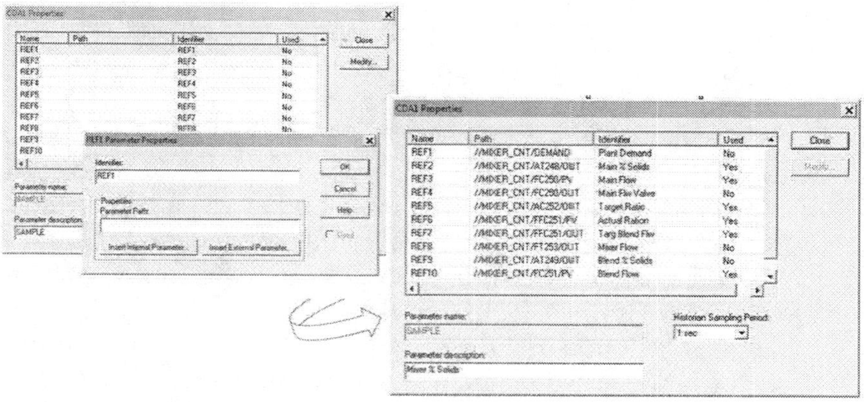

Figure 9-17. Configuring Input for Data Analytics

9.1.2 Model Building

After sufficient historical data has been collected, the data analytics application may be used to create PCA and PLS models for fault detection and fault diagnosis. When the data analytics application is launched, the initial display shows a trend that may be used to plot selected block parameters. Once parameters are assigned to the trends, the user may view the values and select the data that will be used in model development, as illustrated in Figure 9-18.

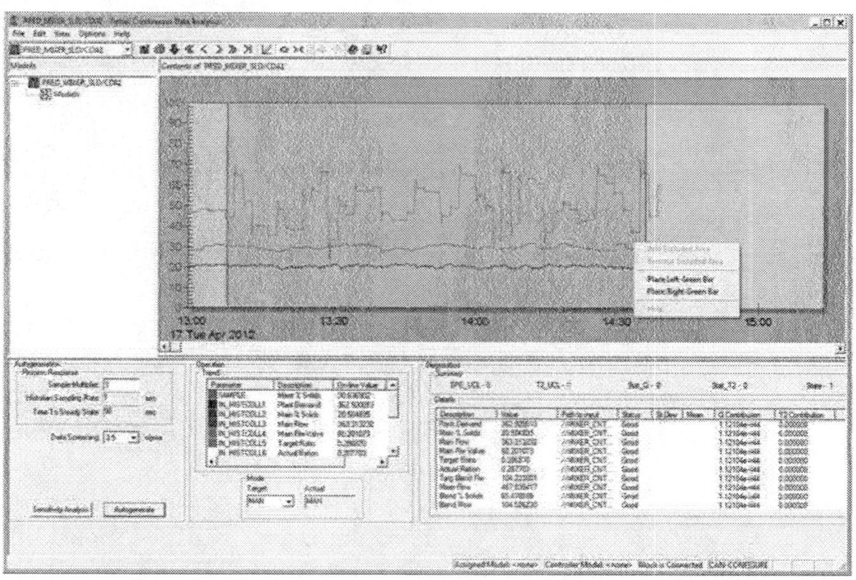

Figure 9-18. Selection of Data for Model Creation

When Sensitivity Analysis (Figure 9-19) is selected, the contribution and delay associated with each input are calculated without generating a model. In this view of the model, the inputs displayed with an X were not selected because they had little or no impact on the quality parameter. The relative contribution of each input and its delay in impacting the quality parameter are noted by the bar height and position within the window. To obtain a more detailed view of this information, the user may double click on the input and see a detailed view of the cross-correlation used to determine the input delay. Also, as shown in Figure 9-19, in the Input detail view the user may view the cross-correlation in the plot, adjust the delay that was automatically determined based on the cross-correlation analysis, and choose whether or not the input is included in the model generation.

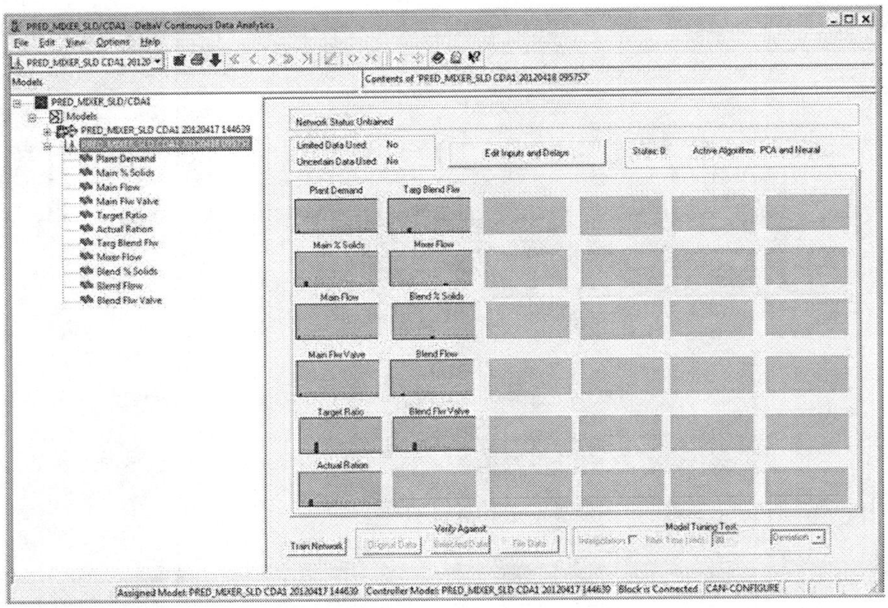

Figure 9-19. Input Sensitivity Analysis and Selection

Once the sensitivity analysis is complete, the user may select to train the models after accepting the sensitivity analysis results. In response, a dialog is presented that allows the user to change the default parameters used in model generation and to select the state parameter, as illustrated in Figure 9-20.

After model generation is complete, the results of the model generation are displayed in a dialog as illustrated in Figure 9-21.

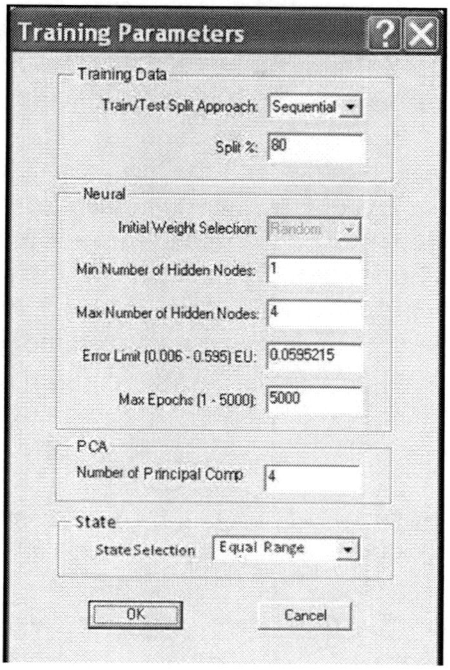

Figure 9-20. Modifying Model Generation Defaults

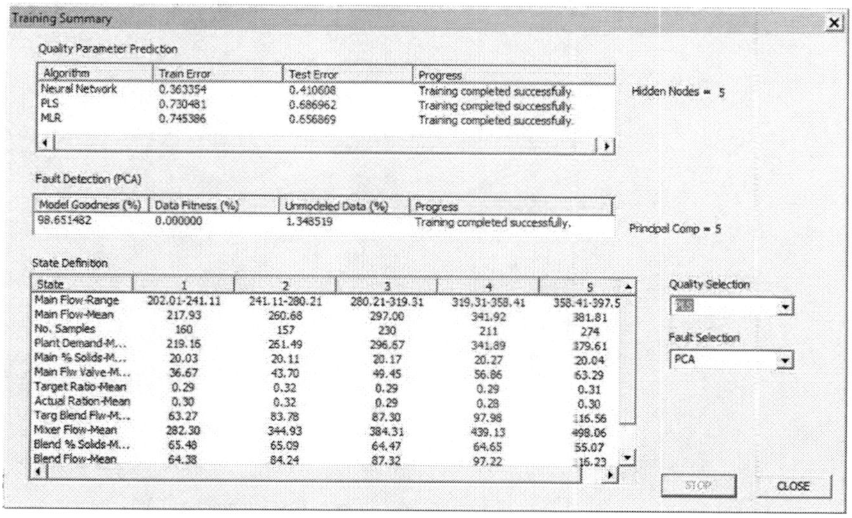

Figure 9-21. Model Generation Dialog

The modeling technique that provides the minimum test error in the quality parameter prediction is highlighted and is automatically selected as the algorithm in the data analytics block used for on-line prediction. The range for state is automatically selected based on the variation of the state parameter in the data used to generate the model. The start of state range and the number of data samples that fell in the state are displayed. For each state, the mean value and standard deviation are automatically calculated for each parameter used in the model development.

As a result of the model generation, a model file in the application database contains the coefficients required for the PLS. The mean value and standard deviation for each parameter is also saved in the model for each state.

If a model has been generated, the user has the option of verifying the model prediction versus the sample input using original data (data used to create the model), data selected by the vertical bars in the overview, or data in a file. When one of these options is selected, the user is presented with a dialog that may be used to compare the accuracy of the predictions using each of the algorithms supported by the data analytics block, as illustrated in Figure 9-22.

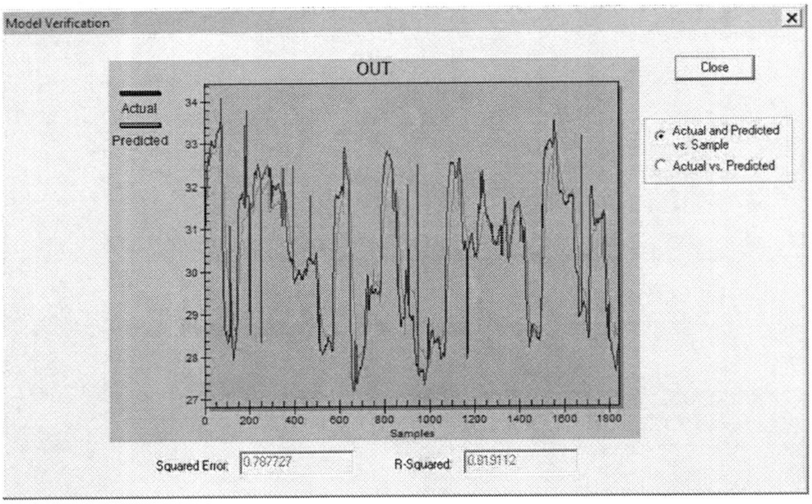

Figure 9-22. Model Verification

Based on the verification the user may elect to deploy the model to enable its use on-line.

9.2 Viewing Data Analytics On-line

After generating the model, the user may download the module that contains the associated data analytics block to view the on-line prediction. When a module contains a data analytics block with a generated model, an icon appears in the left pane of the application display. Multiple models may be generated for a data analytics block and are contained in the model folder for that block. When a model folder is selected, a summary of the models that have been created for the block is displayed as shown in Figure 9-23.

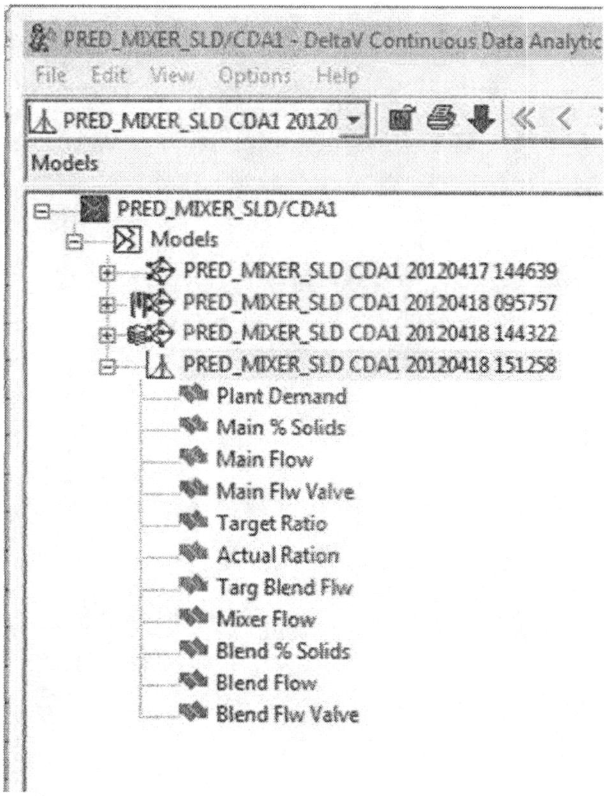

Figure 9-23. Viewing Data Analytics Models

The on-line view of Continuous Data Analytics may be accessed by any work station that has web browser access to the server containing the data analytics application.

The web interface to data analytics is designed to provide information on all data analytics function blocks that have been downloaded in the control system. The top level data analytics view summarizes the prediction and fault detection for all data analytics function blocks. If a fault or prediction alarm is indicated, selecting this block in the overview changes the view to show a trend of the T^2 and Q statistics for that block. In addition, the parameters that most contribute to a fault at a selected time in the trend are shown in the bottom right portion of the display. When a contributing parameter is selected, a trend view of the parameter is displayed along with an indication of its mean value and normal standard deviation from the mean, as determined during model development. By selecting the Quality Parameter tab, a plot of the prediction, future value, and sample input is displayed along with the prediction low and high alarm limits. This web interface is shown in Figure 9-24.

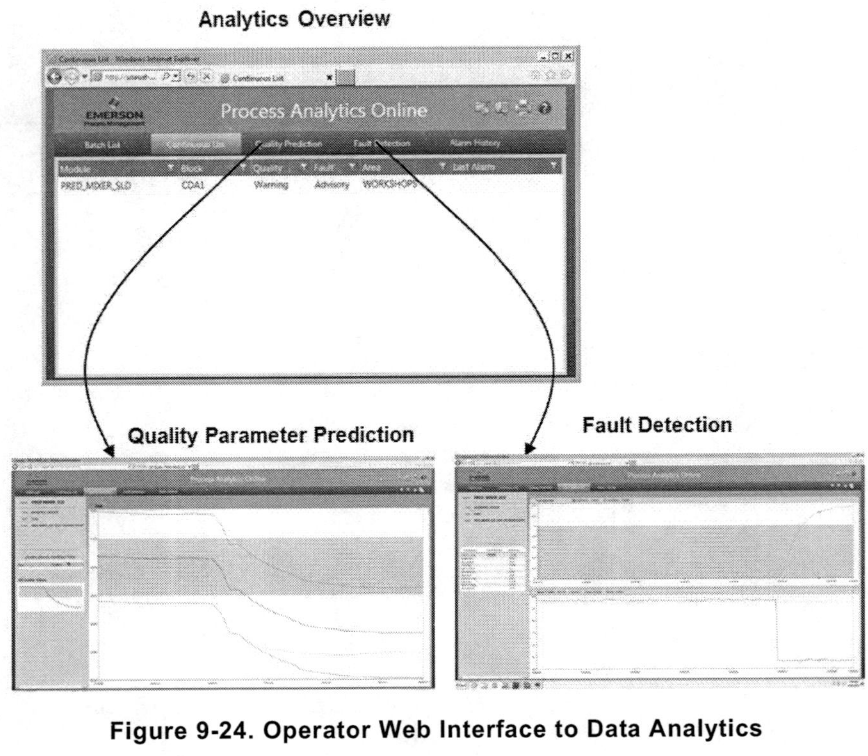

Figure 9-24. Operator Web Interface to Data Analytics

To examine all data analytics alerts that have been detected, the Alarm History tab may be selected to view a list of current and past alerts. By selecting one of the parameters that contributed to the fault, historic information on this parameter at the time of the fault is shown as illustrated in Figure 9-25.

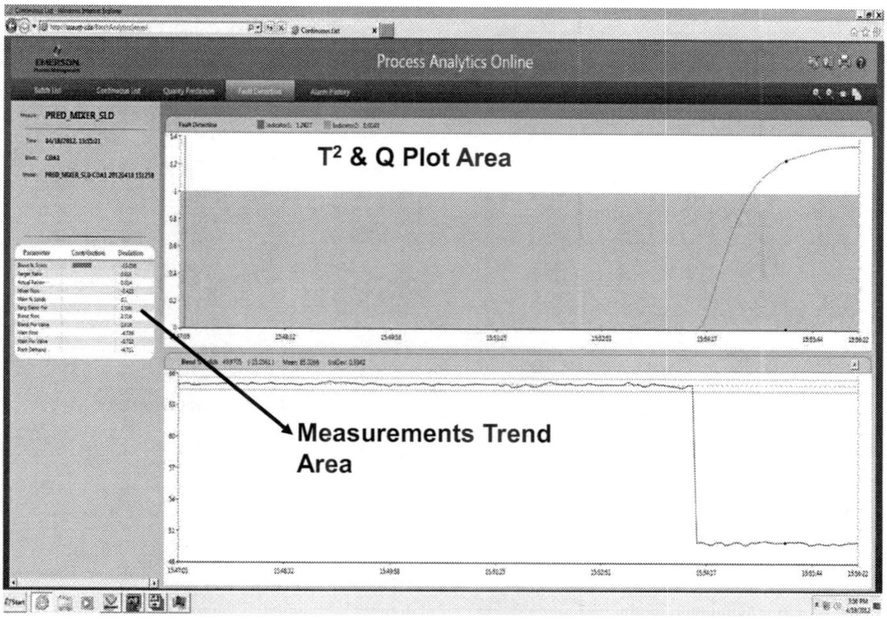

Figure 9-25. Viewing Information Associated with a Process Fault

This ability to easily view all alerts that have occurred in the past and to easily access associated historic information at the time of the alert has many advantages over the traditional alarm summary interface, which does not provide an alert summary and provides no support to display associated historic information at the time of the alert.

9.3 Continuous Data Analytics Workshop

This workshop provides several exercises that may be used to further explore the on-line use of data analytics for fault detection and quality parameter prediction. A simulation of the mixer process (Figure 9-26) is used in this workshop. Access the book's web site http://www.advancedcontrolfoundation.com and select the Solution tab to view the workshop being performed.

Step 1: Open the Data Analytics on-line interface to observe the current status of the process. Select the Alarm History Tab to see the recent faults that have been detected.

Step 2: Access the History Tab view and select the most recent fault and click on the point in the trend that shows variation in T^2 or Q statistics above a value of 1. Examine the parameter contribution and select the parameter that shows that it contributed the most to the fault condition.

Step 3: Based on the trend of the parameter that most contributed to the fault condition, decide which of the process disturbances listed was the source of the fault condition. Select the Quality tab to determine if the fault had an impact on the quality parameter, product % solids.

Step 4: Repeat Steps 2 and 3 for the last four fault conditions. Determine which process disturbance caused each fault condition.

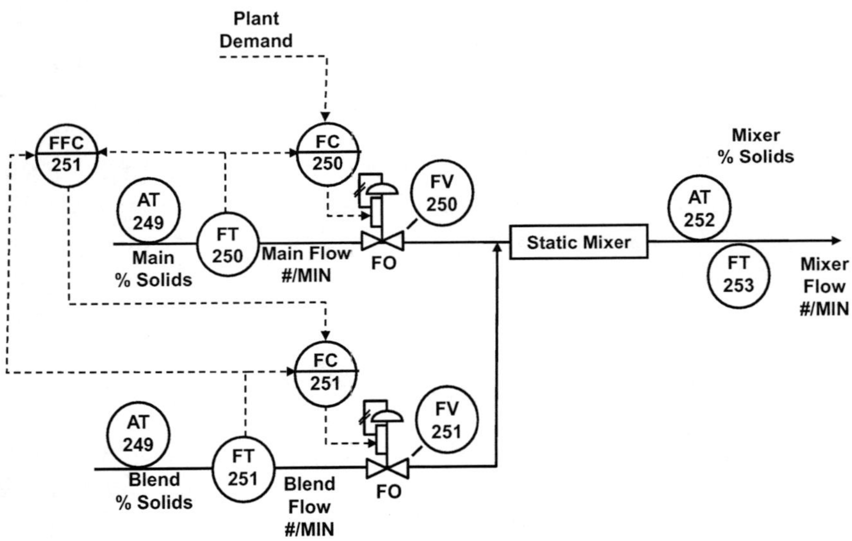

Figure 9-26. Mixer Process

9.4 Technical Basis

Neural Network (NN) technology presented in Chapter 7 has been applied for property inferential prediction for many years. Having many advantages for non-linear modeling, Neural Network suffers some disadvantages. The main flaw of the NN is that the range of data where its model fits is limited to the training data. Predictions are not applicable beyond the training data range. In addition, it is difficult to analyze the NN predictor and develop clear dependencies of the predicted properties from process parameters. Those dependencies are important for on-line property correction. Linear predictors based on PLS may not perform as well for highly non-linear processes, but they perform equally well or better than NN for moderately non-linear processes. In comparison to NN, linear predictors based on PLS are easier to develop and make it easier to better understand and interpret dependencies, thus users often prefer them.

PLS and PCA techniques can be applied jointly for quality prediction, process operation monitoring, and fault detection, the functionality not achievable in Neural Network. In this section the concise theory behind continuous data analytics is examined and addressed in the following topics:

- Data formatting for predictive model development
- Process monitoring and predictive algorithms review
- PCA and PLS model development algorithms details
- Algorithms for on-line fault detection and quality prediction

9.4.1 Data Formatting for Predictive Model Development

A predictive model is typically developed from a data structure such as that shown in Figure 9-27, where J process parameters affect L product properties. To develop a model, process data are collected over K scans (matrix X). It is a good practice to create matrix X by averaging several raw parameter readings.

Product quality or property data comprise matrix Y. Readings of those data are usually available less frequently than the process parameters' scan rate and do not necessarily occur at regular time intervals. To develop a predictive model, the available product properties must be matched with the appropriate process data as shown in Figure 9-28. As it can be seen from Figure 9-28, not all matrix X data are used to develop the predictive model.

Another important step in data formatting is to adjust the process parameters' time delay such that they affect most the property reading, by appropriately shifting columns of the matrix X relative to the property reading. The selected time delay usually differs for most parameters, therefore data matrix X may be shaped as shown in Figure 9-29. As we have seen, one way to estimate the time delay for a process parameter is to define the cross-correlation between the product property and the process parameter for various time delays and use the time delay at which the maximum cross-correlation was achieved for modeling.

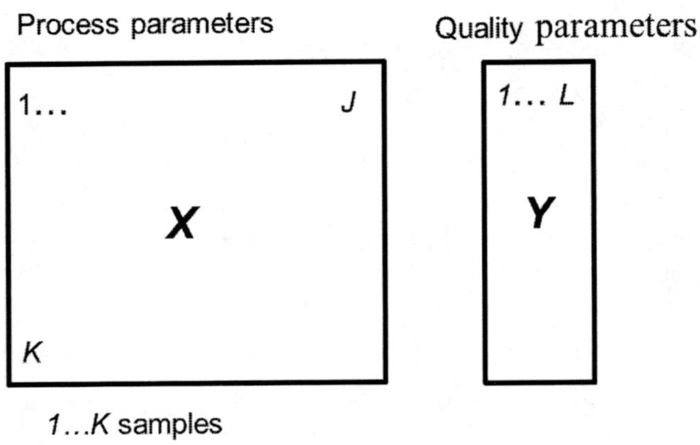

Figure 9-27. Data Matrices Used for Prediction Modeling

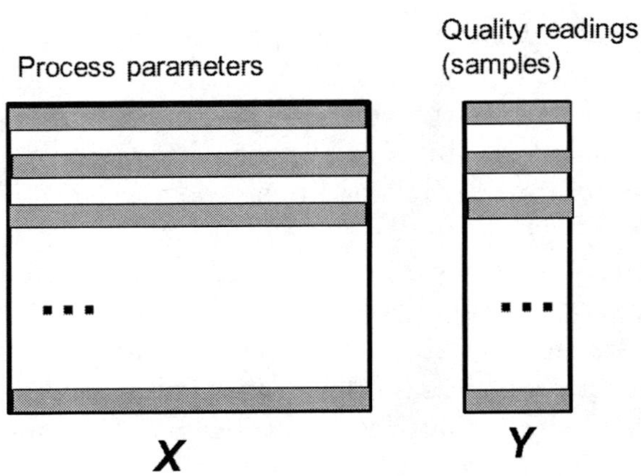

Figure 9-28. Product Property Readings in Matrix Y and Corresponding Process Data in Matrix X

9.4.2 Process Monitoring and Predictive Algorithms Review

Principal Component Analysis (PCA) is used for process monitoring and fault detection. Partial Least Squares (PLS) is a main algorithm applied for quality prediction. For a wider perspective, the following

*Note: **X** data alignment for one property reading shown*

Figure 9-29. Time Delay for Property Y Accounted for in Modeling Data Matrix X

sections in addition to PLS review three other predictive algorithms used in Analytics as specified below:

- Multiple Linear Regression (MLR)
- Principal Components Regression (PCR)
- Partial Least Squares (PLS)
- Partial Least Squares – Discriminant Analysis (PLS-DA)

Principal Component Analysis

Principal Component Analysis extracts from matrix X, principal component matrix T, and residual matrix E from the process data in such a way that matrix T captures maximum correlated variations contained in the original matrix X. The residual, un-correlated variations from matrix X, known as unmodeled variations, are collected in matrix E (Figure 9-30). Matrix T contains modeled variations and is created by applying model matrix P to data matrix X. The decomposition allows the calculation of multivariable and interactive processes' total variability because the principal component vectors are orthogonal (independent), and the variability of each component can simply be added.

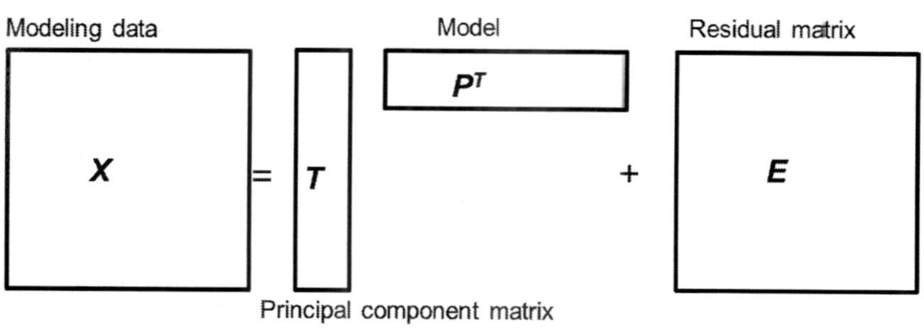

Figure 9-30. PCA Modeling Matrices

Matrix P is in essence a PCA model. It allows the calculation of the principal components' matrix T from data matrix X and performs basic calculations for process monitoring:

$$T = XP \Rightarrow \hat{X} = TP^T \Rightarrow E = X - \hat{X} \tag{9-1}$$

Matrix ***T*** and matrix ***E*** are used to calculate T^2 and Q indexes, as detailed in the following sections.

Multiple Linear Regression

Multiple Linear Regression (MLR) provides product property $\hat{y}$ as the weighted sum of process variables x_i and modeling bias b_0:

$$\hat{y} = b_0 + b_1x_1 + b_2x_2 + \ldots + b_Jx_J + \varepsilon \tag{9-2}$$

Coefficient ε represents bias due to the process disturbances and dynamic operation and can be adjusted by feedback in on-line operation.

The major drawback to MLR is the difficulty in dealing with correlated process parameters.

To develop a predictive MLR model as in Equation 9-2, the least squares method is applied for process data **X** and product property data **y.** For all predictors including MLR, it is assumed that a separate model is built for every property in order to get better predictive results:

$$b = (X^TX)^{-1}X^Ty \tag{9-3}$$

$$b = [b_1, b_2, \ldots, b_J] \tag{9-4}$$

where:

J = number of process variables

Principal Component Regression

Principal Component Regression (PCR) performs significantly better than MLR when data X are collinear. PCR applies PCA first for converting data matrix X into principal components score orthogonal matrix T with a significantly smaller number of columns than matrix X

(T usually has 5–7 columns for any number of X columns). A linear regression is then developed for matrix Y and matrix T (Equation 9-5):

$$\hat{y} = b_0 + b_1 t_1 + b_2 t_2 + \ldots + b_a t_a + \varepsilon \qquad a \ll J \tag{9-5}$$

where:

J	=	number of process parameters in the model
a	=	number of principal components in the PCA model
t_i	=	principal components

Figure 9-31 illustrates PCR modeling.

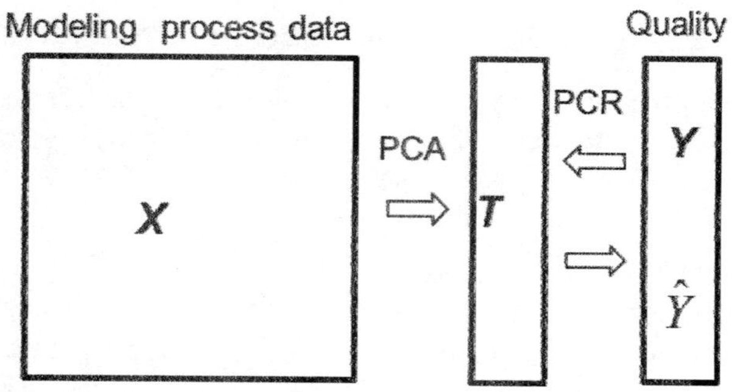

Figure 9-31. PCR Two-Stage Modeling and Prediction

As indicated, PCR is effective for dealing with collinear data. Additionally, the same PCA model used with PCR can be applied (in some cases) for process operation monitoring and fault detection.

Partial Least Squares

PLS is an algorithm which, like PCA, uses principal components. PLS's principal components (matrix T) are orthogonal like PCA and, in addition, are created in such a way as to maximize their covariance both with process data X and product property Y (Figure 9-32).

PLS predictions are better than PCR predictions when a small number (up to four) of principal components are used. When a larger number of principal components are used, the predictions are about the same. PLS is commonly recognized as the preferred analytic prediction algorithm (Geladi, P. and Kowalski, 1986).

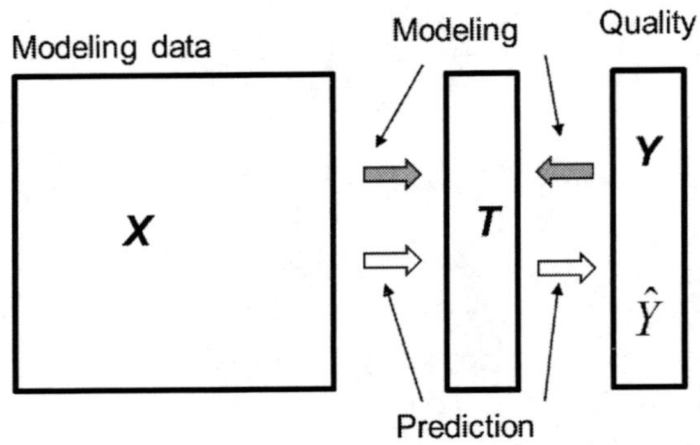

Figure 9-32. PLS Modeling and Prediction

Partial Least Squares – Discriminant Analysis (PLS-DA)

Product property prediction is stochastic in nature. PLS defines a probable property range (confidence interval) and the most probable property value (predicted property value).

Some product properties, on the other hand, are discrete (product grades, for example), or may be discretized for a practical purpose (below low limit, acceptable low, good, acceptable high, and above high limit, for example). PLS-DA is a good modeling technique for identifying such product variations. The technique applies so-called supervised training, which is required to prepare modeling data, as in Figure 9-33. PLS-DA modeling is applied like an original PLS.

Modeling data X, representing property (grade) 1, are on the top of the matrix, with the corresponding property value 1 in column 1 of Y matrix. Property 2 data follows property 1 data with value 1 in column 2 of Y matrix, and so on.

When a model is developed, it predicts certain values in all columns, but in one column the value is closest to that indicating a specific property grade (Figure 9-33 – prediction).

A grade predictor usually delivers values in the range 0.5 to 1.2 when true and -.3 to .4 when false.

9.4.3 Data Preprocessing and Scaling

Among many different data preprocessing methods, data matrices X and Y are scaled into zero mean and unit variance for each column in most process monitoring applications – Figure 9-34.

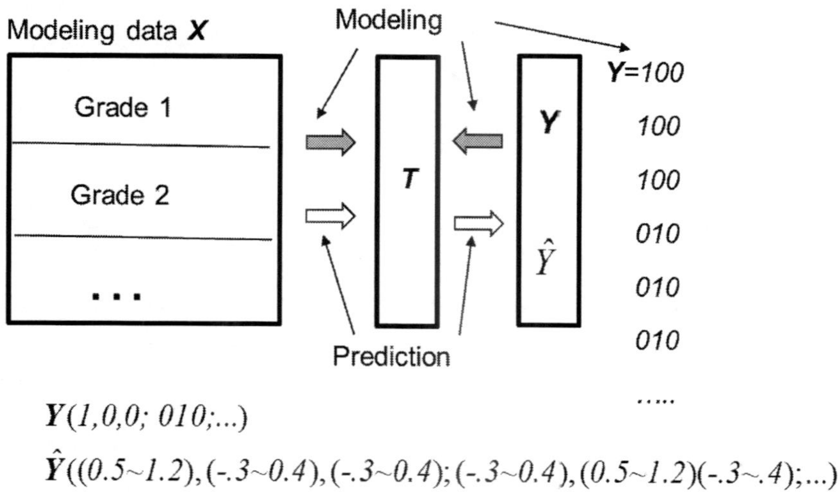

Figure 9-33. PLS–DA Data Structure for Modeling and Prediction

A typical procedure includes the following steps:

1. Reading vector of J process variables x_k^i *(i- reading scan number within statistical scan, k- statistical scan index)* with the data collection scan rate defined in the historical data, accumulating variables vector s times, creating accumulated process variables vector $x_k^{\Sigma} = \sum_{i=1}^{s} x_k^i$ and calculating the vector of averaged process variables within a defined model scan period k: $x_k = \frac{x_k^{\Sigma}}{s} \qquad k = 1,2,\ldots K$

2. Calculating the mean values vector for K statistical scans

$$x_{k\_mean} = \frac{\sum_{1}^{K} x_k}{K}$$

3. Calculating standard deviation estimates for every parameter $j = 1,2,\ldots J$

$$\hat{\sigma}(j) = \sqrt{\frac{\sum_{1}^{K}\left(x_k(j) - x_{k\_mean}(j)\right)^2}{K-1}}$$

4. Scaling/normalizing every parameter $j = 1,2,...J$ for every scan k by its standard deviation

$$x_k(j) = \frac{x_k(j) - x_{k\_mean}(j)}{\hat{\sigma}(j)}$$

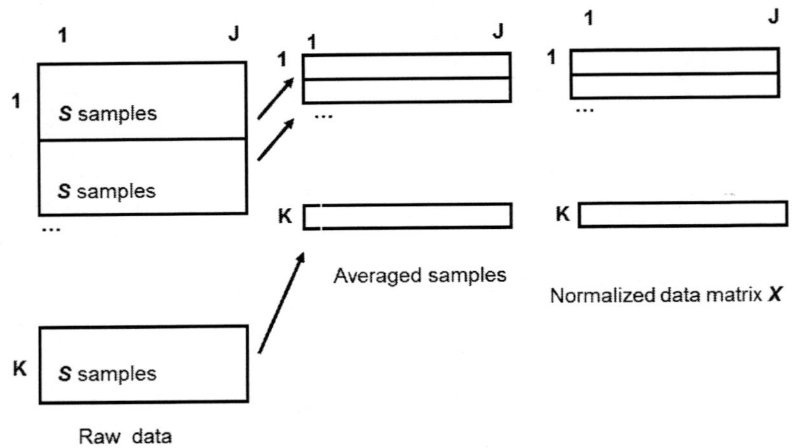

Figure 9-34. Developing Normalized Data for Model Development

9.4.4 PCA Modeling and On-line Fault Detection

The Nonlinear Iterative Partial Least Squares (NIPALS) algorithm is applied for developing a PCA model from data matrix ***X***$(K x J)$ by creating score matrix ***T***$(K x a)$ and load matrix ***P***$(J x a)$ where:

J	=	number of process parameters used in the model
K	=	number of statistical scans in modeling data
a	=	number of principal components

Computationally, the NIPALS algorithm can be carried out by iteration and deflation as follows:

1. Select any column vector x_i from the matrix X as the starting point of t_i .
2. Calculate p_i

$$p_i = X^T t_i (t_i^T t_i)^{-1}$$

3. Normalize p_i

$$p_i^T = \frac{p_i^T}{\|p_i^T\|}$$

4. Calculate t_i

$$t_i = Xp_i$$

5. Compare t_i used in Step 2 and calculated from Step 4. If the difference is smaller than the convergence factor, go to Step 6; otherwise go to Step 2.

6. Deflate X by $X = X - t_i p_i$, set $i = i + 1$, and go to Step 1.

PCA on-line processing consists of the following steps:

1. Scaling process parameter vector

$$x_k = \frac{x_k - x_{k\_mean}}{\sqrt{x_{k\_variance}}} = \frac{x_k - x_{k\_mean}}{\sigma_k}$$

2. Calculating the score vector $t_k^T = x_k^T P$

3. Calculating T^2 statistics for every scan k, $T_k^2 = t_k^T S_k^{-1} t_k$

4. Calculating modeling error vector and SPE/Q statistics $e_k = (I - PP^T)x_k$, $SPE_k = e_k^T e_k$

5. Validating if $T_k^2 < T_{UCL}^2$ and $SPE_k^2 < SPE_{UCL}^2$

6. Developing contribution to T_k^2 for variables $1,\ldots j,\ldots J$

$$c_j(T_k^2) = t_k^T S_k^{-1} [x_{kj} p_j^T]^T$$

where x_{kj} is variable j value at the scan period k

7. Developing contribution for variables $1,\ldots j,\ldots J$ to SPE_k

$$c_j(SPE_k) = e_{kj}$$

where e_{kj} is a component j of the vector e_k.

9.4.5 PLS Modeling and On-line Prediction

The PLS model development algorithm uses normalized data matrices X and Y. The most popular algorithm used for PLS modeling is Non-linear Iterative Partial Least Squares (NIPALS), an extended version of the NIPALS algorithm applied for PCA modeling.

A sound algorithm description can be found in Geladi and Kowalski (1986). This section presents the basic steps of the algorithm followed by a graphic illustration (Figure 9-35).

1. Select initial principal component score ***u*** for the data block **Y**

   Take $u_{start} = y_j - j$ is any **Y** column; larger column variance may speed up convergence

X block processing:

2. Calculate loading vector $w' = u'X/u'u$
3. Normalize loading vector $w' = w'/\|w'\|$
4. Calculate score vector $t = Xw/w'w$

Y block processing:

5. Calculate loading vector $q' = t'Y/t't$
6. Normalize $q' = q'/\|q'\|$
7. Calculate score vector $u = Yq/q'q$

Note: If **Y** has one variable, Steps **5-8** can be omitted by putting ***q* = 1** and no more iterations are necessary.

8. Check convergence:

   IF $\|t_{last} - t_{previous}\| < \varepsilon$ go to Step **9**, else go to Step **2**

   Calculate X loadings and rescale the scores and weights (9-12):

9. $p' = t'X/t't$
10. $p'_{new} = p'_{old}/\|p'_{old}\|$
11. $t_{new} = t_{old}/\|p'_{old}\|$
12. $w'_{new} = w'_{old}/\|p'_{old}\|$

Loadings p', q', w' serve for prediction, scores t, u serve for the diagnostic and classification.

13. Find the regression coefficient for the inner relation $b = u't/t't$

14. Calculate the residuals for the block **X** (for the component ***h***)

$$E_h = E_{h-1} - t_h p'_h \qquad X = E_0$$

15. Calculate the residuals for the block **Y**

$$F_h = F_{h-1} - b_h t_h q'_h \qquad Y = F_0$$

16. Replace matrices **X** and **Y** by their residuals and go to Step 1, to calculate next components (***w***, ***t***, ***p*** for **X** and ***q***, ***u*** for **Y**).

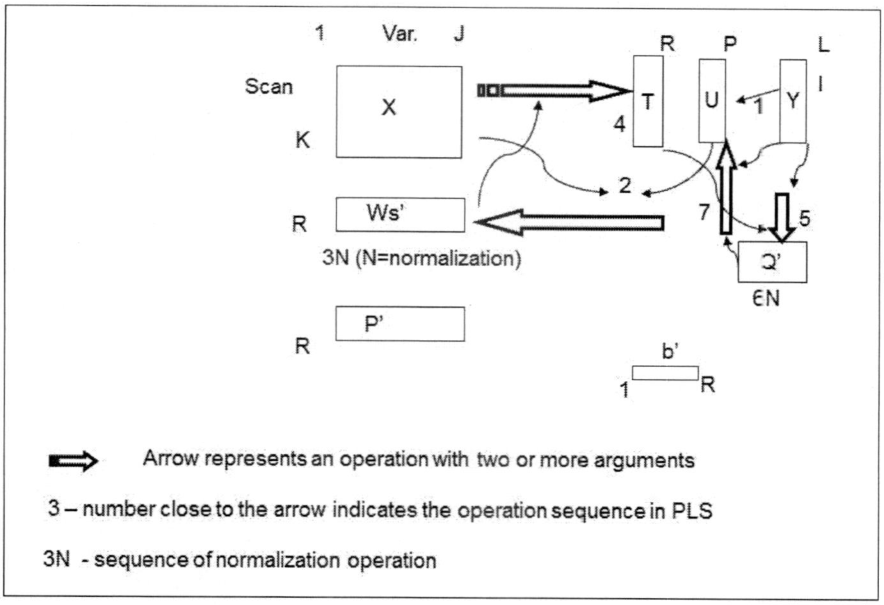

Figure 9-35. Graphic Illustration of PLS NIPALS Algorithm

Developing on-line PLS prediction of product property ***Y*** starts by calculating the actual score matrix ***T***. This is done by successive calculations of p principal components t_i, using model matrices ***W*** and ***P*** - 9-6 and 9-7:

$$t_i = X_{i-1} w_i / (w_i^T w_i) \quad \text{principal component score calculation} \qquad (9\text{-}6)$$

$$X_i = X_{i-1} - t_i P^T \quad \text{matrix } X \text{ deflation} \tag{9-7}$$

Repeating Equations 9-6 and 9-7 for all principal components:

$$T = [t_1 \ldots t_i \ldots t_p] \qquad W = [w_1 \ldots w_i \ldots w_p] \tag{9-8}$$

t_i, w_i = column vectors of the matrices T, W, respectively

Quality is calculated then by using matrices T and Q

$$\hat{Y} = TQ^T \quad \text{property prediction} \tag{9-9}$$

$\hat{Y}$ = predicted property (one or more parameters)

In an on-line application the data matrix X is one row J column vector aligned for the time delay as discussed.

BIBLIOGRAPHY

1. Chiang, L. H., Russell, E. L., and Braatz, R. D. *Fault Detection and Diagnosis in Industrial Systems,* London, Berlin, Heidelberg: Springer-Verlag, 2001.
2. Geladi, P. and Kowalski, B. R. "Partial Least Squares Regression: a Tutorial," *Analitica Chemica Acta* 185, 1986, pp. 1–17.
3. Mason, R. L. and Young, J. C. *Multivariate Statistical Process Control with Industrial Applications,* Philadelphia, PA: SIAM-ASA, 2002.

10 Batch Data Analytics

The trend in process plant design is often towards the use of continuous process operations because this usually minimizes the capital expense required to achieve high rates of production. However, in many industries one or more batch process operations may be used in manufacturing. Batch processing is often required when small lots of material are to be produced or when the product is produced through chemical or biological reactions that take time to complete. For example, the specialty chemical and life science industries depend heavily on batch processing to produce low volume, high value products.

The control logic used in batch processing is often complicated by the fact that multiple products may be produced using the same equipment. The wide range of operating conditions and changes in the process that impact dynamic process behavior are often a challenge from a measurement and control standpoint. In addition, there are the added challenges of assembling, cleaning, and sequencing the necessary data for conducting a thorough data analysis. For these reasons, the application of on-line analytics for fault detection and real-time prediction of the end point value of quality parameters has historically not been adequately addressed, yet it remains a priority for companies that utilize batch processing.

When multiple inputs come together in a batch process, a change in batch operation potentially impacts some or all of the measurements used with that process and that piece of equipment. This interactive nature of the process, combined with the slow integrating response that is characteristic of many batch processes, makes it difficult for an operator to identify abnormal conditions in the batch and to evaluate their impact on final product quality. The fact that quality parameters are often not available as on-line measurements further complicates the operator's job in evaluating the process operation.

Over the last 10 years, significant progress has been made in the development of the technology and understanding needed to successfully apply on-line statistical analytics to batch processes. The integration of such tools into a process control system can provide many benefits. As one moves forward in selecting tools for on-line analytics for batch processes, it is helpful to understand the challenges in applying data analytics to batch processing and the changes in technology that now enable success (Zhang and Edgar, 2007).

10.1 Batch Production Challenges

Regardless of the industry, many plants operate some combination of batch, continuous, and semi-continuous processes, as illustrated in Figure 10-1.

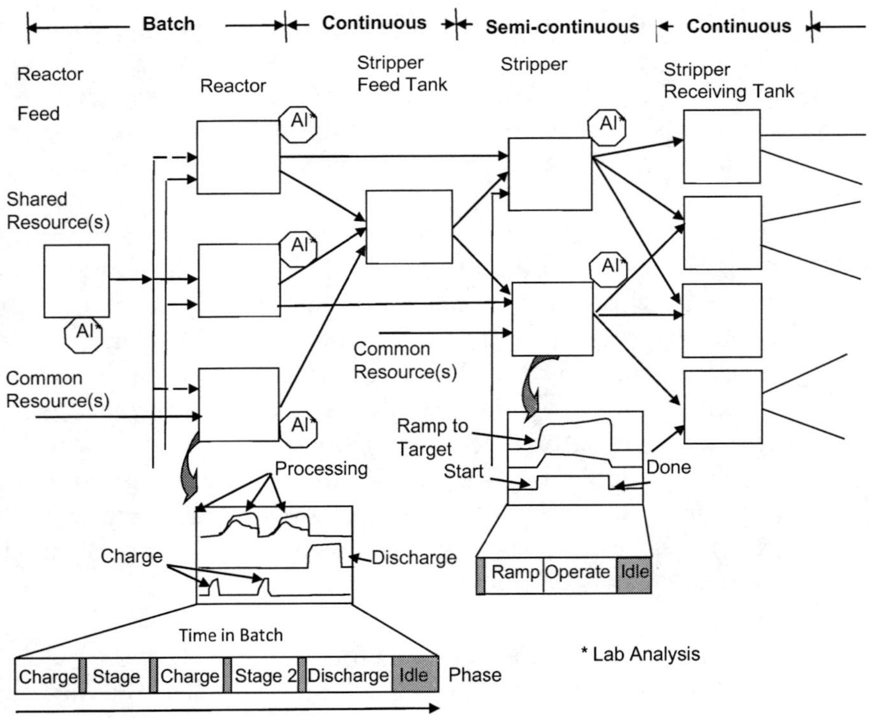

Figure 10-1. Batch, Semi-continuous and Continuous Processes in Plant Operation

The processing associated with a batch unit may be broken down into one or more operations depending on the product that is to be manufactured. In the initial portion of a batch operation, the unit may be charged using discrete addition and/or continuous feed throughout that portion of the operation. Some of the inputs to the batch unit may be shared resources, such as storage tanks, while other inputs may be dedicated to the unit. Feed rates and operating targets are determined by the target product and the operation of the batch.

In some cases, multiple units may be available for processing a batch. Even though the equipment may appear the same physically, the performance of interchangeable production units used for batch processing will often vary because of equipment specification differences, for example, heat exchanger area, capacity, valve characteristics, or measurement sensor location.

Within each batch, the material charged to the unit is retained within the unit and is processed by chemical reaction and/or mechanical means, such as heating, cooling, and agitation. Once all the material associated with a batch has been processed, the product may be discharged to another unit for further processing or to a tank for storage. Lab analysis of some input feed streams may be available and may impact the charge rate and total charge to a unit. Quality parameters associated with the product may be available at the end of the batch or after a batch operation. Such data may be used by the operator or the control system to correct the charge or the operating conditions used in the next phase of the batch operation.

There are significant differences between the operation of a batch and the operation of a continuous process. An interesting fact from a data analysis perspective is that all continuous processes have batch aspects to them and all batch processes have continuous aspects to them. In many cases, analytic tools designed for continuous process analysis do not have the necessary features to effectively address batch process analysis requirements. In particular, some key considerations that must be addressed for successful application of data analytics to a batch process are as follows:

Process Holdups: Operator- and event-initiated processing halts and restarts. Sometimes these halts and restarts are part of the batch process design, such as adding a special ingredient; other times the progression of a batch may be delayed by the need to wait for shared equipment to become available. Regardless of the initiating cause, the time to complete a batch operation may therefore vary. This variation impacts the way data must be processed during analytic model development and the on-line application of analytics.

Access to Lab Data: Due to the nature of the products produced by batch processing, on-line measurement of quality parameters may not be technically feasible or economically justifiable. Thus, it is common practice that at various points in the batch a grab sample is taken and analyzed in the lab. In many plants today, the lab data associated with the processing of grab samples may only reside within the lab system. The lab results may be communicated to the operator by phone or through a lab terminal that is not tied to the control system. To implement on-line analytics, it is necessary that lab results be made available to the on-line analytic toolset.

Variations in Feedstock: The charge to a batch may come from storage tanks that are periodically refilled by upstream processes or by truck or rail shipments from an outside supplier. Changes in the incoming raw material properties may directly impact batch operations and quality parameters. Though the properties associated with each material shipment may be provided by the supplier, these data are often available only to purchasing. If the data are not available for use in on-line analytic tools, the accuracy of the predictions provided by data analytics may be impacted.

Varying Operating Conditions: A batch may be broken into multiple operations. The processing conditions may vary significantly with each batch operation and with the product that is produced by the unit. Thus, the analytic model that is applied to a batch should take into account the product that is being produced and the operation that is active in the processing unit.

10.1.1 Role of Data Analytics in Facing Batch Production Challenges

Advances in data analytics technology over the last 10 years now make it possible to automatically compensate for varying operating conditions and process holdups. It is also now possible to compensate for feedstock property information and sampled lab data of quality parameters, by adjusting process operating targets. However, such technology is not uniformly available in commercial analytic tools. In nearly all cases, the integration of lab data and supplier information on the material properties of truck or railcar shipments is customized to each installation and is therefore not addressed by analytic tools. One of the reasons for this is that all the business systems, lab systems, and distributed control systems available today are nicely designed for their intended use rather than for facilitating data integration and analysis.

With this level of data integration and analysis, within many companies or organizations, the status of batch processing has been judged by a standard set by the so-called "golden" or ideal batch. A golden batch is typically defined as the time-based profile of the measurement values that were recorded for a particular batch that met product quality targets. When this standard is applied, a batch is judged in terms of how closely the golden batch profile can be maintained though the adjustment of process inputs.

One of the biggest weaknesses of this approach is that the conditions indicated by each measurement may impact product quality in a different way. For example, it may be important to tightly control some parameters while other measurements may be allowed to vary significantly without impacting product quality. Through the use of multivariate statistical techniques, the variation within a batch may be used to detect atypical batch events and predict end-of-batch quality characteristics.

One of the important multivariate statistical techniques is Principal Component Analysis, discussed in Chapter 9, as an analytic technique applied for fault detection in a continuous process.

Applied as a tool, **Principal Component Analysis (PCA)** enables the identification and evaluation of material and process variables that may be critical to product quality and performance. Equally important, this tool implemented in DCS workstation may be used to develop an understanding of the interactive relationship of process inputs and measurements and on-, in-, or at-line analysis of final product. When applied on-line, PCA may be used to identify potential failure modes and mechanisms and to quantify their effects on product quality.

At the heart of PCA technology applied to the batch is the concept that a time-based profile for measurement values may be established using a variety of batches that produced good quality product with no significant process upsets. Analysis tools that are designed for batch analysis make it possible to extract, analyze, and use data from multiple batches. For these batches, the measurements' normal variation from this profile is quantified in terms of a PCA model. The model may then be used to develop a better understanding of how multivariate parameters relate to one another and how all those parameters can impact the batch-to-batch costs—energy, waste, time, and so on—of producing a product.

The model structure automatically takes into account that many of the measurements used in the batch operations are related to each other and respond in a similar fashion to a process input change, that is, they are collinear. As a result, a batch operation's multivariable environment can be reduced to just a few simple statistics that the operator can use to assess the batch's progress. These statistics take into account the significance of a parameter's variation from its established profile in detecting a fault.

Through the use of Principal Component Analysis, it is possible to detect abnormal operations resulting from both modeled and unmodeled variations as detailed below.

Modeled Variations: The PCA model captures each process measurement's contributions to the normal modeled process operation. Deviations in the process operation may be quantified through the application of Hotelling's T^2 statistic. Given an assigned level of significance, threshold values can be determined for the detection of an abnormal condition. The T^2 statistic is a multivariable generalization of the Shewhart chart.

Unmodeled Variations: The portion of a process deviation that is not captured by the PCA model reflects changes in unmeasured disturbances or unmodeled process behavior. The Q statistic, also known as the Squared Prediction Error (SPE), is a measure of unmodeled variations in process operation.

Through the use of these two statistics, it is possible to determine fault conditions sooner in the batch and thereby allow corrections to be made to counter the impact of the fault.

There are other two important analytic techniques applied for batch processes: Projection to Latent Structures and Discriminant Analysis. **Projection to Latent Structures (PLS),** sometimes referred to as Partial Least Squares, may be used to analyze the impact of processing conditions on final product quality parameters that are often measured using on-, in-, and at-line analysis of the final product. When PLS is applied in an on-line system, it is possible to provide operators with a continuous prediction of end-of-batch quality parameters. PLS is also effectively used with PCA so that only important process variation is identified in PCA to minimize false alarms of variation that may be present but not meaningful.

Where the objective is to classify the operation results into categories of importance (for example, fault category, good vs. bad batch, etc.), Discriminant Analysis is used in conjunction with PCA and PLS.

Discriminant Analysis (DA) is related to PCA and other statistical analysis methods such as analysis of variance. It is often used in conjunction with PLS as a powerful classification technique. Discriminant analysis (as opposed to a continuous output measurement) is used to relate process variation to categorical or classification criteria. Examples of a categorical classification could be something as simple as the batch being "in specification" or "out of specification," or there could be many different categories representing abnormal situations in the batch or other categorical classifications of quality or economic parameters.

Through the application of data analytics, it is possible for the operator to monitor a batch operation simply by looking at a plot of the PCA statistics (for fault detection) and the PLS predicted end-point value for quality parameters, as illustrated in Figure 10-2.

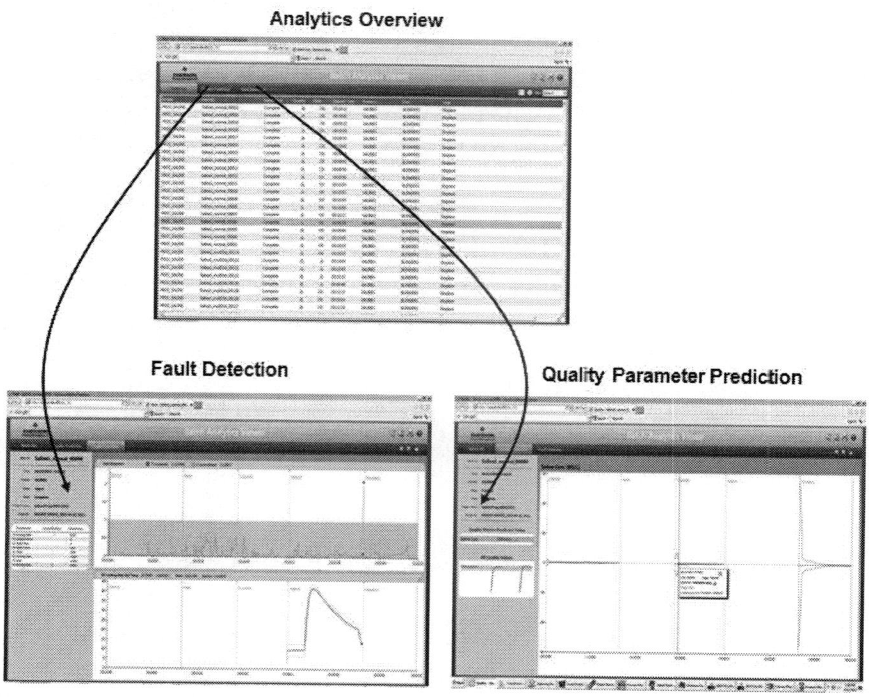

Figure 10-2. Example of a Simple Operator Interface

As illustrated in Figure 10-2, the primary measurements that contribute to a process deviation may be displayed as contribution plots in the left segment of the screen shown. Contribution plots show how much each process variable contributes to the deviation, which can be helpful in quickly determining the source of a measured or unmeasured disturbance. Thus, when the current batch is evolving, if something deviates from acceptable variation relationships derived from the analysis of past batches, the operations staff may drill down into the process variables to understand why, by observing the most contributing parameter trends.

10.1.2 Batch Analytics Overview

The successful application of multivariate statistical techniques such as PCA and PLS depends in part on the toolset selected. Various techniques for PCA and PLS model development are used in commercial products. In some cases, a product may be designed to support the analysis of continuous processes. In such cases, data analysis and model development often assume that a process is maintained at just one operating condition.

To successfully address the requirements of batch processes, it is important that multivariate tools used for model development are designed to address varying process conditions over a wide range of operating conditions. Multiway PCA and PLS algorithms are commonly used as multivariate tools to address the requirements of batch applications.

Multiway PCA (MPCA) is an extension of PCA and provides an effective means of developing an analytic model from aligned batch data. Raw batch data may be associated with different batch duration time and has to be aligned prior to applying MPCA (Nomikos and MacGregor, 1994). Using MPCA, engineers can extract precise information in order to monitor conditions and also relate conditions to faults in batch processes.

Tools that support MPCA algorithms are designed to allow a normal trajectory to be automatically established for each process input and measurement over the period of batch duration. In many cases, the toolset may support only off-line, post-production analysis. There is potentially much greater value in applying analytics on-line so that changes can be made in the batch to correct for detected faults or deviations in the predicted value of key quality parameters. The tools that are selected for implementing on-line analytics should provide effective solutions to address the varying operating conditions and process holdups.

The PCA/PLS techniques applied to batch processing should account for the physical differences between interchangeable batch units, the product that is being produced, and the operations associated with a batch. To provide this capacity, the PCA/PLS model should be stored and organized by batch unit. For each unit, models may be organized by product and stages as illustrated in Figure 10-3.

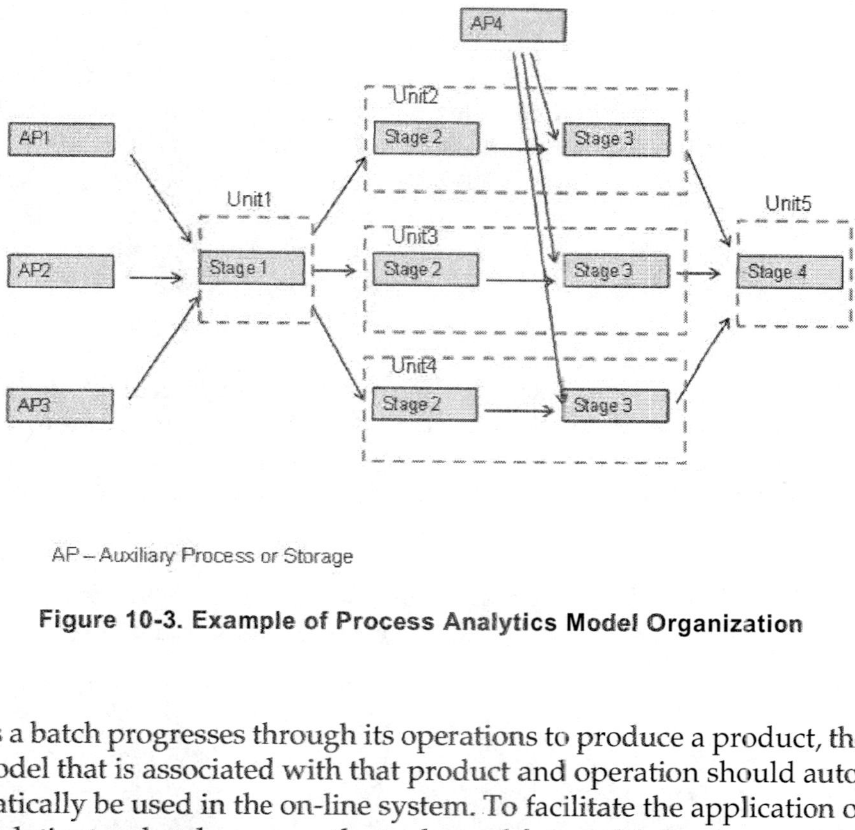

Figure 10-3. Example of Process Analytics Model Organization

As a batch progresses through its operations to produce a product, the model that is associated with that product and operation should automatically be used in the on-line system. To facilitate the application of analytics to a batch process, the tools used for model development should allow historic data for batch operations to be accessed and screened by product and operation in a model development procedure.

In some cases, the collection of process data in a format that can be used by a given analytic tool is one of the greatest challenges in PCA and PLS model development. However, when those tools are integrated into the control system, it becomes possible for a manufacturer to automatically provide data for each batch as required for model development.

Prior to the model development, the data must be appropriately arranged ("unfolded") as required by statistical analysis. Three techniques have traditionally been used to unfold batch data for use in model development: time-wise unfolding, variable-wise unfolding, and batch-wise unfolding. A new approach for PCA data unfolding known as *hybrid unfolding* offers significant technical advantages (Lee, Yoo, and Lee, 2004).

The time required to complete one or more operations associated with a batch may vary because of process holdups or variations in processing conditions. However, the batch data used in model development must be of the same time duration. There are many ways that this may be achieved. For example, to achieve uniform batch length, the data past a certain time in the batch could be simply chopped off, compressed, or expanded in some fashion to achieve the same number of time increments. However, a more effective approach is to apply a relatively new technique known as Dynamic Time Warping (DTW) (Kassidas, MacGregor, and Taylor, 1998).

Dynamic time warping (DTW) algorithms effectively measure the similarities between two sequences that vary in duration and/or speed. The algorithms then adjust for the differences so that data from multiple batches may be combined into one data set for analysis (Figure 10-4).

Dynamic Time Warping allows variations to be addressed by automatically synchronizing batch data using the key characteristics of a reference trajectory.

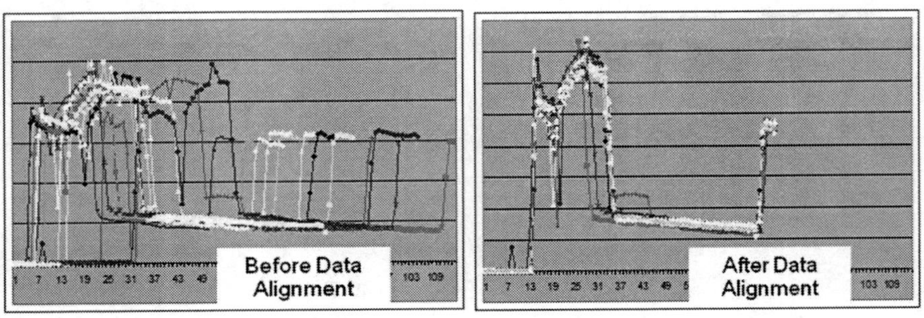

Figure 10-4. Raw Batch Trends and Aligned Trends Used for Analytics Modeling

Once PCA and PLS models have been developed using data from normal batches, their performance in detecting faults and predicting variations in end-of-batch quality parameters may be tested by replaying data collected from abnormal batches. Most commercial modeling programs facilitate testing a model by this means. More important is their performance in detecting faults and predicting variations in end-of-batch quality parameters in real time as in-production batches are evolving.

10.1.3 Application of Batch Analytics to Specialty Chemicals Production

On-line batch analytics has been applied at Lubrizol's Rouen, France, specialty chemicals plant for fault detection and prediction of quality parameters (Wojewodka and Blevins, 2008). This application in the specialty chemicals industry is in many ways representative of the analytics applications used in the chemical industry as a whole. The success of an analytics project, as with any engineering endeavor, depends greatly on the steps taken in applying the technology. To address this application, a multi-disciplinary team was formed that included the toolset provider as well as experts from Lubrizol's plant operations, statistics, MIS/IT, and engineering staffs. The approach that was successfully used at the Rouen plant will be further refined and followed in the future for other applications.

Recommended preparatory steps for application project to apply data analytics in a plant are illustrated in Figure 10-5.

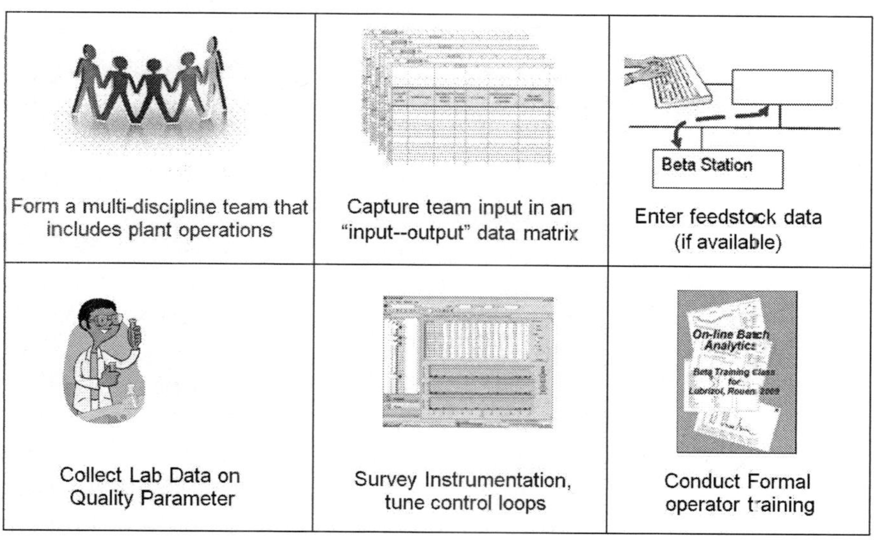

Figure 10-5. Setting the Foundation for the Application of Batch Analytics

Collection of Process Information: The measurements and the manner in which the batch operations are defined for the production of various products are unique to each installation. To apply data analytics to a batch process, it is important that the team has a good understanding of the process, the products produced, and the organization of the batch control. Early in the project, existing documentation on the process and the batch control was distributed to the project team for study. A project meeting was also conducted at the plant to allow operations to provide input and for the team to become more familiar with the process.

Based on this information, listings of the process measurements, lab analysis data and truck data for raw material shipments were created. These data formed the basis of what Lubrizol refers to as the Inputs – Process – Outputs data matrix. This matrix defined the data that was considered in the PCA and PLS model development.

Instrumentation and Control Survey: A basic assumption in the application of analytics to a batch process is that the process operation is repeatable. Any problems associated with process measurement or control tuning and setup should be addressed before data are collected for model development. Thus, in parallel with the initial project meeting, an instrumentation and control survey was conducted for the two batch process areas addressed by the project. Any instrumentation problems found in the survey were corrected. Also, changes in loop tuning were made to provide best process performance. For example, the temperature and pressure loops associated with three reactors were retuned to provide improved, repeatable performance (Blevins and Beall, 2007).

Integration of Lab Data: Key quality parameters associated with the batch operation at the Rouen plant are obtained by lab analysis of a grab sample. The lab analysis data are then entered into their ERP system (SAP in the case of Lubrizol) and are used for quality reporting, generation of certificates of analysis, and process improvement studies. The properties analyses for truck shipments are also entered into SAP. To allow these data to be used in on-line analytics, an interface was created between the SAP system and the process control system.

At the Rouen plant the material properties associated with truck shipments are used to calculate the properties of material drawn from storage—inputs used in the PCA and PLS analysis. As with most processes, for this project it was important to characterize both the quality characteristics of incoming raw materials and the quality characteristics of the end product.

Historian Collection: When the process control system was originally installed at the Rouen plant, all process measurements and critical operation parameters associated with the batch control were set up for historian collection at 1-minute intervals using data compression. However, for the purpose of analytic model development, it is desirable to save data in an uncompressed format. Additional historian collection was therefore defined for the measurements, lab data, and batch operation data. These data were collected using 10-second samples to more accurately capture measurement change and saved in an uncompressed format. Doing this allowed data analysis to be done at a finer time resolution and also to define a more appropriate resolution for future implementation. Ongoing analysis of the data will then define whether the time resolution needs to remain fine or if the sampling rate may be reduced.

Model Development: The tools available to the Rouen plant for model development are designed to allow the user to easily select and organize from the historian a subset of the data associated with parameters that will be used in model development for a specified operation(s) and product. The tools provide the ability to organize and sequence all of the data into a predetermined data file structure that permits data analysis. Once a model has been developed, it may be tested by using playback of data not included in model development. Because the typical batch time is measured in days, this playback may be done faster than real-time. This capability allows the model to be quickly evaluated for a number of batches.

Training: The statistics provided by on-line analytics are primarily used by the plant operator. Thus, operator training was a vital part of commissioning this capability. In addition, separate training classes on the use of the analytic tool were conducted for plant engineering and maintenance.

Evaluation: During the first three months of the application of on-line analytics, operator feedback and data collected on improvements in process operation were used to evaluate the savings attributable to the application of analytics. These were also used to obtain valuable input into improving the user interfaces, the displays and the terminology used in the displays. This allowed the project team to further improve the analysis modules to maximize use and understanding by the operators and engineers.

As illustrated by the project steps outlined above, most of the time required to deploy on-line analytics is associated with collecting process information, performing an instrumentation and control survey, integrating lab data, setting up the historian for data collection, and training.

The on-line prediction of the quality parameter has been reported by the Lubrizol Rouen installation to closely match the end-of-batch lab analysis of that parameter (Wojewodka and Blevins, 2009). A comparison of the lab analysis and the on-line prediction for 50 batches is shown in Figure 10-6.

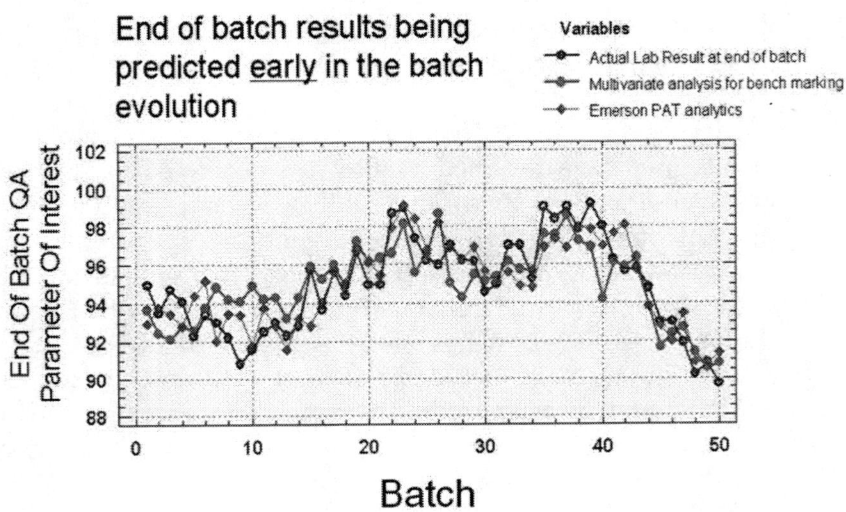

Figure 10-6. End-of-Batch Prediction vs. Lab Analysis

During the initial three months of operation using on-line data analytics, multiple process faults were detected. For example:

- The reactor pump intensity was two times higher than normal during two stages in a batch after one unit's day off. The pump was closely checked, and maintenance was ready to respond in case of a breakdown.
- There was a regularly repeating problem in the reactor heating control loop. This led to revisiting the loop tuning on key process control parameters.

In some other cases, faults were detected before they impacted production. For example, the plant reported that a problem was identified with an upstream boiler that negatively impacted operations. The operations personnel explained, "An equipment failure was discovered in advance, and we avoided losing five hours per batch."

10.2 Data Analytics Application Example – Modeling and On-line Operation

Many of the data analytics products that are marketed for batch processing use multi-way PCA and PLS for fault detection and quality parameter prediction. However, there are significant differences in the way these products address the various challenges associated with batch processing.

For example, variation in the total time required to complete a batch can have a significant impact on the results achieved in off-line model development and in on-line operation. In off-line model development, some products simply remove data after a period of time to achieve equal length batches. Other products compress the data uniformly throughout the batch to achieve equal length batches. The variation in batch time due to contention for resources or holds in a batch is often ignored in the on-line application of PCA and PLS models. Similarly, variation in the operating conditions and the measurements that reflect these conditions may not be considered in off-line model development and in on-line model use.

The recent introduction of dynamic time warping and the development of models by batch *stage* (defined in 10.5.3) can often improve the results achieved in the application of data analytics to batch processes. The product described in this section uses dynamic time warping and data analytic models that are developed and evaluated by batch stage.

The steps involved in developing and deploying the multi-way PCA and PLS models will be shown using a dynamic process simulation of a "Saline" batch process that uses two mixers and two blenders. The equipment associated with this batch process may be used to produce two different products, Saline 1 and Saline 2. The equipment and the on-line measurements and actuators associated with each mixer are illustrated in Figure 10-7.

Based on the recipe associated with the Saline product that is made in this equipment, an initial charge of salt and water is processed in the mixer and then displaced to either available blender for further processing. The equipment and measurements associated with a blender are shown in Figure 10-8.

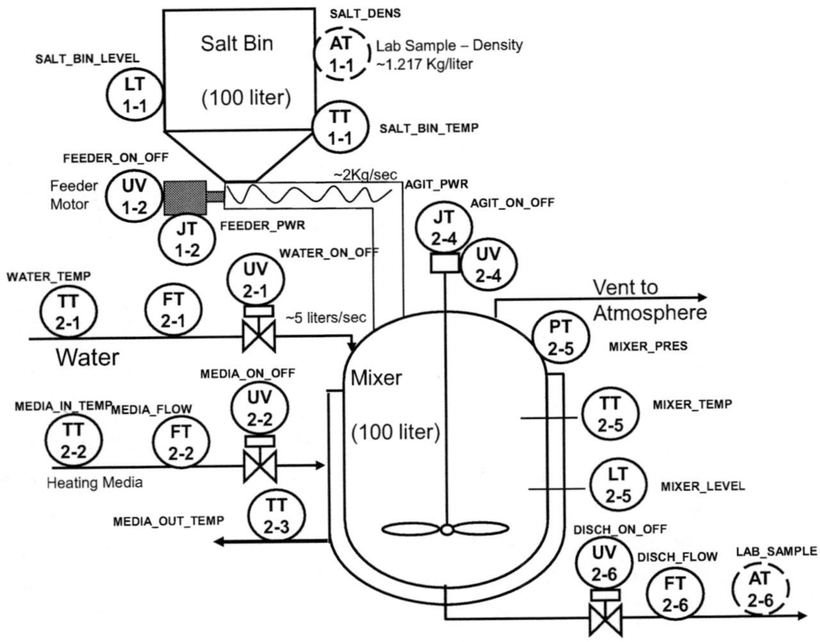

Figure 10-7. Saline Process – Mixer

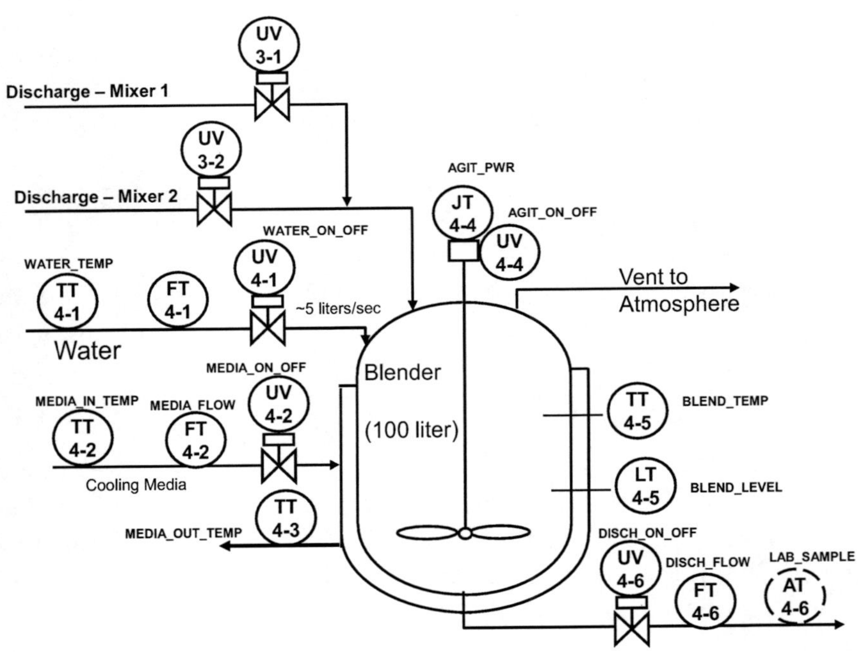

Figure 10-8. Saline Process – Blender

The alternate paths through the process and the stages that the batch may be broken into for the production of a Saline product are illustrated in Figure 10-9.

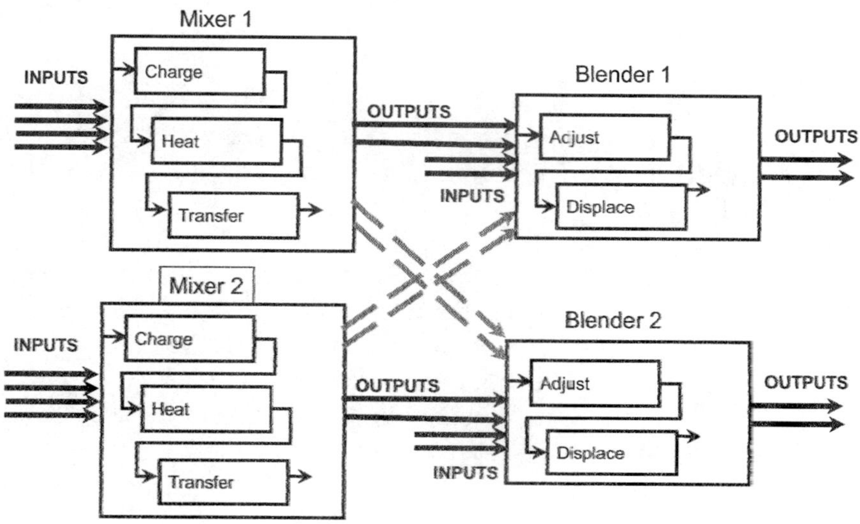

Figure 10-9. Equipment and Stages for Production of Saline Product

As shown in Figure 10-9, the batch is broken into five stages. The first three stages of the batch are executed in the mixer, and the last two stages are executed in the blender. The batch logic automatically selects the first mixer that becomes available and then begins a batch based on the product recipe. The recipe for the Saline 1 and Saline 2 products determines the quantity of water and salt added in the mixer as well as the quantity of water added in the blender and the final product concentration target in the blender.

To allow fault detection to be tested, the Saline batch simulation is designed to allow process faults to be introduced. During the collection of data to train the data analytics models, normal process variations are automatically introduced into the process simulation. Once analytic models have been developed, the process may be operated under normal operating conditions, or a specific disturbance may be introduced to demonstrate how this fault impacts the quality parameter.

10.2.1 Defining Model Input

The mixer and bender process parameters that may be used to develop a PCA model for fault detection and a PLS model for quality parameter prediction are summarized in a data matrix. An example data matrix that identifies the measurements used in each stage of the Saline batch process is shown in Table 10-1.

The Saline process quality parameter is the % solids measurement after the blender. The quantities of salt and water are specified in the product recipe. The lab results are entered into the control system by the lab.

The starting point in data analytics configuration is to identify the product and the associated units that will be used in the production of a Saline product. For example, a project may be created that addresses the production of the Saline 1 product using Mixer 1 and Blender 1. Thus, only these two pieces of equipment would be selected from the plant equipment, as illustrated in Figure 10-10.

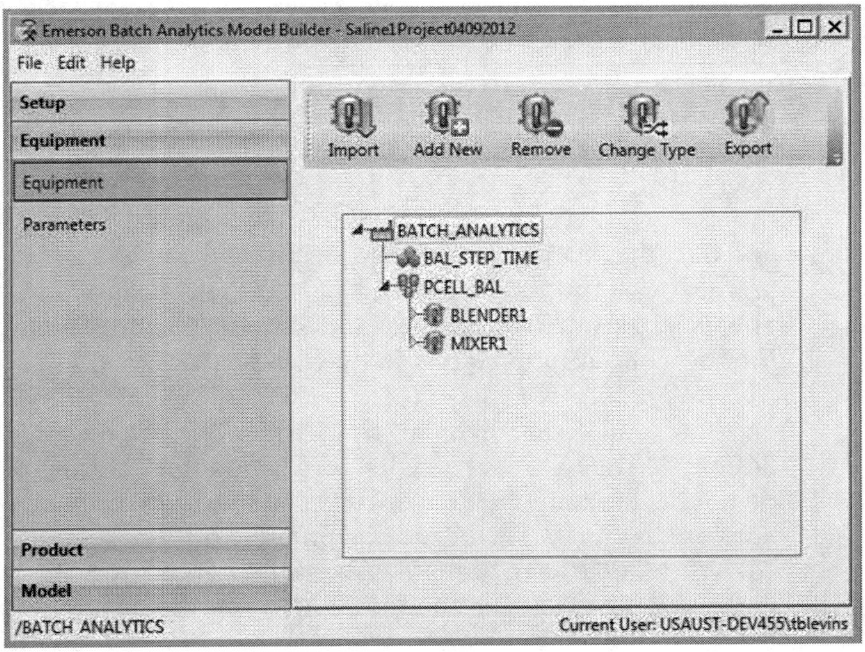

Figure 10-10. Equipment Selection

Table 10-1. Data Matrix for Saline Process

Process Unit	Parameter	Stage				
		Charge	Heat	Transfer	Adjust	Displace
MIXER1 and MIXER2	BATCH_ID.CV	X	X	X		
	PRODUCT.CV	X	X	X		
	UNIT.CV	X	X	X		
	OPERATION.CV	X	X	X		
	STATE.CV	X	X	X		
	AGIT_ON_OFF.CV		X			
	AGIT_PWR.CV		X			
	DISC_ON_OFF.CV			X		
	DISCH_FLOW.CV			X		
	FEEDER_ON_OFF.CV	X				
	FEEDER_PWR.CV	X				
	LAB_PROD_CONC.CV		X	X		
	LAB_SALT_DENS.CV	X	X	X		
	MEDIA_FLOW.CV		X			
	MEDIA_IN_TEMP.CV		X			
	MEDIA_ON_OFF.CV		X			
	MEDIA_OUT_TEMP.CV		X			
	MIXER_LEVEL.CV	X	X	X		
	MIXER_PRES.CV		X	X		
	MIXER_TEMP.CV	X	X	X		
	SALT_BIN_LEVEL.CV	X				
	SALT_BIN_TEMP.CV	X				
	WATER_FLOW.CV	X				
	WATER_ON_OFF.CV	X				
	WATER_TEMP.CV	X				
BLEND1 and BLEND2	BATCH_ID.CV				X	X
	PRODUCT.CV				X	X
	UNIT.CV				X	X
	OPERATION.CV				X	X
	STATE.CV				X	X
	AGIT_ON_OFF.CV				X	
	AGIT_PWR.CV				X	
	BLENDER_LEVEL.CV				X	X
	BLENDER_TEMP.CV				X	X
	CHILLER_FLOW.CV				X	
	CHILLER_ON_OFF.CV				X	
	DISC_ON_OFF.CV					X
	DISCH_FLOW.CV					X
	LAB_PROD_CONC.CV				X	X
	MEDIA_IN_TEMP.CV				X	
	MEDIA_OUT_TEMP.CV				X	
	WATER_FLOW.CV				X	
	WATER_ON_OFF.CV				X	
	WATER_TEMP.CV				X	

After identifying the equipment that will be used to manufacture the product of interest, the inputs associated with those pieces of equipment are automatically extracted from the batch control application that is part of the control system. Based on the data matrix defined for the project, it is possible to quickly select the inputs of interest, as illustrated in Figure 10-11.

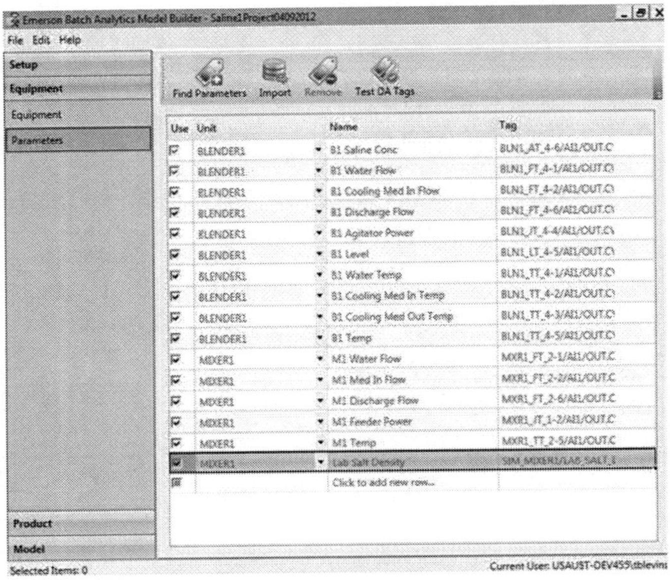

Figure 10-11. Selection of Inputs in Model Building

As part of the input selection, other information needed in model building can also be defined such as input use, that is, measurement or initial condition, and a user-friendly name for the parameter.

Data from the batches that created the product of interest using the selected pieces of equipment are automatically extracted from batch history data and can be selected for model building. Information about the batch parameters may be displayed and used to determine which batches should be selected, as illustrated in Figure 10-12.

Based on the batches selected, the associated lab data for each batch can be configured along with other information such as engineering unit range.

The batch steps, such as starting and stopping a unit operation, are automatically extracted from the batch history and used to define the

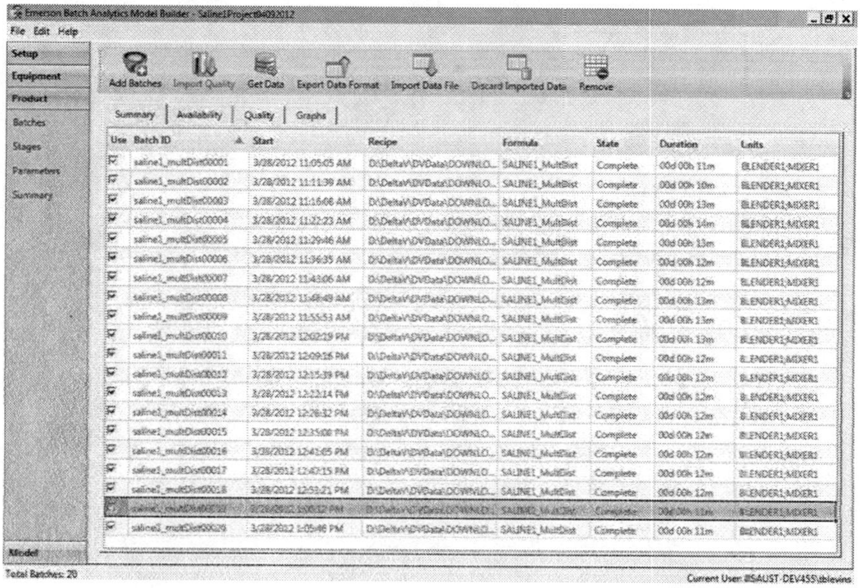

Figure 10-12. Selection of Batches for Model Development

stages of a batch and the steps that defined the beginning and end of each stage as shown in Figure 10-13.

Based on the process input selection and the stages defined for a batch, the inputs that will be monitored and used in each stage may be identified by the user as illustrated in Figure 10-14.

The PCA and PLS models are created automatically for the stages and the associated equipment defined for the project using the batches the user previously selected.

Before initiating model generation, the user may select the batches that will be used to train the model and the batches that will be used to test the model that has been created. By default, a fraction of the batches is automatically selected on a random basis. Information about model goodness for fault detection and quality parameter prediction is as illustrated in Figure 10-15.

Once a model has been generated, it may be put on-line using the Deploy Project selection illustrated in Figure 10-15.

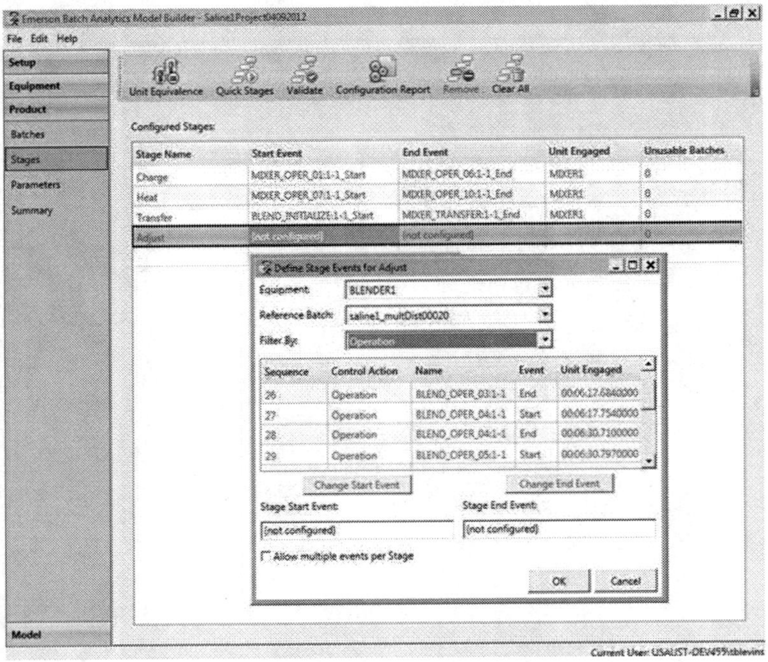

Figure 10-13. Defining Batch Stage

10.2.2 Viewing Data Analytics On-line

After a model has been deployed, the user may access the on-line view of batch data analytics by launching a web browser on any computer that has access to the data analytics server. At any given time, models for multiple products may be deployed. In the on-line view, the current and past batches that produce the products for which models have been deployed are listed.

This top level data analytics view summarizes the prediction and fault detection for all data analytic function blocks as illustrated in Figure 10-16.

If a fault or prediction alarm is indicated for a batch, selecting the batch in the Overview and then selecting Quality Parameter shows a trend of the T^2 and Q statistics for the batch. In addition, the parameters that most contribute to a fault at a selected time in the trend are shown in the bottom left portion of the display. By selecting a contributing parameter, a trend view of the parameter may be displayed along with an indication of its normal trajectory and normal standard deviation from the trajectory as determined during model development, as illustrated in Figure 10-17.

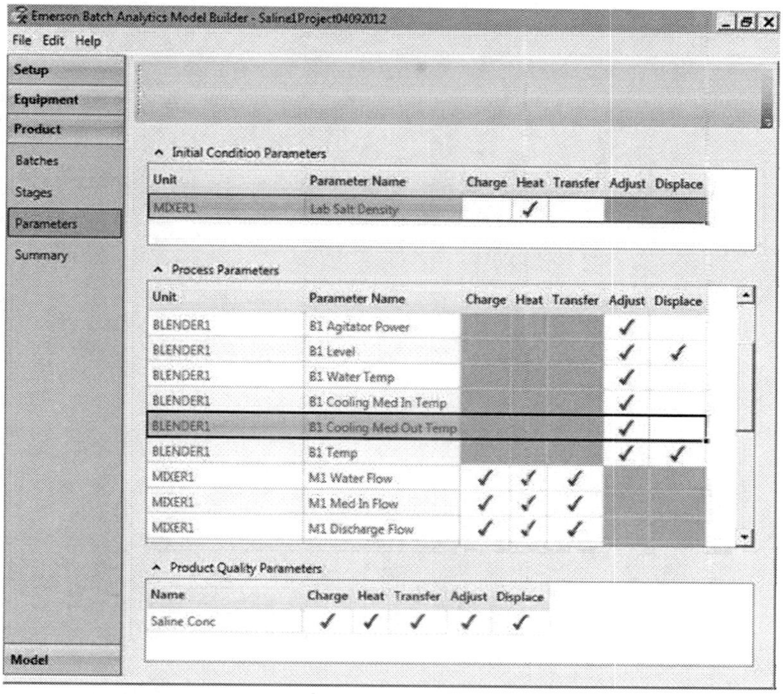

Figure 10-14. Defining Inputs Used in Each Stage

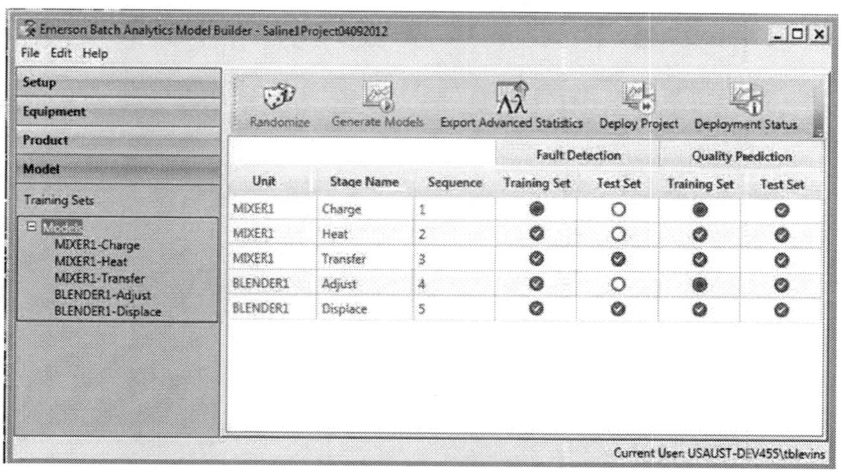

Figure 10-15. Generated Model Goodness

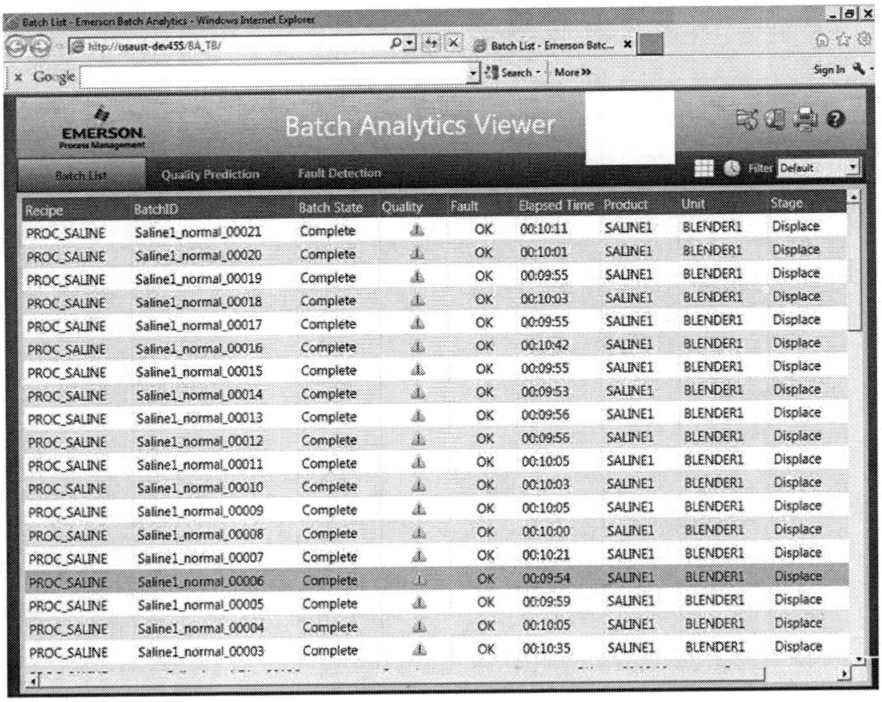

Figure 10-16. Viewing Batch Data Analytics Models

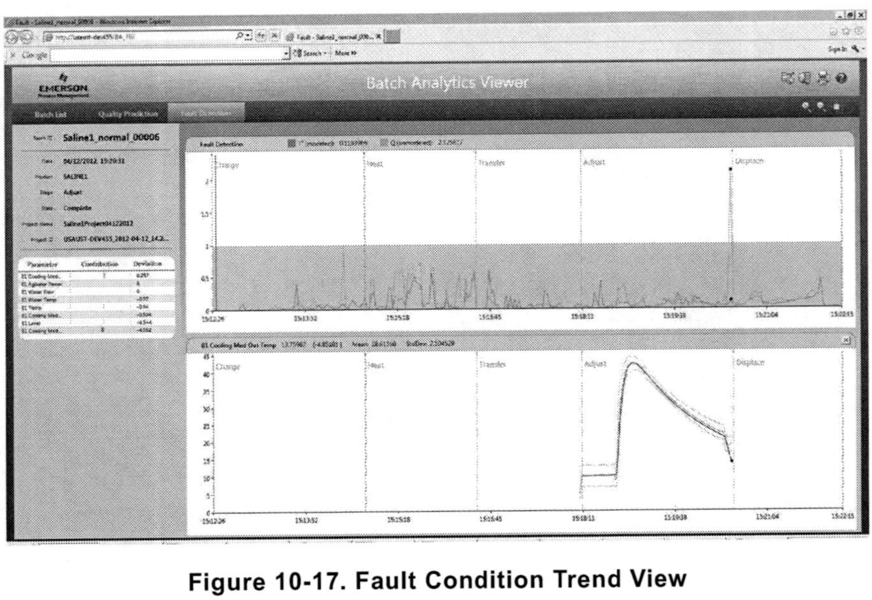

Figure 10-17. Fault Condition Trend View

By selecting the Quality Parameter tab, a plot of the prediction's future value may be displayed along with the configured specification limits for the product, as illustrated in Figure 10-18.

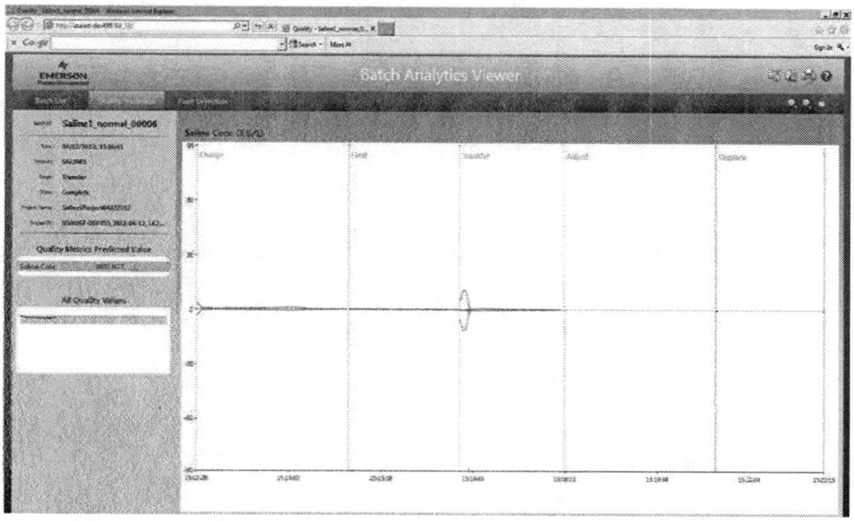

Figure 10-18. Quality Prediction View

This ability to view current and completed batches and to easily access associated historic information at the time a fault was detected may be used to analyze and determine the source of batch faults.

10.3 Batch Data Analytics Workshop

This workshop provides several exercises that may be used to further explore the on-line use of batch data analytics for fault detection and quality parameter prediction. The operation of the mixer and blender in the Saline process (Figure 10-19) will be observed using the data analytics on-line view. Access the book's web site http://www.advancedcontrolfoundation.com and select the Solution tab to view the workshop being performed.

Step 1: Open the Batch Data Analytics web-based interface and select the Overview tab to see the current status of the process.

Step 2: Select a batch that is shown to have a fault condition and then select the Fault Detection tab. Click on the point in the trend that shows maximum variation in T^2 or Q statistics.

Step 3: Examine the parameters that contributed the most to the fault condition. Based on the parameter trend, identify the process disturbance that is the source of the fault condition.

Step 4: Select the Quality tab to determine if the fault had an impact on the quality parameter, product % solids.

Step 5: Repeat steps 2, 3, and 4 for the last four fault conditions shown in the Overview. Identify the process disturbance that caused the fault condition.

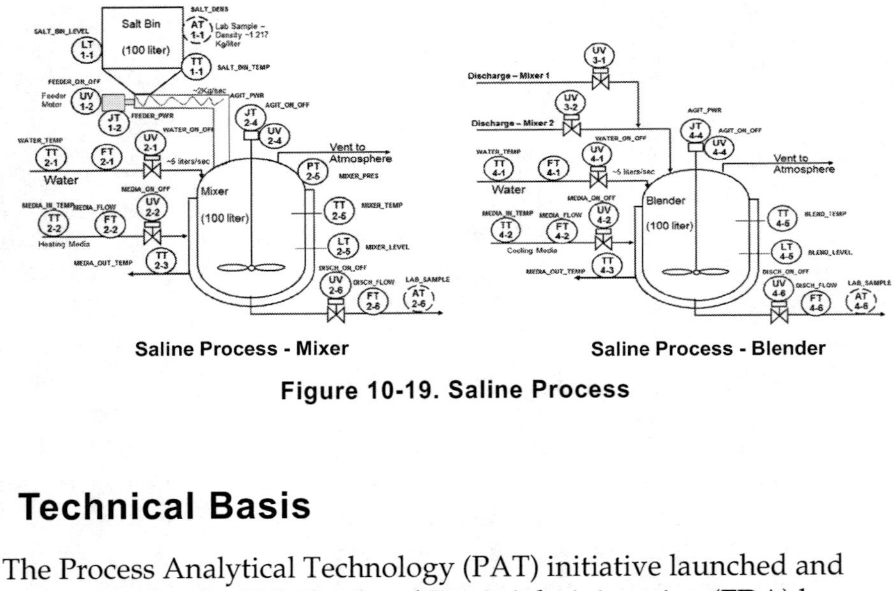

Saline Process - Mixer **Saline Process - Blender**

Figure 10-19. Saline Process

10.4 Technical Basis

The Process Analytical Technology (PAT) initiative launched and supervised by the U.S. Food and Drug Administration (FDA) has boosted research in, and applications of, analytical techniques. It is important to note that the term *analytical* in PAT is applied broadly to include chemical, physical, microbiological, mathematical, and risk analysis that is conducted in an integrated manner. The goal of PAT is to understand and control the manufacturing process, which is consistent with the current FDA approach to quality system: *quality cannot be tested into products; it should be built-in or should be by design* (U.S. FDA, http://www.fda.gov/Cder/OPS/PAT.htm).

Multivariate statistics is based on Principal Component Analysis (PCA) for fault detection and Projection to Latent Structures (PLS) for end-of-batch quality parameters prediction (Mason and Young, 2002). These techniques are pivotal to improving process operation. The on-line batch analytic applications are hard to imagine without recent research advances in batch data unfolding and modeling, feedstock property modeling, and data alignment by using Dynamic Time Warping (DTW) for off-line modeling and on-line operation. To provide the reader with a comprehensive overview of data analytics technology, the following techniques are outlined in this section:

- Feedstock property modeling
- Data preprocessing and validation
- Multi-stage batch modeling
- Batch data alignment and model alignment with on-line process operation
- Batch data arrangement—unfolding
- PCA process modeling and on-line operation
- PLS process modeling and on-line quality prediction

10.4.1 Feedstock Property Modeling

Knowing on-line feedstock properties is important for process monitoring and for quality prediction because batch product quality depends in large part on the properties of the raw materials loaded from storage tanks.

Determining the properties of material pumped from a storage tank is complicated by the varying ways that material may have been pumped into the storage tank. As illustrated in Figure 10-20, it may enter at the top of the tank or be directed by internal piping to the middle or bottom of the tank.

In many cases, storage tanks are not equipped with agitators, and, in general, tank pumps do not run in recirculation mode. Thus, because material exits the tank at the bottom, it may be assumed that the movement of material within the tank can be modeled as plug flow, with slight mixing taking place over time due to concentration differences and to turbulence caused by the addition of new material to the tank.

The tank material properties model is built in a heuristic way by projecting a layer of material in a tank into a zone of the model (Figure 10-21). In this way, a multi-compartment "stack" type model is formed.

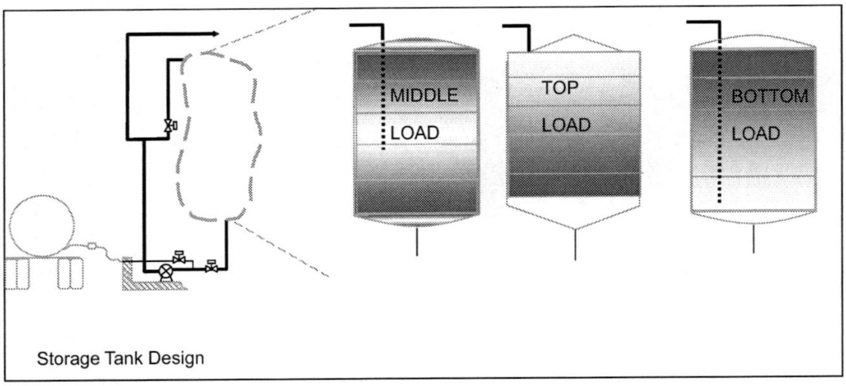

Figure 10-20. Tank Types Named According to Loading Process

Each zone represents one truck, one car, or one batch load. Through the use of zones, it is possible to simulate the more complex behavior associated with plug flow or a short circuiting flow pattern with some degree of mixing. When the new material is loaded, the properties and quantity from the new material will be pushed into a stack as illustrated in Figure 10-21.

The push operation is performed in such a sequence that stack zones are shifted down, starting with the load 1 zone, then the load 2 zone, and finally the new shipment 4 becomes the last load zone. The first row of Figure 10-21 includes the properties' averages and the total material quantity in the model. A complete model operation can be summarized as follows:

- Models the tank loading by shifting current zone properties to the next zone and establishing properties for the first zone from the most recent shipment.
- Establishes current pump-out zone based on the tank level measurement.
- Calculates average tank properties.
- Calculates outlet properties from the current pump-out zone properties and average tank properties.

Every functional component of the calculations can be executed separately. This feature makes the algorithm robust and flexible, and makes it easy to monitor the model update by tracking arrays representing the zones.

Time	Average 1	Average 2	....	Total Quantity
New Shipment 4	Property 1	Property2	....	Quantity
Last Load 3	p_1^3	p_2^3	....	Q^3
2			....	
Oldest Load 1	p_1^1	p_2^1	....	Q^1

Figure 10-21. Tank Zone Blending Model Format

In a default model, one mixing coefficient is applied for all properties. If the blending of various properties differs, the model allows a separate mixing coefficient to be applied for every property.

The model's most important function is calculating outlet properties. It starts with defining the weighted average $\bar{P}_i$ for every property P_i, assuming perfect mixing:

$$\bar{P}_i = \sum_{k=1}^{n} \frac{p_i^k w^k}{W} \tag{10-1}$$

where:

p_i^k = value of property P_i for zone k

w^k = material quantity (weight) in zone k

W = total material weight in the tank with n loaded zones

$$W = \sum_{k=1}^{n} w^k \tag{10-2}$$

The tank outlet property p_i^{outlet} is defined as:

$$p_i^{outlet} = (1 - m)p_i^k + m\bar{P}_i \tag{10-3}$$

k = ***1*** for the bottom load and $\boldsymbol{k = n}$ for the top load, and $1 \le k \le n$ for the middle loading

m = mixing factor $0 \le m \le 1$

When $\boldsymbol{m = 1}$, there is a case of perfect mixing, $p_i^{outlet} = \bar{P}_i$ and when $\boldsymbol{m = 0}$ there is no mixing and $p_i^{outlet} = P_i^k$, that is, outlet material properties are the same as the original properties in the current pump-out zone.

For implementations in which the material quantity for various zones is about the same, Equation 10-1 can be simplified to:

$$\bar{P}_i = \sum_{k=1}^{n} \frac{p_i^k}{n} \tag{10-4}$$

Tank outlet property p_i^{outlet} as defined by Equation 10-3 can be used directly in batch modeling as a process input or initial condition parameter.

10.4.2 Data Preprocessing and Validation

When data analytics are to be applied to a real plant, it must be recognized that the data will not be complete. Simply deleting batches and variables of incomplete data will result in reduced datasets that may not meet the analysis objectives. There is a need to handle missing data, up to some limit, before removing data.

A criterion can be established to use the last known good value for a variable while tracking the percentage of bad values and the number of consecutive bad values per variable per batch. If a specific variable exceeds a percentage of bad values or a maximum number of consecutive bad values, the variable is tagged for reconstruction. A variable is tagged per variable and per batch. If a batch has too many tagged variables, it is thrown out. If a variable is tagged for too many batches, it is thrown out. If the result is a data set with too few batches, the model generation is stopped.

After tagged variables are identified, the data are aligned. This may be done, for example, by the Dynamic Time Warping procedure (DTW), with a weight of zero given to each tagged variable. This means that the tagged variable is ignored with respect to the DTW algorithm.

Once the batches are time warped, each batch is of uniform length. This allows for statistics to be compiled for each observation across all batches. The statistics compiled are the mean and the standard deviation of each variable. This mean is then substituted for the values of tagged variables. In implementation, the substitution is made after normalization such that tagged variable values are simply assigned a zero. The result is a complete data set. A specific example of data validation criteria might be as follows:

- A bad status data value is given the last known good value for that variable.
- If more that 30 consecutive values are bad, the variable is tagged for that batch.
- If more than 20% of the values for a variable are bad, the variable is tagged for that batch.
- If a batch has more than 50% of the variables tagged, the batch is thrown out.
- If a variable is tagged for more than 50% of the batches, the variable is thrown out.
- A tagged variable is "ignored" through DTW, for the specific batch, and then the variable is reconstructed (set to 0 after normalization).

Reconstruction of a variable is done for the variable over the entire duration of the batch in which it was tagged. For multi-stage batches, data validation (as above) is performed and criteria are applied separately for every stage.

10.4.3 Multi-stage Batch Modeling

The on-line use of analytic tools such as PCA and PLS for fault detection and prediction of quality parameters has been limited in many instances to continuous applications in which a single product is produced. In such instances, the process is often treated as a single unit with a fixed set of measurements and lab analysis. For these types of applications, a single PCA and PLS model can be developed and applied in an on-line environment. However, to address the requirements of batch processes, or continuous processes in which multiple products are made using one or more pieces of equipment with their own set of instrumentation and quality parameters, a more general approach must be taken in off-line model development and in the application of on-line analytics.

Some of the challenges in applying on-line analytic tools to continuous and batch process may be summarized as follows:

- In a batch operating environment, product manufacturing uses numerous pieces of equipment that may be run in series, parallel, or a combination of series and parallel configurations. The equipment used and the associated process operating conditions depend on the product being manufactured. Different lab and field measurements may be used at various points in the process.
- A continuous operating environment may involve multiple major pieces of equipment arranged in different configurations. The processing associated with each piece of equipment and the associated measurements and control may vary as the processing conditions change with throughput or with the product that is being processed.

Thus, tools that are designed to support on-line analytics should take into consideration the product, the equipment arrangement(s) that may be used to make the product, and the different operating conditions and associated field and lab measurement needed to manufacture the product. To address these requirements, the analytic model for a batch or continuous manufacturing process may be broken into the different stages of manufacturing required to produce a specific product. In this context, a process *stage* is characterized by:

- The type of equipment required for processing
- The field and lab measurement required to monitor or control the process
- The process operating conditions that must be maintained
- The impact on the product of various process operating conditions

The concept of a stage may be applied to the development of on-line analytic models and their application to both continuous and batch process. Once the different stages associated with a product have been defined, analytic models may be constructed by stage.

A stage of a batch process can be defined based on the product recipe and the active operation(s). For example, when a batch is initiated to manufacture a product, the batch is typically assigned a unique batch identifier (ID). As the batch progresses from one piece of equipment to another in the manufacturing process, this batch ID moves along with the product. If the batch control is structured using SP88 nomenclature and structures, the unit representing each piece of equipment contains parameters for batch ID, the product, and the active operation.

> **ISA88:** a batch standard that uses object-oriented concepts to define terminology and models for batch control.

In general, one or more operations are associated with a single stage. Thus, the on-line analytics application can use this information to determine the stage of processing and to track the product as it moves between pieces of equipment. The progression of a batch and its association with an analytic stage can be illustrated as shown in Figure 10-22.

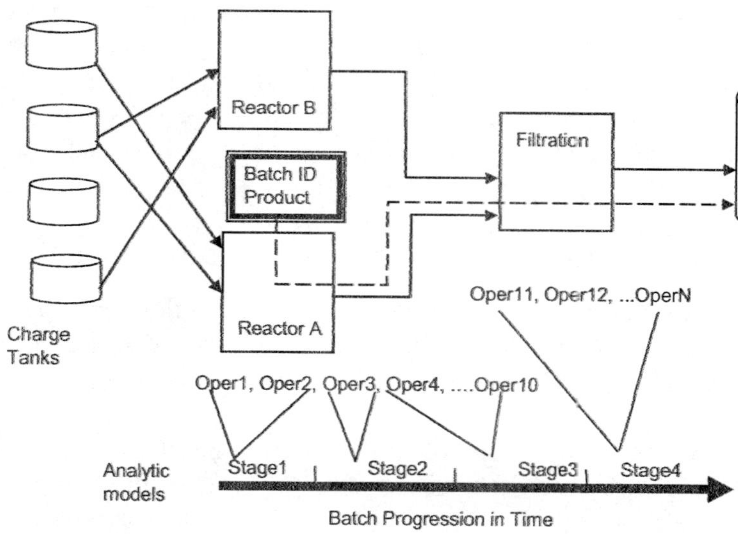

Figure 10-22. Stages in a Batch Process

There are two additional justifications for using stages in analytic modeling:

- Manufacturing time for successive batches differs significantly. Process data from all batches used for model development must be adjusted time-wise. Introducing stages helps to accomplish data synchronization in a more efficient way.

- Analytic models used for on-line process monitoring must be synchronized with process operation. As with model development, dividing a process into stages makes this task more efficient on-line.

Industrial processes are characterized as non-linear and non-stationary. Analytic models assume process linearity and a stationary process. Decomposing a continuous or batch process into stages and developing multi-block models more closely aligns modeling assumptions with the actual process. Applying stages makes the decomposition well-defined based on scheduled operations, acceptable operating conditions, and predictable events.

Multi-block process modeling provides numerous advantages over single-block process modeling, such as flexibility in modeling for successive process modeling and testing (in particular), along with easier on-line deployment and easier model reconfiguration. At the same time, multi-block modeling creates the additional task of transferring modeling results to the successive blocks as processing progresses (Westerhuis, Kourti, and MacGregor, 1998). The predicted end-of-batch or end-of-stage quality itself can be used for transferring prediction from the preceding blocks. Also, if they are available, end-of-stage quality analysis result—instead of predicted end-of-stage quality—can be applied.

10.4.4 Alignment of Batch Data and On-line Model

As mentioned earlier, a substantial obstacle in developing data analytics tools for batch processes is the different time duration of batches and batch stages (Figure 10-23).

The traditional technique used for aligning batch data is to apply an indicator variable that represents the progress of the batch. The indicator variable should be smooth, continuous, and monotonic, and should span the range of all other process variables within the batch data set (Cinar, Parulekar, Undey, and Birol, 2003). For monitoring new batches, data are collected from all process variables and then adjusted with respect to the indicator variable. This alignment technique is illustrated for two variables in Figure 10-24.

Better alignment is delivered by Dynamic Time Warping (DTW), a technology that was borrowed from speech recognition (Kassidas, MacGregor, and Taylor, 1998). As discussed earlier in the chapter, DTW minimizes the distance between respective process trajectories for different batches. In doing so, DTW takes into account all of the variables in the analysis and is an effective approach for batch data

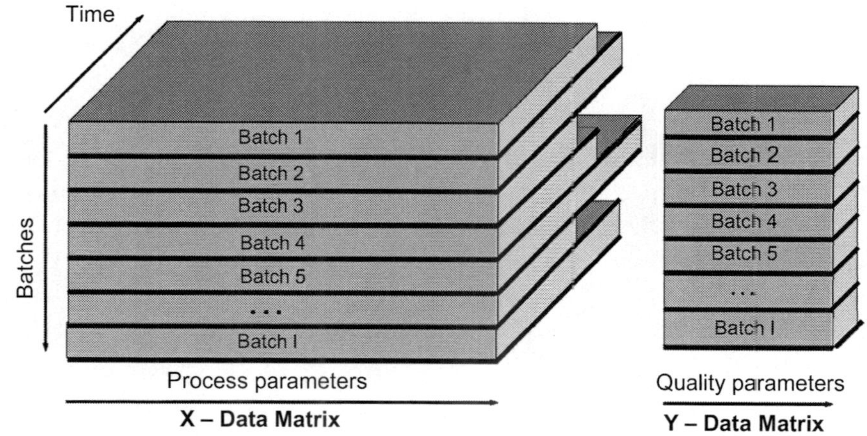

Figure 10-23. The Nature of Batch Data

alignment. DTW can use an indicator variable (described above), or an additional variable can be created and defined as a fraction of the batch (or stage) completion time. This indicator variable can be added to the original process variables set to improve the robustness of the DTW calculation and to prevent convergence to local minima over an extended period of time.

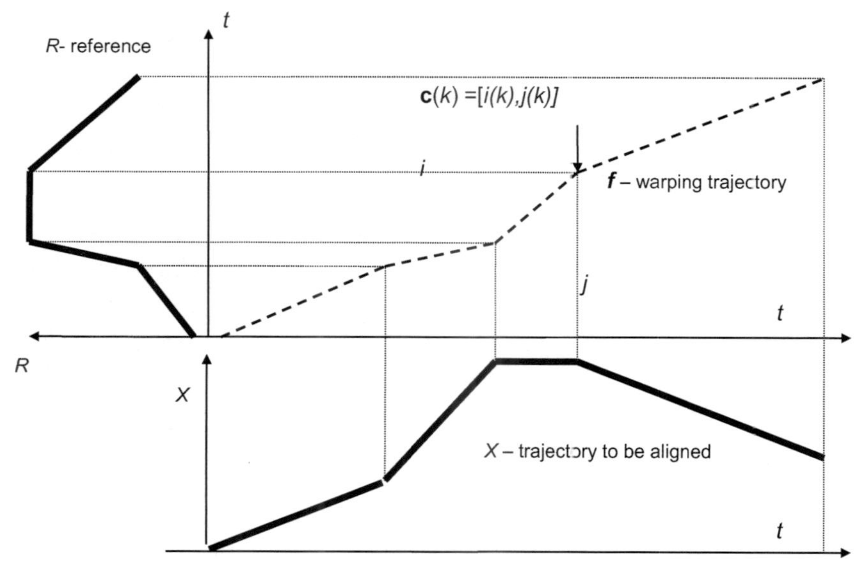

Figure 10-24. Illustration of the Reference and Raw Trajectories R and X and the Aligning Optimal Path c(k)

The basics of using the DTW algorithm to align two sets of batch data with variable time duration trajectories are explained below and illustrated in Figure 10-24.

First, one batch is selected with a reference trajectory R. Normally, when considering batch data training sets, the trajectory of the batch with the median time duration is used as the initial reference trajectory. The distance d between the reference trajectory and the remaining trajectories X at a point k can be defined as:

$$d(i(k), j(k)) = \{R[i(k)] - X[j(k)]\} \cdot W \cdot \{R[i(k)] - X[j(k)]\}^T \tag{10-5}$$

W is a positive weight matrix that reflects the importance of each measured variable; k is a grid point along the path ($k = 1,2,\ldots,K$). The lengths of X and R are t and r.

The total distance $D(t, r)$ between the two trajectory sets is:

$$D(t, r) = \sum_{k=1}^{K} d(i(k), j(k)) \tag{10-6}$$

The optimal path that minimizes $D(t, r)$ is:

$$D^{opt}(t, r) = \min\left\{\sum_{k=1}^{K} d(i(k), j(k))\right\} \tag{10-7}$$

A path can be denoted as: $f = \{c(1), c(2), \ldots c(K)\}$ (10-8)

where $c(k) = [i(k), j(k)]$ represents a grid point connecting i and j. Figure 10-25 is a simple illustration of optimal path development.

Minimization (10-7) is usually subject to the global constraints (Equation 10-9) and local constraints (Equation 10-10), which enforce matching initial and end trajectory points and limit, for the point (i,j), the choice of the predecessor point.

$$c(1) = [1, 1] \qquad c(K) = [t, r] \tag{10-9}$$

$$c(k-1) = (i-1, j), (i-1, j-1) \text{ or } (i, j-1) \tag{10-10}$$

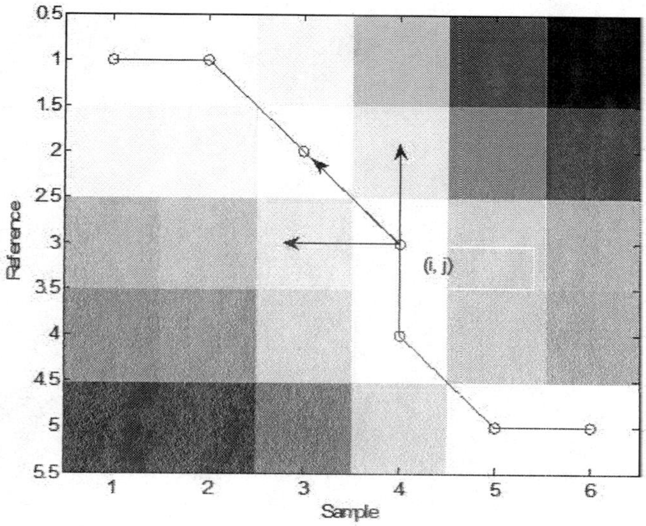

Figure 10-25. Illustration of the Optimal Path c(k) for the Five-Point Reference Trajectory and Six Point Sample Data Trajectory (Boudreau and McMillan, 2007)

The optimal trajectory path is used to build the aligned trajectory, as shown in Figure 10-26. In the figure, points 5 and 6 on the raw trajectory (upper trend) are combined into one point on the aligned trajectory (middle trend) to match point 5 on the reference trajectory (lower trend), and point 8 is expanded on several points to best match points 7, 8, 9, 10, and 11 on the reference trajectory, etc.

As a result of applying DTW, all batch (or stage, if the model is developed for a stage) data sets will have the same length and are ready for creating data structures suitable for model generation.

In on-line operation, PCA or PLS model position is adjusted with the progress of the current batch every modeling scan. After *J* parameters have been read, and the current scan parameter vector has been updated, the parameter vector should be rescaled in the same way as off-line data prior to using DTW.

Figure 10-27 shows the specifics of on-line DTW. The DTW on-line alignment is for one specific modeling scan. The algorithm evaluates the distance from the current batch trajectory at scan *k'* to the model trajectories at the last used model scan and several scans ahead ***(k, k+1, k+2…)*** and selects the model scan with the minimum distance from the current trajectory. The selection criteria may also include the

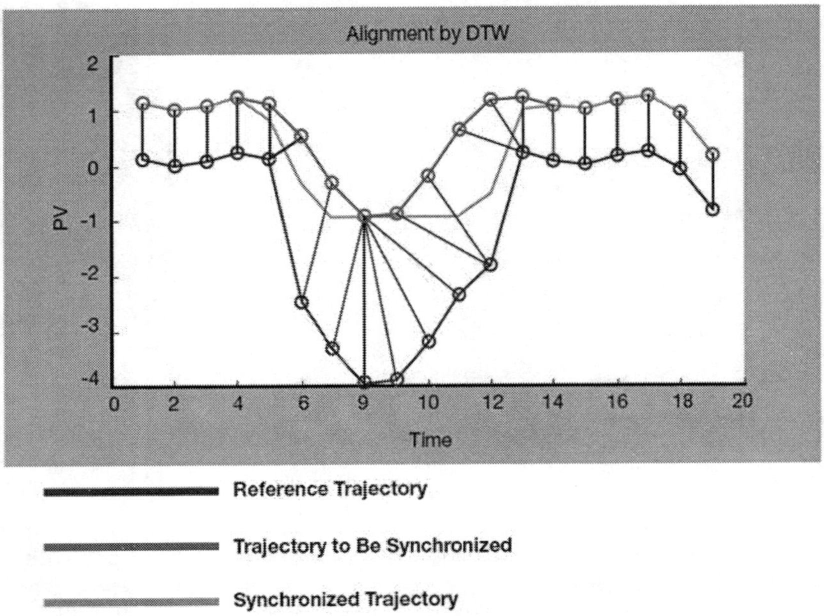

Figure 10-26. Building Aligned Trajectory from the Optimal Path

change in distance between the model trajectories and the current process trajectory since the last scan.

More on DTW can be found in Ramaker, v. Sprang, Westerhuis, and Smilde (2003) and in Pravdova, Walczak, and Massart, 2002.

10.4.5 Data Arrangement – Unfolding

The creation of the batch data structure for modeling is known as data unfolding. A typical use of the batch-wise unfolding of historical data from a number of batches used for model development (training data set) is illustrated in Figure 10-28. Three-dimensional data X(I×J×K) are transformed into two-dimensional space X(I×(J×K)), where J is the number of process parameters used in the model, I is the number of batches, and K is the number of model scans.

For each parameter in the data slice X(I×J), mean values and standard deviations are calculated. These mean values will form the model reference trajectories used in model development and on-line operation. Batch quality data are grouped in the matrix Y(I×L), where L is the number of quality parameters used in the modeling.

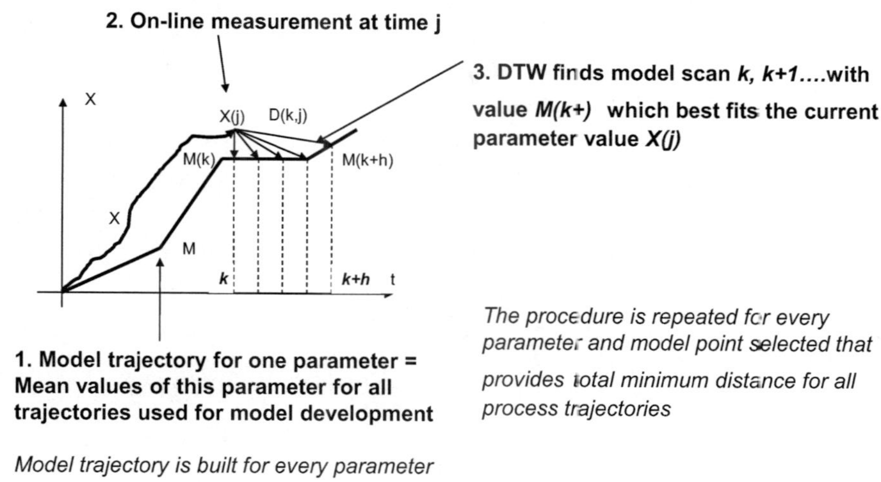

Figure 10-27. Aligning Batch Trajectories On-line

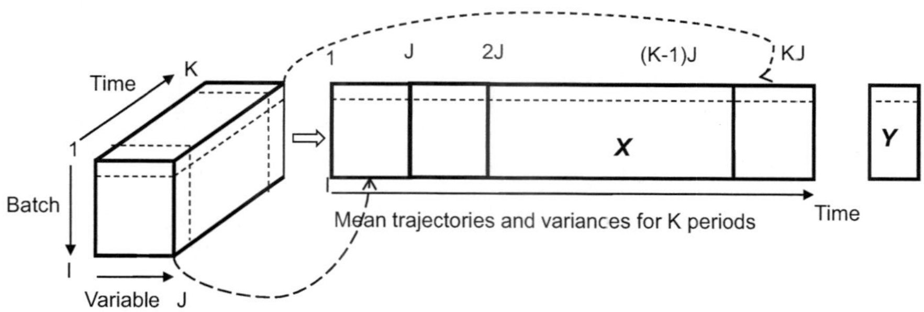

Figure 10-28. Batch-wise Data Unfolding

Batch-wise unfolding is mainly used for PLS modeling. For PCA modeling, variable-wise unfolding or hybrid unfolding is used. Hybrid unfolding combines batch-wise and variable-wise unfolding features and is best suited to PCA modeling.

10.4.6 PCA Modeling and On-line Fault Detection

Figure 10-29 represents the data flow for batch PCA off-line modeling. It is assumed in the figure that K is the number of scans used by DTW for data alignment. The following basic steps are performed to develop a model for on-line deployment:

- A three-dimensional data set X(I×J×K) is batch-wise unfolded into a two-dimensional set X((I×K)×J). Mean value trajectories are calculated, and the data are rearranged in the hybrid unfolded matrix (left bottom matrix).
- The PCA algorithm NIPALS (Nonlinear Iterative Partial Least Squares) delivers a score matrix T((I×K)×R) and a load matrix P(J×R) from the data matrix X((I×K)×J). The score matrix ***T*** is not used on-line; additional matrices applied on-line are inverses of covariance matrices for every model scan—invS$_k$(R×R).
- The Upper Control Limit trajectory for the T^2 statistic is defined
- The Upper Control Limit trajectory for the Q statistic is defined
- The model quality indices are calculated

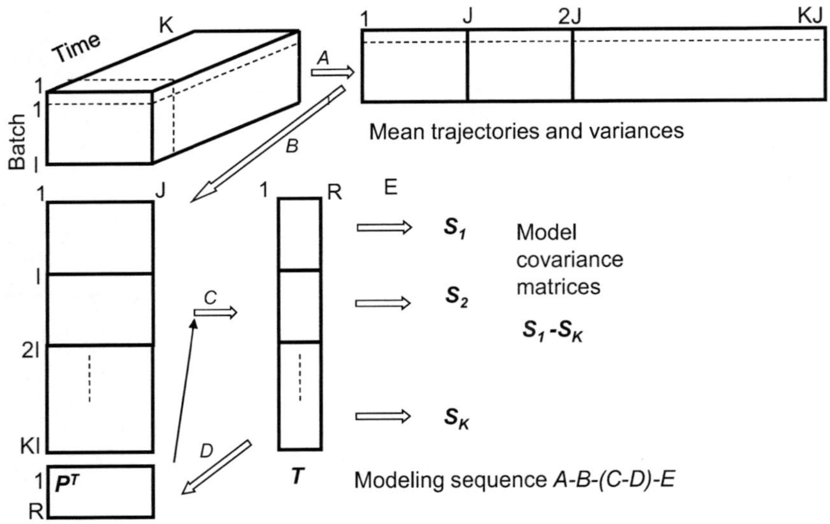

Figure 10-29. PCA Off-line Hybrid Data Unfolding and Model Development Flow Diagram

NIPALS is the core algorithm of analytics, and the algorithm is sketched below – steps 1 to 6. The reader can find more detail in Geladi and Kowalski (1986). Computationally, the algorithm is carried out by iteration and deflation in a few steps:

1. Select a column vector x_j from X as the starting point of t_i.
2. Calculate p_i by:

$$p_i = X^T t_i (t_i^T t_i)^{-1}$$

3. Normalize p_i:

$$p_i^T = \frac{p_i^T}{\|p_i^T\|}$$

4. Calculate t_i:

$$t_i = Xp_i$$

5. Compare t_i used in Step 2 and calculated from Step 3.

If the difference is smaller than a predefined convergence limit, go to Step 6; otherwise go to Step 2.

6. Deflate X by $X = X - t_i p_i$, set $i = i + 1$ and go to Step 1.

Preprocessing of data is crucial to PCA, and different preprocessing criteria will lead to different results. Among many preprocessing methods, X is scaled into zero mean and unit variance for each column in most process monitoring applications. Due to the iterative nature of NIPALS, there might be convergence concerns when very similar eigenvalues exist in the covariance matrix. Luckily, in real applications, which are important for process monitoring, eigenvalues are always different from each other (usually with a different magnitude).

The T^2 limit is estimated from the F-statistic (Fisher distribution). The F-distribution is defined by the degree of freedom of two parameters: the number of batches in the preprocessed data I and the number of selected principal components R (Equation 10-11):

$$T_\alpha^2 = \frac{R(I^2 - I)}{I(I - R)} F_a(R, I - R) \tag{10-11}$$

α defines the statistical threshold for alarming. In common practice, the top 5% of the distribution is considered as the alarming threshold. A T^2 limit is normally used for the PCA model.

Limits for the Q statistic are calculated for every model scan by applying a weighted χ^2 distribution:

$$Q_k = \frac{v_k}{2m_k}\chi^2(2m_k^2, v_k) = \frac{v_k}{2m_k}\chi^2\left(\frac{2m_k^2}{v_k}\right) \tag{10-12}$$

where m_k and v_k are the mean and variance of the $\boldsymbol{Q}$ statistic at time period k.

The chi-square distribution (χ^2) is one of the most widely used probability distributions used in inferential statistics. If Xi are n independent, normally distributed random variables with mean 0 and variance 1, then the random variable

$$Q = \sum_{i=1}^{n} X_i^2$$

is distributed according to the chi-squared distribution.

The model goodness index R_x^2 is a basic tool for model evaluation. This index shows how much data variance is explained by the model:

$$0 < R_x^2 < 1 \tag{10-13}$$

The summary of PCA modeling and operation is illustrated in Figure 10-30. The data block X is decomposed to the modeled part TP^T and the un-modeled part $\boldsymbol{E}$:

$$X = TP^T + E \tag{10-14}$$

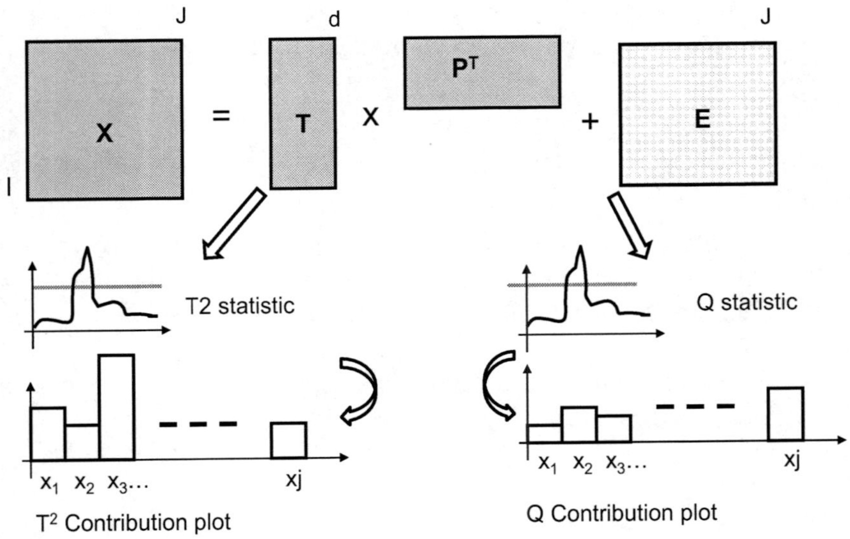

Figure 10-30. Summary of PCA Modeling

On-line PCA performs the following sequence of operations for every modeling scan:

- Reading and rescaling model input parameters
- Calculating the score vector ***t***
- Calculating T^2 statistics for the batch scan *k*
- Calculating modeling error and Q statistics
- Developing contribution plots for T^2 and Q

10.4.7 PLS Modeling and On-line Quality Prediction

PLS off-line model development steps can be summarized as follows:

- PLS algorithm (NIPALS) delivers score matrix ***Ts****(I×R)* and load matrices ***Qs****(L×R),* ***Ws****(JK×R) and* ***Ps****(JK×R)*. The following additional matrices and parameters are created for on-line calculations (Figure 10-31):
- *Inv(Ts(k)'Ts(k))* for calculating confidence interval. *K = 1,2,…,K;* the range of data for every *k* is *[1,…k],* that is, includes all preceding scans.
- *MSE(k)* – mean squared error, actual quality vs. predicted end-of-batch quality for every scan k, for the batches used in model development

Figure 10-31 illustrates the PLS modeling structure.

The key to developing PLS quality prediction $\hat{Y}$ during on-line operation is the calculation of the latent structure score matrix ***T***. This is done by successive calculations of the latent variables matrix ***T*** with *p* principal components, using model matrices ***W***, ***P***, and ***Q*** as shown below:

$$t_i = X_{i-1}w_i/(w_i^T w_i) \tag{10-15}$$

$$X_i = X_{i-1} - t_i P^T \tag{10-16}$$

$$T = [t_1 \ldots t_i \ldots t_p] \qquad W = [w_1 \ldots w_i \ldots w_p] \tag{10-17}$$

t_i, w_i = column vectors of the matrices *T*, *W* respectively

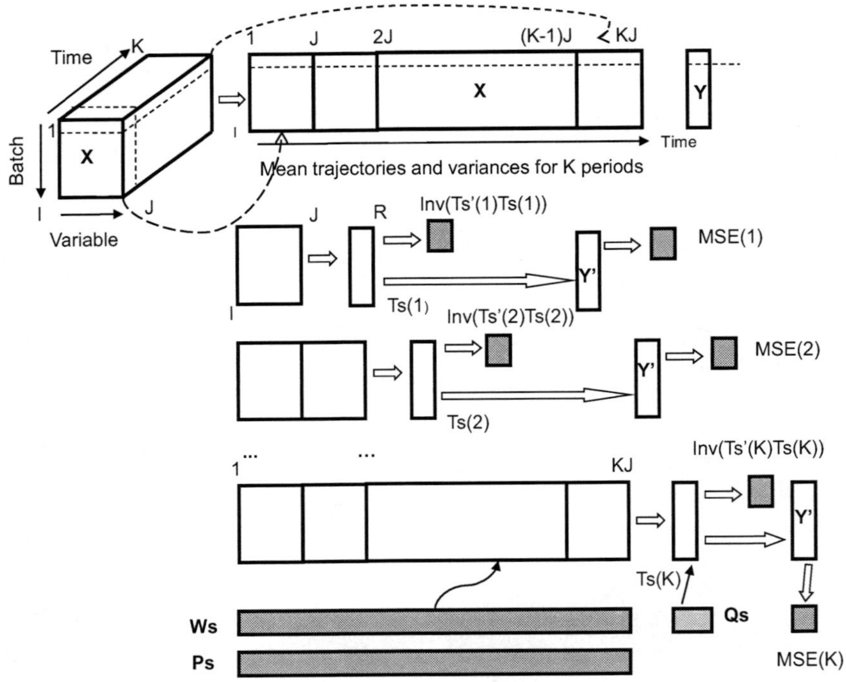

Figure 10-31. Batch-wise Unfolded Data and PLS Modeling Transformations

$$\hat{Y} = TQ^T \tag{10-18}$$

$\hat{Y}$ is a vector of predicted quality parameters change relative to quality mean values for the batches used for PLS model development.

In on-line applications the data matrix $\mathbf{X}$ is a one row matrix. The number of columns of $\mathbf{X}$ increases for every PLS calculation scan on J up to the value KJ at the end of a stage. The data matrix $\mathbf{X}$ carries the variables from the beginning of the batch up to the current time.

The PLS prediction should be accompanied by an on-line confidence interval indicating the accuracy of the prediction:

$$\hat{Y} \pm t_{1-R,\,\alpha/2}\sqrt{MSE(k)}(1 + t_k(T'(k)T(k))^{-1}t'_k) \tag{10-19}$$

All formula parameters are defined in the model, except t_k, which is defined on-line in every scan.

BIBLIOGRAPHY

1. Blevins, T. and Beall, J. "Monitoring and Control Tools for Implementing PAT," *Pharmaceutical Technology,* March 2007.

2. Boudreau, M. A. and McMillan, G. K. *New Directions in Bioprocess Modeling and Control,* Research Triangle Park: ISA, 2007.

3. Cinar, A., Parulekar, S. J., Undey, C., and Birol, G. "*Batch Fermentation – Modeling, Monitoring, and Control,*" New York, NY: Marcel Dekker, 2003.

4. Geladi, P. and Kowalski, B. R. "Partial Least Squares Regression: a Tutorial," *Analitica Chemica Acta,* 185, 1986, pp. 1–17.

5. Kassidas, J. F., MacGregor, J., and Taylor, P. A. "Synchronization of Batch Trajectories Using Dynamic Time Warping," *AIChE Journal,* Vol. 44, 1998, pp. 864–875.

6. Lee, J. M., Yoo, C. K., and Lee, I. B. "Enhanced Process Monitoring of Fed-batch Penicillin Cultivation Using Time-varying and Multivariate Statistical Analysis," *Journal of Biotechnology,* 110, 2004, pp. 119–136.

7. Mason, R. L. and Young, J. C. *Multivariate Statistical Process Control with Industrial Applications,* Philadelphia: ASA-SIAM, 2002.

8. Nomikos, P. and MacGregor, J. F. "Monitoring Batch Processes Using Multiway Principal Component Analysis," *AIChE J.,* 40, 1994, p. 1361.

9. Pravdova, V., Walczak, B,. and Massart, D. L. "A Comparison of Two Algorithms for Warping of Analytical Signals," *Analytica Chimica Acta,* Vol. 456, 2002, pp. 77–92.

10. Ramaker, H. J., v. Sprang, E., Westerhuis, J., and Smilde, A. K. "Dynamic Time Warping of Spectroscopic Batch Data," *Analytica Chimica Acta,* Vol. 498, 2003, pp. 133–153.

11. Reiss, R., Wojsznis, W., and Wojewodka, R. "Partial Least Squares Confidence Interval Calculation for Industrial End-of-batch Quality Prediction," *Chemometrics and Intelligent Laboratory Systems,* 100, 2010, pp. 75–82.

12. U.S. Food and Drug Administration, Office of Pharmaceutical Science web site: http://www.fda.gov/Cder/OPS/PAT.htm.

13. Westerhuis, J. A., Kourti, T., and MacGregor, J. F. "Analysis of Multiblock and Hierarchical PCA and PLS Models," *Journal of Chemometrics,* 12, 1998, pp. 301–321.

14. Wojewodka, R. and Blevins, T., "Data Analytics in Batch Operations," *Control,* May 2008, p.53, www.controlglobal.com/articles/2008/164.html.

15. Wojewodka, R. and Blevins, T., "Benefits Achieved Using On-line Data Analytics," Presented at Emerson Exchange Conference, Orland, FL, 2009.

16. Zhang, Y. and Edgar, T. F. "Multivariate Statistical Process Control," *New Directions in Bioprocess Modeling and Control,* ed. Boudreau, M. and McMillan, G. Research Triangle Park: ISA, 2007.

11
Simple MPC

In 1980, when Cutler and Ramaker first published and presented information on a new control technique known as Dynamic Matrix Control, it caused quite a stir in the process control community (Cutler, Ramaker, 1980). Shell Oil had developed and deployed this technique for the control of large interactive multiple input-multiple output (MIMO) processes such as refinery distillation columns. This work by Shell was the first practical version of what is commonly referred to today as Model Predictive Control (MPC).

Model Predictive Control has been installed in thousands of plants since the initial installations at Shell. Applications have been focused on large and medium size installations in petrochemical, refining, chemical, paper and other industries (Froisy, 1994). MPC is still not common in small installations, despite its many proven advantages over multi-loop PID control techniques in addressing a variety of small applications, as summarized below.

Single or multi-loop control that is characterized by long delay or inverse process response. Because the control action taken by the MPC blocks is based on a process response model, better control can be achieved than that possible with PID feedback control or deadtime compensation techniques such as the Smith Predictor (Shinskey, 1988). Products for implementation of MPC allow the response model to be determined automatically during commissioning

The interaction of two or more control loops impacts process operation, that is, a change in the output of one control loop impacts the controlled parameter of one or more of the other control loops to a significant degree. Such interaction is automatically accounted for when control action is taken by the MPC block. Thus, control may be improved because any correction taken to adjust one control parameter will be balanced to minimize the impact on other parameters.

One or more disturbances to a controlled parameter(s) are measured. By including these measurements in the control, the MPC block automatically compensates for the impact of the disturbances. Thus, changes in the measured disturbance input will have little or no impact on the controlled parameters.

One or more measurable constraints must be observed in the control of the process. The MPC block constantly calculates the impact of a disturbance or control action on constraint outputs. If the future value of a constraint output violates its limit, MPC automatically takes the appropriate action to prevent the constraint limit from being exceeded.

Production is limited by one or more inputs to a process. Control of throughput may be included in the MPC control objectives, along with the control associated with the inputs that limit production. Throughput is automatically adjusted to maintain the process at its input limits, thereby achieving maximum production.

Traditionally, a significant hindrance to small and medium size MPC applications has been the computation power and memory needed to implement model predictive control, which usually required a computer layered on top of an older distributed control system. As a result, the application of MPC has been limited to large, high throughput processes where an increase in throughput could justify the high cost of installing, commissioning and maintaining an MPC control strategy.

In a modern DCS, however, advances in the processors and memory of the controller make it possible to embed an MPC capability in the controller (Thiele 2000, Wojsznis, Blevins, and Nixon 2000). In particular, MPC designed to address small processes characterized by fast dynamics can be used for process applications that have historically been controlled using multi-loop PID techniques.

To focus reader attention on such installations, this chapter illustrates how an MPC capability embedded in a DCS controller can be used to replace traditional control techniques. In addition, it provides a basic understanding of MPC for less experienced users. The two chapters that follow will guide the reader to more advanced MPC products integrated with an optimizer and to more sophisticated MPC applications.

11.1 MPC as a Replacement for PID

Model Predictive Control may be used for the control of single input-single output (SISO) as well as multiple input-multiple output (MIMO) processes. In its simplest form, MPC may be used to control a SISO process. For example, MPC can be used to provide an effective means of controlling a SISO process that is described as "deadtime-dominant" (i.e., when the process deadtime is equal to or greater than the process time constant). The control performance achieved may not be satisfactory when PID feedback control is applied to a deadtime-dominant process because as the ratio of deadtime to time constant increases, the proportional and reset gains must be decreased for stability. As a result, the control response to changes in setpoint and load disturbances is often slower than that required for best process operation. In such cases, control performance can be improved by replacing PID feedback control with model predictive control as illustrated in Figure 11-1.

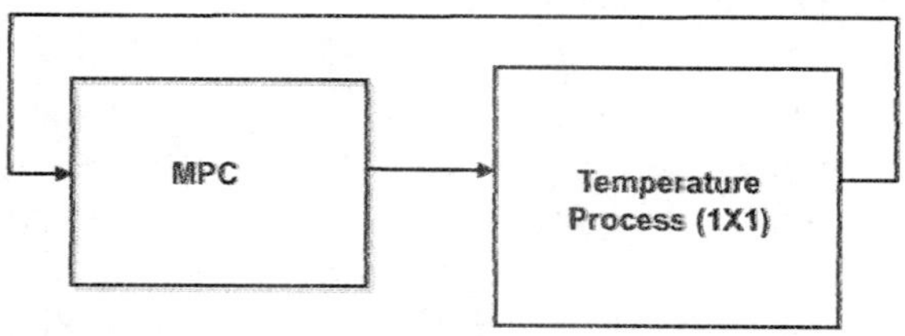

Figure 11-1. MPC as SISO Controller for Addressing Difficult Process Dynamics

As introduced in Chapter 3, this improvement over PID feedback control performance is possible because unlike the PID algorithm, which is predefined, the MPC algorithm is generated based on the process response to a step change in process inputs; this is known as the step response model. When a change is made in an input of a deadtime-dominant process, unlike the PID algorithm the MPC algorithm is aware that this change will not be immediately reflected in the process output. This awareness of the process response allows the MPC algorithm to do a better job of controlling deadtime-dominant processes.

Implementing Model Predictive Control of a SISO process is as simple as using PID control. The MPC block simply replaces the PID block in the control module, as illustrated in Figure 11-2. The connection of the analog input and output blocks to the MPC block is done in exactly the same way. However, when using MPC, the controller tuning is approached differently. For example, the MPC block does not contain parameters for proportional gain, integral gain and derivative gain. Instead, the identified step response (the step response model) is used to generate the MPC control matrix.

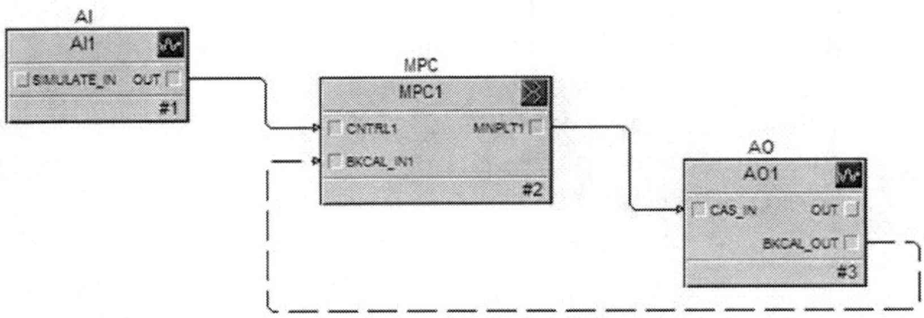

Figure 11-2. MPC Configuration for Single Loop Control

11.2 Commissioning MPC

When MPC is supported in a control system, a software application is provided to automatically identify the process step response model. Similar to the tuning applications provided to tune the PID, the MPC application is designed to allow the process to be automatically tested. With the push of a Test button, the process is tested, the step response model is automatically identified, and the MPC algorithm is generated and automatically transferred to the MPC block and used for control. The identified step response may be viewed as illustrated in Figure 11-3.

Before initiating testing, the user may specify how much the manipulated parameter is to be changed during testing. It is also necessary for the user to input their best guess, based on observation, of how long it takes the process to fully respond to a change in the manipulated parameter, that is, the process time to steady state as illustrated in Figure 11-4.

For example, if the process were to fully respond to a change in the manipulated parameter in 120 seconds, the user would enter 120 seconds as the time to steady state. The duration of the automated test is based on the estimated time to steady state and determines the maximum time duration of pulses generated during testing. When testing is initiated, the controller output is automatically changed to generate a series of pulses, the duration of which changes in a pseudo-random

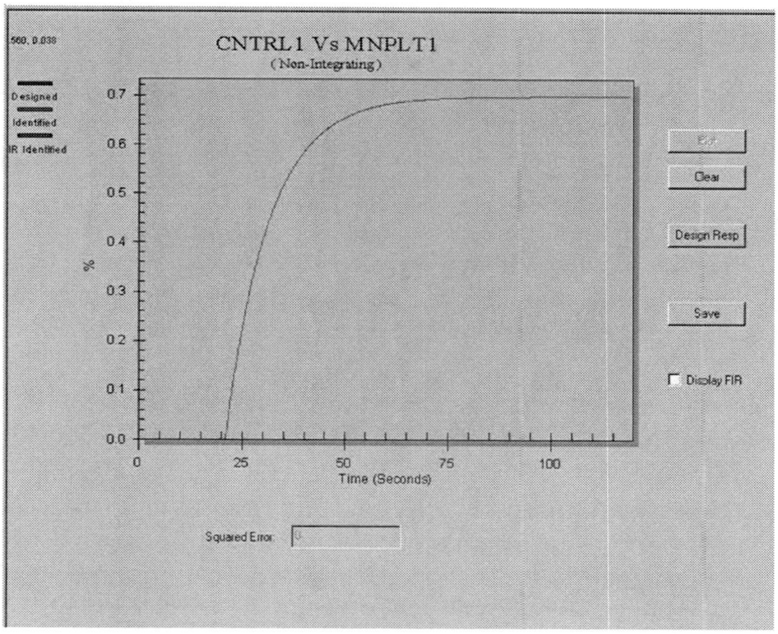

Figure 11-3. Identified Process Step Response

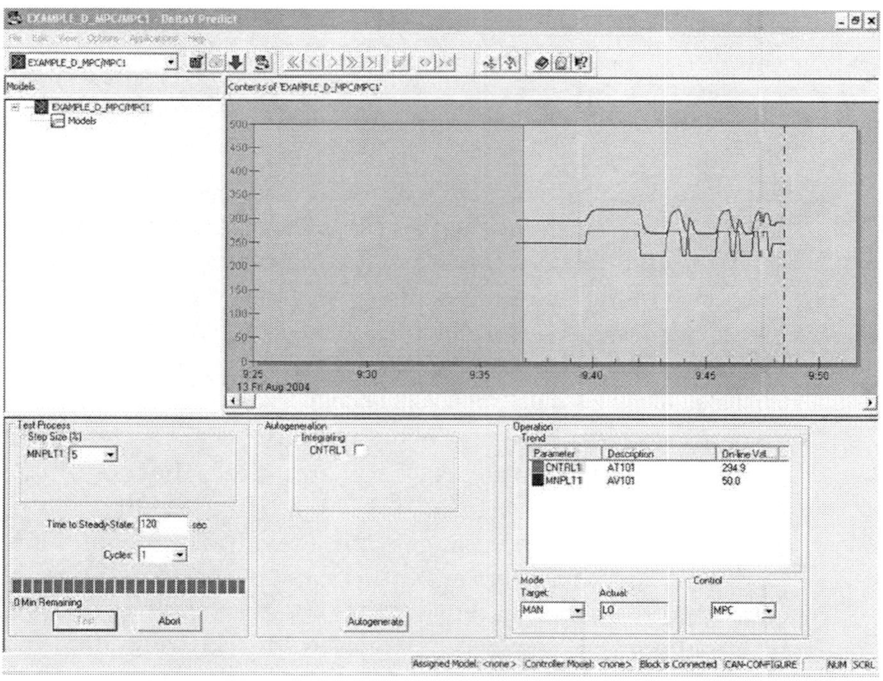

Figure 11-4. Automated Process Testing to Identify the Process Model

fashion, to allow the process response to be observed and collected for model identification.

> **Pseudo-random** - A series of changes that appear to be random but are in fact generated according to some prearranged sequence.

After testing is complete and the user selects the Autogenerate button, the manner in which the process input and output varied during the test are automatically analyzed, and the step response model and MPC controller are generated.

MPC is most often used to control processes having multiple inputs and multiple outputs. When MPC is applied to a MIMO process, the automated testing introduces changes in all manipulated process inputs, and data are collected for all process inputs and outputs to support model identification. The display of the step response model is designed to show the step response of each output to a step change in each input when all other inputs are held constant. For example, if the MPC block supports a process with as many as four manipulated and four disturbance inputs and four controlled parameters and four constraint outputs, the identified step response display for a single input and a single output might appear as illustrated in Figure 11-5.

The user can see a larger view of a specific step response by clicking on the step response.

When the MPC block is configured for multiple process inputs and/or outputs, the application shows the impact of a 1% change for each manipulated or disturbance input (with all other inputs constant) on each process output. When an MPC block is used to control a process that has eight process inputs and eight process outputs, the step response model would contain 64 step responses. In such a case, the process might be referred to as an 8×8 process to indicate (the number of process inputs) × (the number of process outputs).

When a step response model has been identified by testing the process, a natural question to ask is, "How accurate is the identified step response model prediction?" To validate the identified step response model prediction, the changes in the manipulated and disturbance input that occurred during testing can be passed through the model. The response calculated using the model can then be plotted against the actual process output response as illustrated in Figure 11-6.

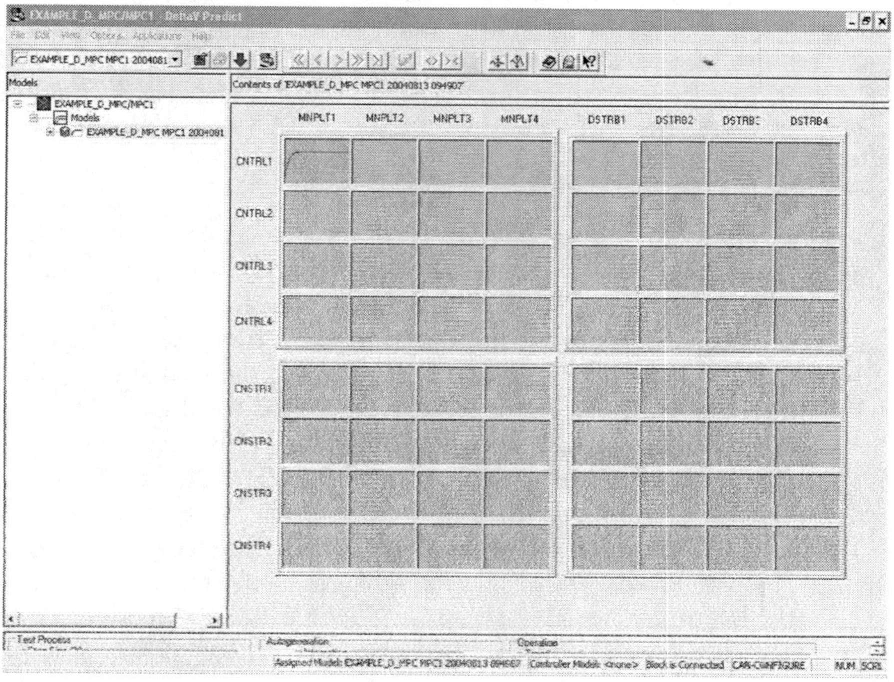

Figure 11-5. Step Response Model Display

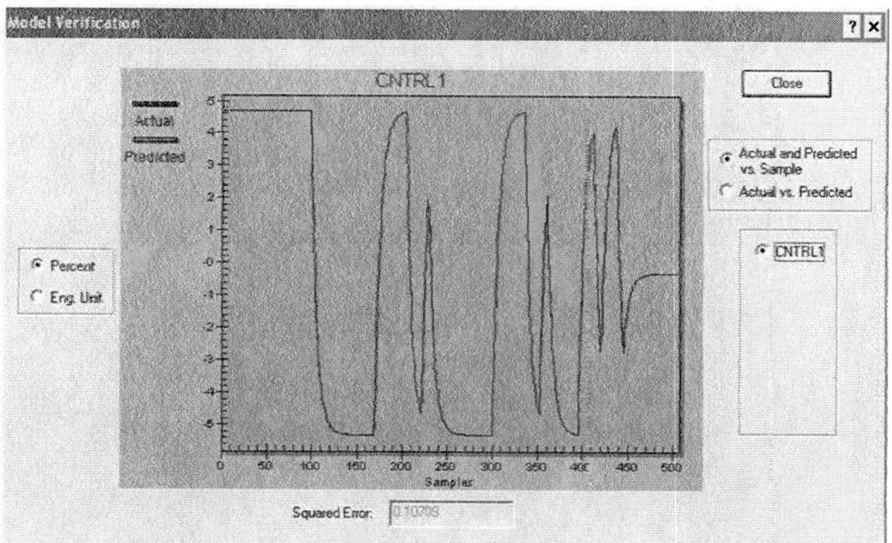

Figure 11-6. Verification of Identified Model

In this example, for the process output selected in the right-hand pane of the display, there is little noticeable difference in the calculated and the actual value of the process output. This indicates the model identified by the MPC application accurately predicts the process output response to changes in the process input.

Similar to the PID tuning application, the MPC application can be designed to allow the control response to be observed before downloading the module that contains the MPC block with the identified step response model. By selecting Simulate, it is possible to view how the process will respond to a setpoint change. One of the unique features of model predictive control is using the model to predict process outputs internally in the MPC block based on changes in the process inputs. This predicted response may be shown in the operator interface to an MPC block and is also shown when the MPC block is viewed in the simulation environment. Such a prediction is especially valuable for operator guidance when one or more manipulated variables are in MAN (Manual), that is, not being manipulated by MPC. In the example shown in Figure 11-7, the predicted response of the process is shown in a lighter area on the right side of the trend plot.

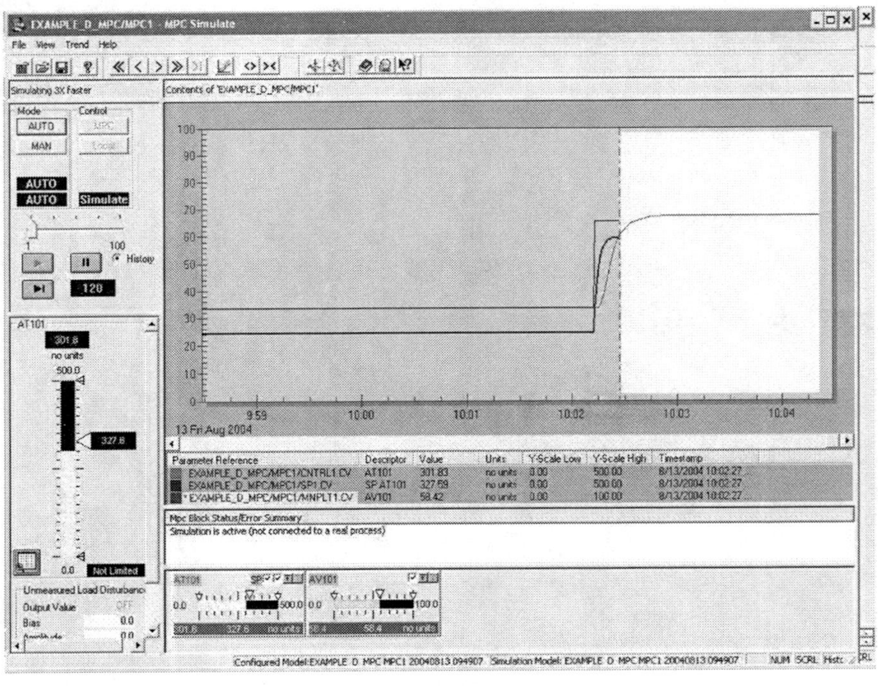

Figure 11-7. Testing of Control Using Simulated Environment

If the process time constant and deadtime are very large, a change in a process input may not be reflected in the process response for many minutes or even, in some cases, for many hours. Thus, the MPC simulation environment may be designed to allow the process response to be displayed much faster than in real time. In the example application shown in Figure 11-7, the slider in the upper left corner of the screen may be used to view the process response as much as 180 times faster than in real time.

To enable users to view and work with MPC control in exactly the same way as PID control, some manufacturers provide pre-built dynamos (dynamic elements) that can be included in the operator display. From the operator's perspective, single loop MPC control may be used in the exact manner as single loop PID control. Also, to support an engineer who is working with an MPC application, manufacturers may supply pre-built dynamos that are very similar to the interface provided for simulation with the exception that the speedup capability is removed, as illustrated in Figure 11-8.

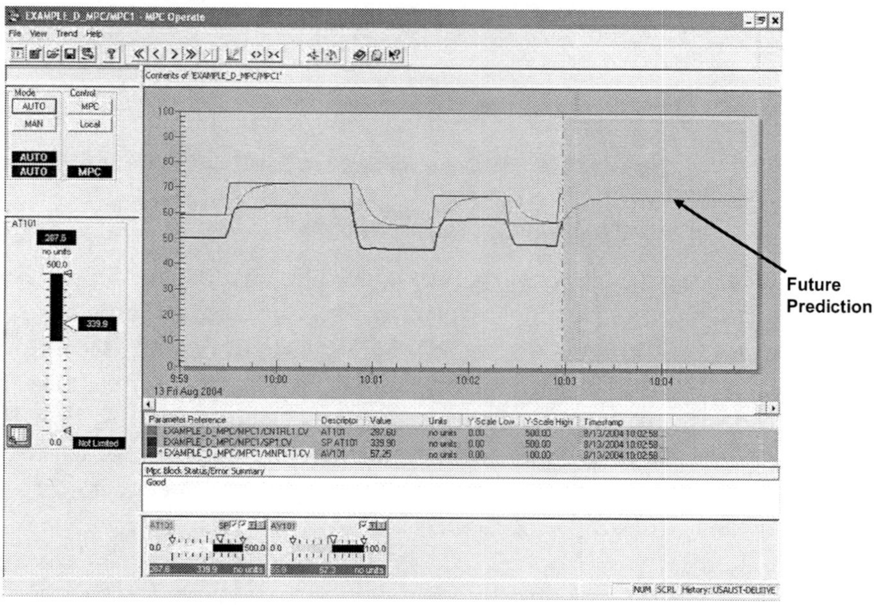

Figure 11-8. Engineer's Interface to MPC

An engineer who is accustomed to working in the MPC simulation environment may use this on-line interface to view and work with an MPC block that is controlling the process.

11.3 MPC Replacement for PID with Feedforward

If a disturbance to a process can be measured, this input may be incorporated into the MPC application as illustrated in Figure 11-9. In this way, MPC control may be used to address applications that are traditionally addressed by adding a feedforward input to the PID. When a measured disturbance is connected as an input to the MCP block, the MPC application may be used to automatically identify the impact of the disturbance on the process output. By including the measured disturbance as an input, the MPC block can automatically compensate for the disturbance and thus better maintain the controlled parameter at setpoint.

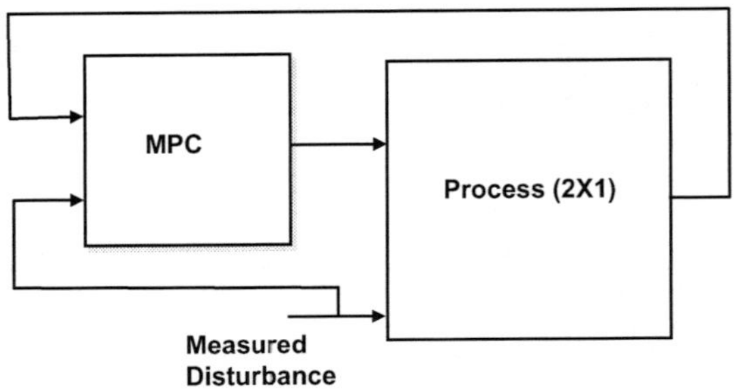

Figure 11-9. MPC Replacement for PID with Feedforward

When a process is characterized by one or more measured disturbance inputs, the MPC block inputs are expanded to include a disturbance input by right clicking on the MPC block and selecting Extensible Parameters. The same procedure is used to add other types of inputs and outputs to the MPC block. In response, the block view is automatically updated to reflect a connection point on the MPC block. An example of where one disturbance input has been added to the MPC block is illustrated in Figure 11-10.

The default configuration information associated with MPC block inputs or outputs may be viewed or changed by right clicking on the block and choosing Properties. In response, a dialog box is presented for viewing and changing the associated configuration; for example, Setpoint high and low limit or Output low and high limit.

MPC configuration typically involves significantly fewer parameters than PID configuration of an equivalent control strategy. Because the MPC block is generated from the step response model, corrections are

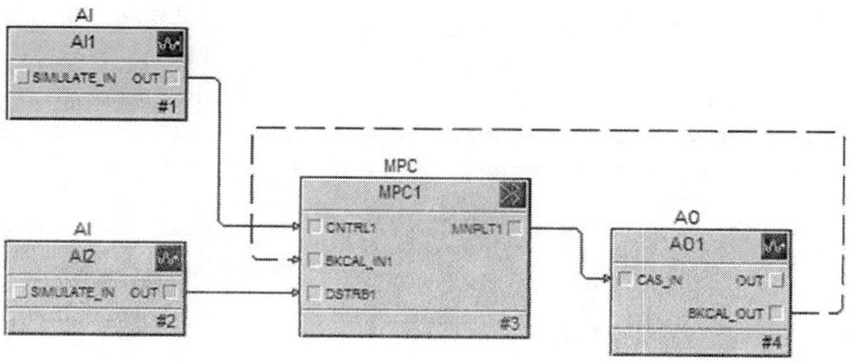

Figure 11-10. Example MPC Implementation for One Measured Disturbance Input

automatically made for any differences in the controlled parameter response to changes in the disturbance inputs and the manipulated process inputs.

11.4 MPC Replacement for PID Override

When a process output is a constraint variable in PID feedback control, the constraint variable measurement is simply added as an input to the MPC block as illustrated in Figure 11-11. The design and implementation of an override control strategy using an MPC block are thus much simpler than implementing override control using two PID blocks and a control selector block.

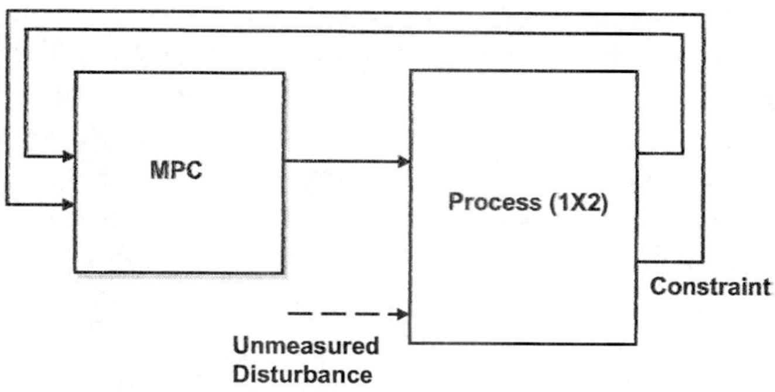

Figure 11-11. MPC Replacement for PID Override

When a measured constraint is included as an input to the MPC block, the step response of the constraint variable to changes in the process input(s) is automatically identified when the MPC application is used to test the process and to generate the step response model. A constraint variable is added to the MPC block by right clicking on the block and selecting Extensible Parameters. The view of the block is updated to include the constraint input as illustrated in Figure 11-12.

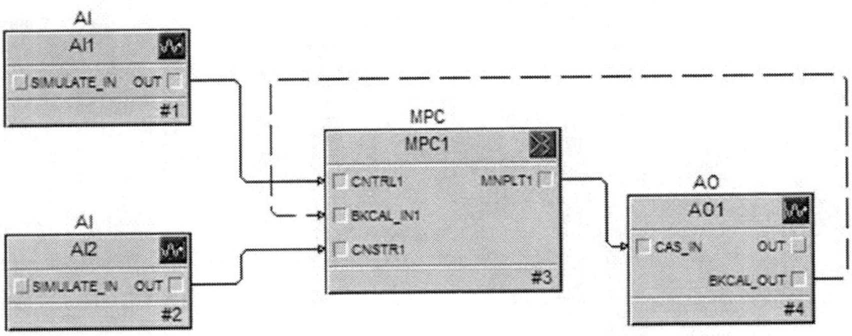

Figure 11-12. MPC Constraint Control

Because the process step response is used to generate the MPC controller, the responses of constraint variables and controlled parameters to changes in the process inputs are automatically taken into account to minimize variations from setpoint while observing the constraint limits.

11.5 Using MPC to Address Process Interactions

When a process is characterized by multiple manipulated process inputs and multiple controlled process outputs, there is potential for process interaction, that is, a change in one manipulated input may impact multiple controlled outputs. As we have seen, detuning the PID controller is the most common way to deal with significant interaction between control loops. When MPC is used, the interaction of the manipulated inputs and controlled outputs may be addressed by one MPC block as illustrated in Figure 11-13.

Because the impact of each manipulated input parameter on each controlled output parameter is identified by the step response model used in MPC block generation, the MPC block automatically compensates for any interactions.

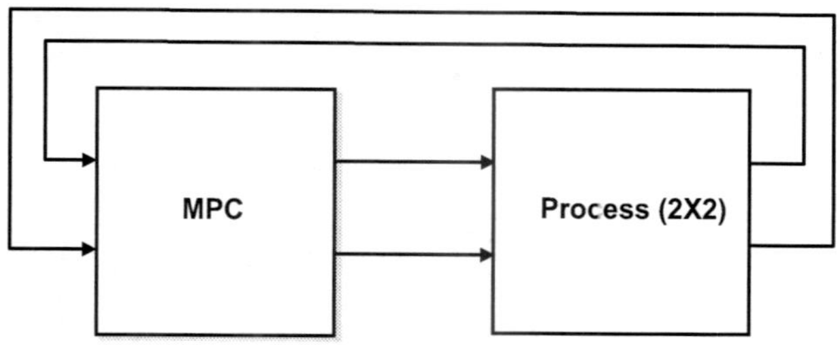

Figure 11-13. Addressing Process Interaction Using MPC

Manipulated process inputs and controlled process outputs are added to an MPC block by right clicking on the block and selecting Extensible Parameters. In response, the view of the block will be updated to show the added inputs and outputs. When two manipulated inputs and two controlled outputs have been configured, the MPC block and its connections to associated input and output blocks appear as illustrated in Figure 11-14.

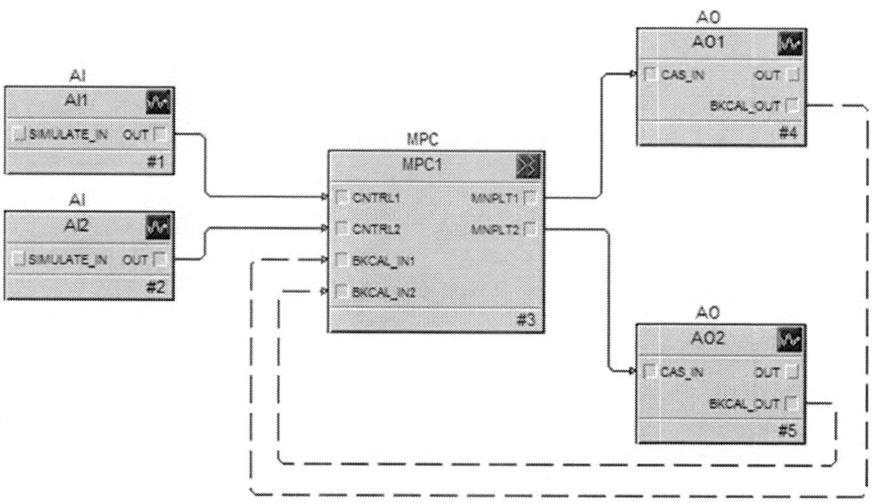

Figure 11-14. MPC Implementation for Interactive Processes

A typical step response model for a process with two manipulated inputs and two controlled outputs is illustrated in Figure 11-15. In this example, manipulated input 1 has a much larger impact on controlled output 1 than on controlled output 2, while manipulated input 2 has a larger impact on controlled output 2. The step response model shows that a change in either input has an impact on both controlled outputs, therefore significant process interaction exists.

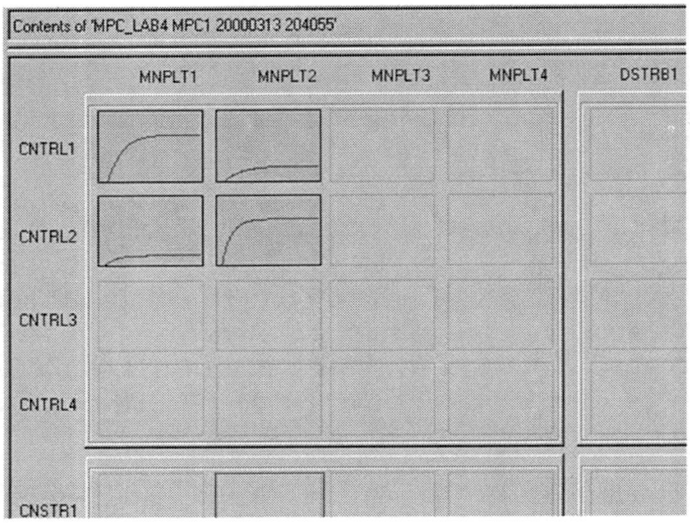

Figure 11-15. Process Model for Interactive Process

Because the step response model is used to generate the MPC block, these interactions are automatically taken into account. Thus, each controlled parameter may be independently maintained at setpoint without interaction.

11.6 Application Examples

Model Predictive Control has been applied to many different processes. Table 11-1 lists a small sample of processes in which simple MPC has been applied to replace traditional control strategies based on PID.

Table 11-1. Examples of Simple MPC Applications

Control Application	Industry
Evaporator	Chemical
Lime Kiln	Pulp & Paper
Pipeline Gas Blending	Power
Bleach Plant	Pulp & Paper
pH Control	Pulp & Paper
Distillation Column	Pharmaceutical

As noted, the process industry has been slow to adopt MPC to replace PID in smaller process applications. In addition to the reasons cited earlier, one reason may be there is often not enough time in a project to try something new. Also, only the latest distributed control systems support MPC at the controller level. However, little by little, plant by plant, the use of MPC for small applications is gaining momentum. When MPC is available as a function block in the control system, implementation is much easier and less costly than using a computer layered on top of the control system.

11.6.1 Evaporator Control

Model Predictive Control has been effectively used in the control of evaporators. Evaporators are used for the efficient concentration of liquids in the food, chemical and pharmaceutical industries. Most frequently used are falling film evaporators, which are particularly suitable for heat sensitive products. The heating of the evaporator is typically done by live steam or by thermal vapor recompression. The operation of a single effect evaporator that employs one stage of evaporation is depicted in Figure 11-16.

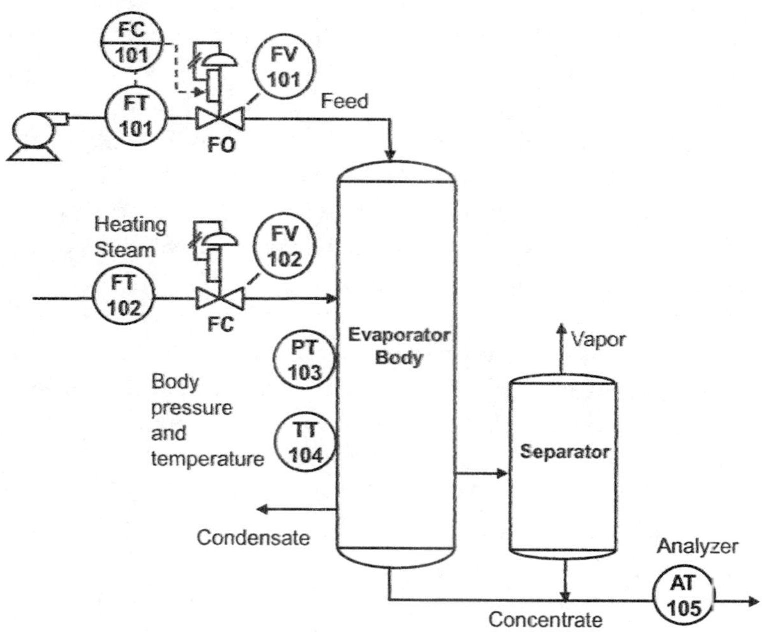

Figure 11-16. Single Effect Evaporator

Liquid to be concentrated is preheated to its boiling temperature while flowing downward. This evaporator process includes product density as a controlled variable and temperature and pressure as constraint variables. The conventional solution was based on three PID controllers and posed problems with interactions, tuning and decoupling. A simple MPC controller replaced the PID controllers. The MPC controller proved to be easier to develop and operate and the application of MPC improved product uniformity and reduced energy consumption.

11.6.2 Dryer Control

Spray dryers are used in the chemical, pharmaceutical, food and dairy industries to manufacture a wide variety of products. In a spray dryer process, a slurry feed is sprayed into the dryer and the liquid component is removed by a hot gas stream to create a powdered product. The dryer design and the type of spray nozzles (or atomizers) are selected based on the properties of the slurry to be dried and the desired characteristics of the powder.

The slurry feed rate is normally set by the operator. However, if the spray nozzles begin to clog, the pressure to the nozzles will increase. Under this condition, the feed flow rate may be automatically reduced using pressure override control to help maintain consistent particle size in the spray.

Any reduction in slurry flow rate will impact the air temperature required to dry the material. The air flow rate to the dryer is based on the target slurry feed flow rate and the target product moisture. The temperature of the air to the dryer is automatically controlled by regulating the air heater fuel input.

Any significant change in the slurry concentration, slurry feed rate and spray pressure will impact the residual moisture content of the product. If the product's residual moisture content is measured on-line, moisture control may be cascaded to the air heater temperature loop as illustrated in Figure 11-17. The air temperature is automatically adjusted to bring the moisture content back to setpoint.

When the moisture content is only available from the lab, the lab measurement may be used by the operator to manually change the air heater temperature setpoint.

In some facilities, Model Predictive Control has been successfully used to automatically account for the impact of these changes on product residual moisture as shown in Figure 11-18.

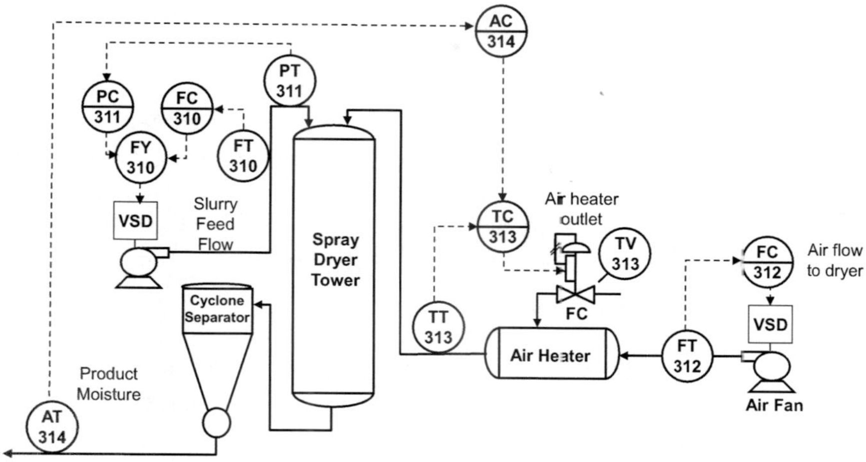

Figure 11-17. Typical Spray Dryer

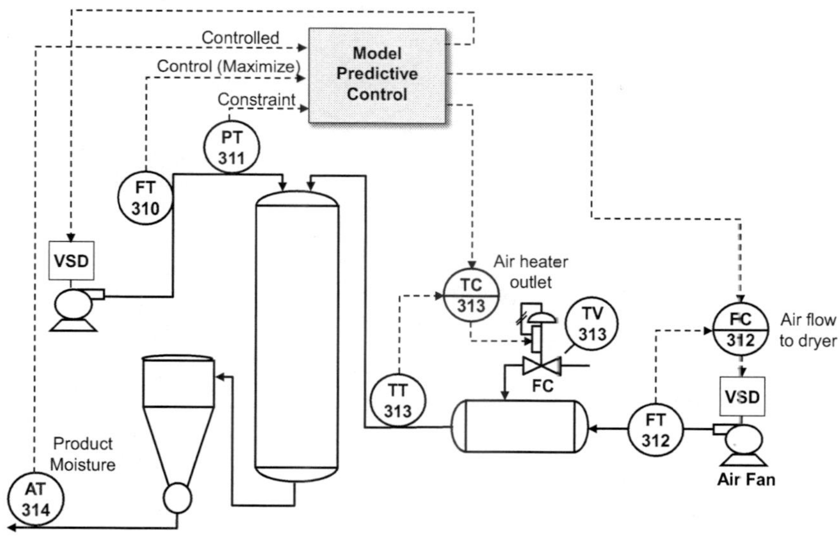

Figure 11-18. MPC Configuration for Dryer

Through the use of Model Predictive Control, it is also possible to maximize dryer throughput by maintaining the feed rate at a value that maintains the dryer at its spray pressure constraint. In addition, any delay between a change in a process input and when it is reflected in product moisture content is automatically compensated for when Model Predictive Control is used to automate the spray dryer.

11.6.3 Rotary Kiln Control

An example of a medium-size MPC application is rotary kiln control. Rotary kilns (as illustrated in Figure 11-19) are used widely in the thermal processing of solids for cement production, calcination and the reduction of metal ores. They are essentially refractory-lined steel tubes that rotate slowly, typically at 0.5 to 2 revolutions per minute. The kiln is fired by fuel and air, which are usually injected at the lower end. The flame travels up the kiln, counter to the solids, heating them directly by radiation and conduction and indirectly through contact with the refractory lining.

For calcination, the feed consists of a source rich in calcium such as limestone. For the cement process, the feed also contains silica. The processing occurs in three zones:

> **Drying zone,** where the water evaporates from the raw material slurry.
>
> **Calcining zone,** with temperature up to 2200 °F. The intense heat causes a reaction that produces lime:
>
> $$CaCO_3 \longrightarrow CaO + CaO_2$$
>
> **Burning zone,** where calcination, or in the cement process, clinker formation is completed.

Gas or oil may be burned in the kiln. A forced draft (FD) fan is used to provide some of the air required for combustion and to shape the flame within the kiln. However, most of the air for combustion is normally pulled into the kiln by the ID (Induced Draft) fan and is preheated as it travels over the hot material leaving the kiln. The ID fan also must maintain a negative draft in the kiln to prevent hot gases from blowing back through access ports in the firing end of the kiln. A two-color pyrometer is used to measure the material temperature within the lower end of the kiln. The exit gas temperature and oxygen content are measured as illustrated in Figure 11-19.

To maintain the correct flame shape as fuel input changes to maintain kiln temperatures with varying slurry feed rate, the FD fan speed may be characterized as a function of fuel input using a characterizer block, FY430.

For best conversion, the temperature of the material leaving the kiln must be maintained above what is required for the calcination process. For example, the temperature must be maintained above the dissocia-

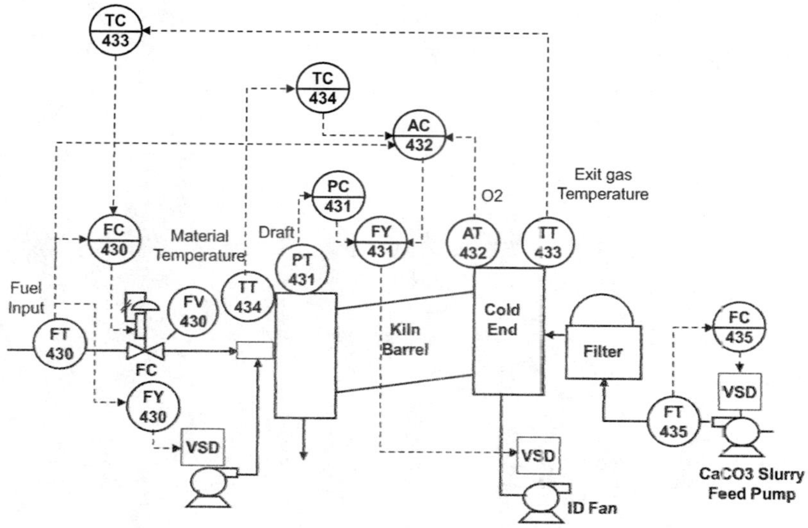

Figure 11-19. Rotary Kiln Process

tion temperature of the carbonates in the lime mud, which is between 1436 °F and 2444 °F. Any imbalance in the amount of energy supplied by the fuel and the energy required for drying and to support calcination is reflected in the temperature of the gas leaving the kiln. The fuel input is regulated to maintain the exit gas temperature at the value that is required to dry and process the slurry feed. Because the total air flow in the kiln is not measured, the ID fan is regulated to maintain a target O_2 level in the gas leaving the kiln. To ensure a negative pressure is always maintained in the kiln, actions by the O_2 controller may be overridden by the pressure controller. The lime kiln inputs and outputs are shown in Figure 11-20.

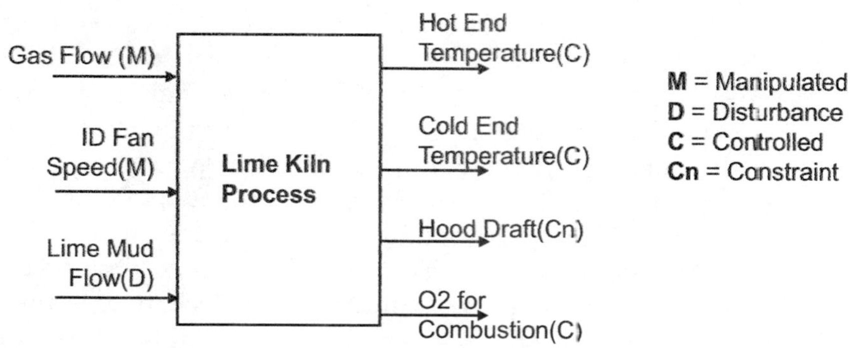

Figure 11-20. Lime Kiln Process Inputs and Outputs

Because the flame temperature is a function of the excess air level, the material temperature is maintained at setpoint indirectly by regulating the O_2 setpoint over a narrow range that ensures complete combustion. The process responses for step changes in the process input are shown in Figure 11-21.

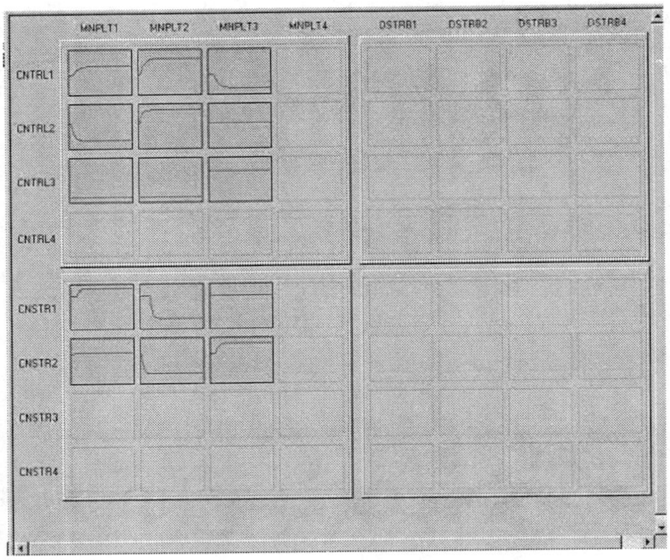

Figure 11-21. Lime Kiln Step Responses

Because of the interactions and operating constraints associated with the lime kiln process, it is often possible to increase throughput and improve product quality and operating efficiency through the use of Model Predictive Control.

However, as shown in the step response, the lime kiln process is characterized by the process outputs responding in very different time frames to changes in the process inputs. Because the MPC controller must be generated based on the slowest responding output, the associated MPC execution rate may not be fast enough to address the requirements associated with the fastest changing output. In such cases, it is possible to break the control into two MPC blocks, one addressing the fast responding outputs and the other addressing the slow responding outputs. For example, the lime kiln process hot end temperature responds much more slowly to changes in fuel and air flow than the cold end temperature, kiln O_2 and draft pressure. Because the O_2 and hot end temperature are highly correlated, it is possible to use the O_2 measurement as a controlled parameter of an inner loop MPC that is manipulated by an outer loop MPC, as illustrated in Figure 11-22.

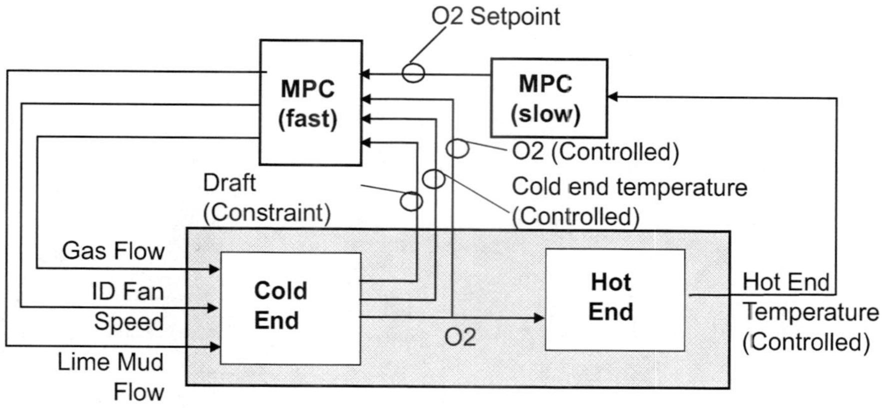

Figure 11-22. Cascade MPC Kiln Control

Through the application of MPC to the lime kiln process, the following benefits have been reported (Chmelyk, 2002):

- Increase in throughput/capacity
- Decrease in energy consumption
- Decrease in environmental emissions
- Improved product quality through a uniform residual alkali level
- Increase in lime availability, active calcium oxide weight ratio of total lime

11.6.4 Pulp Brightness Control

In most MPC implementations, it is assumed the process may be characterized as linear and time invariant. If these assumptions are true, the control generated from a process model identified during normal plant operations will perform well. However, if the process dynamics change from those captured by the model, these changes may degrade the control that is based on the model.

In some cases, the process feed rate may change the delay time required for a process output to reflect a change in a process input. Such variations in the process response can often be traced to the effect of process feed rate on the transport delay through the process. For example, the time required for pulp stock to move through a tower in the wood pulp bleaching process is strictly a function of the feed rate, assuming plug flow through the tower. Thus, the time required for the brightness sensor (located after the tower) to indicate changes in chemical addition that take place ahead of the tower varies with the pulp stock flow rate as illustrated in Figure 11-23.

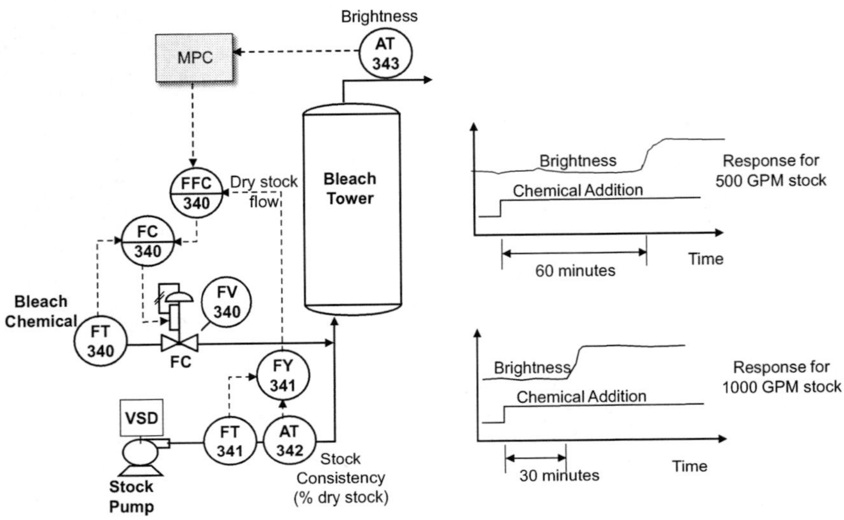

Figure 11-23. Impact of Throughput–Bleach Tower Example

For this example, the brightness control strategy using MPC might be generated using the step response model identified at a stock flow rate of 500 GPM. Brightness control would be degraded at flow rates other than 500 GPM because the model used to generate the control matches the process only when the flow rate is 500 GPM.

11.6.5 MPC Control at Changing Production Rate

The primary objective of many traditional MPC installations within the process industry is to maximize the throughput of a large unit operation, for example, a distillation column. Using the ability of MPC to maintain a process at its operating constraint, it is often possible to achieve a 1–3 percent increase in throughput in such installations. For these operating conditions, the throughput to the process is expected to vary over a very small range. Thus, the impact of throughput changes on the process response is often negligible during normal operation for a particular product. An MPC controller generated from a process model identified at normal throughput could provide good control when the feed rate is varied over a narrow range. However, for startup and upset conditions and product changeovers, feed rates can change drastically.

Drastic changes in feed rate can occur because the instantaneous production rate of different unit operations usually does not match up, and the surge volumes (usually in the form of a surge tank) are often too small. As a result, problems anywhere in the plant can ripple through as feed rate changes. There are also many non-traditional installations, such as brightness control in the pulp and paper industry (as just discussed), paper machine MD (Machine Direction) control, H/N (hydrogen/nitrogen) ratio in the ammonia industry, and MD control for webs and sheets that are characterized by large changes in throughput (Butler, Cameron, Brown and McMillan, 2001). These types of applications have large changes in the process transport delay as a function of changes in process throughput. Thus, to provide best control under all operating conditions, it is necessary to modify MPC control to match this varying delay caused by large changes in throughput.

Many MPC products address the changes in the process response by providing the ability to switch controllers. To use this technique, a process model is identified for each operating condition. A controller is then generated for each of these process models. As processing conditions change, the MPC control is designed to switch controllers to match the current process condition, as illustrated in Figure 11-24.

Controller switching is a general approach that allows any nonlinearity that is a function of a measured process input to be addressed. However, to apply this technique, multiple process models must be identified to generate control at different operating conditions. In addition, the introduction of multiple models and control definitions increases the complexity of commissioning and maintaining the MPC operation.

When changes in time delay as a function of throughput are the primary variation in the process response, it is possible to avoid the complexity of control model switching. By comparing the step responses of such a process at different throughputs, some insight may be gained into how this may be accomplished, as shown in Figure 11-23. By examining the two step responses, it may be seen that the only difference is the response is spread over a different time to steady state, that is, the coefficients of the step response are the same, but the time difference between the time samples is inversely proportional to the feed rate. Therefore, the control matrix generated for models created at dif-

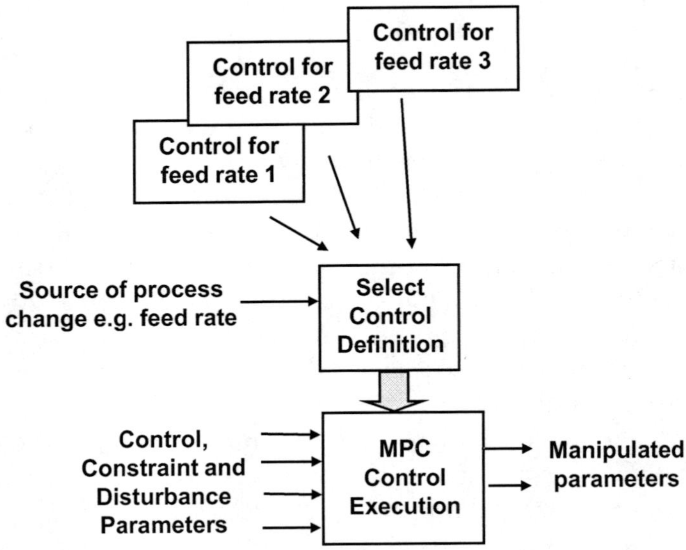

Figure 11-24. Controller Switching to Match Changes in Process Condition

ferent throughputs would be the same except for the fact that the control execution rate would be different (because the model time to steady state is different). Thus, for this type of process, adjusting the control execution period compensates for the variation in transport delay with the variation in throughput–Equation 11-1a.

$$\text{Execution Period} = (\text{block execution period}) \times (\text{execution modifier})$$
$$= (Tss/\ Nc) \times (Fn/Fa) \qquad (11\text{-}1a)$$

where:

Tss	=	Time to steady state of the model used to generate the control
Nc	=	Number of coefficients used in the model
Fn	=	Throughput of process when data was collected to generate the model
Fa	=	Current throughput to the process

In any practical implementation, the calculated execution period must be limited and restricted to values supported by the control system. In addition, the current throughput value would need to be heavily filtered to eliminate the effects of process noise.

To allow the user to control processes that are characterized by a variable deadtime, an execution modifier block can be provided with the MPC block. Figure 11-25 shows an example of how this execution modifier block is designed to automatically change the MPC function block execution period as a function of a throughput input.

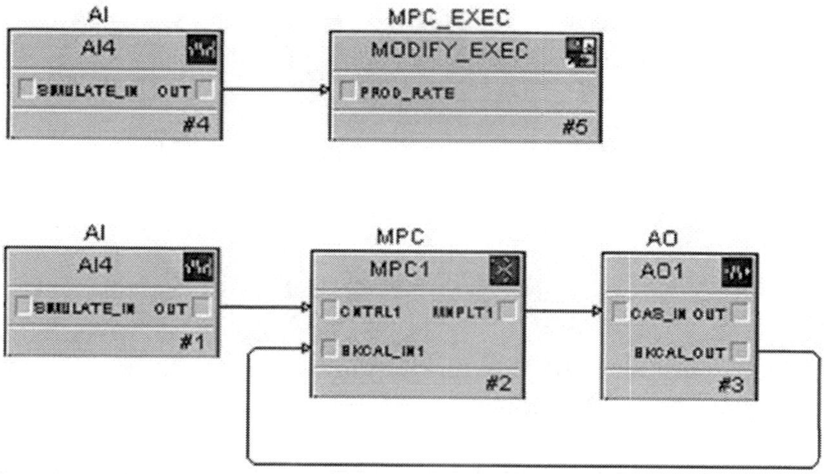

Figure 11-25. Example MPC Implementation that Addresses Variable Deadtime

If the adopted MPC technology does not allow model swapping, online deadtime adjustment or controller execution rate adjustment, and the model is known to cover a range of deadtime, a more robust MPC controller can be designed by overestimating the deadtime. While an MPC controller with a modeled deadtime that is longer than the actual process deadtime will be sluggish in performance, a controller with an underestimated model deadtime can create cycling and stability problems.

11.6.6 Batch Reactor Control

Batch chemical reactors may be used to manufacture a variety of products. In a typical application, the reactor is initially charged (filled) with one or more chemicals that do not react with each other, the contents of the reactor are heated to a target temperature and then a metered quantity of one or more reactive chemicals is added to the batch. The chemical reaction may be exothermic, so cooling may be required to maintain the reactor charge temperature at the target temperature. A different reaction may be endothermic, and heating will be required. In other dedicated reactors, in order to support both heating and cooling, temperature control is implemented as a split-range control loop that adjusts heating and cooling media flow to the reactor shell. The reactor pressure is maintained at setpoint by venting vapors and gases generated by the chemical reaction.

With an exothermic reaction, when the flow rate of the reactive chemical(s) to the reactor increases, more cooling is required. If the cooling requirements exceed what the cooling media can provide with the cooling valve wide open, the batch temperature will exceed the temperature controller setpoint—TC211 in Figure 11-26. If the temperature continues to rise, when the temperature reaches the setpoint of the override controller (TC211A), the feed flow rate to the reactor will be reduced to prevent the temperature from exceeding the override temperature setpoint.

If the production of gas and vapor generated by the reaction exceeds the capacity of the vent system, then when the pressure reaches the setpoint of the override controller (PC212), the feed flow rate to the reactor will be reduced to prevent the pressure from exceeding the override pressure setpoint. The control loops and measurements commonly found on a batch reactor are shown in Figure 11-26.

The feedstock tank and valve manifold used to initially charge the reactor are often shared among multiple reactors. Thus, the timing of the reactor charge must be coordinated by the batch control software. The sequence of feed addition to such a batch process and potential for reducing batch time may be illustrated as shown by arrows in Figure 11-27.

The feed rate of exothermically reactive chemical(s) to a batch chemical reactor is often set by the operator to a value that ensures that the capability of the cooling and vent systems will not be exceeded, that is, the pressure and temperature overrides will remain inactive throughout the batch. Thus, at times a higher feed and throughput rate could be achieved because the capacity of the cooling and vent systems, which support multiple reactors, may vary with plant operating conditions, for example, cooling media temperature. Using MPC, it is possible to operate at the temperature and pressure operating constraints to achieve the maximum possible feed rate and reduce the batch cycle time as illustrated by arrows in Figure 11-27. The configuration of MPC for batch reactor control is shown in Figure 11-28.

The capability of MPC to actively maintain the process at a constraint is often provided in small MPC applications through an approach

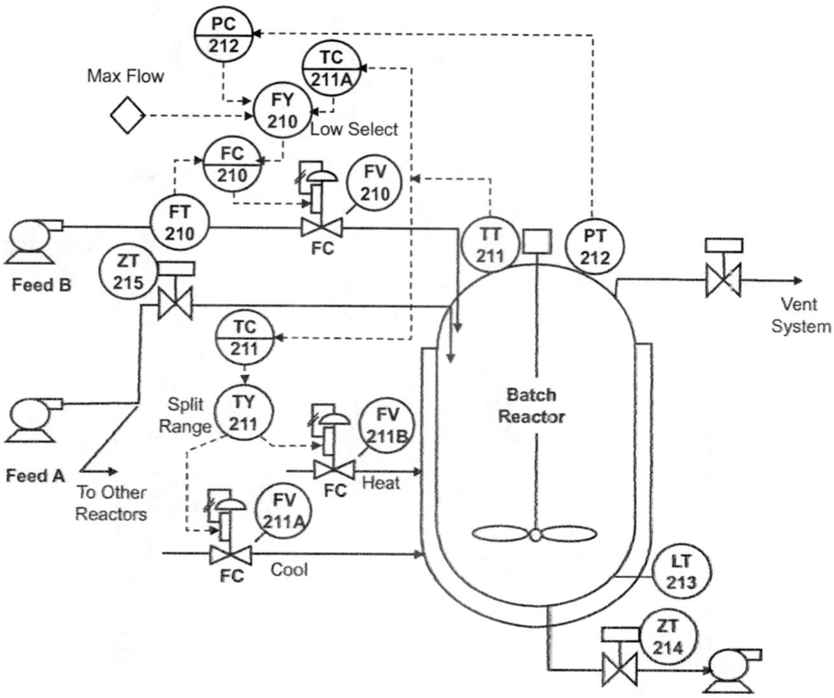

Figure 11-26. Traditional Approach to Batch Reactor Control

known as a *pusher*. For example, in the case of the reactor, the throughput would be selected during MPC configuration as the parameter to be optimized (in this example, maximized). When the operator increases the target throughput rate, the MPC controller will automatically increase throughput ("push") until the target is achieved or is at a value that maintains the process at one or more operating constraints.

11.7 MPC Application Development Procedure

Based on the MPC applications presented in sections 11.1 – 11.6, the procedure for developing typical small and medium-size applications may be generalized as outlined in Figure 11-29.

The procedure for MPC application development consists of the following steps:

- Process analysis, MPC configuration design and creating MPC module
- Process testing using automatic test signal generator
- Process model identification

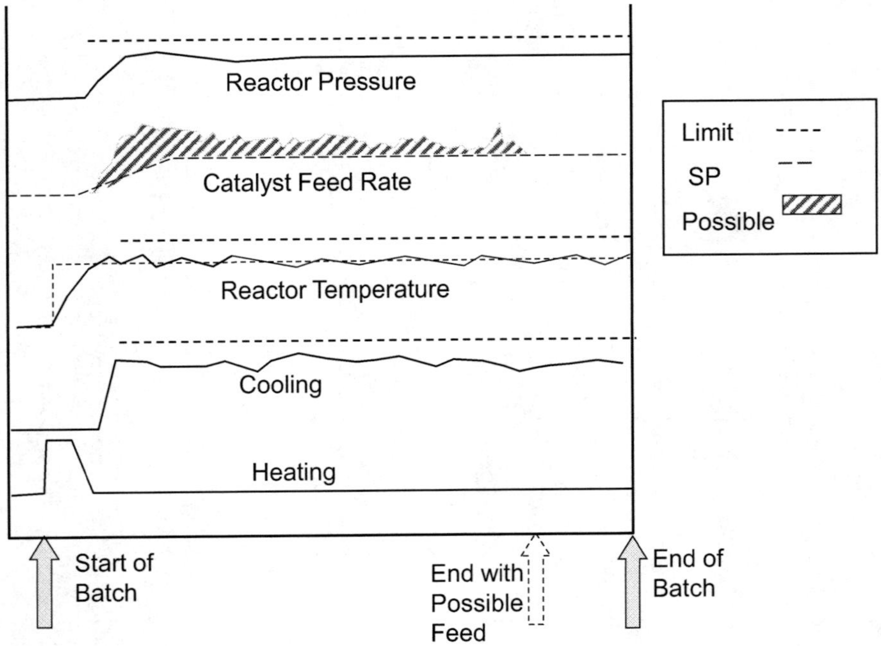

Figure 11-27. Potential for Reducing Batch Time by Applying MPC

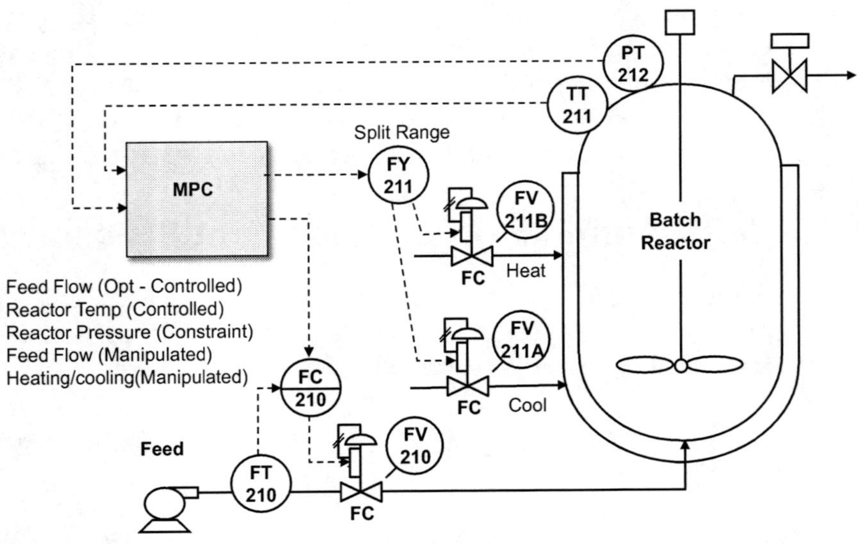

Figure 11-28. MPC Configuration for Batch Reactor Control

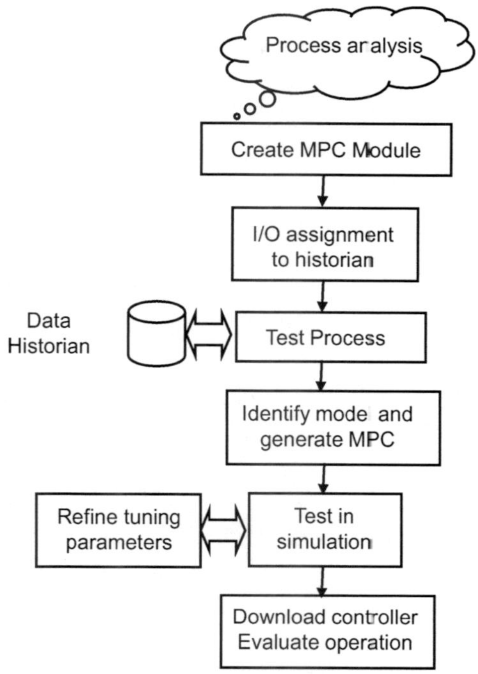

Figure 11-29. A Typical Procedure for Developing an MPC Application

- Controller generation
- MPC simulation and tuning validation and refining
- MPC control evaluation and tuning adjustment

The details of the procedure are discussed below.

11.7.1 Process Analysis and MPC Configuration Design

Process analysis should deliver a clear formulation of process control objectives and limitations as well as a means for achieving those objectives in MPC configuration design.

The process inputs and outputs are grouped into five different categories based on how they are used in the control of a process:

> *Manipulated Variable (MV)* – Process input that is adjusted to maintain a controlled output at setpoint.
>
> *Controlled Variable (CV)* – Process output that is maintained at a specific value (for example, the setpoint).

Disturbance Variable (DV) – Process input that might also affect the value of controlled outputs.

Constraint Variable (AV) – Process output that must be maintained within an operating range by restricting the adjustment of manipulated inputs.

Optimized Variable – Process input that decisively affects the unit production rate, coupled with a "shadow" controlled variable. The optimized variable is a subcategory of manipulated and controlled variables.

When MPC is used to replace the primary loop of a cascade strategy, the PV of a primary cascade loop becomes an MPC controlled variable. Therefore, the SP of a secondary loop would be a manipulated variable of the MPC configuration. Otherwise when interfacing to a single loop MPC directly manipulates the loop SP. This function block mode and block structure facilitates an easy switch between MPC and PID cascade control.

There are typically more process inputs than controlled variables. These extra process inputs are considered to be disturbances (measured or unmeasured). If possible, all inputs that affect the controlled outputs should be measured and included in the model. If there are extra inputs that can be manipulated (i.e., changed by the control system), it is possible to select only those inputs that most affect the production rate and couple these inputs with a shadow controlled variable called an optimized variable. When a constraint variable is defined, it should be associated with a controlled variable. The associated controlled variable setpoint will be managed to drive the constraint variable out of limits.

A review of the existing regulatory controls and instrumentation is also an important part of the process analysis step, because they are the foundation of any MPC control strategy. Any shortcomings in design or tuning must be addressed before the testing stage. All regulatory loops, whether or not they are included as manipulated variables, should be checked and tuned. The selected tunings will affect the response of the process and are therefore considered part of the process from the MPC perspective.

11.7.2 Process Testing

During the test, the manipulated and measured disturbance inputs to the process must be changed. Traditionally, testing of the process has been performed using manual manipulation of these process inputs.

Most MPC products generate pseudo-random testing sequences automatically, with only one parameter (process time to steady state) set by the user (Figure 11-30).

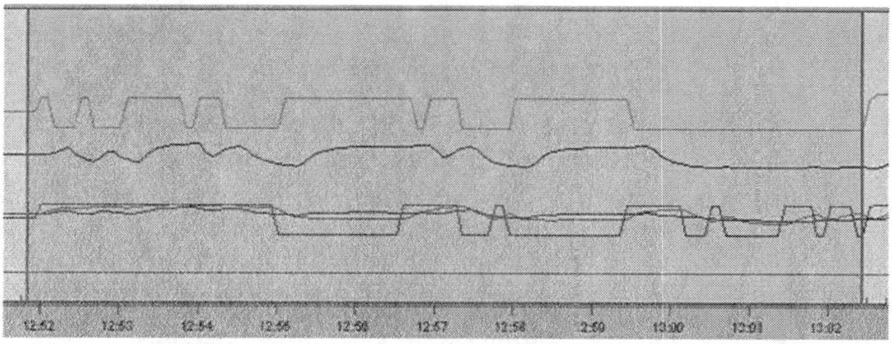

Figure 11-30. Testing a Process Applying Pseudo Random Binary Sequence (PRBS) Changes

11.7.3 Process Model Generation

Once data are collected, it is important to review the values of the data and remove data from any periods of time that do not reflect normal operating conditions. Then, the process model and control can be generated automatically. It is a good practice to examine the process model that was identified and validate it using practical knowledge of the process.

The process model is displayed as process step responses. The step response for each process input shows the impact of a unit process input change on the process outputs (with all other inputs maintained constant). If the process input does not change sufficiently during the time that data was collected, the gain or dynamics identified for that input might not be correct. Process model accuracy may be examined using the verification features of the MPC product (for example, a comparison of the step responses calculated using two different techniques, Finite Impulse Response (FIR) and Auto Regressive with eXternal inputs (ARX). Comparable results for these different techniques are to be expected when the model is accurate.

In addition, the MPC identification tool provides a comparison of the measured process output values to those calculated based on the process inputs. If a significant error is identified in the model during its verification, the easiest solution may be to collect more process data and perform model identification and verification again. However, as in the case of disturbance inputs, further collection of data may not lead to a better result. Therefore, correcting the associated response models based on process knowledge would be a practical solution. The MPC software provides options for editing the step response graphically or numerically by entering the step response parameters (e.g., gain, lags, and deadtime).

11.7.4 Controller Generation

The MPC controller is generated using the verified process model. Both the speed of MPC control response and controller robustness are influenced by the parameters used in controller generation. During controller generation, these parameters can be set automatically based on the process model, and there is no need for user involvement. To satisfy specific requirements, the user may adjust the following controller generation parameters:

> *Penalty on Move:* Control sensitivity to changes in a process is determined by the controller robustness. The Penalty on Move (PM) parameter used in controller generation has the most impact on robustness. The PM defines how much the MPC controller is penalized for a change in the manipulated output (MV). The PM parameter is typically defined independently for every MV. High PM values result in a slow controller with a wide stability margin. With such settings, control is relatively insensitive to changes in the process parameters over time or to model errors. Low PM values result in a fast controller with a narrow stability margin.
>
> The PM value most affects the controller performance when the model does not match the real process.
>
> *Penalty on Error:* The Penalty on Error (PE) factor allows more importance to be placed on a specific controlled variable. Typically, the default value for the PE gives equal weight to all controlled parameters and provides good control for most applications. If the control strategy clearly indicates that one of the controlled variables should be of lower priority, the associated PE can be decreased. However, in general the PE should not be used to change overall control performance. In most cases, it is recommended to change the PE only after adjusting the MPC controller using the PM and then testing MPC control in simulation. The primary crite-

rion for adjusting the PE is acceptable variability on a specific controlled parameter.

In addition to the parameters used during controller generation, the user can influence control behavior and robustness during on-line operation. One way is to apply a reference trajectory (i.e., the filtered predicted setpoint value), which can also be used for funnel or range control (Figure 11-41). In funnel or range control, the process predicted CV values that are within an acceptable setpoint range or within the funnel are not penalized (B, C area), thus PE is applied to the values outside the funnel.

11.7.5 MPC Simulation and Tuning Validation

Before the new controller is tried on the real process, the user must observe how well the control responds to setpoint changes and load disturbances in simulation. The model used in simulation is the one identified for the process. However, by using a modified model, the user can test the control response to changes in the process.

The ability to introduce disturbances and test MPC control for changes in process gain or dynamics is extremely valuable for testing the robustness of the control. If the process response is too aggressive, it is easy to improve the results by modifying the default values for the Penalty on Error and the Penalty on Move.

If the process responds slowly to input changes, it may take many hours in an operating plant to see the full control response. In a simulated environment, the process and control simulation can support faster than real-time execution to allow the response of even the slowest processes to be seen in a few minutes. Similarly, the ability to simulate the control and process response at slower than real time can be valuable when working with a very fast process. The ability to adjust the speed of execution allows the control response to be verified quickly for a variety of operating conditions.

11.7.6 MPC Control Evaluation and Tuning Adjustment

The MPC operator interface is configured automatically for the MPC module. The trending display serves as the main means for evaluating MPC operation and performance. After the MPC controller is turned on, the user should observe its operations for various conditions: setpoint changes, constrained control, disturbances compensation and optimization. If controller performance is not satisfactory after applying a setpoint filter, the user should regenerate the controller with new controller settings. In such cases, thorough model revision is advised as well.

11.8 Simple MPC Workshop

This workshop provides several exercises that are used to explore simple MPC. A batch reactor process (Figure 11-31) will be used to demonstrate how feed rate may be maximized within pressure constraints and the capacity of the cooling system for temperature control. Access the book's web site http://www.advancedcontrolfoundation.com and select the Solution tab to view the workshop being performed.

> Step 1: Increase the feed flow rate setpoint in small steps until the temperature constraint is reached.
>
> Step 2: Introduce a disturbance that reduces the cooling capacity and observe how MPC automatically reduces the feed rate to keep the temperature from exceeding setpoint.
>
> Step 3: Increase the cooling capacity and observe how the feed rate is automatically increased and returns to the original setpoint.
>
> Step 4: Introduce a 20% increase in the feed rate and observe how the feed rate is limited to avoid exceeding the reactor pressure setpoint.
>
> Step 5: Reduce the cooling capacity and observe how the feed rate is automatically reduced to keep the temperature from exceeding setpoint.

11.9 Technical Basis

There have been several good books published on MPC technology in recent years, including Camacho and Bordens (1995), Maciejowski (2002), Tatjewski (2007) and Rossiter (2003). In addition, an excellent MPC applications overview by Qin and Badgwell (1997) provides an outline of MPC technology and future technology trends.

It is evident, therefore, that there is a fair amount of literature on the background of MPC technology. However, much less attention is paid in the literature to MPC implementation. MPC embedded into a DCS controller was presented in a book by Blevins, McMillan, Wojsznis and

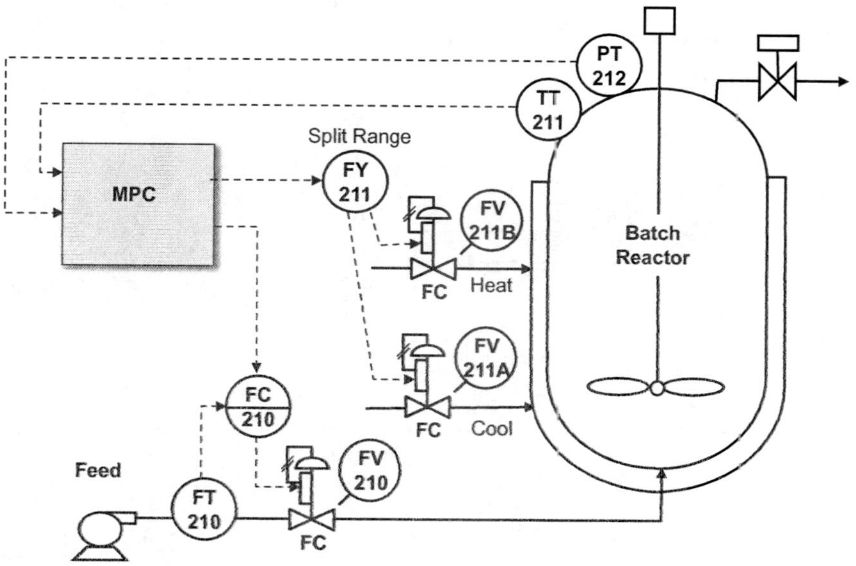

Figure 11-31. Batch Reactor Process

Brown (2003), and since then, there have been no significant updates to the topic.

The Technical Basis sections in this chapter and in Chapter 12 will focus in particular on the industrial MPC control techniques embedded into a DCS controller.

These Technical Basis sections are structured in such a way as to fill the gap in the literature and to thereby benefit the control engineer who wants to refresh his or her knowledge of MPC or has not yet had a chance to learn about MPC and wants to understand MPC basics. It also opens the door for potential users of easy-to-use MPC products, designed to be applied by a broad spectrum of process control engineers (Wojsznis, Blevins, Nixon, 2000).

For an engineer, a good way to be introduced to MPC is by showing an analogy with PID or with a feedback controller in general. Model Predictive Control uses modeling and a specific type of feedback that makes it distinctive. As opposed to a traditional control loop where the controller uses a difference (error) between the setpoint and the recent values of the measurement as its input, the MPC controller applies the difference between the future trajectory of the set point (whether a constant or a known function of time) and the predicted trajectory of the controlled variable as its input. The difference is expressed not by a single value as in a traditional feedback loop, but as a vector of error values from setpoint from the present time to a set time in the future, usually defined to cover the settling time of the process.

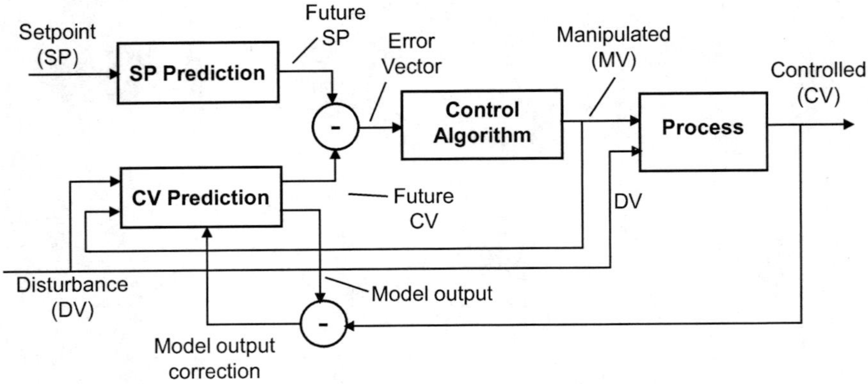

Figure 11-32. MPC Controller Operation Principle

Figure 11-32 shows an MPC controller for a process with two inputs and one output, in a form that allows one to see the analogy with a typical feedback control loop. The process has a manipulated variable (MV) and a disturbance variable (DV) on the input and a controlled variable (CV) on the output. A simple MPC controller used in this configuration has three basic components:

- A process model that predicts the process output for 120 or more scans ahead. (For consistency, we will use a prediction horizon of 120 scans throughout this discussion).
- A future trajectory of the setpoint for the same number of scans as the trajectory of the predicted process output.
- A control algorithm for computing a control action based on the error vector as the difference between the future trajectories of the setpoint and the predicted process output.

The controller output is a manipulated variable (MV) applied to the inputs of the process and the process model. A measured load upset to the process input is also applied as a disturbance variable (DV) to the model input. The process model computes a predicted trajectory of the controlled variable (CV) that is the process output. After this trajectory is corrected for any mismatch between the predicted value and an actual measured value of the controlled variable, the predicted trajectory is subtracted from the future trajectory of the setpoint to form an error vector as shown in Figure 11-33.

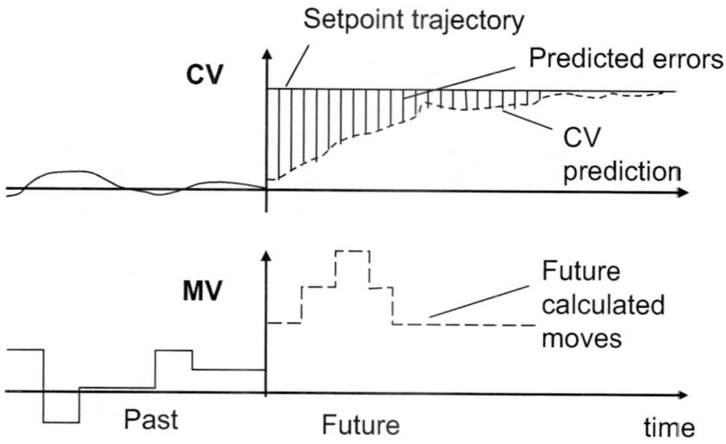

Figure 11-33. Illustration of MPC Controller Operation

An MPC controller is developed from the dynamic model of the future trajectory of the process formed from step responses. A specific gain is used for every element of the error vector. For a simple process with just one MV and one CV, the MPC matrix has 120 gain coefficients. In the MPC algorithm, the predicted errors are multiplied by the appropriate gains, and all the terms are summed to compute an increment of the controller output. The prediction is adjusted every scan by an actual measurement.

The major differences between the operation of an MPC controller and that of a classic PID feedback controller are:

- A predicted error vector is applied to the MPC controller algorithm instead of the scalar values of recent errors that are used in a classic PID feedback controller.
- The error vector is computed as the corrected model prediction subtracted from the future setpoint values. The scalar values of errors applied in a classical PID feedback controller are obtained from the measurement value subtracted from the setpoint value for the current and previous scans.
- The increments in the controller output required to bring the process output trajectory as close as possible to the setpoint trajectory are limited and are spread over several moves into the

future. Only the first move is implemented, however, and the computation procedure is corrected and repeated for the next scan.

- Disturbances with the proper dynamics based on identified models from process step responses are included in the prediction of the process output. In comparison, the feedforward signal in a basic control system provides a single current value. Some simple dynamic compensation may be applied to this value, but the lead and lag times and time delays are often adjusted on-line by trial and error. Because a simple feedforward signal added to a PID controller output cannot predict where the process is headed in the future, it may take the wrong action.

The operation of an MPC controller relies on a good process model.

The advantages of MPC are most evident when it is used as a multivariable controller. Therefore, the acronym MPC often means *Multivariable Predictive Control*. In applications to date, MPC has been used predominantly as a multivariable controller. However, for the sake of consistency, this book uses the original meaning of the acronym MPC: *Model Predictive Control*.

A generic multivariable process controlled by MPC is presented as a black box in Figure 11-34.

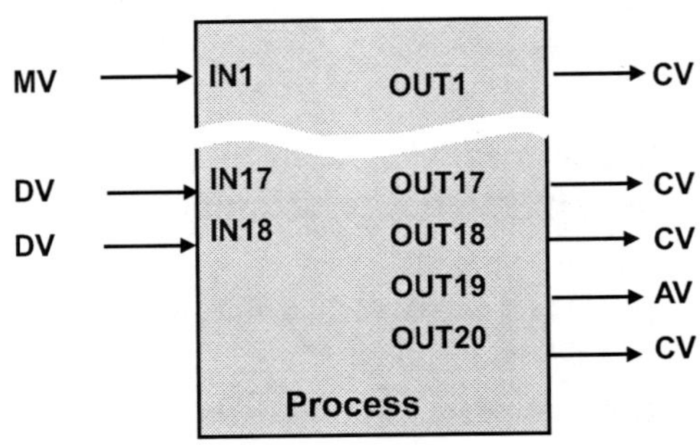

Figure 11-34. Multivariable MPC Controlled Generic Process Configuration

To review and expand upon earlier discussions in this chapter, in MPC terminology the process inputs are manipulated variables (MV) and measured disturbances (DV). The process outputs are controlled variables (CV) and auxiliary or constraint variables (AV).

Manipulated variables are managed by MPC controller outputs, predominantly by the manipulation of setpoints of the fast loops that stabilize the inputs to the process. In calculating the optimal solution, an MPC controller assumes in its calculations that it can calculate several moves of the manipulated variables ahead. Only the first move is implemented. The procedure for an optimal solution is repeated every scan. The *control horizon* is the number of manipulated variable moves ahead taken into account in developing the MPC optimal solution.

Manipulated variables have *hard constraints*, which means no violations of the limits are allowed at any time. Both *absolute* and *rate* or *incremental* constraints are set for manipulated variables.

Measured disturbances are also inputs to the process, but they are not managed by MPC.

Controlled variables are process outputs kept at specific setpoints (targets) or within specified ranges. Process models predict the future values of controlled variables for a number of scans ahead. The *prediction horizon* is the range in scans of process output prediction.

Finally, *constraint variables* are a type of controlled variable with range control only and no setpoints. Constraint variables have so-called *soft constraints*. Violations of this type of constraints may happen temporarily to satisfy other criteria or constraints of the system.

Some writers and practitioners use the term "controlled variables" for controlled and constraint variables, differentiating between them by using "controlled variables with setpoints" for controlled variables and "controlled variables with limits" for constraint variables.

A dynamic matrix is a matrix built from step responses to predict the changes in the process outputs that result from moves of the manipulated variables over the control horizon. *Dynamic Matrix Control* (DMC) has historically been the most successful MPC implementation approach, based on the dynamic matrix (Cutler and Ramaker, 1980).

11.9.1 The Basics of Process Modeling

The process model is the basis of MPC technology. Most MPC implementations to date use step response models proven in DMC applications. Step response models provide a prediction of process outputs. The prediction is used for computing a predicted error vector as an input to an MPC controller as shown in Figure 11-35 and explained below.

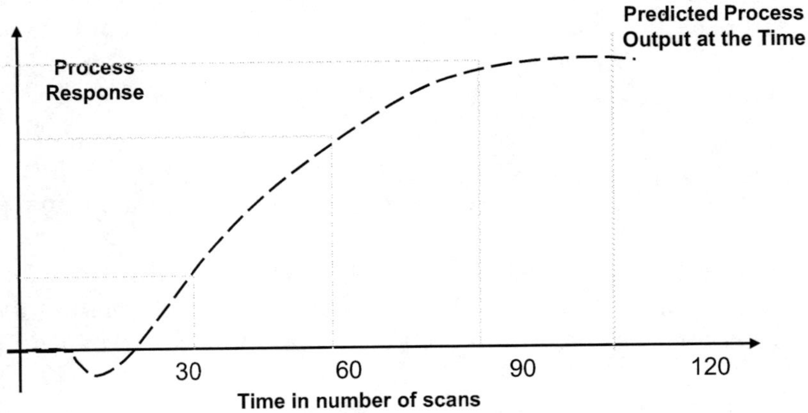

Figure 11-35. Step Response with 120 Coefficients

The step response is represented by a number of coefficients (120 coefficients in our example). Every coefficient corresponds to the value of the model step response at a specific time instance.

In MPC terms we see the step response as a prediction of the process output for up to 120 scans ahead, for a unit step input that was applied at scan zero. The value of the prediction at a particular instant in the future is represented by a specific coefficient. Figure 11-35 marks coefficients representing the process output at 30, 60, 90 and 120 scans ahead after the application of a step value on the input.

An MPC controller uses an *incremental model*. In such a model, real process input and output values are assigned to the model at the time of model initialization, and afterward, the controller accounts only for increments on the inputs and calculates increments of the model outputs.

An MPC application assumes a linear process model. Linear processes and systems possess the two properties of *superposition* and *homogeneity*. A system that produces an output ***y1*** for an input ***x1***, an output ***y2*** for an input ***x2*** and an output ***y1+y2*** for an input ***x1+x2*** possesses the property of superposition. Similarly, a system that produces an output *y* for an input *x* and produces an output ***ay*** for an input ***ax*** possesses the property of homogeneity. These properties are used to build multivariable models as in Figure 11-36. Every process output is a sum of input actions.

The superposition principle is illustrated graphically in Figure 11-37.

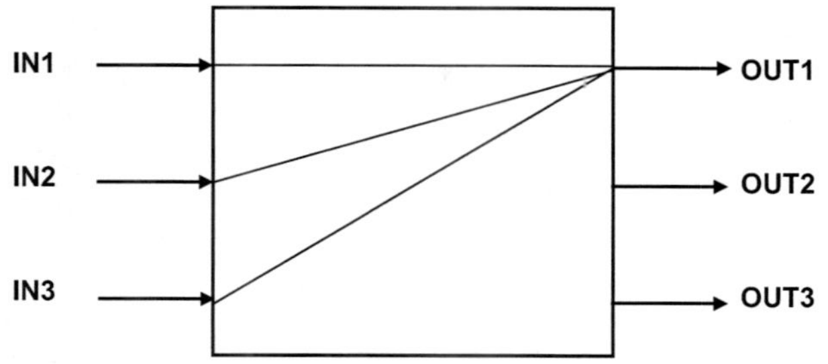

Figure 11-36. Superposition and Homogeneity Are Applied on Every Output of an MPC Process Model

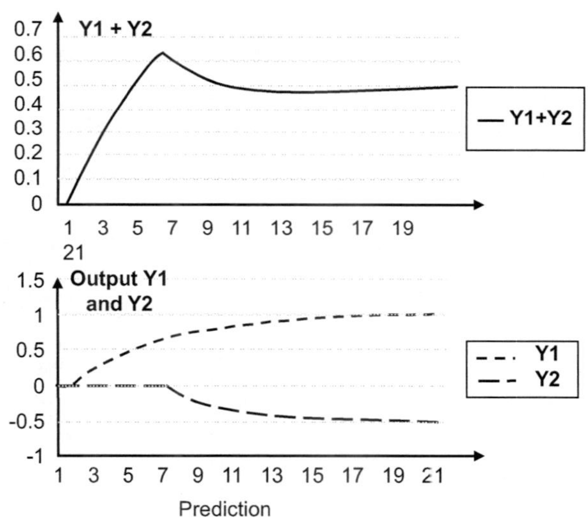

Figure 11-37. Superposition Principle in Linear Models

Mathematically, we can consider a predicted output trajectory of a process as a process state and use a modified state space form for process modeling (Lee, Morari and Garcia 1994). For single input, single output (SISO) process prediction, the equations are in this form:

$$\begin{aligned} x_{k+1} &= Ax_k + B\Delta u_k + Fw_k \\ y_0 &= Cx_{k+1} \end{aligned} \tag{11-1}$$

where:

$x_k = [y_0, y_1, \ldots, y_i, \ldots, y_{p-1}]^T$ is a vector of process output prediction at a time k, 0,1,2,...p-1 steps ahead

A is a shift operator, defined as $Ax_k = [y_1, y_2, \ldots, y_i, \ldots, y_{p-1}, y_{p-1}]^T$

$B = [b_0, b_1, \ldots, b_i, \ldots, b_{p-1}]^T$ is a vector of p step response coefficients

$\Delta u_k = u_k - u_{k-1}$ is the change on the process input/controller output

$w_k = y^p - y^m$ is the process output measurement minus the model output (the mismatch between the process and the model that results from noise, unmeasured disturbances and model inaccuracy)

$F = [f_0, f_1, \ldots, f_i, \ldots, f_{p-1}]$ $0 < f_i \le 1$ Filter is applied to the mismatch between the process and model outputs

C is the operator for selecting the current model output, defined as $y_0 = Cx_{k+1}$

For a process with n outputs and m inputs, vector x_k has the dimension $n{\times}p$ and vector $\boldsymbol{B}$ becomes a matrix with dimensions $n{\times}p$ rows and m columns. The graphical illustration of Equation 11-1 in Figure 11-38 explains the prediction principles.

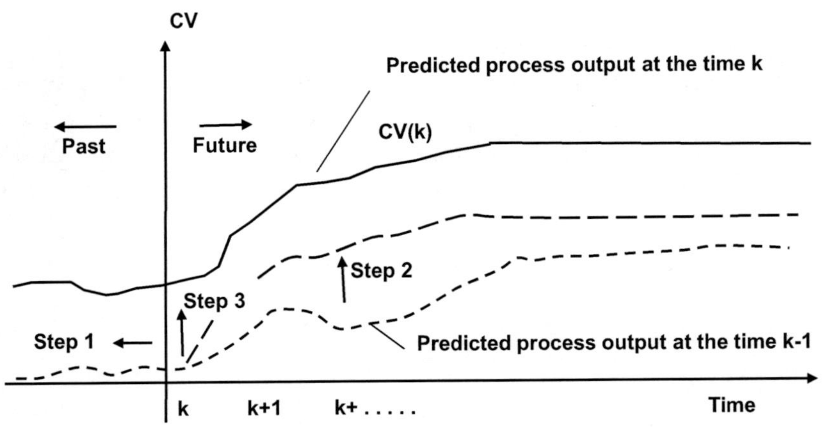

Figure 11-38. Graphical Illustration of Linear Process Modeling

At any time instance, the *k* process output prediction (bottom curve) is updated in three steps:

1. The prediction made at time *k-1* (the bottom dotted curve) is shifted one scan to the left.
2. A step response, scaled by the current change on the process input, is added to the output prediction.
3. The prediction curve is moved to the point to match the current measured process output, for a filter coefficient = 1, or in general the prediction shift is Fw_k.

11.9.2 Process Model Identification

As has been mentioned, MPC is used predominantly for multivariable and highly interactive processes. Therefore, there is little chance that the application of a single step change to a specific process input will produce outputs that would provide a good process model. Multivariable process outputs are influenced by other inputs, disturbances and noise. Better results are obtained by applying a special pulse test sequence instead of a single step and then using mathematics to process test data and build a process model. An example of testing for the identification of a 3×3 process is shown in Figure 11-30.

Using test data, we can build as many equations as the number of collected output samples for every output. The number of collected output samples in a test data set is usually significantly higher than the number of unknown model coefficients. Such equations are solved using the least squares technique. This technique finds the coefficients that may not fit perfectly in any equation, but fit optimally for all equations, that is, the total squared error for all equations is minimal.

The form of the equations used for process modeling is defined by the modeling technique. A number of modeling techniques are used. The most common modeling techniques are Finite Impulse Response (FIR) and Auto Regressive with eXternal inputs (ARX) – (Box, Jenkins and Reinsel, 1994). FIR identifies pulse response coefficients, as in Equation 11-2 for an SISO process:

$$\Delta y_k = \sum_{i=1}^{p} h_i \Delta u_{k-i} \tag{11-2}$$

where:

p = prediction horizon, with a typical default value for an MPC model of 120

Δy_k = change in the process output at the time k
Δu_{k-i} = change in the process input at the time k-i
h_i = pulse response coefficients of the model

Step response coefficients are calculated directly from the pulse response as follows:

$$b_k = \sum_{i=1}^{k} h_i \qquad k = 1, 2, \ldots, p \tag{11-3}$$

An advantage of FIR is it does not require any preliminary knowledge about the process. However, identifying step responses with a full prediction horizon with as many as 120 coefficients results in low confidence levels for identified coefficient values. Therefore, a shorter horizon with 60 points is more suitable for an FIR model. An FIR model with a shorter horizon provides the initial part of the step response, which is adequate for defining process deadtimes using a heuristic approach.

On the other hand, the ARX model for an SISO process (Equation 11-4) has fewer coefficients, which are defined with higher confidence provided the process deadtimes are known:

$$y_k = \sum_{i=1}^{V} a_i y_{k-i} + \sum_{i=1}^{A} b_i u_{k-d-i} \tag{11-4}$$

where:

A, V = autoregressive and moving average orders of the ARX model. $A = 4$, $V = 4$ satisfy most applications
a_i, b_i = moving average and autoregressive coefficients of the ARX model
d = deadtime in scans

Applying FIR first for defining deadtimes and then applying those deadtimes to ARX provides the best model identification results.

Finally, having an ARX model and applying a unit step change to the input, step responses for any prediction horizon may be calculated directly from Equation 11-4.

For MIMO processes, process superposition is applied from all inputs to every output in FIR and ARX models.

Combining FIR and ARX model identification techniques, and properly testing the process, delivers a high quality process model. An example of an identified 1×1 process model is shown in Figure 11-5. For this example, errors for process steady state gains were less than 5%, and the deadtime error was one scan.

The identified model should be validated by applying real input data to the model inputs and comparing the model output with real output. A validation plot for the output of the cited example is shown in Figure 11-6.

The validation procedure may also include statistical techniques for calculating *confidence intervals.* A confidence interval forms an area around a nominally defined step response (Figure 11-39).

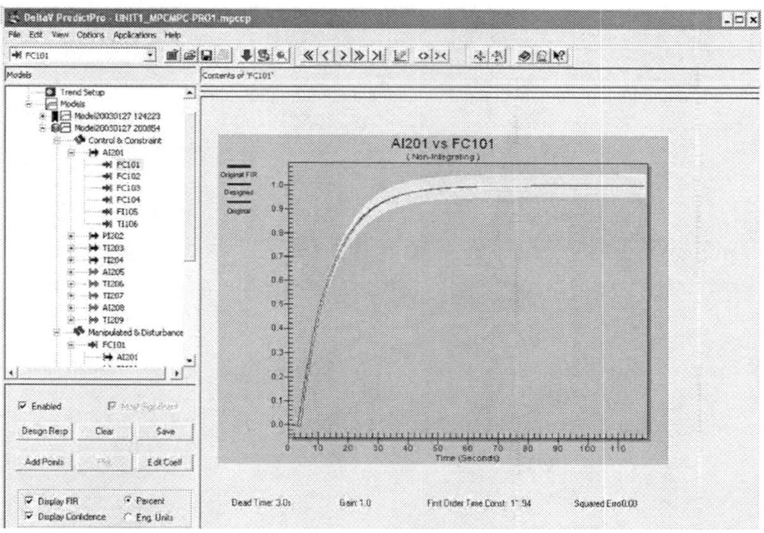

Figure 11-39. Step Response with Marked Confidence Intervals

The true value of the step response coefficients is found within confidence intervals with a predefined confidence (or probability). Normally, 95% confidence intervals are used.

11.9.3 Unconstrained Model Predictive Control

The unconstrained MPC algorithm is a core function of MPC process control. Therefore, it is useful to present some details on deriving the basic dependencies of this algorithm (Morari, Ricker and Zafiriou, 1992). The derivation follows the DMC approach and assumes, for the sake of simplicity, an SISO process and a control horizon = 1.

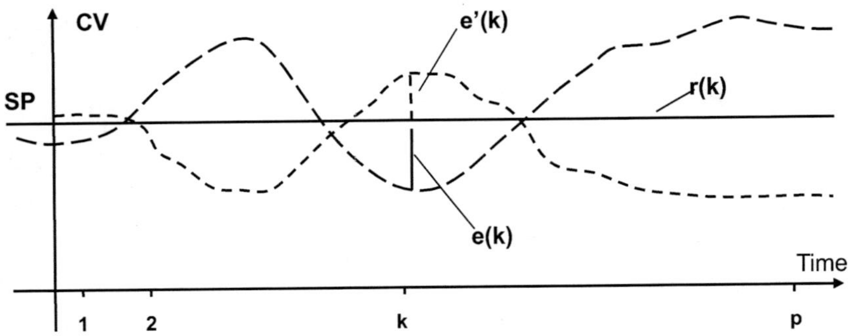

Figure 11-40. Compensating Prediction Errors *e(k)*

Let's assume that as a result of a setpoint change or disturbance, a control error occurred (Figure 11-40). Applying the setpoint trajectory $r(k)$ and a prediction of the process output $y(k)$, the predicted errors $e(k)$ are expressed as:

$$e(k) = r(k) - y(k) \qquad k = 1, 2, \ldots, p \tag{11-5}$$

To compensate for the errors, the change in the MPC controller output should cause a change in the process output with predicted errors that are the mirror image of the existing errors relative to the setpoint trajectory:

$$e'(k) = -e(k) \quad \text{for} \quad k = 1, 2, \ldots, p \tag{11-6}$$

$e'(k)$ is calculated as $e'(k) = b(k) \times \Delta u = -e(k)$, $k = 1, 2, \ldots, p$ (11-7)

b_k are step response coefficients $k = 1, 2, \ldots, p$

To design a controller, we have to solve p equations, finding one controller move Δu. This is called an over-determined problem; that is, the solution cannot satisfy all of the equations perfectly, so we need to find the solution that best satisfies the equations. This is a least squares solution. Using matrix notations we define the MPC controller design problem as finding Δu that minimizes the total squared error $e'(k) - e(k)$ for all $k = 1, 2, \ldots, p$.

$$\min\{B \times \Delta u - E\}$$

E = error vector of p dimension

B = step response vector of p dimension

From linear algebra the solution of the least squares problem is known as:

$$\Delta u = (B^T \times B)^{-1} \times B^T \times E \tag{11-8}$$

Designing a controller with a control horizon greater than 1 requires that the number of moves ahead over the whole control horizon is accounted for.

The compensating error that results from a number of control output moves is found in a similar way as shown in Equation 11-7. A dynamic matrix is required to apply a multi-step control horizon.

The dynamic matrix is formed from the step response coefficients in such a way that the first column of the matrix consists of coefficients of the original step response $b_k \times k = 1, 2, \ldots, p$. The second column of the matrix consists of coefficients of the same step response shifted to the right, with the first coefficient equal to zero. The next columns are built in a similar way from the preceding columns. The number of columns in the dynamic matrix is equal to the control horizon. Equation 11-9 is an example dynamic matrix control for a control horizon equal to two and with a five coefficients step response. Multiplication of the matrix over the vector of the two sequential control output moves develops a prediction of process output change resulting from the two moves.

$$\begin{bmatrix} \Delta out1 \\ \Delta out2 \\ \Delta out3 \\ \Delta out4 \end{bmatrix} = \begin{bmatrix} b_1 0 \\ b_2 b_1 \\ b_3 b_2 \\ b_4 b_3 \\ b_5 b_4 \end{bmatrix} \times \begin{bmatrix} \Delta u1 \\ \Delta u2 \end{bmatrix} \tag{11-9}$$

The solution of the MPC design problem for two control moves is analogous to Equation 11-8, except that instead of step response B, dynamic matrix $\mathbf{S}$ is applied.

A more general formulation of the MPC controller includes minimization of both the squared sum of predicted control error and the squared sum of the calculated controller moves. To get the desired MPC controller performance, two tuning parameters called *penalty on moves* and *penalty on error* (defined in 11.7.4) are used in the formulation.

The MPC controller objective for minimizing the squared error of the controlled variable includes the penalty on error over the prediction horizon and the objective for minimizing the squared changes in the controller output over the control horizon includes the penalty on moves in the following way:

$$\min_{\Delta U(k)} \left\{ [\Gamma^y [X(k) - R(k)]^2] + [\Gamma^u \Delta U(k)]^2 \right\}$$

where:

$X(k)$	=	controlled output, p = step ahead prediction vector
$R(k)$	=	p is the step-ahead setpoint vector
$\Delta U(k)$	=	m is the step-ahead incremental controller output moves vector
Γ^y	=	$\text{diag}\{\Gamma_1^y, \ldots, \Gamma_p^y\}$ is the penalty matrix on the output error
Γ^u	=	$\text{diag}\{\Gamma_1^u, \ldots, \Gamma_m^u\}$ is the penalty matrix on the control moves

The solution is in the form:

$$\Delta U(k) = (S^{uT}\Gamma^{yT}\Gamma^y S^u + \Gamma^{uT}\Gamma^u)^{-1} S^{uT}\Gamma^{yT}\Gamma^y E_p(k) \qquad (11\text{-}10)$$

where:

S^u	=	a $p \times m$ process dynamic matrix
$E_p(k)$	=	the error vector over the prediction horizon

For a multivariable $n \times n$ process, the dimensions of the S^u matrix would be $(n \times p) \times (n \times m)$. The dimensions of Γ^y would be $n \times p$, $\Gamma^u - n \times m$ and $E_p(k) - n \times p$.

The performance of the control algorithm may be modified by the adjustable parameters p, m, Γ^u and Γ^y. From an implementation point of view, p and m are not convenient to use as tuning parameters; instead, Γ^u and Γ^y are applied as scalars. The control error is multiplied by the scalar γ^y over the whole prediction horizon, and the controller moves over the control horizon are multiplied by the scalar γ^u.

By making Γ^y and Γ^u coefficients dependent on the control and prediction horizon, controller robustness can be improved. Various functions, such as linear or exponential, for shaping the coefficients of Γ^y can be used (Wojsznis and Wojsznis, 1993), (Muskie and Rowlings, 1993).

Γ^u is a basic controller tuning parameter at the controller generation phase. Increasing Γ^u reduces and slows the control response and action; decreasing Γ^u makes the control action more aggressive and the control response faster.

It follows from experience that the deadtime should be accounted for as a major factor in setting the penalty on moves (Wojsznis, Gudaz, Mehta and Blevins, 2003). Equation 11-11 defines a penalty on moves for MV_i that provides stable and responsive MPC operation for a model error of up to about 50%:

$$\Gamma^{ui} = 3\left(1 + \frac{6DT_i}{p} + \frac{3G_iDT_i}{p}\right) \tag{11-11}$$

where:

DT_i	=	the deadtime in MPC scans for an MV_i - CV_i pair
G_i	=	gain (no units) for the MV_i - CV_i pair
p	=	the prediction horizon

MV and CV denote the manipulated variable and the controlled variable, respectively.

In addition to shaping the controller structure at the generation phase, a number of features can be added to the unconstrained MPC controller to manage controller behavior and robustness on-line.

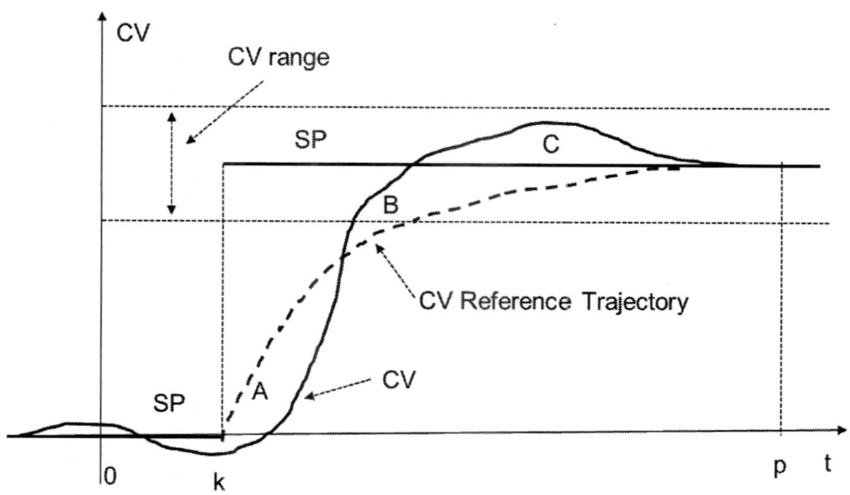

Figure 11-41. Using Setpoint Trajectory for MPC On-line Tuning

A primary, intuitively understandable on-line tuning parameter is the setpoint filter or reference trajectory (Figure 11-41). One trajectory design (Wojsznis, Blevins and Nixon, 2000) acts in such a way that instead of any departure from the trajectory being penalized, only those deviations that are below the trajectory or above the setpoint value are penalized (Area A and area C only if range = 0). In addition, the control error is considered zero if the controlled variable is within range (Area C with range wider than error value). These features constitute a funnel control (which was introduced in 11.7.4), which is shaped on-line by changing the setpoint filter time constant and range, further enhancing controller robustness and flexibility.

Presenting the MPC control equation in the form:

$$\Delta U_k = K_{mpc} E_p \tag{11-12}$$

where:

$$K_{mpc} = (S^{uT}\Gamma^{yT}\Gamma^{y}S^{u} + \Gamma^{uT}\Gamma^{u})^{-1}S^{uT}\Gamma^{yT}\Gamma^{y} \tag{11-13}$$

is the MPC controller gain, we can see that MPC uses integral action. This, however, does not mean that we can draw a complete analogy between an MPC controller and a conventional integral-only feedback controller. The major difference is the MPC controller uses predicted errors and integrates over the prediction horizon.

11.9.4 Integrating Constraints Handling, Optimization and Model Predictive Control

Simple Constraints Handling

A typical MPC configuration includes controlled variables and constrained variables. The unconstrained MPC controller does not manage constrained variables directly. A supervisory constraints handling algorithm or optimizer performs the task. First, let's review the simple solution with the supervisory algorithm.

The constraints handling algorithm uses two basic assumptions:

- MPC predicts future outputs for control and constraint variables
- There is an interactive gain between the control and constraint variables

The algorithm becomes active when a predicted constraint variable violates a constraint limit in the future. To prevent constraint violations, the algorithm adjusts the working setpoint of a selected controlled variable on the value:

$$\Delta SP_{cv} = -r \times G_{CV-AV} \times \Delta AV \tag{11-14}$$

where:

ΔAV = the magnitude of the predicted steady-state constraint violation

G_{CV-AV}= the gain relationship between AV and CV

r = the relaxation factor, $r < 1$

ΔSP_{cv} = is the requested working setpoint change for the controlled variable

The association between a constrained variable and a controlled variable is established during configuration, thus the gain relationship is defined from the process model as follows:

$$G_{CV-AV} = \frac{G_{CV-MV}}{G_{AV-MV}} \tag{11-15}$$

G_{CV-AV} gain tells how much we need to move the CV's setpoint to compensate for a unit constraints violation.

The working setpoint is adjusted every scan according to constraint violations. If a constraint violation ends, the working setpoint gradually returns to the original setpoint value. Figure 11-42 illustrates the concept.

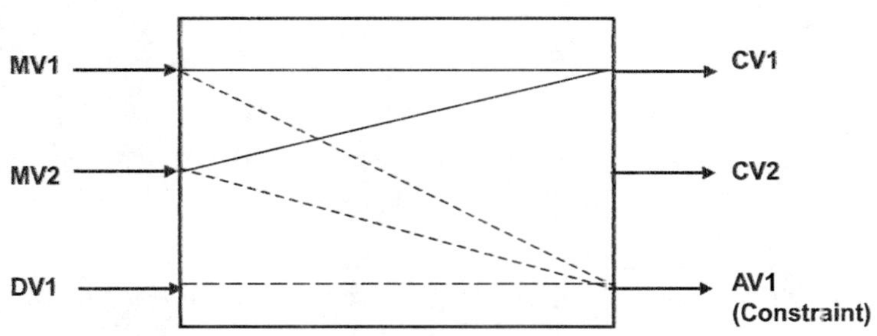

Figure 11-42. Relationship between Controlled Variable CV and Constrained Variable AV

Referring to Figure 11-42, it may be seen that both CV1 and AV1 depend on the same manipulated variables MV1 and MV2. If AV1 violates constraint limits, changing the setpoint of CV1 in the proper direction will cause MV1 and MV2 moves that result in the return of AV1 to within its desired limits.

Simple Optimization

The objectives of optimization are usually to maximize product value by maximizing throughput and minimize raw material cost. A simple optimization is applied in the cases where a priori knowledge defines optimization targets as minimizing or maximizing one of the process parameters. In traditional applications, this type of solution is known as a *pusher*. To perform optimization in an MPC installation, the number of MVs should be greater than the number of CVs. A simple solution consists of defining one or more of the excess MVs as optimizing MVs. For example, in a configuration with three CVs, up to four AVs, and up to four DVs, three MVs are required for CV control and at least one more MV for optimization. An MV used for optimization should be a process input that directly affects production rate or raw material cost. Introducing an optimized CV as a shadow of an optimized MV makes it possible to generate an MPC $n \times n$ optimizing controller. By setting the setpoint of an optimizing CV as the maximum (or minimum) achievable value of an optimizing MV, the MPC controller drives the optimizing MV toward an optimal value, that is, toward the setpoint of the optimizing CV, until any of the manipulated variables, or more exactly their predicted values, violate their own limits. If that happens, the optimizing MV moves out of an optimal value until other MVs are out of the limit. For managing MV limits, a similar approach is applied as with AV constraints. The setpoint change for an optimizing CV is calculated as:

$$\Delta SP_{cv}^{opt} = -r \times G_{OPTMV-MV} \times \Delta MV \tag{11-16}$$

where:

ΔMV = magnitude of the predicted MV limit violation

$G_{OPTMV-MV}$ = gain relationship between optimized MV and MV with limits being violated

r = relaxation factor, $r < 1$

ΔSP_{cv}^{opt} = requested working set point change for optimized CV

The gain relationship between MVs can be established using Equation 11-17:

$$G_{optMV-MV} = \frac{\Delta MV^{opt}}{\Delta MV} \text{ with MPC closed loop control} \tag{11-17}$$

$G_{optMV-MV}$ gain; tells how much we need to move an optimal CV's setpoint to compensate for a unit MV limit violation.

If the constraint limit violation ends, an optimal CV setpoint returns in a stepwise manner to the original setpoint value. Figure 11-43 illustrates the concept.

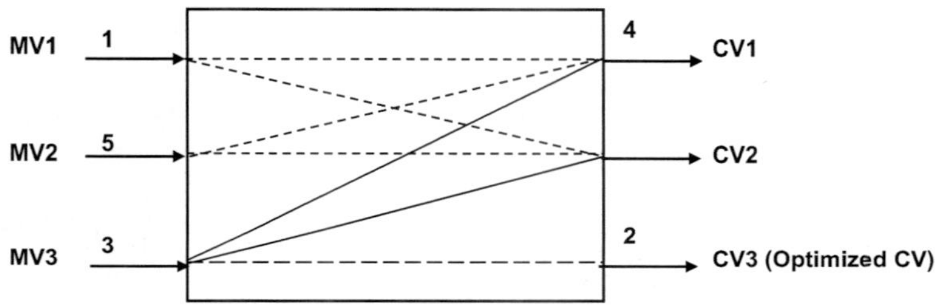

Figure 11-43. MPC Configuration for Simple Optimization

In Figure 11-43, MV3 is a manipulated variable to set the production rate or production cost for the MPC controlled unit. MV3 adds a degree of freedom to the system, which is required for optimization. CV3 is a virtual parameter; it is the shadow of MV3 that was created to integrate optimization into MPC control.

The CV3 setpoint is set as a maximum limit of MV3 if it represents production rate, or as a minimum limit of MV3 if it represents production cost. When the predicted MV3 = CV3 stays within limits, the CV3 setpoint is achieved and is kept as a normal MPC setpoint. However, when the predicted value of one or more MVs is out of limits, the optimized working setpoint value is changed according to Equation 11-16 to keep MVs within limits, effectively retaining CV control. The following sequence of events happens in Figure 11-43:

1. Predicted MV1 value is out of limits.
2. CV3 setpoint is moved out of maximal or minimal value, sacrificing optimal control to retain controllability.
3. Achieving the new CV3 setpoint requires that MV3 move.

4. The MV3 move causes a change in CV1 and CV2 as they move out of their respective setpoints.
5. Predicted MV1 moves in an out-of-limit direction, pushing CV1 and CV2 back toward their respective setpoints.

The techniques for constraints handling and optimization feature simplicity, which makes it possible to easily implement these techniques in the system controller as a part of an integrated MPC algorithm. Dealing with constraints by managing working setpoints as presented above, instead of by controlling constraints variables directly, has another advantage, that of avoiding *ill conditioned* control configurations. Ill conditioning happens when a controller needs excessive output movements in order to control changes in process outputs. Because the configuration of the MPC controller with this approach does not change during operation, controller conditioning can be properly set at the time of controller generation.

Introduction to Integrated Constraints Handling, Optimization and Model Predictive Control

For processes with an objective function dependent on several manipulated or controlled variables, optimization techniques are an essential component of Model Predictive Control technology. The proven approach applies linear programming (LP) or quadratic programming (QP) with steady state models.

Linear programming is a mathematical technique for solving a set of linear equations and inequalities in such a way as to maximize or minimize a certain additional function. This additional function is called an *objective function*. Objective functions usually express economic value, such as cost or profit.

MPC optimization, on the other hand, uses incremental MV values at the present time or the sum of MV increments over the control horizon and incremental CV values at the end of the prediction horizon, instead of positional current values as in typical LP applications. The LP technique uses steady state models, therefore steady state conditions are required for its application. With the prediction horizon normally used in MPC a future steady state for self-regulating processes is guaranteed.

Optimal MV values are applied to the MPC algorithm as target MV values to be achieved within the control horizon. Therefore, the MPC algorithm working with the optimizer has two main objectives:

- Minimize CV control error with minimal MV moves
- Achieve optimal steady state MV values set up by optimizer

The original MPC algorithm is extended to include the MV targets into the least square solution.

More details on optimizer and MPC functionality are provided in the next two chapters.

BIBLIOGRAPHY

1. Blevins, T., McMillan, G., Wojsznis, W. and Brown, M. *Advanced Control Unleashed: Plant Performance Management for Optimum Benefit*, Research Triangle Park: ISA, 2003; ISBN 978-1-55617-815-3.
2. Box, G. E. P., Jenkins, G. M. and Reinsel, G. C. *Time Series Analysis, Forecasting and Control*, Third Edition, Prentice Hall, Hoboken, NJ, 1994.
3. Butler, D. L., Cameron, R. A., Brown, M. W. and McMillan, G. K. "Constrained Multivariable Predictive Control of Plastic Sheets," ISA Technical Conference, Paper 1022, Houston, TX, 2001.
4. Camacho, E. F. and Bordens, C. "Model Predictive Control in the Process Industry," London: Springer-Verlag, 1995, ISBN 3-540-19924-1.
5. Chmelyk, T. "Lime Kiln Model Predictive Control with a Residual Carbonate Soft Sensor," *ACC*, WP01, Anchorage, Alaska, 2002.
6. *Cutler*, C. and *Ramaker*, *B*. *"Dynamic Matrix Control—A Computer Control Algorithm," Proceedings of Joint Automatic Control Conference*, San Francisco, CA, 1980.
7. Froisy, J. B. "Model Predictive Control: Past, Present and Future" *ISA Transactions*, 33, pp. 235–243, 1994.
8. Lee, J. H., Morari, M. and Garcia, C. E. "State-space Interpretation of Model Predictive Control," *Automatica*, Vol. 30, No. 4, pp. 707–717, 1994.
9. Maciejowski, J. M. *Predictive Control with Constraints*, Prentice Hall, Harlow, England, 2002.
10. Morari, M., Ricker, N. L. and Zafiriou, E. "Model Predictive Control," Workshop No. 4, *American Control Conference*, June 1992, Chicago.
11. Muskie, K. R. and Rowlings, J. B. "Model Predictive Control with Linear Models," *AIChE*, 39(2), pp. 262–287, 1993.
12. Qin, S. J. and Badgwell, T. A. "An Overview of Industrial Model Predictive Control Technology," *Proceedings of Fifth International Conference on Chemical Process Control*, pp. 232–256, AIChE and CACHE, 1997.
13. Rossiter, J. A. *Model-based Predictive Control: a Practical Approach*, Boca Raton, London, New York, Washington, D.C., CRC Press LLC, 2003.
14. Shinskey, F. G. *Process Control Systems: Application, Design, and Adjustment*, 3rd edition, New York: McGraw-Hill, 1988.

15. Tatjewski, P. *Advanced Control of Industrial Processes,* London, Springer, 2007.

16. Thiele, D. "Benefits and Challenges of Implementing Model Predictive Control as a Function Block," Paper presented at ISA Technical Conference, New Orleans, 2000.

17. Wojsznis, W. K. and Wojsznis, P. W. "Robust Predictive Controller in Object-Oriented Implementation," *Advances in Instrumentation and Control,* Vol. 48, Pt. 1, p. 521, 1993.

18. Wojsznis, K.W., Blevins, L. T. and Nixon, M. "Easy Robust Optimal Predictive Controller," *Advances in Instrumentation and Control,* Proceedings of ISA Technical Conference, New Orleans, 2000.

19. Wojsznis, W., Gudaz, J., Mehta, A. and Blevins, T. "Practical Approach to Tuning MPC," *ISA Transactions,* Vol. 42, No. 1, pp. 149-162, January 2003.

12

MPC Integrated with Optimization

The simple Model Predictive Control (MPC) capability detailed in Chapter 11 can be effectively used to address a wide variety of small applications that are common to the process industry. Successful examples of such implementations are presented in Chmelyk (2002) and Blevins et al (2003), wherein simple MPC has been implemented at the controller level. Significant advantages can be obtained by using a simple MPC capability that is embedded within the control system.

However, there are often limits on the size and complexity of the process that can be addressed using a simple MPC capability. Such products, for example, may be designed for a specific maximum process size (e.g., eight inputs and eight outputs). In addition, the embedded optimizer included with simple MPC may be designed to maximize or minimize only one selected process input (feed rate, for example). For larger, more complex applications that are used in batch or continuous processing, the plant operation objective(s) may be best met using MPC integrated with optimization. Such applications are often characterized by numerous operating constraints and the need to address broader operating objectives, such as maximizing throughput while minimizing production costs (Qin and Badgwell, 1997).

Optimizers are an inherent part of MPC functionality, wherein they are applied toward the economic optimization and constraint handling objectives (*Cutler* and *Ramaker, 1980).* Under normal operating conditions, an optimizer provides an economically optimal solution within predefined acceptable ranges and limits. If a solution does not exist within the predefined ranges and limits, the optimizer should have the means to recover from the infeasibility.

The optimization that is achieved through the application of MPC integrated with optimization is related to the means by which process constraints are addressed. An added degree of flexibility is achieved in optimization by allowing the user to easily define the plant operating objectives. The process is then dynamically controlled to satisfy an optimal solution for predicted process steady-state values while accounting for constraints.

One of the proven approaches uses linear programming (LP) with steady-state models. Linear programming is a mathematical technique for solving a set of linear equations and inequalities that maximizes or minimizes a certain additional function. This additional function is called the *objective function,* which was introduced in Chapter 11. Objective functions usually express an economic value, such as cost or profit. MPC optimization considers incremental manipulated variable (MV) values at the present time or the sum of increments of the MV over the control horizon, and incremental values of controlled and constrained values (CV) at the end of prediction horizon, instead of positional current values as in typical LP applications. With the prediction horizon normally used in MPC design, future process steady-state is guaranteed for self-regulating processes.

The optimal solution determined by the linear program is always located at one of the vertices of the region of feasible solutions. To find this solution, the LP algorithm calculates the objective function for an initial vertex and improves the solution with every step until it determines the vertex with the maximum (or minimum) value of the objective function as the optimal solution. In operation, the optimizer sets up and updates the steady-state targets for the MPC unconstrained controller at every scan; thus, the MPC controller executes the unconstrained algorithm. Because the targets are set in a manner that accounts for constraints, as long as a feasible solution exists, the controller works within constraint limits. This is illustrated in Figure 12-1 for a two-dimensional problem.

MPC with integrated optimization has been successfully used to address the control requirements of a wide variety of complex processes. In this chapter, some examples are discussed that illustrate the application of, and benefits achieved through, the installation of MPC with integrated optimization.

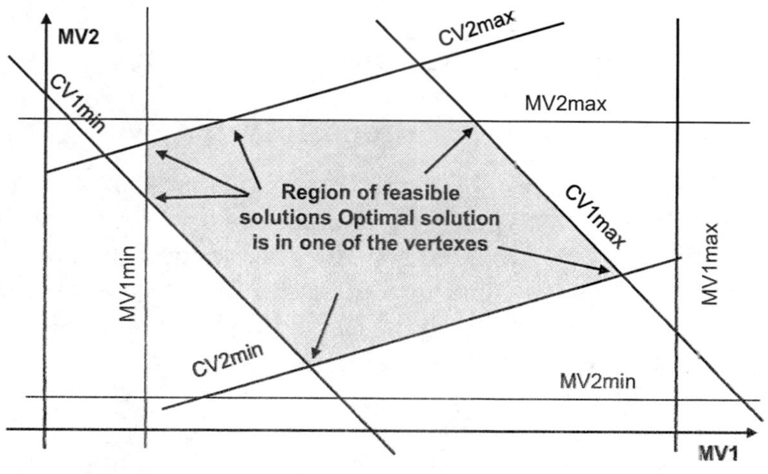

Figure 12-1. Optimization Problem of a Two-dimensional System

12.1 Application Example – Multiple Effect Evaporator

(Section co-authors: Terrance Chmelyk and Saul Mtakula, Spartan Controls, Burnaby, BC, Barry Hirtz, Canfor Pulp Limited Partnership, Prince George, BC)

The Multiple Effect Evaporators (MEE) are a key unit operation in the chemical recovery cycle of a kraft pulp mill because of the overall impact of the MEE on mill production, energy use, water use and potentially the environment. This paper presents a Model Predictive Control (MPC) strategy that was deployed on a set of Multiple Effect Evaporators at Canfor Pulp Limited Partnership's (CPLP) Prince George Pulp and Paper mill in northern British Columbia. A hybrid approach utilizing MPC, energy balances and abnormal situation management techniques produced a robust control solution for enhancing the control of the MEE. As a result of this control improvement project, which was initiated in the spring of 2004, the mill has realized several ongoing operational benefits that will be presented, including:

- Consistent and stable operations across all crews
- Reduced strong black liquor solids variability
- Optimized liquor throughput
- Significantly reduced contamination of condensate

- Decreased fouling caused by high solids excursions
- Improved startup times
- Optimized energy consumption (steam, water).

12.1.1 Evaporator Process Overview

The Prince George Pulp and Paper Mill first started up in 1966 as a single line mill with production of around 600 admt/d (air-dried metric tons per day) of fully bleached softwood kraft pulp and unbleached sack kraft paper. Continual efforts to de-bottleneck the mill have increased production to the current daily target of 811 admt/d. The optimization project presented in this section was focused around improving the operation of the #1 Multiple Effect Evaporators (MEE). At the Prince George Pulp and Paper Mill, the #1 Multiple Effect Evaporators consist of five long tube vertical rising film vessels and one tubular type falling film vessel, making up a six-effect set. The feed flow rate of the MEE is approximately 80 l/s of weak black liquor (WBL) at 18% solids. They evaporate enough water to produce 53% solids strong black liquor (SBL) at an economy of 5.3 kg vaporized / kg steam input. The SBL is then fed to a concentrator set before being fired in the recovery boiler at 75% solids. Figure 12-2 provides a general overview of the process.

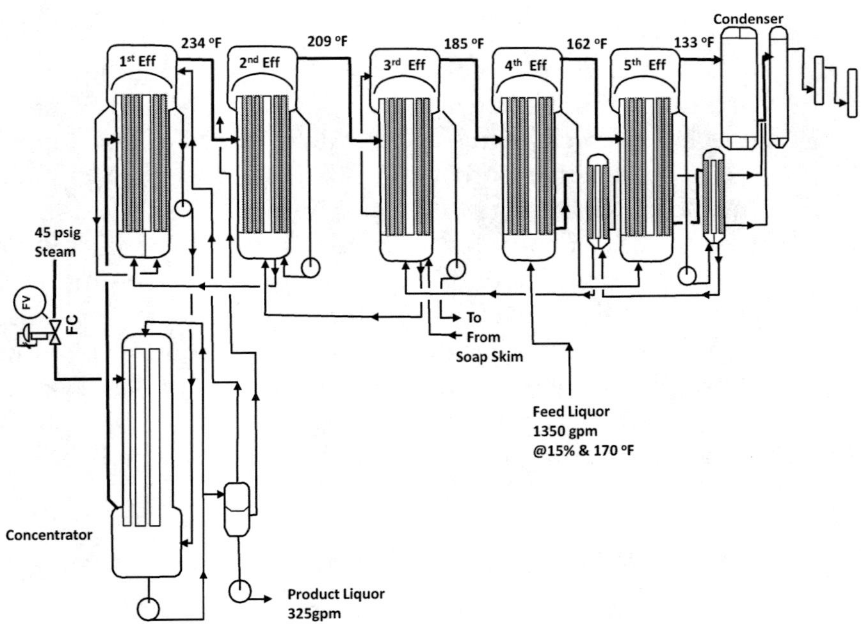

Figure 12-2. Process Flow Diagram for the #1 Evaporators

12.1.2 Project Motivation and Design Considerations

Prior to the optimization project, the MEE was equipped with basic DCS regulatory (single loop) controls that were manually coordinated and set by the operator. When the process was disturbed, because of a large number of interacting variables and significant deadtimes it was difficult to maintain consistent operation and meet product targets. As a result, the MEE ran with a high standard deviation in product solids of 4.5%. This created a number of operating problems including:

- Fouling the tube sheets at high solids levels
- Throughput limitations
- Contaminated condensate resulting in wasted energy through re-evaporation
- Recovery boiler efficiency loss at low solids levels
- Wasted energy through over-evaporation and then dilution at cyclones to lower solids

It was decided that a more sophisticated control scheme was required to deal with the coordination of the system, handle the difficult dynamics of the process and minimize the impact of frequent disturbances. Research into the application of MPC to other pulp and paper processes (Chmelyk, 2002) led the authors to investigate the suitability of applying the technology to the MEE problem. Some initial modeling and analysis determined that the development of an MPC-based control solution for the MEE could potentially yield some good improvements in the operation.

12.1.3 Why Use Model Predictive Control?

MPC has achieved the greatest benefits when applied to processes that have long process delays, multiple constraints, interactions among variables and the need for some optimization (Chemlyk, 2002; Blevins et al, 2003 and Cutler and Ramaker, 1980). In general, the MEE met several of the key criteria used to assess the suitability of the system to a Model Predictive Control based strategy (Blevins et al, 2004):

- *Interactive process:* Variables are highly correlated.
- *Long process delay:* Although process delay was not extreme, the time to steady state from changes in weak black liquor density to product density changes was roughly 45 minutes.
- *Multiple constraints:* Condensate conductivities, pressures and temperatures must all be maintained within limits.
- *Multiple disturbances:* There are changes in feed flow, WBL density and temperature.

- *Optimization:* Goals consisted of maximization of liquor solids and throughput as well as energy use considerations.

From a statistical and operational analysis perspective, the historical operating data on the MEE confirmed that a "comfort margin" was normally maintained by the operators on the product solids target and main constraints, because there was significant fear of contaminating and fouling i.e. accumulation of unwanted material on solid surfaces which decreases heat transfer capability. The authors concluded that the application of MPC would reduce the variability and allow the process to be optimized to constraint limits, as this functionality would be easily incorporated into the MPC design.

12.1.4 Model Predictive Control Strategy Development

Figure 12-3 shows the configuration of the MEE assembly and the main process variables. Weak (low concentration) black liquor is introduced into the system at one end of the system and flows counter current to steam introduced at the opposite (high concentration) end. The liquor enters the bottom of a series of vertically mounted heating tubes. As the weak black liquor is indirectly heated by the steam supplied to the evaporator, the escaping vapor forces the liquor to quickly pass up the tube wall, with the mixture of liquor and vapor released into the vapor dome at the top of each rising film evaporator. Steam is supplied to the first effect (evaporator), and the evaporated vapor from the first effect is used to provide heat energy to the second effect. Evaporated vapor from the second effect then provides heat energy to the third effect, and so on.

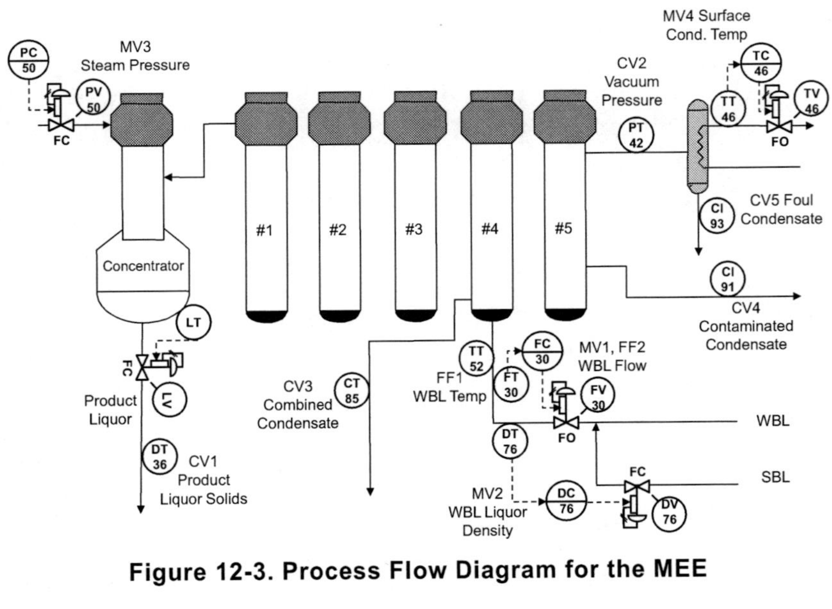

Figure 12-3. Process Flow Diagram for the MEE

The variables of interest for the design of the MPC controller classified as shown Table 12-1 to fit into the MPC strategy design.

Table 12-1. MPC Controller Design

Controlled Variables	Constrained Variables	Manipulated Variables	Disturbance Variables
D36-Solids Density (CV1)	C93-Combined Condensate Conductivity(AV1)	F30-Weak Black Liquor Flow (MV3)	T52-Weak Black Liquor Temperature (DV1)
F30-Production Rate (CV2) Optimized	C91-Contaminated Condensate Conductivity (AV2)	P50-Steam Pressure (MV2)	D76 Weak Black Liquor Density (DV2)
	C85-Foul Condensate Conductivity (AV3)	T46-Surface Condensate Temperature (MV1)	P42 Condenser Vacuum Pressure
	T27 #1 Effect Boiling Point Rise		L70 Soap Skimmer Effect (DV4)

The conceptual basis for the design as summarized in Table 12-1 was motivated by physical plant behavior, desired control outcomes and environmental constraints. For example, because conductivities must be constrained within safe ranges for trouble-free operation, defining them as constraints was quite natural. By the same token, understanding that steam pressure and the effective pressure difference across all of the evaporators influence evaporation led to the designation of the steam pressure and condensate flow as manipulated variables.

The other manipulated variable, weak black liquor flow, was defined as an optimized control variable, in that the intention in throughput control is to maximize that throughput while maintaining good solids control, the solids density being the most important controlled variable. The optimized variable, weak black liquor flow, might not achieve exact regulation; in fact, in transient instances it may be changing significantly in order to maintain solids on target.

The variables classified as disturbance variables are variables that would invite feedforward compensation in the normal scheme of things. One unexpected disturbance variable turned out to be the valve output of the soap skimmer, which by its behavior turned out to be decoupling the front end from the back end of the MEE. This means that depending on that valve output, the influence of a weak black liquor inflow increase or decrease might not necessarily register at the product outflow point, with the soap skimmer acting as a buffer. The feedforward action of the predictive controller was able to take that into account, resulting in better control.

12.1.5 Model Development and Verification

Model development and system identification are an integral part of the MPC application that was used for the project. The process was automatically tested using a series of pseudo-random bump tests that allowed the process to supply the necessary modeling information. During the system identification process, the predicted process output responses were compared to the actual process output responses for the same series of process input data. This verification was used to ensure the process models were accurate. A screenshot of the verification analysis of the final product liquor density is shown in Figure 12-4.

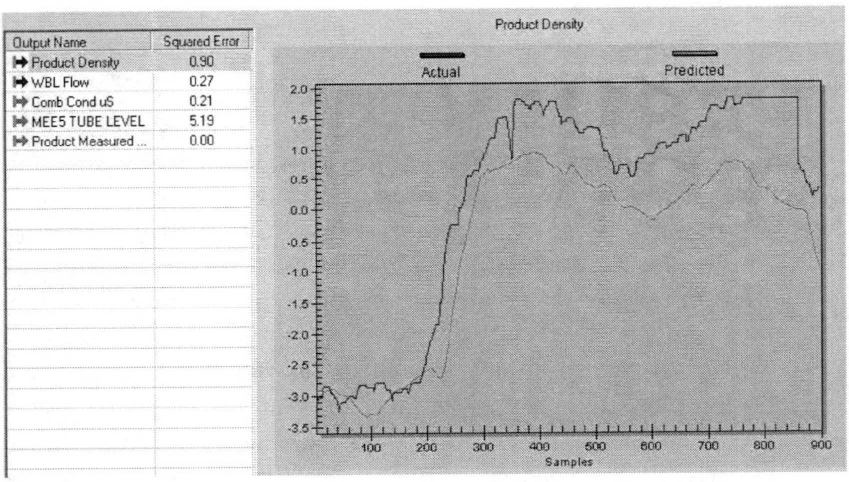

Figure 12-4. Verification Analysis of the Product Density Process Model

Once the final process models were verified, the controller configuration was commissioned, and the MPC controller was placed on-line. The models generated from the available data are shown in Table 12-2, where K = process gain, τ = process time constant and T_d = process deadtime.

Table 12-2. Evaporator Process Model

		Manipulated			Disturbance			
	Parameter	T46	P50	F30	T52	D76	T42	L70
Controlled	D36	$K = 0.05$ $\tau = 162$ $T_d = 300$	$K = 0.69$ $\tau = 697$ $T_d = 310$	$K = -1.16$ $\tau = 831$ $T_d = 530$	$K = 0.19$ $\tau = 762$ $T_d = 720$	$K = 0.03$ $\tau = 715$ $T_d = 30$	$K = -0.25$ $\tau = 600$ $T_d = 210$	$K = 2.27$ $\tau = 1233$ $T_d = 240$
	F30	$K = 0$	$K = 0$	$K = 0.96$ $\tau = 69$ $T_d = 30$	$K = 0$	$K = 0$	$K = 0$	$K = 0$
Constraint	C85	$K = -0.05$ $\tau = 208$ $T_d = 150$	$K = -0.2$ $\tau = 323$ $T_d = 450$	$K = 0.35$ $\tau = 485$ $T_d = 150$	$K = -0.12$ $\tau = 1015$ $T_d = 330$	$K = -0.1$ $\tau = 508$ $T_d = 210$	$K = 0$	$K = 0.39$ $\tau = 817$ $T_d = 1440$
	C91	$K = 0$	$K = -0.12$ $\tau = 462$ $T_d = 50$	$K = 0.4$ $\tau = 208$ $T_d = 120$	$K = 0$	$K = 0$	$K = 0.2$ $\tau = 462$ $T_d = 120$	$K = 0.8$ $\tau = 462$ $T_d = 120$
	C93	$K = 0$	$K = -0.4$ $\tau = 23$ $T_d = 29$	$K = 0.1$ $\tau = 323$ $T_d = 120$	$K = 0$	$K = 0.1$ $\tau = 138$ $T_d = 20$	$K = 0$	$K = 0.4$ $\tau = 323$ $T_d = 30$
	T27	$K = 0$	$K = 0.5$ $\tau = 600$ $T_d = 600$	$K = -1.3$ $\tau = 831$ $T_d = 630$	$K = 0.2$ $\tau = 785$ $T_d = 210$	$K = 0.3$ $\tau = 603$ $T_d = 950$	$K = -0.3$ $\tau = 600$ $T_d = 210$	$K = 1.6$ $\tau = 623$ $T_d = 420$

A point to note is that while the linear model identifier yielded some linear process parameters for the condensate conductivity variables, the reality was the behavior of those variables was highly non-linear. In the end, viable linear models were never obtained, even when feedback linearization transformations were attempted. The non-linear behavior of the condensate conductivities was dealt with via external logic (Section 12.1.7) with quite good results.

12.1.6 MPC Operation, Tuning and Optimization

Although the actual MPC control application was not native to the existing DCS, an OPC communication link provided a reasonably robust interface. As a result, the MPC configuration is tightly integrated with the regulatory controls through mode-switching logic, and the operators are not aware of the system integration complexity. To engage MPC, the operator simply has to place the controller into SUPV ("supervisory") mode. Once the MPC controller is engaged, the operator may make setpoint changes to product density or production rate only. Alternately, the system can automatically optimize product solids and throughput in a coordinated fashion. Figure 12-5 shows the primary operator interface graphic.

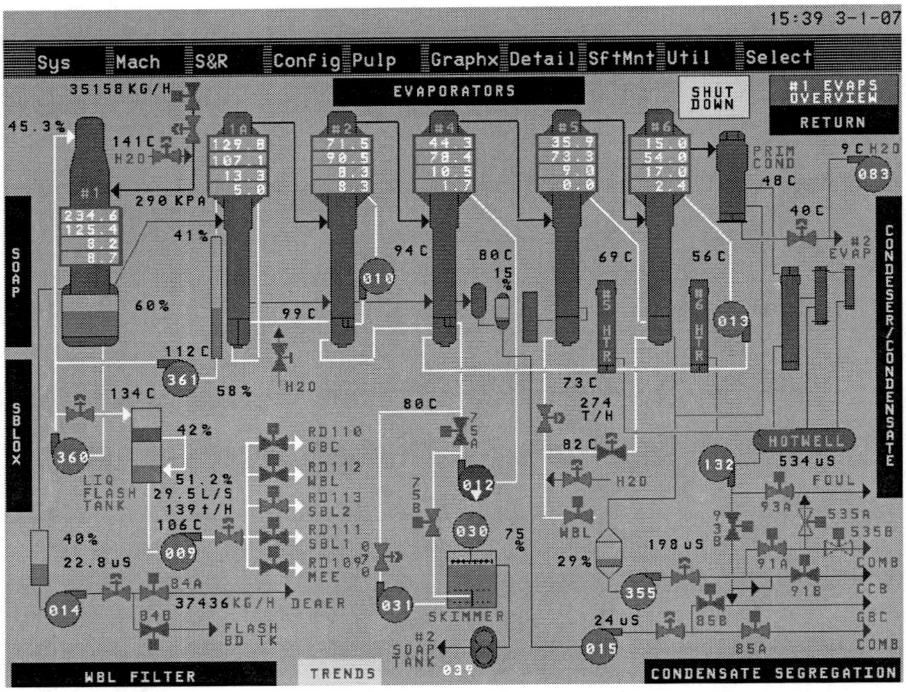

Figure 12-5. Evaporator DCS Operator Interface Graphic

12.1.7 Dealing with Major Process Constraints

During commissioning, the authors were able to successfully model the relationships between feed solids, feed temperature, condenser temperature, first effect steam pressure, soap skimmer level and outlet solids. However, significant difficulty was encountered in modeling and predicting the major process upsets of contamination of the condensate flows. Not only were the conductivity measurements highly non-linear, the conditions that seemed to cause the upsets were not repeatable. This seemed to change each day due to some unmeasured variable that proved impossible to fully detect.

To handle this unique event, an external rule-based controller or "kicker" algorithm was programmed to override the setpoint to the weak black liquor flow controller. This logic would cut the feed flow rate rapidly as the conductivity of the condensate increased and then ramp the flow back up slowly when the conductivity fell below a set limit. With this logic it became possible to ramp up the evaporator feed flow rate until the conductivity started to climb and then ramp down about 2% until the conductivity fell, and then ramp up again. This allowed the evaporators to be run within 1% or 2% of maximum rate without diverting or having to reprocess any condensate. Figure 12-6 shows how the Weak Black Liquor feed cuts back quickly to handle the increasing conductivity while the MPC maintains solids at target value and then, after the conductivity falls, the feed rate is slowly increased.

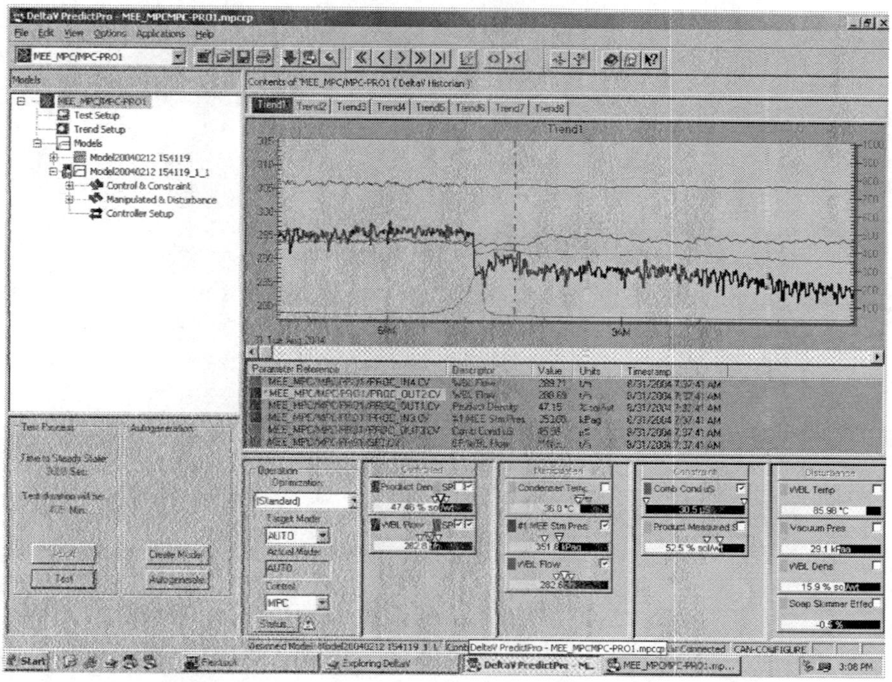

Figure 12-6. Operator Display Showing Response to Conductivity Constraint Violation

12.1.8 Evaporator Flush Control

When the MPC was first put in place, the steam pressure, inlet flow and product solids targets were restricted to about 50% of range, but as time has passed and confidence has increased, these limits have been removed. The operator is now allowed to decrease the product solids target to 30%, which is low enough to wash down the first effect.

This has greatly simplified the procedure for doing a MEE weak black liquor flush and has reduced the time and energy required. The operator simply enters a solids target of 30%, waits 1 hour and then sets the target back to 53%.

12.1.9 Solids Measurement

At the beginning of this project, it was found that the MPC model and the measured solids of the SBL were drifting day to day by as much as 2%. Through further investigation and manual tests, the authors found that a calculated solids (Gullichsen and Fogelholm, 1999) using Boiling Point Rise (BPR) was more repeatable than the solids meter. Over the last two years, it has proven possible to stop using the online solids meter and rely solely on the BPR calculation. The calculated solids is used directly in the MPC model and has provided excellent control results. The calculation being used is shown below:

$$\text{Calculated Solids} = \text{BPR} \times 3.7879 + (\text{Sqrt(BPR)} \times 0.0529) + 2.1 \qquad (12\text{-}1)$$

12.1.10 Operational Results

Since the commissioning of the MPC system in March 2004, operator acceptance has been excellent. The mill is very satisfied with the three year average uptime of 94% for the MPC supervisory control. One of the key benefits of the project has been consistent and stable operation across all crews. As a result of the increased stability and uniformity of operation, the average product solids from the evaporators has increased by at least 3% solids. Table 12-3 provides a summary of the results of the evaporator control optimization.

Table 12-3. Results of Evaporator Control Optimization

	2004	2006	% Change
Average Product Solids	49%	52%	+3% Solids
Average Error From Solids Target	2.43	0.68	-72%
% Time Product Solids > 54% High Limit	1.79%	0.01%	-99%
% Time Combined Condensate Diversion	2.51%	0.38%	-85%
% Time Contaminated Condensate Diversion	0.36%	0.04%	-89%
% Time Foul Condensate Diversion	0.72%	0.19%	-74%

The table shows the average increase in products solids, the percentage of time that the condensates were diverted due to contamination episodes, and the percentage of time that the final product solids was above the 54% high solids limit. As can be seen in this comparison analysis, the improvement in solids variability (72% reduction) has significantly stabilized the evaporator operation. The improvements have reduced the contamination of condensate and diversion time by 70 to 90%. Even more apparent is the fact that the product solids now virtually never exceed the 54% high solids limit. The result of this improvement is that fouling of the effects has been reduced dramatically. There has been no boil-out requirement in the last two years because of the reduction in effect fouling.

12.1.11 Summary

Model Predictive Control (MPC) has proven to be an effective approach to optimizing multiple-effect evaporators in the chemical recovery cycle of a kraft pulpmill. The embedded constraint handling and optimization algorithms allow the safe operation of the MEE at near optimum efficiency, maximizing solids and minimizing times of constraint violation. As a result of the control improvement project, the mill has realized many ongoing operational benefits, including:

- Consistent and stable operation across all crews
- Reduced strong black liquor solids variability by 72%
- Increased strong black liquor solids from 49% to 52%
- Optimized liquor throughput
- Significantly reduced contamination of condensate and diversion time by 70 to 90%
- Decreased fouling due to high solids excursions (no boil-out requirements in the last two years)
- Improved startup times
- Improved and automatic weak black liquor flush control

The constraint control to prevent condensate diversion was extremely important to the success of the project. Some rule-based control was necessary to handle severe upsets due to unmeasured disturbances and system non-linearity. Future work is aimed at looking into modeling and incorporating a better prediction for environmental constraints. The hope is this work may further minimize or even possibly eliminate the contamination events if the controls can be made to respond sooner.

12.1.12 Recent Work and Future Opportunities

Recent work in the last year has been focused on utilizing the economic optimizer that is embedded in the MPC application for further energy optimization. The authors have incorporated an objective function that increases the surface condenser temperature target (Figure

12-3, MV4) to a maximum value when an appropriate mass and energy balance can be obtained that satisfies all the primary control conditions. As a result, any increase in hot water temperature that is achieved at the surface condenser has the potential of displacing low pressure steam that is used in process heat exchangers. Initial results show that it has been possible to raise the temperature of the hot water by as much as 4°C.

12.1.13 Acknowledgments

The authors would like to acknowledge the fantastic support and contributions from the operators at the PGPP Pulpmill.

12.2 Application Example – CTMP Refiner

(Section co-authors Manny Sidhu and Carl Sheehan, Spartan Controls and Stewart McLeod, Catalyst Paper, Port Alberni, BC.)

In order to produce high quality chemi-thermo-mechanical pulp (CTMP), the refining process must be under tight control. Closed loop control of a mechanical refining system is one of the most complex and challenging control problems in a pulp mill. The process is inherently multivariable and exhibits strong interactions. The refining process is also complicated by non-stationary process dynamics due to wear of the refiner plates (Du, 1998).

A significant amount of work has been done to improve TMP refiner control (Alsip, 1981; Roche et al, 1996). All of these control strategies use a single-loop PID based decentralized control architecture. The choice of decentralized architecture was appealing as it was easy to understand and simple to implement using the available distributed control systems of the past. Consequently, in the mid 1990s, a decentralized control strategy based on PID controllers and discrete logic was implemented at Catalyst Paper Port Alberni. This control strategy is referred to as TMPCON.

The system works well in maintaining a constant motor load and attempts to attenuate consistency disturbances after the refiner. However, the multivariable nature of the process was not taken into account, and pulp quality feedback was not considered as part of the design.

In addition to the multivariable nature of the process, process constraints such as limits around the pulp quality and maximum motor load must also be considered. Constrained model based predictive control (MPC) is a natural candidate in the process industry to control such systems. MPC provides a unified framework to efficiently handle a highly interactive process with constraints. MPC technology has also become mainstream in the process industry, and microprocessor based embedded control algorithms are available to simplify the deployment process. A DCS with embedded MPC product was chosen for the project.

In this application example, a fully centralized control strategy for a single CTMP refiner line is presented. The MPC control strategy is defined to maintain the key process variables within their respective constraints while the total applied electrical energy is deliberately minimized. After an overview of the CTMP process at Port Alberni, process modeling and analysis are presented, followed by a discussion of the control strategy design and the use of economic optimization. Finally, the results are presented.

12.2.1 Process Description

Catalyst Paper Port Alberni uses a two-stage high consistency CTMP process to produce about 175,000 tons of pulp every year. The process topology includes one refiner line followed by screening, rejects refining and single-stage peroxide bleaching. The pulp is utilized by paper machines to produce high quality lightweight paper grades and telephone directory grades.

The CTMP mainline refiner line was commissioned in 1987 with two Sprout-Bauer Twin-60 refiners (refer to Figure 12-7). Each refiner is rated for 27,000 hp. The use of an Impressafiner screw press manufactured by Andritz Inc., (Figure 12-7) greatly reduces the bulk density and moisture variations in the feedstock to the refiner. Chemical addition in the impressafiner, known as pre-sulfonation, is used to enhance pulp quality and final pulp brightness. Following the primary refiner (PR) is a pressure cyclone that separates the semi-refined pulp and steam from the primary blow-line. The semi-refined pulp is then fed into the secondary refiner (SR) for further fiber development. At the exit of the secondary refiner is another pressure cyclone, which empties into the latency chest. The steam from the pressure cyclones is reclaimed for mill use.

The process is well instrumented, as depicted in Figure 12-7. The PR blow-line consistency is measured by an in-line optical sensor. The SR blow-line consistency is estimated based on a mass and energy balance around the refiner (Tessier et al, 1997).

At the exit of the latency chest is a pulp quality analyzer that periodically samples the pulp from the latency chest and provides an analysis

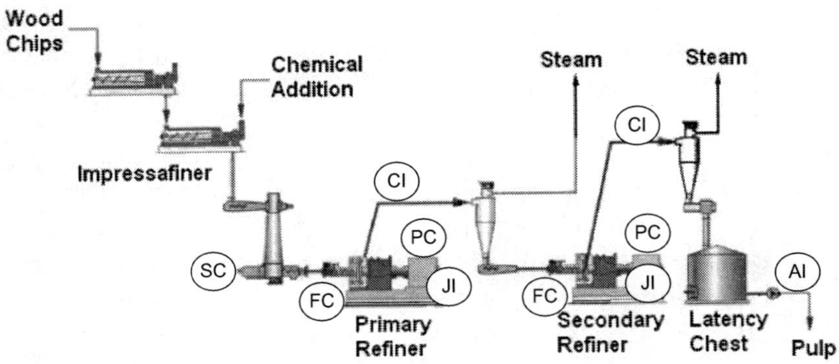

SC - screw speed controller
FC - dilution flow controller
PC - plate pressure controller
JI - motor load indicator
CI - consistency indicator
AI - pulp quality analyzer

Figure 12-7. CTMP Process Schematic

of the pulp quality. In particular, the pulp freeness (the rate at which a dilute suspension of pulp may be drained) as defined by Canadian Standard Freeness (CSF) is measured.

12.2.2 Electrical Energy Consumption Model

The energy consumption for the mainline refiners is modeled using linear regression. Figure 12-8 shows the relationship between specific energy and production rate. The specific energy decreases as production rate increases. This inverse relationship is attributed to the fact that as the production rate increases, the required refiner plate gap decreases. As the production rate increases, the motor load also increases, but the increase in motor load is not sufficient to maintain the necessary energy per unit mass, specific energy, needed for the refining process and therefore plate gap must be decreased to increase the motor load. The decrease in plate gap has a disproportionately high impact on the pulp quality and as result the original pulp quality target must be achieved by decreasing motor load and hence at lower specific energy.

There is a direct relationship between specific energy (kWh/Tonne) and pulp quality (freeness). This relationship is constant provided the furnish composition (wood chip mixture, bulk density, moisture, etc.) is constant for a given production rate. The spread around the trend line illustrates the changes in specific energy requirements, which vary as a function of wood chip furnish, to achieve a given pulp quality target. Therefore, a control system that holds the specific energy target constant is unable to provide optimal pulp quality control and is unable to minimize energy consumption.

Based on historical plant data, the specific energy consumption as a function of production is defined as:

$$\dot{e} = -1.522\dot{m} + 2409.2 \qquad (12\text{-}2)$$

where:

$\dot{e}$	=	specific energy (kWh/Tonne)
$\dot{m}$	=	production rate (Tonnes/day)

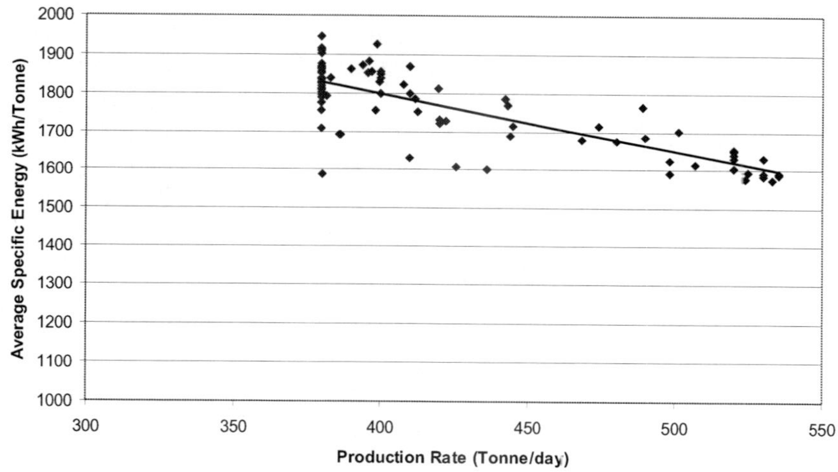

Figure 12-8. Specific Energy versus Production Rate

12.2.3 Process Model Identification

A considerable amount of work has been done on modeling the CTMP process (Allison et al, 1995; Rosenqvist et al, 2002; Ruscio, 1993; McQueen et al, 1999; Berg, et al 2003). The CTMP process exhibits strong interactions that necessitate constructing a multivariable model. For example, a change in the feed screw speed will affect the PR motor load and blow-line consistency. The increase in the feed screw speed will also affect the operation of the SR. However, the changes in secondary refiner (SR) dilution and SR plate gap do not impact the runability of the primary refiner (PR).

To model the refining process, bump tests were performed on each manipulated variable (MV) and the resulting response of all the control/constraint variables (CV) was recorded. The CVs and MVs are listed in Table 12-4:

Table 12-4. Controlled (CVs) and Manipulated (CVs) selected for the MPC Controller

MVs	CVs
Plug screw speed	Primary motor load
Primary plate pressure	Secondary motor load
Secondary plate pressure	Canadian Standard Freeness (CSF)
Primary dilution flows (DE and TE)	Primary blow-line consistency
Secondary dilution flows (DE and TE)	Secondary blow-line consistency

The bump tests were designed to excite the process sufficiently to allow for the development of a rich data set for model identification and verification. The duration and magnitude of the bump tests were adjusted to ensure a high signal-to-noise ratio. Once the necessary data were obtained, built-in model identification subroutines were used to identify, validate and simulate the process models. Each CV (output channel) was modeled with respect to each MV (input channel).

Figure 12-9 shows the step response of the various process sub-models.

12.2.4 MPC Control Strategy

Definition of the control objectives and priorities is a key step in defining the MPC control strategy. The main objectives of the control strategy are:

- Attenuate wood chip density variations
- Control pulp quality (freeness)
- Minimize total applied electrical load
- Minimize refiner motor loads
- Control blow-line consistency
- Maximize production rate

The primary objective of the MPC controller is to minimize refiner motor loads while maintaining the freeness within operator-specified targets. The operators may specify the freeness target between 135 and 145 ml (milliliter), while other variable ranges are be production rate dependent. Table 12-5 summarizes the optimization and control objectives and priorities.

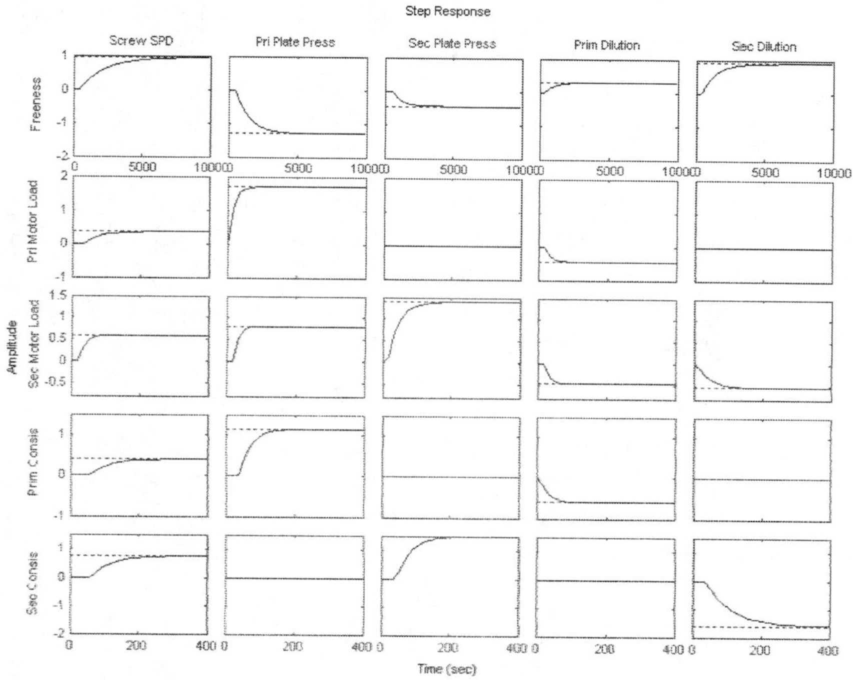

Figure 12-9. CTMP Refiner Process Step Response

Table 12-5. MPC Optimization and Control Objectives and Priorities

CVs	MPC Objective and Priorities
Primary motor load	Minimize
Secondary motor load	Minimize
Canadian Standard Freeness (CSF)	High priority constraint (keep within operator specified target)
Primary blow-line consistency	Medium priority constraint (keep within operator specified target)
Secondary blow-line consistency	Medium priority constraint (keep within operator specified target)
MVs	
Plug screw speed	Maximize (maximize production rate provided sufficient degrees of freedom exist)
Primary plate pressure	Hard constraint (keep within operator-specified limits)
Secondary plate pressure	Hard constraint (keep within operator-specified limits)
Primary dilution flows (DE and TE)	Hard constraint (keep within operator-specified limits)
Secondary dilution flows (DE and TE)	Hard constraint (keep within operator-specified limits)

The MPC algorithm embedded into DCS controller was applied to implement the above control objectives and priorities (Blevins et al, 2003). The algorithm had two control objectives:

1. Minimize CV control error with minimal MV moves within operational constraints.
2. Achieve optimal steady state values set up by the optimizer and the target CV value calculated directly from MV values.

In operation, the optimizer sets up and updates the steady state targets for the MPC unconstrained controller at every scan, thus the MPC controller executes the unconstrained algorithm. Because the targets are set in a manner that accounts for constraints, as long as a feasible solution exists the controller works within constraint limits. Optimization, therefore, is an integral part of the MPC controller. The integrated MPC controller and optimizer performs the following sequence of operations each scan:

1. Update CV measurements
2. Update CV prediction
3. Determine optimal MV steady state targets and calculate the CV targets
4. Provide MV outputs accounting for CV and MV targets
5. Update MV

Figure 12-10 shows communication between the optimizer and MPC algorithm. A typical situation frequently occurs when the optimizer cannot find the optimal solution because constraint limits are too tight, there are too many setpoints or the disturbances are too severe. In such cases, the system constraints are relaxed by changing setpoints to range control and/or abandoning some constraints. This can be done by minimizing the squared error sequentially.

Poor model conditioning, resulted from the model colinearity (defined in Chapter 9), is another common problem an optimizer has to handle gracefully. As with the MPC algorithm, poor conditioning manifests itself as excessive changes in calculated MV targets, even for minor corrections of constraints. In this example application of MPC, poor conditioning is removed dynamically by changing the configuration of active constraints (abandoning some constraints) or by removing the association between constraint or controlled variables and manipulated variables that have excessively big moves.

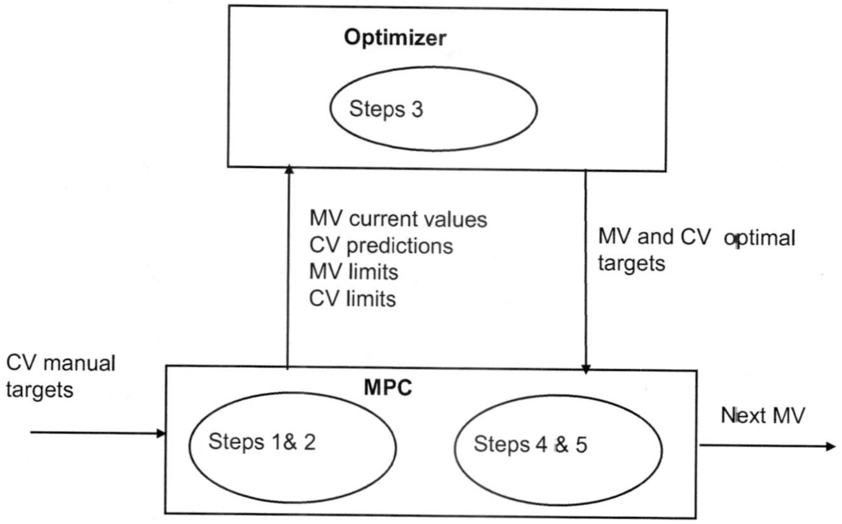

Figure 12-10. MPC and Optimizer Communication Schematic

12.2.5 Results

MPC was successful in reducing the specific energy consumption and improving the pulp quality. Table 12-6 compares the specific energy consumption at a production rate of 390 tonnes/day.

Table 12-6. Specific Energy Comparison

	Mean	Standard Deviation
MPC Off	1776.86	34.67
MPC On	1746.68	61.42

On average, there was a decrease of 30 kWh/Tonne in electrical consumption, and the estimated savings are $303k per year.

The MPC controller has also been able to improve refiner line operation at various production rates. The electrical consumption rate, which relates specific energy and production rate, after MPC commissioning is:

$$\dot{e} = -2.6911\dot{m} + 2806.7 \tag{12-3}$$

This equation can be compared with Equation 12-2 to estimate the specific energy savings at various production rates. As stated previously, the MPC controller was also able to improve the pulp quality as defined by the variations in pulp freeness. Table 12-7 shows the before and after results and shows a 25% reduction in freeness variability. Qualitatively, it can also be seen in Figure 12-11 by the improvement in freeness under MPC control.

Table 12-7. Pulp Freeness Results

	Mean	Standard Deviation
MPC Off	134.7	14.96
MPC On	140.9	11.11

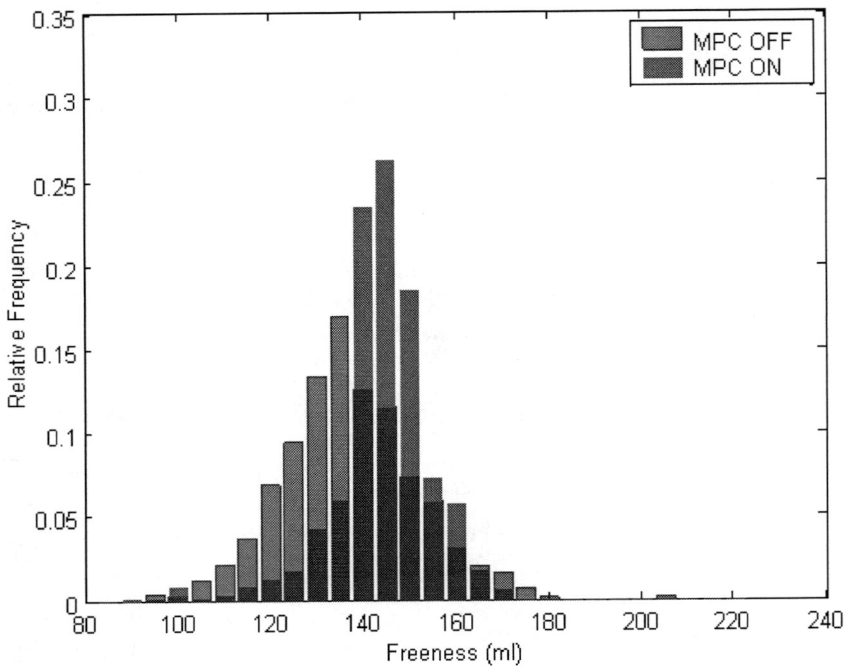

Figure 12-11. Pulp Freeness Distribution

12.2.6 Summary

The mainline refiners consume a significant amount of electrical energy to reduce wood chips to pulp. In fact, CTMP refiners are the single biggest consumer of electrical energy in the mill and hence present the largest opportunity for energy and cost reductions.

Based on historical data, a process energy consumption model was derived. A fully centralized control strategy based on MPC using on-line linear programming optimization was developed. The benefits of using the MPC based control strategy resulted in savings of $303k per year and a reduction of 25% in freeness variability. Future plans will consider the use of a measure of fiber length, mean-fiber length, and chemical addition as part of the control strategy.

12.2.7 Acknowledgments

The authors would like to thank Catalyst Paper management for their financial support. The authors would also like to acknowledge BC Hydro Power Smart (Lester Dyck, Kevin Wallace, Alex Adams, Markus Zeller and Daniel Ouellet) for technical discussions and for providing partial funding to conduct an energy savings study in the CTMP plant. The study was instrumental in uncovering electrical energy saving opportunities in the plant. Finally, the authors would like to thank Port Alberni Division Maintenance and E&I (Dale Shimell, Rick Skelton and Ken Jones) and Operations (all operators, Jason Seabrook and Kelly Sasaki) for their support.

12.3 Application Example – Heavy Oil Fractionator

(Section co-author: Michael Boudreau, Consultant)

The Shell Heavy Oil Fractionator (HOF) problem originally proposed by Shell may be used to illustrate the steps required to configure and commission MPC. In addition, a simulation of the process defined by Shell may be used to evaluate the performance of MPC products.

The Heavy Oil Fractionator problem was originally presented at a Shell Process Control Workshop (Prett and Morari, 1987). The problem was devised by Shell to contain all of the relevant elements present in typical control problems in the petrochemical industry, with the objective of providing academia with a benchmark problem to which to apply new control methods (Ansari et al, 2000; Maciejowski, 2002a). In the problem, it is assumed the heat required for the column enters with the feed.

The column that was the focus of this problem (Figure 12-12) has three product draws and three side circulating reflux loops that remove heat to achieve the desired product separation. The heat duty (heat transfer rate) associated with the bottom loop can be adjusted. The heat duty of the other recirculating loops can be measured but is determined by other parts of the plant. The specifications for the top and side draws are determined by economics and operating requirements. There is no product specification for the bottom draw, but there is an operating constraint on the temperature in the lower part of the column.

In the simulated process, constraint violation is achieved by applying high-level disturbances or by setting low/high limits on the inputs and outputs that are violated by their actual values. Figure 12-12 shows the process proposed by Shell.

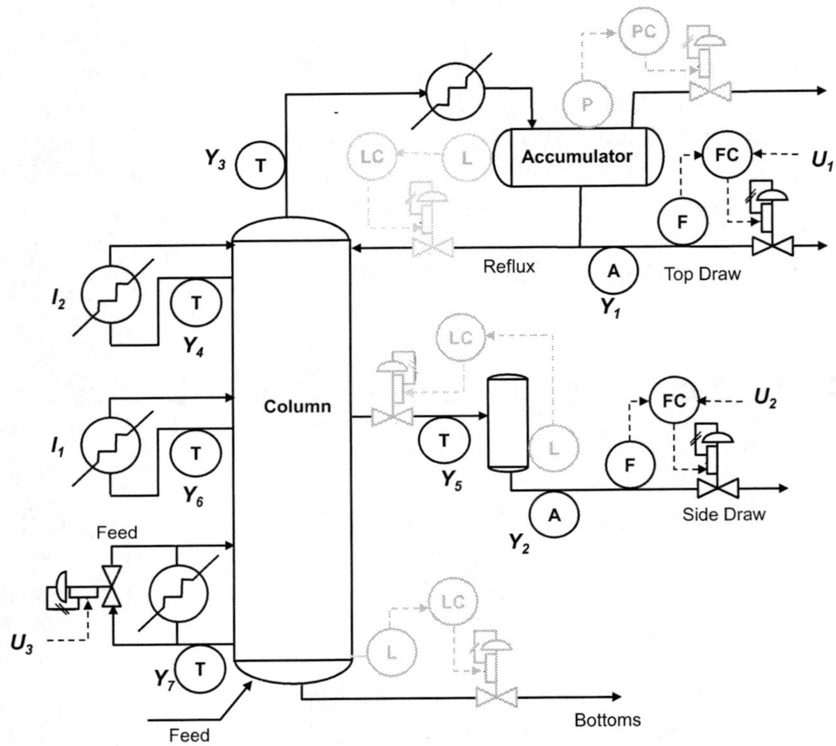

Figure 12-12. Shell Heavy Oil Fractionator

From a control perspective, the process inputs and outputs can be visualized as shown in Figure 12-13 based on the process design, available measurements and operating requirements.

This problem examines a multivariable control strategy's ability to track setpoints, handle constraints, reject disturbances and meet economic objectives in the presence of long deadtimes, model gain uncertainty, multiple constraints, and mixed fast and slow responses.

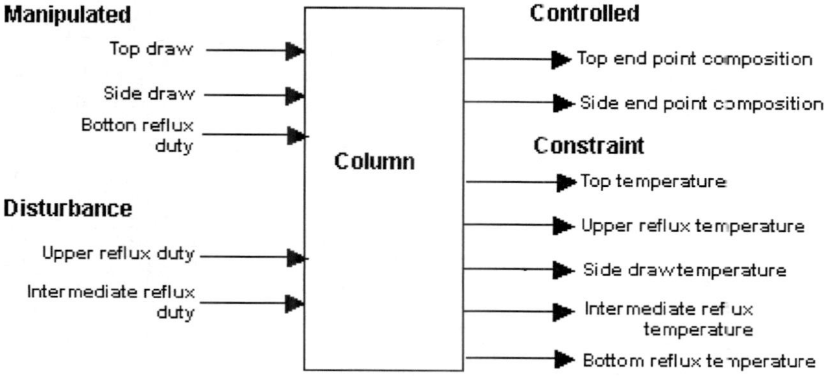

Figure 12-13. Fractionator Inputs and Outputs

The process is difficult to control because it is close to being collinear. The relationship between a change in a process input (with all other inputs constant) and its impact on each process output is modeled as first order-plus-deadtime that is characterized by gain (K), deadtime (T_d) and time constant (τ) expressed in minutes (Table 12-8).

Thus, as addressed in Chapter 14, a dynamic simulation of the process can easily be created within the DCS to allow MPC control of this process to be tested under different operating conditions. As part of the problem proposed by Shell, an uncertainty model gain was defined that allows the control to be tested for different operating conditions. The uncertainties in process gain were defined as shown in Table 12-9 where: $-1 \le \varepsilon_i \le 1$; $i = 1,2,3,4,5$.

Table 12-8. Process Model

		Manipulated			Disturbance	
	Parameter	**Top Draw**	**Side Draw**	**Bottom Reflux Duty**	**Intermediate Reflux Duty**	**Upper Reflux Duty**
Controlled	Top End Point	$K = 4.05$ $\tau = 50$ $T_d = 27$	$K = 1.77$ $\tau = 60$ $T_d = 28$	$K = 5.88$ $\tau = 50$ $T_d = 27$	$K = 1.20$ $\tau = 45$ $T_d = 27$	$K = 1.44$ $\tau = 40$ $T_d = 27$
	Side End Point	$K = 5.39$ $\tau = 50$ $T_d = 18$	$K = 5.72$ $\tau = 60$ $T_d = 14$	$K = 6.90$ $\tau = 40$ $T_d = 15$	$K = 1.52$ $\tau = 25$ $T_d = 15$	$K = 1.83$ $\tau = 20$ $T_d = 15$
Constraint	Top Temperature	$K = 3.66$ $\tau = 9$ $T_d = 2$	$K = 1.65$ $\tau = 30$ $T_d = 20$	$K = 5.53$ $\tau = 40$ $T_d = 2$	$K = 1.16$ $\tau = 11$ $T_d = 0$	$K = 1.27$ $\tau = 6$ $T_d = 0$
	Upper Reflux Temperature	$K = 5.92$ $\tau = 12$ $T_d = 11$	$K = 2.54$ $\tau = 27$ $T_d = 12$	$K = 8.10$ $\tau = 20$ $T_d = 2$	$K = 1.73$ $\tau = 5$ $T_d = 0$	$K = 1.79$ $\tau = 19$ $T_d = 0$
	Side Draw Temperature	$K = 4.13$ $\tau = 8$ $T_d = 5$	$K = 2.38$ $\tau = 19$ $T_d = 7$	$K = 6.23$ $\tau = 10$ $T_d = 2$	$K = 1.31$ $\tau = 2$ $T_d = 0$	$K = 1.26$ $\tau = 22$ $T_d = 0$
	Intermediate Reflux Temperature	$K = 4.06$ $\tau = 13$ $T_d = 8$	$K = 4.18$ $\tau = 33$ $T_d = 4$	$K = 6.53$ $\tau = 9$ $T_d = 1$	$K = 1.19$ $\tau = 19$ $T_d = 0$	$K = 1.17$ $\tau = 24$ $T_d = 0$
	Bottom Reflux Temperature	$K = 4.38$ $\tau = 33$ $T_d = 20$	$K = 4.42$ $\tau = 44$ $T_d = 22$	$K = 7.20$ $\tau = 19$ $T_d = 0$	$K = 1.14$ $\tau = 27$ $T_d = 0$	$K = 1.26\ \tau$ $= 32$ $T_d = 0$

Table 12-9. Uncertainties in Model Gains

		Manipulated			Disturbance	
	Parameter	Top Draw	Side Draw	Bottom Reflux Duty	Intermediate Reflux Duty	Upper Reflux Duty
Controlle d	Top End Point	4.05+2.11ϵ**1**	1.77+0.39ϵ**2**	5.88+0.59ϵ**3**	1.20+0.12ϵ**4**	1.44+0.16ϵ**5**
	Side End Point	5.39+3.29ϵ**1**	5.72+0.57ϵ**2**	6.90+0.89ϵ**3**	1.52+0.13ϵ**4**	1.83+0.13ϵ**5**
Constraint	Top Temperature	3.66+2.29ϵ**1**	1.65+0.35ϵ**2**	5.53+0.67ϵ**3**	1.16+0.39ϵ**4**	1.27+0.08ϵ**5**
	Upper Reflux Temperature	5.92+2.34ϵ**1**	2.54+0.24ϵ**2**	8.10+0.32ϵ**3**	1.73+0.02ϵ**4**	1.79+0.04ϵ**5**
	Side Draw Temperature	4.13+2.39ϵ**1**	2.38+0.93ϵ**2**	6.23+0.30ϵ**3**	1.31+0.03ϵ**4**	1.26+0.02ϵ**5**
	Intermediate Reflux Temperature	4.06+2.39ϵ**1**	4.18+0.35ϵ**2**	6.53+0.72ϵ**3**	1.19+0.08ϵ**4**	1.17+0.01ϵ**5**
	Bottom Reflux Temperature	4.38+3.11ϵ**1**	4.42+0.73ϵ**2**	7.20+1.33ϵ**3**	1.14+0.18ϵ**4**	1.26+0.18ϵ**5**

The test cases proposed by Shell were to demonstrate through simulation that the applied MPC embedded into DCS controller satisfies the control objectives without violating the control constraints (assume all inputs and outputs are initially at zero; the magnitude for the upper and intermediate reflux duty set changes are as indicated below):

$\epsilon_1 = \epsilon_2 = \epsilon_3 = \epsilon_4 = \epsilon_5 = 0$. Upper reflux duty = 0.5, intermediate reflux duty = 0.5.

$\epsilon_1 = \epsilon_2 = \epsilon_3 = -1$; $\epsilon_4 = \epsilon_5 = 1$. Upper reflux duty = - 0.5, intermediate reflux duty = - 0.5.

$\epsilon_1 = \epsilon_3 = \epsilon_4 = \epsilon_5 = 1$. $\epsilon_2 = -1$. Upper reflux duty = - 0.5, intermediate reflux duty = - 0.5.

$\epsilon_1 = \epsilon_2 = \epsilon_3 = \epsilon_4 = \epsilon_5 = 1$. Upper reflux duty = - 0.5, intermediate reflux duty = -0.5.

$\epsilon_1 = -1$, $\epsilon_2 = 1$, $\epsilon_3 = \epsilon_4 = \epsilon_5 = 0$. Upper reflux duty = - 0.5, intermediate reflux duty = 0.5.

When addressing a large multivariable control application, such as that proposed by Shell, the first step in configuration is to define the inputs and outputs that are included in the control problem. If MPC is implemented in the DCS as a function block, then a module may be created that only contains that block. By creating a module that is independent of the modules that contain measurement and lower level control, it is possible at any time to download the module that contains the MPC block. The MPC block may be designed to allow external references to be configured that address the measurements and control loops in the other modules. For simplicity, in this example, the MPC block has been included in the same modules as the measurements and loops it references, as shown in Figure 12-14.

The complete configuration of the MPC block is shown in the following figures. The setpoint limits are based on the normalized range that is used in the Shell control problem. In an actual application, these limits should be set in engineering units based on the actual range of the measurements and actuators (Figure 12-15).

After the module containing the MPC block is downloaded, the application that is used to commission and test the control may be opened. Figure 12-16 shows the main screen of the application as launched from the example MPC block.

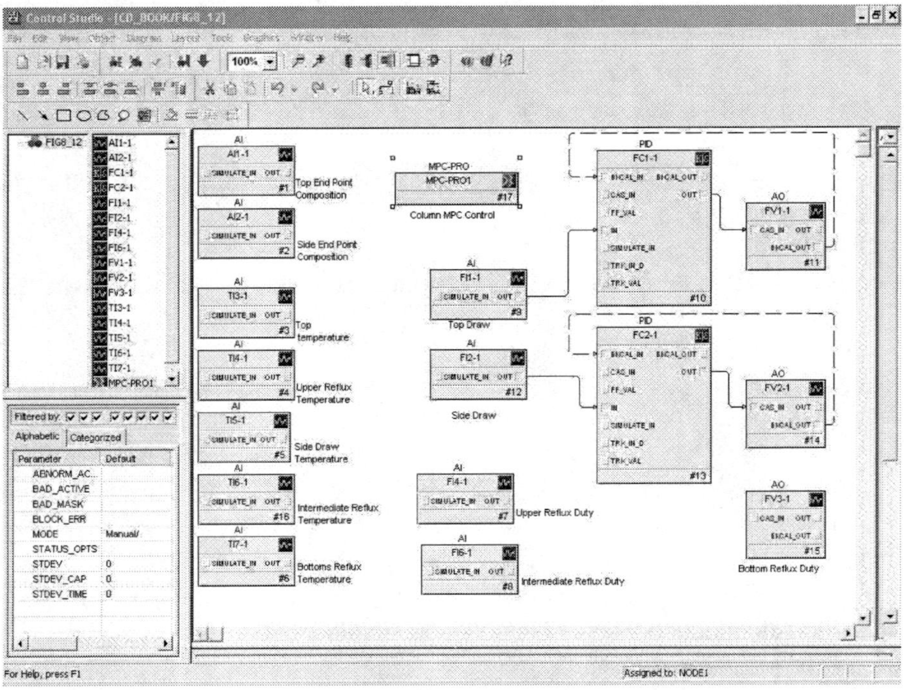

Figure 12-14. Module Containing the MPC block

Controlled | Manipulated | Disturbance | Constraint | Deleted Items

Description	Track SP	SP Filter Time (secs)	SP Low Limit	SP High Limit	Path
AI1-1	Yes	0	0	100	^/AI1-1/OUT
AI2-1	Yes	0	0	100	^/AI2-1/OUT

Controlled | Manipulated | Disturbance | Constraint | Deleted Items

Description	Low Limit	High Limit	Maximum R...	Path
FC1-1	0	100	10	^/FC1-1/RCAS_IN
FC2-1	0	100	10	^/FC2-1/RCAS_IN
FV3-1	0	100	10	^/FV3-1/RCAS_IN

Controlled | Manipulated | Disturbance | Constraint | Deleted Items

Description	Path
FI4-1	^/FI4-1/OUT
FI6-1	^/FI6-1/OUT

Controlled | Manipulated | Disturbance | Constraint | Deleted Items

Description	Low Limit	High Limit	Path
TI3-1	0	100	^/TI3-1/OUT
TI4-1	0	100	^/TI4-1/OUT
TI5-1	0	100	^/TI5-1/OUT
TI6-1	0	100	^/TI6-1/OUT
TI7-1	0	100	^/TI7-1/OUT

Figure 12-15. Configuration of MPC block

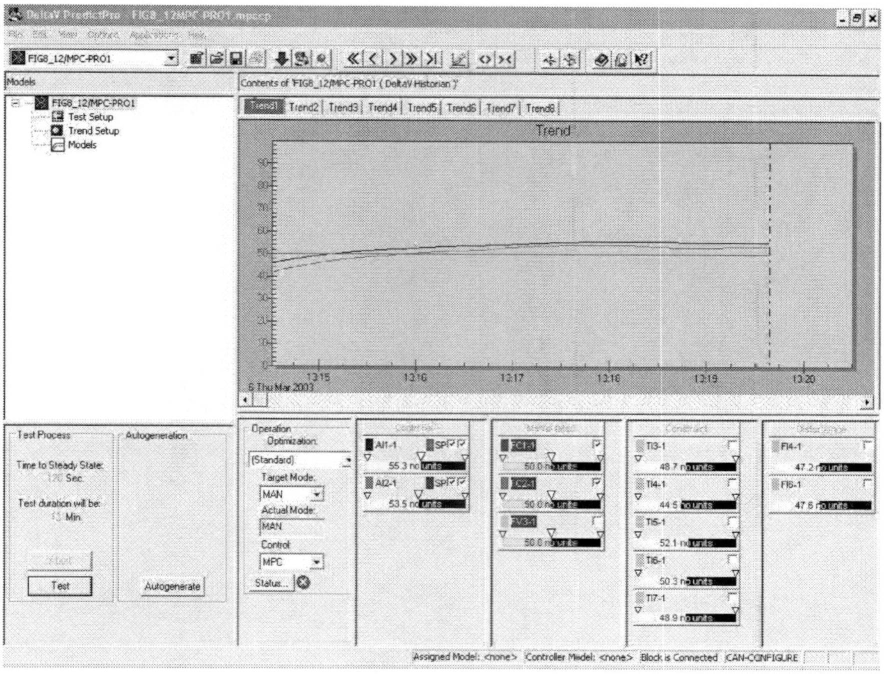

Figure 12-16. MPC Model Development Screen

Based on the MPCPro block configuration, the Controlled, Manipulated, Constraint, and Disturbance parameters are automatically displayed. Notice that the Description that was configured for each of these parameters is used in the faceplate representation. For this example, the Control selection has already been changed from Local to MPC, which has forced the downstream blocks to automatically change their target mode to RCAS. Thus, the mode of the MPCPro block is MAN. Notice also that some parameters are being trended. This is accomplished by clicking in the faceplate check box for the parameter to be plotted.

By selecting the setup button, the Manipulated parameters that are included in the test can be selected, as shown in Figure 12-17.

In this example, all the Manipulated parameters have been chosen for use in the automated test. The Select All feature is used to include all parameters in the process test. The default step size of 5% will be used in the test. Because all process outputs and disturbances are to be identified (the default selection), then nothing more must be done in the setup. The next step is to return to the application's top level view and click the Test button to run the test. In response, the Manipulated

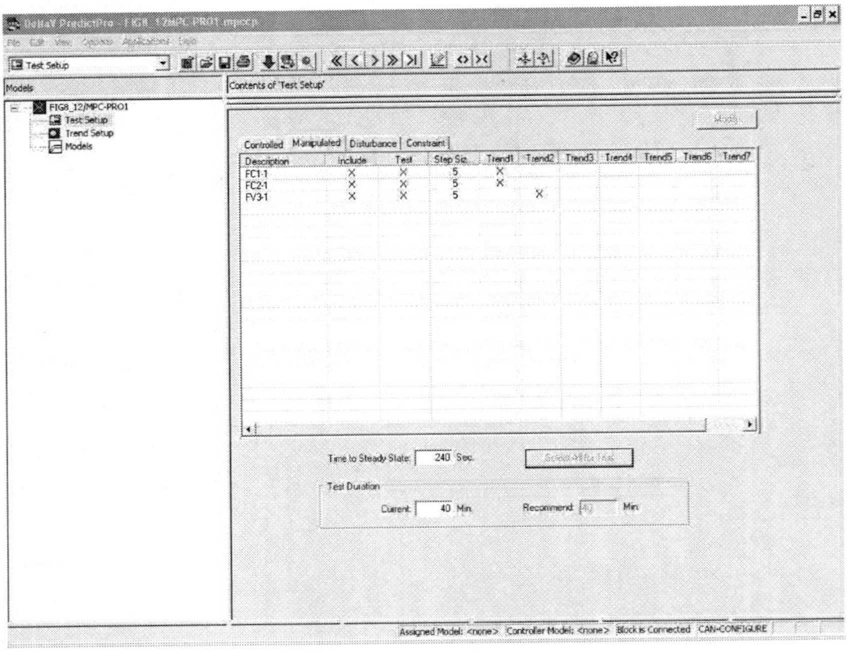

Figure 12-17. MPC Test Setup

parameters will be automatically changed in a pseudo-random manner as shown in Figure 12-18.

When the automated test is complete, the data for testing period is automatically selected as shown in Figure 12-19.

Because process operation during the time of the test was normal, all the test data will be used to generate the model. Thus, by selecting the Autogeneration button, the model and controller for the process may be generated. Verification of the set response is shown in Figure 12-20. For the purposes of this example, the process response has been scaled by a factor of 100.

Based on the squared error indicated in the verification, the model has been accurately identified. Further verification of the model can be done by examining each step response on the screen and comparing it to your knowledge of the process. Example of the Top End Point Composition step responses is shown in Figure 12-21.

Also, further verification of the model can be done by comparing the FIR response to the ARX response and examining the confidence inter-

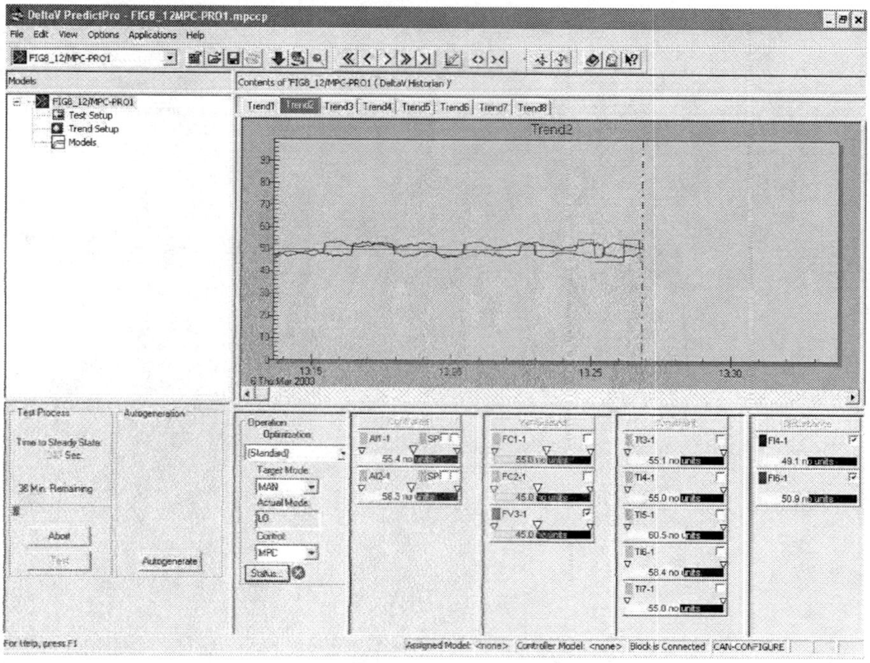

Figure 12-18. Testing Process for Model Identification

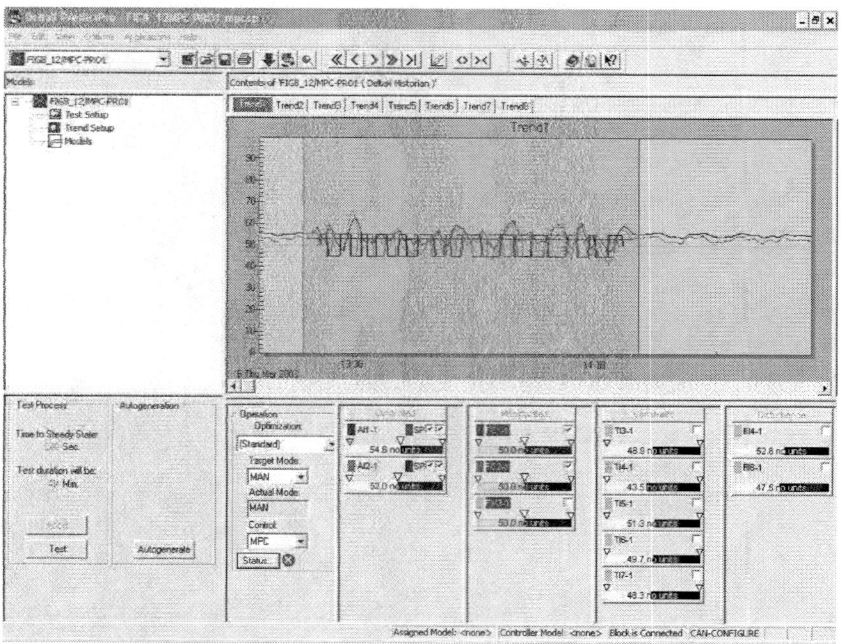

Figure 12-19. Selection of Data for Model Development

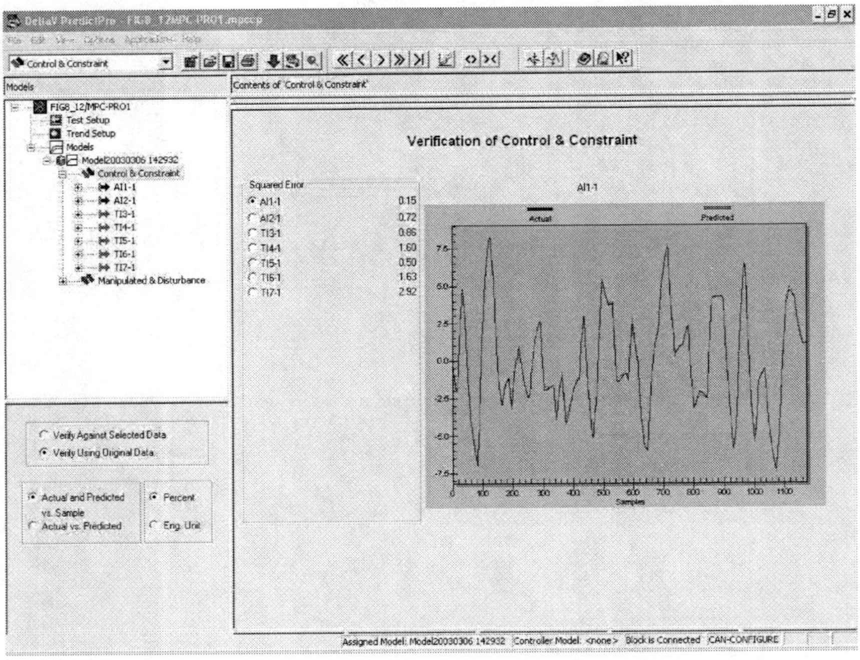

Figure 12-20. Verification of Model

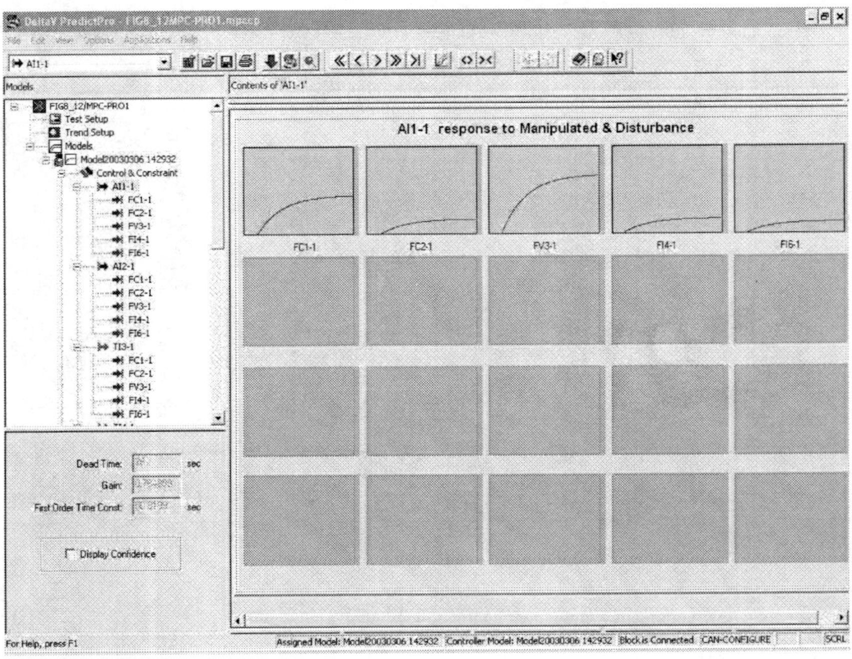

Figure 12-21. User Interface for Examining Step Response

val for one of the step responses as shown in Figure 12-22. Options | Expert must be chosen to use this command.

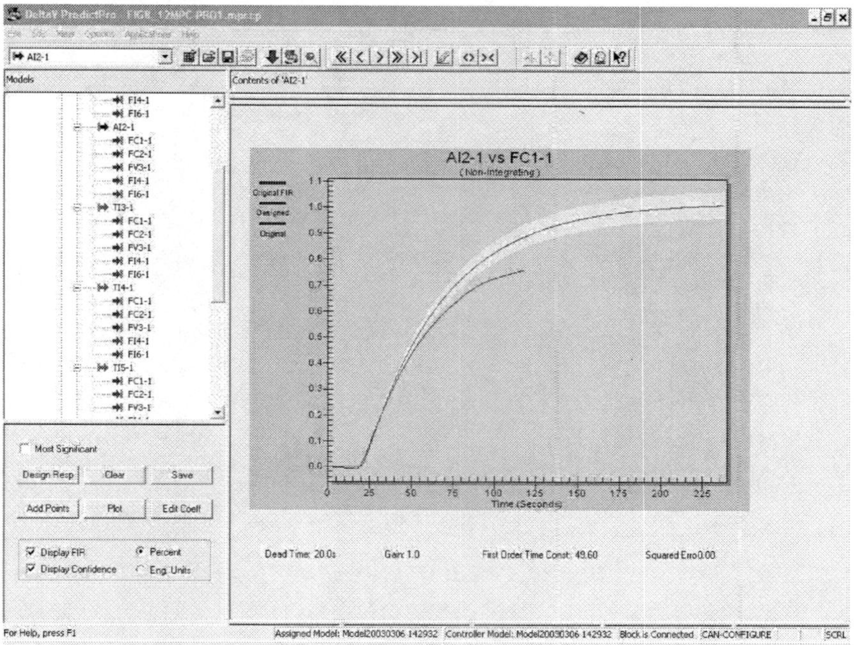

Figure 12-22. Comparison for Identified FIR and ARX Model

The Controlled or Constraint parameter that best reflects changes in the Manipulated parameter is automatically selected for use in controller generation. The Controlled or Constraint parameters that have been selected are represented in different colors on the screen in the step response for each Manipulated parameter. For example, the selection for the Top Draw Flow is shown by the darker background color in Figure 12-23.

The selection of the parameters used in control is based on the parameters' impact on the condition number associated with the control matrix that is generated. Also, the deadtime and gain associated with a parameter should be considered in this selection. Preference is given to Controlled parameters if the performance is nearly the same as that given for Constraint parameters. The Controlled and Constraint parameters selected for the control generation may be examined by selecting Controller Setup in the left pane. Option | Expert must be chosen to see this option. For this example, the information displayed is as shown in Figure 12-24.

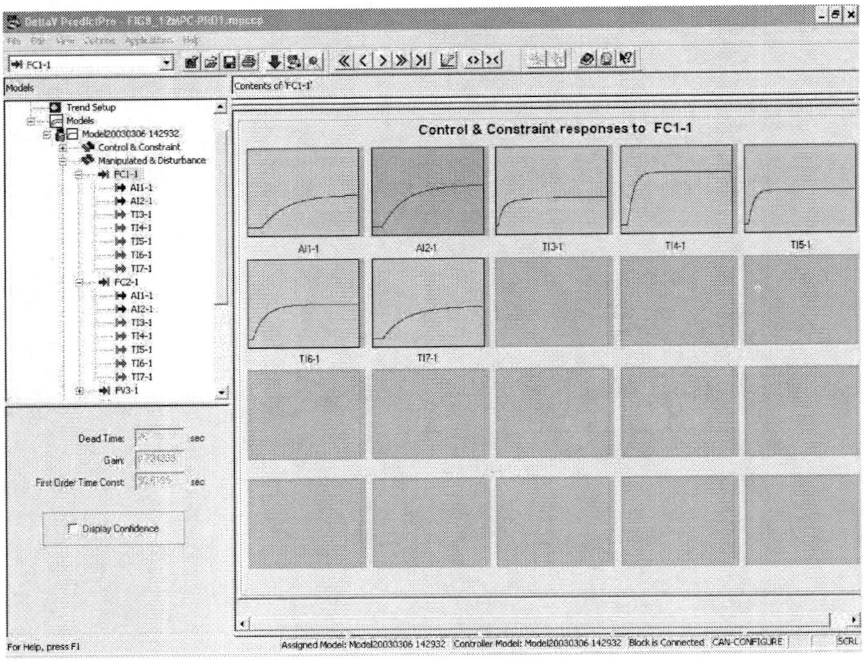

Figure 12-23. Overview of Step Response Model

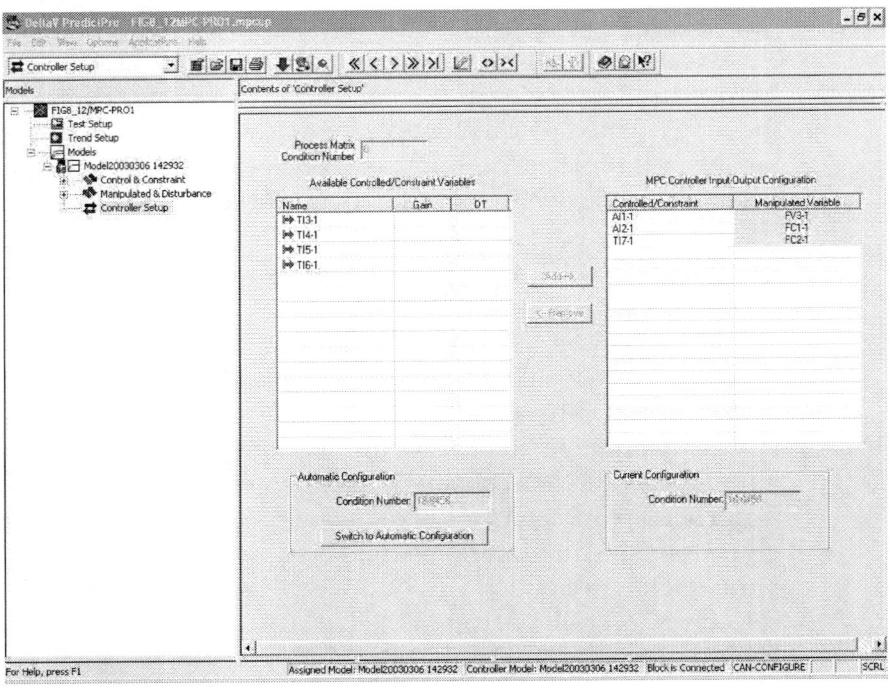

Figure 12-24. MPC Controller Generation

For this example, the automatic choice of parameters gives a condition number of 18. If the Controlled and Constraint parameters are highly collinear (i.e., reflecting the same information), it is possible that Controlled and Constraint parameters cannot be selected for some of the Manipulated parameters. Also, under certain conditions, the user can choose to force this condition by removing a parameter in the selection list as shown above—if this leads to a lower condition number. In general, better control will be provided by a lower condition number.

The Standard objective function provided as the default for the MPC block can be used without modification for many applications. This default objective function is designed to provide control action that maintains the Controlled parameters at setpoint within the operating constraints. When there are extra degrees of freedom, for example when the number of Manipulated parameters exceeds the number of Controlled parameters as in this example, the Manipulated parameters will be minimized while maintaining the Controlled parameters at setpoint.

When it is predicted that a Constraint parameter will violate one of its limits, the working setpoint of the Controlled parameters can be automatically modified within their specified range. If there is a conflict in satisfying the constraints and operating range, then the Controlled or Constraint parameters that are at their limit and are of lowest priority will be given up so that the remaining Controlled and Constraint parameters can be maintained within limits. For this application, the Standard objective function can be used. However, under certain market conditions, it may be desirable to maximize the top draw while maintaining the composition. Under these conditions, a second objective function can be defined as shown in Figure 12-25.

In the case of maximizing top draw, the dollar profit associated with a percent change in the Top and Side End Point composition and the Top Draw flow are not known or can change daily. Thus, the unit cost that is provided is used to indicate the relative value of maintaining composition versus maximizing top draw. For this objective, an objective name of MaxTopDraw is defined. The Operator Selectable checkbox must be clicked for this objective to be available to the operator. Once any objectives have been defined, it is important to download the module to use these new objective functions in simulation and in on-line operation.

To test the column control off-line, it is necessary to select the model in the left pane and click the Simulate button. This command is available if Options | Expert has been selected. Figure 12-26 shows the Simulate

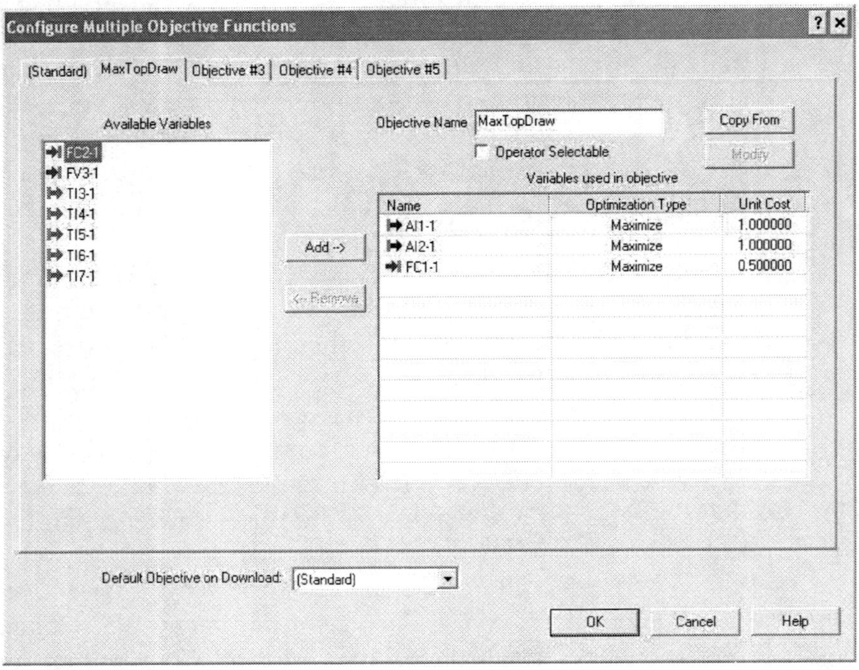

Figure 12-25. Configuration of Second Objective Function

interface for the example MPC block when the block is in Automatic mode and the Standard objective function has been selected.

The control responses may be observed by changing the setpoint. The control objective may also be changed to examine the impact on the process. For example, the impact of changing the objective from Standard to MaxTopDraw is reflected in the trend shown in Figure 12-27.

Once the MPCPro control has been verified using the simulation environment available with the MPC product, the module containing the MPC block can be put into service. The predefined dynamos for PredictPro can be included in the operator interface. Figure 12-28 shows an MPC Operate view for this example.

In the tests defined by Shell, the fractionator process deadtimes and time constants do not change, but in four of the five proposed test cases, the gains are changed in a correlated manner. The MPC algorithm under study is tested for how it meets the control objectives in one test case with no model mismatch and in four test cases with different gain settings.

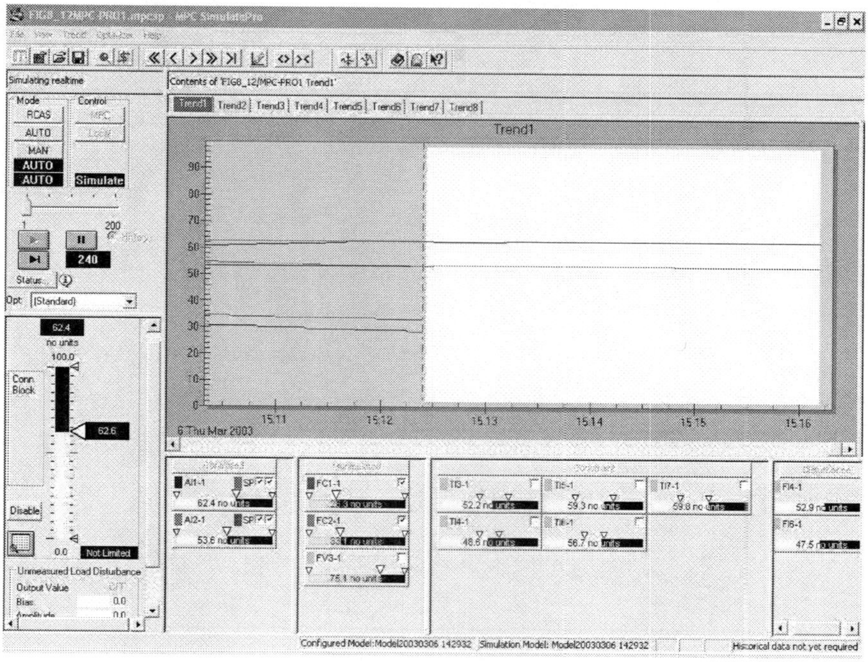

Figure 12-26. Simulation Environment for Testing MPC

The MPCPro formulation combines a squared model used for MPC controller generation (number of CV equal number of MV) with soft constraints (Mehta et al, 2003). The controller is not reconfigured on-line to meet hard constraints (Wojsznis et al, 2002). Instead, it is automatically configured prior to controller generation so that manipulated variables are matched to controlled or constraint variables. Selection of the controlled and constraint variables used in control is based on the impact on condition number of the control matrix and the gain and deadtime. Controlled variables are preferred over constraint variables that give the same performance.

The Model Predictive Controller has been tested (Boudreau, 2003) with feedforward of the measured disturbances (Cuthrell et al, 1990). As previously shown, the controller is configured with End Points as controlled variables and the Bottoms Reflux Temperature as a constraint variable. Controlled variables can be configured with a "Setpoint Delta Range" and are maximized or minimized within the range. Controlled variable setpoints are maintained as long as constraint variables are projected to be within their limits. If a constraint violation is predicted, then the working setpoints of controlled variables are

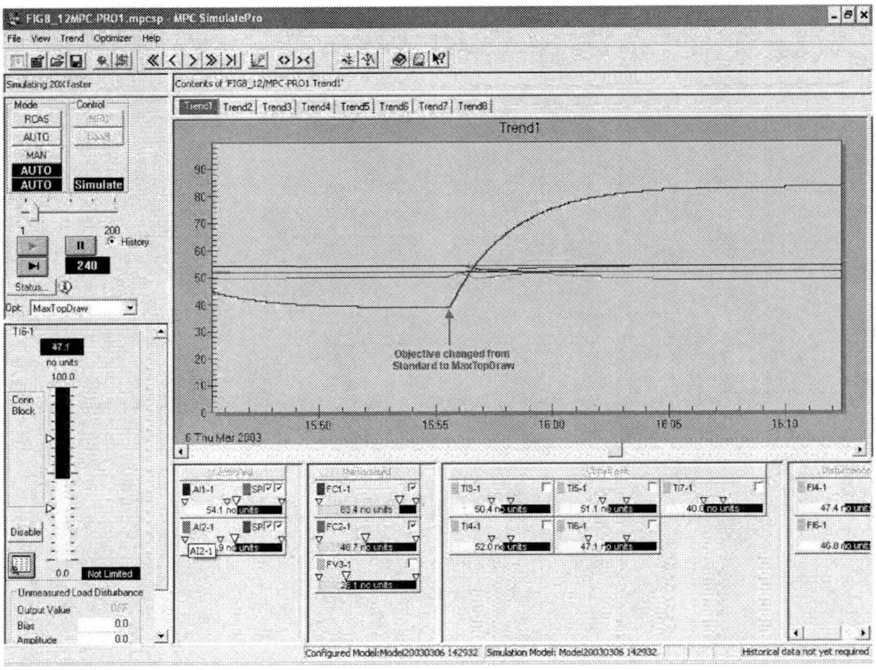

Figure 12-27. Testing Impact of a Change in Objective Function

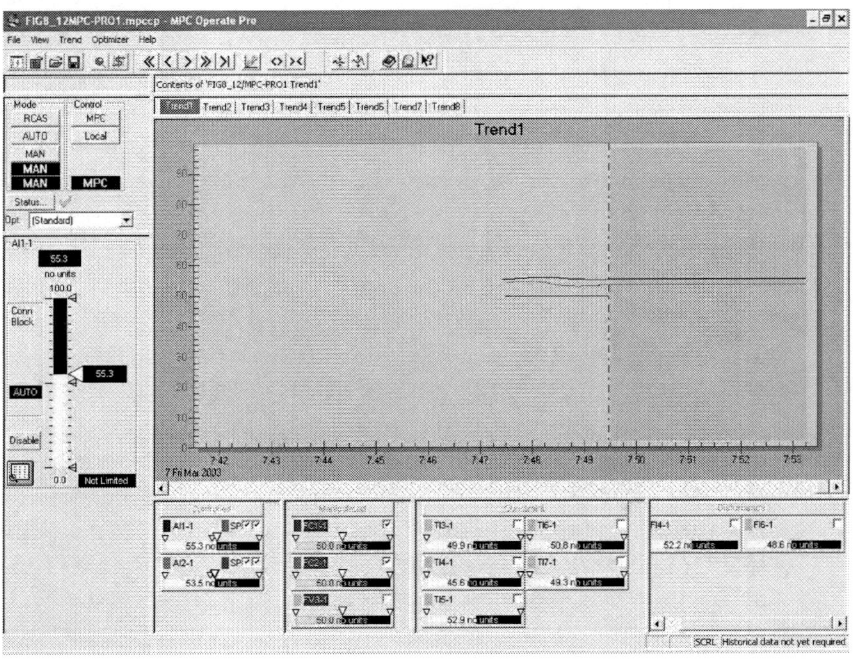

Figure 12-28. Operator Interface to MPC

moved within the Setpoint Delta Range to prevent the constraint violations.

To meet the steady-state requirement of the Shell problem, the Setpoint Delta Range is set to zero for both controlled variables. When the optimizer cannot find a solution that satisfies constraints while maintaining controlled variables in their Setpoint Delta Range, constraint variables are allowed to deviate beyond their constraints.

To find the best solution, constraint and controlled variables are prioritized. The variable with the highest priority has the smallest priority number assigned to it. Variables with the lowest priority are removed from the equation, one after another, until a solution is found. When constraint violations are present, the optimizer searches for a solution that keeps all controlled variables within their Setpoint Delta Range while minimizing the weighted error of variables that exceed their range. To meet the Shell Top End Point constraint requirement, it is the highest priority. Bottoms Reflux Temperature is the lowest priority.

The Bottoms Reflux Duty is minimized. By decreasing the heat returned to the Fractionator, steam production is increased. The MPC's hard manipulated variable constraints meet the Shell problem's requirement. To meet the maximum move size requirement of 0.05 per minute, all "Max MV Move Per Sec" parameter settings are 0.0008333 per second. The recommended prediction horizon in this MPC formulation is based on the settling time. Three times the longest Fractionator time constant plus deadtime gives a settling time of 12,480 seconds.

The controller executes 120 times over the prediction horizon, so the period between controller executions is 104 seconds. However, the open loop time constant between Bottoms Reflux Temperature and Bottoms Reflux Duty is only 19 minutes. There is no deadtime. To increase resolution here, the controller execution is set at the fastest sampling time permitted in the Shell problem. The minimal computational requirement of this MPC formulation allows sampling times of much less than one minute. However, the resulting reduction in controller gains in the longest open loop step responses would effectively de-tune the controller.

In the test proposed by Shell, the MPC is expected to reject two simultaneous step disturbances and at the same time perform an economic optimization in five test cases. Gain uncertainties exist in four of the five test cases. The presenters at the Second Shell Process Control Workshop voted Test Case 2 as the worst (Prett et al, 1990). Test Case 5 came second in that category. Gain changes in the process outputs related to Top Draw vary from 40% to 70%. Side Draw gain changes vary from 8% to 22%. In Test Case 2, Top and Side Draw gains are decreased. Gains related to disturbances are increased, but on average the increase is less than 10%. The significant decrease in process gains effectively increases controller gain. Top Draw related gains are decreased in Test Case 5, but Side Draw related gains are increased so the effect on the controller is not as pronounced. Process gains are increased in Test Cases 3 and 4, and thus move the controller in the direction of stability. Disturbances in opposite directions make Test Case 4 the easiest to satisfy test requirements.

The results for all responses with feedforward (Boudreau, 2003) can be seen in Figures 12-29 through 12-39. Disturbances are available to the controller and shown on the trends. All test cases met the Shell performance criteria with the Penalty on Move and Penalty on Error settings calculated automatically on controller generation. No additional manual tuning was done. All test case responses exhibit integral action and satisfy the designated constraints. Settling times vary from 217 to 358 minutes.

Table 12-10. Settling Times of Controller with Feedforward Control

Test Case	1	2	3	4	5
Settling Time, min.	217	294	358	345	213

In Test Case 1 (Figures 12-29, 12-30), the optimization is good. The Bottoms Reflux Temperature rests at its lower constraint of -0.5. The oscillations that occur when initially reaching the constraint may be reduced by increasing the Penalty on Move. Since the Bottom Reflux Duty is optimized, its action does not have to be so aggressive. The optimizer solution is found at different constraints with the negative load direction in Test Cases 2, 3 and 5. In all these cases, the Top Draw is at its 0.5 constraint value. Note that it is a hard constraint and is not violated in any of the tests.

Test Cases 2 and 3 have the same loads, but the model errors are in different directions. Test Case 2 lives up to expectations and demonstrates the worst disturbance rejection performance. The Side End Point violates its lower constraint briefly. Test Case 3 shows better disturbance rejection performance, but Manipulated Variables migrate further from their origin.

Negative load direction puts Test Case 5 in the same category as Test Cases 2 and 3. Recall that the largest gain uncertainties are in opposite directions. Those uncertainties related to the Top Draw are increased while those of the Side Draw group of gains decrease. The disturbance rejection and response time performance is much better in Test Case 5 than in Test Cases 2 and 3. The Top Draw saturates as quickly as it did in Test Case 2, but Manipulated Variable movement is minimal.

This soft constraint MPC formulation disagrees with the Second Shell Process Control Workshop consensus and votes Test Case 3 as the second worst case. With increased process gains and opposing disturbances, the optimizer in Test Case 4 comes to the same constraint as it did in Test Case 1. Controlled variable deviation from setpoint during disturbance rejection is the smallest in this test case. However, manipulated variable movement is larger and settling time is longer here than in Test Case 1.

One of the control objectives of the Shell problem is to reject disturbances even when one or both end point analyzers fail. If a sensor has a failed status or an analyzer status is constant, this MPC formulation allows the model value from the last execution to be passed to the MPC algorithm as a measured input. When the MPC is so configured, a timer is set at the prediction horizon and begins to count down. If the sensor or analyzer status has not returned to "Good" by this time, the MPC status changes from "Simulated" to "Failed."

This simulation feature is most useful in avoiding nuisance MPC controller shutdowns on analyzer quality or calibration cycles. A bump in the process measurement is not likely on an analyzer outage of short duration. However, there is the possibility of a difference between predicted and actual measurements on an optimized parameter in the presence of model mismatch (Rawlings, 1990). Load rejection performance while in Simulation mode as well as the controller's ability to handle the transition from simulated to measured data after a long period were examined with the worst case model when both end points are simulated for a period slightly less than the prediction horizon (Figure 12-39).

Both end point analyzers were put in Manual mode soon after the start of this test. Manual mode caused the endpoint measurement status to become "Uncertain Constant" and caused the switch to simulation. Just over an hour later, as the end points were approaching steady state, the analyzers were placed back into Automatic mode. On sensing a good signal, simulation was switched off, and the measured value was sent to the controller. This MPC algorithm does not filter the transition to real data. Although the bump was significant, it was only a small fraction of the measured disturbance. The controller remained stable.

In conclusion, an MPC formulation containing an LP optimizer based on steady state models met the Shell HOF control problem criteria in all test cases. A prudent controller configuration that included feedforward rejected measured disturbances in the presence of model mismatch and end point analyzer failure. After the original reduction of the prediction horizon, no further user tuning was required.

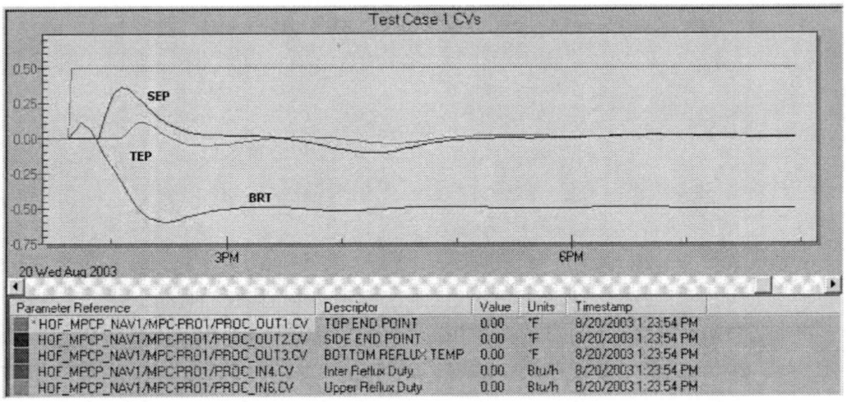

Figure 12-29. Controlled Variable Responses in Test Case 1

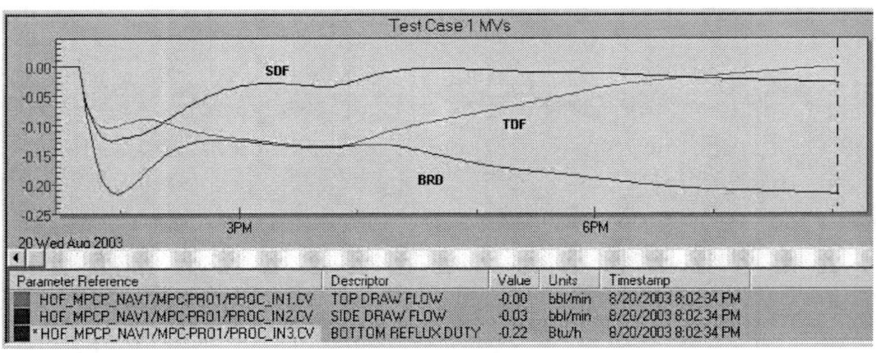

Figure 12-30. Manipulated Variable Moves in Test Case 1

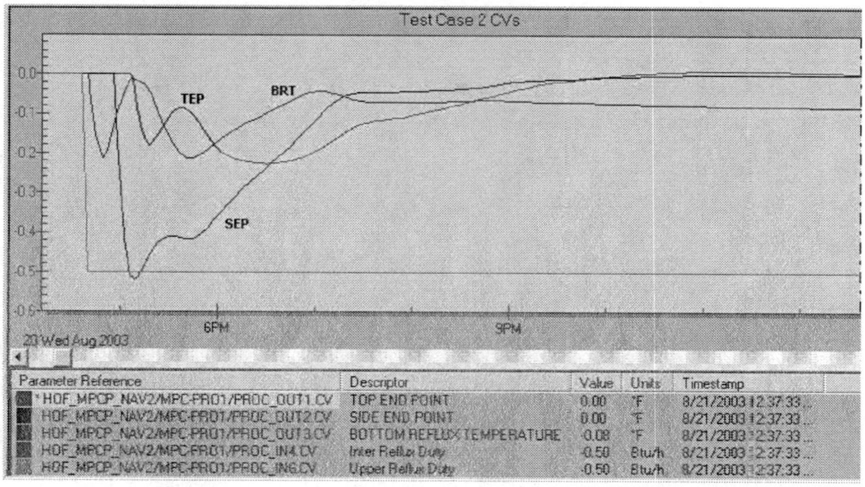

Figure 12-31. Controlled Variable Responses in Test Case 2

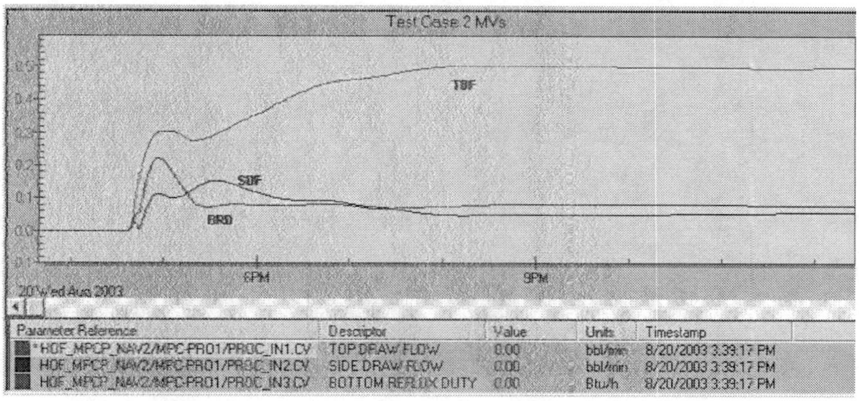

Figure 12-32. Manipulated Variable Moves in Test Case 2

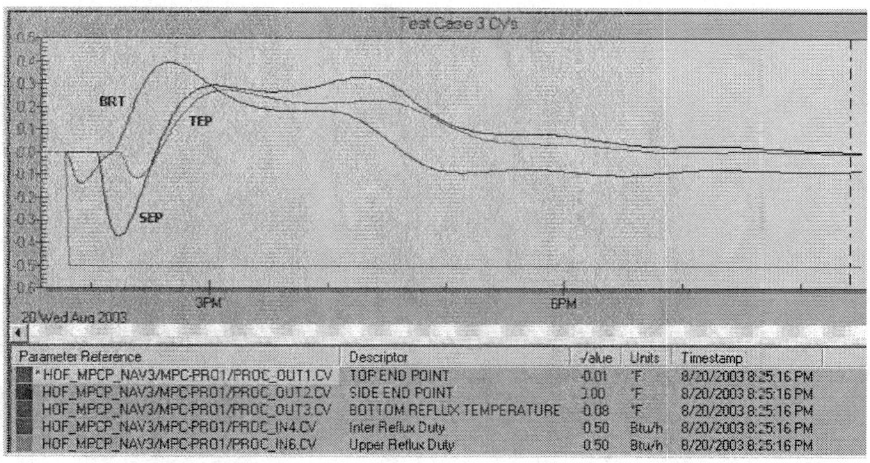

Figure 12-33. Controlled Variable Responses in Test Case 3

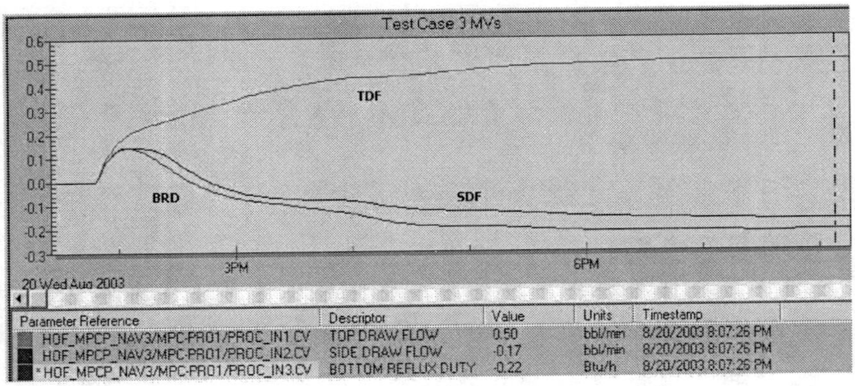

Figure 12-34. Manipulated Variable Moves in Test Case 3

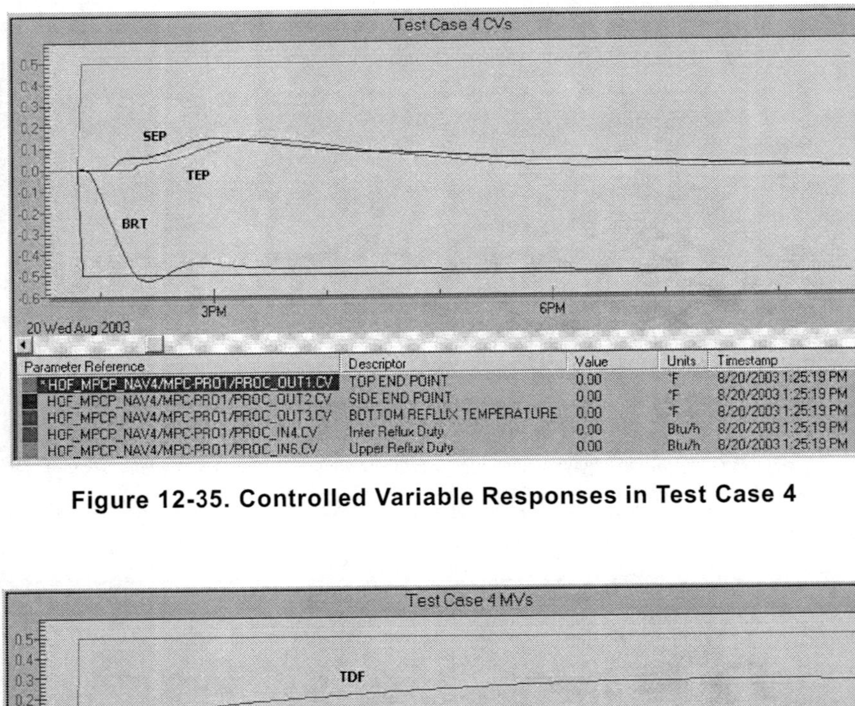

Figure 12-35. Controlled Variable Responses in Test Case 4

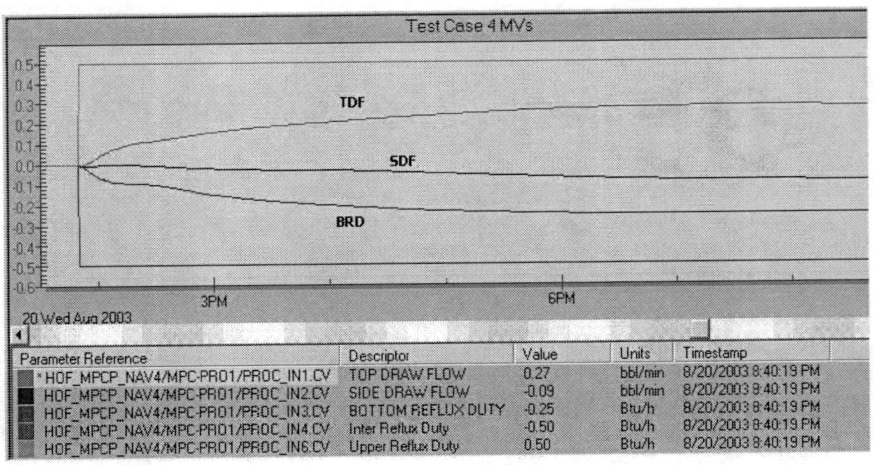

Figure 12-36. Manipulated Variable Moves in Test Case 4

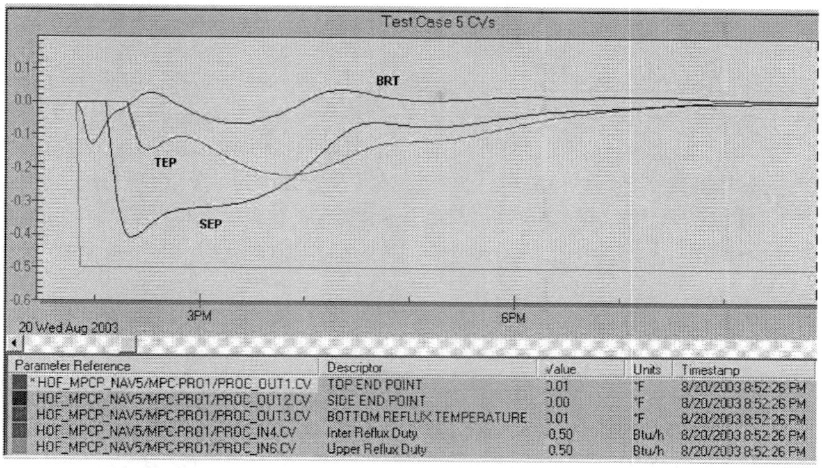

Figure 12-37. Controlled Variable Responses in Test Case 5

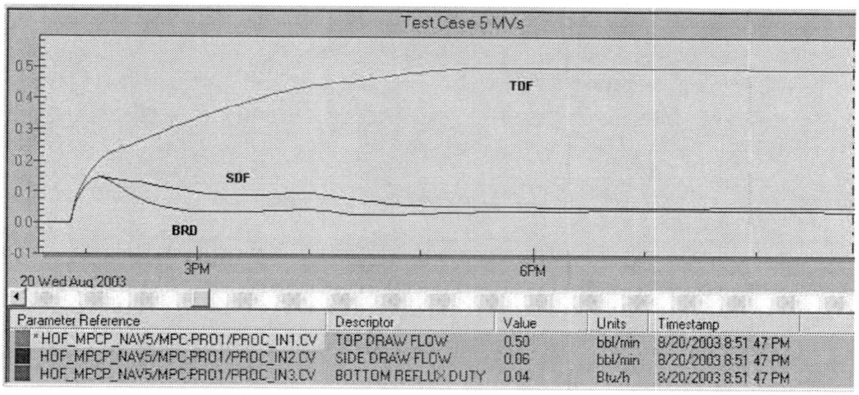

Figure 12-38. Manipulated Variable Moves in Test Case 5

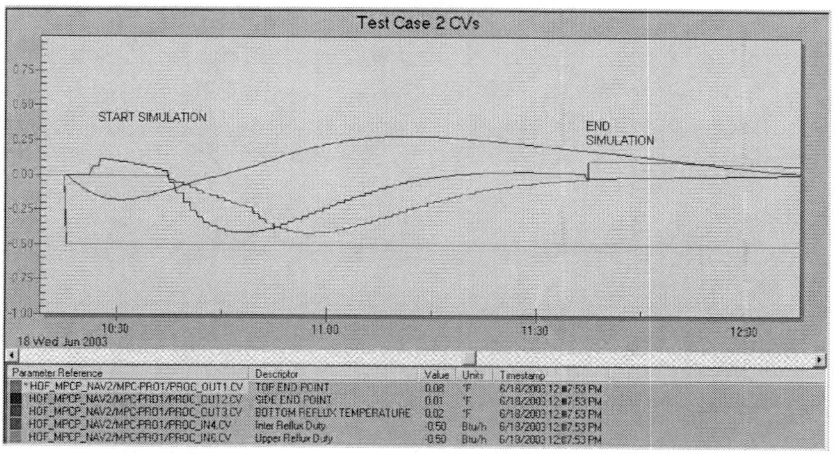

Figure 12-39. Rejection of Disturbance when Both End Point Analyzers Have Failed

12.4 MPC Integrated with Optimization Workshop

This workshop provides several exercises that may be used to further explore the MPC capability that was addressed in the Heavy Oil Fractionator application example. The fractionators process simulation is based on the process dynamics defined by Shell but running faster than real-time. Access the book's web site http://www.advancedcontrolfoundation.com, and select the Solution tab to view the workshop being performed.

Step 1. Observe the MPC configuration and the step response identified during MPC commissioning.

Step 2. Using MPC with the simulated process, observe how the control automatically compensates for a step changes in the top and intermediate reflux duty to maintain the top end and side end point composition at setpoint.

Step3. When the top end composition setpoint is changed, observe how the MPC works to achieve this new setpoint.

Step 4. Note the impact that a change in control objective has on the column operation.

12.5 Technical Basis

A basic feature of the most industrial MPC controllers is tight integration with an optimizer. A major benefit of MPC comes from process optimization, which is inherently related to MPC process constraints handling. MPC control itself can be defined as an optimization problem (Maciejowski, 2002). Quadratic programming, and more current techniques like the interior point method (Rao, Wright, and Rawlings, 1998) have been proposed as an efficient solution of this problem.

To run control by an optimizer, the optimization solution is required every scan, and it should provide MPC controller moves, taking into account process dynamics, current constraints and optimization objectives. However, the computational burden of this approach is tremendous. Therefore, the most common approach in industry is to solve the problem in two steps:

1. Find the optimal solution for predicted process steady state values, accounting for constraints.

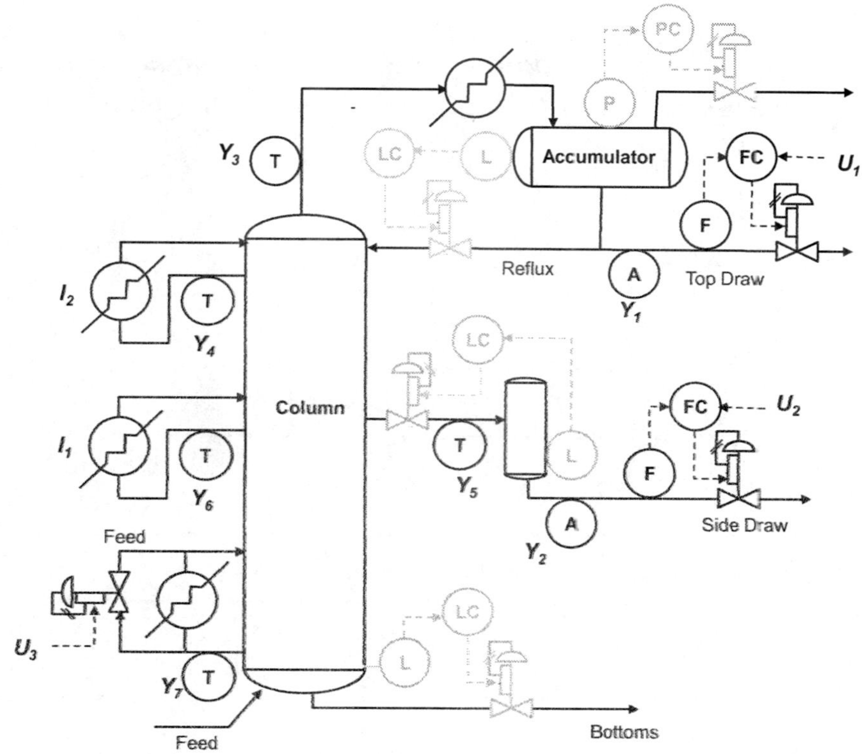

Figure 12-40. Shell Heavy Oil Fractionator

2. Control the process dynamically, satisfying steady state targets defined by the optimizer.

Significant advantages can be obtained by integrating MPC with the control system infrastructure. As a result, it can be configured, tested and commissioned with much less effort, delivering reliable operation as well as simplicity in use and maintenance. Some initial material on the subject has been already provided in Chapter 11. The following, more advanced topics will be presented in this section:

- MPC controller integrated with optimizer operation overview
- On-line optimizer structure and functionality
- MPC unconstrained controller integrated with optimizer operation
- MPC constrained controller integrated with optimizer operation

12.5.1 MPC Operation Overview

A general MPC diagram (Figure 12-41) shows the basic components of the constrained MPC control structure: on-line steady state optimizer, MPC dynamic controller, and DCS measurements and regulatory control.

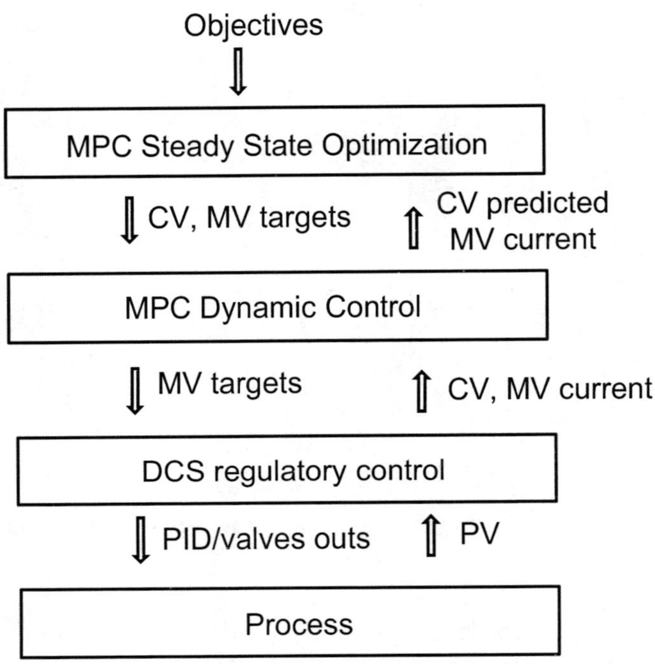

Figure 12-41. Basic Components of the Constrained MPC Control Structure

For processes with several manipulated or controlled variables, optimization techniques are essential components of Model Predictive Control technology. One of the proven approaches, as discussed earlier in this chapter, uses linear programming (LP) with steady state models. Linear programming is a mathematical technique for solving a set of linear equations and inequalities that maximizes or minimizes a certain additional function. This additional function is called the *objective function*. Objective functions usually express an economic value like cost or profit.

Specifically, (and also by way of review) MPC optimization considers incremental Manipulated Variable (MV) values at the present time or the sum of increments of the MV over the control horizon, and incremental values of Controlled and Constrained Variables (CV) at the end of the prediction horizon, instead of positional current values as in typical applications of LP. The LP technique uses steady state model, and therefore a steady state condition is required for its application. With the prediction horizon normally used in MPC design, for predicted process outputs at the end of the prediction horizon steady state is guaranteed for self-regulating processes.

The predicted process steady state equation for an *m* by *n* input-output process, with prediction horizon *p* and control horizon *c*, in the incremental form is:

$$\Delta CV(t+p) = A \times \Delta MV(t+c) \tag{12-4}$$

where:

$\Delta CV(t+p) = \begin{bmatrix} \Delta cv_1 \\ \dots \\ \Delta cv_n \end{bmatrix}$ = vector of predicted changes in *n* outputs at the end of the prediction horizon

$A = \begin{bmatrix} a_{11} \dots a_{1m} \\ \dots\dots\dots \\ a_{n1} \quad a_{nm} \end{bmatrix}$ = the process steady state $m \times n$ gains matrix

$\Delta MV(t+c) = \begin{bmatrix} \Delta mv_1 \\ \dots \\ \Delta mv_m \end{bmatrix}$ = vector of changes in *m* manipulating variables at the end of the control horizon

Vector $\Delta MV(t+c)$ represents the sum of the incremental changes over the control horizon made by every controller output mv_i:

$$\Delta mv_i = \sum_{j=1}^{c} \Delta mv_i(t+j) \qquad i = 1, 2, \dots, m$$

The changes should satisfy limits on both MVs and CVs:

$$MV_{\min} \le MV_{current} + \Delta MV(t+c) \le MV_{\max} \tag{12-5}$$

$$CV_{\min} \le CV_{predicted} + \Delta CV((t+p) \le CV_{\max}) \tag{12-6}$$

The objective function for maximizing product value and minimizing raw material cost can be defined jointly in the following way:

$$\underset{\min}{Q} = -P^T \times \Delta CV(t+p) + D^T \times \Delta MV(t+c) \tag{12-7}$$

where P - cost vector for a one unit change in CV process value and D - cost vector for a one unit change in MV process value. Applying Equation 12-1, the objective function expressed in terms of MV only is:

$$\underset{\min}{Q} = -P^T \times A \times \Delta MV(t+c) + D^T \times \Delta MV(t+c) \tag{12-8}$$

The LP solution is always located at one of the vertexes of the region of feasible solutions. This is illustrated in Figure 12-1 for a two dimensional problem. The region of feasible solutions for two controlled/constrained variables and two manipulated variables is an area contained within MV1 and MV2 limits, represented by the vertical and horizontal lines, and CV1 and CV2 limits, represented by the straight lines (Equations 12-9 and 12-10).

$$CV1_{\min} = a_{11}MV1 + a_{12}MV2 \qquad CV1_{\max} = a_{11}MV1 + a_{12}MV2 \tag{12-9}$$

$$CV2_{\min} = a_{21}MV1 + a_{22}MV2 \qquad CV2_{\max} = a_{21}MV1 + a_{22}MV2 \tag{12-10}$$

As has been mentioned, the optimal solution is located at one of the vertexes marked by arrows. To find this solution, the LP algorithm calculates the objective function for an initial vertex and improves the solution every step until it determines the vertex with the maximum (or minimum) value of the objective function as the optimal solution.

Optimal MV values are applied to the MPC control as the target MV values to be achieved within the control horizon. If the MPC controller is squared, that is, the number of MVs is equal to the number of CVs, then MV targets can be effectively achieved by change in CV value:

$$\Delta CV = A \times \Delta MVT \tag{12-11}$$

$\Delta MVT =$ optimal target change of MV

$\Delta CV =$ CV target change to achieve optimal MV.

A CV target change is implemented by managing CV setpoints. As we have seen, the MPC algorithm working with an optimizer therefore has two main objectives:

- Minimize CV control error with minimal MV moves, within operational constraints.
- Achieve optimal steady state MV values set up by optimizer and target CV values, calculated directly from MV values.

To satisfy these objectives in a better way than managing only target CV, the original unconstrained MPC algorithm has been extended to include MV targets into the least squares solution. The Objective function for this MPC controller is:

$$\min_{\Delta MV(k)} \left\{ \left\| \Gamma^y [CV(k) - R(k)] \right\|^2 + \left\| \Gamma^u \Delta MV(k) \right\|^2 + \left\| \Gamma^o \left[\sum \Delta MV(k) - \Delta MVT \right] \right\|^2 \right\} \tag{12-12}$$

where:

$CV(k)$	=	the controlled output p-step ahead prediction vector
$R(k)$	=	the p-step ahead reference trajectory (setpoint) vector
$\Delta MV(k)$	=	the c-step ahead incremental control moves vector
Γ^y	=	$diag\{\Gamma^y_1,\ldots,\Gamma^y_p\}$ = a penalty matrix on the controlled output error
Γ^u	=	$diag\{\Gamma^u_1,\ldots,\Gamma^u_c\}$ = a penalty matrix on the control moves
p	=	the prediction horizon (number of scans)
c	=	the control horizon (number of MV steps)
Γ^o	=	a penalty on error of the sum of controller output moves over the control horizon relative to the target optimal change of MV defined by the optimizer.

For notation simplicity, the objective function (Equation 12-12) is shown for SISO control.

The first two terms are the objective function (Equation 12-12) for an unconstrained MPC controller. The third term sets up an additional condition, which is to make the sum of the controller output moves equal to the optimal targets. In other words, the first two terms set up objectives for controller dynamic operation, the third term sets up steady state optimization objectives. Optimized MPC objectives are graphically illustrated in Figure 12-42.

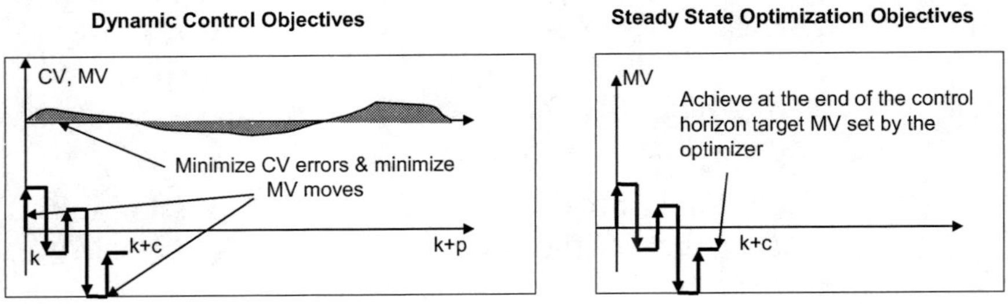

Figure 12-42. Optimized MPC Controller Objectives

The general unconstrained solution for this controller, similar to that for the unconstrained MPC controller, can be expressed as:

$$\Delta MV = (S^{u^T}\Gamma^T\Gamma S^u + \Gamma^{u^T}\Gamma^u)^{-1} S^{u^T}\Gamma^T\Gamma E_{p+1}(k) = K_{ompc}E_{p+1}(k) \qquad (12\text{-}13)$$

where:

$\Delta MV = [\Delta MV(k), \Delta MV(k+1), \ldots, \Delta MV(k+c)]$ = incremental move plan for the MPC controller output at the time k

K_{ompc} = optimized MPC controller gain matrix

S^u = the process dynamic matrix built from the step responses of dimension $p \times c$ for an SISO model and $p{*}n \times c{*}m$ for an MIMO model with m manipulated inputs and n controlled outputs. For the optimized MPC, the dynamic matrix is extended to the size $(p+1) \times m$ for an SISO model and $(p+m){*}n \times c{*}m$ for an MIMO model, to accommodate the MV error.

$E_{p+1}(k)$ = combines the CV error vector over the prediction horizon, and the error of the sum of controller output moves over the control horizon relative to the target optimal change of MV. Matrix Γ combines matrixes Γ^y and Γ^o. It is a square matrix of dimension *(p+1)* for an SISO controller and *[n(p+m)]* for a multivariable (MIMO) controller. Superscript *T* denotes a transposed matrix.

In operation, the optimizer sets up and updates the steady state targets for the MPC unconstrained controller at every scan, thus the MPC controller executes the unconstrained algorithm. Because the targets are set in a manner that accounts for constraints, as long as a feasible solution exists, the controller works within constraint limits. The integrated MPC controller and optimizer performs the following sequence of operations each scan:

1. Update CV and DV measurements
2. Update CV predictions
3. Determine optimal MV steady state targets and calculate CV targets
4. Provide MV outputs accounting for CV and MV targets
5. Update MV

Figure 12-43 illustrates the communication and integrated operation of the optimizer and the MPC algorithm. There is a typical situation in which the optimizer cannot find an optimal solution because of too-tight constraint limits, too many setpoints or excessively severe disturbances. In such cases, the system constraints are relaxed by changing setpoints to range control and/or abandoning some constraints. This can be done by minimizing the squared error over a number of constraints or by sequentially abandoning the constraints with the lowest priority.

Ill conditioning is another typical problem an optimizer has to deal with. As with the MPC algorithm, ill conditioning manifests itself as excessive changes in calculated MV targets, even for minor corrections of constraints. In this approach, ill conditioning is removed dynamically by changing the configuration of active constraints (abandoning some constraints) or by removing the association between constraint or controlled variables and manipulated variables that have excessive moves.

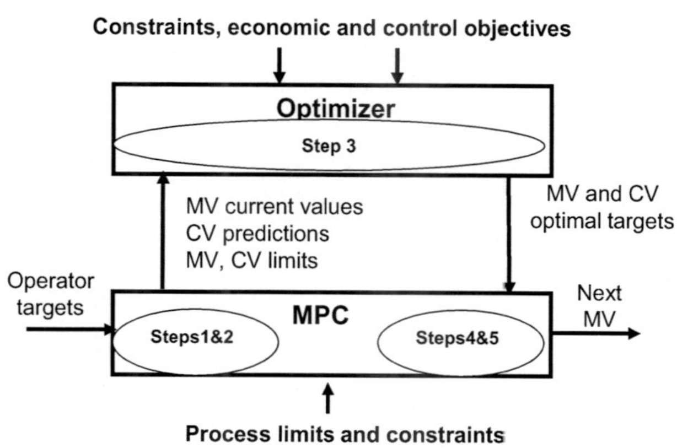

Figure 12-43. MPC and Optimizer Communication

12.5.2 On-line Multi-objective Optimizer Functionality Overview

As mentioned earlier in the chapter, optimizers are an inherent part of MPC defined by an objective function, provides an economically optimal solution while keeping the process variables within acceptable ranges and limits. In the original linear programming (LP) formulation (Press, Teukolsky, Vetterling and Flannery, 1997), an objective function comprises terms that include some or all the independent variables. However, in process control applications the independent variables (process inputs/manipulated variables) and the dependent variables (process outputs) need to be optimized, requiring superposition of the objectives. An objective function term for any dependent variable may then be effectively projected on some or all the independent variables, based on the process model gain relationships. Therefore, setting an objective for a dependent variable is equivalent to setting an objective function composed of some or all the independent variables (Ehrgott, Gandibleux, 2002).

Because the LP solution lies at the constraints boundary, the optimal steady state targets tend to bounce around. In order to prevent such behavior, QP (quadratic programming) optimization is applied, substantially increasing the complexity of the implementation.

In addition, the objective function is applicable only when the solution exists within limits. The original optimization objectives, however, need to be redefined if a solution does not exist within the predefined limits, and in such cases the optimizer should have the means to recover from the infeasibility. The existing recovery techniques are based on the priorities of the constrained and controlled variables. There are several known approaches to this problem. Tyler and Morari (1997) used integer variables to cope with prioritization. The minimization of the size of the violation is performed by solving a sequence of mixed integer optimization problems. Vada, Slupphaung and Foss (1999) proposed an algorithm that solves a sequence of LP or QP problems in instances of infeasibility. This algorithm, like the previous one of Tyler and Morari (1997), minimizes the number of violations of those constraints that cannot be fulfilled. However, both approaches are computationally intensive and are therefore inadequate, particularly for fast real-time applications.

Another algorithm has been developed for off-line LP weights, designed in such a way that the computed constraint violations are optimal (Vada, Slupphaung, Johansen, 2002). While it takes the burden of excess computations off-line, it presents additional off-line optimization problems.

In the remainder of this section, a multi-objective optimization technique is presented whose application simultaneously satisfies the three major requirements of MPC control and optimization, namely (Wojsznis, Mehta, Wojsznis, Thiele and Blevins, 2007):

- Constraint handling: The solution does not violate the defined process variable hard limits and minimizes possible violations of the soft limits.
- Economic optimization: The solution maximizes or minimizes the economic criteria based on user specified economic values
- Control functionality: The solution establishes targets for the dynamic behavior of the process inputs (manipulated variables) and outputs (controlled and constraint variables).

In its standard form, and with built-in default objectives, the optimizer addresses normal control operation. Constraint and economic objectives are addressed by extending the LP functionality, using penalized slack variables, which are added new variables, defined in such a way that they are non-zero and heavily penalized in the objective function only if the constraints are violated. In such a way the optimizer is suitable for on-line implementation. This technique provides the functionality delivered by the QP optimizer with an insignificant increase in LP complexity.

12.5.3 Multi-objective MPC Optimization Background

As follows from Equation 12-8, the objective function for maximizing product value and minimizing raw material cost can be defined jointly in the following way:

$$\underset{\min}{Q} = P^T \times A \times \Delta MV(t+c) + D^T \times \Delta MV(t+c) \qquad (12\text{-}14)$$

where P denotes the cost vector for a unit change in CV process value, and D denotes the cost vector for a unit change in MV process value.

In essence, the objective defined for every output is projected onto the independent variables as separate objective functions. It follows, then, that a composite objective function that includes the objective functions for the independent variables would be a superposition of $n+1$ objective functions, or:

$$\underset{\min}{Q} = P^T \times A \times \Delta MV(t+c) + D^T \times \Delta MV(t+c)$$

$$= \sum_{1}^{n} q_i(\Delta MV) + q_{n+1}(\Delta MV) = \sum_{1}^{n+1} q_i(\Delta MV) \qquad (12\text{-}15)$$

where:

$$q_i(\Delta MV) = [p_i a_{i1}, p_i a_{i2}, \ldots, p_i a_{im}]\Delta MV$$
$$= p_i a_{i1} \Delta mv_1 + p_i a_{i2} \Delta mv_2 + \ldots + p_i a_{im} \Delta mv_m \qquad (12\text{-}16)$$

is the objective function defined for a dependent control variable cv_i projected onto independent manipulated variables $mv_i \quad i = 1,\ldots,m$.

The form (Equation 12-16) of the objective function corresponds to the non-normalized weighted sum multi-objective function. The objective function, after summation, is a function of the independent manipulated variables mv_i or their increments Δmv_i:

$$Q = e_1 mv_1 + e_2 mv_2 + \ldots + e_m mv_m \qquad (12\text{-}17)$$

where ε_i is a system composite cost or profit for unit mv_i change, projected from all control variables, $cv_i \quad i = 1,2,\ldots,n$.

The objective function (Equation 12-15 or 12-17) is composed of multiple objective functions, such as in Equation 12-16, with diverse and possibly contradictory objectives. This poses a significant challenge in creating practically meaningful objective functions and understanding optimized solutions. The MPC optimizer needs to satisfy the economic objective function such that the changes in process variable values do not violate the predefined limits on both *MVs* and *CVs*, that is, additionally satisfy the constraint handling objectives:

$$MV_{min} \leq MV_{current} + \Delta MV(t+c) \leq MV_{max} \qquad (12\text{-}18)$$

$$CV_{min} \leq CV_{predicted} + \Delta CV(t+p) = CV_{predicted} + A \times \Delta MV(t+c) \leq CV_{max} \qquad (12\text{-}19)$$

The LP solution is always located at one of the vertices of the region of feasible solutions set by the inequalities in Equations 12-18 and 12-19. The optimal *MV* values are applied to the MPC control as the target *MV* values to be achieved within the control horizon.

12.5.4 Multi-objective Optimization Function for Infeasibility Handling

In normal control operation, the target setpoints are within acceptable ranges, and the constraint variables are within limits. However, when disturbances are too severe to be compensated for, the optimizer may not find a solution within limits. Because the basic requirement for on-line optimization is to have an optimal solution in all circumstances, an effective mechanism is needed for handling infeasibilities when process variables are predicted to be out of their limits.

The extension of the objective function for such cases is based on a special use of parameters called slack variables. In linear programming, slack variables or in the vector notation slack vectors $S_{\max} \geq 0$ and $S_{\min} \geq 0$ are used to transform the inequalities in Equation 12-19 into the following equalities:

$$CV_{predicted} + A \times \Delta MV(t+c) = CV_{min} + S_{min} \qquad (12\text{-}20)$$

$$CV_{predicted} + A \times \Delta MV(t+c) = CV_{max} - S_{max} \qquad (12\text{-}21)$$

The equality is required for the linear programming model, thus slack variables serve only as formal parameters with no specific application meaning. However, adding slack variables does serve to positively increment the degrees of freedom. Therefore, adding slack variable for each equation will increase the degrees of freedom and will allow finding a solution in situations where the original solution violates the limits. Slack variables may be applied to extend the range limits, with the slack vector $S^{+} \geq 0$ for the high limit violation and the slack vector $S^{-} \geq 0$ for the low limit violation, as below:

$$CV_{predicted} + A \times \Delta MV(t+c) = CV_{min} + S_{min} - S^{-} \qquad (12\text{-}22)$$

$$CV_{predicted} + A \times \Delta MV(t+c) = CV_{max} - S_{max} + S^{+} \qquad (12\text{-}23)$$

Figure 12-44 illustrates this concept of slack variables for constraint variables (process outputs with no setpoints). $S(i)$ denotes the component i of the relevant slack vector. Note that there will be a pair of equations (12-17, 12-18) for each value of the CV prediction. Depending on the actual prediction value, some of the slack variables will be zero. Only the non-zero slack variables applied for various objectives are shown in Figures 12-44 to 12-47.

To find an LP solution that is within the process variable limits or violates the limits, the new slack variables are penalized, such that these

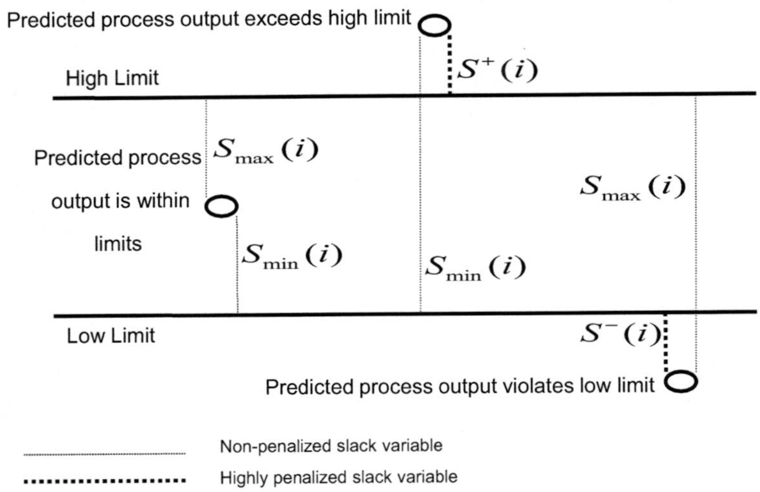

Figure 12-44. Application of Slack Variables for Constraints Handling with No Objectives

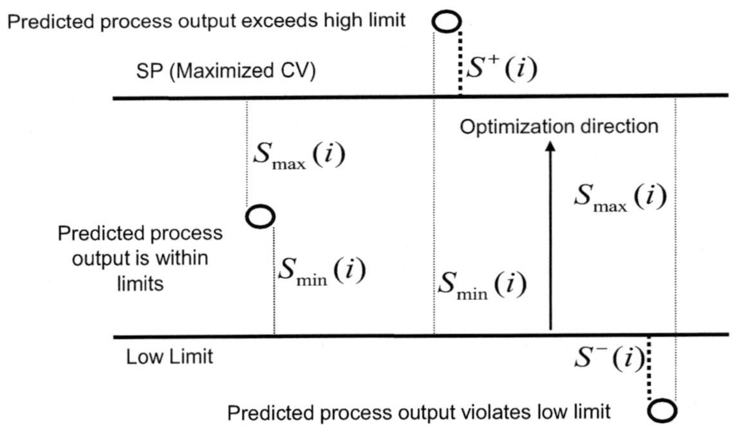

Figure 12-45. Slack Variables Application to Maximized CV Control

penalties are significantly higher than the terms that denote economic costs or profits. Therefore, the objective function in Equation 12-24 needs to be extended by adding the ***constraint*** penalizing terms $P_{C-}^T \times S^-$ and $P_{C+}^T \times S^+$.

$$\underset{min}{Q} = P^T \times A \times \Delta MV(t+c) + D^T \times \Delta MV(t+c) + P_{C-}^T \times S^- + P_{C+}^T \times S^+ \quad (12\text{-}24)$$

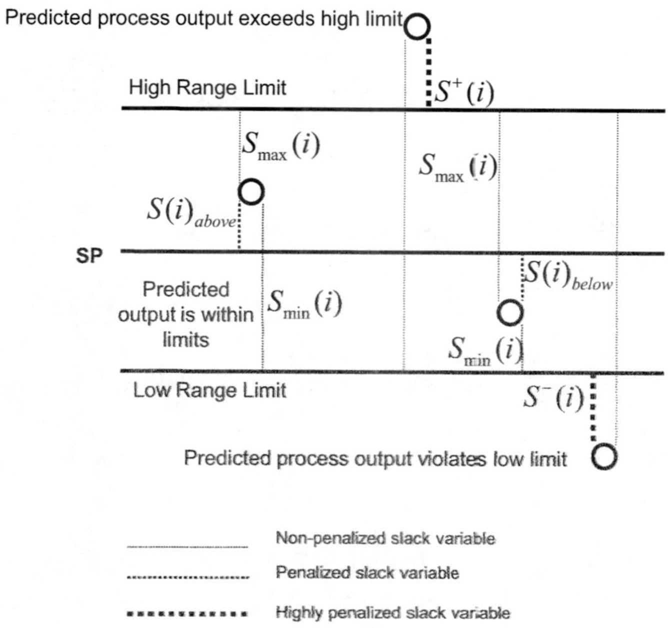

Figure 12-46. Slack Variables Application to the Target Control with the Two-Sided Range

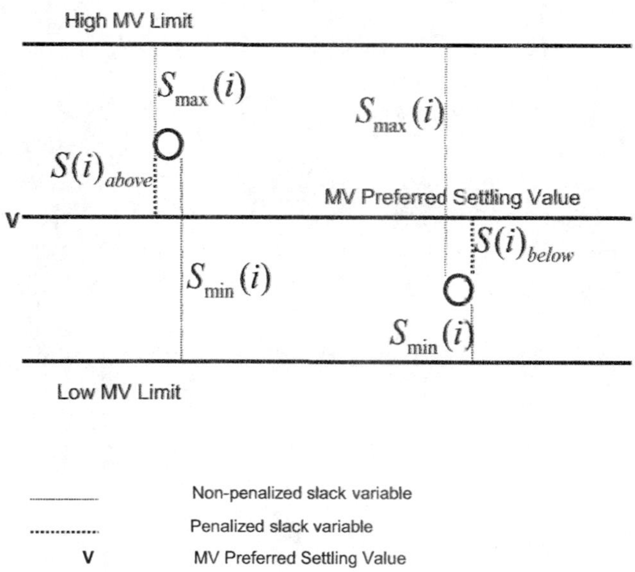

Figure 12-47. Using Penalized Slack Variables to Account for the Preferred Settling Value

where:

P_{C-}^T = the penalty vector for violating low constraint limits
P_{C+}^T = the penalty vector for violating high constraint limits

$$P_{C-} \gg P \text{ and } P_{C+} \gg P \text{, also } P_{C-} \gg D \text{ and } P_{C+} \gg D \quad (12\text{-}25)$$

Simplifying notations, we assume:

$$P_C \geq P_{C-} \text{ and } P_C \geq P_{C+} \quad (12\text{-}26)$$

All components of the vector P_C should be significantly larger than the economic cost/profit vectors. Generally, it is reasonable to assume that the smallest component of the vector P_C is greater than the largest component of the vector P.

12.5.5 Multi-objective Optimization Function for Extending LP Functionality

It is common engineering practice to approximate a range around the setpoint of a controlled variable, within which no further control action needs to be taken. Because slack variables lend themselves naturally to such a representation, an extension of the multi-objective optimizer is the use of slack variables to define acceptable ranges for setpoint optimization. The ranges are defined around the setpoint value, and are necessarily within the low/high setpoint limits. Setpoint ranges can be single sided or two sided.

Single sided ranges are associated with CV minimization and maximization (Figure 12-45) objective functions. Two-sided ranges (Figure 12-46) have no economic objective other than realizing the optimal solution as close as possible to the setpoint within the defined ranges, as extended by the penalized slack variables. A range equal to zero indicates a setpoint with highly penalized slack variables around it.

The equations for setpoint control with two-sided ranges for each CV prediction are of the form:

$$CV_{predicted} + A \times \Delta MV(t+c) = SP - S_{below} + S_{above} \quad (12\text{-}27)$$

$$CV_{predicted} + A \times \Delta MV(t+c) = CV_{min} + S_{min} - S^- \quad (12\text{-}28)$$

$$CV_{predicted} + A \times \Delta MV(t+c) = CV_{max} - S_{max} + S^+ \quad (12\text{-}29)$$

S_{below} and S_{above} are additional vectors of slack variables for the solutions below and above the setpoints. Therefore, an additional term: $P^T{}_{below} \times S_{below} + P^T{}_{above} \times S_{above}$ should be added to the objective function, where:

$P^T{}_{below}$ = the unit penalty for the solution below the setpoint, and
$P^T{}_{above}$ = the unit penalty for the solution above the setpoint.

It is reasonable to assume that $P^T_{below} = P^T_{above} = P^T_S$.

By applying penalized slack variables in this way, the optimizer can always find a solution at its first execution, even if the solution is outside the output limits. This approach also allows the handling of process inputs in a more flexible way. Inputs (*MVs*) have only hard constraints. It is also possible to define soft constraints, contained within hard constraints, for some of the inputs. By introducing penalized slack variables for the soft constraints, it is easy to define a penalized range for the *MVs*. For doing this, the following equations should be included along with Equation 12- 25:

$$MV_{min}{}^{soft} - S^{-}{}_{softmin} = MV_{current} + \Delta MV(t-c) \qquad (12\text{-}30)$$

$$MV_{max}{}^{soft} + S^{+}{}_{softmax} = MV_{current} + \Delta MV(t+c) \qquad (12\text{-}31)$$

The same approach may be used to define the preferred settling value (*PSV*) or *V*, the configured settling value of the *MV*. If there are no active conditions that require a change to the *MV* value, the *MV* will tend towards its *PSV* value. *PSV* is either set by the user or it may be the last value of the *MV*.

The equations for an *MV* with a *PSV* (Figure 12-47) are:

$$V - S_{below} + S_{above} = MV_{current} + \Delta MV(t+c) \qquad (12\text{-}32)$$

$$MV_{min} + S_{min} = MV_{current} + \Delta MV(t+c) \qquad (12\text{-}33)$$

$$MV_{max} - S_{max} = MV_{current} + \Delta MV(t+c) \qquad (12\text{-}34)$$

PSV is the lowest priority control objective, to be achieved after the other objectives (economic, constraint and/or control) are satisfied. Objective function terms for the penalized *MV* slack vectors S_{below}, S_{above} are set in a similar way as those for the *CV* setpoint slack vectors.

The three optimization objectives (economic, constraint and/or control), using the penalized slack variables S_C^+ and S_C^- for constraint and S_{SP}^+, S_{SP}^-, M^+ and M^- for control, can then be combined to result in the following general form of the multi-objective optimization function:

$$Q = (P^T \times A + D^T)MV^t - (P^T \times A + D^T)MV^{t-1} \tag{12-35}$$

$$- P_C^T \times S_C^+ - P_C^T \times S_C^- - P_S^T \times S_{SP}^+ - P_S^T \times S_{SP}^- - D_M^T \times M^+ - D_M^T \times M^-$$

12.5.6 System Functionality for Multi-objective Optimization

In an industrial MPC implementation, the multi-objective LP optimizer can be integrated with the MPC controller and implemented as a control function block executing in the controller. The key requirements of a viable engineering implementation are ease of use and clear interpretation of the required settings. As will be seen, the three optimization objectives, in order of priority: constraint handling, economic optimization and control functionality, have a well-defined setup. The economic objective is satisfied prior to the control objectives unless limits are exceeded, in which case constraint handling takes the highest priority.

As we have seen, the three process variable types that the optimizer operates on are:

a. *Controlled variables*: Process outputs that need to be driven to their configured setpoint values. In addition, an acceptable range around the setpoint value may be defined

b. *Constraint variables*: Process outputs that, unlike control variables do not have setpoints, but need to be maintained within configured high/low limits

c. *Manipulated variables*: Process inputs that need be maintained within high/low limits, with the additional requirement that these are hard limits

In addition, some of these process variables have an associated economic value (cost/profit). Therefore, the optimizer needs to maintain the controlled variables at their setpoints while keeping all the variables within their limits, such that the final solution (target steady state values of the process variables) maximizes economic value. The optimization objectives for the different process variables of the MPC configuration are set up as follows (Figures 12-48 to 12-50 illustrate some of the parameter settings):

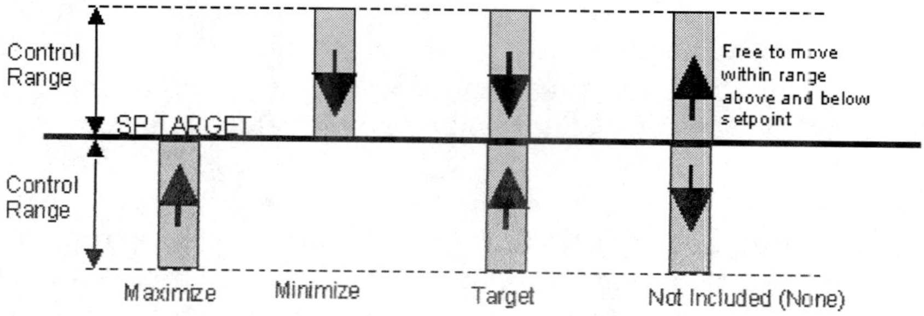

Figure 12-48. Target Control with the Two-Sided Ranges with Shown Direction Driven by the Optimizer

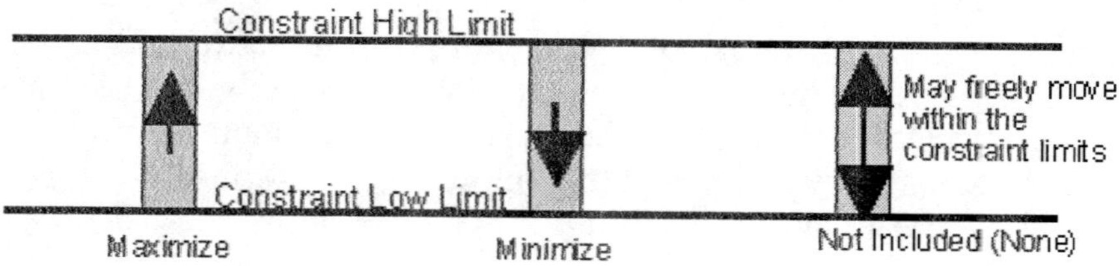

Figure 12-49. Constraint Parameter —Range and Direction Driven by the Optimizer

Constraint handling is set up by assigning priorities to the process outputs (control and constraint variables) of the MPC configuration. Input limits (MV hard constraints) are included in the problem formulation and are inherently accounted for in the LP solution. The optimizer operates such that if a constraint (or controlled) variable is predicted to exceed its limits (or its control range), the *working* target values of other constraint (or controlled) variables are adjusted to bring the process output within limits (or control range). In case processing conditions do not allow all the constraint and controlled variables to be maintained within their constraint limits and control ranges, the lower priority constraint or controlled variables are allowed to violate the limits or ranges in order to satisfy the higher priority constraints. The MPC function block allows five priorities for the process outputs: *high, higher than normal, normal, lower than normal,* and *low*. By default, controlled variables have *normal* priority, while constraint variables have *higher than normal* priority. Priorities should always be reviewed during the configuration setup of the MPC controller. In addition, the priorities may have to be changed on-line in response to actual process conditions.

Economic optimization is configured by setting a value, for example in units of *dollar value per percent* for some (or all) the controlled, constraint and manipulated variables included in the MPC block. An economic objective with a defined value outweighs all related control-only objectives that have no economic value (value per percent is zero).

Control objectives define the multivariable control functionality at normal operating conditions, that is, when all process outputs are within their control ranges and constraint limits and the economic optimum has been achieved. At that stage, the control functionality establishes the dynamic behavior of the remaining controlled, constraint and manipulated variables, including those not included in the objective function.

Controlled variables have well defined behavior around setpoints and the corresponding control range. As a rule, controlled variables defined in the objective function are driven towards their setpoint values. However, a controlled variable may deviate from its setpoint, depending upon one of the following user defined optimization types (Figure 12-48):

- *Maximize*: Controlled variable moves below the setpoint within the control range, if necessary, to achieve constraint handling or economic objectives. When other objectives are achieved, it then moves towards the setpoint by maximizing the controlled variable.
- *Minimize*: Controlled variable moves above the setpoint within the control range, if necessary, to achieve constraint handling or economic objectives. When other objectives are achieved, it then moves towards the setpoint by minimizing the controlled variable.
- *Target*: Controlled variable moves above or below the setpoint within the control range, if necessary, to achieve constraint handling or economic objectives. When other objectives are achieved, it moves towards the setpoint.
- *None*: Controlled variable not included in the objective function is maintained within the control range around the setpoint, with no attraction towards the setpoint.

Constraint variables included in the objective function may be driven towards their high (low) limit based on a *maximize* (*minimize*) definition. When a constraint variable is not included in the objective function (*default: none*), the optimizer makes changes that cause it to move freely within its upper and lower limits, as shown in Figure 12-49.

Manipulated variables included in the objective function may be driven towards their high (low) limit based on the associated *maximize* (*minimize*) setting. If a manipulated parameter is not included in the objective function (*default: none*), the optimizer attempts to maintain the current value if other conditions do not require the manipulated variable to be moved. In addition, the following MV optimization types are available (Figure 12-11):

- *Preferred Settling Value (PSV)*: Manipulated variable moves towards the configured preferred settling value if active conditions do not require a change in MV value.
- *Equalize*: Two or more manipulated variables set to equalize are maintained at their average value, provided all other objectives are satisfied.

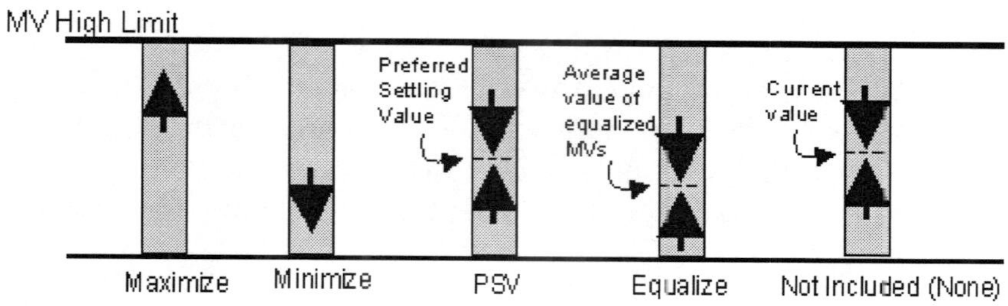

Figure 12-50. Manipulated Parameter—Range and Direction Driven by the Optimizer

The optimization objective, at both the individual variable and the functional level, needs to be configured in an off-line application. The default objective function is to maintain the controlled variables at their setpoints and to maintain all variables within their limits. If the configuration contains only constraint variables, the last MV values are maintained, thereby preserving the stability of the configuration. The active objective function may be defined by including process variables and then setting their individual optimization type (such as maximize, minimize, target, PSV, equalize) and, if required, their economic value. This is illustrated in Figure 12-51.

The on-line view (Figure 12-52) provides all the requisite information to an operator, such as current values, prediction at steady state, opti-

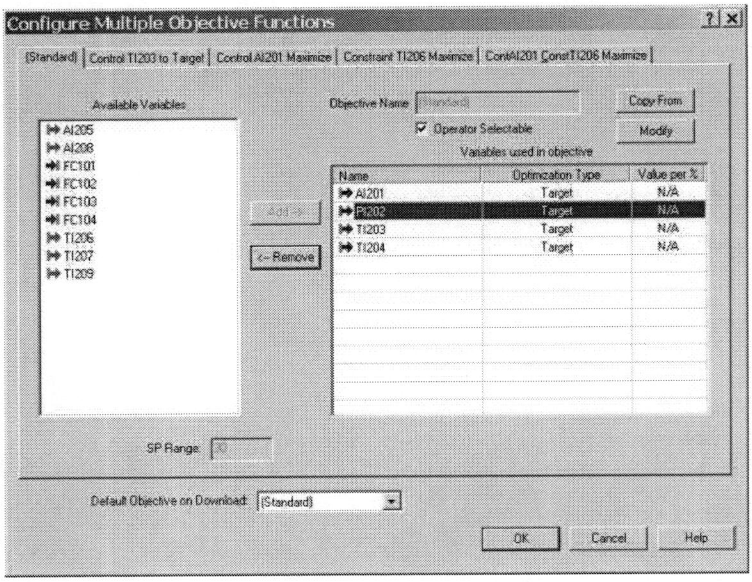

Figure 12-51. Dialog for Defining Control and Economic Objectives

mal targets, ranges/limits, the current optimization parameters (value per percent, optimization type, variable priority, etc.) An operator may also change these at run time, for example in response to price changes of feedstock, so that the results of the optimization objective remain valid.

One interesting consequence of the use of economic values is illustrated via Figure 12-53. Except for the economic objectives that are defined explicitly in the "Value per %" for MVs, the "Objective Function per %" column for each MV displays the effective value per percent. This column includes the value defined for the MV itself, in addition to values projected from the other control and constraint variables associated with this MV. These projected values may outweigh the original MV objective, and the optimizer would drive the MV in a direction opposite to that which was intended. Therefore, it is essential to consider the values in the "Objective Function per %" column in the initial evaluation of the optimizer's functionality.

There may often be special requirements for multiple optimization objectives because the overall objective, such as maximize throughput, minimize energy, maximize quality, etc., may change in response to feedstock cost, product value, utilization or plant conditions. This lends itself to the concept of configuring multiple objective functions for the same MPC controller, as shown in Figure 12-53. During on-line

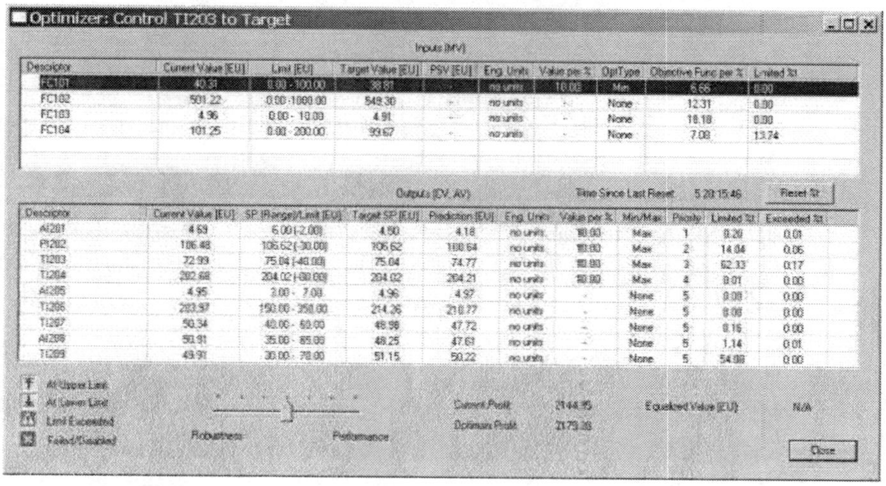

Figure 12-52. The Optimizer Window for the Operator

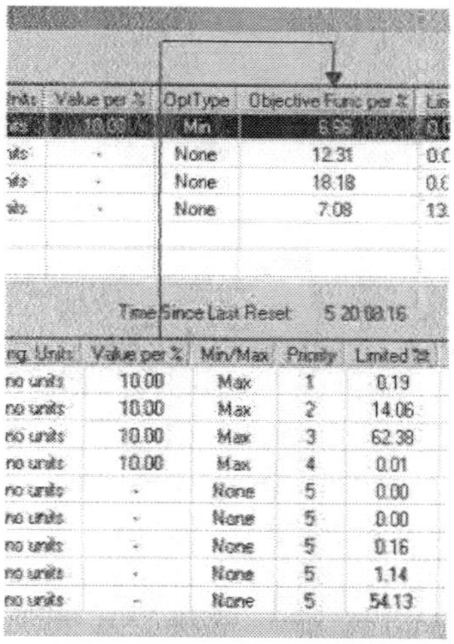

Figure 12-53. Relationship of the Economic Objectives Defined for Controlled, Constraint and Manipulated Variables and Effective Economic Objective Projected on the MVs

operation of the MPC controller, the operator may switch between the various configured optimization objectives depending on current operating conditions.

In summary, an optimizer with a multi-objective optimization function designed for the three operational levels of economic, control and constraint handling realizes flexibility and robustness of the multivariable control configuration required for on-line operation. Assigning targets for the Manipulated Variables, which may be desired pre-defined values or the actual values, is an extremely useful application feature, in particular when economic objectives have not been defined for some or all the MVs. The full functionality, including handling process outputs violating their limits and having a guaranteed solution for any condition, is achieved by using slack variables. In essence, the approach extends LP functionality to match QP features without the additional complexity. The technique is successfully embedded into an industrial controller as a standard Function Block, providing MPC integrated with optimizer functionality.

12.5.7 Unconstrained MPC Controller Supervised by the Optimizer

The optimizer solution provides steady state MV and CV targets for the MPC controller (Figure 12-43). The MPC controller's task is to achieve those targets while accounting for the process dynamics, limits and constraints. Two approaches are discussed in this and the following section:

1. Applying targets and constraints as the settings for constrained controller development. The controller should be generated on-line, effectively satisfying targets and constraints – section 12.5.8.
2. Applying targets to the unconstrained MPC controller and forcing it to comply with the constraints. Because the targets were calculated by the optimizer for the constrained configuration, complying with those targets should deliver controller operation equivalent to that of the constrained MPC operation – section 12.5.7.

In this section, the unconstrained controller is presented. The unconstrained controller is generated off-line from the process model and tuning parameters (Figure 12-54).

The most important procedure in unconstrained controller generation is the creation of a dynamic matrix from the step responses (Cutler and Ramaker, 1980). The dynamic matrix principle was discussed in Chapter 11; the general form of dynamic matrix for an SISO process is shown in Equation 12-36.

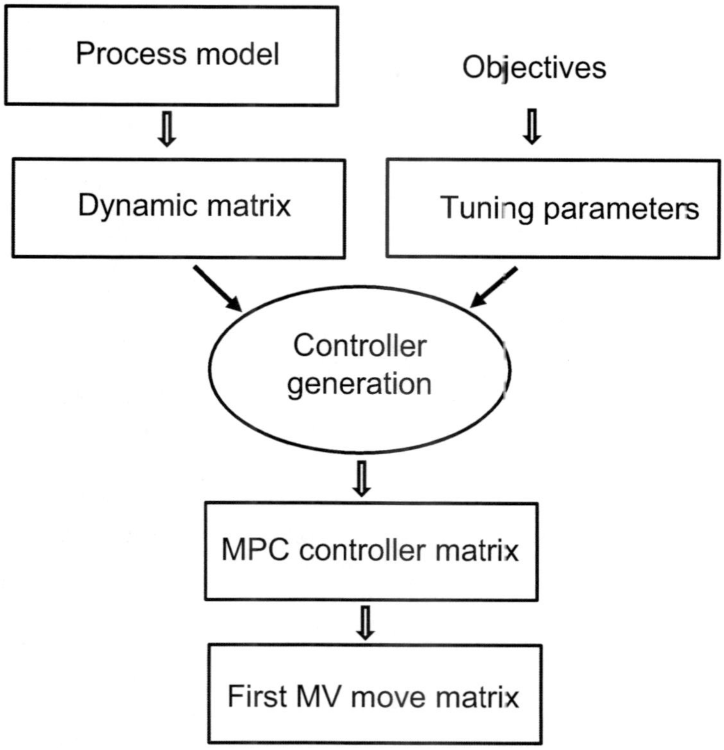

Figure 12-54 Unconstrained Controller Generation

$$A_{j,k} = \begin{pmatrix} r_{j,k}^{1} & 0 & 0 & 0 & \cdots & 0 \\ r_{j,k}^{2} & r_{j,k}^{1} & 0 & 0 & \cdots & 0 \\ r_{j,k}^{3} & r_{j,k}^{2} & r_{j,k}^{1} & 0 & \cdots & 0 \\ r_{j,k}^{4} & r_{j,k}^{3} & r_{j,k}^{2} & r_{j,k}^{1} & \ddots & 0 \\ \vdots & \vdots & \vdots & \vdots & \ddots & 0 \\ \vdots & \vdots & \vdots & \vdots & & r_{j,k}^{1} \\ \vdots & \vdots & \vdots & \vdots & & \vdots \\ r_{j,k}^{nPH-1} & r_{j,k}^{nPH-2} & r_{j,k}^{nPH-3} & r_{j,k}^{nPH-4} & \cdots & r_{j,k}^{nPH-nCH-1} \\ r_{j,k}^{nPH} & r_{j,k}^{nPH-1} & r_{j,k}^{nPH-2} & r_{j,k}^{nPH-3} & \cdots & r_{j,k}^{nPH-nCH} \end{pmatrix} \quad (12\text{-}36)$$

where $r_{j,k}^{l}$ is l^{th} step response coefficient between the jth CV and k^{th} MV.

For the multivariable $n \times n$ controller the dynamic matrix takes the form:

$$\begin{pmatrix} A1_{1,1} & A1_{1,2} & \cdots & A1_{1,nMV} \\ A1_{2,1} & A1_{2,2} & \cdots & A1_{2,nMV} \\ \vdots & \vdots & \ddots & 0 \\ A1_{nCV,1} & A1_{nCV,2} & \cdots & A1_{nCV,nMV} \end{pmatrix} \qquad (12\text{-}37)$$

The dynamic matrix (Equation 12-37) is used to minimize dynamic error between setpoint trajectory and controlled variable future trajectory. The second objective for MPC working with the optimizer is to minimize future error *MV*, that is, to achieve optimal steady state *MV* values set up by the optimizer.

To generate such an MPC controller, the dynamic process matrix should be extended in the following way:

$$\textit{Extended dynamic matrix} \quad \begin{pmatrix} A1_{1,1} & A1_{1,2} & \cdots & A1_{1,nMV} \\ A1_{2,1} & A1_{2,2} & \cdots & A1_{2,nMV} \\ \vdots & \vdots & \ddots & 0 \\ A1_{nCV,1} & A1_{nCV,2} & \cdots & A1_{nCV,nMV} \\ (11...)_{nCH} & (00...)_{nCH} & \cdots & (00...)_{nCH} \\ (00...)_{nCH} & (11...)_{nCH} & \cdots & (00...)_{nCH} \\ \vdots & \vdots & \ddots & \vdots \\ (00...)_{nCH} & (00...)_{nCH} & \cdots & (11...)_{nCH} \end{pmatrix} \qquad (12\text{-}38)$$

The number of ones (1) or zeroes (0) in a column is equal to a number of controller *MV* future moves, that is, the control horizon. The number of columns with 1s or 0s in a row is equal to the number of *MVs*. The penalty on the error Γ matrix should match the size of the modified dynamic process matrix and should include an optimization error penalty $\Gamma^o = (po_1, po_2, ..., po_{nMV})$ as a generation parameter.

Controller generation delivers the MPC controller matrix (Equation 12-13). For on-line operation, only part of the matrix K_{ompc} is used for calculating the first move $\Delta MV(k)$ by performing at every scan the simple operation $K_{mpc1}E_{p+1}(k)$. Error vector $E_{p+1}(k)$ is the *CV* error vector over the prediction horizon and the error of the sum of controller output moves over the control horizon relative to the target optimal change of *MV*.

Though the MPC controller is generated off-line, there are several parameters available for on-line controller tuning. The best adjustment of controller performance can be achieved by the reference trajectory (setpoint filter). In addition, the error vector can be multiplied by a factor greater than 1, to speed up controller response, or multiplied by a factor smaller than 1 to slow down the response.

The major advantage of an unconstrained MPC controller guided by an optimizer is its simplicity and short on-line execution time.

12.5.8 Constrained MPC Controller Integrated with the Optimizer

Compared to an unconstrained pre-generated controller, a constrained MPC controller requires significantly more computing power for on-line operation. On-line operation, however, is becoming feasible even for controllers with a 1 second scan, because computer speeds and memory can now deliver more than adequate performance.

A constrained controller has two advantageous features (Johnston, 2010). First, the control matrix is generated on-line at each control cycle. Differentiating between control and constraints is no longer required. All dependent variables are now included in the dynamic matrix in the move calculation. The Penalty on Error (PE) for each dependent variable is adjusted at each control cycle. If a CV is far from the limits, its PE can be set to zero to effectively remove it from the dynamic control problem. Conversely, if a CV is near a limit, the full value of the PE may be used. By this means, the dynamic move calculation at each control cycle includes only those CVs that are near their limits, and excludes those that are far from their limits.

Second is the enforcement of MV constraints on the future move plan. This prevents the controller from planning moves that cannot be implemented. The general approach is to first calculate the unconstrained MV solution. The second step is to find the first MV to become constrained in time (find the earliest MV constraint violation). The allowable portion of the move plan for this MV is then imposed on the problem. Next, the unconstrained solution using the remaining MVs is calculated, and the process is repeated until either all MVs are constrained or the unconstrained solution does not violate any constraints.

To achieve a feasible MPC controller generation, a number of algorithmic techniques are applied at every scan. The major computation savings are achieved by the following procedures:

- MV move blocking, which generates the controller with a subset of the future calculated moves within the control horizon. The columns of the dynamic matrix that corresponds to the blocked moves are removed. The original dynamic matrix with 40 or so columns can be reduced to a matrix with 10 or fewer columns.
- CV blocking, which causes the successive equations of the dynamic matrix to be added together to speed up calculations. This is essentially equivalent to defining future prediction intervals of unequal duration. CV blocking can reduce the number of dynamic matrix rows to less than one-fourth of the original number.
- The MPC constrained controller is implemented in the function control block and can work in the application workstation. More extensive discussion of this topic exceeds the scope of this book. The reader interested in the topic can find more details in the referenced monograph books (Tatjewski 2007, Macjewski, 2002 and Rossiter, 2003) or in the product manuals (e.g., DeltaV BOL).

BIBLIOGRAPHY

1. Allison, B. J., Isaksson, A. J., and Karlström, A. "Grey-box Identification of a TMP Refiner," *Proceedings of IMPC*, TAPPI, 1995.
2. Alsip, W. P. "Digital Control of a TMP Operation," *Pulp & Paper Canada*, 82(3): T76-T79, 1981.
3. Ansari, R. and Ansari, T. "Non-linear Model Based Multivariable Control of a Crude Distillation Process," *Non-linear Model Based Process Control*, London: Springer-Verlag, 2000, pp. 79-101.
4. Berg, D. and Karlström, A. Gustavsson, M., "Deterministic Consistency Estimation in Refining Processes," *Proceedings of IMPC*, TAPPI, 2003.
5. Boudreau, M. "Squared Model Predictive Controller Performance on the Shell Standard Control Problem," *Proceedings of ISA Conference*, Houston, October 2003.
6. Butler, D. L., Cameron, R. A., Brown, M. W., and McMillan, G. K. "Constrained Multivariable Predictive Control of Plastic Sheets," ISA Technical Conference, Paper 1022, Houston, TX, 2001.
7. Camacho, E. F. and Bordens, C. "Model Predictive Control in the Process Industry," London: Springer-Verlag, 1995; ISBN 3-540-19924-1.
8. Chmelyk, T. "Lime Kiln Model Predictive Control with a Residual Carbonate Soft Sensor," *ACC*, WP01, Anchorage, Alaska, 2002.
9. Chmelyk, T. "An Integrated Approach to Model Predictive Control of an Industrial Lime Kiln," *Proceedings of the 2002 IEEE APC Conference*, Vancouver, BC 2002.

10. Cuthrell, J. E., Rivera, D. E., Schmidt, W. J. and Vegeais, J. A. "Solution to the Shell Standard Control Problem," *Proceedings of The Second Shell Process Control Workshop,* Stoneham: Butterworth Publishers, 1990, pp.27-58.

11. Cutler, C. and Ramaker, B. "Dynamic Matrix Control—A Computer Control Algorithm," *Proc. Joint Automatic Control Conference,* San Francisco, CA, 1980.

12. Du, H. "Multivariable Predictive Control of a TMP Plant," *Ph.D. dissertation*, UBC, Vancouver, BC, Canada, 1998.

13. Edgar, T., Himmelblau, D., and Lasdon, L. *Optimization of Chemical Processes,* New York, McGraw-Hill, 2001.

14. Ehrgott, M. and Gandibleux, X. *Multiple Criteria Optimization,* Boston, Dordrecht, London, Kluwer's International Series, 2002.

15. Illikainen, M. et al. "Power Consumption and Fibre Development in TMP Refiner Plate Gap," IMPC, 2007.

16. Johnston, C. "MPC Documentation," *Emerson technical document*, Emerson Process Control, 2010.

17. Maciejowski, J.M. *Predictive Control with Constraints,* Harlow: Pearson Education Limited - Prentice Hall, 2002.

18. Maciejowski, J. M. "Two Case Studies," *Predictive Control with Constraints,* Harlow: Pearson Education Limited, 2002a, pp. 248-269.

19. Mehta, A., Wojsznis, W., Thiele, D., and Blevins, T. "Constraints Handling in Multivariable System by Managing MPC Squared Controller," *Proceedings of ISA Conference*, Houston, October 2003.

20. Press, W. H., Teukolsky, S. A., Vetterling, W. T. and Flannery, B. P. *Numerical Recipes in C,* Cambridge: Cambridge University Press, 1997.

21. Prett, D., García, C. and Ramaker, B. "Workshop Papers Discussion Sessions" *Proceedings of The Second Shell Process Control Workshop,* Stoneham: Butterworth Publishers, 1990, pp. 375-398.

22. Prett, D. and Morari, M. "Shell Control Problem," *Proceedings of Shell Process Control Workshop,* Stoneham: Butterworth Publishers, 1987, pp. 355-360.

23. Qin, S. J. and Badgwell, T. A. "An Overview of Industrial Model Predictive Control Technology," *Proceedings of Fifth International Conference on Chemical Process Control,* pp. 232-256, AIChE and CACHE, 1997.

24. Rao, C., Wright, S., and Rawlings, J. "Application of Interior-Point Methods to Model Predictive Control," *J. Optim. Theory Appl*, 99:723-757, 1998.

25. Rawlings, J. and Eaton, J. "Optimal Control and Model Identification Applied to the Shell Standard Control Problem" *Proceedings of The Second Shell Process Control Workshop,* Stoneham: Butterworth Publishers, 1990, pp. 209-240.

26. Roche, A., Owen, J., Miles, K., and Harrison, R. "A Practical Approach to the Control of TMP Refiners," *Control Systems*, 1996, p. 129.

27. Rosenqvist, F., Berg, D., Karlstrom, A., Eriksson, K. and Breitholtz, C. "Internal Interconnections in TMP Refining," *Proceedings of IEEE Control Systems Society Conference on Control Applications*, 2002, Glasgow, UK.

28. Rossiter, J. A. *Model-based Predictive Control: a Practical Approach*, Boca Raton, London, New York, Washington, D.C., CRC Press LLC, 2003.

29. Ruscio, D. D. "Topics in Model Based Control with Application to the Thermo Mechanical Pulping Process," *Ph.D. dissertation*, Norwegian Institute of Technology, Trondheim, Norway, 1993.

30. Sidhu, M. S. et al. "Stabilization and Optimization of a TMP Process," *Control Systems*, 2004.

31. Gullichsen and Fogelholm editors, *Chemical Pulping*, Vol. 5 of Papermaking Science and Technology Series, TAPPI Press, 1999.

32. Tatjewski, P. *Advanced Control of Industrial Processes*, London, Springer, 2007.

33. Tessier P., Broderick, G., Desrochers, C. and Bussiere, S. "Industrial Implementation of Motor Load and Freeness Control of Chemimechanical Pulp Refiners," *TAPPI Journal*, 80(12): 135-142, 1997.

34. Tyler, M. L. and Morari, M. "Propositional Logic in Control and Monitoring Problems," in *Proceedings of European Control Conference '97*, pp. 623-628, Bruxelles, Belgium, June 1997.

35. Vada, J., Slupphaug, O., and Foss, B.A. "Infeasibility Handling in Linear MPC Subject to Prioritized Constraints," in *Preprints IFAC'99 14th World Congress*, Beijing, China, July 1999.

36. Vada, J., Slupphaug, O., and Johansen, T. A. "Efficient Infeasibility Handling in Linear MPC Subject to Prioritized Constraints," in *ACC 2002 Proceedings*, Anchorage, Alaska, May 2002.

37. Wojsznis, K. W., Blevins, L. T., and Nixon, M. "Easy Robust Optimal Predictive Controller," *Advances in Instrumentation and Control, Proceedings of ISA Technical Conference*, New Orleans, 2000.

38. Wojsznis, W., Mehta, A., Wojsznis, P., Thiele, D. and Blevins, T. "Multi-objective Optimization for Model Predictive Control," *ISA Transactions*, Vol. 46, No. 3, pp. 351-361, 2007.

39. Wojsznis, W., T., Thiele, D., Wojsznis, P., and Mehta, A. "Integration of Real Time Process Optimizer with a Model Predictive Function Block," *Proceedings of ISA Conference*, Chicago, October 2002.

13

On-line Optimization

Optimization has been commonly applied to various industries for nearly 50 years. Since the introduction of MPC, process optimization has played a pivotal role in achieving benefits from MPC applications. The early optimization applications were mainly for off-line optimal planning or scheduling. More benefits come from the tight integration of the optimizer with the process and by providing by optimizer on-line solution accounting for the current measurements and process constraint operation. In some processes, such as power generation, on-line optimization can be effectively applied using only steady-state process models. This approach simplifies application development, in particular process modeling.

This chapter provides examples of on-line LP (linear programming) optimizer to minimize the cost of power generation. On-line operation was achieved by using MPC with an integrated optimizer; however, the MPC functionality was disabled. The economic benefits achieved from the known on-line optimization applications indicate use of on-line optimization will be expanding in the coming years.

The included workshop and the section on the basics of on-line optimization should provide the reader with a good idea of the functionality, achievable benefits and design approach of on-line optimization applications.

13.1 Why Optimization – A Look at Boiler Load Allocation

(Section co-authors: Chris Hawkins, George Buchanan, Andrew Riley – Emerson Process Management, UK)

A common problem in process plants is the necessity to allocate a quantity of feed, energy or production between two or more similar items of equipment or processing trains (Cohen and Sherkat 1987), (Arroyo and Conejo, 2000). Consider the simple system shown in 3-1: two steam generators (boilers), both burning the same fuel and supplying steam through a common header. The approach may be extended to more boilers, and is applicable to sets of turbines, compressors or any other system that may be adequately described by linear equations. The example also dispels the myth that load should always be allocated to the more/most efficient equipment or train.

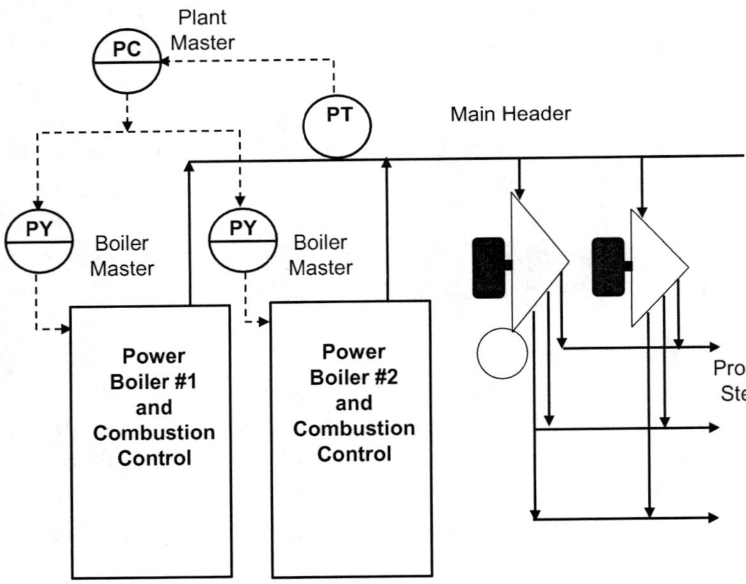

Figure 13-1. Two Boiler Example

The quantity of fuel energy F, required to generate a quantity of steam S, is given by:

$$F = F_0 + \frac{S\lambda}{\eta}$$

where λ is the latent heat of steam, η is the efficiency of the conversion from chemical energy in the fuel to thermal energy in the steam and F_0 is the flow quantity of fuel energy consumed but does not contribute to steam generation because of heat losses associated with the boiler.

Other equipment, such as steam turbines, exhibits a similar relationship in which a finite quantity of steam is required to compensate for heat loss and mechanical efficiency. Let's introduce some real numbers for our two boilers, that are characterized by linear models (Table 13-1).

Table 13-1. Example Boiler Operation

	Boiler 1	Boiler 2
λ	2400 MJ/Ton	
η	0.7	0.8
F_0	10 MW	40 MW

If η and λ were constant, then the boiler response would be linear. In reality, η is a function of load, and λ varies with changes in boiler feed water and steam conditions, resulting in a non-linear response. Figure 13-2 compares the responses for constant (linear) and varying η at fixed feedwater and steam conditions.

Process knowledge and the analysis of operating data should indicate the linearity of the process. Although linear programming requires linear response models, we will later show that mild non-linearities can be handled by linear programming without resort to more advanced methods such as mixed-integer arithmetic.

Let's look at the overall efficiency of our two boilers. This is generally expressed as:

$$Efficiency = \frac{Output}{Input} \times 100$$

This definition yields many different values for efficiency depending upon the value of Input and Output and whether the higher or lower heating value of the fuel is used. The ASME Code for Fired Steam Generators (ASME PTC 4-1998) recognizes two definitions of steam generator efficiency:

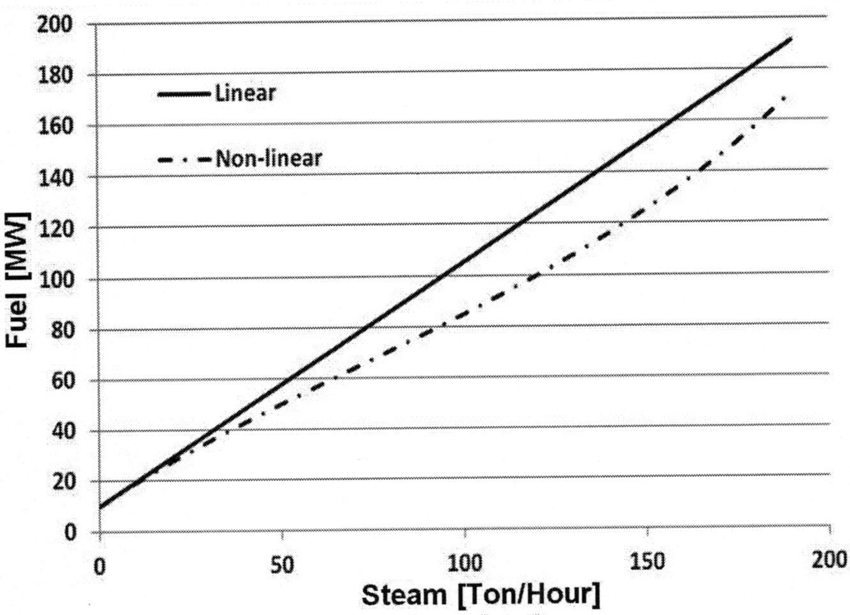

Figure 13-2. Boiler Response to Fuel Demand

- Fuel efficiency includes all energy absorbed by the steam output but counts only the chemical energy of the fuel as input; usually the higher heating value.
- Gross efficiency also includes all energy absorbed by the steam as output and counts all energy entering the steam generator as input. Gross efficiency is usually less than or equal to fuel efficiency.

Regardless of which definition is used, there are two generally accepted methods for determining the efficiency of a steam generator:

- Input/Output
- Energy balance

Each method has its pros and cons, and although the energy balance method is arguably the more rigorous, the input/output method is generally adopted in practical implementations.

In our illustration, we'll use simple input/output fuel efficiency:

$$\eta_{i/o} = 100\frac{S\lambda}{F}$$

Figure 13-3 shows the overall efficiency for our two "linear" boilers. Note that Boiler 2 appears less efficient than Boiler 1 despite the higher value of η . This is a consequence of Boiler 2 requiring significantly more fuel to offset its greater heat losses.

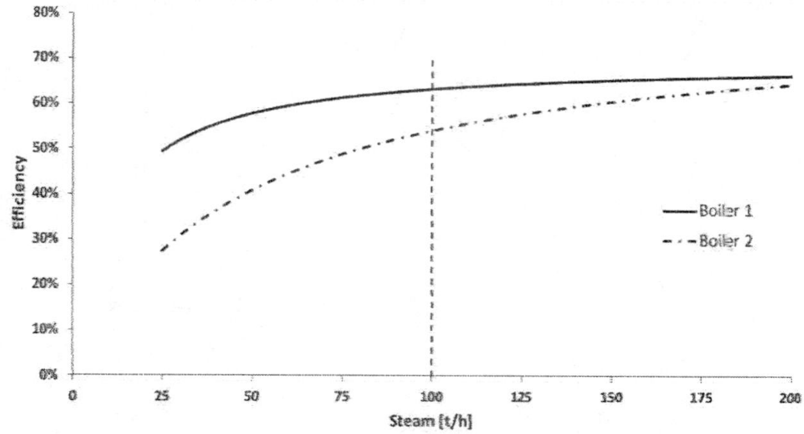

Figure 13-3. Change in Efficiency with Steam Generation

Assume that the total steam demand of 200 t/h is met equally by each boiler. If the steam demand increases to 210 t/h, on (that is, using) which boiler should we generate the additional 10 t/h? The answer is surely obvious: Boiler 1 since as shown in Figure 13-3 Boiler 1 is the more efficient as measured by the percentage of the total absorption heating value of outlet steam in the total supply heating value.

However, unless Boiler 2 is constrained, this answer is incorrect and illustrates the myth that the additional load should be met by the more efficient boiler.

To understand why this is so, we'll look beyond the fuel efficiency and return to the fuel-steam relationship depicted in Figure 13-2. Restricting ourselves to the linear case, shows the response for the two boilers.

Table 13-2 shows that raising the additional 10 t/h steam on the *more efficient* Boiler 1 requires 1.2 MW more fuel than on the *less efficient* Boiler 2. Why is this? Boiler 1 has the higher *fuel* efficiency because of its lower zero-load fuel demand. Hence the total amount of fuel required to generate 100 t/h or indeed 110 t/h of steam is less than that required for Boiler 2. However, Boiler 1 has lower *incremental* effi-

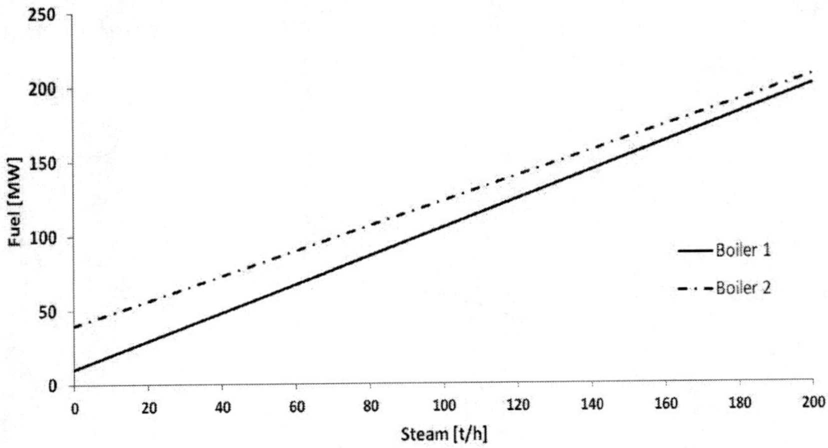

Figure 13-4. Linear Boiler Response

ciency, in that the additional fuel required to raise an additional 10 t/h on Boiler 1 exceeds that required on Boiler 2. This may be attributed to poorer combustion and/or heat transfer within the boiler.

Table 13-2. Boiler Incremental Efficiency

	Boiler 1		Boiler 2		
Total Steam	Steam	Fuel	Steam	Fuel	Total Fuel
[t/h]	[t/h]	[MW]	[t/h]	[MW]	[MW]
200	100	105.24	100	123.33	228.57
210	110	114.76	100	123.33	238.10
210	100	105.24	110	131.67	236.90

Regardless of the number of boilers or the linearity of the describing relationships, the boiler with the highest incremental efficiency will always consume the least additional fuel for a given quantity of additional steam. Similarly, should steam demand be reduced, steam should be taken from the boiler with the lowest incremental efficiency.

We can solve this problem for an arbitrary number of boilers of differing capacity and performance using linear programming. Moreover, we can implement the application on-line to ensure that the allocation remains optimal with changing circumstances.

The standard form for linear programming (Huang et al 1992), (Lahdelma et al 2003) is:

$$\text{Minimize:} \quad F(x) = c_1x_1 + c_2x_2 + \ldots + c_nx_n$$

$$\text{Subject to:} \quad a_{11}x_1 + a_{12}x_2 + \ldots + a_{1n}x_n = b_1$$

$$a_{21}x_1 + a_{22}x_2 + \ldots + a_{2n}x_n = b_2$$

$$\vdots$$

$$a_{m1}x_1 + a_{m2}x_2 + \ldots + a_{mn}x_n = b_m$$

$$x_i \geq 0 \quad i = 1, n$$

$$b_j \geq 0 \quad \mathrm{j} = 1, m$$

In more compact form:

$$\text{Min} \quad \mathbf{f} = c^T x$$

$$\text{Subject to:} \quad Ax = b$$

$$x \geq 0, b \geq 0$$

For our two boiler examples, the optimization variables, x, are the flow of fuel to each boiler. (Alternately, we could invert the problem and define the quantity of steam generated on each boiler as the optimization variables.)

Because we have a single fuel, its cost is not a factor, so we set $c_1 = c_2 = 1$ such that our objective function is:

$$\min f(x) = x_1 + x_2$$

The constraints are the steam header balance and the total quantity of fuel consumed, that is:

$$1.05x_1 + 1.2x_2 = \text{steam demand}$$

$$x_1 + x_2 \leq \text{maximum fuel available}$$

In the example, each boiler was producing 100 t/h of steam. But is this the optimum allocation for a total of 200 t/h steam with both boilers operating? Almost certainly not, the optimum is to produce all 200 t/h of steam on Boiler 2—the least efficient of the pair (Figure 13-4)!

Can this be so? 13-3 shows that raising the 200 t/h on Boiler 2 alone uses almost 24 MW less fuel. In fact, using Boiler 1 alone would use more fuel than any other allocation alternatives:

Table 13-3. Impact of Allocation on Fuel Consumption

	Boiler 1		Boiler 2		
Total Steam	**Steam**	**Fuel**	**Steam**	**Fuel**	**Total Fuel**
[t/h]	[t/h]	[MW]	[t/h]	[MW]	[MW]
200	100	105.24	100	123.33	228.57
200	200	200.48	0	40	240.48
200	0	10	200	206.67	216.67

But it would appear that Boiler 1 is running at minimum (zero) load, while Boiler 2 is at maximum load. So should we turn Boiler 1 off?

No; as Table 13-4 shows, we should turn off Boiler 2!

Table 13-4. Impact of Turning One Boiler Off

	Boiler 1		Boiler 2		
Total Steam	**Steam**	**Fuel**	**Steam**	**Fuel**	**Total Fuel**
[t/h]	[t/h]	[MW]	[t/h]	[MW]	[MW]
200	200	200.48	0	0	200.48
200	0	0	200	206.67	206.67

Why is this? Because when we no longer have to consume fuel simply to overcome the heat losses, it is better to turn off Boiler 2 as it has the higher zero-load demand for fuel.

Selecting which boilers to operate is a mixed integer optimization problem. In practice, such decisions are rarely made solely on fuel use minimization. Industrial boilers take a finite time to start up and shut down, and it is often an operational requirement that a certain "instantaneous" spare steam generating capacity is maintained.

The astute reader may now realize that given *n* boilers, each with differing but linear steam-fuel responses, the optimal load will always comprise n-1 boilers at the minimum or maximum load, and 1 boiler providing the balancing steam flow.

So far we've assumed boilers with linear steam-fuel response. While this is often a reasonable assumption, when requirement on linearity is relaxed, the optimal loading can change markedly. Can we solve the problem using basic linear programming? We can if we exploit the fact that the optimum loading occurs at the steam flows which total 200 t/h and which have equal gradients on the steam-fuel curves. This rule applies regardless of the number of boilers and the form of their performance equations. However, if we assume that the response curves are adequately represented as quadratic equations, for example:

$$S = aF^2 + bF + c$$

where a, b, and c would generally be determined by fitting the performance relationship to operating data, the first derivative of the quadratic equation is the linear equation:

$$\frac{dS}{dF} = 2aF + b$$

So our linear programming problem becomes:

$$\min f(x) = x_1 + x_2$$

With the steam header constraint replaced with the constraint that the derivatives of each response curve are the same:

$$2(a_1x_1 - a_2x_2) + (b_1 - b_2) = 0$$

While retaining the total fuel constraint:

$$x_1 + x_2 \leq \text{maximum fuel available}$$

However, we can no longer include the steam demand constraint as an explicit variable in the LP formulation. Instead, we must introduce a simple updating algorithm for the total quantity of fuel to be allocated between boilers based upon the ratio between the steam demand and the quantity of steam generated:

$$F_{\Sigma}^{k+1} = F_{\Sigma}^{k}\frac{S_{SP}}{S^k}$$

The example uses the performance curves shown in Table 13-5:

Table 13-5. Performance Curve Coefficients

	Boiler 1	Boiler 2
a	-2.678×19^{-3}	-3.195×10^{-3}
b	1.711	2.131
c	–25.17	–89.22

With the same total steam demand of 200 t/h, the optimum loading is 77 t/h and 123 t/h steam on Boilers 1 and 2 respectively. Clearly this is a less obvious solution, partly because the majority of the steam is raised on the less fuel efficient boiler.

13.2 Energy Optimization in a Pulp and Paper Mill

(Section co-authors: Terrance Chmelyk, Carl Sheehan – Spartan Controls, Burnaby, BC, Devin Marshman – University of British Columbia, Vancouver, BC)

Pulp and paper companies have long recognized that energy is much more than a cost center. In fact, energy management and cogeneration can be looked upon as a separate business that can have a tremendous impact on the mill's overall profitability. Many mills have the ability to cogenerate power and export their electrical energy to the marketplace and, as a result, are entering into longer term, often complex buy/sell agreements with the local utilities (Gruhl et al 1975). This presents a significant challenge in making critical decisions in a timely manner and requires a great deal of real-time economic information in conjunction with the process operating parameters (Ferrari-Trecate et al, 2005). As a result, the actual achievable profit is sometimes not optimized for all operating conditions. Significant manual intervention occurs sometimes as operations personnel are multitasking and trying to make critical process decisions that include:

- Adjusting steam supply to meet process demands
- Selecting various possible fuels based on cost and availability
- Loading the available boilers based on cost and constraints
- Deciding on the amount of electricity to purchase versus generate

- Loading the turbines based on cost, steam availability and constraints
- Responding to disruptions and equipment breakdowns

In this section, we present an integrated real-time energy management optimization system that is used to reduce the total cost of energy in an integrated pulp and paper mill by automating critical decision-making and process adjustment. The Energy Management System (EMS) employs facility-specific business models to continually manage the energy contract with the local utility while determining the optimum operating parameters to maximize profitability.

13.2.1 Introduction to Cogeneration

The production of wood chips for pulp production can yield up to approximately 50% waste material, also known as "hog," depending on the type of wood and the pulping process used. Rather than disposing of this excess material, the pulp and paper industry often exploits it by burning the waste in multi-fuel power boilers to produce steam. In addition to being a much cheaper source of fuel, the scrap product is considered to be a "green" fuel with significant environmental benefits when compared to conventional fuels such as natural gas or coal (Sampson et al, 1993). Once the hog fuel is combusted in the boilers, the resulting steam is used as a source of heat for various unit operations within the pulp mill or is passed through a series of turbines to generate electrical power. Such systems are referred to as combined heat and power, powerhouse or cogeneration systems because they generate steam and electricity, two commodities required by many industrial processes including pulp production.

The generation of power and steam by a single system, as opposed to by two separate systems, is economically advantageous. Combining the two processes leads to savings in the range of 10% to 40% through reduced fuel costs due to higher overall efficiencies (Madlener and Schmid, 2003). Increased efficiency results in an equivalent reduction in harmful emissions, making cogeneration beneficial from an environmental standpoint regardless of the fuel used. The combustion of cleaner fuels such as natural gas generates fewer emissions that combustion of coal or hog fuel for the same quantity of steam generated. The cost of steam generation using natural gas has recently been significantly reduced through the use of hydraulic fracturing in gas exploration. The differences in boiler efficiencies and fuel costs may be further exploited because cogeneration systems in many mills are capable of producing power at rates well in excess of plant demand, making the sale of power to regional electricity providers at contracted rates an additional source of income for the mill.

Cogeneration systems are usually complex, each consisting of several generators, boilers, turbines, condensers, relief valves, vents and pressure headers. Utilizing a cogeneration system at its maximum potential requires stable, accurate and quick control over each of the individual units as well as an effective strategy that coordinates all units towards a common objective such as maximum profitability (Thorin et al, 2005). An energy management system (EMS) is the generic name for a tool or strategy capable of satisfying these demands.

13.2.2 Introduction to Energy Management

Successful management of a cogeneration facility is by no means straightforward, and remains a critical component of recovering profits in today's pulp and paper industry. Elaborate plant configurations result in complex, constantly changing unit co-dependencies. As mentioned earlier, complex contractual agreements may exist between a mill and a regional electricity provider. The sale price of electricity may change based on the time of day, the month of year, the market price of power, and/or the current rate of power production. Other complicating factors include variable fuel costs and availability, dynamic unit efficiencies, fluctuating mill heat requirements and many production rate limiting process constraints. Most cogeneration facilities have therefore developed and/or employed optimization algorithms to aid in the process of plant energy management (Vasebi et al, 2007).

These algorithms, often referred to as EMS, are responsible for the optimal scheduling of power and heat generation through efficient resource and unit operation management (Williams et al, 2005). It is important to note that these energy management systems are often distinct from individual unit controllers. An EMS often dictates optimal unit setpoints, whereas the unit controllers achieve and maintain those setpoints.

An energy management system is designed to drive the operation of a plant, and the individual unit operations within that plant, toward a common goal through the use of mathematical models and optimization strategies. A typical EMS consists of two essential components: a unit coordination strategy and a business strategy. The unit coordination strategy uses current unit efficiencies and potential mill configurations to determine the relative cost of steam and electrical energy production. The unit coordination strategy also handles process constraints such as unit operation limitations or trips, steam consumption requirements and environmental constraints. The business strategy uses real-time energy pricing, fuel cost/availability and power contract information to develop a current and accurate relationship between power production rate and overall profitability. By effectively combining and implementing these components of energy management, a cogeneration facility can reliably maximize profitability.

13.2.3 An Integrated Real-time Energy Management System

As discussed in previous sections, it is highly desirable to effectively combine the unit coordination and business strategy (models) of an energy management system in order to achieve the maximum profitability and reliability from the system. In the past, energy management systems have been offered to industry in various forms that typically used a "layered" architecture, with the EMS functions being distinctly separate from the regulatory control systems. These "layered" systems were often limited in scope, lacked robustness and were often difficult to maintain by mill technical resources.

A recent advance in commercially available control system technology has provided the framework upon which a fully integrated EMS could be developed. Application technology, such as Model Predictive Control (MPC), linear programming (LP), and advanced energy metering functionality, are now standard within some control systems. This allows for tight integration between the regulatory control layer and the economic optimization required in the business strategy models.

In developing the integrated EMS, several key attributes and capabilities were considered as a basis for the design. These included:

- Using standard real-time open architecture hardware and software components so that the system would be stable and maintainable
- Ensuring the base unit regulatory controls were designed as an integral part of the EMS
- Ensuring the EMS was designed to perform full automatic control, but could operate in either Advisory or Direct Control mode. This flexibility would allow operators to make system-suggested changes and would also allow the system to make setpoint changes itself in conjunction with the unit process control systems.
- Ensuring the integrated EMS was scalable and expandable to various configurations of cogeneration plants.

The integrated EMS that was developed uses the DeltaV Digital Automation System embedded advanced control functionality to coordinate and optimize the steam headers, steam turbine generator(s), and boiler combustion controls to maximize overall power generation and provide stable operation of the various steam headers through upset conditions. The following control and optimization functionality was incorporated into the integrated design:

- *Header Pressure Management:* Maintains steam header pressures within specific limits during upset conditions and, when necessary, slides certain low-priority header pressure setpoints to maintain high-priority header pressures.
- *Boiler Load Allocation*: Determines the optimal steam load allocation between the various multifuel boilers based upon boiler availability, efficiency, and constraints, thereby maximizing steam production from the most efficient available boiler.
- *Steam Turbine Power Generation Optimization:* Determines the optimal steam load allocation and maximizes STG power generation based upon available steam and mill constraints.
- *Contract Base Load /Generator Base Load Tracking:* Continuously tracks power use and generation against the contract baselines.
- *Energy Cost and Efficiency Calculations:* Determines the cost of producing steam and generating power on a per unit basis.
- *Real-time Power Contract Calculations:* Determines the incremental price for generated power at any time of the day and year.
- *Data Validation and Error Handling:* Ensures that all process data used for control and economic decision making is "good" data.
- *Enhanced operator interface displays:* Provide good insight into EMS operation.

To illustrate how the integrated EMS was designed and how it functions, a case study of the application will be described in detail in the following section.

13.2.4 Integrated EMS Case Study

The EMS described in the following case study was designed for an integrated cogeneration steam plant that consisted of (Figure 13-5):

- One recovery boiler

- Two power boilers burning hog and natural gas
- One 3-stage condensing steam turbine generator (STG) with 1150 kPa steam extraction and 450 kPa steam extraction
- Interconnecting steam header systems at pressures of 4000 Kpa, 1150 Kpa, 450 Kpa and 65 Kpa.

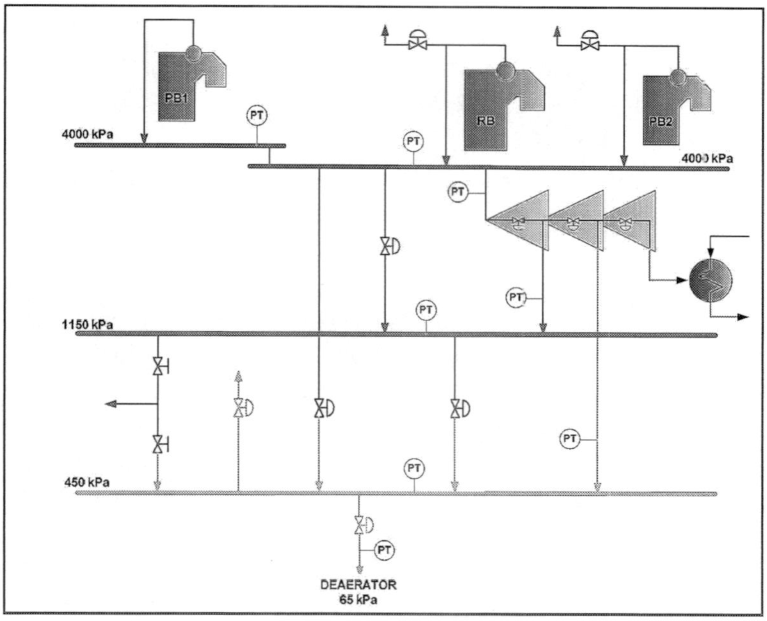

Figure 13-5. Process Overview Schematic for the Case Study Mill Cogeneration Facility

The integrated real-time EMS contains the following core components:

- Regulatory controls
- Real-time economic analysis of steam production and power generation
- Model Predictive Control and Linear Programming (LP) optimization that incorporates process constraints

Regulatory Controls

To maintain the fundamental operation of the steam plant (that is, the safe generation and delivery of steam to the process), a layer of regulatory controls is required. The regulatory controls are designed specifically to maintain the operation of the steam plant while also providing the EMS with specific means to coordinate the unit operations that make up the steam plant. In general, the regulatory controls are designed to satisfy the following control objectives:

- Provide stable pressure control of the high pressure (4000 kPa) steam header and intermediate-pressure steam header(s) in the steam plant
- Deliver all incremental steam produced in the boilers to the process through the steam turbine generator (STG) if possible
- Provide a predictable and repeatable path for incremental steam produced in the boilers if any stage of the STG is limited in its capacity
- Optimize the boiler combustion processes and STG power generation

Figure 13-6 provides an overview of the normal header pressure control scheme with no constraint control. Pressure control of the high pressure (4000 kPa) steam header is achieved by controlling the combustion rate of hog and/or natural gas in the power boilers. Individual header pressure controllers are used to control the firing rate of each fuel. Both pressure controllers maintain the same steam header pressure, but the hog fuel header pressure controller is given a higher setpoint than the natural gas header pressure controller. Doing this ensures that hog is the primary fuel used to produce steam. Natural gas is only used if the hog fuel header pressure controller cannot add enough hog fuel to maintain the pressure in the high-pressure steam header; as the steam header pressure drops below the natural gas header pressure controller's setpoint, natural gas is added to maintain the steam header pressure.

Coordination of the combustion rates in each power boiler is achieved by the EMS, as explained in later sections of this paper. The combustion rate of black liquor in the recovery boiler is dictated by pulping and recausticizing rates in the kraft mill. Therefore, the recovery boiler is not used to maintain the high-pressure steam header. The power boilers maintain steam header pressure when the firing rate in the recovery boiler is adjusted.

Control of the 1150 kPa steam header and the 450 kPa steam header is maintained by the STG extraction valves. The 1150 kPa steam header is controlled by the STG's 2nd stage extraction valve and the 450 kPa steam header pressure is controlled by the STG's 3rd stage extraction valve (3rd stage extraction is to the condenser). The STG's inlet valve is used to maintain STG inlet pressure. Individual pressure controllers

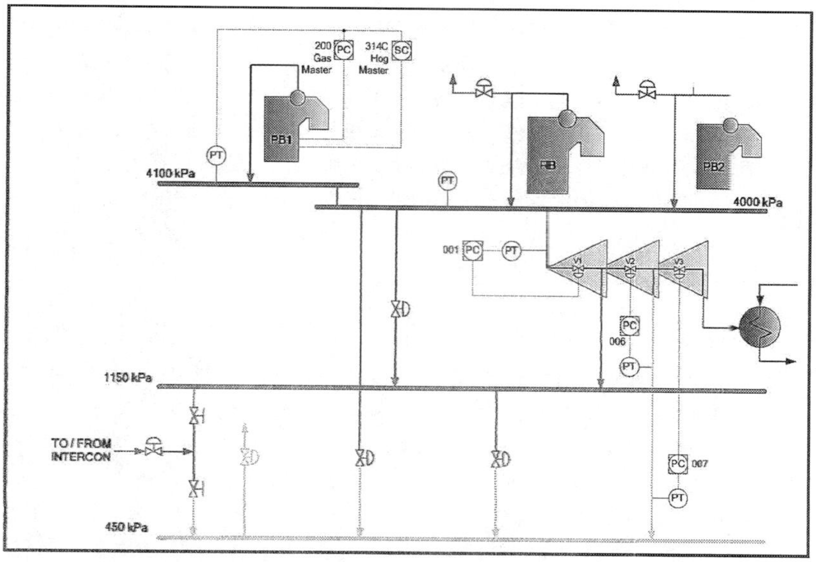

Figure 13-6. General Header Pressure Control with No Constraints

are used to manipulate these valves (i.e., the STG inlet pressure controller manipulates the STG inlet valve, the 1150 kPa header pressure controller manipulates the STG 2nd stage extraction valve, etc.)

By using boiler combustion rates to control the high pressure steam header and STG inlet/extraction valves to control the intermediate-pressure headers, incremental steam produced will always flow through the STG before going to the process, provided that none of the STG valves are limited. To illustrate, consider the effect on the steam system as an extra ton of steam is generated in a power boiler. As the steam enters the high-pressure steam header, the STG inlet pressure increases. As a result, the STG inlet pressure controller opens the STG inlet valve to relieve the pressure at the STG inlet and the extra ton of steam passes through the 1st stage of the STG. This ton of steam now causes the 1st stage extraction pressure to increase, causing the 1150 kPa header pressure controller to open the 2nd Stage extraction valve; the ton of steam now moves through the 2nd Stage of the STG, and so on.

When any of the STG inlet/extraction valves become limited with respect to their flow capacity (that is, increased valve position does not result in increased flow through the valve), header pressure control is compromised for the valve that is limited. To mitigate this, separate pressure controllers for each intermediate header are configured that manipulate the inter-header pressure reducing valves (PRVs). An example is the high pressure (4000 kPa) header over-pressure controller that manipulates the 4000/1150 kPa header PRV (Figure 13-7). This "over-pressure" controller is given a higher pressure setpoint than the STG inlet pressure controller (which manipulates the STG inlet valve). When the STG inlet valve is limited and cannot pass any more flow, the STG inlet pressure increases. Once the STG inlet pressure increases over the setpoint of the over-pressure controller, the over-pressure controller opens the 4000/1150 kPa PRV to relieve the STG inlet pressure.

Over-pressure controllers are configured for each of the PRVs in the steam plant. If any of the STG inlet/extraction valves become limited, the PRVs will move incremental steam through the system. However, while maintaining steam header pressures, the over-pressure controllers effectively bypass steam from specific stages of the STG. Using the previous example, as incremental steam flows through the 4000/1150 kPa PRV when the STG inlet valve is flow-limited, the 1st stage of the STG is bypassed. Obviously, if incremental steam bypasses the 1st stage of the STG, less power is produced by that incremental steam, thereby decreasing the profit that can be gained from that steam. Therefore, the position of each STG inlet/extraction valve must be accounted for by the EMS.

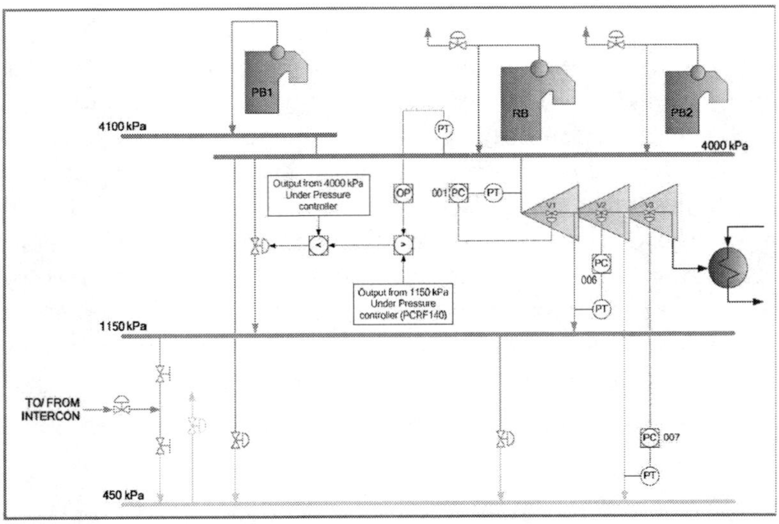

Figure 13-7. Header Pressure Control with 4000 Kpa Over-Pressure Constraint

Real-Time Economic Analysis

The EMS conducts a real-time economic analysis of steam generation and electrical power production in a specific order. Ultimately, the EMS must determine the profitability of producing electrical power from each boiler/fuel combination that it manipulates. Profitability is simply the difference between the cost to produce electrical power and the price at which electrical power can be sold. Figure 13-8 showing the EMS graphic with some of the key economic data that are calculated for each unit.

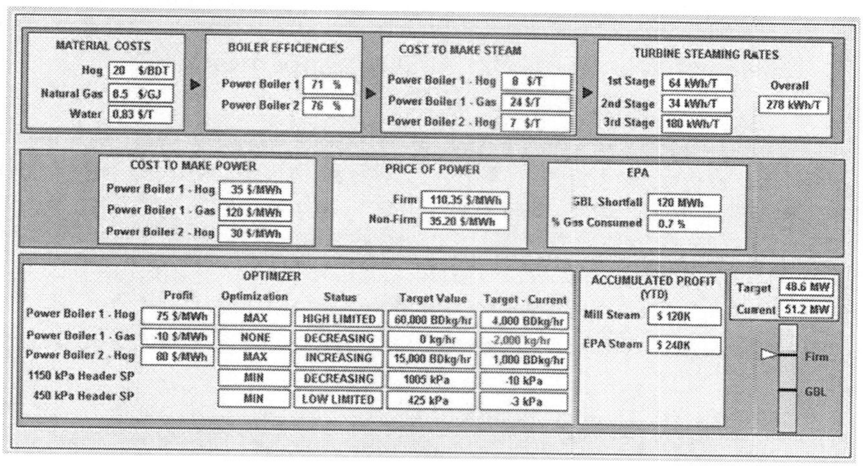

Figure 13-8. Graphic Showing Typical EMS Economic Data

First, the cost to produce steam from each boiler/fuel combination is determined. The incremental cost of hog ($/BDT Hog, (BDT) Bone Dry Tons) is manually entered into the system as the price of hog changes with different outside suppliers of hog fuel. The cost of natural gas ($/GJ Gas, (GJ) Giga Joule) is a real-time value that is automatically updated through a series of secure connections between the EMS and the Internet. The cost to chemically treat raw water ($/Ton Water) is another manually entered value. This cost is added to the overall cost of steam only when the steam will be vented to atmosphere. The EMS determines the path that incremental steam will take through the steam plant and then adds the extra cost of raw water treatment to the cost to produce steam if the incremental steam will be vented.

Boiler efficiencies are calculated to determine the incremental amount of steam that can be produced from each fuel, a value called the Steam Factor. The Steam Factor is expressed as mass of steam per unit of fuel.

Thus, the Steam Factor converts the cost of fuel into the cost of steam. Knowing the Steam Factor for each boiler and the raw material cost of fuel for each boiler, the cost to produce steam ($/Ton Steam) from each boiler/fuel combination can be determined.

In the next step of the economic analysis, the cost of steam must be corrected for the benefit of incremental steam production through desuperheating. High pressure, superheated steam from the boilers is desuperheated to the saturation point before it is used in the process. The boiler feedwater that is used to desuperheat the high-pressure steam is, itself, converted to steam. Thus, to account for the added energy needed to superheat steam, the cost to generate saturated steam for use in the process is reduced by the amount of the extra steam that is produced through desuperheating that steam.

Desuperheating occurs at each extraction point of the STG and downstream of all PRVs. At each of these points, the amount of steam produced through desuperheating is different based on the amount of energy that is retained by the steam as it passes through each of these points. Thus, the EMS must account for the path that incremental steam will take through the steam plant when determining the overall amount of desuperheating that will occur.

Now that the cost of steam ($/Ton Steam) has been determined and corrected for desuperheating, the amount of power that can be generated from the steam (MWh/Ton Steam) is calculated to convert the cost of steam into the cost of producing power. For this, an analysis of the operation of the STG is conducted. This analysis is conducted for each stage of the STG because incremental steam produced in the boilers will only flow through specific stages of the STG, depending on the operation of the steam plant. For each stage of the STG, the efficiency and Steaming Rate are calculated. The Steaming Rate is the amount of electrical power produced by steam flow through the turbine (MWh/Ton Steam) and is determined based on steam enthalpy at each inlet/extraction point of the STG.

As with the correction for desuperheating, the path that incremental steam will flow through the steam plant will determine which stages of the STG will contribute to overall electrical power production. Using one of the examples from the *Regulatory Controls* section, if an increase in steam production causes the high-pressure header overpressure controller to open the 4000/1150 kPa PRV, the 1st stage of the STG is essentially bypassed by that extra steam. Therefore, electrical power is not produced in that stage of the STG.

With the result of the analysis of operation of the STG, the cost of steam ($/Ton Steam) for each boiler/fuel combination is now converted to a cost of producing power ($/MWh) for each boiler/fuel combination. These costs can now be subtracted from the price of electrical power ($/MWh) to determine the profit for each boiler/fuel combination.

The excess electrical power produced by the STG is sold to the local power utility through complex buy/sell agreements. These agreements typically consist of a fixed price portion and an open-market price portion for exported power, based on how much power is produced. The fixed price of power is adjusted based on demand; the time of day (off-peak, peak, or super-peak) and month of the year impact the fixed price of power. For instance, power produced at 5:00 pm (super-peak) in January may be double the price of power produced at 3:00 am (off-peak) in July. The open-market price typically follows demand as well. The EMS manages the buy/sell agreement in real-time and tracks the annual contract. Figure 13-9 shows an example of an EMS display for managing the utility contract agreement in real time.

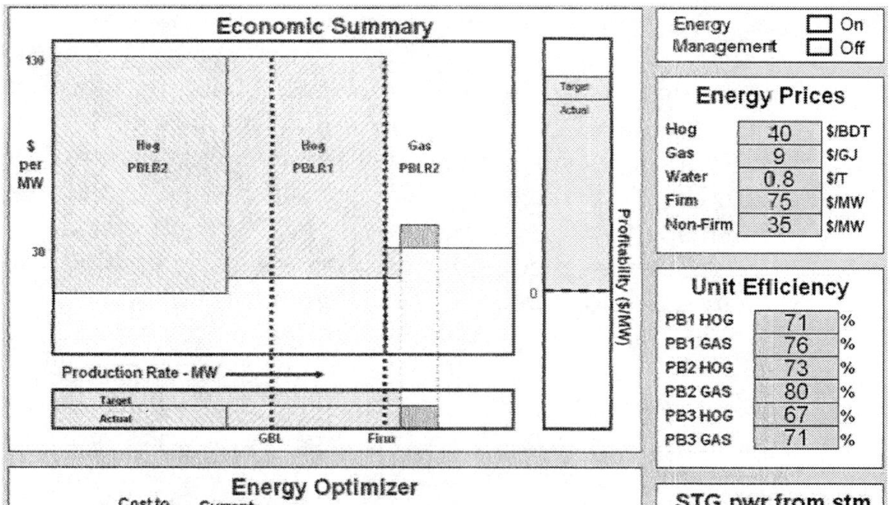

Figure 13-9. Graphic Showing Example EMS Electrical Contract Management

Other factors must be accounted for as well in the contract agreements. The fixed price of power, which may be scheduled monthly for many years in the future at the time the contract is negotiated, must be adjusted for inflation, as quantified by the Consumer Price Index (CPI). The open-market price of power is typically indexed in United States currency ($US) and must therefore be converted to Canadian currency ($CDN) where applicable.

The EMS calculates the real-time price of power based on the applicable contract agreement for power. To determine many of the real-time price factors (open market price of power, currency exchange rate, CPI), the EMS receives information securely from various Internet-based services. For instance, a service may be established with the Dow Jones Index to update the EMS with daily spreadsheets containing current on-peak and off-peak open-market prices for electrical power.

LP Optimizer

Optimization of real-time power generation in the STG is achieved using an embedded Linear Programming (LP) Optimizer that is part of the control system's MPC. The LP Optimizer contains a series of process dynamic models that correlate steam production from each boiler/fuel combination to power production in the STG. The LP Optimizer achieves the most profitable operating point by maximizing or minimizing steam production from each boiler/fuel combination while maintaining the process within constraint limits. Thus, for each boiler/fuel combination that the Optimizer manipulates, the Optimizer can predict how much incremental power can be generated from that boiler/fuel combination. The Optimizer contains another series of process models that correlate steam production from each boiler/fuel combination to specific process constraints (for example, high-pressure steam header pressure, boiler stack opacity, etc.) Thus, the Optimizer can predict the maximum combustion rates for each boiler/fuel combination before a process constraint is violated. The LP Optimizer uses all of the process dynamic models to determine the most profitable combination of the various steam producers that will maintain the process within its various constraint limits. The Optimizer manipulates the combustion rate of each fuel in each boiler (for example hog in Power Boiler 1, natural gas in Power Boiler 1, etc.) In doing this, the Optimizer overrides the high-pressure steam header controls that manipulate the combustion rate of each fuel in each boiler. Therefore, high-pressure header pressure control is maintained by the STG inlet pressure controller, which manipulates the STG inlet valve. Thus, as the Optimizer manipulates the combustion rates of the boilers, the STG inlet pressure controller ensures that incremental steam moves through the STG. When the STG inlet valve becomes flow-limited, the high-pressure header overpressure controller manipulates the 4000/1150 kPa header PRV to move the steam from the high-pressure header. Over-pressure controllers on the remaining steam headers ensure incremental steam continues to move through as many stages of the STG as possible.

Once the economic analysis determines which fuels are to be maximized in which boilers, the Optimizer seeks to maximize steam production from the most profitable boiler/fuel combination. Using the process dynamic models, the Optimizer calculates the maximum steam generation from that boiler/fuel steam producer that will keep the process within its constraints and keep the total power generated in the STG within its limits. If there is still power capacity in the STG with this boiler/fuel combination at its calculated maximum, the Optimizer seeks to maximize steam production from the next most profitable boiler/fuel combination. This series of calculations is carried out until the STG power is maximized or all process constraints are reached.

The economic analysis described previously determines the net profit ($/MWh) from each boiler/fuel combination that is manipulated by the Optimizer. If a boiler/fuel combination has a positive profit, the Optimizer maximizes the combustion of that fuel in that boiler; when maximized, the combustion of that fuel in that boiler is pushed to its high limit. Conversely, if a boiler/fuel combination has a negative profit (loses money), the Optimizer minimizes the combustion of that fuel in that boiler by pushing the combustion rate of that fuel to its low limit. Each profit calculation passes through a hysteresis function so that if a net profit is marginally positive or negative, the Optimizer does not oscillate between "maximize" and "minimize."

Figure 13-10 shows graphically how electric and steam generation are optimized based upon economics and constraint limits.

The optimization solution is calculated using the process dynamic models before any combustion rates in any boilers are adjusted. Once the solution has been calculated, the combustion rates for each fuel in each boiler are set to their optimal values through ramp functions.

The optimization solution is calculated and adjusted every five seconds. Measured and unmeasured process disturbances that will affect the optimization solution are continuously evaluated and adjusted for. For example: The Optimizer determines that the optimal hog feed rate in a boiler is 30 T/h and begins to ramp the hog feed rate to that point.

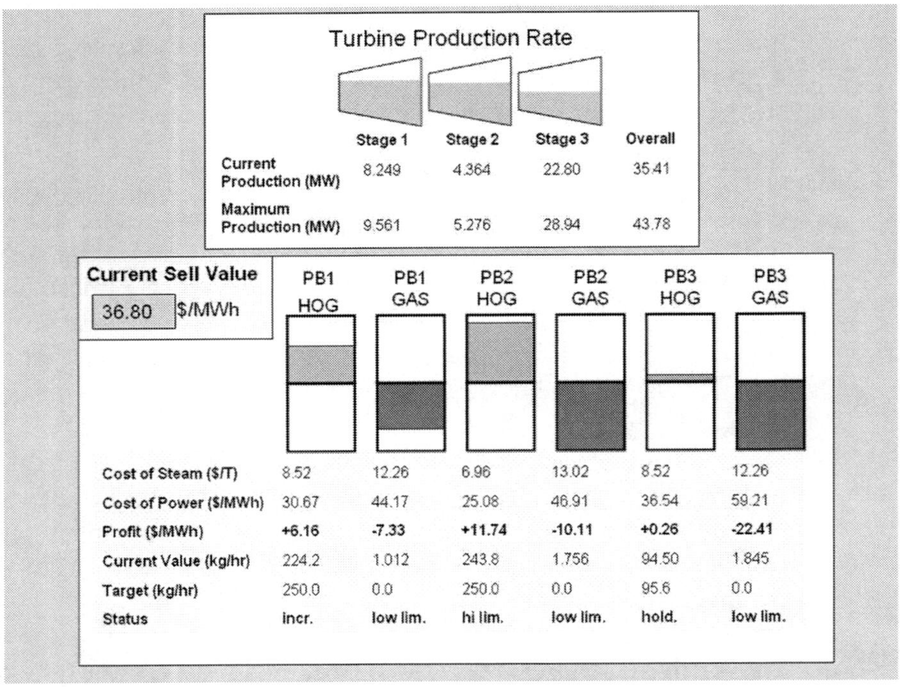

Figure 13-10. Graphic Example of Unit Level Optimization

At 20 T/h, the hog feeder is limited due to a problem with hog hanging up in a hog bin. At this point, the Optimizer recalculates an increase in steam production from the next most profitable fuel to compensate for the power not obtained from the 10 T/h of hog fuel that was not burned. When the hog feeder problem is eliminated, the Optimizer increases the hog feed rate to 30 T/h and decreases steam production from the next most profitable fuel.

By continuously calculating the optimization solution and accounting for real-time disturbances, the EMS is effective and less sensitive to modeling errors.

13.2.5 Summary of Results

To illustrate the results achieved from the integrated real-time EMS, a summary of case study results is presented in this section. The example case study contrasts pre-existing operational strategies that were used to try to optimize the profitability of the cogeneration plant against the performance of the EMS. The baseline assumes plant management methods that try to optimize the plant based upon steam turbine capacity and some rules around the electrical contract agreement.

The following case study assumptions form the basis for the analysis:

- Pulp mill with integrated cogeneration and 55 MW total steam turbine capacity (3 stage condensing)
- Three hog fired powerboilers with varying operating efficiencies
- Electrical contract agreement with the following general characteristics:
  - Fixed power prices up to 40 MW are $75/MWh (peak) and $55/MWh (off-peak)
  - Non-fixed power price ranging from ~$30–$60/MWh
- Incremental hog fuel available at $2.05/GJ
- Baseline of existing baseline operating strategies with rules established based on turbine capacity:
  - Run constantly at 100% turbine stage 3 (STG S3) capacity (condensing)
  - Run at 100% stage 3 condensing capacity during peak, 50% stage 3 capacity during off-peak
  - Hold power production at 40 MW (fixed-price limited)
- Integrated EMS system is operated in closed loop mode to achieve optimized results

A brief summary of the results is presented below. Figure 13-11 shows a plot of profitability ($/hr) over time for the integrated EMS approach versus the three baseline operating strategies. The chart shows that the plant profitability varies throughout the 48 hour period shown, and that the integrated EMS maximizes this profit during peak and non-peak times of day. Table 13-6 compares the improvement in profitability against the base operating strategies.

Table 13-6.

Operating Mode	Profit ($/day)	Profit Improvement ($/day)	Profit Improvement vs. Fixed Mode (%)
Integrated EMS	$47,814	$3,087	6.46%
100% S3 Capacity	$44,824	$97	0.22%
50/100% S3 Capacity	$45,438	$711	1.56%
Fixed	$44,727	n/a	n/a

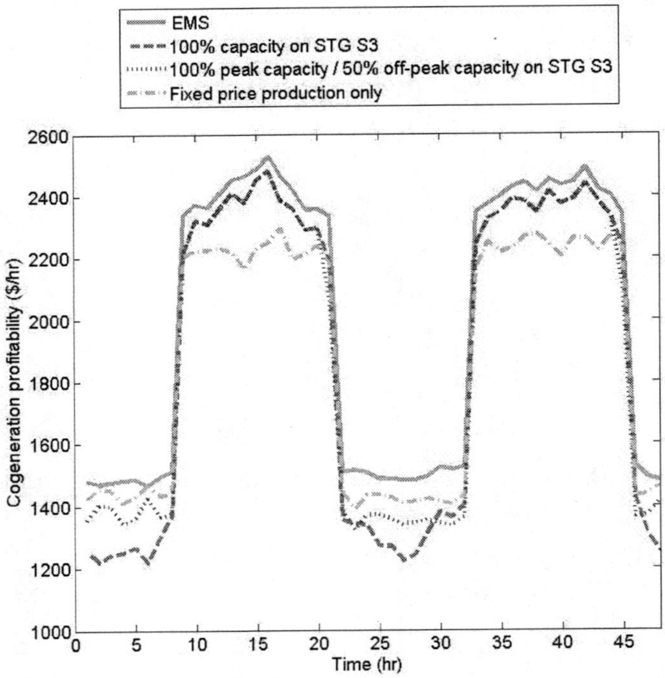

Figure 13-11. Plot of Plant Profitability ($/hr) over Time for the Integrated EMS System Approach versus the Three Baseline Operating Strategies

The integrated EMS was able to achieve an almost 6.5% improvement in overall cogeneration profitability, or about $3,000/day. Based upon a typical 350-day operating year for the plant, we would expect that the accumulated gain in profits will be over $1MM/year due to having a real-time optimized plant.

The explanation for the improvement in profitability comes from the real-time manipulation of the unit operations by the economic optimizer. The open-loop operational strategies that pre-existed saw the human operators make infrequent changes to unit processes and did not consider the changing efficiencies or economic models. This is typical of most cogeneration operations, because the complexity and the number of combinations are extremely difficult for a human operator to process on a real-time basis.

In contrast, the integrated EMS with closed-loop control of the STG, in conjunction with control of the boilers and PRVs, is able to make real-time adjustments to find the optimal combinations minute by minute. The break-even point in generation vs purchase is a function of total steam generation as shown in Figure 13-12. In this figure, the location of the break-even point is shown using a sample plot of the real-time power price and real-time steam cost curves.

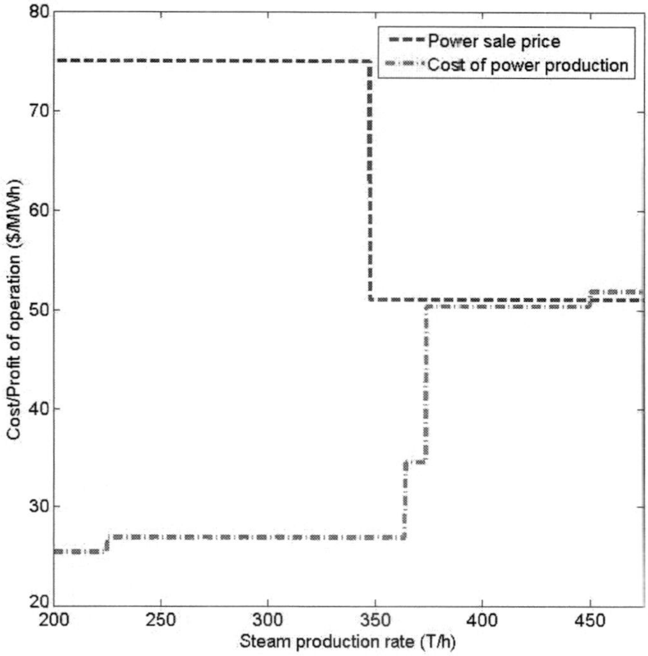

Figure 13-12. Sample Plot of the Real-time Power Price and Steam Cost Curves Demonstrating Location of Break-even Point

Figures 13-13, 13-14 and 13-15 show sample plots of how the integrated EMS varies the unit targets based upon real-time economic calculations.

13.2.6 Summary

This section has discussed a real-time Energy Management System that can be used to maximize the profitability of cogeneration facilities in a kraft pulp mill. Recent advances in commercially available control system technology have provided the framework from which a fully integrated EMS can be deployed that effectively combines regulatory control, unit coordination and economic models in order to achieve the maximum performance and reliability from the EMS. In the past, energy management functions have been offered to industry in various forms that typically involved a "layered" architecture with the EMS functions being distinctly separate from the regulatory control systems. These systems were often limited in scope, lacked robustness and were often difficult to maintain by mill technical resources.

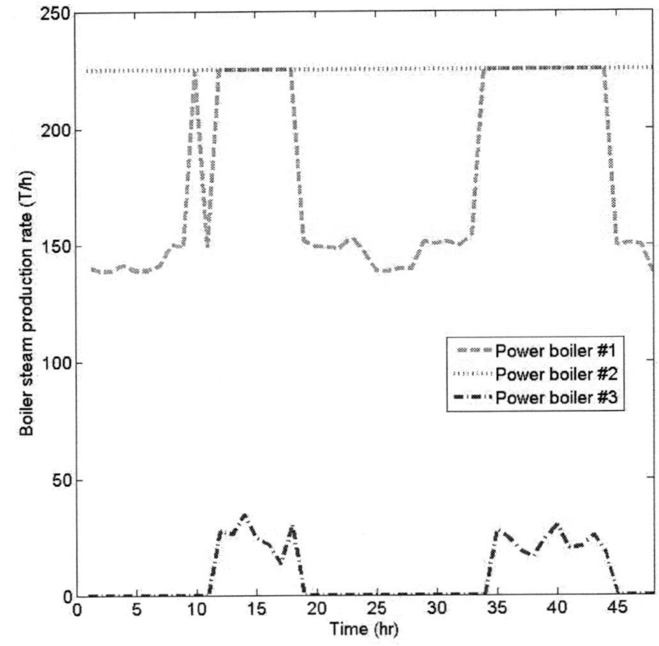

Figure 13-13. Plot of Integrated EMS Optimization of Power Boiler Production Targets vs. Time

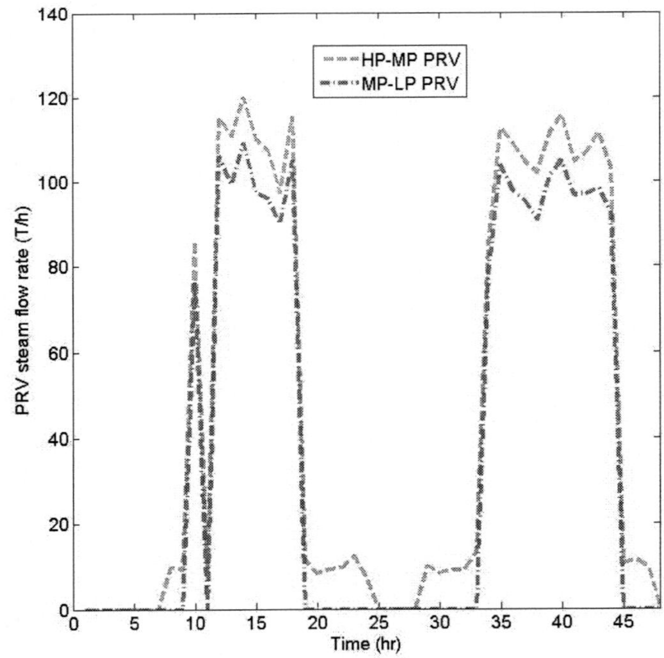

Figure 13-14. Plot of Integrated EMS Optimization of Pressure Reducing Valve Steam Flow Rates vs. Time

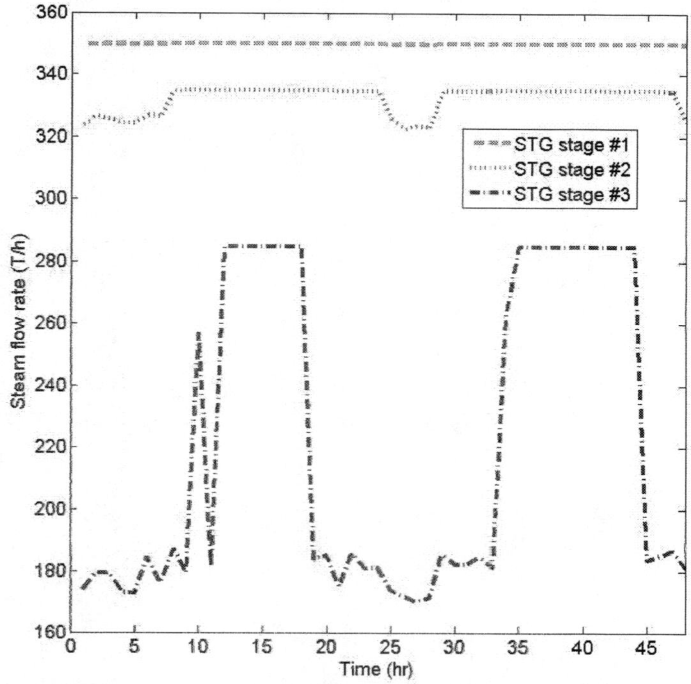

Figure 13-15. Plot of Integrated EMS Optimization of Steam Turbine Generator Extraction Stage Steam Flow Rates vs. Time

A case study was presented based upon applying an integrated EMS approach, and the results indicate there is significant opportunity to manage cogeneration plant profitability in real-time. When compared to existing open-loop operational strategies, the integrated EMS system was able to achieve as much as a 6.5% improvement in plant profitability. Based upon a typical 350 day operating year for this case study plant, we would expect that the accumulated gain in profits will be over $1MM/year due to the real-time decision making provided by the integrated EMS system.

13.3 On-line Optimization Workshop

This workshop exercise is designed to show how real-time optimization may be used to minimize cost in steam generation when the steam load is to be allocated to two boilers with different operating efficiencies. Access the book's web site http://www.advancedcontrolfoundation.com, and select the Solution tab to view the workshop being performed.

Step 1: Observe the impact of boiler load allocation on fuel consumption for a constant steam demand of 200 T/h.

Step 2: Examine the cost of generating 200 T/h of steam when only Boiler 1 or Boiler 2 is in service.

Step 3: Change the steam demand in steps from 100 to 300 T/h and observe how the steam load is allocated by the optimizer to minimize the cost of steam generation.

The interface shown in Figure 13-16 will be used to examine the impact of boiler load allocation on the cost of steam generation.

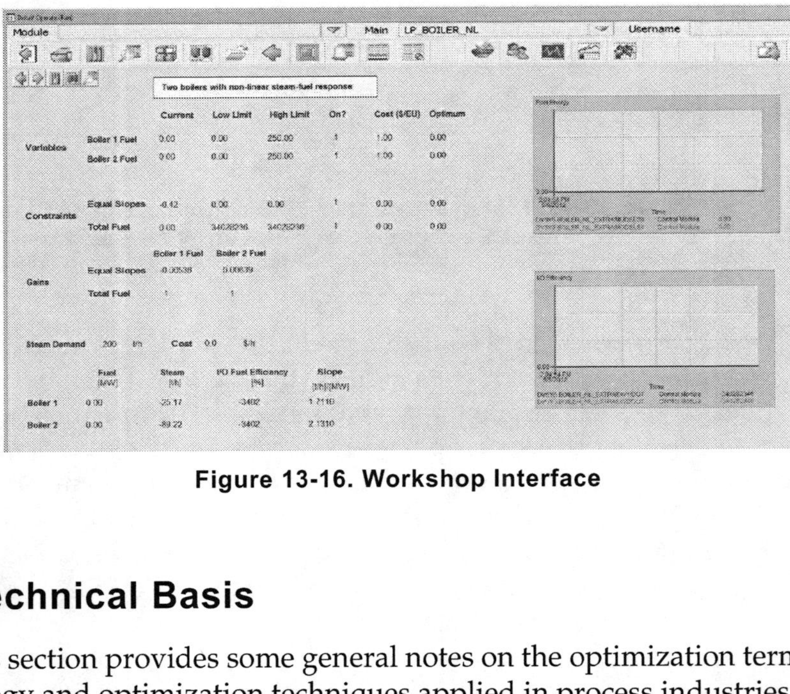

Figure 13-16. Workshop Interface

13.4 Technical Basis

This section provides some general notes on the optimization terminology and optimization techniques applied in process industries. In general, optimization can be applied to diverse industry problems such as designing equipment, scheduling operations, minimizing inventory or allocating resources. In this chapter, we focused on on-line process optimization. This can be seen as optimization when all process optimization model parameters are available on-line and optimization results can be transferred to the process, affecting its operation.

In general, an optimization software application consists of three components:

- The process model
- The operation constraints
- The objective function

Optimization with a linear process model and a linear objective function is presented in Chapter 12. Such optimization is known as Linear Programming (LP). LP is the most widely used and perhaps the most effective technique. When the objective function contains quadratic terms, optimization then requires Quadratic Programming (QP). As was illustrated in Chapter 12, LP always provides the solutions on the intersection of constraints, while a QP solution can be off constraints, inside the feasible region, which is the area limited by lines CV1min, CV2max, MV2max, CV1max and CV2min (Figure 13-17). If the optimal operating point is inside the feasible region, the optimal QP solution may not change with the constraints. This allows for more robust optimization.

When the process model uses non-linear equations, nonlinear optimization is applied, called Nonlinear Programming (NP). A Successive Linear Programming (SLP) technique is commonly used for solving NP. SLP techniques solve a sequence of linear programming approximations for a nonlinear programming problem.

Many industrial problems involve integer values as well. The optimizer solution should answer questions such as which equipment to operate or install or how to develop an optimal schedule. Such optimization problems are known as Mixed Integer Programming (MIP) and are solved with Branch and Bound (BB) methods.

For more complex optimization problems, Real Time Optimization (RTO) is applied. RTO is a complex, rigorous model based system that optimizes plant operations. A typical RTO involves thousands of variables and equations and uses the linear and nonlinear optimization techniques most suitable for the problem. Providing details on any of the above optimization techniques is far beyond the scope of this book. More information is available in the referred literature, in particular in Edgar, Himmelblau and Lasdon (2001) and Maciejowski (2002).

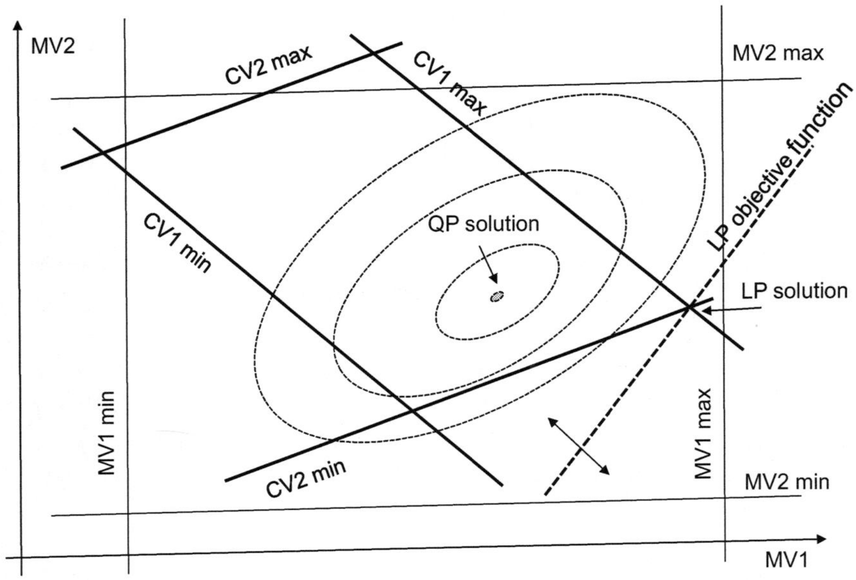

Figure 13-17. LP and QP Optimization Solutions

BIBLIOGRAPHY

1. Arroyo, J. and Conejo, J. "Optimal Response of a Thermal Unit to an Electricity Spot Market," *IEEE Transactions on Power Systems* 15(3):1098–1104, August 2000.

2. Cohen, A. and Sherkat, V. "Optimization-based Methods for Operations Scheduling," *Proceedings of the IEEE,* 75(12):1574–1591, December 1987.

3. Edgar, T., Himmelblau, D., and Lasdon, L. *Optimization of Chemical Processes,* 2nd edition, Boston, McGraw-Hill, 2001.

4. Ferrari-Trecate, G., Gallestey, E., Letizia, P., Spedicato, M., Morari, M. and Antoine, M. "Modeling and Control of Co-generation Power Plants: A Hybrid System Approach," *IEEE Transactions on Control Systems Technology* 12(5):694–705, September 2004.

5. Gruhl, J., Schweppe, F., and Ruane, M. "Unit Commitment Scheduling of Electric Power Systems," in *System Engineering for Power: Status and Prospects,* Springfield, VA, US DOE Report NTIS-CONF-750867, 1975, pp. 116-129.

6. Huang, G., Baetz, B., and Patry, G. "A Grey Linear Programming Approach for Municipal Solid Waste Management Planning under Uncertainty," *Civil Engineering and Environmental Systems* 9(4):319–335, 1992.

7. Lahdelma, R. and Hakonen, H. "An Efficient Linear Programming Algorithm for Combined Heat and Power Production," *European Journal of Operational Research* 148(1):141–151, July 2003.

8. Maciejowski, J. M. *Predictive Control with Constraints,* Harlow: Pearson Education Limited, 2002.

9. Madlener, R. and Schmid, C. "Combined Heat and Power Generation in Liberalised Markets and a Carbon-constrained World," *GAIA – Ecological Perspectives for Science and Society* 12(2):114–120, June 2003.

10. Sampson, R., Wright, L., Winjum, J., Kinsman, J., Benneman, J., Kürsten, E., and Scurlock, J. "Biomass Management and Energy," *Water, Air, & Soil Pollution* 70(1- 4):139–159, October 1993.

11. Thorin, E., Brand, H. and Webe, C. "Long-term Optimization of Cogeneration Systems in a Competitive Market Environment," *Applied Energy* 81(2):152–169, June 2005.

12. Vasebi, A., Fesanghary, M., and Bathaee, S. "Combined Heat and Power Economic Dispatch by Harmony Search Algorithm," *International Journal of Electrical Power & Energy Systems* 29(10):713–719, December 2007.

13. Williams, J., Huff, F. and Francino, P. "Islands of Optimization: a Web-based Economic Optimization Tool," White paper, Emerson Process Management, 2005.

14 Process Simulation

Dynamic process simulations may be used to demonstrate and test monitoring and control functions within the control system, as illustrated in many of the workshop exercises. In most cases, the control system's features may be used to create process simulations. Using this capability to create the process simulation as a module, process simulations may be added to a control system to check out or demonstrate a control strategy without modifying the modules configured for monitoring, calculation and control. Comprehensive operator training systems may also be created using this capability.

Because a process simulation can be such a useful tool when working with advanced control techniques such as model predictive control and with the advanced monitoring functions provided by continuous and batch data analytics, this chapter addresses how to add process simulations to a control system using tools that exist in most modern control systems.

14.1 Process Simulation Techniques

In a control module, the analog output block is used to manipulate final control elements such as valves, dampers or variable speed drives. Because the analog output block's OUT parameter represents the current or digital signal to these final control elements, this parameter is connected as an input to the process simulation. Similarly, the OUT_D parameter of discrete output blocks may be added as inputs to the process simulation. Measured and unmeasured disturbances may be included in the simulation as adjustable discrete or analog parameters. When included, these parameters are also connected as inputs to the process simulation.

The process conditions that result from the manipulated and disturbance inputs, for example, monitored measurements of process conditions such as flow, pressure, temperature, level, composition or the discrete state of a limit switch, are reflected in the process simulation's outputs and may be easily included in the simulation. Some of these process conditions may be measured in the field using transmitters or switches, and thus the simulated values can be provided to the modules that access these field values using analog and discrete input blocks. The current or digital outputs of these transmitters and switches are normally accessed in a control system using the analog and discrete input blocks. However, the normal processing of analog and discrete input blocks can be altered by enabling the Simulate parameter, as illustrated in Figure 14-1.

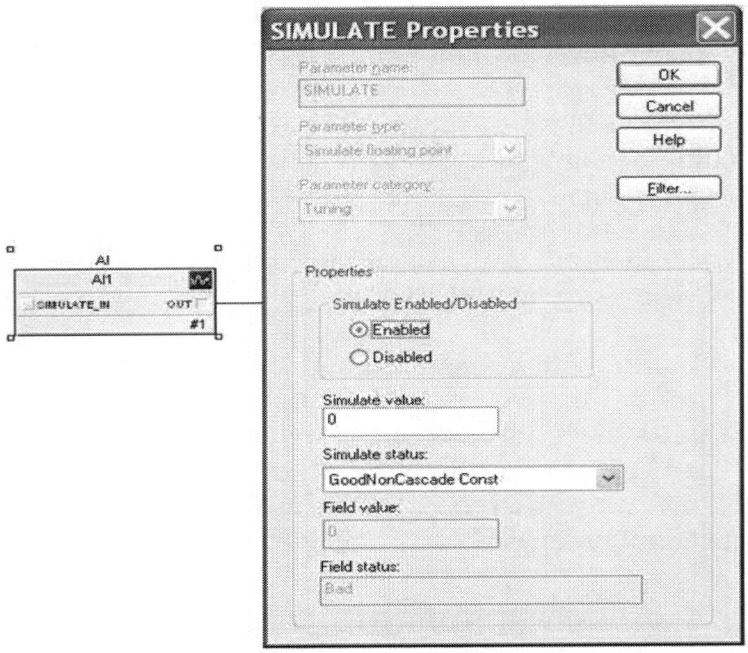

Figure 14-1. Simulate Parameter – Analog Input Block

When simulation is enabled in an input block, a Simulate parameter value is used in place of the field measurement. Thus, the simulated process output values may be used by analog and discrete input blocks in place of field measurements. Process simulation may be incorporated into the control system via this basic control system capability.

The most common means of adding process simulation capability to a control system is to create new modules that contain only the process simulation. These simulation modules may be added without modifying existing modules, because most modern control systems include the capability for a module to read and write parameters contained in another module. Using this capability, simulation modules may read the output value of analog and discrete output function blocks that are part of a control module. The results of the simulation modules may then be written back to the simulation input parameters of the analog and discrete input blocks of the existing modules. An example of this implementation is illustrated in Figure 14-2.

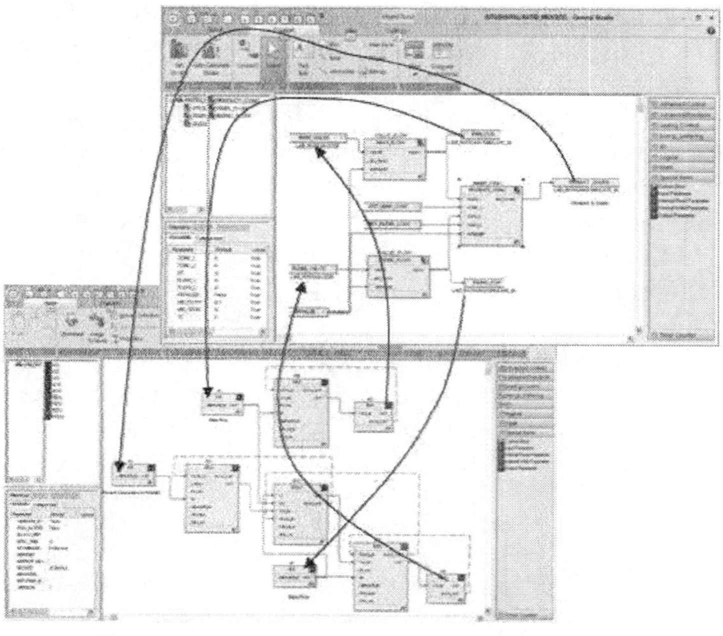

Figure 14-2. Transfer of Simulation Parameters

In the existing modules, the I/O blocks' Simulate parameter can be enabled to use the process simulation. When simulation is enabled, the normal operator control displays and engineering tools may be used to view the monitoring and control functions performed by the existing control system modules.

14.2 Developing a Process Simulation from the P&ID

The Piping and Instrumentation Diagram (P&ID) often serves as the starting point in the development of a process simulation for control system checkout or operator training. The major pieces of equipment and their connections are shown on the plant P&ID. In addition, the final control elements and transmitters that are accessed by the process control system are identified on the P&ID. In most cases, one process simulation module should be developed for the modules that support control or monitoring functions associated with a piece of equipment. Thus, a first step in developing a process simulation is to break the equipment and piping shown on the P&ID into multiple processes that may be simulated in one or more simulation modules.

As a general rule, pieces of equipment or piping that contain transmitters and final control elements used in a regulatory or advanced control strategy or monitoring function should be represented as one or more processes in the same simulation module. For example, the actuator position implied by an analog output block in a control module would be an input to the simulated process. Any measurements in the flow path between the actuator(s) and transmitters used in control or monitoring functions would be outputs of the simulated process. Unmeasured disturbances may be included as parameters of the simulation module. These disturbance parameters may be changed when testing a control strategy or during operator training to demonstrate the impact of a disturbance on the process.

An example of how the equipment and piping shown on the P&ID may be broken into smaller sets of equipment that represent processes is illustrated in the Recycle Tank example shown in Figure 14-3.

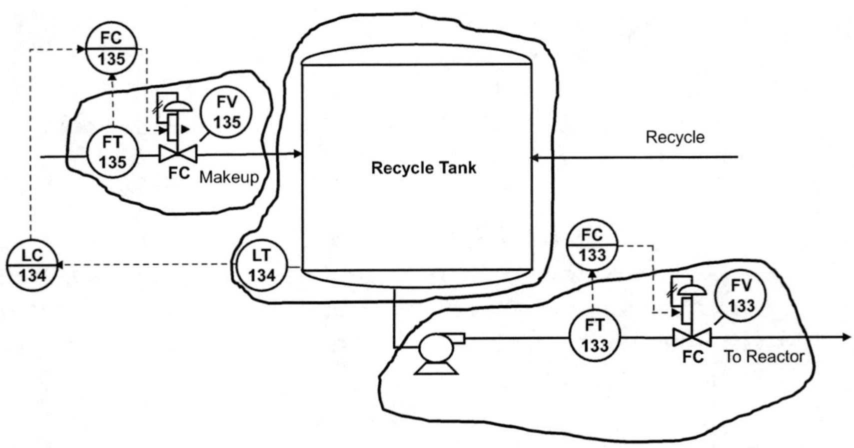

Figure 14-3. Breaking a P&ID into Small Processes

In this example, the circled areas of the P&ID will be simulated processes that are interconnected and can be documented in a simulation diagram as illustrated in Figure 14-4.

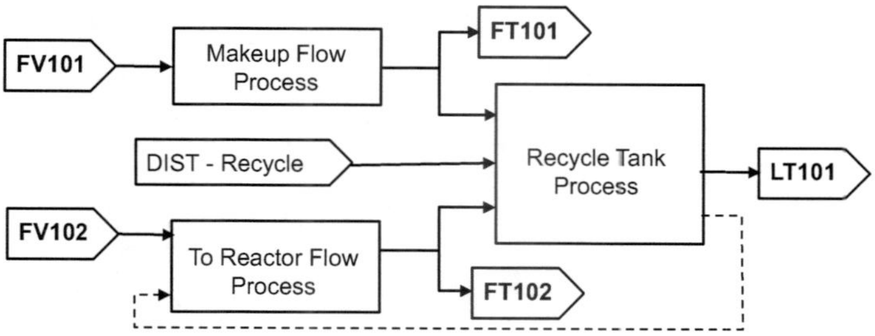

Figure 14-4. Simulation Diagram for Recycle Tank Example

Once the P&ID has been analyzed in this manner, the simulation diagram may be implemented as a simulation module that references the existing control modules. The processes on the simulation diagram are implemented as composite blocks in the simulation module. Process inputs from final control elements are added as external references to the setpoints of the analog output blocks in the existing control modules. Unmeasured inputs to a process should be added as parameters. Outputs of the process simulation that represent field transmitter measurements are then implemented as external references that write to the Simulate parameter of the existing modules' analog input blocks.

The simulation module for the recycle tank example is shown in Figure 14-5.

For the recycle tank example, tank level control LC134 and makeup flow FC135 are implemented in one control module, LIC134, as illustrated in Figure 14-6.

The flow loop to the reactor in the recycle tank example is implemented as one module, FIC133, as illustrated in Figure 14-7.

When a control system is commissioned, the step response of a self-regulating process is often characterized as first-order-plus-deadtime. Similarly, the step response of an integrating process is often characterized by the deadtime and the integrating gain.

It is appropriate to design a process simulation to duplicate the step response (Cutler, 2005), (Blevins et al, 2011). For example, a filter

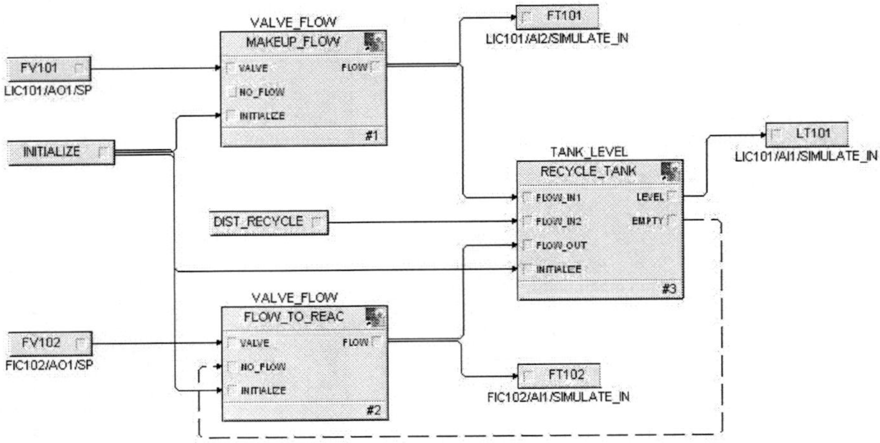

Figure 14-5. Simulation Module for Recycle Tank Example

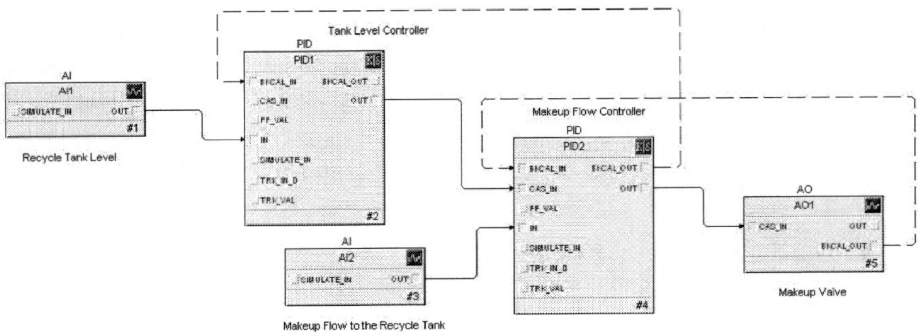

Figure 14-6. Module LIC134 – Recycle Tank Example

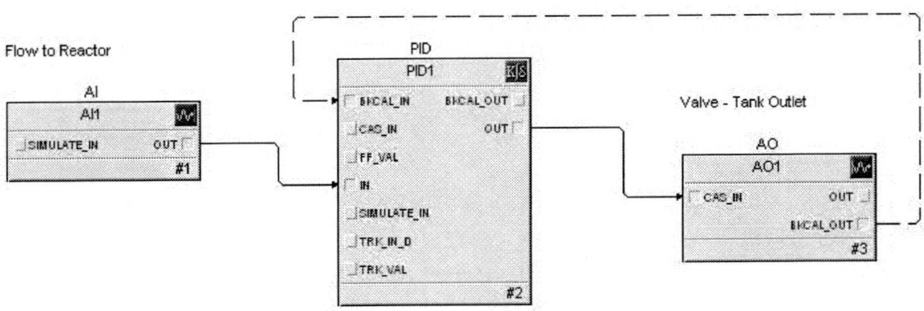

Figure 14-7. Module FIC133 – Recycle Tank Example

block, a deadtime block and a multiplier block may be used to simulate the process deadtime, time constant, and gain of a single input-single output (SISO) *self-regulating* process. Process noise may be added to the simulated process outputs using a signal generator block. For example, the FLOW_VALVE composite illustrated in Figure 14-8 uses this approach to simulate the makeup and reactor flow processes in the recycle tank example. This composite was created using these standard function blocks (filter block, deadtime block and multiplier block and a signal generator block to simulate process noise).

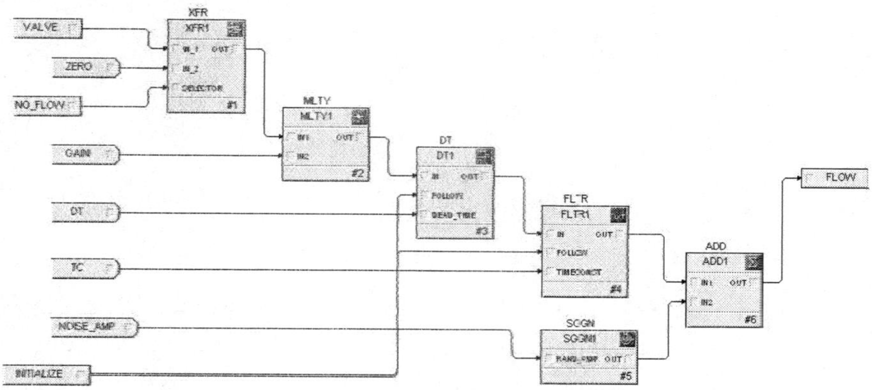

Figure 14-8. FLOW_VALVE Composite

Additional inputs may be included in the process simulation to address abnormal operating conditions. As illustrated in the FLOW_VALVE composite, a NO_FLOW input is included to allow the simulation to accurately reflect conditions that prevent flow.

Outputs may also be included in the simulation to reflect abnormal conditions. For example, the LEVEL_TANK composite used to simulate the recycle tank includes an EMPTY output. In the level simulation, this Boolean parameter transitions from a value of 0 to 1 when the tank is empty. As illustrated in the simulation module for the recycle tank shown in Figure 14-5, this EMPTY condition is used in the simulation of the flow to the reactor.

The simulation of an *integrating process* should take into account the module execution period, and should support a varying number of inputs. In addition, the calculated value for the level should be limited to the container/vessel height. A combination of standard function blocks and a calculation block may be used to implement the recycle tank level simulation as illustrated in Figure 14-9.

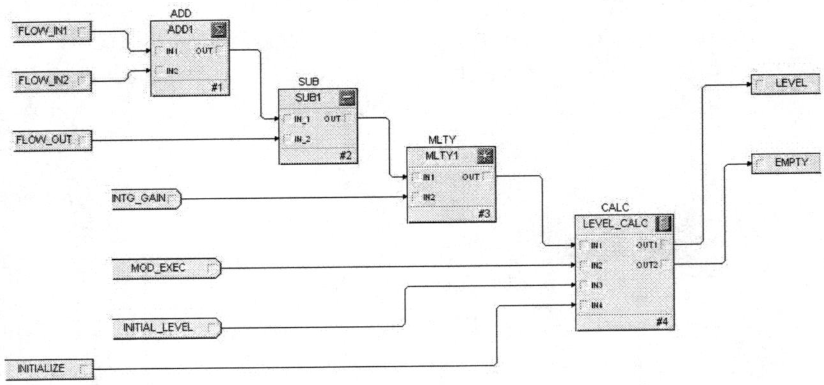

Figure 14-9. Recycle Tank Level Simulation

The real-time dynamic response of the simulated recycle tank and associated control modules will accurately reflect changes in the flow to the reactor and in the recycle tank inlet flow disturbance input that is defined in the recycle tank simulation module.

A simulated process may have multiple inputs and outputs. In the recycle tank level simulation, an identical change in inlet and outlet flow rates could be safely assumed to have the same dynamic impact on the tank level. However, in general, each process input may be expected to impact each process output differently.

As stated in Chapters 11 and 12, the impact of each process input on each process output may be characterized by the output response to a step change in the input. Process simulations can be designed to duplicate the process output response to step changes in the process input. For example, the simulation module diagram for the batch reactor process, Figure 14-10, shows the batch reactor process has inputs that impact each of the three process outputs, as illustrated in Figure 14-11.

Examining the process simulation diagram shows that the three process inputs to the batch reactor process impact the temperature and the pressure outputs. The feedrate only impact the reactor level. The design of the reactor simulation must take into account the impact of each of the three process inputs on the three outputs. The implementation using a composite block that supports a process as large as 3×3 is

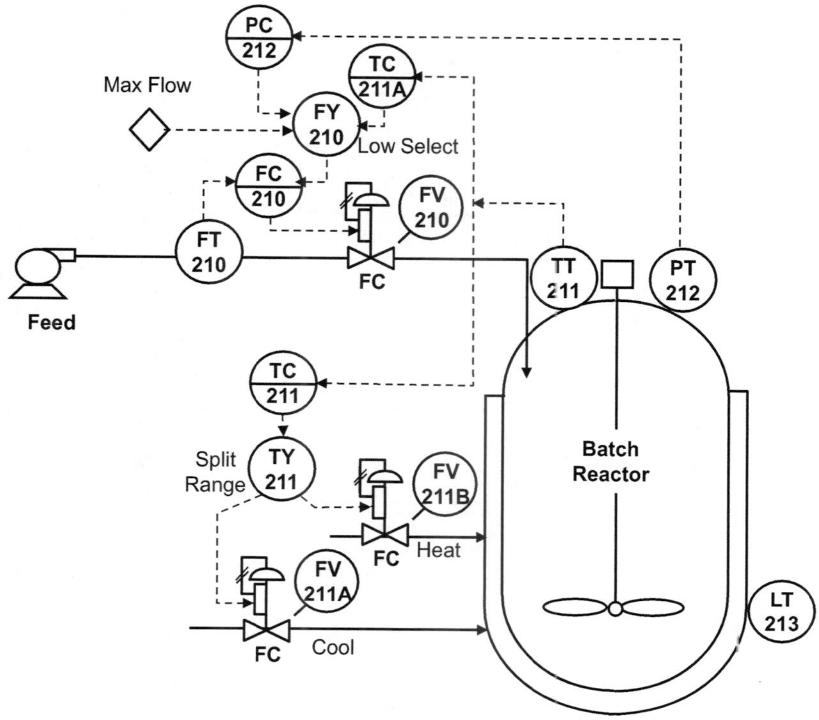

Figure 14-10. Batch Reactor Process

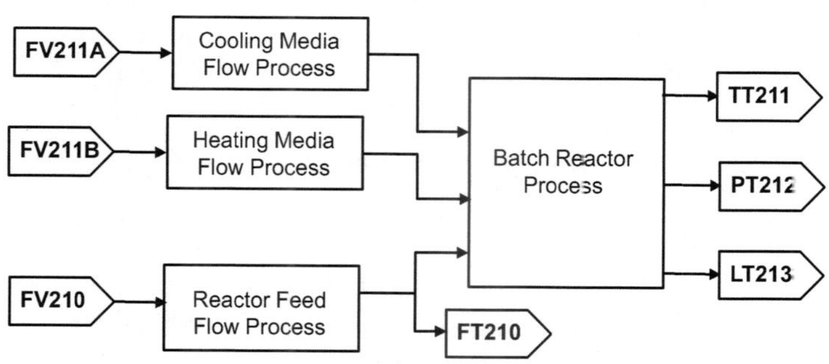

Figure 14-11. Batch Reactor Process Simulation Diagram

shown in Figure 14-12. Only standard function blocks are required to implement this composite block.

The composite for simulating a 3×3 process may be easily expanded to simulate much larger processes, such as distillation columns, that are characterized by a large number of process inputs that impact multiple

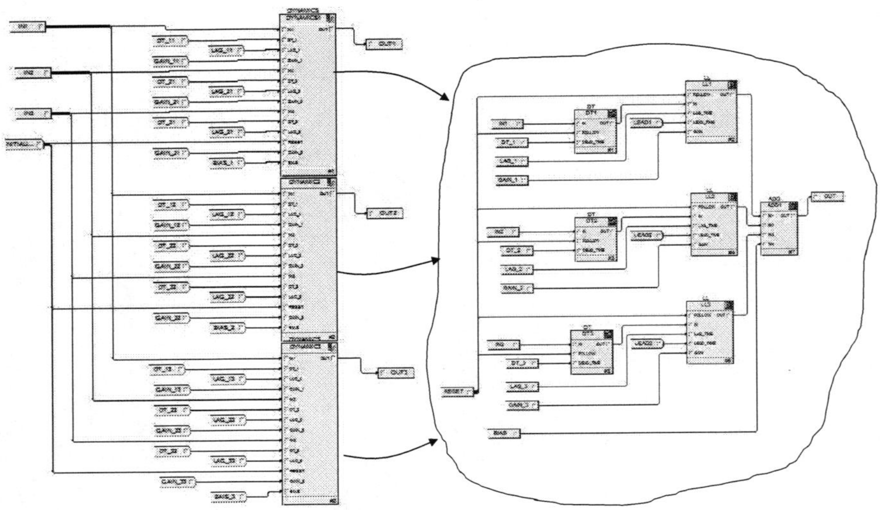

Figure 14-12. Composite Block for 3X3 Process Simulation

process outputs as illustrated in some of the neural network and model predictive control workshops in this book.

14.3 Simulating Process Non-linearity

Simulations based on duplicating the process step response may be effectively used to calculate a variety of self-regulating and integrating process measurements. However, this simulation technique assumes the process behaves in a linear fashion. When the simulation is used over a wide operating range, the step response may not accurately show the impact of process non-linearity.

A non-linear installed valve characteristic may be simulated by substituting a characterizer block for the gain multiplier block in the FLOW_VALVE composite. The non-linearity introduced by an oddly-shaped tank may be simulated by including a characterizer block in the LEVEL_TNK composite to calculate the integrating gain as a function of level. In a similar fashion, the non-linear response often seen in analytic measurements, such as concentration after blending two streams or heater outlet temperature for varying inlet flow rates, may be accounted for in the process simulation.

The necessary corrections in process gain must be based on input/output relationships determined by performing an energy and/or mass balance for steady-state operation. The inputs and outputs of a process are known from the simulation diagram.

An energy or mass balance is based on the fact that in general what goes in plus accumulation or loss within the process must be equal to the process output. However, under steady-state operation, what goes into the process must come out (i.e., there is a balance). For example, when two liquid flow streams are blended using a mixer, the concentration of material dissolved or suspended in the mixer outlet stream (expressed as the weight percent of the outlet stream) varies in a non-linear fashion with the inlet flow rates and inlet stream concentrations.

The simulation diagram for a mixer process is shown in Figure 14-13.

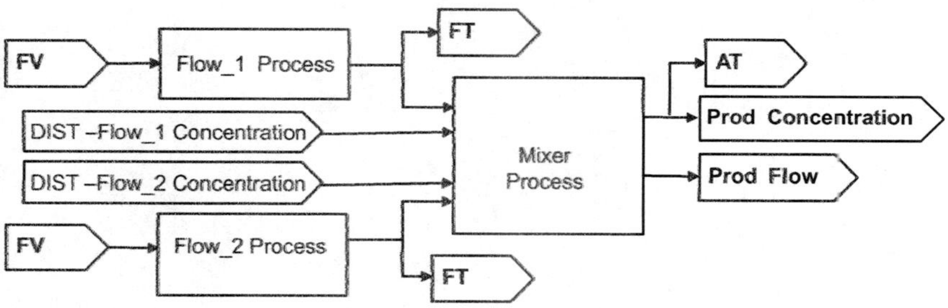

Figure 14-13. Simulation Diagram for a Mixer Process

The outlet stream concentration may be calculated based on a mass balance around the mixer, and blocks may be added to provide a dynamic response to input changes. The outlet concentration may be calculated as follows:

$$Solids_{in} = Solids_{out} = F1 \times C1 + F2 \times C2$$

$$Mass_{in} = Mass_{out} = F1 + F2$$

$$Outlet\ Concentration = \frac{Solids_{out}}{Mass_{out}} = \left(\frac{F1 \times C1 + F2 \times C2}{F1 + F2}\right)$$

where:

$F1$	=	Inlet mass Flow_1
$F2$	=	Inlet mass Flow_2
$C1$	=	Solids concentration in Flow_1
$C2$	=	Solids concentration in Flow_2

As shown in this calculation, the process gain (i.e., the change in mixer outlet concentration for a change in the flow rate of inlet flow to the mixer) for steady-state operation before and after the input change depends on the flow rate and the concentration of both streams. This nonlinear response can be included in the process simulation by calculating the outlet concentration based on the steady-state mass balance. Any transport delay or lag due to mixing may be accounted for by using a deadtime and filter block in combination with this calculation, as illustrated in the mixer simulation composite shown in Figure 14-14.

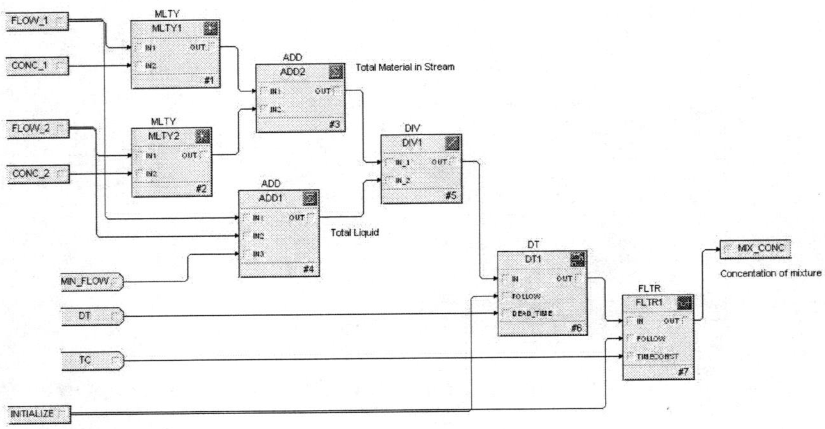

Figure 14-14. Mixer Outlet Concentration Simulation

When a combustion process such as a heater or boiler is to be simulated, it is often necessary to simulate the oxygen concentration in the flue gases. Over a wide operating range, the oxygen concentration in the flue gases, as measured by an O_2 analyzer, varies in a non-linear fashion with the fuel input rate and total air flow. This is another example of how the simulation range may be extended by accounting for this non-linear response in the simulation. The simulation diagram for a combustion process is illustrated in Figure 14-15.

To accurately simulate the flue gas O_2 concentration over a wide operating range, the calculation of O_2 using a mass balance may be combined with the lag and deadtime identified in the process step response. The oxygen concentration in air is approximately 21 percent by volume. Thus, the steady-state O_2 concentration after combustion may be calculated as follows:

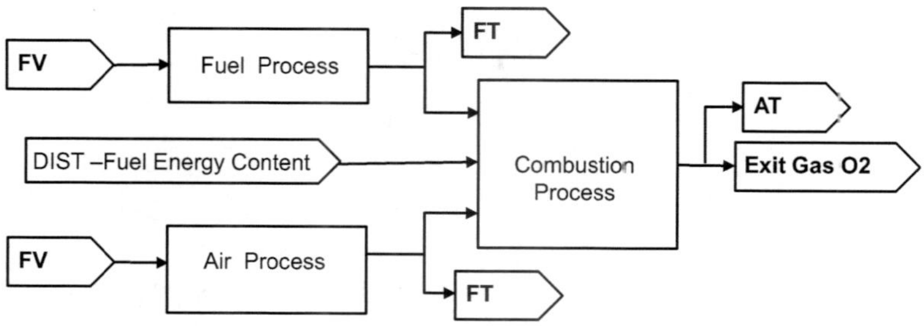

Figure 14-15. Combustion Process Simulation Diagram

$$Consumed\ Air = Fuel\ Flow \times K = Air\ Flow \times \left(1 - \frac{\%O_2}{21}\right)$$

where:

K = Air required per unit of fuel for complete combustion

Solving for O_2, the following is obtained:

$$O_2 = 21 \times \left(1 - \frac{K \times Fuel\ Flow}{Air\ Flow}\right)$$

As mentioned above, the impact of a change in fuel or air flow rate on the flue gas O_2 concentration varies with the fuel and air flow rates. To account for this non-linear response, the combustion process O_2 simulation may be based on the steady-state relationship between O_2, fuel flow and air flow. The calculated value must be limited to 0–21% to accommodate abnormal situations such as zero air flow. A deadtime and filter block may be used to account for transport delay and the mixing of gases in the combustion process as illustrated in Figure 14-16.

The simulations for the processes included in the neural network and model predictive control workshops demonstrate how non-linear process response may be accurately simulated by combining the results of steady-state analysis with the process step response.

14.4 Other Considerations

When process simulation is used for control system checkout or operator training, it may be desirable to speed up the process response. A

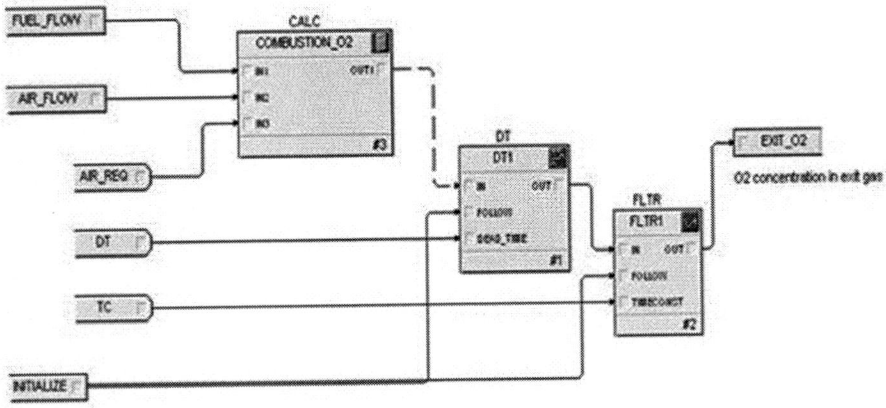

Figure 14-16. Combustion Process O_2 Simulation Composite

simulated process may be designed to provide faster than real-time response. This can be achieved by reducing the time constants and deadtime of a self-regulating process. For an integrating process, the integrating gain can be increased and the deadtime reduced. When the process response is modified in this way, it is often necessary to re-tune the PID control to match the modified process response in the simulation.

Some control systems provide simulation environment that allows process simulation modules and control modules to provide faster than real-time response thus eliminating the need to modify simulation parameters and control tuning to achieve this behavior. Such tools may also allow the input block's Simulate parameter to be enabled or disabled in all modules or in selected modules with one click of a button. An example of the interface provided for such a simulation environment is shown in Figure 14-17.

Another feature of the Simulation interface is the ability to save and restore simulations. This capability may be incorporated into the training simulator and used to capture and restore state information in preparation for running specific scenarios. By building these features into the actual operator interface, it is possible to use same operator displays for both plant operation and training.

Many excellent commercial simulation products may be purchased for process design. The simulation tools supported by these products are based on first principle analysis of process equipment.

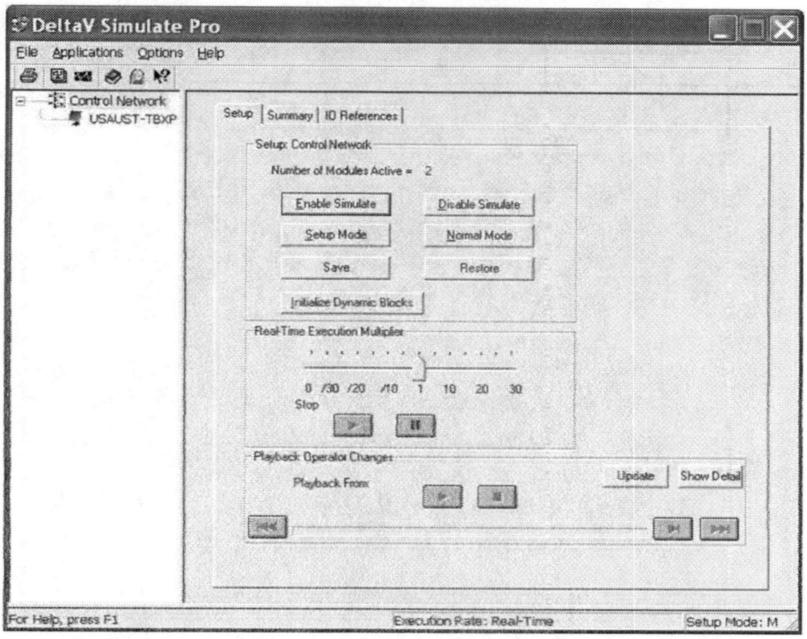

Figure 14-17. Example Simulation Interface

> **First Principle Analysis** – Analysis based on established laws of physics and chemistry that does not make assumptions such as empirical model and fitting parameters.

These process simulations typically provide better results over a wider operating range than is possible using a step response simulation model. Though such tools have been successfully used for operator training (Blevins et al, 2000), the cost of engineering a simulation is often prohibitive if the simulation is to be used only for control system checkout or operator training. Also, the expertise needed to update these types of simulations to reflect changes made at plant startup is often not available in process plants. It is also very difficult to keep the first principle models up-to-date with the actual process. A process simulation based on step responses is sufficient to meet the requirements for control checkout and operator training. In most cases, the added cost associated with the purchase, engineering and maintenance of a simulation based on a first principle analysis is not justified by the difference in the simulated process response.

14.5 Process Simulation Workshop

Spray dryers are used in the process industry to manufacture a variety of products. As described in Chapter 11, the drying mechanism within the spray tower is quite complex and is impacted by the slurry flow rate, the spray pressure, the air flow and the air temperature. This workshop exercise is designed to show how the spray dryer process (Figure 14-18) may be simulated. Access the book's web site http://www.advancedcontrolfoundation.com, and select the Solution tab to view the workshop being performed.

Step 1: Identify the small processes associated with the spray dryer process by circling these on a copy of the P&ID shown in the Process tab.

Step 2: Create a simulation diagram that represents the small processes identified in Step 1.

Step 3: Examine the control and process simulation modules that were created based on the spray dryer process simulation diagram.

Step 4: View the dynamic response of the spray dryer simulation when a change is made in the slurry flow setpoint in the control module. Note the impact of this change on the simulated slurry flow, heater outlet temperature and product moisture.

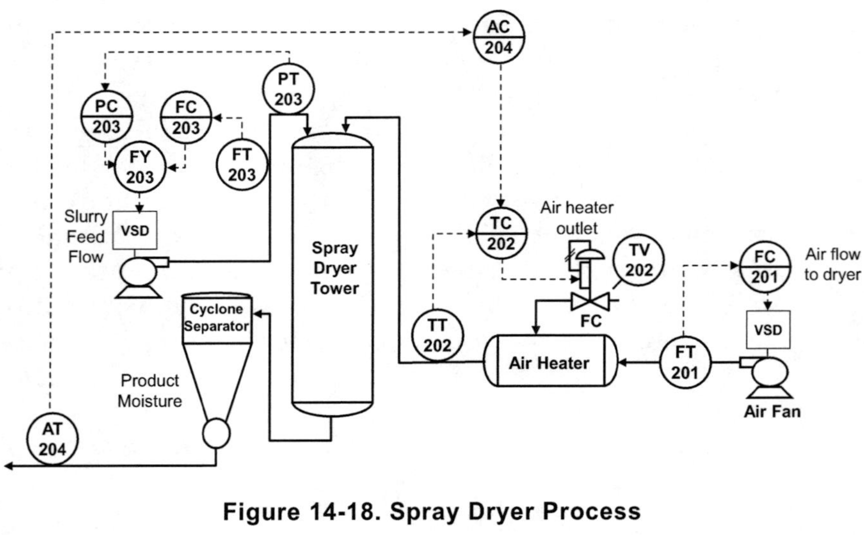

Figure 14-18. Spray Dryer Process

Workshop Discussion

The equipment associated with the spray dryer process can be broken into five logical processes that can each be modeled based on step response:

1. Slurry Flow Process
2. Spray Pressure Process
3. Air Flow Process
4. Air Heater Process
5. Tower Process

The simulation diagram based on this selection is shown in Figure 14-19. The slurry and air flow processes could be simulated in the same manner as the VALVE_FLOW composite. The air heater and spray pressure processes and tower process could be simulated using a general step response model, that is, a 2×1 and 4×2 process, respectively, implemented in a similar way to the 3×3 process presented earlier in this chapter.

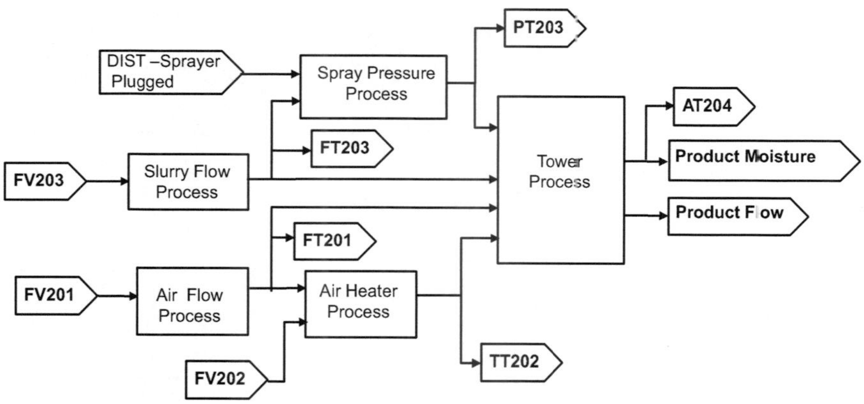

Figure 14-19. Simulation Diagram for Spray Dryer Process

The spray pressure process is non-linear over a wide operating range. To illustrate this in the simulation, a characterizer block could be used to define the process gain as a function of the spray pressure process flow rate.

14.6 Theory – Simulation Based on Step Response

Many of the basic thermal and physical mechanisms reflected in the design of process equipment—heat transfer, the blending of gas or liquids—tend to provide a first order response to input changes. Also, because of physical equipment size, some transportation delay is often seen in the output response. In such equipment, the output response to a change in a process input may be described as a first order-plus-deadtime response. The dynamic response of a first order-plus-deadtime process is fully captured by identifying the process gain, deadtime and time constant.

In such a process, when a step change is made in a process input, a response is not immediately seen in the process output, as illustrated in Figure 14-20.

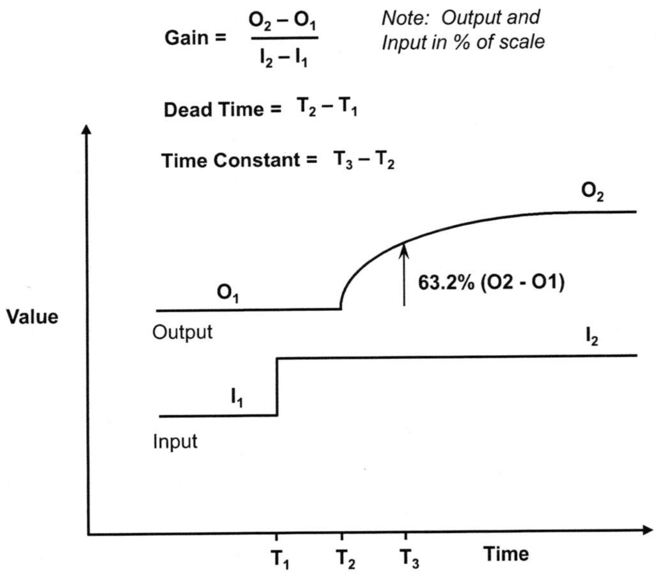

Figure 14-20. First-Order-Plus-Deadtime Process

The process deadtime is the time between a step input change and the first detectable change in output. The process time constant is the time from the first detectable change in output until the output reaches 63% of its final change in value (not 63% of its final value). The process gain is the (percent) change in the process output divided by the (percent) change in the process input after the process has fully responded.

For analytical and control system design purposes, the operation of the process equipment that makes up a plant can usually be broken into processes that are characterized as having a first-order-plus-deadtime response. However, a close examination of the process equipment will show that the process output response is the end result of multiple mechanisms working together and interacting: heat transfer, friction, inertia, air capacitance in an actuator and so on. While the combined effect of these different mechanisms constitutes a higher order process response, it often closely (and, for most purposes, adequately) approximates a first-order-plus-deadtime response.

One example of a process that may be characterized as having a first-order-plus-deadtime response is a feed stream heater, as illustrated in Figure 14-21.

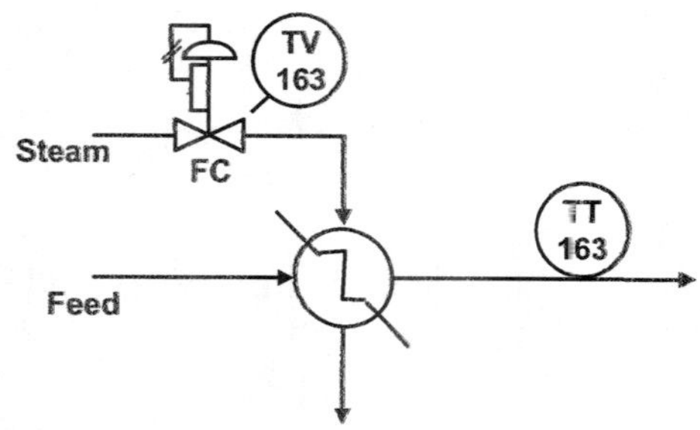

Figure 14-21. Heater Example – First-Order-Plus-Deadtime Process

When a step change is made in the heater steam inlet valve position, a delay will be observed in the heater outlet temperature response. Depending on the feed flow rate, it takes some time for the liquid feed flowing through the heater tube bundle to reach the temperature transmitter. Once the temperature starts to change, it exhibits a first-order response. This is because of the thermal mass of the tube bundle that makes up the heater and the heat transfer mechanism between the liquid feed inside the tube bundle and the steam that is outside the tube bundle. The combined impact of transport delay and heat transfer provides a first-order-plus-deadtime response.

As mentioned above, multiple mechanisms working together within a process often combine to provide what may be approximated as a first order-plus-deadtime response, even when no transport or measurement delay is involved. Figure 14-22 illustrates a process that is made up of three lags in series (a third-order system) but can be approximated as a first order-plus-deadtime process.

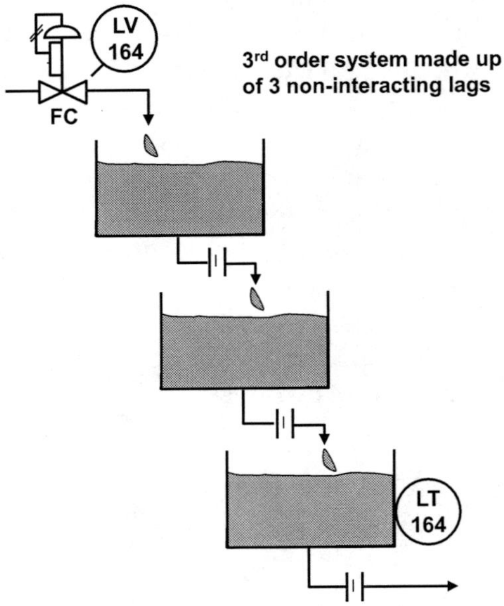

Figure 14-22. Process Made up of 3 Non-interacting Lags

In this example, three pure lag processes (often simply called *lag processes*) are combined in series to create a third-order process. The manipulated input is the inlet flow to the first lag process. The output of the first lag process becomes the input of the second lag process, and its output becomes the input of the third lag process. The level in the third lag process is the controlled parameter of the combined third-order process. A change in the output of a lag process begins immediately when the process input changes. However, as shown in Figure 14-23, the combined impact of the three lag processes in series gives a level output response that closely approximates the response of a first-order-plus-deadtime process.

In some cases, a step change in a process input will causes the process output to change continuously. A process that exhibits such a response is known as an integrating process. There are similar examples of integrating processes in the process industry, such as level control in storage tanks and pressure control in a gas or steam header.

When an input to an integrating process changes, the output may not begin to respond for a period of time because of measurement or transport delay. Thus, one parameter that may be used to characterize an integrating process is the process deadtime. Once the process output begins to change in response to a change in input, the observed rate of change in the output is used to characterize an integrating process.

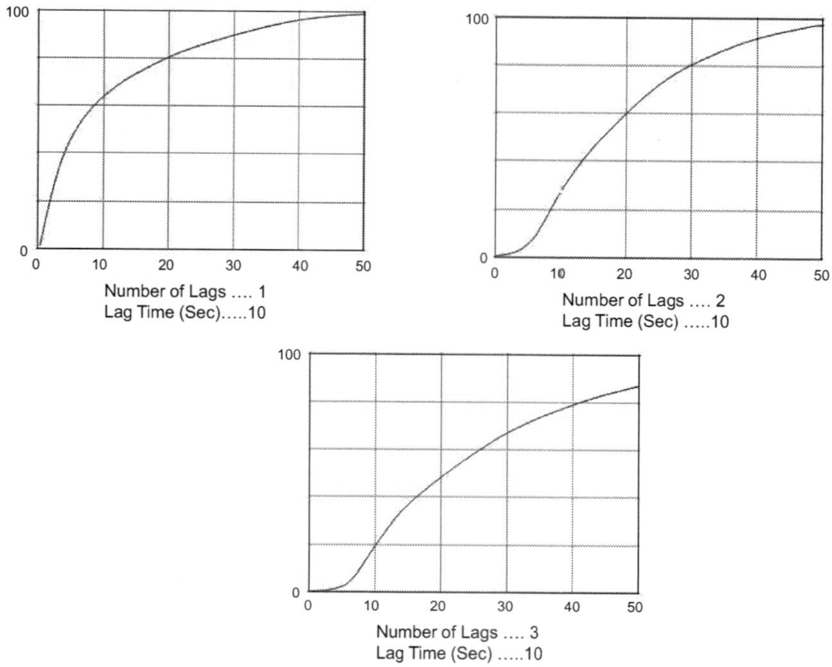

Figure 14-23. Approximating Higher Order Systems

Specifically, the rate of change in output per percent change in input per second is defined as the integrating gain. Thus, the response of an integrating process in general is described by the process integrating gain and deadtime, although the deadtime may be zero.

> **Integrating gain** – In an integrating process, the rate of change in output per percent change in input per second, as illustrated in Figure 14-24.

The response of an integrating process and the calculation of the integrating gain are shown in Figure 14-24.

Integrating processes are also described as non-self-regulating processes. An example of an integrating process is a tank in which the flow into the tank is the manipulated input, the tank level is the controlled output, and the discharge from the tank is a gear pump as illustrated in Figure 14-25. In this process, the outlet flow rate is determined by the speed of the gear pump and thus is not affected by the tank level.

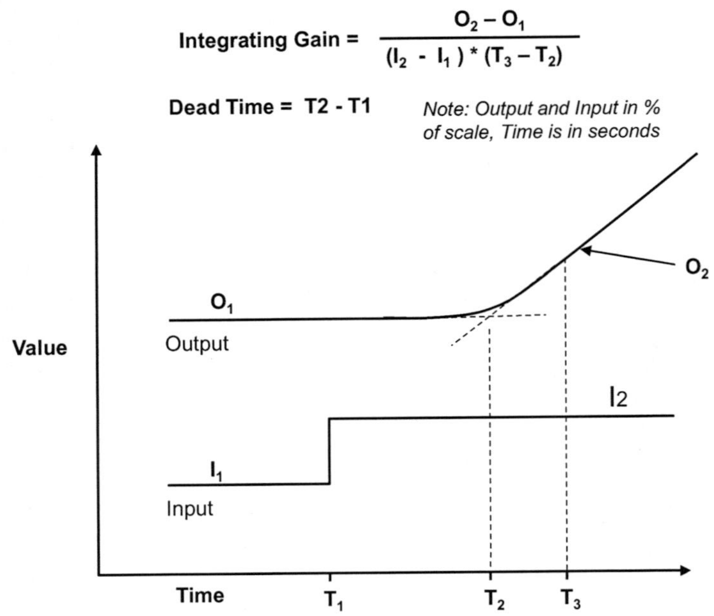

Figure 14-24. Integrating Process Response

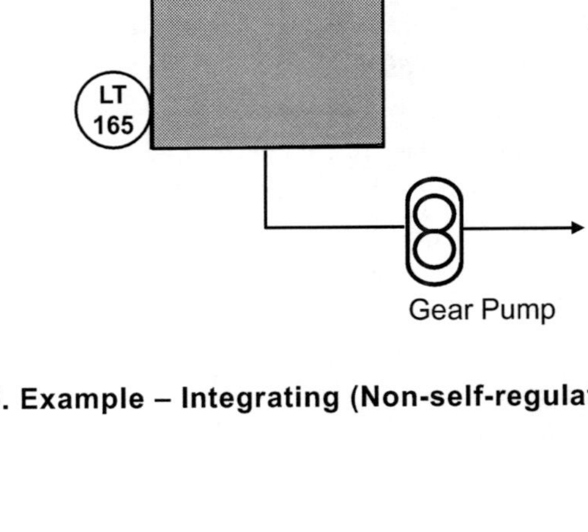

Figure 14-25. Example – Integrating (Non-self-regulating) Process

In this tank level control example, if the inlet flow does not match the outlet flow, then the level will constantly change. If the process continues for a long enough time, the tank will overflow or go dry. This example and the previous tank level example demonstrate that tank level response is determined by the process design.

When the response of a process that has multiple inputs and/or outputs is addressed, the assumption made in simulating the process by step responses is that the process is linear and possesses the two properties of superposition and homogeneity (discussed in more detail in Chapter 11). A process that produces an output ***y1*** for an input ***x1*** and an output *y2* for an input *x2* and an output ***y1+y2*** for an input ***x1+x2*** possesses the property of superposition. Similarly, a system that produces an output *y* for an input *x* and produces an output ***ay*** for an input ***ax*** possesses the property of homogeneity. These properties are the basis on which the process output of a multivariable process may be represented by the sum of the step responses for a change in the process inputs, as illustrated in Figure 14-26. Every process output is the sum of input actions.

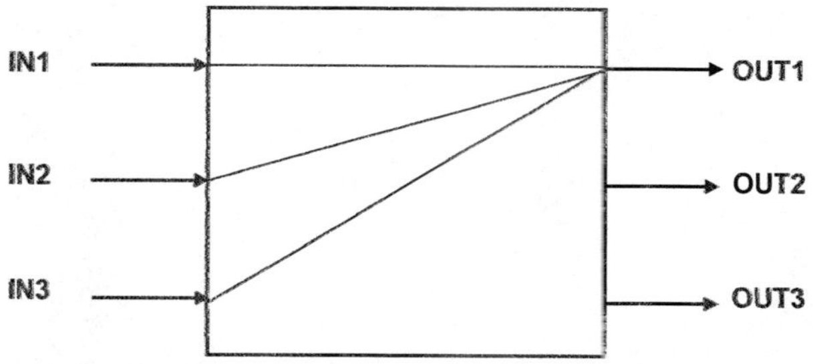

Figure 14-26. Simulation of a Multivariate Process

BIBLIOGRAPHY

1. Blevins, T., Nixon, M., *Control Loop Foundation – Batch and Continuous Processes*, pp. 327–346, International Society of Automation, Research Triangle Park, NC, 2011.
2. Blevins, T., McMillan, G. K., and Wheatley, R. "The Benefits of Combining High Fidelity Process and Control System Simulation," *Proc. ISA EXPO 2000*, Chicago, Oct. 2000.
3. Culter, C., "Operator training simulation," *World Oil*, p. 3, April 1, 2005.
4. Seborg, D., Edgar, T. F., and Mellichamp, D. A. *Process Dynamics and Control*, John Wiley & Sons, Hoboken, NJ, 2004.

15 Integrating Advanced Control into a DCS

Plant production and operation efficiency may often be improved by applying the advanced control techniques addressed in the previous chapters. In some cases, support for advanced control is embedded in the distributed control system (DCS). This approach provides the plant operator with a single window interface to all functions associated with running the process. Related process information may be included with advanced control in the operator displays. The operator is also provided with consistent presentation of process alarming, trending and system interaction for initiating changes in control setpoints and mode of operation. In addition, embedded advanced control makes it possible to provide a single log-in and span of control.

If the DCS does not support advanced control, then the advanced control applications must be layered onto the DCS. Several approaches may be taken for layering advanced control onto an existing distributed control system. The approach that is selected depends on the type of advanced control application and the support provided by the DCS for layered applications.

When advanced control is layered onto an existing DCS, the operators who work with the advanced control may need to see and work with other information from the existing DCS. Similarly, the operators of the existing DCS may need to have feedback from the advanced control in making decisions about how best to run the portion of the plant that is not under advanced control. In most cases, only a small fraction of the information in the advanced control application and the existing DCS must be communicated to support such coordination with the operators. The technology that has most often been successfully used to support the exchange of information between the DCS and a workstation that contains the advanced control application is illustrated in Figure 15-1.

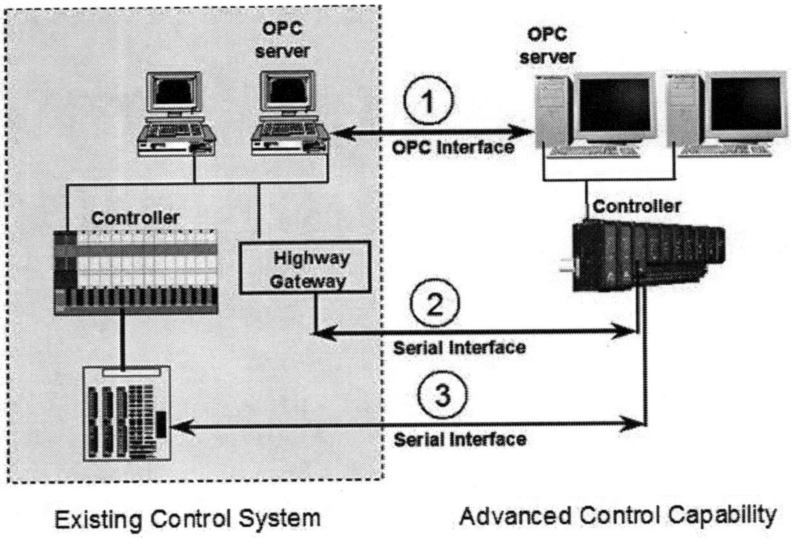

Figure 15-1. Advanced Control Integration into an Existing DCS

As is illustrated in Figure 15-1, when the Distributed Control System supports an OPC server capability then advanced control applications may use OPC capability to access information in the DCS (Rohjans et al, 2011), (Gîrbea et al, 2011), (Hollender et al, 2012). Through this OPC interface, it is possible to access real-time and historic information from the DCS. The addition of an OPC server is an option that is supported by many older DCS and by many business applications. The workstation that contains the advanced control capability may be structured as an OPC client that is configured to directly access information from the OPC server in the existing DCS. If OPC is not supported by the existing system, then it may be possible to interface to the DCS through a supported serial interface.

In some cases, the workstation that contains the advanced control capability may serve only as an OPC server. In such cases, it is necessary to use another application, such as OPC Mirror, that may be configured to transfer information from the DCS server to the workstation containing the advanced control capability. Products such as OPC Mirror may be installed on the DCS server, on the advanced control server or on a separate station as illustrated in Figure 15-2.

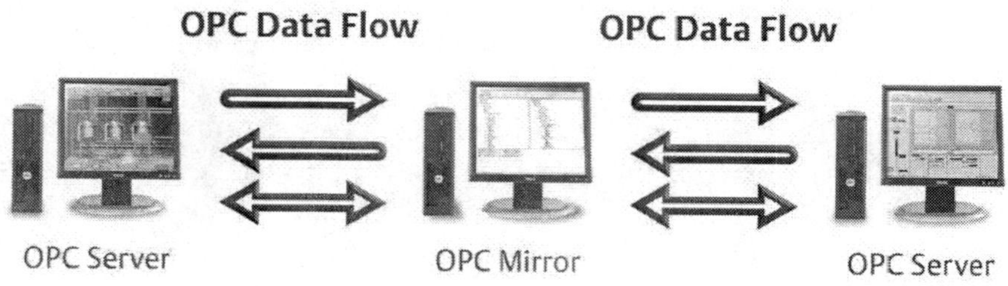

Figure 15-2. Transfer of Data Using OPC

If an OPC server must be added to an existing DCS, this can increase the total cost of implementing an advanced control project. Depending on the level of OPC support provided by the advanced control station, some amount of project time should be budgeted to configure the OPC interface because doing this often requires duplicate configuration to map parameters from one system to another.

In addition, such an interface is often not redundant. In such cases, the DCS interface should be designed to take appropriate action on temporary loss of communications; for example, to automatically switch the mode of downstream control from remote to local control.

In many cases, the data flows from the distributed control system OPC server to the advanced control server will be through networks that are separated by firewalls. These firewalls often contain features that determine the legitimacy of sessions, filter at the network level and evaluate the content of the packets. These firewalls execute algorithms to inspect at the packet layer. They also perform the mapping of addresses from one system into another. Setting up these firewalls requires specialized skills. Since firewalls are commonly used it is important to allocate time for configuring network settings to enable traffic flows.

Another challenge that may be encountered when integrating advanced control into an existing DCS is the fact that control system manufacturers often represent control parameters differently. Take, for example, the representation of mode. Some control systems implement mode as two parameters, that is, Local/Remote and Auto/Man while other systems represent mode as a single parameter, that is, Man/Auto/Cascade/Remote Cascade/Remote Out. Where such differences exist, the basic calculation features of the controller may be used to create a proxy in the new controller to map these parameters to their equivalent representation in the new control system.

In the last few years, several OPC improvements have been made that make advanced control integration easier. The latest OPC standardization efforts are yielding important improvements to the data model, integrating data sources such as alarms and supporting a much more capable web service style interface. The recommended practice for distributed programming is now moving towards service-oriented architecture with payloads encoded in XML, JSON or binary XML. The OPC Foundation has recently introduced new standards that combine existing data models with new service frameworks.

For example, real-time data, alarms and events and history models are being combined. (Mahnke et al, 2011), (Gîrbea et al, 2012), (Gîrbea et al, 2011). These newer models also address problems related to consistent naming and consistent use of status/quality attributes. Where it has been implemented, consistent/standardized naming allows clients to reconcile the naming differences between systems—the same OPC path always refers to the same data.

The standardization of OPC by the OPC Foundation also defines how status is assigned. OPC servers use a consistent status when communicating and receiving data from devices. Uniformity in the assignment of status by OPC servers improves the consistency of clients. This change allows a single client to browse to locate data of interest, subscribe, and then retrieve real-time data, alarms and events, as well as history. Mapping modules may be created in an OPC server to serve as place holders for information transferred to an OPC server, as illustrated in Figure 15-3.

Some distributed control systems offer an embedded advanced control capability in which advanced control functions, such as model predictive control are represented as function blocks. Performance monitoring and tools for commissioning advanced control are supported as applications within the control system. When this level of support for advanced control is embedded in the DCS, the implementation and commissioning of advanced control is much easier. Even so, in many cases, the use of advanced control tools such as MPC in place of PID control is a new concept to a control engineer. The skills and knowledge needed to design and commission advanced control are quite different from those associated with traditional control strategies based on PID.

An easy way to initially learn about MPC and to gain experience in commissioning MPC blocks is to layer MPC blocks on top of PID-based traditional control strategies within a control module, as illustrated in Figure 15-4. The output of the MPC block is connected to the

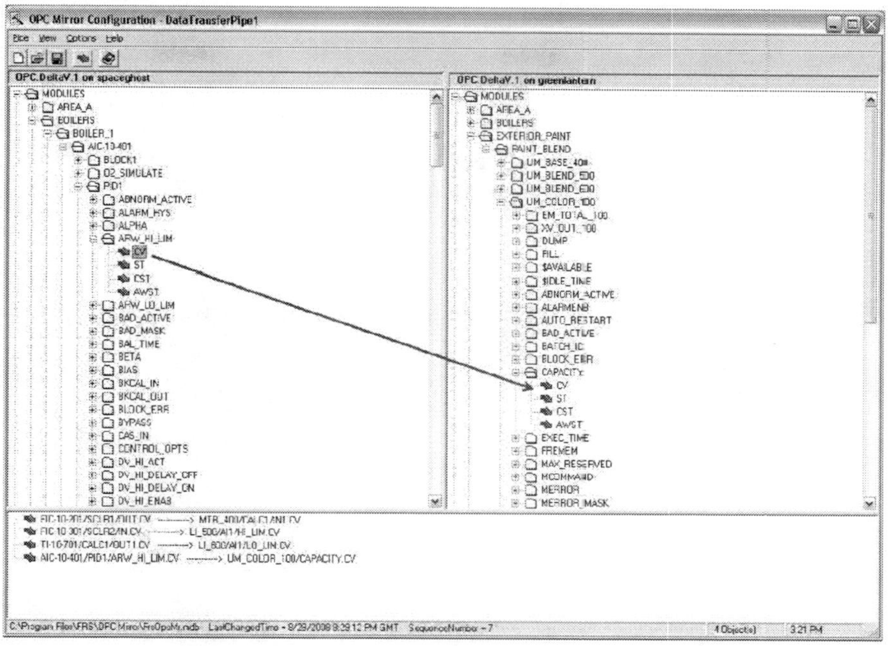

Figure 15-3. Layering Advanced Control on an Existing DCS

RCAS_IN of the analog output block. The RCAS_OUT of the analog output block is connected to the back calculation input of the MPC block.

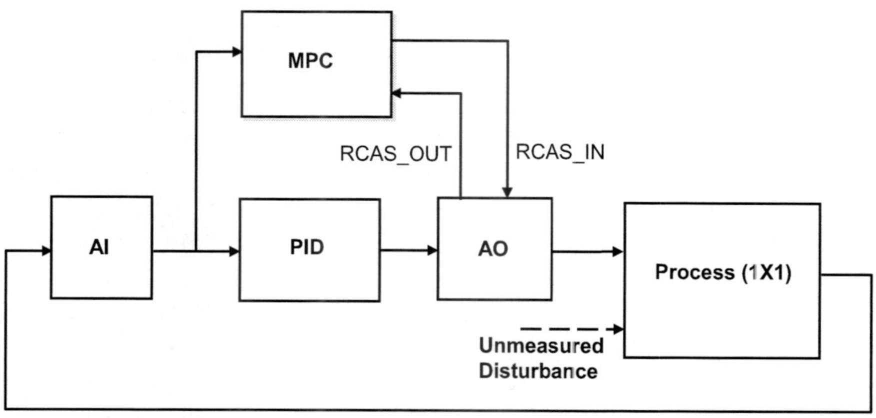

Figure 15-4. Layering MPC onto an Existing Strategy

In this example, the controlled parameter is wired as an input to both the PID block and the MPC block by exposing the RCAS_IN and RCAS_OUT parameters of the AO block.

Once the RCAS_IN and RCAS_OUT parameters of the AO block are exposed, they may be used for connections to the MPC block output and back calculation input as illustrated in Figure 15-5.

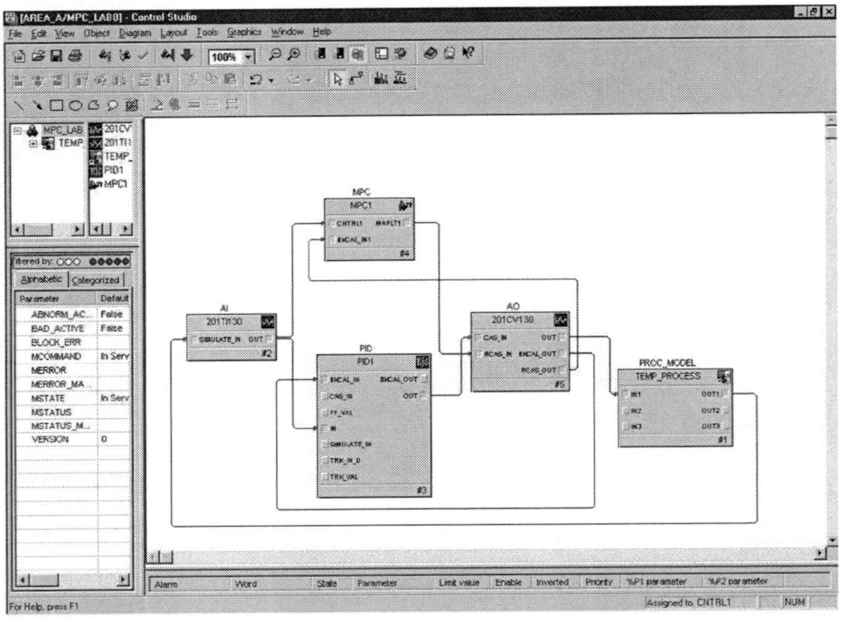

Figure 15-5. Implementing MPC onto an Existing Strategy

By changing the AO block mode to RCAS, the engineer can allow the MPC block to control the process. When the AO mode is switched back to Cascade, the PID block may be used to control the process. Switching between Cascade and RCAS mode is bumpless because the PID and MPC blocks have back calculation connections.

The advanced control application may initially be commissioned using PID control because this is what the engineer and operator typically have the most experience with. However, as time permits, the MPC block may be commissioned and gradually used more and more in the control of the process. When the distributed control system supports the layering of MPC onto an existing strategy, then either option, PID or MPC, is always available to operate the plant.

As mentioned earlier in this chapter, MPC may be used to address situations where control performance using PID is limited. For example, if the process deadtime is significant compared to the process time constant, then better control performance may be achieved using MPC. If there are strong process interactions, this is another reason to use MPC rather than PID. In addition, if a process is characterized by many constraint variables, then the implementation and commissioning of the control system will be much faster and easier using MPC rather than multiple PID blocks and one or more control selectors.

15.1 Integrating with Plant Systems

In an increasing number of organizations, costs are being reduced by integrating external and business systems. For example, lab data and production scheduling systems may be integrated with the DCS and used in the advanced control strategies. The overall plant architecture provides means for data sharing, data acquisition, peer-to-peer data exchange and viewing information through web-enabled applications. Using networking in this way to enhance productivity and reduce costs may lead to vulnerabilities that can expose mission-critical DCS-related systems to security threats. For this reason most organizations take great care in securing their networks through the use of authorization, authentication, and firewall devices.

The information infrastructures in process plants use networking technology to support data sharing, peer-to-peer data exchanges and web-enabled applications. These infrastructures support the communication of data for control applications and operator interfaces. As an example, lab data may be communicated from the lab entry system to the DCS and integrated into data analytics or NN (neural network) applications. In other cases, the operator may use the data to coordinate operations or help make decisions about the overall health of the process.

Two techniques often used to communicate data are MODBUS TCP/IP and OPC. In other cases, data may be communicated via web pages over sockets or via higher level frameworks such as Windows Communication Foundation. Regardless of whether the data are being communicated value-by-value using a protocol such as OPC or communicated by web pages written in HTML5, plants secure these communications by using very carefully designed networks. Figure 15-6 provides a logical view of a network in a process plant.

Network integration provides the capability to move data throughout the organization. The operating characteristics of the overall network must ensure that there is no unacceptable risk of loss of communications that results in operating conditions in that plant which endanger the safety of personnel or could cause damage to the environment or plant assets. For any potentially disrupting condition, contingencies must be in place for maintaining safe operations, including emergency shutdown and startup. To meet these needs, networks are integrated into the plant that are isolated by function and are designed to meet specific requirements depending on how a specific network is used.

The overall design principles for network integration are summarized below:

- **Security:** Security is the level of protection of the production facilities against malicious or inadvertent external intrusions that would threaten quality, integrity or availability. Individual networks must be secured.
- **Standards:** International standards and de facto industry standards provide the foundation for much of the plant's network infrastructure.
- **Availability:** Availability is the percentage of time that a production facility functions under normal operating conditions.
- **Integrity:** Integrity is the capability of the plant's systems to continuously supply measurements and apply control functions under specified quality, delivery and time constraints.
- **Quality:** Quality is defined in this context as the percentage of time that measured values and measurement frequency are assured to be within the design specification.
- **Scalability:** Network integration must allow for different physical implementations supporting different plant sizes, asset maturity and practical constraints (for example, regulatory, environmental, commercial and financial, etc.)

Architecturally, the overall systems at plant sites are generally subdivided into three environments: the Office Environment, the Plant Environment and the Internet. These environments are shown in Figure 15-6. The Office Environment is where planning, forecasting, asset management, modeling, optimal setpoint computation and some diagnostic and maintenance activities are performed. The Plant Environment is where all field operations such as real-time process monitoring and control, safety systems, validation of setpoints, equipment diagnostics and maintenance of the producing facilities are performed. The Internet is where authorized users can perform

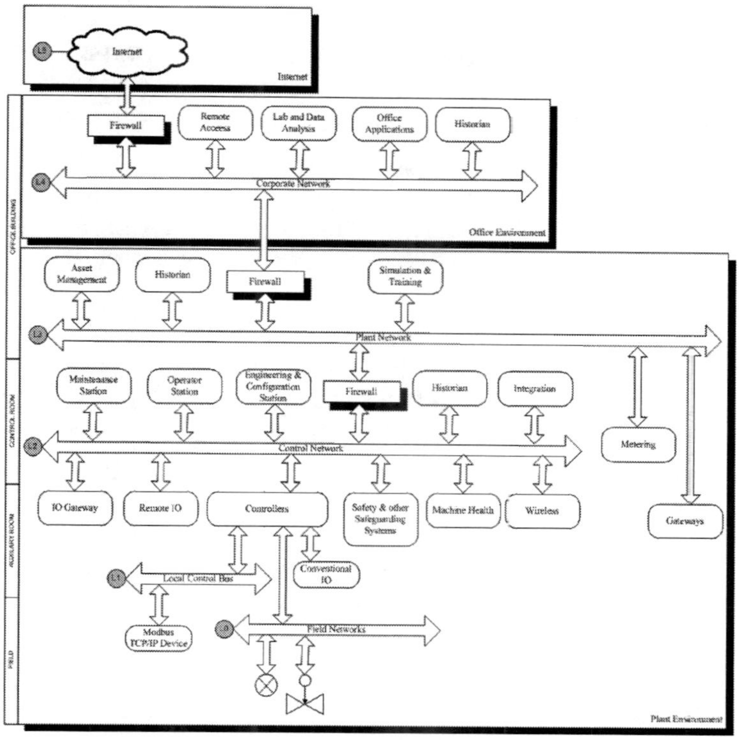

Figure 15-6. Network Integration

activities such as remote diagnostics, data acquisition and office activities as required (direct access to the Internet is never allowed from networks within the Plant Environment). Different organizations customize this partitioning for their specific needs.

The interconnection of the Internet, the Office Environment and the Plant Environment exposes the overall plant to significant business risks. These risks can result in common mode failures of control and safeguarding systems. For example, failure to secure the Plant Environment can result in computer virus, worm or hacker infiltration of the network, resulting in personnel injury or death, environmental damage, asset damage and/or deferred production.

Consequently, all connections between the Office Environment and the Plant Network (including read-only links for data historians) must be firewalled and properly secured. There are many standards covering security including NERC CIP (North American Electricity Reliability Corporation (NERC) Critical Infrastructure Protection (CIP) Cyber Security Standards.), NIST (National Institute of Standards and Technology), the NISCC (National Infrastructure Security Coordination Center Good Practice Guide on Firewall Deployment for SCADA and Process Control Networks Rev 1.1) and ISA publications: ISA-TR99.00.01-2004 (Security Technologies for Manufacturing and Control Systems) and ISA-TR99.00.02-2004 (Integrating Electronic Security into the Manufacturing and Control Systems Environment). NIST provides an extensive set of recommendations that can be easily adopted by system suppliers to lock down security settings for workstations.

Referring again to Figure 15-6, the top-level network in the Plant Environment is the Plant Network. The Plant Network interconnects all plant systems, such as the electricity metering, the Control Network, and gateways with the Office Environment. The Control Network provides a backbone for maintenance stations, operator stations, engineering and configuration stations, historians, controllers, machine health, I/O systems and gateways. In most cases, the Plant Network and the Control Network are connected through firewalls. More recent systems also connect all workstations into the Control Network through smart switches with integrated firewall capabilities.

Process control devices are interconnected by three specialized communication networks:

- Fieldbuses and conventional I/O
- Control buses
- Control network

Fieldbuses and conventional I/O connect field devices through DCS controllers to the rest of the plant control system. Conventional I/O is point-to-point wiring—this has been the standard practice in plants for many years. Fieldbuses such as Fieldbus H1 eliminate a significant portion of the wiring, enable extensive device diagnostics and enable control in the field. Most fieldbuses also provide power to the devices, which eliminates additional wiring.

Recent advances in wireless and conventional I/O also reduce the amount of wiring in the field. In the case of wireless devices powered by batteries or other power sources, signal information is transmitted using wireless technologies. In the case of conventional I/O, electronic marshaling fundamentally changes work practices for wiring, engineering, and commissioning plants.

A local control bus is a local backbone that interconnects the active control modules with other parts of the Plant Environment, such as motor control centers, packaging, and material handling. In many cases, local control buses also include integrated firewalls. These firewalls plug directly in line with device level traffic on a control network.

Most organizations employ a set of best practices when utilizing multiple networking devices. These include firewall devices, routers with configured access control lists (ACLs), configured switches, static routing and routing tables and completely separate physical media. In organizations where a single log-in is required, domains are organized into a single structure or trusts are set up between domains.

15.2 Network and System Setup

Setting up a system to transfer data and web content between networks requires setting up an application such as OPC to transfer data between systems. In many cases, OPC is used to transfer data such as real-time data, lab data, or historical values from a source system to a target system. In many cases, also, mapping modules must be created to land values on the target system. Mapping modules provide a location to land values in the target system so they can be referenced by MPC, NN and other applications. Mapping modules were described earlier in this chapter and are illustrated in Figure 15-3.

Setting up OPC requires defining a pipe to transfer data through and then configuring the parameters that are to be accessed. Because pipes must be authenticated, it is important to have username and password information available for the source and the target system before starting the installation. It is also necessary to know the port numbers that will be used for data transfer. In many cases, it will be necessary to have IT open ports in the firewall for data transfers. An example of an interface for setting up pipe properties is shown in Figure 15-7.

Once the OPC pipe has been configured, the next step is to configure data values to be transferred between systems. It is important to have the link direction properly set up. Configuring OPC links in one commercial application is shown in Figure 15-8.

When configuring links it also important to have the data types correctly configured. Configuring data type information is shown in Figure 15-9.

Most utilities for transferring data values also have capabilities for validating that the data are being transferred correctly. It is important to validate that data values are being correctly transferred and are updating at the correct update rate.

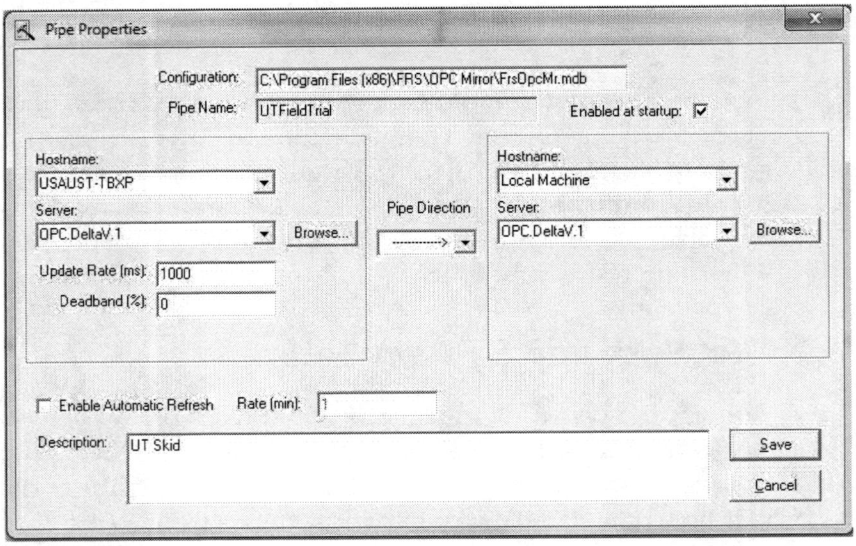

Figure 15-7. Setting Up OPC Pipe Properties

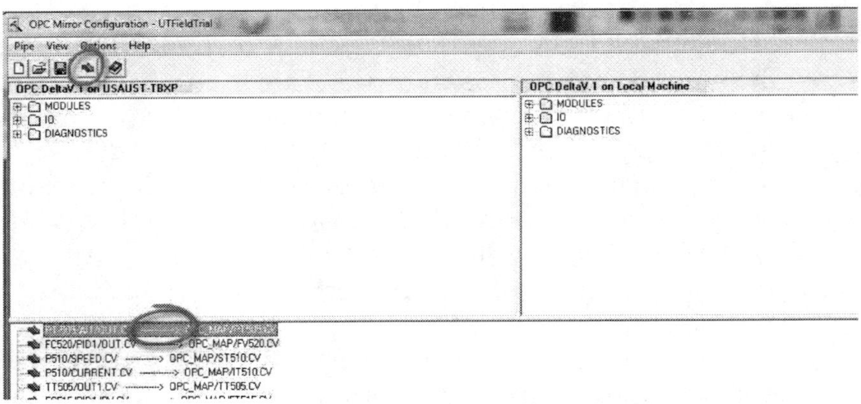

Figure 15-8. Configuring OPC Links

The second step in setting up network connections between systems usually consists of setting up the networking to support transferring web content and supporting data updates between the server and the client browsers. In many cases, the web pages will be written in HTML5 with Java Script used for providing updates and behaviors. For example, values that are being updated on the web page must be first collected by the web page Java Script and then used by additional Java Scripts to update values on the screen. A good example of this

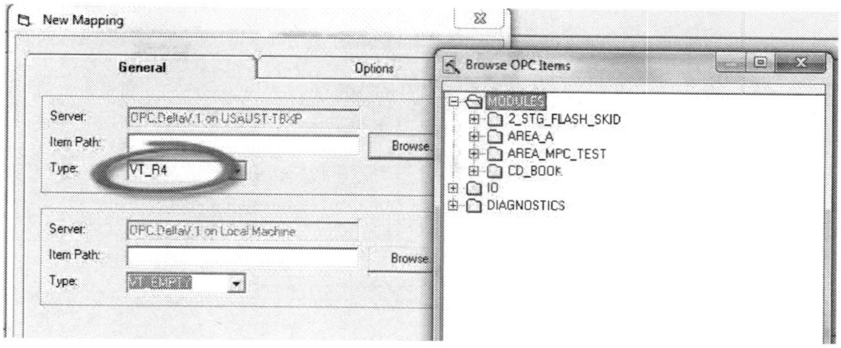

Figure 15-9. Configuring OPC Data Types

type of interaction can be seen in the workspace area provided for workshops and applications at www.controlloopfoundation.com.

To allow Web Pages hosted by an advanced control server to be accessed by a plant operator, IT department will need to open two or three pipes for information transfer. In some cases, it will be also necessary for the user viewing web content to provide some level of authentication. When this is required, the authentication should default to the currently logged-in user.

One problem that sometimes arises is the user at the Plant Network level or in the Office Network will be authenticated through a corporate login that is different from the control system login. There are a number of ways to get around this: the domains can be set up with a trust, the same account information can be used by the server hosting the web pages and the client running the browser or security can be disabled at the web page level.

15.3 Application Example

At a plant site, a server supporting continuous data analytics is to be layered onto an existing DCS. For this example, it is assumed the DCS contains an OPC server. OPC Mirror has been configured to transfer information from the DCS server to the OPC server that is supported by the continuous data analytics applications. An on-line web interface is used by the operator, process engineers and plant management to access the quality parameter prediction and fault detection provided by continuous data analytics. Although not fully representative of the plant in this example, a typical approach to addressing these installation requirements is shown in Figure 15-10.

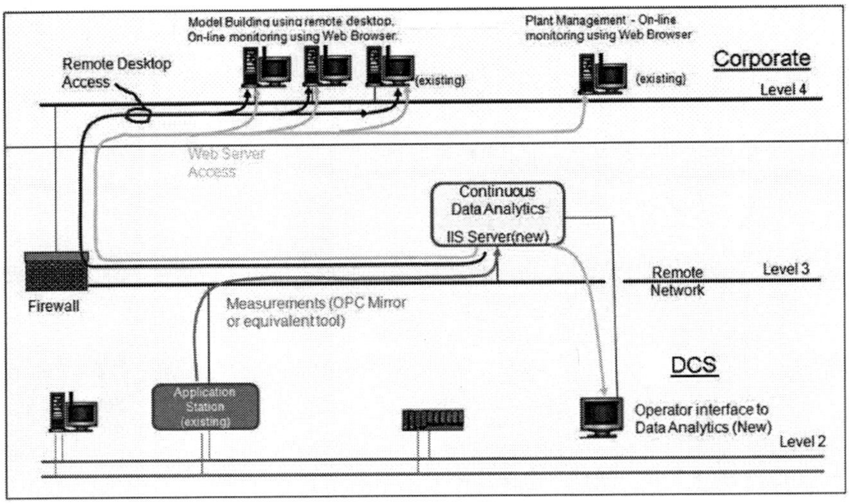

Figure 15-10. Typical Advanced Control Installation

In this application example, the DCS networks are firewalled from the business network. Only traffic that is required to pass in either direction is allowed to do so. Hosts on the DCS network are never allowed to reach the Internet.

All access to DCS protected networks is done from a system that is also protected in a Demilitarized Zone (DMZ). Remote access is done through a firewalled Citrix server that is part of the business network domain, but has limited access. Communication from the end user to the DCS through the Citrix remote access architecture is encrypted end-to-end, and authentication is done using two-factor authentication. This prevents access by someone who steals the password of someone who does have access.

A Microsoft Internet Explorer component used by the data analytics web-enabled application is installed on the Citrix system.

Using a limited number of dedicated hosts to access the DCS allows the plant site to maintain static firewall rules and eliminates the need to enable paths from individual end user systems that would require static IP addresses. Applications that need to run on end user computers must be compatible with Windows 7, and are certified and packaged prior to use to ensure they install correctly, do not require administrator rights to run, are installed in a consistent manner and comply with licensing requirements.

The IT security requirements at a plant site will often influence the manner in which advanced control is integrated into the plant network and the rules by which information within the control system is accessed by the advanced control workstation. For example, in the discussed installation the pipes opened for information transfer were assigned by the plant's IT department.

BIBLIOGRAPHY

1. *OPC Mirror, Product Data Sheet*, www.EmersonProcess.com/DeltaV.
2. Mahnke, W., Leitner, S., Damm, M., *OPC Unified Architecture*, ISBN 978-3-540-68898-3, Springer, 2009.
3. Hollender, M., *Collaborative Process Automation Systems*, ISBN 978-1-936007, International Society of Automation, Research Triangle Park, NC, 2012.
4. Gîrbea, A., Nechifor, S., Sisak, F., Perniu, L., "Efficient address space generation for an OPC UA server," *Software: Practice and Experience*, Vol. 42, Issue 5, pp. 543–557, May 2012.
5. Gîrbea, A. Nechifor, S.; Sisak, F.; Perniu, L. "Design and implementation of an OLE for process control unified architecture aggregating server for a group of flexible manufacturing systems," *Software, IET*, Vol 5, Issue 4, pp. 406–414, August 2011.
6. Rohjans, S., Piech, K., "Standardized Smart Grid Semantics using OPC UA for Communication," *IBIS Interoperability in Business Information Systems*, Issue 1 (6), 2011.

Appendix A

A web site has been established to enable you to view the workshop solutions for evaluating system performance, on-demand and adaptive tuning, fuzzy logic control, neural networks, intelligent PID, continuous and batch data analytics, MPC, real-time optimization and process simulation. In addition, some information is provided on the web site that may be helpful in exploring basic and advanced control techniques.

All you need to access and use the web site is a web browser. The following section contains information on how to access the web site and find the workshops.

A.1 Accessing the Web Site

Using your web browser, go to http://www.advancedcontrolfoundation.com. The workshops for Advanced Control Foundation are displayed on the home page, as shown in Figure A-1.

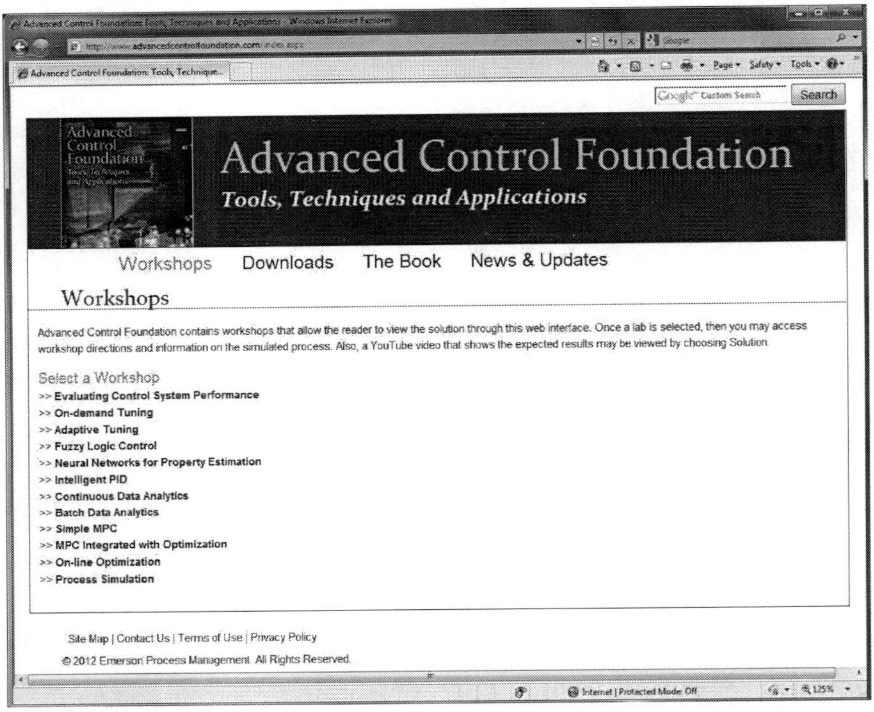

Figure A-1. Web Site Home Page

Selecting a workshop from the Workshops list takes you to the selected workshop. For example, selecting Evaluating System Performance under Workshops opens the interface for this workshop as shown in Figure A-2.

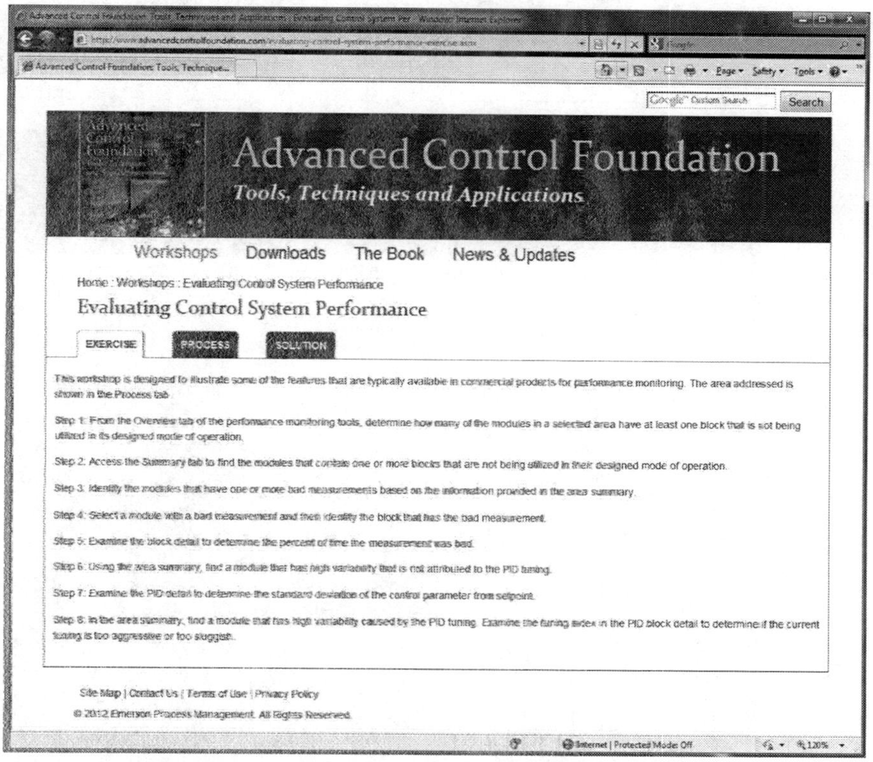

Figure A-2. Evaluating System Performance Workshop

The Exercise tab is shown by default and contains a duplicate of the workshop description that is provided in the book.

Select the Process tab to see the process that is simulated in a workshop. In response, the bottom portion of the screen changes to show a picture of the process, as illustrated in Figure A-3.

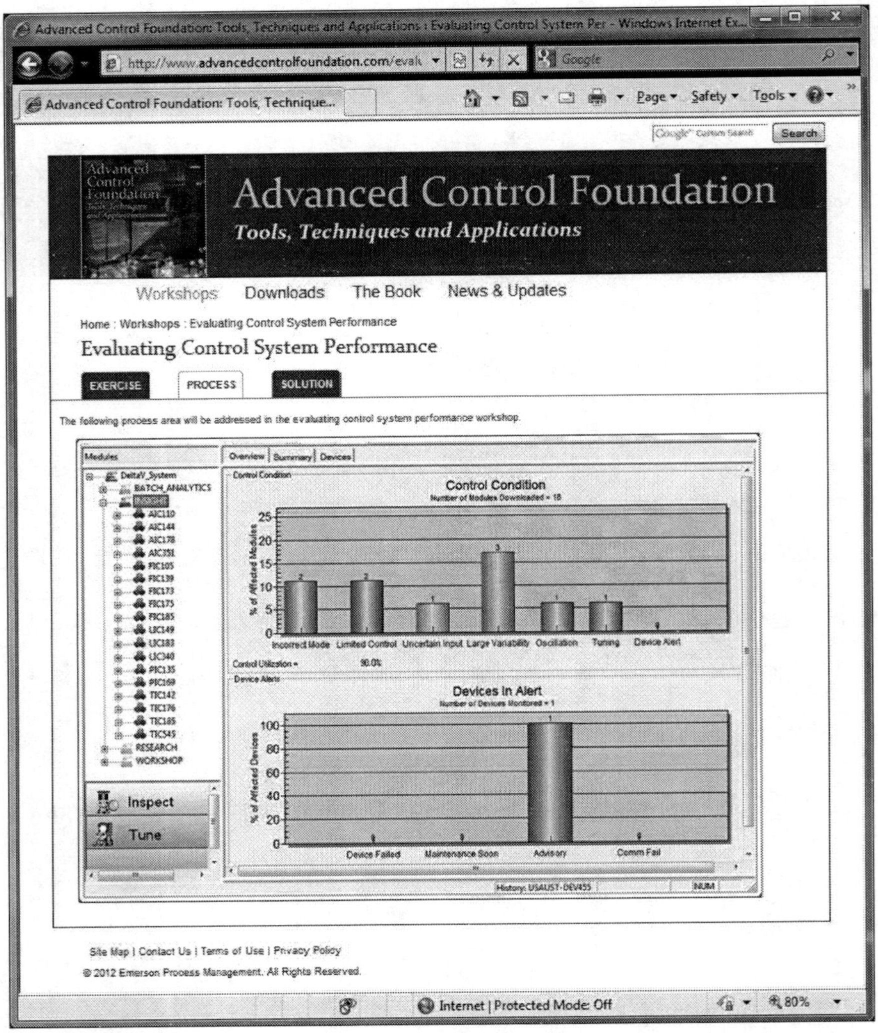

Figure A-3. Process – Evaluating System Performance

Selecting the Solution tab opens a video in the lower portion of the screen that shows the exercise solution, as shown in Figure A-4. In the video, each step of the exercise is shown and discussed.

Figure A-4. Solution – Evaluating System Performance

Using the controls provided in the video area, it is possible to regulate the sound level and to start, stop, and replay the video.

A.2 Downloads Selection

On the web site home page, a Downloads selection is provided for downloading configuration files for the process simulation and the control modules associated with each workshop. These .fhx files may be imported and used with a DeltaV control system or used with the DeltaV Simulate product.

A.3 Book Selection

Information on this Advanced Control Foundation book and its authors is provided through The Book selection on the home page. Also, links are provided to allow the book to be purchased over the web.

A.4 News and Updates Selection

This section of the web site contains information on scheduled education classes that utilize the book *Advanced Control Foundation*. Also, errata will be posted in this section when mistakes are discovered in the book text or figures.

Glossary of Terms

Accuracy – The ability of a measurement to match the actual value being measured.

Actual Mode – The mode of operation that may be achieved based on the target mode and the status of the PID block inputs.

Adaptive Control – A control method in which one or more parameters are sensed and used to modify its operation to achieve the best mode of operation.

Autotuning – The capability to automatically compute the tuning parameters of a PID connected to a plant.

Automatic mode – PID controller mode to enable its set point to be locally adjusted by the operator (also known as LOCAL and AUTO mode).

Batch Process – Process that receives materials and processes them into an intermediate or final product using a series of discrete steps.

Calibrating a Transmitter – Procedure of adjusting a transmitter to accurately reflect reference input values or a known differential pressure or other process measurement.

Cascade Inner, Secondary, or Slave Loop – PID in a cascade strategy that manipulates the process input.

Cascade mode – Secondary PID controller mode to enable its set point to be remotely adjusted by the output of a primary controller (also known as REMOTE and CAS mode).

Cascade Outer, Primary, or Master Loop – PID associated with the final output of the processes making up a cascade.

Cascade Intermediate Loops – PIDs positioned between the outer and inner loops of a cascade.

Closed Loop Control – Automatic regulation of a process input based on a measurement of process output.

Commissioning – The work required to bring a system to the point where it can be used.

Constraint Limit – Value that a constraint parameter must not exceed for proper operation of the process.

Constraint Output (constraint parameter) – Process output that must be maintained within an operating range.

Continuous Communications – Wireless communication in which the device wakes up at a configured update period, senses the measurement and then communicates the value.

Continuous Process – A *process* that continuously receives raw materials and continuously processes them into an intermediate or final product.

Control Loop – One segment of a process control system.

Control Module – Container of measurement, calculations, and control implemented as function blocks. A module may contain other control modules.

Control Utilization – The percent time a control loop is used in its designed mode of operation.

Controlled parameter – A process output to be kept at a set point (e.g. concentration, flow, pH, pressure, temperature, or level).

Control Robustness – A measure of how well control copes with differences between the true process gain and dynamics and that assumed in setting PID tuning.

Controlled Output (controlled parameter) – Process output that is to be maintained at a desired value by adjustment of process input(s).

Control System – A component, or system of components functioning as a unit, which is activated either manually or automatically to establish or maintain process performance within specification limits.

Critically Damped Response – Controlled parameter is restored in minimum time without overshooting the setpoint value. While minimizing response time, unstable control may be observed with changing operating conditions that impact process gain or response time.

Data Analytics – The process of inspecting, cleaning, transforming, and modeling data with the purpose of drawing conclusions about that information.

Discriminant Analysis (DA) is related to PCA and other statistical analysis methods such as analysis of variance. It is often used in conjunction with PLS as a powerful technique. Discriminant analysis as opposed to a continuous output measurement is used to relate process variation to categorical or classification criteria.

Disturbance Input – Process input, other than the manipulated input, which affects the controlled parameter.

Distributed Control System – Control system made up of components interconnected through digital communication networks.

Dynamic Compensation– Network included in the feedforward path to account for differences between the process response to changes in a disturbance input and changes in the manipulated input.

Dynamic time warping (DTW) algorithms effectively measure the similarities between two sequences which may vary in time and/or speed. The method then adjusts for these differences so that multiple batches may be combined together into one data set so that the batches may be analyzed.

Dynamic Process Simulation – Modeling and simulation of an industrial process behavior over time.

Estimator – Online calculation (intelligent sensor or inferential measurement) of a process variable (e.g. composition or quality) corrected by a field or lab measurement.

Execution period – Time interval between successive executions of control algorithm.

External Reset – Calculation of the PID integral contribution based on a measurement of the manipulated process input.

Function Block – A logical processing unit of software consisting of one or more input and output parameters.

Fuzzy Logic Control – Control employing a range of values from "true" to "false" that is used in decision-making with imprecise data.

Gain Margin – Change in process gain that would be required for an unstable closed loop response.

Grab Sample – Manually acquired sample of process product or feed material taken for lab analysis.

Graphic Display – Display containing a pictorial representation of process equipment and piping along with measurement values at the point they are made.

Hysteresis – The difference in the input signals required to produce the same output during a single cycle of input signal when cycled at a rate below that at which dynamic effects are important.

Initialization Manual (IMAN) – Actual mode that indicates the forward path to a physical output is broken and the output is tracking the downstream block.

Installed Characteristics – Manner in which flow rate changes with valve stem position as installed in the process.

Instrumentation and Control Survey – Survey done to identify issues associated with the process measurement or control tuning and setup.

Integrating Gain – Rate of change in the output of an integrating process per percent change in input per second.

Intelligent PID – PID with standard options and algorithm extensions for wireless/event-driven control, loop tuning, adaptive control, loop performance evaluation.

Manipulated input (manipulated parameter) – Process input that is adjusted to maintain the controlled parameter at the setpoint.

Manual Control – Plant operator adjustment of a process input.

Media – A liquid or gel designed to support the growth of microorganisms or cells.

MISO Controller – Control based on the use of multiple process inputs and outputs to determine a manipulated process input.

Model Predictive Control (MPC) – An advanced method of process control that is designed to address large interactive multiple input multiple output (MIMO) processes characterized by long dynamics, multiple constraints, and interactions among variables.

Monotonic – Either entirely non-increasing or non-decreasing.

Multiway PCA (MPCA) is an extension of PCA and provides a very effective means of accounting for and aligning batch data (i.e., different batch lengths).

Multiple Linear Regression (MLR) Model – A linear statistical technique that uses measured process conditions to predict a quality parameter prediction. Can be used to extrapolate outside the data range used in model development.

Neural Network (NN) Model – A non-linear modeling technique for quality parameter prediction based on measured process conditions. Does not extrapolate outside the data range used in model development.

Nonlinear system – Gain, time constant, or time delay and hence controller tuning settings are not constant but a function of time, direction, operating point, or load.

Normal Mode – The mode the PID is expected to be in for normal operation.

On-demand Tuning – Automatic establishment of PID tuning using on-demand testing of the process.

OPC (OLE for Process Control) – An industrial connectivity standard that enables the transfer of automation data between automation hardware and software.

Over-Damped Response – Controlled parameter is gradually restored without overshooting the setpoint value. In most cases, an over-damped response provides best response for varying operating conditions that impact process gain and response time.

Override (LO) – Actual mode that indicates Track or Auto-tuning is active and in control of the output value.

Partial Least Squares (PLS) – Linear technique for prediction of quality parameters based on measured process conditions. This technique can be used with co-linear data and can extrapolate outside the data range used in model development.

Penalty on Error – Model predictive control tuning parameter that allows more importance to be placed on a specific controlled variable.

Penalty on Move – Model predictive control tuning parameter that impacts has most impact on robustness. Model Predictive Control sensitivity to changes in a process is determined by the controller robustness.

Permitted Mode – The mode(s) available to the operator for a given application.

Phase Margin – Added phase shift through the process that would be required for an unstable closed loop response.

PID Tuning Parameters – Parameters of the PID algorithm that may be used to compensate for process gain and dynamic response and to tailor the control response to changes in setpoint and to process disturbance inputs.

Piping and Instrumentation Diagram – Drawing that shows the instrumentation and piping details for plant equipment.

Plant Operator – Person responsible for the minute-to-minute operation of one or more process areas within a plant.

Positioner – Controller mounted on the valve whose set point is desired position and whose controlled variable is shaft position.

Pressure Head – Pressure induced by a given height (or depth) of a liquid column at its bottom end.

Principal component analysis (PCA) – Mathematical technique that enables the identification and evaluation of product and process variables that may be critical to product quality and performance.

Process – Specific equipment configuration (in a manufacturing plant) which acts upon inputs to produce outputs.

Process Analysis – The examination of a process, its elements and their relationships.

Process Area – A functional grouping of equipment within a plant.

Process Flow Diagram – Drawing that shows the general flow between major pieces of equipment of a plant and the expected operating conditions at target production rate.

Process Deadtime – For a step change in the process input, process deadtime is defined as the time measured from the input change until when the first effects of the change are detected in the process output.

Process Diagnostics – The investigation of the cause or nature of a process condition or problem.

Process Gain – For a step change in a process input, the process gain is defined as the change in process output divided by the change in process input. The change in output should be determined after the process has fully responded to the input change.

Process Holdups – Operator and event initiated processing halts and restarts. Sometimes these halts and restarts are part of the batch process design, such as adding a special ingredient.

Process Time Constant – For a step change in process input, the time measured from the first detectable change in output until it reaches 63% of its final change in value.

Process Units – Pieces of equipment in a process area that are similar in construction and function.

Projection to Latent Structures (PLS), sometimes referred to as Partial Least Squares, may be used to analyze the impact of processing conditions on final product quality parameters that are often measured using lab analysis of final product.

Pseudo Random – A series of changes that appears to be random but are in fact generated according to some prearranged sequence.

Rangeability – The ratio of the maximum full-scale range to the minimum full scale range of the flowmeter.

Real Time Optimization – A closed loop online optimization technique..

Remote Seal – Flush-mounted diaphragm installed at the process vessel and connected to the differential pressure transmitter using a capillary tube filled with fluid.

Reset Windup – Continued PID accumulation of reset contribution when further changes in the PID output will have no impact on the process.

Resolution – The ability to detect and faithfully indicate small changes.

Saturation – Controller output is at either low or high limits that are normally set to match the valve being shut and wide open, respectively (controller is effectively disabled).

Self-Regulating – The ability or tendency of the process to maintain internal equilibrium.

Setpoint – Value at which the controlled parameter is to be maintained by the control system.

Single-Loop Control – The manual or automatic adjustment of a single manipulated parameter to maintain the control parameter at a target value.

Slurry – A suspension of insoluble particles in a liquid.

Span – The difference between the minimum and maximum output signals of a pressure sensor.

Step Change – A sudden change from one steady-state value to another.

Superheated Steam – Steam whose temperature, at any given pressure, is higher than water's boiling point.

Target Mode – The mode of operation requested by the operator.

Transmitter – Field device consisting of a housing, electronics, and sensing element that is used to measure process operating conditions and to communicate these values to the process control system.

Transport Delay – Time required for a liquid, gas, or solid material flow to move from one point to another through the process.

Ultimate Gain – Proportional gain required to achieve sustained oscillation of the controlled and manipulated parameter (with integral and derivative gain set to zero).

Ultimate Period – Period of oscillation achieved for sustained oscillation when the proportional gain is set to the ultimate gain (with integral and derivative gain set to zero).

Underdamped Response – Controlled parameter overshoots the setpoint value but eventually settles at the setpoint value. May minimize time to get back to setpoint but at the expense of stability and overshoot to setpoint and load disturbances.

Unstable Response – Controlled and manipulated parameters exhibit an oscillatory response that builds in amplitude until the manipulated parameter reaches its output limits.

Valve Characteristics – Manner in which flow through the valve changes as a function of stem position for a constant differential pressure across the valve.

Valve Positioner – Device used to regulate the valve actuator to maintain the stem position at a target value.

Window Communication – Wireless communication in which the device wakes up at a configured update period, senses the measurement and then communicates the measurement if the specified trigger value is exceeded.

Index

Q

R

Numerical Solution and Applications of Fractional Differential Equations, 2nd Edition

Numerical Solution and Applications of Fractional Differential Equations, 2nd Edition

Guest Editors

Libo Feng
Yang Liu
Lin Liu

Basel • Beijing • Wuhan • Barcelona • Belgrade • Novi Sad • Cluj • Manchester

Guest Editors
Libo Feng
School of Mathematical Sciences
Queensland University of Technology
Brisbane
Australia

Yang Liu
School of Mathematical Sciences
Inner Mongolia University
Hohhot
China

Lin Liu
School of Mathematics and Physics
University of Science and Technology Beijing
Beijing
China

Editorial Office
MDPI AG
Grosspeteranlage 5
4052 Basel, Switzerland

This is a reprint of the Special Issue, published open access by the journal *Fractal and Fractional* (ISSN 2504-3110), freely accessible at: https://www.mdpi.com/journal/fractalfract/special_issues/NSAFDE_II.

For citation purposes, cite each article independently as indicated on the article page online and as indicated below:

Lastname, A.A.; Lastname, B.B. Article Title. *Journal Name* **Year**, *Volume Number*, Page Range.

ISBN 978-3-7258-4111-0 (Hbk)
ISBN 978-3-7258-4112-7 (PDF)
https://doi.org/10.3390/books978-3-7258-4112-7

Contents

Preface

Over the past few decades, tremendous interest has been paid to the field of fractional calculus, which finds wide applications in physics, biology, chemistry, finance, signal and image processing, hydrology, non-Newtonian fluids, etc. Given that the analytical solution to fractional models is extremely complicated in terms of transcendental functions and analytically intractable in complex cases, numerical methods become a powerful and useful tool to handle fractional differential equations. This is the motivation for the Special Issue "Numerical Solution and Applications of Fractional Differential Equations" (NSAFDE) in the Journal of *Fractal and Fractional*.

This issue is the 2nd Edition of NSAFDE. It received 48 submissions and produced 12 accepted and published papers, all of which were subject to a rigorous review process. This reprint is a result of these 12 collections, which cover the topics of analytical techniques, numerical methods, and the applications for fractional differential equations.

Regarding the analytical techniques, the complete discriminant system method of polynomials, the traveling wave transformation, the Strang splitting algorithm, the spectral analysis method, the cones fixed point theorem, Filippov's approach, the natural transform, the residual power series method, and the Tikhonov Regularization have been considered.

The numerical methods in the collection include the mixed finite element method, the artificial boundary method, the Chebyshev spectral method, the meshless technique, the compact difference scheme, and the fast Alikhanov method.

Finally, some interesting applications have been presented, such as the second-grade fluid over a semi-infinite plate, biological population dynamics, the liquid crystal model, the Hopfield neural network, the preloaded compliance system and the fractional Cauchy problem in a circular annular region.

All the Guest Editors of this Special Issue are grateful to the authors for their quality contributions, to the reviewers for their valuable comments and advice, and to the administrative staff of MDPI for their support in completing this Special Issue. Special thanks go to the Section Managing Editor, Mr Ethan Zhang, for his excellent collaboration and valuable assistance.

Libo Feng, Yang Liu, and Lin Liu
Guest Editors

fractal and fractional

MDPI

Article

A Dynamical Analysis and New Traveling Wave Solution of the Fractional Coupled Konopelchenko–Dubrovsky Model

Jin Wang and Zhao Li *

College of Computer Science, Chengdu University, Chengdu 610106, China; wj19882009@163.com
* Correspondence: clizhao10.26@163.com

Abstract: The main object of this paper is to study the traveling wave solutions of the fractional coupled Konopelchenko–Dubrovsky model by using the complete discriminant system method of polynomials. Firstly, the fractional coupled Konopelchenko–Dubrovsky model is simplified into nonlinear ordinary differential equations by using the traveling wave transformation. Secondly, the trigonometric function solutions, rational function solutions, solitary wave solutions and the elliptic function solutions of the fractional coupled Konopelchenko–Dubrovsky model are derived by means of the polynomial complete discriminant system method. Moreover, a two-dimensional phase portrait is drawn. Finally, a 3D-diagram and a 2D-diagram of the fractional coupled Konopelchenko–Dubrovsky model are plotted in Maple 2022 software.

Keywords: Konopelchenko–Dubrovsky model; complete discriminant system; traveling wave solution; phase portrait

Citation: Wang, J.; Li, Z. A Dynamical Analysis and New Traveling Wave Solution of the Fractional Coupled Konopelchenko–Dubrovsky Model. *Fractal Fract.* **2024**, *8*, 341. https://doi.org/10.3390/fractalfract8060341

Academic Editor: Nikolay A. Kudryashov

Received: 7 May 2024
Revised: 29 May 2024
Accepted: 4 June 2024
Published: 6 June 2024

1. Introduction

In the present era, nonlinear evolution equations (NLEEs) [1–4] are employed in a number of areas like physics, chemistry, biology, fluid dynamics, engineering, optical fibers, plasma, and hydrodynamics. The analytical solutions of NLEEs can be applied to control complex behavior and difficult phenomena when the system displays [5–23]. While there is no unified method to obtain the exact solution of nonlinear evolution equations, most of the time, NLEEs can be converted into an ordinary differential equation by taken the traveling wave transformation. Based on the efforts of many predecessors, various methods have been imposed to solve NLEEs precisely and analyze various wave phenomena. He and Wu [24] proposed the first-time Exp-function method to seek solitary solutions, periodic solutions and compacton-like solutions of the KdV equation and Dodd–Bullough–Mikhailov equation. By using Hirota's bilinear transformation method, Ma proved the existence of N-soliton solutions of the (2+1)-dimensional KdV equation, the Kadomtsev–Petviashvili equation, the (2+1)-dimensional Hirota–Satsuma–Ito equation, and a combined pKP–BKP equation [25,26], respectively. Li et al. presented the (w/g)-expansion method [27]. Later, Arafat and his collaborators applied the customized (w/g)-expansion method to found the optical soliton solutions of the paraxial nonlinear Schrödinger equation and fractional Biswas–Arshed model [28,29], respectively. Wazwaz has derived the solitons and periodic solution of the Dodd–Bullough–Mikhailov and the Tzitzeica–Dodd–Bullough equations via the tanh-function approach [30]. Seadawy and Iqbal analyzed the nonlinear damped Korteweg–de Vries equation in an unmagnetized collisional dusty plasma via the direct algebraic approach [31]. Arafat also investigated scores of broad-spectral soliton solutions to the stated system via the auxiliary equation technique [32]. However, research on the traveling wave solutions of more complex fractional order NLEEs is still ongoing, and there are still a large number of open issues that need to be addressed by researchers.

In this paper, we consider the fractional coupled Konopelchenko–Dubrovsky model [33]

$$\begin{cases} D_t^\kappa u - u_{xxx} - 6auu_x + \frac{3}{2}b^2u^2u_x - 3v_y + 3bu_xv = 0, \\ v_x = u_y, \end{cases} \tag{1}$$

where $D_t^\kappa u$ is the conformable fractional derivative. $u = u(x,y,t)$ and $v = v(x,y,t)$ represent the velocity components along the horizontal and vertical axes, respectively. a and b stand for the amplitude of the wave. When $\kappa = 1$, Equation (1) become the well-known integer-order Konopelchenko–Dubrovsky model [34]. The main object of this paper is to study the traveling wave solution of the fractional coupled Konopelchenko–Dubrovsky model by using the complete discriminant system method of polynomials. On the one hand, the main effort of this article is to focus on constructing the traveling wave solution of Equation (1). On the other hand, without solving Equation (1), its dynamic branch will be analyzed.

The conformable fractional derivative was first proposed by Khalil et al. [35]. Compared with traditional fractional derivatives, the conformable fractional derivative has a more intuitive physical meaning. At present, it has been widely used in the construction of infectious disease dynamics models, nonlinear system modeling, and thermal science fields. Its definition is usually described as follows.

Definition 1 ([36]). *Let $f : [0,\infty) \to \mathbf{R}$. Then, the conformable derivative of f of order κ is defined as*

$$D_t^\kappa f(t) = \lim_{\varepsilon \to 0} \frac{f(t + \varepsilon t^{1-\kappa}) - f(t)}{\varepsilon}, \forall t \in (0, +\infty), \kappa \in (0,1], \tag{2}$$

and the function f is κ-conformable differentiable at a point t if the limit in Equation (2) exists.

The remaining sections of this article are arranged as follows: In Section 2, the traveling wave solutions of Equation (1) are constructed by using the complete discriminant system method. Moreover, a two-dimensional phase portrait is drawn. In Section 3, the three-dimensional, two-dimensional, and density plots to some obtained solutions of Equation (1) are plotted. Finally, a brief summary is presented.

2. Dynamical Analysis and Traveling Wave Solutions of Equation (1)

2.1. Traveling Wave Transformation

In this section, we first consider the wave transformation

$$u(x,y,t) = U(\xi), v(x,y,t) = V(\xi), \xi = kx + ly + \mu\frac{t^\kappa}{\kappa}, \tag{3}$$

Substituting Equation (3) into Equation (1), we have

$$\begin{cases} \mu U' - k^3U''' - 6akUU' + \frac{3}{2}kb^2U^2U' - 3lV' + 3kbU'V = 0, \\ kV' = lU'. \end{cases} \tag{4}$$

Integrating the second equation of Equation (4), we obtain

$$V = \frac{l}{k}U. \tag{5}$$

Substituting Equation (5) into the first equation of Equation (4), we have

$$-k^3U'' + \frac{1}{2}kb^2U^3 + (\frac{3lb}{2} - 3ak)U^2 + (\mu - \frac{3l^2}{k})U = d_1, \tag{6}$$

where d_1 is the integral constant.

2.2. Dynamical Analysis

Here, we consider the planar dynamic system of Equation (6) when $a_1 = 0$

$$\begin{cases} \frac{dU}{d\xi} = z, \\ \frac{dz}{d\xi} = \ell_3 U^3 + \ell_2 U^2 + \ell_1 U, \end{cases} \tag{7}$$

where $\ell_3 = \frac{b^2}{2k^2}$, $\ell_2 = \frac{1}{k^3}(\frac{3lb}{2} - 3ak)$, $\ell_1 = \frac{1}{k^3}(\mu - \frac{3l^2}{k})$.

The first integration of Equation (7) is

$$H(U,z) = \frac{1}{2}z^2 - \frac{\ell_3}{4}U^4 - \frac{\ell_2}{3}U^3 - \frac{\ell_1}{2}U^2 = h. \tag{8}$$

By setting the parameter values of fixed Equation (7), we draw the planar phase portrait of Equation (7), as shown in Figure 1.

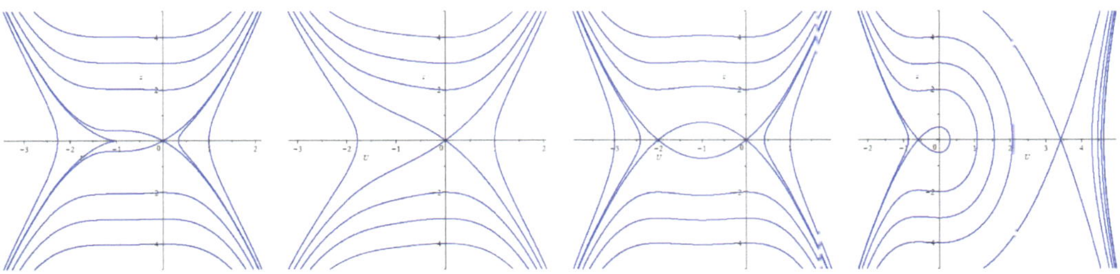

(a) $\ell_3 = 1, \ell_2 = 2, \ell_1 = 1$ **(b)** $\ell_3 = 1, \ell_2 = 2, \ell_1 = 2$ **(c)** $\ell_3 = 1, \ell_2 = 3, \ell_1 = 1$ **(d)** $\ell_3 = 1, \ell_2 = -2\sqrt{2}, \ell_1 = -2$

Figure 1. Phase portrait of Equation (7).

2.3. Traveling Wave Solutions of Equation (1)

Multiplying both sides of Equation (6) by U' simultaneously and integrating it yields

$$(U')^2 = b_4 U^4 + b_3 U^3 + b_2 U^2 + b_1 U + b_0, \tag{9}$$

where $b_4 = \frac{b^2}{4k^2}$, $b_3 = \frac{lb}{k^3} - \frac{2a}{k^2}$, $b_2 = \frac{\mu}{k^3} - \frac{3l^2}{k^4}$, $b_1 = -\frac{2d_1}{k^3}$, $b_0 = 2d_2$; here, d_2 is the integral constant.

Here, we make a transformation:

$$\begin{cases} w = (b_4)^{\frac{1}{4}}(U + \frac{b_3}{4b_4}), \\ \chi = (b_4)^{\frac{1}{4}}\xi. \end{cases} \tag{10}$$

Substituting Equation (10) into Equation (9), we obtain:

$$w_\chi^2 = w^4 + c_2 w^2 + c_1 w + c_0, \tag{11}$$

where $c_2 = \frac{b_2}{\sqrt{b_4}}$, $c_1 = (\frac{b_3^3}{8b_4^2} - \frac{b_2 b_3}{2b_4} + b_1)(b_4)^{-\frac{1}{4}}$, $c_0 = \frac{-3b_3^4}{256b_4^3} + \frac{b_2 b_3^2}{16b_4^2} - \frac{b_1 b_3}{4b_4} + b_0$.

By integrating Equation (11) once, one has:

$$\pm(\chi - \chi_0) = \int \frac{dw}{\sqrt{w^4 + c_2 w^2 + c_1 w + c_0}}, \tag{12}$$

where χ_0 denotes an integrating constant.

Denote that $G(w) = w^4 + c_2 w^2 + c_1 w + c_0$. We derive its complete discrimination system in the following form:

$$D_1 = 4, D_2 = -c_2, D_3 = -2c_2^3 + 8c_2c_0 - 9c_1^2, D_4 = -c_2^3c_1^2 + 4c_2^4c_0 + 36c_2c_1^2c_0 - 32c_2^2c_0^2 - \frac{27}{4}c_1^4 + 64c_0^3,$$
$$E_2 = c_1c_2^2 - 32c_2c_0. \tag{13}$$

According to the root-classifications of Equation (11), we will discuss the traveling wave solutions of Equation (1) under nine cases.

Case 1: $D_2 < 0, D_3 = 0, D_4 = 0, G(w) = [w^2 + q^2]^2$.

By combining Equation (9) with Equation (3), the traveling wave solutions of Equation (1) can be expressed as below:

$$u_1(x,y,t) = \frac{2ak - lb}{kb^2} + \frac{\sqrt{2(\mu k - 3l^2)}}{bk} \tan\left[\frac{\sqrt{2(\mu k - 3l^2)}}{bk^3}\left(\sqrt{\frac{b}{2k}}(kx + ly + \mu\frac{t^\kappa}{\kappa}) - \chi_0\right)\right]. \tag{14}$$

$$v_1(x,y,t) = \frac{2akl - l^2b}{k^2b^2} + \frac{l\sqrt{2(\mu k - 3l^2)}}{bk^2} \tan\left[\frac{\sqrt{2(\mu k - 3l^2)}}{bk^3}\left(\sqrt{\frac{b}{2k}}(kx + ly + \mu\frac{t^\kappa}{\kappa}) - \chi_0\right)\right]. \tag{15}$$

Case 2: $D_2 = 0, D_3 = 0, D_4 = 0$, $G(w) = w^4$. In this situation, the traveling wave solutions of Equation (1) take the form:

$$u_2(x,y,t) = \frac{2ak - lb}{kb^2} - \sqrt{\frac{2k}{b}}\left(\sqrt{\frac{b}{2k}}(kx + ly + \mu\frac{t^\kappa}{\kappa}) - \chi_0\right)^{-1}. \tag{16}$$

$$v_2(x,y,t) = \frac{2akl - l^2b}{k^2b^2} - l\sqrt{\frac{2}{bk}}\left(\sqrt{\frac{b}{2k}}(kx + ly + \mu\frac{t^\kappa}{\kappa}) - \chi_0\right)^{-1}. \tag{17}$$

Case 3: $D_2 > 0, D_3 = 0, D_4 = 0$, $G(w) = (w - \alpha)^2(w - \beta)^2$, where α, β are real number, and $\alpha > \beta$.

(i) If $w > \alpha$ or $w < \beta$, the traveling wave solutions of Equation (1) take the form:

$$u_{3,1}(x,y,t) = \alpha\sqrt{\frac{k}{2b}} - \frac{lb - 2ak}{kb^2} + (\beta - \alpha)\sqrt{\frac{k}{2b}} \coth\frac{(\alpha - \beta)\left(\sqrt{\frac{b}{2k}}(kx + ly + \mu\frac{t^\kappa}{\kappa}) - \chi_0\right)}{2}. \tag{18}$$

$$v_{3,1}(x,y,t) = \frac{l\alpha}{\sqrt{2bk}} - \frac{l^2b - 2akl}{k^2b^2} + \frac{l(\beta - \alpha)}{\sqrt{2bk}} \coth\frac{(\alpha - \beta)\left(\sqrt{\frac{b}{2k}}(kx + ly + \mu\frac{t^\kappa}{\kappa}) - \chi_0\right)}{2}. \tag{19}$$

(ii) If $\beta < w < \alpha$, the traveling wave solutions of Equation (1) take the form:

$$u_{3,2}(x,y,t) = \alpha\sqrt{\frac{k}{2b}} - \frac{lb - 2ak}{kb^2} + (\beta - \alpha)\sqrt{\frac{k}{2b}} \tanh\frac{(\alpha - \beta)\left(\sqrt{\frac{b}{2k}}(kx + ly + \mu\frac{t^\kappa}{\kappa}) - \chi_0\right)}{2}. \tag{20}$$

$$v_{3,2}(x,y,t) = \frac{l\alpha}{\sqrt{2bk}} - \frac{l^2b - 2akl}{k^2b^2} + \frac{l(\beta - \alpha)}{\sqrt{2bk}} \tanh\frac{(\alpha - \beta)\left(\sqrt{\frac{b}{2k}}(kx + ly + \mu\frac{t^\kappa}{\kappa}) - \chi_0\right)}{2}. \tag{21}$$

Case 4: $D_2 > 0, D_3 > 0, D_4 = 0$, $G(w) = (w - \alpha)^2(w - \beta)(w - \gamma)$, where α, β and γ are real number, $\beta > \gamma$.

(i) If $\alpha > \beta$ and $w > \beta$, or $\alpha < \gamma$ and $w < \gamma$, the implicit traveling wave solutions of Equation (1) can be expressed as below:

$$\pm\left(\sqrt{\frac{b}{2k}}(kx+ly+\mu\frac{t^{\kappa}}{\kappa})-\chi_0\right)=\frac{1}{(\alpha-\beta)(\alpha-\gamma)}$$
$$\ln\frac{[\sqrt{(kb^2u_{4,1}+lb-2ak-\sqrt{2}(kb)^{\frac{3}{2}}\beta)(\alpha-\gamma)}-\sqrt{(\alpha-\beta)(kb^2u_{4,1}+lb-2ak-\sqrt{2}(kb)^{\frac{3}{2}}\gamma)}]^2}{|kb^2u_{4,1}+lb-2ak-\sqrt{2}(kb)^{\frac{3}{2}}\alpha|}. \tag{22}$$

$$\pm\left(\sqrt{\frac{b}{2k}}(kx+ly+\mu\frac{t^{\kappa}}{\kappa})-\chi_0\right)=\frac{1}{(\alpha-\beta)(\alpha-\gamma)}$$
$$\ln\frac{[\sqrt{(k^2b^2v_{4,1}+l^2b-2alk-\sqrt{2}(kb)^{\frac{3}{2}}l\beta)(\alpha-\gamma)}-\sqrt{(\alpha-\beta)(k^2b^2v_{4,1}+l^2b-2alk-\sqrt{2}(kb)^{\frac{3}{2}}l\gamma)}]^2}{|k^2b^2v_{4,1}+l^2b-2alk-\sqrt{2}(kb)^{\frac{3}{2}}l\alpha|}. \tag{23}$$

(ii) If $\alpha>\beta$ and $w<\gamma$, or $\alpha<\gamma$ and $w<\beta$, the implicit traveling wave solutions of Equation (1) can be expressed as below:

$$\pm\left(\sqrt{\frac{b}{2k}}(kx+ly+\mu\frac{t^{\kappa}}{\kappa})-\chi_0\right)=\frac{1}{(\alpha-\beta)(\alpha-\gamma)}$$
$$\ln\frac{[\sqrt{(kb^2u_{4,2}+lb-2ak-\sqrt{2}(kb)^{\frac{3}{2}}\beta)(\gamma-\alpha)}-\sqrt{(\beta-\alpha)(kb^2u_{4,2}+lb-2ak-\sqrt{2}(kb)^{\frac{3}{2}}\gamma)}]^2}{|kb^2u_{4,2}+lb-2ak-\sqrt{2}(kb)^{\frac{3}{2}}\alpha|}. \tag{24}$$

$$\pm\left(\sqrt{\frac{b}{2k}}(kx+ly+\mu\frac{t^{\kappa}}{\kappa})-\chi_0\right)=\frac{1}{(\alpha-\beta)(\alpha-\gamma)}$$
$$\ln\frac{[\sqrt{(k^2b^2v_{4,2}+l^2b-2alk-\sqrt{2}(kb)^{\frac{3}{2}}l\beta)(\gamma-\alpha)}-\sqrt{(\beta-\alpha)(k^2b^2v_{4,2}+l^2b-2alk-\sqrt{2}(kb)^{\frac{3}{2}}l\gamma)}]^2}{|k^2b^2v_{4,2}+l^2b-2alk-\sqrt{2}(kb)^{\frac{3}{2}}l\alpha|}. \tag{25}$$

(iii) If $\beta>\alpha>\gamma$, the implicit traveling wave solutions of Equation (1) can be expressed as below:

$$\pm\left(\sqrt{\frac{b}{2k}}(kx+ly+\mu\frac{t^{\kappa}}{\kappa})-\chi_0\right)=\frac{1}{(\beta-\alpha)(\alpha-\gamma)}$$
$$\arcsin\frac{(kb^2u_{4,3}+lb-2ak-\sqrt{2}(kb)^{\frac{3}{2}}\beta)(\alpha-\gamma)+(\alpha-\beta)(kb^2u_{4,3}+lb-2ak-\sqrt{2}(kb)^{\frac{3}{2}}\gamma)}{|(kb^2u_{4,3}+lb-2ak-\sqrt{2}(kb)^{\frac{3}{2}}\alpha)(\beta-\gamma)|}. \tag{26}$$

$$\pm\left(\sqrt{\frac{b}{2k}}(kx+ly+\mu\frac{t^{\kappa}}{\kappa})-\chi_0\right)=\frac{1}{(\beta-\alpha)(\alpha-\gamma)}$$
$$\arcsin\frac{(k^2b^2v_{4,3}+l^2b-2alk-\sqrt{2}(kb)^{\frac{3}{2}}l\beta)(\alpha-\gamma)+(\alpha-\beta)(k^2b^2v_{4,3}+l^2b-2alk-\sqrt{2}(kb)^{\frac{3}{2}}l\gamma)}{|(k^2b^2v_{4,3}+l^2b-2alk-\sqrt{2}(kb)^{\frac{3}{2}}l\alpha)(\beta-\gamma)|}. \tag{27}$$

Case 5: $D_2>0, D_3=0, D_4=0, E_2=0, G(w)=(w-\alpha)^3(w-\beta)$, where α,β are real numbers.

When $w > \alpha$ and $w > \beta$, or $w < \alpha$ and $w < \beta$, the traveling wave solutions of Equation (1) take the form:

$$u_5(x,y,t) = \frac{8\sqrt{2}(\alpha-\beta)k^{\frac{3}{2}}}{\sqrt{b}[(\beta-\alpha)^2(\sqrt{b}(kx+ly+\mu\frac{t^\kappa}{\kappa})-\sqrt{2k}\chi_0)^2-4]} + \frac{\sqrt{2}\alpha(kb)^{\frac{3}{2}}+2ak-lb}{kb^2} \tag{28}$$

$$v_5(x,y,t) = \frac{8\sqrt{2}(\alpha-\beta)lk^{\frac{1}{2}}}{\sqrt{b}[(\beta-\alpha)^2(\sqrt{b}(kx+ly+\mu\frac{t^\kappa}{\kappa})-\sqrt{2k}\chi_0)^2-4]} + \frac{\sqrt{2}l\alpha(kb)^{\frac{3}{2}}+2akl-l^2b}{k^2b^2} \tag{29}$$

Case 6: $D_4 = 0, D_2D_3 < 0$, $G(w) = (w-\alpha)^2\left[(w-l_1)^2+s_1^2\right]$. The traveling wave solutions of Equation (1) take the form:

$$u_6(x,y,t) = \frac{2ak-lb}{kb^2} + \sqrt{\frac{2k}{b}}\frac{\left[e^{\pm\sqrt{(\alpha-l_1)^2+s_1^2}(\sqrt{\frac{b}{2k}}(kx+ly+\mu\frac{t^\kappa}{\kappa})-\chi_0)}-\gamma\right]+\sqrt{(\alpha-l_1)^2+s_1^2}(2-\gamma)}{\left[e^{\pm\sqrt{(\alpha-l_1)^2+s_1^2}(\sqrt{\frac{b}{2k}}(kx+ly+\mu\frac{t^\kappa}{\kappa})-\chi_0)}-\gamma\right]^2-1}, \tag{30}$$

$$v_6(x,y,t) = \frac{2akl-l^2b}{k^2b^2} + l\sqrt{\frac{2}{kb}}\frac{\left[e^{\pm\sqrt{(\alpha-l_1)^2+s_1^2}(\sqrt{\frac{b}{2k}}(kx+ly+\mu\frac{t^\kappa}{\kappa})-\chi_0)}-\gamma\right]+\sqrt{(\alpha-l_1)^2+s_1^2}(2-\gamma)}{\left[e^{\pm\sqrt{(\alpha-l_1)^2+s_1^2}(\sqrt{\frac{b}{2k}}(kx+ly+\mu\frac{t^\kappa}{\kappa})-\chi_0)}-\gamma\right]^2-1}, \tag{31}$$

where $\gamma = \frac{\alpha-2l_1}{\sqrt{(\alpha-l_1)^2+s_1^2}}$.

Case 7: $D_4 > 0, D_3 > 0, D_1 > 0, G(w) = (w-\alpha_1)(w-\alpha_2)(w-\alpha_3)(w-\alpha_4)$, in which $\alpha_1, \alpha_2, \alpha_3, \alpha_4$ is real number and $\alpha_1 > \alpha_2 > \alpha_3 > \alpha_4$.

When $w > \alpha_1$ or $w < \alpha_4$, the traveling wave solutions of Equation (1) take the form:

$$\begin{aligned}
u_{7,1}(x,y,t) &= \frac{2ak-lb}{kb^2} \\
&+\sqrt{\frac{2k}{b}}\frac{\alpha_2(\alpha_1-\alpha_4)\,\mathrm{sn}^2\left(\frac{\sqrt{(\alpha_1-\alpha_3)(\alpha_2-\alpha_4)}}{2}\left(\sqrt{\frac{b}{2k}}(kx+ly+\mu\frac{t^\kappa}{\kappa})-\chi_0\right),m\right)-\alpha_1(\alpha_2-\alpha_4)}{(\alpha_1-\alpha_4)\,\mathrm{sn}^2\left(\frac{\sqrt{(\alpha_1-\alpha_3)(\alpha_2-\alpha_4)}}{2}\left(\sqrt{\frac{b}{2k}}(kx+ly+\mu\frac{t^\kappa}{\kappa})-\chi_0\right),m\right)-(\alpha_2-\alpha_4)}, \\
v_{7,1}(x,y,t) &= \frac{2akl-l^2b}{k^2b^2} \\
&+l\sqrt{\frac{2}{kb}}\frac{\alpha_2(\alpha_1-\alpha_4)\,\mathrm{sn}^2\left(\frac{\sqrt{(\alpha_1-\alpha_3)(\alpha_2-\alpha_4)}}{2}\left(\sqrt{\frac{b}{2k}}(kx+ly+\mu\frac{t^\kappa}{\kappa})-\chi_0\right),m\right)-\alpha_1(\alpha_2-\alpha_4)}{(\alpha_1-\alpha_4)\,\mathrm{sn}^2\left(\frac{\sqrt{(\alpha_1-\alpha_3)(\alpha_2-\alpha_4)}}{2}\left(\sqrt{\frac{b}{2k}}(kx+ly+\mu\frac{t^\kappa}{\kappa})-\chi_0\right),m\right)-(\alpha_2-\alpha_4)},
\end{aligned} \tag{32}$$

$$
\begin{aligned}
u_{7,2}(x,y,t) = & \frac{2ak-lb}{kb^2} \\
& + \sqrt{\frac{2k}{b}} \frac{\alpha_4(\alpha_2-\alpha_3)\,\mathrm{sn}^2\left(\frac{\sqrt{(\alpha_1-\alpha_3)(\alpha_2-\alpha_4)}}{2}\left(\sqrt{\frac{b}{2k}}(kx+ly+\mu\frac{t^\kappa}{\kappa})-\chi_0\right),m\right)-\alpha_3(\alpha_2-\alpha_4)}{(\alpha_2-\alpha_3)\,\mathrm{sn}^2\left(\frac{\sqrt{(\alpha_1-\alpha_3)(\alpha_2-\alpha_4)}}{2}\left(\sqrt{\frac{b}{2k}}(kx+ly+\mu\frac{t^\kappa}{\kappa})-\chi_0\right),m\right)-(\alpha_2-\alpha_4)}, \\
v_{7,2}(x,y,t) = & \frac{2akl-l^2b}{k^2b^2} \\
& + l\sqrt{\frac{2}{kb}} \frac{\alpha_4(\alpha_2-\alpha_3)\,\mathrm{sn}^2\left(\frac{\sqrt{(\alpha_1-\alpha_3)(\alpha_2-\alpha_4)}}{2}\left(\sqrt{\frac{b}{2k}}(kx+ly+\mu\frac{t^\kappa}{\kappa})-\chi_0\right),m\right)-\alpha_3(\alpha_2-\alpha_4)}{(\alpha_2-\alpha_3)\,\mathrm{sn}^2\left(\frac{\sqrt{(\alpha_1-\alpha_3)(\alpha_2-\alpha_4)}}{2}\left(\sqrt{\frac{b}{2k}}(kx+ly+\mu\frac{t^\kappa}{\kappa})-\chi_0\right),m\right)-(\alpha_2-\alpha_4)},
\end{aligned} \tag{33}
$$

in which $m^2 = \frac{(\alpha_1-\alpha_4)(\alpha_2-\alpha_3)}{(\alpha_1-\alpha_3)(\alpha_2-\alpha_4)}$.

Case 8: $D_4 < 0, D_2D_3 \geqslant 0$, then $G(w) = (w-\alpha)(w-\beta)\left[(w-l_1)^2+s_1^2\right]$, where real number $\alpha > \beta, l_1, s_1 > 0$.

The traveling wave solutions of Equation (1) take the form:

$$
u_8(x,y,t) = \frac{2ak-lb}{kb^2} + \sqrt{\frac{2k}{b}} \frac{a\,\mathrm{cn}\left(\frac{\sqrt{-2s_1m_1(\alpha-\beta)}}{2mm_1}\left(\frac{b}{2k}(kx+ly+\mu\frac{t^\kappa}{\kappa})-\chi_0\right),m\right)+e_2}{c\,\mathrm{cn}\left(\frac{\sqrt{-2s_1m_1(\alpha-\beta)}}{2mm_1}\left(\frac{b}{2k}(kx+ly+\mu\frac{t^\kappa}{\kappa})-\chi_0\right),m\right)+e_4}, \tag{34}
$$

$$
v_8(x,y,t) = \frac{2akl-l^2b}{k^2b^2} + l\sqrt{\frac{2}{kb}} \frac{a\,\mathrm{cn}\left(\frac{\sqrt{-2s_1m_1(\alpha-\beta)}}{2mm_1}\left(\frac{b}{2k}(kx+ly+\mu\frac{t^\kappa}{\kappa})-\chi_0\right),m\right)+e_2}{c\,\mathrm{cn}\left(\frac{\sqrt{-2s_1m_1(\alpha-\beta)}}{2mm_1}\left(\frac{b}{2k}(kx+ly+\mu\frac{t^\kappa}{\kappa})-\chi_0\right),m\right)+e_4}, \tag{35}
$$

in which $e_1 = \frac{1}{2}(\alpha+\beta)e_3 - \frac{1}{2}(\alpha-\beta)e_4, e_2 = \frac{1}{2}(\alpha+\beta)e_4 - \frac{1}{2}(\alpha-\beta)e_3, e_3 = \alpha - l_1 - \frac{s_1}{m_1}$, $e_4 = \alpha - l_1 - s_1m_1, E = \frac{s_1^2+(\alpha-l_1)(\beta-l_1)}{s_1(\alpha-\beta)}, m_1 = E - \sqrt{E^2+1}, m^2 = \frac{1}{1+m_1^2}$.

Case 9: $D_4 > 0, D_2D_3 \leqslant 0$, then $G(w) = \left[(w-l_1)^2+s_1^2\right]\left[(w-l_2)^2+s_2^2\right]$, where l_1, l_2, s_1, s_2 are real and $s_1 \geqslant s_2 > 0$. The traveling wave solutions of Equation (1) take the form:

$$
\begin{aligned}
u_9(x,y,t) = & \frac{2ak-lb}{kb^2} \\
& + \sqrt{\frac{2k}{b}} \cdot \frac{e_1\,\mathrm{sn}\left(\eta\left(\frac{b}{2k}(kx+ly+\mu\frac{t^\kappa}{\kappa})-\chi_0\right),m\right)+e_2\,\mathrm{cn}\left(\eta\left(\frac{b}{2k}(kx+ly+\mu\frac{t^\kappa}{\kappa})-\chi_0\right),m\right)}{e_3\,\mathrm{sn}\left(\eta\left(\frac{b}{2k}(kx+ly+\mu\frac{t^\kappa}{\kappa})-\chi_0\right),m\right)+e_4\,\mathrm{cn}\left(\eta\left(\frac{b}{2k}(kx+ly+\mu\frac{t^\kappa}{\kappa})-\chi_0\right),m\right)},
\end{aligned} \tag{36}
$$

$$
\begin{aligned}
v_9(x,y,t) = & \frac{2akl-l^2b}{k^2b^2} \\
& + l\sqrt{\frac{2}{kb}} \cdot \frac{e_1\,\mathrm{sn}\left(\eta\left(\frac{b}{2k}(kx+ly+\mu\frac{t^\kappa}{\kappa})-\chi_0\right),m\right)+e_2\,\mathrm{cn}\left(\eta\left(\frac{b}{2k}(kx+ly+\mu\frac{t^\kappa}{\kappa})-\chi_0\right),m\right)}{e_3\,\mathrm{sn}\left(\eta\left(\frac{b}{2k}(kx+ly+\mu\frac{t^\kappa}{\kappa})-\chi_0\right),m\right)+e_4\,\mathrm{cn}\left(\eta\left(\frac{b}{2k}(kx+ly+\mu\frac{t^\kappa}{\kappa})-\chi_0\right),m\right)},
\end{aligned} \tag{37}
$$

in which $e_1 = l_1e_3 + s_1e_4, e_2 = l_1e_4 - s_1e_3, e_3 = -s_1 - \frac{s_2}{m_1}, e_4 = l_1 - l_2, E = \frac{(l_1-l_2)^2+s_1^2+s_2^2}{2s_1s_2}$, $m_1 = E + \sqrt{E^2-1}, m^2 = \frac{m_1^2-1}{m_1^2}, \eta = \frac{s_2\sqrt{(e_3^2+e_4^2)(m_1^2e_3^2+e_4^2)}}{e_3^2+e_4^2}$.

3. Numerical Simulation

In order to understand the dynamical processes and mechanisms of complex phenomena of the fractional coupled Konopelchenko–Dubrovsky model, numerical simulations of the obtained soliton solutions are given in this section. As is vividly shown in Figures 2a, 3a and 4a, $u_1(x,y,t), u_2(x,y,t)$ and $u_3(x,y,t)$ stand for the tangent function solutions, the rational function solutions and the hyperbolic function solutions, respectively. Figures 2b, 3b and 4b denote the level curve at time $t = 1$. Furthermore, Figures 2c, 3c and 4c represent the density plots. Figures 2d, 3d and 4d stand for the contour plots.

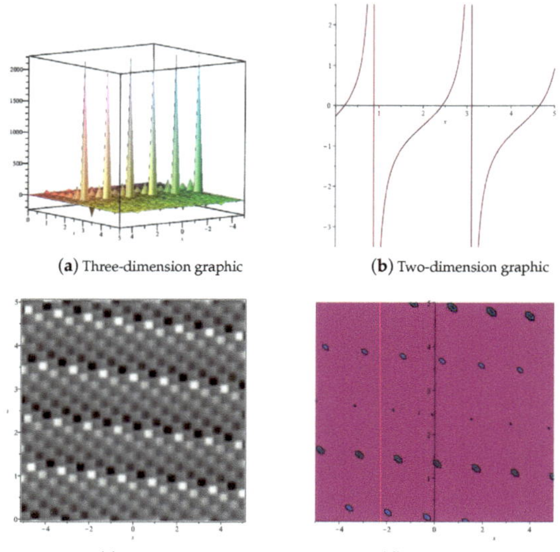

(a) Three-dimension graphic (b) Two-dimension graphic

(c) Density plot (d) Contour plot

Figure 2. Equation (14) for $a = 3, b = 4, k = 1, l = 2, \mu = 16$.

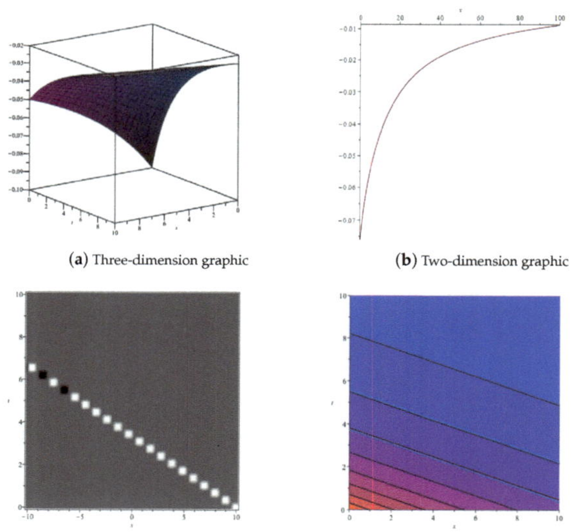

(a) Three-dimension graphic (b) Two-dimension graphic

(c) Density plot (d) Contour plot

Figure 3. Equation (16) for $a = k = l = 1, b = 2, \mu = 3$.

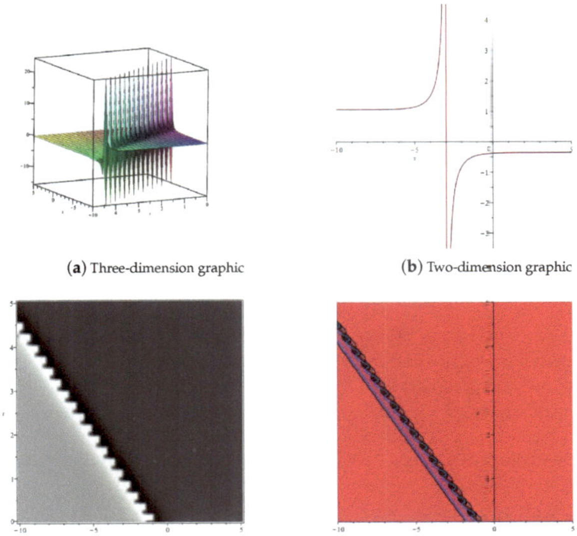

(**a**) Three-dimension graphic (**b**) Two-dimension graphic

(**c**) Density plot (**d**) Contour plot

Figure 4. Equation (18) for $a = k = l = 1, b = 2, \mu = 2, \alpha = \frac{1}{\sqrt{2}}, \beta = \frac{-1}{\sqrt{2}}$.

4. Conclusions

In this article, we have presented the traveling wave solutions of Equation (1) via the complete discriminant system method, which is one of the most useful tools in solving NLEEs. The trigonometric function solutions, the rational function solutions, the hyperbolic function solutions, the exponential function solutions, and the Jacobian elliptic function solutions are obtained. Lastly, in order to understand the mechanisms of physical phenomena for Equation (1), we have also depicted two-dimensional and three-dimensional diagrams. In future work, we will focus on the traveling wave solutions and dynamic behavior of more complex NLEEs. Furthermore, we will also use the Darboux transformations to discuss the N-soliton solutions of more complex NLEEs.

Author Contributions: Software, Z.L.; Writing—original draft, J.W. All authors have read and agreed to the published version of the manuscript.

Funding: This research received no external funding.

Data Availability Statement: No new data were created or analyzed in this study. Data sharing is not applicable to this article.

Conflicts of Interest: The authors declare no conflicts of interest.

References

1. Gu, M.; Peng, C.; Li, Z. Traveling wave solution of (3+1)-dimensional negative-order KdV-Calogero-Bogoyavlenskii-Schiff equation. *Aims Math.* **2024**, *9*, 6699–6708. [CrossRef]
2. Wu, J.; Huang, Y. Boundedness of solutions for an attraction-repulsion model with indirect signal production. *Mathematics* **2024**, *12*, 1143. [CrossRef]
3. Liu, C.; Li, Z. The dynamical behavior analysis and the traveling wave solutions of the stochastic Sasa-Satsuma Equation. *Qual. Theory Dyn. Syst.* **2024**, *23*, 157. [CrossRef]
4. Wu, J.; Yang, Z. Global existence and boundedness of chemotaxis-fluid equations to the coupled Solow-Swan model. *Aims Math.* **2023**, *8*, 17914–17942. [CrossRef]
5. Tang, C.; Li, X.; Wang, Q. Mean-field stochastic linear quadratic optimal control for jump-diffusion systems with hybrid disturbances. *Symmetry* **2024**, *16*, 642. [CrossRef]
6. Chen, T.; Chen, J.; Zhuang, B. Stabilization of coupled delayed nonlinear time fractional reaction diffusion systems using sampled-in-space sensing and actuation. *Int. J. Adapt. Control. Signal Process.* **2024**. [CrossRef]

7. Arfaoui, H.; Makhlouf, A.B. Stability of a time fractional advection-diffusion system. *Chaos Solitons Fractals* **2022**, *157*, 111949. [CrossRef]
8. Khater, M.M.A. In solid physics equations, accurate and novel soliton wave structures for heating a single crystal of sodium fluoride. *Int. J. Mod. Phys.* **2023**, *37*, 2350068. [CrossRef]
9. Li, Z.; Liu, C. Chaotic pattern and traveling wave solution of the perturbed stochastic nonlinear Schrödinger equation with generalized anti-cubic law nonlinearity and spatio-temporal dispersion. *Results Phys.* **2024**, *56*, 107305. [CrossRef]
10. Arafat, S.M.Y.; Fatema, K.; Islam, M.E.; Akbar, M.A. Promulgation on various genres soliton of Maccari system in nonlinear optics. *Opt. Quantum Electron.* **2022**, *54*, 206. [CrossRef]
11. Khater, M.M.A.; Alfalqi, S.H.; Alzaidi, J.F.; Attia, R.A.M. Analytically and numerically, dispersive, weakly nonlinear wave packets are presented in a quasi-monochromaticmedium. *Results Phys.* **2023**, *46*, 106312. [CrossRef]
12. Akinyemi, L.; Rezazadeh, H.; Shi, Q.H.; Inc, M.; Khater, M.M.; Ahmad, H.; Jhangeer, A.; Akbar, M.A. New optical solitons of perturbed nonlinear Schrodinger-Hirota equation with spatio-temporal dispersion. *Results Phys.* **2021**, *29*, 104656. [CrossRef]
13. Li, Z.; Peng, C. Bifurcation, phase portrait and traveling wave solution of time-fractional thin-film ferroelectric material equation with beta fractional derivative. *Phys. Lett.* **2023**, *484*, 129080. [CrossRef]
14. Zhang, K.; Li, Z.; Cao, J. Qualitative analysis and modulation instability for the extended (3+1)-dimensional nonlinear Schrödinger equation with conformable derivative. *Results Phys.* **2024**, *61*, 107713. [CrossRef]
15. Ma, H.; Gao, Y.; Deng, A. Nonlinear superposition of the (2+1)-dimensional generalized Konopelchenko-Dubrovsky-Kaup-Kupershmidt equation. *Nonlinear Dyn.* **2023**, *111*, 619–632. [CrossRef]
16. Khater, M.M.A. Novel computational simulation of the propagation of pulses in optical fibers regarding the dispersion effect. *Int. J. Mod. Phys.* **2023**, *37*, 2350083. [CrossRef]
17. Feng, W.; Chen, L.; Ma, G.; Zhou, Q. Study on weakening optical soliton interaction in nonlinear optics. *Nonlinear Dyn.* **2022**, *108*, 2483–2488. [CrossRef]
18. Ma, H.; Gao, Y.; Deng, A. Fission and fusion solutions of the (2+1)-dimensional generalized Konopelchenko-Dubrovsky-Kaup-Kupershmidt equation: Case of fluid mechanics and plasma physics. *Nonlinear Dyn.* **2022**, *108*, 4123–4137. [CrossRef]
19. Han, T.; Jiang, Y.; Lyu, J. Chaotic behavior and optical soliton for the concatenated model arising in optical communication. *Results Phys.* **2024**, *58*, 107467. [CrossRef]
20. Khater, M.M.A. A hybrid analytical and numerical analysis of ultra-short pulse phase shifts. *Chaos Solitons Fractals* **2023**, *169*, 113232. [CrossRef]
21. Horvath, D.X.; Calabrese, P.; Castro-Alvaredo, O.A. Branch point twist field form factors in the sine-Gordon model II: Composite twist fields and symmetry resolved entanglement. *Sci Post Phys.* **2022**, *12*, 88. [CrossRef]
22. Ismael, H.F.; Bulut, H.; Park, C.; Osman, M.S. M-lump, N-soliton solutions, and the collision phenomena for the (2+1)-dimensional Date-Jimbo-Kashiwara-Miwa equation. *Results Phys.* **2020**, *19*, 103329. [CrossRef]
23. Akbulut, A.; Kaplan, M.; Bekir, A. Auxiliary equation method for fractional differential equations with modified Riemann-Liouville derivative. *Int. J. Nonlinear Sci. Numer. Simul.* **2016**, *17*, 413–420. [CrossRef]
24. He, J.; Wu, X. Exp-function method for nonlinear wave equations. *Chaos Solitons Fractals* **2006**, *30*, 700–708. [CrossRef]
25. Ma, W. N-soliton solution and the Hirota condition of a (2+1)-dimensional combined equation. *Math. Comput. Simul.* **2021**, *190*, 270–279. [CrossRef]
26. Ma, W. N-soliton solution of a combined pKP-BKP equation. *J. Geom. Phys.* **2021**, *165*, 104191. [CrossRef]
27. Li, W.; Chen, H.; Zhang, G. The (w/g)-expansion method and its application to Vakhnenko equation. *Chin. Phys.* **2009**, *18*, 400.
28. Arafat, S.M.Y.; Khan, K.; Islam, S.M.R.; Rahman, M.M. Parametric effects on paraxial nonlinear Schrodinger equation in Kerr media. *Chin. J. Phys.* **2023**, *83*, 361–378. [CrossRef]
29. Arafat, S.M.Y.; Islam, S.M.R.; Rahman, M.M.; Saklayen, M.A. On nonlinear optical solitons of fractional Biswas-Arshed Model with beta derivative. *Results Phys.* **2023**, *48*, 106426. [CrossRef]
30. Wazwaz, A.-M. The tanh method: Solitons and periodic solutions for the Dodd-Bullough-Mikhailov and the Tzitzeica-Dodd-Bullough equations. *Chaos Solitons Fractals* **2005**, *25*, 55–63. [CrossRef]
31. Seadawy, A.R.; Iqbal, M. Propagation of the nonlinear damped Korteweg-de Vries equation in an unmagnetized collisional dusty plasma via analytical mathematical methods. *Math. Methods Appl. Sci.* **2021**, *44*, 737–748. [CrossRef]
32. Yiasir Arafat, S.M.; Fatema, K.; Rayhanul Islam, S.M.; Islam, M.E.; Ali Akbar, M.; Osman, M.S. The mathematical and wave profile analysis of the Maccari system in nonlinear physical phenomena. *Opt. Quantum Electron.* **2023**, *55*, 136. [CrossRef]
33. Aslam, A.; Majeed, A.; Kamran, M.; Inc, M.; Alqahtani, T.R. Dynamical behavior of the fractional coupled Konopelchenko-Dubrovsky and (3+1)-dimensional modified Korteweg-de Vries-Zakharov-Kuznestsov equations. *Opt. Quantum Electron.* **2023**, *55*, 543. [CrossRef]
34. Arafat, S.M.Y.; Rahman, M.M.; Karim, M.F.; Amin, M.R. Wave profile analysis of the (2+1)-dimensional Konopelchenko-Dubrovsky model in mathematical physics. *Partial. Differ. Equ. Appl. Math.* **2023**, *8*, 100573. [CrossRef]
35. Khalil, R.R.; Horani, M.A.H.H.; Yousef, A.; Sababheh, M. A new definition of fractional derivative. *J. Comput. Appl. Math.* **2014**, *264*, 65–70. [CrossRef]

36. Li, Z.; Han, T.; Huang, C. Exact single traveling wave solutions for generalized fractional Gardner equation. *Math. Probl. Eng.* **2020**, *2020*, 8842496. [CrossRef]

fractal and fractional

Article

Dynamic Behavior and Optical Soliton for the M-Truncated Fractional Paraxial Wave Equation Arising in a Liquid Crystal Model

Jie Luo [1] and Zhao Li [2,*]

[1] School of Electronic Information and Electrical Engineering, Chengdu University, Chengdu 610106, China; luojie01@cdu.edu.cn
[2] College of Computer Science, Chengdu University, Chengdu 610106, China
* Correspondence: lizhao@cdu.edu.cn; Tel.: +86-183-8207-1390

Abstract: The main purpose of this article is to investigate the dynamic behavior and optical soliton for the M-truncated fractional paraxial wave equation arising in a liquid crystal model, which is usually used to design camera lenses for high-quality photography. The traveling wave transformation is applied to the M-truncated fractional paraxial wave equation. Moreover, a two-dimensional dynamical system and its disturbance system are obtained. The phase portraits of the two-dimensional dynamic system and Poincaré sections and a bifurcation portrait of its perturbation system are drawn. The obtained three-dimensional graphs of soliton solutions, two-dimensional graphs of soliton solutions, and contour graphs of the M-truncated fractional paraxial wave equation arising in a liquid crystal model are drawn.

Keywords: paraxial model; bifurcation; phase portrait; chaos behavior; M-truncated fractional derivative

Citation: Luo, J.; Li, Z. Dynamic Behavior and Optical Soliton for the M-Truncated Fractional Paraxial Wave Equation Arising in a Liquid Crystal Model. *Fractal Fract.* **2024**, *8*, 348. https://doi.org/10.3390/fractalfract8060348

Academic Editor: Carlo Cattani

Received: 22 May 2024
Revised: 3 June 2024
Accepted: 9 June 2024
Published: 12 June 2024

1. Introduction

Fractional partial differential equations (FPDEs) are partial differential equations [1,2] that involve fractional derivatives and are commonly used to describe complex systems with memory effects. FPDEs have wide applications in fluid mechanics, biology, signal processing, and financial mathematics [3]. In general, due to the widespread application of fractional derivatives in multiple fields [4–8], many researchers have proposed many different fractional derivatives from different perspectives. For example, the Riemann–Liouville fractional derivative, the conformable fractional derivative, the Caputo fractional derivative, and the Grünwald–Letnikov fractional derivative. Therefore, with the continuous development of fractional derivative theory, researchers have also delved deeper into the theory of fractional derivatives. Additionally, many different types of FPDEs have been proposed. Due to the complexity and diversity of fractional derivatives, on the one hand, many experts use the finite element method, the spectral method, and the finite difference method to solve the numerical solutions of these equations. On the other hand, many researchers use mathematical analysis methods to construct exact solutions [9,10] to these equations. The main purpose of this article is to study the dynamic behavior and soliton solutions of a very important class of FPDEs.

In this study, the M-truncated fractional paraxial wave equation arising in a liquid crystal model is presented as follows [11]:

$$i\frac{\partial\psi}{\partial y} + \frac{a_1}{2}\,{}_{\kappa}\mathbb{D}^{2\alpha,d}_{M,t}\psi + \frac{a_2}{2}\frac{\partial^2\psi}{\partial x^2} + a_3|\psi|^2\psi = 0, \tag{1}$$

where a_1, a_2, a_3 are real constants, which represent the coefficients of the dispersal effect, the Kerr nonlinearity effect, and the diffraction effect, respectively. ${}_{\kappa}\mathbb{D}^{2\alpha,d}_{M,t}$ stands for the

M-fractional derivative, which was proposed by Cliveira and Sousa [12]. $\psi = \psi(x,y,t)$, where the variables x, y, t represent the transverse, longitudinal, and temporal propagation, respectively. In [13,14], Hamood, et al. used the ϕ^6 model expansion technique and the Sardar subequation method to study the optical solitons of the paraxial wave model, respectively. In [11], Mannaf et al. studied the optical soliton solutions of Equation (1) by using the extended tanh method and the modified extended tanh method, respectively. However, despite our best efforts, there is still an insufficient body of literature on the dynamic behavior and soliton solutions of Equation (1). Solitons are a special wave phenomenon that occurs in nonlinear physics. Solitons were first proposed in the study of shallow water waves, but were later widely applied in fields such as optics, acoustics, and quantum physics. The optical soliton solution is usually used to describe the mathematical solution of the propagation of optical solitons under specific conditions, for example, the solution to the Schrödinger equation. This article will conduct research from two aspects. On the one hand, by using the method of planar dynamical systems, the dynamic behavior of two-dimensional dynamical systems and their disturbance systems are studied. On the other hand, the optical soliton solution of Equation (1) is constructed using the planar dynamical system method.

The remaining part of this article is arranged as follows: In Section 2, the phase portrait of the two-dimensional dynamical system and its perturbed system are discussed. In Section 3, the optical soliton solutions of Equation (1) are constructed. Finally, a brief conclusion is given.

2. Bifurcation and Chaotic Behaviors

2.1. Preliminary

Definition 1 (M-truncated fractional derivative [15]). *For $\alpha \in (0,1]$, the M-truncated fractional derivative of $f : [0,+\infty) \to (-\infty,+\infty)$ is defined as*

$$ {}_{\kappa}\mathbb{D}_{M,t}^{2\alpha,d}(f) = \lim_{h\to 0} \frac{f(t_{\kappa}E_d(ht^{\alpha})) - f(t)}{h}, \quad 0 < \alpha < 1, d > 0 $$

In Definition 1, $E_d(z)$ represents the truncated Mittag-Leffler function of one parameter, which is defined as

$$ {}_{\kappa}E_d(z) = \sum_{j=0}^{\kappa} \frac{z^j}{\Gamma(dj+1)}, \; z \in [0,+\infty). $$

M-truncated fractional derivative has very important properties, and relevant conclusions can be referenced in reference [15].

2.2. Mathematical Derivation

Firstly, let us introduce the the wave transformation

$$ \psi(x,y,t) = \Psi(\xi)e^{i\eta}, \xi = m_1x + m_2y + \frac{\Gamma(d+1)}{\alpha}\omega t^{\alpha}, \eta = r_1x + r_2y + \frac{\Gamma(d+1)}{\alpha}\tau t^{\alpha} + \delta. \quad (2) $$

Inserting Equation (2) into Equation (1), we obtain the real and imaginary components of the resultant expression

$$ \begin{cases} (a_1\omega^2 + a_2m_1^2)\Psi''(\xi) - 2a_3\Psi^3(\xi) - (a_1\tau^2 + a_2r_1^2 + 2r_2)\Psi(\xi) = 0, \\ (2a_1\tau\omega + 2a_2m_1r_1 + 2m_2)\Psi'(\xi) = 0, \end{cases} \quad (3) $$

where $\Psi'(\xi) \neq 0$.

From the second equation of Equation (3), we have

$$ m_2 = -(a_1\tau\omega + a_2m_1r_1). \quad (4) $$

In order to further analyze the dynamic behavior and soliton solutions of Equation (1), under the conditions satisfied by Equation (3), we can transform the first equation of Equation (3) into the following ordinary differential equation:

$$\Psi'' - \Im_1\Psi^3 - \Im_2\Psi = 0, \tag{5}$$

Here, $\Im_1 = \frac{2a_3}{a_1\omega^2 + a_2 m_1^2}$ and $\Im_2 = \frac{a_1\tau^2 + a_2 r_1^2 + 2r_2}{a_1\omega^2 + a_2 m_1^2}$, where $a_1\omega^2 + a_2 m_1^2 \neq 0$.

2.3. *Qualitative Analysis*

The two-dimensional dynamic system of Equation (5) can be described as follows:

$$\begin{cases} \frac{d\Psi}{d\xi} = u, \\ \frac{du}{d\xi} = \Im_1\Psi^3 + \Im_2\Psi, \end{cases} \tag{6}$$

with its first integral

$$H(\Psi, u) = \frac{1}{2}u^2 - \frac{\Im_1}{4}\Psi^4 - \frac{\Im_2}{2}\Psi^2 = h. \tag{7}$$

Let $F(\Psi_j) = 0$ ($j = 0, 1, 2$) be the abscissa of the equilibrium point, where $F(\Psi_j) = \Im_1\Psi_j^3 + \Im_2\Psi_j$. Assume that $\mathbf{M}(\Psi_j, 0) = \begin{pmatrix} 0 & 1 \\ 3\Im_1\Psi_j^2 + \Im_2 & 0 \end{pmatrix}$ is the coefficient matrix of (6) at the equilibrium point. Then, we obtain

$$\mathbf{det}(\mathbf{M}(\Psi_j, 0)) = -F'(\Psi_j), j = 0, 1, 2. \tag{8}$$

If $\Im_1\Im_2 > 0$, system (6) has one equilibrium point $(0, 0)$ (see Figure 1a,b). If $\Im_1\Im_2 < 0$, the system (6) has three equilibrium points: $(0, 0)$, $(\sqrt{-\frac{\Im_2}{\Im_1}}, 0)$, and $(-\sqrt{-\frac{\Im_2}{\Im_1}}, 0)$ (see Figure 1c,d).

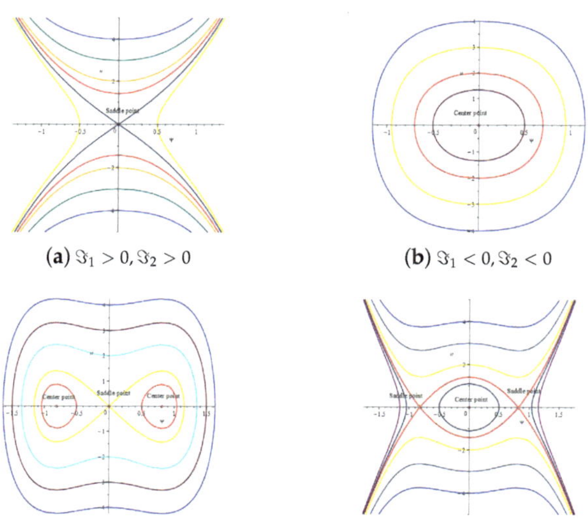

(**a**) $\Im_1 > 0, \Im_2 > 0$ (**b**) $\Im_1 < 0, \Im_2 < 0$

(**c**) $\Im_1 < 0, \Im_2 > 0$ (**d**) $\Im_1 > 0, \Im_2 < 0$

Figure 1. Two-dimensional phase portrait of system (7).

2.4. Qualitative Analysis with Perturbation Term

In this section, we add the following small perturbation term to system (6):

$$\begin{cases} \frac{d\Psi}{d\xi} = u, \\ \frac{du}{d\xi} = \Im_1 \Psi^3 + \Im_2 \Psi + f(\xi), \end{cases} \tag{9}$$

where $f(\xi) = A\sin(\varpi\xi)$ and $f(\xi) = Ae^{-0.05\xi}$ are the perturbed terms. A stands for the amplitude of system (9). ϖ is the frequency of system (9).

By fixing the parameters $\Im_1, \Im_2, A, \varpi$, we have drawn the two-dimensional, three-dimensional, and Paincaré section diagrams of system (9), as shown in Figures 2–5. Specifically, when drawing two-dimensional and three-dimensional phase diagrams, we consider the graphs under different initial values. In Figure 6, we plotted the branch phase diagram of system (9) when ϖ takes different values. Obviously, it can be seen from Figure 6 that when A reaches a critical point, the phase diagram of system (9) exhibits chaotic behavior.

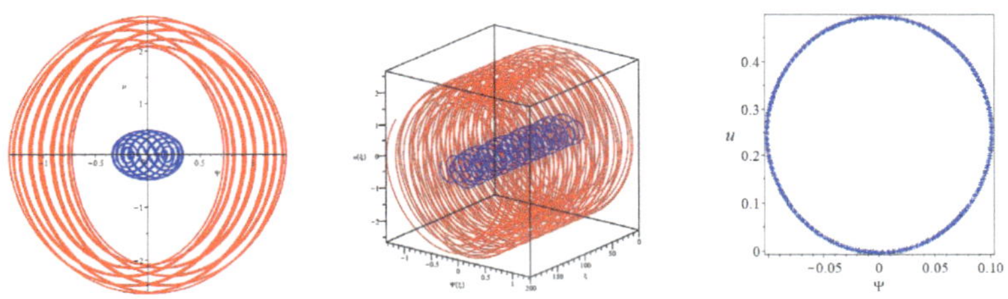

(**a**) 2D phase portrait (**b**) 3D phase portrait (**c**) Paincaré section

Figure 2. The chaotic behaviors of system (9) with $\Im_1 = 1, \Im_2 = -6, A = 1.2, \varpi = 1$.

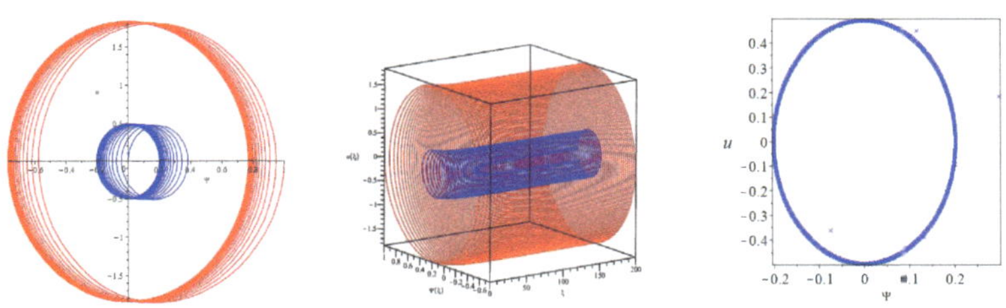

(**a**) 2D phase portrait (**b**) 3D phase portrait (**c**) Paincaré section

Figure 3. The chaotic behaviors of system (9) with $\Im_1 = 1, \Im_2 = -6, A = 1.2$.

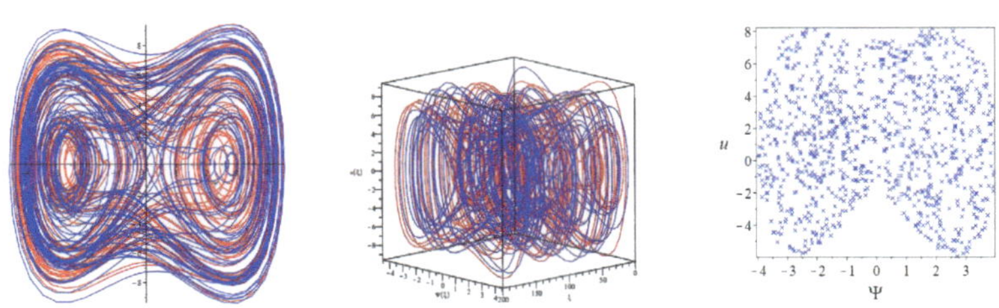

(**a**) 2D phase portrait (**b**) 3D phase portrait (**c**) Paincaré section

Figure 4. The chaotic behaviors of system (9) with $\Im_1 = -1, \Im_2 = 6, A = 6, \varpi = 1$.

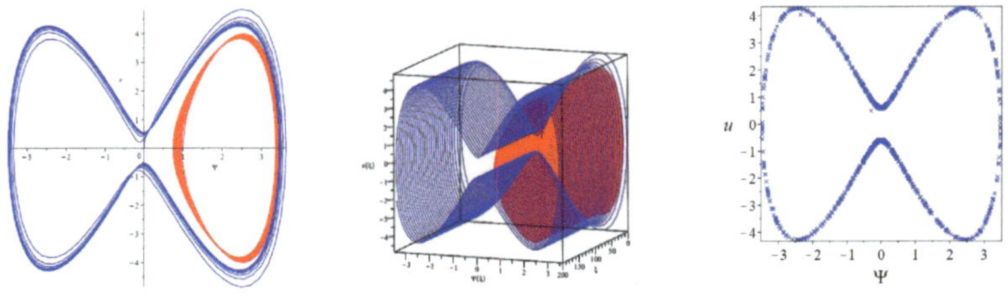

(**a**) 2D phase portrait (**b**) 3D phase portrait (**c**) Paincaré section

Figure 5. The chaotic behaviors of system (9) with $\Im_1 = -1, \Im_2 = 6, A = 1.2$.

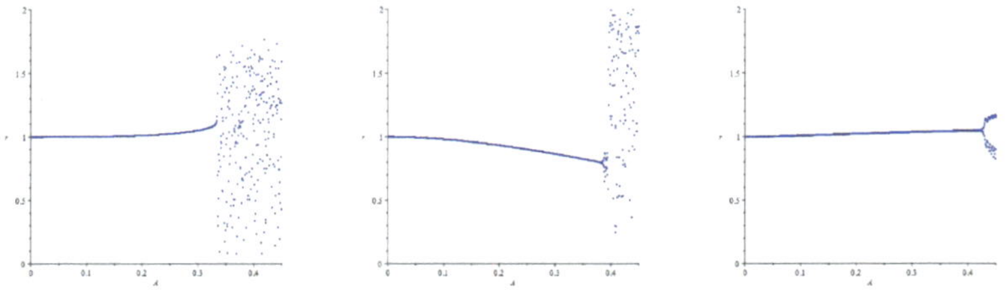

(**a**) $\varpi = \frac{\pi}{8}$ (**b**) $\varpi = \frac{\pi}{4}$ (**c**) $\varpi = \frac{\pi}{2}$

Figure 6. The bifurcation portraits of system (9) with $\Im_1 = -1, \Im_2 = 1, r = \sqrt{u^2 + \Psi^2}$.

3. Optical Soliton Solution of Equation (1)

Let be $h_0 = H(0,0) = 0, h_1 = H(\pm\sqrt{-\frac{\Im_2}{\Im_1}}, 0) = \frac{\Im_2^2}{4\Im_1}$.

3.1. $\Im_1 > 0, \Im_2 < 0, 0 < h < \frac{\Im_2^2}{4\Im_1}$

Then, system (7) becomes

$$u^2 = \frac{\Im_1}{2}(\Psi^4 + \frac{2\Im_2}{\Im_1}\Psi^2 + \frac{4h}{\Im_1}) = \frac{\Im_1}{2}(\varrho_{1h}^2 - \Psi^2)(\varrho_{2h}^2 - \Psi^2), \tag{10}$$

where $\varrho_{1h}^2 = \frac{-\Im_2 + \sqrt{\Im_2^2 - 4\Im_1 h}}{\Im_1}$ and $\varrho_{2h}^2 = \frac{-\Im_2 - \sqrt{\Im_2^2 - 4\Im_1 h}}{\Im_1}$.

Substituting (10) into $\frac{d\Phi}{d\xi} = u$ and integrating it, we can present the Jacobian function solutions

$$\psi_1(x,y,t) = \pm\varrho_{1h}\mathbf{sn}(\varrho_{2h}\sqrt{\frac{\Im_1}{2}}(m_1 x + m_2 y + \frac{\Gamma(d+1)}{\alpha}\omega t^\alpha), \frac{\varrho_{1h}}{\varrho_{2h}})\mathbf{e}^{i(r_1 x + r_2 y + \frac{\Gamma(d+1)}{\alpha}\tau t^\alpha + \delta)}. \tag{11}$$

3.2. $\Im_1 > 0, \Im_2 < 0, h = \frac{\Im_2^2}{4\Im_1}$

When $\varrho_{1h}^2 = \varrho_{2h}^2 = -\frac{\Im_2}{\Im_1}$, we can obtain the soliton solution of (1)

$$\psi_2(x,y,t) = \pm\sqrt{-\frac{\Im_2}{\Im_1}}\mathbf{tanh}(\sqrt{-\frac{\Im_2}{2}}(m_1 x + m_2 y + \frac{\Gamma(d+1)}{\alpha}\omega t^\alpha))\mathbf{e}^{i(r_1 x + r_2 y + \frac{\Gamma(d+1)}{\alpha}\tau t^\alpha + \delta)}. \tag{12}$$

3.3. $\Im_1 < 0, \Im_2 > 0, -\frac{\Im_2^2}{4\Im_1} < h < 0$

Then, system (7) becomes

$$u^2 = -\frac{\Im_1}{2}(-\Psi^4 - \frac{2\Im_2}{\Im_1}\Psi^2 - \frac{4h}{\Im_1}) = -\frac{\Im_1}{2}(\Psi^2 - \varrho_{3h}^2)(\varrho_{4h}^2 - \Psi^2), \tag{13}$$

where $\varrho_{3h}^2 = \frac{-\Im_2 + \sqrt{\Im_2^2 - 4\Im_1 h}}{\Im_1}$ and $\varrho_{4h}^2 = \frac{-\Im_2 - \sqrt{\Im_2^2 - 4\Im_1 h}}{\Im_1}$.

Substituting (13) into $\frac{d\Phi}{d\xi} = u$ and integrating it, we can present the Jacobian function solutions

$$\psi_3(x,y,t) = \pm\varrho_{4h}\mathbf{dn}\Big(\varrho_{4h}\sqrt{-\frac{\Im_1}{2}}\Big(m_1x + m_2y + \frac{\Gamma(d+1)}{\alpha}\omega t^{\alpha}\Big), \frac{\sqrt{\varrho_{4h}^2-\varrho_{3h}^2}}{\varrho_{4h}}\Big) \mathbf{e}^{i(r_1x+r_2y+\frac{\Gamma(d+1)}{\alpha}\tau t^{\alpha}+\delta)}. \tag{14}$$

3.4. $\Im_1 < 0, \Im_2 > 0, h = 0$

When $\varrho_{3h}^2 = 0, \varrho_{4h}^2 == -\frac{2\Im_2}{\Im_1}$, we can obtain the soliton solution of (1)

$$\psi_4(x,y,t) = \pm\sqrt{-\frac{2\Im_2}{\Im_1}}\mathbf{sech}\Big(\sqrt{\Im_2}\Big(m_1x + m_2y + \frac{\Gamma(d+1)}{\alpha}\omega t^{\alpha}\Big)\Big)\mathbf{e}^{i(r_1x+r_2y+\frac{\Gamma(d+1)}{\alpha}\tau t^{\alpha}+\delta)}. \tag{15}$$

3.5. $\Im_1 < 0, \Im_2 > 0, h > 0$

Then, system (7) becomes

$$u^2 = -\frac{\Im_1}{2}\Big(-\Psi^4 - \frac{2\Im_2}{\Im_1}\Psi^2 - \frac{4h}{\Im_1}\Big) = -\frac{\Im_1}{2}(\varrho_{5h}^2 + \Psi^2)(\varrho_{6h}^2 - \Psi^2), \tag{16}$$

where $\varrho_{5h}^2 = \frac{\Im_2-\sqrt{\Im_2^2-4\Im_1 h}}{\Im_1}$ and $\varrho_{6h}^2 = \frac{-\Im_2-\sqrt{\Im_2^2-4\Im_1 h}}{\Im_1}$.

Substituting (16) into $\frac{d\Phi}{d\xi} = u$ and integrating it, we can present the Jacobian function solutions

$$\psi_5(x,y,t) = \pm\varrho_{6h}\mathbf{cn}\Big(\sqrt{-\frac{\Im_1(\varrho_{5h}^2+\varrho_{6h}^2)}{2}}\Big(m_1x + m_2y + \frac{\Gamma(d+1)}{\alpha}\omega t^{\alpha}\Big), \frac{\varrho_{6h}}{\sqrt{\varrho_{5h}^2+\varrho_{6h}^2}}\Big) \mathbf{e}^{i(r_1x+r_2y+\frac{\Gamma(d+1)}{\alpha}\tau t^{\alpha}+\delta)}. \tag{17}$$

3.6. *Numerical Simulations*

In this section, we plotted the solutions $\psi_1(x,y,t)$, including the three-dimensional graph, two-dimensional graph, and counter graph, when $a_1 = 1, a_2 = 1, \omega = 1, m_1 = 1, m_2 = 6, \tau = -7, r_1 = 1, r_2 = -\frac{3}{2}, \alpha = \frac{1}{2}, d = 1, h = \frac{3}{16}$ as shown in Figure 7. Obviously, the solution $\psi_1(x,y,t)$ of Equation (1) is a periodic function solution. We plotted the solutions $\psi_2(x,y,t)$, including the three-dimensional graph, two-dimensional graph, and counter graph, when $a_1 = 1, a_2 = 1, \omega = 1, m_1 = 1, m_2 = 6, \tau = -7, r_1 = 1, r_2 = 1, \alpha = \frac{1}{2}$, as shown in Figure 8. Obviously, the solution $\psi_2(x,y,t)$ of Equation (1) is a kink-like soliton. Moreover, we also plot solutions $\psi_3(x,y,t)$, $\psi_4(x,y,t)$, and $\psi_5(x,y,t)$ of Equation (1), as shown in Figures 9–11.

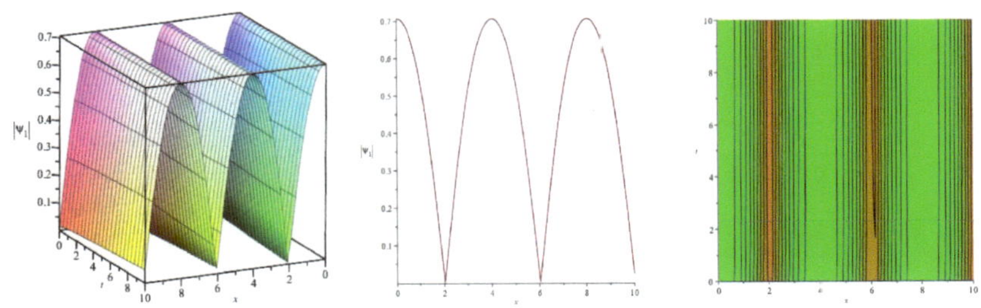

(a) 3D graph **(b)** 2D graph **(c)** Contour graph

Figure 7. The solution $\psi_1(x,y,t)$ with $a_1 = 1, a_2 = 1, a_3 = 1, \omega = 1, m_1 = 1, m_2 = 6, \tau = -7, r_1 = 1, r_2 = -25, \alpha = \frac{1}{2}, d = 1, h = \frac{3}{16}$.

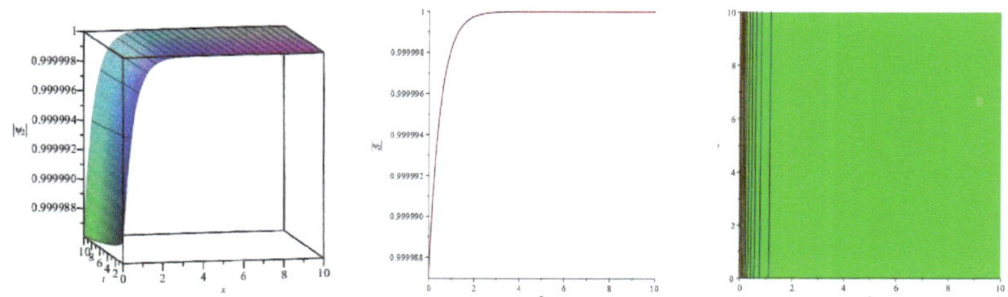

(a) 3D graph **(b)** 2D graph **(c)** Contour graph

Figure 8. The solution $\psi_2(x,y,t)$ with $a_1 = 1, a_2 = 1, a_3 = 1, \omega = 1, m_1 = 1, m_2 = 6, \tau = -7, r_1 = 1, r_2 = 1, d = 1, \alpha = \frac{1}{2}, h = \frac{1}{4}$.

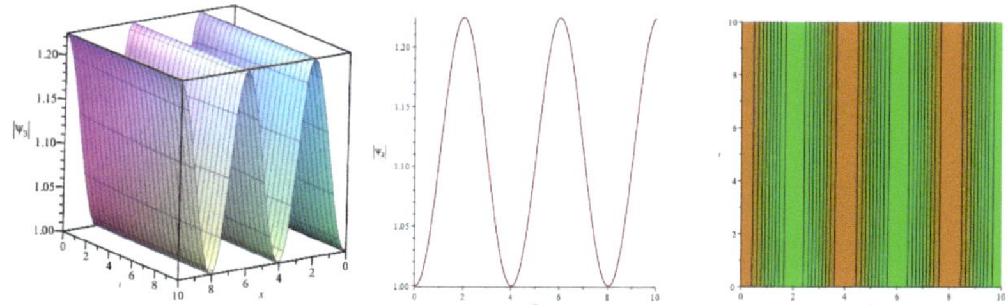

(a) 3D graph **(b)** 2D graph **(c)** Contour graph

Figure 9. The solution $\psi_3(x,y,t)$ with $a_1 = 1, a_2 = 1, a_3 = -1, \omega = 1, m_1 = 1, m_2 = 6, \tau = -7, r_1 = 1, r_2 = -24, d = 1, \alpha = \frac{1}{2}, h = -\frac{3}{16}$.

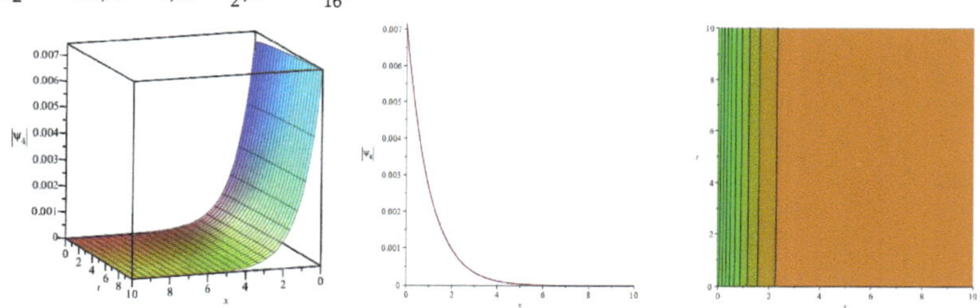

(a) 3D graph **(b)** 2D graph **(c)** Contour graph

Figure 10. The solution $\psi_4(x,y,t)$ with $a_1 = 1, a_2 = 1, a_3 = -1, \omega = 1, m_1 = 1, m_2 = 6, \tau = -7, r_1 = 1, r_2 = -24, d = 1, \alpha = \frac{1}{2}, h = 0$.

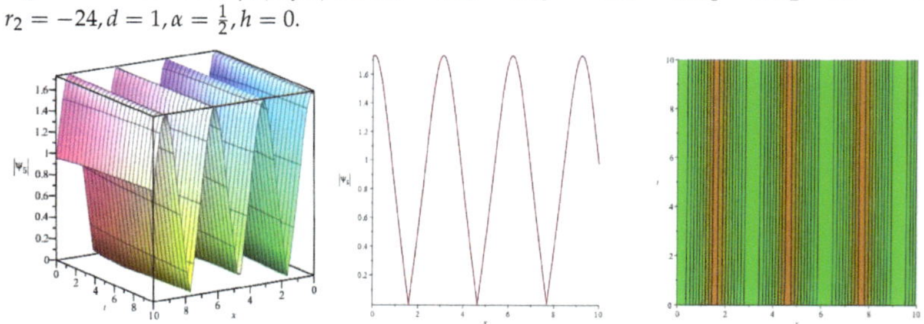

(a) 3D graph **(b)** 2D graph **(c)** Contour graph

Figure 11. The solution $\psi_5(x,y,t)$ with $a_1 = 1, a_2 = 1, a_3 = -1, \omega = 1, m_1 = 1, m_2 = 6, \tau = -7, r_1 = 1, r_2 = -\frac{49}{2}, d = 1, \alpha = \frac{1}{2}, h = \frac{3}{4}$.

4. Conclusions

In this article, we use the theory of dynamical systems to study the dynamic behavior and optical soliton for Equation (1) in a liquid crystal model. Furthermore, we used mathematical software to draw a planar phase diagram of a two-dimensional dynamical system, and we can easily obtain the characteristics of some equilibrium points of the planar dynamical system from the planar phase diagram. And by adding small perturbations, the dynamic behavior of the two-dimensional system is analyzed. Based on different initial values, we have drawn planar phase diagrams using red and blue colors in the same coordinate system. From the perspective of plane dynamics theory, we have drawn a bifurcation phase diagram and Poincaré sections of a disturbance system. And we separately considered the dynamic behavior under periodic and small disturbances. In future research, we will consider the dynamic behavior and optical soliton solutions of more complex FPDEs.

Author Contributions: Software, Z.L.; writing—original draft preparation, J.L.; writing—review and editing, Z.L. All authors have read and agreed to the published version of the manuscript.

Funding: This research received no external funding.

Data Availability Statement: Data are contained within the article.

Conflicts of Interest: The authors declare no conflicts of interest.

References

1. Wu, J.; Yang, Z. Global existence and boundedness of chemotaxis-fluid equations to the coupled Solow-Swan model. *AIMS Math.* **2023**, *8*, 17914–17942. [CrossRef]
2. Wu, J.; Huang, Y.J. Boundedness of solutions for an attraction-repulsion model with indirect signal production. *Mathematics* **2024**, *12*, 1143. [CrossRef]
3. Tang, L. Dynamical behavior and multiple optical solitons for the fractional Ginzburg-Landau equation with β-derivative in optical fibers. *Opt. Quant. Electron.* **2024**, *56*, 175. [CrossRef]
4. Wang, Y.; Qian, Z. Regularizing a two-dimensional time-fractional inverse heat conduction problem by a fractional Landweber iteration method. *Comput. Math. Appl.* **2024**, *164*, 104–115. [CrossRef]
5. Jornet, M. On the Cauchy-Kovalevskaya theorem for Caputo fractional differential equations. *Physica D* **2024**, *462*, 134139. [CrossRef]
6. Yu, J.C.; Feng, Y.Q. On the generalized time fractional reaction-diffusion equation: Lie symmetries, exact solutions and conservation laws. *Chaos Solitons Fractals* **2024**, *182*, 114855. [CrossRef]
7. Espinosa-Paredes, G.; Cruz-López, G.A. A new compartmental fractional neutron point kinetic equations with different fractional orders. *Nucl. Eng. Des.* **2024**, *423*, 113184. [CrossRef]
8. Lu, Y.S.; Hu, Y.Z.; Qiao, Y.; Yuan, M.J.; Xu, W. Sparse least squares via fractional function group fractional function penalty for the identification of nonlinear dynamical systems. *Chaos. Soliton. Fract.* **2024**, *182*, 114733. [CrossRef]
9. Liu, C.Y.; Li, Z. The dynamical behavior analysis and the traveling wave solutions of the stochastic Sasa-Satsuma Equation. *Qual. Theor. Dyn. Syst.* **2024**, *23*, 157. [CrossRef]
10. Gu, M.S.; Chen Peng, C.; Li, Z. Traveling wave solution of (3+1)-dimensional negative-order KdV-Calogero-Bogoyavlenskii-Schiff equation. *AIMS Math.* **2023**, *9*, 6699–6708. [CrossRef]
11. Mannaf, M.A.; Islam, M.E.; Bashar, H.; Basak, U.S.; Akbar, M.A. Dynamical behavior of optical self-control solion in a liquid crystal model. *Results Phys.* **2024**, *57*, 107324. [CrossRef]
12. Usman, Y.; Abdulkadir, S.T.; Ren, J.L. Propagation of M-truncated optical pulses in nonlinear optics. *Opt. Quant. Electron.* **2023**, *55*, 102.
13. Rehman, H.U.; Awan, A.U.; Allahyani, S.A.; Tag-ElDin, E.M.; Binyamin, M.A.; Yasin, S. Exact solution of paraxial wave dynamical model with kerr media by using ϕ^6 model expansion technique. *Results Phys.* **2022**, *42*, 105975. [CrossRef]
14. Rehman, H.U.; Seadawy, A.R.; Younis, M.; Yasi, S.; Raza, S.T.R.; Althobaiti, S. Monochromatic optical beam propagation of paraxial dynamical model in kerr media. *Results Phys.* **2021**, *31*, 105015. [CrossRef]
15. Roshid, M.M.; Uddin, M.; Mostafa, G. Dynamical structure of optical solution for M-fractional paraxial wave equation by using unified technique. *Results Phys.* **2023**, *51*, 106632. [CrossRef]

fractal and fractional

MDPI

Article

Positive Solutions and Their Existence of a Nonlinear Hadamard Fractional-Order Differential Equation with a Singular Source Item Using Spectral Analysis

Cheng Li [1] and Limin Guo [2,*]

[1] School of Automotive Engineering, Changzhou Institute of Technology, Changzhou 213032, China; licheng@czu.cn
[2] School of Science, Changzhou Institute of Technology, Changzhou 213032, China
* Correspondence: guolm@czu.cn

Abstract: Based on the spectral analysis method, Gelfand's formula, and the cones fixed point theorem, some positive solutions with their existence of a nonlinear infinite-point Hadamard fractional-order differential equation is achieved on the interval $[a, b]$ under some conditions, and particularly the nonlinear term allows singularities for time and spatial parameters in the present study. Finally, an analysis case is carried out to reveal the principal results.

Keywords: hadamard; fractional-order differential; positive solution; infinite-point; specral analysis

MSC: 70K50; 34B18

Citation: Li, C.; Guo, L. Positive Solutions and Their Existence of a Nonlinear Hadamard Fractional -Order Differential Equation with a Singular Source Item Using Spectral Analysis. *Fractal Fract.* **2024**, *8*, 377. https://doi.org/10.3390/fractalfract8070377

Academic Editor: Riccardo Caponetto

Received: 27 May 2024
Revised: 23 June 2024
Accepted: 25 June 2024
Published: 26 June 2024

1. Introduction

The fractional-order differential model can exactly depict the physical and mechanical process containing historical memories and thus establish spatial global relationship briefly. Accordingly, the fractional-order differential model is concise in form and the parameter meaning is obvious, so it becomes one of the effective approaches for modeling complex mechanical and physical behaviors. Boulham et al. [1] proposed an adaptive monitor for projective chaotic synchronization of general class fractional-order systems with chaotic uncertainty affected by unknown input nonlinear factors, and the closed-loop stability was strictly verified by two simulation examples and related comparative studies. Wu et al. [2] used a new nonlinear fractional-order damaged pattern with viscosity, elasticity and plasticity for rock and soil materials to characterize three-stage creep behaviors, and their pattern fully calculated and predicted the deformation with hysteresis derived from the rapid creep in tunnel engineering. Ma et al. [3] developed a fractional-order model termed as SEIR type and subsequently the beingness, uniqueness, boundness of such a fractional-order system were determined. The research conclusions indicated that it is difficult for humans to avoid coexistence with the spread of the COVID-19 epidemic. Yang et al. [4] designed a fractional controller of EV hybrid energy storage system that can significantly improve the control performance with good robustness. In addition, fractional-order models have been found to be extensively applied in photovoltaic panels [5], electromagnetic waves [6], electric circuits [7], viscoelastic materials [8], chemical reactions [9] and biological systems [10,11].

Owing to its so many advantages and fine results while simulating the system, the fractional-order differential model has received increasingly research attentions during the past several years. How to get the existence result of positive solution for the fractional-order differential model as well as its related nonlinear dynamics becomes a hot research direction in the field of fractional-order differential models, and many meaningful results are achieved in recent years [12–17]. Xu et al. [18] depicted the positive solutions of a kind of fractional-order differential models as follows

$$^{H}D^{\alpha}_{a+}x(t) + \hbar(t, x(t)) = 0, t \in (a, b),$$

with the boundary value condition

$$x(a) = x'(a) = 0, x(b) = \int_a^b \chi(t)x(t)\frac{dt}{t},$$

where α, a, b are real positive mumbers, $2 < \alpha < 3, a < b < +\infty$, ${}^{H}D^{\alpha}_{a+}$ is the Hadamard fractional derivative of order α, $\chi : [a, b] \to \mathbb{R}^+$ with $\chi(t) \not\equiv 0, t \in [a, b]$, and $\int_a^b \chi(t)(\ln \frac{t}{a})^{\alpha-1}\frac{dt}{t} \in [0, (\ln \frac{b}{a})^{\alpha-1})$, $\hbar \in C([a, b] \times \mathbb{R}^+, \mathbb{R}^+)$. Arul and Karthikeyan [19] discussed the existence and uniqueness of solutions with some integral boundary conditions of implicit Hadamard differential equation

$$-^{H} D^{\kappa}_{a+}v(t) = g(t, v(t), ^{H} D^{\kappa}_{a+}v(t)),\ a < t < b,$$

with boundary value condition

$$v(a) = 0, v(b) = \varsigma \int_0^{\sigma} v(s)ds, a < \sigma < b, \varsigma \in \mathbb{R},$$

where ${}^{H}D^{\kappa}_{a+}v$ is the standard Hadamard fractional derivative of order $1 < \kappa \leq 2$, Arul and Karthikeyan [19] investigated the existence and uniqueness based on Banach and Schauder's fixed point theorem. Wang et al. [20] studied a fractional-order iterative functional differential equation containing parameter on the interval $[a, b]$, showed as

$$^{c}D^{\alpha}_{a+}u(t) = f(t, u(t), u(u^{\nu}(t))) + m, t \in [a, b], \nu \in R\backslash\{0\}, q \in (0, 1), \lambda \in R,$$

with boundary conditions

$$u(t) = \varphi(t), t \in [a_1, a], u(t) = \psi(t), t \in [b, b_1],$$

where ${}^{c}D^{q}_{a}$ is the Caputo fractional derivative of order q with the oower limit a, $a_1 \leq a < b \leq b_1, a_1 \leq a_1^{v}, b_1^{v} \leq b_1$, $f \in C([a, b] \times [a_1, b_1]^2, R), \varphi \in C([a_1, a], [a_1, b_1])$ and $\psi \in C([b, b_1], [a_1, b_1])$. The existence theorem is established by the method of Schauder's fixed point theorems [20], and Wang et al. [20] obtained data dependence of solutions and related parameters.

In this article, an infinite-point Hadamard fractional-order differential equation is taken into account, shown as

$$-^{H} D^{\flat}_{a+}\mathscr{V}(t) + \varrho(t)\ell\left(t, \mathscr{V}(t), ^{H} D^{\mu}_{a+}\mathscr{V}(t)\right) = 0,\ a < t < b, \tag{1.1}$$

with boundary value conditions

$$^{H}D^{\mu}_{a+}\mathscr{V}(a) = ^{H} D^{\mu+1}_{a+}\mathscr{V}(a) = 0, ^{H} D^{\mu}_{a+}\mathscr{V}(b) = \sum_{j=1}^{\infty} \kappa_j {}^{H}D^{\mu}_{a+}\mathscr{V}(\varsigma_j), \tag{1.2}$$

where $2.5 < \flat \leq 3$, $\kappa_j \geq 0$, $0 < \mu < \frac{1}{2}$, $1 < \varsigma_1 < \varsigma_2 < \cdots < \varsigma_{j-1} < \varsigma_j < \cdots < e (j = 1, 2 \cdots)$, $\sum_{j=1}^{\infty} \kappa_j \left(\frac{\ln \varsigma_j - \ln s}{\ln b - \ln s}\right)^{\flat-1} < 1$ and $\sum_{j=1}^{\infty} \kappa_j (\ln \frac{\varsigma_j}{a})^{\flat-1} < (\ln \frac{b}{a})^{\flat-1}$, $\varrho(t)$ is singular at $t = a$ or/and $t = b$, $\ell : [a, b] \times \mathbb{R} \times \mathbb{R} \to \mathbb{R}$ is a given function and $\ell(t, x, y)$ is continuous, and ${}^{H}D^{\flat}_{1+}$ is the standard Hadamard derivative.

Compared with [18,21], the present study contains the the derivative term in the nonlinear term of the equation and we will deal with this difficulty, moreover, and involved infinite points in the boundary conditions. Compared with [19], the nonlinear term is singular in this article, and the method we used is spectral analysis and infinite-points are contained in the boundary conditions. Compared with [22,23], the interval we discussed is arbitrarily closed interval $[a,b]$, the interval $[1,T]$ or $[0,1]$ in [22,23] are special cases of interval $[a,b]$, considering the solution of the equation on this interval will bring about a series of problems, and we overcome this difficulty.

2. Preliminaries And Lemmas

Some important definitions and lemmas which may be adopted during the certification in the present study, could be reviewed in the newly published papers, e.g., see [24,25], we introduce some of them herein.

Definition 1 ([24,25]). *Let $a>0$, then the Hadamard-type fractional left integral of order $\beta>0$ of a function $h:[a,\infty)\to\mathbb{R}$ is defined by*

$$ {}^{H}I^{\beta}_{a^+}h(t)=\frac{1}{\Gamma(\beta)}\int_a^t(\ln\frac{t}{\varsigma})^{\beta-1}h(\varsigma)\frac{d\varsigma}{\varsigma},t\geq a. $$

Definition 2 ([24,25]). *Let $a>0$, $h:[a,\infty)\to\mathbb{R}, t^{n-1}h^{(n-1)}(t)\in AC[a,\infty), n\in\mathbb{N}$, $\beta\in(n-1,n)$, then the Hadamard fractional left derivative of from $[a,+\infty)$ is defined by*

$$ {}^{H}D^{\beta}_{a^+}h(t)=\frac{1}{\Gamma(n-\beta)}(t\frac{d}{dt})^n\int_a^t(\ln\frac{t}{\varsigma})^{n-\beta-1}h(\varsigma)\frac{d\varsigma}{\varsigma},t>a. $$

Lemma 1 ([24,25]). *For $\beta\in(n-1,n), n\in\mathbb{N}, h\in L[a,\infty), a>0$, the fractional equation ${}^{H}D^{\beta}_{a^+}h(t)+\omega(t)=0, t>a$ has expression of solution*

$$ x(t)=\sum_{i=1}^{n}\chi_i(\ln\frac{t}{a})^{\beta-i}-\frac{1}{\Gamma(\beta)}\int_a^t(\ln\frac{t}{\varsigma})^{\beta-1}\omega(\varsigma)\frac{d\varsigma}{\varsigma},t\geq a, $$

where $\chi_i\in\mathbb{R}, k=1,2,\ldots,n$.

Let $\mathscr{U}(t)={}^{H}D^{\mu}_{a^+}\mathscr{V}(t), \mathscr{V}(t)\in C[a,b]$, then the BVP (1.1,1.2) can be reduced to a modified model

$$ -{}^{H}D^{\flat-\mu}_{a^+}\mathscr{U}(t)+\varrho(t)\ell\Big(t,{}^{H}I^{\mu}_{a^+}\mathscr{U}(t),\mathscr{U}(t)\Big)=0,\ a<t<b, \tag{2.1} $$

with boundary value condition

$$ \mathscr{U}(a)=\mathscr{U}'(a)=0,\mathscr{U}(b)=\sum_{j=1}^{\infty}\kappa_j\mathscr{U}(\varsigma_j). \tag{2.2} $$

Lemma 2. *Given $\mathscr{A}\in L^1(a,b)$, then the model*

$$ {}^{H}D^{\flat-\mu}_{1^+}\mathscr{U}(t)+\mathscr{A}(t)=0,\ a<t<b, \tag{2.3} $$

with boundary value condition (2.2) can be expressed by

$$ \mathscr{U}(t)=\int_a^b\mathscr{H}(t,s)\mathscr{A}(s)\frac{ds}{s},\ t\in[a,b], \tag{2.4} $$

where

$$ \mathscr{H}(t,s)=\mathscr{G}(t,s)+\frac{(\ln\frac{t}{a})^{\flat-\mu-1}}{\Delta}\sum_{j=1}^{\infty}\kappa_j\mathscr{G}(\varsigma_j,s), $$

$$ \mathscr{G}(t,s)=\frac{1}{(\ln\frac{b}{a})^{\flat-\mu-1}\Gamma(\flat-\mu)}\begin{cases}(\ln t-\ln a)^{\flat-\mu-1}(\ln b-\ln s)^{\flat-\mu-1}\\ -(\ln t-\ln s))^{\flat-\mu-1}(\ln\frac{b}{a})^{\flat-\mu-1},\ a\leq s\leq t\leq b,\\ (\ln t-\ln a)^{\flat-\mu-1}(\ln b-\ln s)^{\flat-\mu-1},\ a\leq t\leq s\leq b,\end{cases} \tag{2.5} $$

in which $\Delta = (\ln \frac{b}{a})^{\flat-\mu-1} - \sum_{j=1}^{\infty} \kappa_j (\ln \frac{\varsigma_j}{a})^{\flat-\mu-1}$.

Proof. By Lemma 1, Formula (2.3) is changed into an equivalent integral equation

$$\mathscr{U}(t) = -^{H}I_{a^+}^{\flat-\mu}\mathscr{A}(t) + d_1(\ln\frac{t}{a})^{\flat-\mu-1} + d_2(\ln\frac{t}{a})^{\flat-\mu-2} + d_3(\ln\frac{t}{a})^{\flat-\mu-3}.$$

From $u(a) = 0$, we have $d_3 = 0$, then

$$\mathscr{U}'(t) = -^{H}I_{a^+}^{\flat-\mu-1}\mathscr{A}(t) + d_1(\flat-\mu-1)(\ln\frac{t}{a})^{\flat-\mu-2}\cdot\frac{a}{t} + d_2(\flat-\mu-2)(\ln\frac{t}{a})^{\flat-\mu-3}\frac{a}{t},$$

by $\mathscr{U}'(a) = 0$, we have $d_2 = 0$. Hence, we get

$$\begin{aligned}\mathscr{U}(t) =& d_1(\ln\frac{t}{a})^{\flat-\mu-1} -^{H} I_{a^+}^{\flat-\mu}\mathscr{A}(t)\\ =& -\frac{1}{\Gamma(\flat-\mu)}\int_a^t(\ln t - \ln s)^{\flat-\mu-1}\mathscr{A}(s)\frac{ds}{s} + d_1(\ln\frac{t}{a})^{\flat-\mu-1},\end{aligned}$$

therefore, $\mathscr{U}(\xi_j) = -\frac{1}{\Gamma(\flat-\mu)}\int_a^{\varsigma_j}(\ln\xi_j - \ln s)^{\flat-\mu-1}\mathscr{A}(s)\frac{ds}{s} + d_1(\ln\frac{\varsigma_j}{a})^{\flat-\mu-1}$. Moreover, by $\mathscr{U}(b) = \sum_{j=1}^{\infty}\kappa_j\mathscr{U}(\varsigma_j)$, we have

$$\begin{aligned}\mathscr{U}(b) =& -\frac{1}{\Gamma(\flat-\mu)}\int_a^b(\ln b - \ln s)^{\flat-\mu-1}\mathscr{A}(s)\frac{ds}{s} + d_1(\ln\frac{b}{a})^{\flat-\mu-1}\\ =& \sum_{j=1}^{\infty}\kappa_j\Big(-\frac{1}{\Gamma(\flat-\mu)}\int_a^{\varsigma_j}(\ln\varsigma_j - \ln s)^{\flat-\mu-1}\mathscr{A}(s)\frac{ds}{s} + d_1(\ln\frac{\varsigma_j}{a})^{\flat-\mu-1}\Big).\end{aligned}$$

Then, we get

$$d_1\Bigg((\ln\frac{b}{a})^{\flat-\mu-1} - \sum_{j=1}^{\infty}\kappa_j(\ln\frac{\varsigma_j}{a})^{\flat-\mu-1}\Bigg) = \frac{1}{\Gamma(\flat-\mu)}\int_a^b(\ln b - \ln s)^{\flat-\mu-1}\mathscr{A}(s)\frac{ds}{s}$$
$$-\frac{1}{\Gamma(\flat-\mu)}\sum_{j=1}^{\infty}\kappa_j\int_a^{\varsigma_j}(\ln\varsigma_j - \ln s)^{\flat-\mu-1}\mathscr{A}(s)\frac{ds}{s},$$

thus, we have

$$d_1 = \frac{1}{\Gamma(\flat-\mu)\Delta}\int_a^b(\ln b - \ln s)^{\flat-\mu-1}\mathscr{A}(s)\frac{ds}{s} - \frac{1}{\Gamma(\flat-\mu)\Delta}\sum_{j=1}^{\infty}\kappa_j\int_a^{\varsigma_j}(\ln\varsigma_j - \ln s)^{\flat-\mu-1}\mathscr{A}(s)\frac{ds}{s},$$

hence,

$$
\begin{aligned}
\mathscr{U}(t) =& \frac{(\ln\frac{t}{a})^{\flat-\mu-1}}{\Gamma(\flat-\mu)\Delta}\int_a^b(\ln b-\ln s)^{\flat-\mu-1}\mathscr{A}(s)\frac{ds}{s}-\frac{(\ln\frac{t}{a})^{\flat-\mu-1}}{\Gamma(\flat-\mu)\Delta}\sum_{j=1}^{\infty}\kappa_j\int_a^{\varsigma_j}(\ln\varsigma_j-\ln s)^{\flat-\mu-1}\mathscr{A}(s)\frac{ds}{s}\\
&-{}^{H}I_{a^+}^{\flat-\mu}\mathscr{A}(t)\\
=& \frac{(\ln\frac{t}{a})^{\flat-\mu-1}}{\Gamma(\flat-\mu)\Delta}\int_a^b(\ln b-\ln s)^{\flat-\mu-1}\mathscr{A}(s)\frac{ds}{s}-\frac{(\ln\frac{t}{a})^{\flat-\mu-1}}{\Gamma(\flat-\mu)\Delta}\sum_{j=1}^{\infty}\kappa_j\int_a^{\varsigma_j}(\ln\varsigma_j-\ln s)^{\flat-\mu-1}\mathscr{A}(s)\frac{ds}{s}\\
&-\frac{1}{\Gamma(\flat-\mu)}\int_a^t(\ln t-\ln s)^{\flat-\mu-1}\mathscr{A}(s)\frac{ds}{s}\\
&+\frac{1}{\Gamma(\flat-\mu)(\ln\frac{b}{a})^{\flat-\mu-1}}\int_a^b(\ln\frac{t}{a})^{\flat-\mu-1}(\ln\frac{b}{s})^{\flat-\mu-1}\mathscr{A}(s)\frac{ds}{s}\\
&-\frac{1}{\Gamma(\flat-\mu)(\ln\frac{b}{a})^{\flat-\mu-1}}\int_a^b(\ln\frac{t}{a})^{\flat-\mu-1}(\ln\frac{b}{s})^{\flat-\mu-1}\mathscr{A}(s)\frac{ds}{s}\\
=&\int_a^b\mathscr{G}(t,s)\mathscr{A}(s)\frac{ds}{s}+\frac{1}{\Gamma(\flat-\mu)\Delta}\int_a^b(\ln\frac{t}{a})^{\flat-\mu-1}(\ln b-\ln s)^{\flat-\mu-1}\mathscr{A}(s)\frac{ds}{s}\\
&-\frac{1}{\Gamma(\flat-\mu)(\ln\frac{b}{a})^{\flat-\mu-1}}\int_a^b(\ln\frac{t}{a})^{\flat-\mu-1}(\ln\frac{b}{s})^{\flat-\mu-1}\mathscr{A}(s)\frac{ds}{s}\\
&-\frac{(\ln\frac{t}{a})^{\flat-\mu-1}}{\Gamma(\flat-\mu)\Delta}\sum_{j=1}^{\infty}\kappa_j\int_a^{\varsigma_j}(\ln\varsigma_j-\ln s)^{\flat-\mu-1}\mathscr{A}(s)\frac{ds}{s}\\
=&\int_a^b\mathscr{G}(t,s)\mathscr{A}(s)\frac{ds}{s}+\frac{(\ln\frac{b}{a})^{\flat-\mu-1}-\Delta}{\Delta\Gamma(\flat-\mu)(\ln\frac{b}{a})^{\flat-\mu-1}}\int_a^b(\ln\frac{t}{a})^{\flat-\mu-1}(\ln\frac{b}{s})^{\flat-\mu-1}\mathscr{A}(s)\frac{ds}{s}\\
&-\frac{(\ln\frac{t}{a})^{\flat-\mu-1}}{\Gamma(\flat-\mu)\Delta}\sum_{j=1}^{\infty}\kappa_j\int_a^{\varsigma_j}(\ln\varsigma_j-\ln s)^{\flat-\mu-1}\mathscr{A}(s)\frac{ds}{s}\\
=&\int_a^b\mathscr{G}(t,s)\mathscr{A}(s)\frac{ds}{s}\\
&+\frac{(\ln\frac{t}{a})^{\flat-\mu-1}}{\Delta}\Bigg(\frac{1}{\Gamma(\flat-\mu)(\ln\frac{b}{a})^{\flat-\mu-1}}\sum_{j=1}^{\infty}\kappa_j\int_a^b(\ln\frac{\varsigma_j}{a})^{\flat-\mu-1}(\ln\frac{b}{s})^{\flat-\mu-1}\mathscr{A}(s)\frac{ds}{s}\\
&-\frac{1}{\Gamma(\flat-\mu)}\sum_{j=1}^{\infty}\kappa_j\int_a^{\varsigma_j}(\ln\varsigma_j-\ln s)^{\flat-\mu-1}\mathscr{A}(s)\frac{ds}{s}\Bigg)\\
=&\int_a^b\mathscr{G}(t,s)\mathscr{A}(s)\frac{ds}{s}\\
&+\frac{(\ln\frac{t}{a})^{\flat-\mu-1}}{\Delta}\sum_{j=1}^{\infty}\kappa_j\Bigg(\frac{1}{\Gamma(\flat-\mu)(\ln\frac{b}{a})^{\flat-\mu-1}}\int_a^b(\ln\frac{\varsigma_j}{a})^{\flat-\mu-1}(\ln\frac{b}{s})^{\flat-\mu-1}\mathscr{A}(s)\frac{ds}{s}\\
&-\frac{1}{\Gamma(\flat-\mu)}\int_a^{\varsigma_j}(\ln\varsigma_j-\ln s)^{\flat-\mu-1}\mathscr{A}(s)\frac{ds}{s}\Bigg)\\
=&\int_a^b\mathscr{H}(t,s)\mathscr{A}(s)\frac{ds}{s}.
\end{aligned}
$$

Therefore, the expression (2.4) is ritht. □

Lemma 3. *The properties of the Green function (2.4) are as follows:*

(i) $\mathscr{H}(t,s)\leq\frac{\varpi(s)}{\Gamma(\flat-\mu)(\ln\frac{b}{a})}w;$

(ii) $\mathscr{H}(t,s)\geq\frac{\vartheta(t)\varpi(s)}{\Gamma(\flat-\mu)(\ln\frac{b}{a})^{\flat-\mu+1}},$

where

$$\vartheta(t) = (\ln\frac{t}{a})^{\flat-\mu-1}(\ln\frac{b}{t}), \varpi(t) = (\ln\frac{t}{a})(\ln\frac{b}{t})^{\flat-\mu-1}, w = 1 + \frac{(\ln\frac{b}{a})^{\flat-\mu-1}}{\Delta}\sum_{j=1}^{\infty}\kappa_j. \quad (2.6)$$

Proof. From Lemma 5 of [26], we get

$$\mathscr{G}(t,s) \le \frac{\varpi(s)}{\Gamma(\flat-\mu)(\ln\frac{b}{a})}, \mathscr{G}(t,s) \ge \frac{\vartheta(t)\varpi(s)}{\Gamma(\flat-\mu)(\ln\frac{b}{a})^{\flat-\mu+1}}, \quad (2.7)$$

where $\vartheta(t), \varpi(t)$ is as Lemma 3, $\mathscr{G}(t,s)$ is from (2.5). By (2.7), we get

$$\begin{aligned}
\mathscr{H}(t,s) =& \mathscr{G}(t,s) + \frac{(\ln\frac{t}{a})^{\flat-\mu-1}}{\Delta}\sum_{j=1}^{\infty}\kappa_j\mathscr{G}(\xi_j,s) \\
\le& \frac{\varpi(s)}{\Gamma(\flat-\mu)(\ln\frac{b}{a})} + \frac{(\ln\frac{b}{a})^{\flat-\mu-1}}{\Delta}\sum_{j=1}^{\infty}\kappa_j\frac{\varpi(s)}{\Gamma(\flat-\mu)(\ln\frac{b}{a})} \\
\le& \frac{\varpi(s)}{\Gamma(\flat-\mu)(\ln\frac{b}{a})}\left(1 + \frac{(\ln\frac{b}{a})^{\flat-\mu-1}}{\Delta}\sum_{j=1}^{\infty}\kappa_j\right) \\
=& \frac{\varpi(s)}{\Gamma(\flat-\mu)(\ln\frac{b}{a})}w.
\end{aligned}$$

On the other hand, we can calculate that

$$\mathscr{H}(t,s) \ge \mathscr{G}(t,s) \ge \frac{\vartheta(t)\varpi(s)}{\Gamma(\flat-\mu)(\ln\frac{b}{a})^{\flat-\mu+1}},$$

then we have (ii) of Lemma 3. As a consequent, we complete the the proof of Lemma 3. □

Let $E = C[a,b]$, $\|\mathscr{U}\| = \max_{a\le t\le b}|\mathscr{U}(t)|$, then $(\mathscr{E}, \|.\|)$ is a Banach space. In this paper,

$$\mathscr{P} = \{u \in E : \mathscr{U}(t) \ge 0,\ t \in [a,b]\},$$

$$\mathscr{K} = \left\{\mathscr{U} \in P : \mathscr{U}(t) \ge \frac{\vartheta(t)}{(\ln\frac{b}{a})^{\flat-\mu}w}\|\mathscr{U}\|,\ t \in [a,b]\right\},$$

where $w = 1 + \frac{(\ln\frac{b}{a})^{\flat-\mu-1}}{\Delta}\sum_{j=1}^{\infty}\kappa_j$. Apparently, $\mathscr{K}$ is a subcone of P, and $(\mathscr{E},\mathscr{K})$ has an ordering relation. Let $\mathscr{K}_r = \{\mathscr{U} \in \mathscr{K} : \|\mathscr{U}\| < r\}$, $\partial\mathscr{K}_r = \{\mathscr{U} \in \mathscr{K} : \|\mathscr{U}\| = r\}$ and $\overline{\mathscr{K}}_r = \{\mathscr{U} \in \mathscr{K} : \|\mathscr{U}\| \le r\}$.

Now the following conditions is listed which will be used later.

(A_1) $\varrho : (a,b) \to \mathbb{R}_+^1$ is nonnegative, $\varrho(t) \not\equiv 0$ and $\varrho(t)$may be singular at $t = a$, $t = b$, and

$$\int_a^b \varpi(s)\varrho(s)\frac{ds}{s} < +\infty.$$

(A_2) $\ell : [a,b] \times (0,+\infty) \times (0,+\infty) \to \mathbb{R}_+^1$ is continuous, and for any $0 < \imath < \jmath < +\infty$,

$$\begin{aligned}
&\limsup_{m\to+\infty}\left\{\sup\int_{e(m)}\varpi(s)\varrho(s)\ell(s,x_1(s),x_2(s))\frac{ds}{s}\right|, \\
&x_1 \in \overline{\mathscr{K}}_{\bar{\jmath}} \setminus \mathscr{K}_\imath,\ x_2 \in \overline{\mathscr{K}}_\jmath \setminus \mathscr{K}_\imath\} = 0.
\end{aligned}$$

where $e(m) = [0, \frac{1}{m}] \cup [\frac{m-1}{m}, 1]$, $\bar{\jmath} = \frac{1}{\Gamma(\mu+1)}(\ln\frac{b}{a})^{\mu}\jmath$.

Nonlinear operator $A : \mathscr{K} \setminus \{0\} \to \mathscr{P}$ and linear operator $\mathscr{T} : \mathscr{E} \to \mathscr{E}$ are defined, shown as

$$(A\mathscr{U})(t) = \int_a^b \mathscr{H}(t,s)\varrho(s)\ell\Big(s, {}^H I^{\mu}_{a^+}\mathscr{U}(s), \mathscr{U}(s)\Big)\frac{ds}{s},\ t \in [a,b], \tag{2.8}$$

$$(\mathscr{T}\mathscr{U})(t) = \int_a^b \mathscr{H}(t,s)\varrho(s)\mathscr{U}(s)\frac{ds}{s},\ t \in [a,b]. \tag{2.9}$$

Lemma 4 ([27]). *Assume $\mathscr{T} : \mathscr{E} \to \mathscr{E}$, and $\mathscr{T}$ be a linear operator and is continuous, $\mathscr{T}(\mathscr{K}) \subset \mathscr{K}$, where $\mathscr{K}$ is a total cone. If there exist a positive constant d and $\zeta \in \mathscr{E} \setminus (-\mathscr{K})$ that makes $d\mathscr{T}(\zeta) \geq \zeta$, then the spectral radius of $\mathscr{T}$ be greater than 0, and which has a positive eigenfunction in regard to its the first eigenvalue $\lambda = r(\mathscr{T})^{-1}$. This formula is often called Krein-Rutmann's theorem.*

Lemma 5 ([27]). *The spectral radius of $\mathscr{T}$ meets*

$$r(\mathscr{T}) = \lim_{n \to +\infty} \|\mathscr{T}^n\|^{\frac{1}{n}},$$

where $\mathscr{T}$ is a linear bounded operator, and $\|.\|$ is the norm of operator. This formula is often called Gelfand's formula.

Lemma 6. *$\mathscr{T} : \mathscr{K} \to \mathscr{K}$ defined by (2.9) is a linear operator with complete continuity under the condition (A_1), and the spectral radius $r(\mathscr{T})$ of $\mathscr{T}$ is unequal zero, furthermore, $\mathscr{T}$ exists a positive eigenfunction ζ in regard to its first eigenvalue $\lambda_1 = (r(\mathscr{T}))^{-1}$.*

Proof. For $\forall\, \mathscr{U} \in \mathscr{K}$, from Lemma 3, one arrives

$$\begin{aligned} \|\mathscr{T}\mathscr{U}\| &= \max_{t\in[a,b]} \int_a^b \mathscr{H}(t,s)\varrho(s)\mathscr{U}(s)\frac{ds}{s} \\ &\leq \frac{1}{\Gamma(\flat-\mu)(\ln\frac{b}{a})}\left(1 + \frac{(\ln\frac{b}{a})^{\flat-\mu-1}}{\Delta}\sum_{j=1}^{\infty}\kappa_j\right)\int_a^b \varpi(s)\varrho(s)\mathscr{U}(s)\frac{ds}{s} \\ &\leq \frac{w}{\Gamma(\flat-\mu)(\ln\frac{b}{a})}\int_a^b \varpi(s)\varrho(s)\mathscr{U}(s)\frac{ds}{s} \end{aligned} \tag{2.10}$$

Moreover, by Lemma 3, the following result can be derived

$$\mathscr{T}\mathscr{U}(t) \geq \frac{\vartheta(t)}{\Gamma(\flat-\mu)(\ln\frac{b}{a})^{\flat-\mu+1}}\int_a^b \varpi(s)\varrho(s)\mathscr{U}(s)\frac{ds}{s}, t \in [a,b]. \tag{2.11}$$

Formula (2.10) and Formula (2.11) mean $\mathscr{T}$ from $\mathscr{K}$ to $\mathscr{K}$. One finds that $\mathscr{T}$ has completely continuous property from $\mathscr{K}$ to $\mathscr{K}$ by combing (A_2) and the uniform continuous property of $\mathscr{H}(t,s)$ on $[a,b] \times [a,b]$.

Next, based on Krein-Rutmann's theorem, we prove that $\mathscr{T}$ has the first eigenvalue λ_1 and $\lambda_1 > 0$. In reality, by process of proof Lemma 3, there exists $t_0 \in (a,b)$ that makes $\mathscr{H}(t_0,t_0) > 0$. Therefore, there exists $[m,n] \subset (a,b)$ that makes $t_0 \in (m,n)$ and $\mathscr{H}(t,s) > 0$ for $\forall\, t,s \in [m,n]$. Select $\mathscr{U} \in K$ that makes $\mathscr{U}(t_0) > 0$ and $\mathscr{U}(t) = 0$ for all $t \notin [m,n]$. After that, for $\forall\, t \in [m,n]$, we have

$$(\mathscr{T}\mathscr{U})(t) = \int_a^b \mathscr{H}(t,s)\varrho(s)\mathscr{U}(s)\frac{ds}{s} \geq \int_m^n \mathscr{H}(t,s)\varrho(s)\mathscr{U}(s)\frac{ds}{s} > 0.$$

Hence, there is $d > 0$ such that $d(\mathscr{T}\mathscr{U})(t) \geq \mathscr{U}(t)$ for $t \in [a,b]$, then the spectral radius $r(\mathscr{T}) \neq 0$ according to Lemma 4, furthermore, by Lemma 6, for the given first eigenvalue $\lambda_1 = (r(\mathscr{T}))^{-1}$, there exists a positive eigenfunction $\varphi^\star$ for A that makes

$$\lambda_1 \mathscr{T}\varphi^\star = \varphi^\star.$$

The proof is over. □

Lemma 7. *In case of $(A_1), (A_2)$ hold, then A acts as an operator with complete continuity from $\overline{\mathscr{K}_\jmath} \setminus \mathscr{K}_\iota$ to $\mathscr{K}$.*

Proof. First of all, we show $A(\overline{\mathscr{K}_\jmath} \setminus \mathscr{K}_r) \subset \mathscr{K}$. Actually, for $\forall\, u \in \overline{\mathscr{K}_\jmath} \setminus \mathscr{K}_\iota, t \in [a,b]$, and combing Lemma 3, one has

$$\begin{aligned}(A\mathscr{U})(t) &= \int_a^b \mathscr{H}(t,s)\varrho(s)\ell\Big(s, {}^H I^\mu_{a+}\mathscr{U}(s), \mathscr{U}(s)\Big)\frac{ds}{s} \\ &\leq \frac{w}{\Gamma(\flat-\mu)(\ln\frac{b}{a})}\int_a^b \varpi(s)\varrho(s)\ell\Big(s, {}^H I^\mu_{a+}\mathscr{U}(s), \mathscr{U}(s)\Big)\frac{ds}{s},\end{aligned}$$

Then

$$\|(A\mathscr{U})\| \leq \frac{w}{\Gamma(\flat-\mu)(\ln\frac{b}{a})}\int_a^b \varpi(s)\varrho(s)\ell\Big(s, {}^H I^\mu_{a+}u(s), u(s)\Big)\frac{ds}{s}.$$

On the other side, the following result is obtained from Lemma 3

$$\begin{aligned}(A\mathscr{U})(t) &= \int_a^b \mathscr{H}(t,s)\varrho(s)\ell\Big(s, {}^H I^\mu_{a+}\mathscr{U}(s), \mathscr{U}(s)\Big)\frac{ds}{s} \\ &\geq \frac{\vartheta(t)}{\Gamma(\flat-\mu)(\ln\frac{b}{a})^{\flat-\mu+1}}\int_a^b \varpi(s)\varrho(s)\ell\Big(s, {}^H I^\mu_{a+}\mathscr{U}(s), \mathscr{U}(s)\Big)\frac{ds}{s} \\ &\geq \frac{\vartheta(t)}{(\ln\frac{b}{a})^{\flat-\mu}w}\|A\mathscr{U}\|, t \in [a,b].\end{aligned}$$

Consequently, $A(\overline{\mathscr{K}_\jmath} \setminus \mathscr{K}_\iota) \subset \mathscr{K}$. Next, for $\forall\, \iota > \complement$, we prove

$$\sup_{\mathscr{U}\in\overline{\mathscr{K}_\jmath}\setminus\mathscr{K}_\iota}\int_a^b \varpi(s)\varrho(s)\ell\Big(s, {}^H I^\mu_{a+}\mathscr{U}(s), \mathscr{U}(s)\Big)\frac{ds}{s} < +\infty, \tag{2.12}$$

that is to say, $A : \overline{\mathscr{K}_\jmath} \setminus \mathscr{K}_\iota \to \mathscr{K}$ is well defined. As a matter of fact, for $\forall \mathscr{U} \in \overline{\mathscr{K}_\jmath} \setminus \mathscr{K}_\iota$, we have

$$\begin{aligned}{}^H I^\mu_{a+}\mathscr{U}(t) &= \frac{1}{\Gamma(\mu)}\int_a^t (\ln t - \ln s)^{\mu-1}\mathscr{U}(s)\frac{ds}{s} \\ &\leq -\frac{1}{\Gamma(\mu)}\|\mathscr{U}\|\int_a^t (\ln t - \ln s)^{\mu-1}d(\ln t - \ln s) \\ &= \frac{1}{\Gamma(\mu+1)}\|\mathscr{U}\|(\ln t - \ln s)^\mu|_a^t = \frac{1}{\Gamma(\mu+1)}(\ln\frac{b}{a})^\mu\|\mathscr{U}\| \\ &\leq \frac{1}{\Gamma(\mu+1)}(\ln\frac{b}{a})^\mu \jmath, t \in [a,b],\end{aligned} \tag{2.13}$$

Then $\|^{H}I^{\mu}_{a^+}\mathscr{U}\| \le \frac{1}{\Gamma(\mu+1)}(\ln\frac{b}{a})^{\mu}\jmath$, hence, by (A_2), there must be a non-negative integer $v_0 > 1$ such that

$$\sup_{\mathscr{U}\in\overline{K}_\jmath\setminus K_\iota}\int_{e(v_0)} w(s)\varrho(s)\ell(s,{}^{H}I^{\mu}_{a^+}\mathscr{U}(s),\mathscr{U}(s))\frac{ds}{s} < \frac{\Gamma(\flat-\mu)(\ln\frac{b}{a})}{w}. \tag{2.14}$$

where

$$w = 1 + \frac{(\ln\frac{b}{a})^{\flat-\mu-1}}{\Delta}\sum_{j=1}^{\infty}\kappa_j.$$

Choosing

$$\varrho = \min\left\{\frac{1}{(\ln\frac{b}{a})^{\flat-\mu-1}w}, \frac{(\ln a)^{\flat-\mu}(-1)^{\flat-\mu-1}}{\Gamma(\mu)\gamma(\ln\frac{b}{a})^{\flat-\mu}w}, 1\right\},$$

$$\overline{\varrho} = \max\left\{1, \frac{(\ln\frac{b}{a})^{\mu}}{\Gamma(\mu+1)}\right\}.$$

Hence, for $\forall\,\mathscr{U}\in\overline{\mathscr{K}}_\jmath\setminus\mathscr{K}_\iota$, we get

$$u(t) \le \|\mathscr{U}\| \le \jmath \le \overline{\varrho}\jmath, \tag{2.15}$$

$$\begin{aligned}\mathscr{U}(t) \ge& \frac{\vartheta(t)\|\mathscr{U}\|}{(\ln\frac{b}{a})^{\flat-\mu}w} = \frac{(\ln\frac{t}{a})^{\flat-\mu-1}(\ln\frac{b}{t})\|\mathscr{U}\|}{(\ln\frac{b}{a})^{\flat-\mu}w}\\ \ge& \frac{(\ln t)^{\flat-\mu-1}}{(\ln\frac{b}{a})^{\flat-\mu-1}w}\|\mathscr{U}\| \ge (\ln t)^{\flat-\mu-1}\varrho\|\mathscr{U}\|,\end{aligned} \tag{2.16}$$

by (2.13), we have

$${}^{H}I^{\mu}_{a^+}\mathscr{U}(t) \le \frac{1}{\Gamma(\mu+1)}(\ln\frac{b}{a})^{\mu}\|\mathscr{U}\| \le \overline{\varrho}\|\mathscr{U}\|, t\in[a,b], \tag{2.17}$$

$$\begin{aligned}{}^{H}I^{\mu}_{a^+}\mathscr{U}(t) =& \frac{1}{\Gamma(\mu)}\int_a^t(\ln t-\ln s)^{\mu-1}\mathscr{U}(s)\frac{ds}{s}\\ \ge& \frac{1}{\Gamma(\mu)}\int_a^t(\ln t-\ln s)^{\mu-1}\frac{(\ln\frac{s}{a})^{\flat-\mu-1}(\ln\frac{b}{s})}{(\ln\frac{b}{a})^{\flat-\mu}w}\|\mathscr{U}\|\frac{ds}{s}\\ =& \frac{\|\mathscr{U}\|}{\Gamma(\mu)(\ln\frac{b}{a})^{\flat-\mu}w}\int_a^t(\ln t-\ln s)^{\mu-1}(\ln s-\ln a)^{\flat-\mu-1}(\ln b-\ln s)d(\ln s),\\ & t\in[a,b].\end{aligned} \tag{2.18}$$

If $\ln s = \Lambda \ln t$, one gets

$$\begin{aligned}
&\frac{\|\mathscr{U}\|}{\Gamma(\mu)(\ln\frac{b}{a})^{\flat-\mu}w}\int_a^t(\ln t-\ln s)^{\mu-1}(\ln s-\ln a)^{\flat-\mu-1}(\ln b-\ln s)d(\ln s)\\
&=\frac{\|\mathscr{U}\|}{\Gamma(\mu)(\ln\frac{b}{a})^{\flat-\mu}w}\int_{\frac{\ln a}{\ln t}}^1(\ln t-\Lambda\ln t)^{\mu-1}(\iota\ln t-\ln a)^{\flat-\mu-1}(\ln b-\iota\ln t)d\tau\\
&=\frac{\|\mathscr{U}\|}{\Gamma(\mu)(\ln\frac{b}{a})^{\flat-\mu}w}(\ln t)^{\mu-1}\int_{\frac{\ln a}{\ln t}}^1(1-\Lambda)^{\mu-1}(\iota\ln t-\ln a)^{\flat-\mu-1}(\ln b-\iota\ln t)d\tau\\
&\geq\frac{\|\mathscr{U}\|}{\Gamma(\mu)(\ln\frac{b}{a})^{\flat-\mu}w}(\ln t)^{\mu-1}\int_0^1(1-\Lambda)^{\mu-1}(\Lambda\ln b-\ln a)^{\flat-\mu-1}(\ln b-\Lambda\ln a)d\tau\\
&\geq\frac{\|\mathscr{U}\|(\ln a)^{\flat-\mu-1}}{\Gamma(\mu)(\ln\frac{b}{a})^{\flat-\mu}w}(\ln t)^{\mu-1}\int_0^1(1-\Lambda)^{\mu-1}(\Lambda-1)^{\flat-\mu-1}\ln a(1-\Lambda)d\tau\\
&=\frac{\|\mathscr{U}\|(\ln a)^{\flat-\mu}}{\Gamma(\mu)(\ln\frac{b}{a})^{\flat-\mu}w}(\ln t)^{\mu-1}\int_0^1(1-\Lambda)^{\mu-1}(\Lambda-1)^{\flat-\mu-1}d\tau\\
&=\frac{\|\mathscr{U}\|(\ln a)^{\flat-\mu}}{\Gamma(\mu)(\ln\frac{b}{a})^{\flat-\mu}w}(\ln t)^{\mu-1}(-1)^{\flat-\mu-1}\int_0^1(1-\Lambda)^{\flat-1}\\
&=\frac{\|\mathscr{U}\|(\ln a)^{\flat-\mu}}{\gamma\Gamma(\mu)(\ln\frac{b}{a})^{\flat-\mu}w}(\ln t)^{\mu-1}(-1)^{\flat-\mu-1}\\
&\geq\|\mathscr{U}\|(\ln t)^{\mu-1}\varrho\geq\|u\|(\ln t)^{\flat-\mu-1}\varrho, t\in[a,b].
\end{aligned}\tag{2.19}$$

Thus, for $\forall\, a+\frac{1}{v_0}\leq t\leq b-\frac{1}{v_0}$, by (2.15)–(2.19), we obtain

$$r\varrho(\ln(a+\frac{1}{v_0}))^{\flat-\mu-1}\leq\mathscr{U}(t),{}^H I_{a^-}^{\mu}\mathscr{U}(t)\leq\overline{\varrho}_J.\tag{2.20}$$

Combing (2.14) and (2.20), we get

$$\begin{aligned}
&\sup_{\mathscr{U}\in\overline{\mathscr{K}}_J\setminus\mathscr{K}_i}\int_a^b\frac{w}{\Gamma(\flat-\mu)(\ln\frac{b}{a})}\varpi(s)\varrho(s)\ell\Big(s,{}^H I_{a^+}^{\mu}\mathscr{U}(s),\mathscr{U}(s)\Big)\frac{ds}{s}\\
&\leq\sup_{\mathscr{U}\in\overline{\mathscr{K}}_J\setminus\mathscr{K}_i}\int_{e(v_0)}\frac{w}{\Gamma(\flat-\mu)(\ln\frac{b}{a})}\varpi(s)\varrho(s)\ell\Big(s,{}^H I_{a^+}^{\mu}\mathscr{U}(s),\mathscr{U}(s)\Big)\frac{ds}{s}\\
&+\sup_{\mathscr{U}\in\overline{\mathscr{K}}_J\setminus\mathscr{K}_i}\int_{a+\frac{1}{v_0}}^{b-\frac{1}{v_0}}\frac{w}{\Gamma(\flat-\mu)(\ln\frac{b}{a})}\varpi(s)\varrho(s)\ell\Big(s,{}^H I_{a^+}^{\mu}\mathscr{U}(s),\mathscr{U}(s)\Big)\frac{ds}{s}\\
&\leq 1+D_1\int_a^b\frac{w}{\Gamma(\flat-\mu)(\ln\frac{b}{a})}\varpi(s)\varrho(s)\frac{ds}{s}<+\infty,
\end{aligned}\tag{2.21}$$

where

$$w=1+\frac{(\ln\frac{b}{a})^{\flat-\mu-1}}{\Delta}\sum_{j=1}^{\infty}r_j,$$

$$D_1=\max\left\{\ell(t,\varsigma_1,\varsigma_2):(t,\varsigma_1,\varsigma_2)\in\left(\frac{1}{v_0},1-\frac{1}{v_0}\right)\times\left(\iota\varrho(\ln(a+\frac{1}{v_0}))^{\flat-\mu-1},\overline{\varrho}_J\right)^2\right\}.$$

Thus, (2.12) is valid and indicating that A has the uniformly bounded property on any bounded set.

For $\forall\,\epsilon>0$, by (A_2), there must be a non-negative integer $v_0>1$ that makes

$$\sup_{x_1\in\overline{\mathscr{K}}_J\setminus\mathscr{K}_i,x_2\in\overline{\mathscr{K}}_J\setminus\mathscr{K}_i}\int_{e(v_0)}w(s)\varrho(s)\ell(s,x_1(s),x_2(s))\frac{ds}{s}<\frac{\varepsilon\Gamma(\flat-\mu)(\ln\frac{b}{a})}{4w}.\tag{2.22}$$

Then, we show that A is continuous from $\overline{\mathscr{K}}_J \setminus \mathscr{K}_\iota$ to $\mathscr{K}$. Let $\mathscr{U}_k, \mathscr{U}_0 \in \overline{\mathscr{K}}_J \setminus \mathscr{K}_\iota$ and $\|\mathscr{U}_k - \mathscr{U}_0\| \to 0$ $(k \to \infty)$. On account of $\ell(t, x_1, x_2)$ has uniformly continuous property on

$$\left(a + \frac{1}{v_0}, b - \frac{1}{v_0}\right) \times \left(r\varrho(\ln(a + \frac{1}{v_0}))^{\flat-\mu-1}, \overline{\varrho}R\right)^2,$$

we have

$$\lim_{k\to+\infty}\left|\ell\left(s,^H I^\mu_{a^+}\mathscr{U}_k(s), \mathscr{U}_k(s)\right) - \ell\left(s,^H I^\mu_{a^+}\mathscr{U}_0(s), \mathscr{U}_0(s)\right)\right| = 0$$

Uniformly holds for $t \in \left[a + \frac{1}{v_0}, b - \frac{1}{v_0}\right]$. Then, by Lebesgue control convergence theorem and (A_2), we have

$$\int_{a+\frac{1}{v_0}}^{b-\frac{1}{v_0}} w(s)\varrho(s)\left|\ell\left(s,^H I^\mu_{a^+}\mathscr{U}_k(s), \mathscr{U}_k(s)\right) - \ell\left(s,^H I^\mu_{a^+}\mathscr{U}_0(s), \mathscr{U}_0(s)\right)\right|\frac{ds}{s} \to 0, \text{ as } k \to \infty.$$

Thus, for the ϵ shown as in (2.18), there is a natural number N, for $k > N$, that makes

$$\begin{aligned}&\int_{a+\frac{1}{v_0}}^{b-\frac{1}{v_0}} w(s)\varrho(s)\left|\ell\left(s,^H I^\mu_{a^+}\mathscr{U}_k(s), \mathscr{U}_k(s)\right) - \ell\left(s,^H I^\mu_{a^+}\mathscr{U}_0(s), \mathscr{U}_0(s)\right)\right|\frac{ds}{s}\\ &< \frac{\epsilon\Gamma(\flat-\mu)(\ln\frac{b}{a})}{2w}.\end{aligned} \tag{2.23}$$

For $k > N$, by (2.22) and (2.23), one has

$$\begin{aligned}&\|A\mathscr{U}_k - A\mathscr{U}_0\|\\ &\leq \sup_{\mathscr{U}_k\in\overline{\mathscr{K}}_J\setminus\mathscr{K}_\iota}\int_{e(v_0)}\frac{w}{\Gamma(\flat-\mu)(\ln\frac{b}{a})}\varpi(s)\varrho(s)\ell\left(s,^H I^\mu_{a^+}\mathscr{U}_k(s), \mathscr{U}_k(s)\right)\frac{ds}{s}\\ &\quad+\sup_{\mathscr{U}_k\in\overline{\mathscr{K}}_J\setminus\mathscr{K}_\iota}\int_{e(v_0)}\frac{w}{\Gamma(\flat-\mu)(\ln\frac{b}{a})}\varpi(s)\varrho(s)\ell\left(s,^H I^\mu_{a^+}\mathscr{U}_k(s), \mathscr{U}_k(s)\right)\frac{ds}{s}\\ &\quad+\int_{a+\frac{1}{v_0}}^{b-\frac{1}{v_0}}\frac{w}{\Gamma(\flat-\mu)(\ln\frac{b}{a})}w(s)\varrho(s)\left|\ell\left(s,^H I^\mu_{a^+}\mathscr{U}_k(s), \mathscr{U}_k(s)\right)\frac{ds}{s}\right.\\ &\quad-\ell\left(s,^H I^\mu_{a^+}\mathscr{U}_0(s), \mathscr{U}_0(s)\right)\frac{ds}{s}\left|\frac{ds}{s}\right.\\ &<2\left(\frac{w}{\Gamma(\flat-\mu)(\ln\frac{b}{a})}\right)\times\frac{\varepsilon}{4\left(\frac{w}{\Gamma(\flat-\mu)(\ln\frac{b}{a})}\right)}+\left(\frac{w}{\Gamma(\flat-\mu)(\ln\frac{b}{a})}\right)\frac{\varepsilon}{2\left(\frac{w}{\Gamma(\flat-\mu)(\ln\frac{b}{a})}\right)}=\varepsilon.\end{aligned}$$

Thus, A is continuous from $\overline{\mathscr{K}}_J \setminus \mathscr{K}_\iota$ to K.

For $\forall$ bounded set Ω and $\Omega \subset \overline{\mathscr{K}}_J \setminus \mathscr{K}_\iota$, we show that $A(\Omega)$ is equicontinuous. As a matter of fact, by (A_2), for $\forall\, \epsilon > 0$, there is a natural number $\kappa_0 > 1$ so that

$$\sup_{x_2\in\overline{\mathscr{K}}_J\setminus\mathscr{K}_\iota}\int_{e(\kappa_0)} w(s)\varrho(s)\ell(s, x_1(s), x_2(s))\frac{ds}{s} < \frac{\varepsilon}{4\left(\frac{w}{\Gamma(\flat-\mu)(\ln\frac{b}{a})}\right)}.$$

Taking

$$D_2 = \max\left\{\ell(t, x_1, x_2) : (t, x_1, x_2) \in \left(a + \frac{1}{\kappa_0}, b - \frac{1}{\kappa_0}\right) \times \left(r\varrho(\ln(a + \frac{1}{n_0}))^{\flat-\mu-1}, \overline{\varrho}J\right)^2\right\}.$$

On account of $\mathscr{H}(t,s)$ is uniformly continuous on $[a,b]\times[a,b]$, for the ϵ which is defined by above, there is $\delta>0$, for $\forall\, s\in[a+\frac{1}{\kappa_0}, b-\frac{1}{\kappa_0}]$, we have

$$|\mathscr{H}(t,s)-\mathscr{H}(t',s)|\leq\frac{\varepsilon}{2}\left(D_2\int_{a+\frac{1}{\kappa_0}}^{b-\frac{1}{\kappa_0}}\varrho(s)\frac{ds}{s}\right)^{-1},$$

for $|t-t'|<\delta, t,t'\in[a,b]$. Thus, for $\forall\,|t-t'|<\delta, t,t'\in[a,b]$ and $\mathscr{U}\in\Omega$, we obtain

$$\begin{aligned}&\|A\mathscr{U}(t)-A\mathscr{U}(t')\|\\&\leq 2\sup_{\mathscr{U}\in\overline{\mathscr{K}}_j\backslash\mathscr{K}_i}\left(\frac{w}{\Gamma(b-\mu)(\ln\frac{b}{a})}\right)\int_{e(\kappa_0)}w(s)\varrho(s)\ell\left(s,{}^H I^{\mu}_{a+}\mathscr{U}(s),\mathscr{U}(s)\right)\frac{ds}{s}\\&+\sup_{\mathscr{U}\in\overline{\mathscr{K}}_j\backslash\mathscr{K}_i}\int_{a+\frac{1}{\kappa_0}}^{b-\frac{1}{\kappa_0}}|\mathscr{H}(t,s)-\mathscr{H}(t',s)|\varrho(s)\ell\left(s,{}^H I^{\mu}_{a+}\mathscr{U}(s),\mathscr{U}(s)\right)\frac{ds}{s}\\&<2\left(\frac{w}{\Gamma(b-\mu)(\ln\frac{b}{a})}\right)\times\frac{\varepsilon}{4(\frac{w}{\Gamma(b-\mu)(\ln\frac{b}{a})})}+\frac{\varepsilon}{2}=\varepsilon,\end{aligned}$$

which demonstrates that $A(\Omega)$ is equicontinuous. By the Arzela-Ascoli theorem, A is fully continuous from $\overline{\mathscr{K}}_j\setminus\mathscr{K}_i$ to $\mathscr{K}$, then the verification is finished. □

3. Principal Results

In this section, the main results will be provided according to these preliminaries and the following lemmas.

Lemma 8 ([27]). *Suppose $\mathscr{K}$ is a cone within Banach space $\mathscr{E}$. Let $A:\overline{\mathscr{K}}_r\to\mathscr{K}$ is a fully continuous operator. In case of $\mathscr{U}_0\in\mathscr{K}\backslash\{\theta\}$ so that $\mathscr{U}-A\mathscr{U}\neq\mu\mathscr{U}_0$ for an arbitrary $\mathscr{U}\in\partial\mathscr{K}_r$ and $\mu\geq0$, thus $i(A,\mathscr{K}_r,\mathscr{K})=0$.*

Lemma 9 ([27]). *Suppose $\mathscr{K}$ is a cone within Banach space E. Let $A:\overline{\mathscr{K}}_r\to\mathscr{K}$ is a fully continuous operator. In case of $Au\neq\mu\mathscr{U}$ for an arbitrary $\mathscr{U}\in\partial\mathscr{K}_r$ and $\mu\geq1$, thus $i(A,\mathscr{K}_r,\mathscr{K})=1$.*

Theorem 1. *Suppose the conditions $(A_1$–$A_2)$ are established, and*

$$\liminf_{\substack{x_i\to0^+\\ i=1,2}}\min_{t\in[a,b]}\frac{\ell(t,x_1,x_2)}{x_1+x_2}>\lambda_1, \tag{3.1}$$

$$\limsup_{x_2\to+\infty}\max_{t\in[a,b]}\frac{\ell(t,x_1,x_2)}{x_2}<\lambda_1 \tag{3.2}$$

Hold uniformly for $x_i\in[0,+\infty)$ $(i=1,2)$, and λ_1 is the first eigenvalue of $\mathscr{T}$ which is defined by (2.9), then there's at least one positive solution of the BVP (1.1,1.2).

Proof. By (3.1), there exists $\iota>0$, that makes

$$\ell(t,x_1,x_2)\geq\lambda_1(x_1+x_2), 0<x_i\leq\iota, i=1,2,\ t\in[a,b]. \tag{3.3}$$

Choosing $\iota_0=\min\{\iota,\frac{\iota}{\bar\varrho}\}$ and for $\forall\,\mathscr{U}\in\partial\mathscr{K}_{\iota_0}$, one has

$$0<{}^H I^{\mu}_{a+}\mathscr{U}(s)\leq\bar\varrho\iota_0\leq\iota, 0<\mathscr{U}(s)\leq\iota,\ s\in[a,b]. \tag{3.4}$$

Accordingly, one can arrive at according to (3.3) and (3.4)

$$\begin{aligned}(A\mathscr{U})(t) &= \int_a^b \mathscr{H}(t,s)\varrho(s)\ell(s,{}^H I_{a^+}^{\mu}\mathscr{U}(s),\mathscr{U}(s))\frac{ds}{s} \\ &\geq \lambda_1 \int_a^b \mathscr{H}(t,s)\varrho(s)\left({}^H I_{a^+}^{\mu}\mathscr{U}(s)+\mathscr{U}(s)\right)\frac{ds}{s} \\ &\geq \lambda_1(\mathscr{T}\mathscr{U})(t), t\in[a,b].\end{aligned} \tag{3.5}$$

By Lemma 6, $\mathscr{T}$ has a positive eigenfunction ζ in regard to λ_1, i.e., $\zeta = \lambda_1\mathscr{T}\zeta$. Then we will show

$$\mathscr{U} - A\mathscr{U} \neq d\zeta,\ \mathscr{U}\in\partial\mathscr{K}_{\iota_0},\ d\geq 0. \tag{3.6}$$

Otherwise, there is $\mathscr{U}_0\in\partial\mathscr{K}_{\iota_0}$ and $d_0\geq 0$ that makes $\mathscr{U}_0 - A\mathscr{U}_0 = d_0\zeta$, then $d_0 > 0$ and $\mathscr{U}_0 = A\mathscr{U}_0 + d_0\zeta \geq d_0\zeta$. Assume $\bar{d} = \sup\{d|\mathscr{U}_0 \geq d\zeta\}$, then $\bar{d}\geq d_0, \mathscr{U}_0 \geq \bar{d}\zeta, \lambda_1\mathscr{T}\mathscr{U}_0 \geq \lambda_1\bar{d}\mathscr{T}\zeta = \bar{d}\zeta$. Thus, from (3.5), we get

$$\mathscr{U}_0 = A\mathscr{U}_0 + d_0\zeta \geq \lambda_1\mathscr{T}\mathscr{U}_0 + d_0\zeta \geq \bar{d}\zeta + d_0\zeta = (\bar{d}+d_0)\zeta,$$

which is a contradiction with the definition of $\bar{d}$. So (3.6) holds and combing Lemma 8, we have

$$i(A,\mathscr{K}_{\iota_0},\mathscr{K}) = 0. \tag{3.7}$$

Now taking a constant $0<\sigma<1$ makes

$$\limsup_{x_2\to+\infty}\max_{t\in[a,b]}\frac{\ell(t,x_1,x_2)}{x_2} < \omega\lambda_1, \tag{3.8}$$

and a linear operator $\widetilde{\mathscr{T}}\mathscr{U} = \omega\lambda_1\mathscr{T}\mathscr{U}$ is defined, then subsequently $\widetilde{\mathscr{T}}$ is a bounded linear operator from $\mathscr{E}$ to $\mathscr{E}$, and $\widetilde{\mathscr{T}}(\mathscr{K})\subset\mathscr{K}$. Additionally, $\widetilde{\mathscr{T}}\zeta = \omega\lambda_1\mathscr{T}\zeta = \omega\zeta$, and so the spectral radius of $\widetilde{\mathscr{T}}$ is $r(\widetilde{\mathscr{T}}) = \omega$ and $\widetilde{\mathscr{T}}$ also has the first eigenvalue $r^{-1}(\widetilde{\mathscr{T}}) = \omega^{-1} > 1$. Adopting Gelfand's formula, one can obtain

$$\omega = \lim_{n\to+\infty}\|\widetilde{\mathscr{T}}^n\|^{\frac{1}{n}}. \tag{3.9}$$

Let $\varepsilon_0 = \frac{1}{2}(1-\omega)$ and by (3.9), there is a large enough natural number N_0, and when $n\geq N$, we have $\|\widetilde{\mathscr{T}}^n\| \leq [\omega+\epsilon_0]^n$. For $\forall\,\mathscr{U}\in\mathscr{E}$, we define

$$\|\mathscr{U}\|^\star = \sum_{i=1}^N [\omega+\epsilon_0]^{N-i}\|\widetilde{\mathscr{T}}^{i-1}\mathscr{U}\|, \tag{3.10}$$

where $\widetilde{\mathscr{T}}^0 = I$ is the unit operator. Apparently, $\|.\|^\star$ is another norm of $\mathscr{E}$.

In addition, according to (3.2), there is $\jmath_1 > \iota$, we have

$$\ell(t,x_1,x_2) \leq \omega\lambda_1 x_2,\ \text{for } x_2\geq \jmath_1, x_i\geq 0,\ i=1,2,\ t\in[a,b]. \tag{3.11}$$

Choosing

$$\jmath > \max\left\{\jmath_1, \frac{2(\omega+\varepsilon_0^{N-1})^{-1}}{\varepsilon_0}C^\star\right\},$$

where $C^\star = \|C\|^\star$ and

$$\begin{aligned}C = &\sup_{\mathscr{U}\in\mathscr{K}_{\jmath_1}}\int_a^b \frac{w}{\Gamma(\flat-\mu)(\ln\frac{b}{a})}\varpi(s)\varrho(s)\ell(s,{}^H I_{a^+}^{\mu}\mathscr{U}(s),\mathscr{U}(s))\frac{ds}{s} \\ &< +\infty \text{ (by (2.11))}.\end{aligned} \tag{3.12}$$

Then we prove that

$$A\mathscr{U} \neq \mu\mathscr{U}, \mathscr{U}\in\partial\mathscr{K}_{\jmath},\ \mu\geq 1. \tag{3.13}$$

If not, then there is $\mathscr{U}_1 \in \partial\mathscr{K}_\jmath$ and $d_1 \geq 1$ that makes $A\mathscr{U}_1 = d_1\mathscr{U}_1$. Assume $\widetilde{\mathscr{U}}(t) = \min\{\mathscr{U}_1(t), \jmath_1\}$ and $D(\mathscr{U}_1) = \{t \in [a,b] : \mathscr{U}_1(t) > \jmath_1\}$. Assume

$$\xi(t) = \frac{\Gamma(\flat - \mu)(\ln \frac{b}{a})}{w}\vartheta(t).$$

According to $\widetilde{\mathscr{U}} \in C([a,b],[0,+\infty)), \xi(t)\jmath \leq \mathscr{U}_1(t) \leq \|\mathscr{U}_1\| = \jmath$ and $A\mathscr{U}_1 = \mu_1\mathscr{U}_1$, we have $\mathscr{U}_1$ meets boundary conditions, Thus, we get $u_1(0) = 0$, hence, there is $0 < t_0 \leq 1$ that makes $u_1(t_0) = R$. Hence, $\widetilde{\mathscr{U}}(t) = \min\{\mathscr{U}_1(t), \jmath_1\} \leq \min\{\jmath, \jmath_1\} = \jmath_1$ for $t \in [a,b]$, and $\widetilde{\mathscr{U}}(t_0) = \min\{\mathscr{U}_1(t_0), \jmath_1\} = \min\{\jmath, \jmath_1\} = \jmath_1$, then we get $\|\widetilde{\mathscr{U}}\| = \jmath_1$. Because

$$\widetilde{\mathscr{U}}(t) = \min\{\mathscr{U}_1(t), \jmath_1\} \geq \min\{\xi(t)\jmath, \jmath_1\} \geq \jmath_1\xi(t),\ t \in [a,b],$$

Thus, $\tilde{u} \in \partial\mathscr{K}_{\jmath_1}$. For $t \in D(\mathscr{U}_1)$, $\mathscr{U}_1(t) \geq \jmath_1, {}^H I^{\mu}_{a^+}\mathscr{U}_1(t) \geq 0$, by (3.11) and Lemma 2, one gets

$$\begin{aligned}(A\mathscr{U}_1)(t) &= \int_a^b \mathscr{H}(t,s)\varrho(s)\ell\Big(s, {}^H I^{\mu}_{a^+}\mathscr{U}_1(s), \mathscr{U}_1(s)\Big)\frac{ds}{s} \\ &\leq \int_{D(\mathscr{U}_1)} \mathscr{H}(t,s)\varrho(s)\ell\Big(s, {}^H I^{\mu}_{a^+}\mathscr{U}_1(s), \mathscr{U}_1(s)\Big)\frac{ds}{s} \\ &+ \int_{[a,b]\backslash D(u_1)} \mathscr{H}(t,s)\varrho(s)\ell\Big(s, {}^H I^{\mu}_{a^+}\mathscr{U}_1(s), \mathscr{U}_1(s)\Big)\frac{ds}{s} \\ &\leq \sigma\lambda_1 \int_a^b \mathscr{H}(t,s)\varrho(s)\mathscr{U}_1(s)\frac{ds}{s} \\ &+ \int_a^b \frac{w}{\Gamma(\flat - \mu)(\ln\frac{b}{a})}\varpi(s)\varrho(s)\ell\Big(s, {}^H I^{\mu}_{a^+}\widetilde{\mathscr{U}}(s), \widetilde{\mathscr{U}}(s)\Big)\frac{ds}{s} \\ &\leq (\mathscr{T}\mathscr{U}_1)(t) + C,\ t \in [a,b].\end{aligned} \tag{3.14}$$

It is noted that $\mathscr{T}$ is a bounded linear operator and from $\mathscr{K}$ to $\mathscr{K}$, from (3.14) one arrives

$$0 \leq (\mathscr{T}^j(A\mathscr{U}_1))(t) \leq (\mathscr{T}^j(\mathscr{T}\mathscr{U}_1 + C))(t),\ j = 0,1,2,\cdots,n-1,\ t \in [a,b]. \tag{3.15}$$

Then, by (3.15), we have

$$\|(\mathscr{T}^j(A\mathscr{U}_1))\| \leq \|(\mathscr{T}^j(\mathscr{T}\mathscr{U}_1 + C))\|,\ j = 0,1,2,\cdots,n-1,$$

that gives rise to

$$\begin{aligned}\|A\mathscr{U}_1\|^\star &= \sum_{i=1}^n [\omega + \varepsilon_0]^{n-i}\|\mathscr{T}^{i-1}(A\mathscr{U}_1)\| \\ &\leq \sum_{i=1}^n [\omega + \varepsilon_0]^{n-i}\|\mathscr{T}^{i-1}(\mathscr{T}\mathscr{U}_1 + C)\| = \|\mathscr{T}\mathscr{U}_1 + C\|^\star.\end{aligned} \tag{3.16}$$

By $\mathscr{U}_1 \in \partial\mathscr{K}_\jmath$, $\|\mathscr{U}_1\| = \jmath$, and (3.10), the following formula can be developed

$$\|\mathscr{U}_1\|^\star > [\omega + \varepsilon_0]^{n-1}\|\mathscr{U}_1\| = [\omega + \varepsilon_0]^{n-1}R > \frac{2}{\varepsilon_0}C^\star,$$

which results in

$$C^\star < \frac{\epsilon_0}{2}\|\mathscr{U}_1\|^\star. \tag{3.17}$$

According to (3.10), (3.16) and (3.17), one obtains

$$\begin{aligned}
\mu_1\|\mathscr{U}_1\|^\star &= \|A\mathscr{U}_1\|^\star \le \|\widetilde{\mathscr{T}}\mathscr{U}_1\|^\star + C^\star = \sum_{i=1}^{n}[\sigma+\epsilon_0]^{n-i}\|\widetilde{\mathscr{T}}^i\mathscr{U}_1\| + C^\star \\
&= [\sigma+\varepsilon_0]\sum_{i=1}^{n-1}[\sigma+\varepsilon_0]^{n-i-1}\|\widetilde{\mathscr{T}}^i\mathscr{U}_1\| + \|\widetilde{\mathscr{T}}^N\mathscr{U}_1\| + C^\star \\
&\le [\sigma+\varepsilon_0]\sum_{i=1}^{N-1}[\sigma+\varepsilon_0]^{n-i-1}\|\widetilde{\mathscr{T}}^i\mathscr{U}_1\| + [\sigma+\epsilon_0]^N\|\mathscr{U}_1\| + C^\star \\
&= [\sigma+\varepsilon_0]\sum_{i=1}^{N}[\sigma+\varepsilon_0]^{n-i}\|\widetilde{\mathscr{T}}^{i-1}\mathscr{U}_1\| + C^\star \\
&= [\sigma+\varepsilon_0]\|u_1\|^\star + C^\star \le [\sigma+\varepsilon_0]\|\mathscr{U}_1\|^\star + \frac{\varepsilon_0}{2}\|\mathscr{U}_1\|^\star \\
&= \left[\frac{1}{4}\sigma + \frac{3}{4}\right]\|\mathscr{U}_1\|^\star.
\end{aligned}$$

Taking care of $d_1 \ge 1$, we have $\frac{1}{4}\omega + \frac{3}{4} \ge 1$, and thus $\omega \ge 1$, which contradict with $0 < \omega < 1$. Thus (3.13) is right, and by Lemma 9 , we have

$$i(A, \mathscr{K}_J, \mathscr{K}) = 1. \tag{3.18}$$

Combing (3.7) with (3.18), we have

$$i(A, \mathscr{K}_J \setminus \overline{\mathscr{K}}_{\iota_0}, J) = i(A, \mathscr{K}_J, \mathscr{K}) - i(A, \mathscr{K}_{\iota_0}, \mathscr{K}) = 1.$$

Hence, A has at least one fixed point in $K_R\setminus\overline{K}_{r_0}$, that is to say, the BVP (2.1, 2.2) has at least one positive solution, which implies that BVP (1.1, 1.2) also has at least one positive solution. Another case for Equations (1.1) and (1.2) will be discussed herein. For this purpose, for $\forall$ small enough $0 < \varepsilon < 1$, we define a linear operator $\mathscr{T}_\varepsilon$

$$(\mathscr{T}_\varepsilon\mathscr{U})(t) = \int_{a+\varepsilon}^{b-\varepsilon} \mathscr{H}(t,s)\varrho(s)\mathscr{U}(s)\frac{ds}{s},\ t \in [a,b].$$

According to Lemma 7, we have$\mathscr{T}_\varepsilon$ form $\mathscr{K}$ to $\mathscr{K}$ denotes a linear operator with complete continuity, too, and the spentral radius $r(\mathscr{T}_\varepsilon)$ of $\mathscr{T}_\varepsilon)$ unequal to 0, and furthermore, $\mathscr{T}_\varepsilon$ has a positive eigenfunction ζ_ε in regard to its first eigenvalue $\lambda_\varepsilon = (r(\mathscr{T}_\varepsilon))^{-1}$. □

Lemma 10. *Assume that* (A_1) *holds, then* $\mathscr{T}$ *has an eigenvalue* $\widetilde{\lambda}_1$ *that makes*

$$\lim_{\varepsilon\to 0^+} \lambda_\varepsilon = \widetilde{\lambda}_1.$$

Proof. Choose $\ldots \le \varepsilon_n \le \ldots \le \varepsilon_2 \le \varepsilon_1$ and ε_n tend to 0 as $n \to +\infty$. Then for $\forall\ n < m$ and $\zeta \in \mathscr{E}$, we get

$$(\mathscr{T}_{\varepsilon_n}\zeta)(t) \le (\mathscr{T}_{\varepsilon_m}\zeta)(t) \le (\mathscr{T}\zeta)(t),\ t \in [a,b],$$

and

$$(\mathscr{T}_{\varepsilon_n}^k\zeta)(t) \le (\mathscr{T}_{\varepsilon_m}^k\zeta)(t) \le (\mathscr{T}^k\zeta)(t),\ t \in [a,b],\ k = 2,3,\cdots,$$

where $\mathscr{T}_{\varepsilon_n}^k = \mathscr{T}(\mathscr{T}_{\varepsilon_n}^{k-1})$ $(k = 2,3,\cdots)$. Thus, $\|\mathscr{T}_{\varepsilon_n}^k\| \le \|\mathscr{T}_{\varepsilon_m}^k\| \le \|\mathscr{T}^k\|$ $(k = 1,2,\cdots)$. By Gelfand's formula, we obtain $\lambda_1 \le \lambda_{\varepsilon_m} \le \lambda_{\varepsilon_n}$, and λ_1 is the first eigenvalue of $\mathscr{T}$. According to $\{\lambda_{\varepsilon_n}\}$ monotonous property with lower boundedness λ_1, assume

$$\lambda_{\varepsilon_n} \to \widetilde{\lambda}_1(n \to +\infty).$$

Now we prove that $\widetilde{\lambda}_1$ is one of eigenvalue of $\mathscr{T}$. Assume ζ_{ε_n} is one of positive eigenfunction of $\mathscr{T}_{\varepsilon_n}$ in regard to λ_{ε_n} with $\|\zeta_{\varepsilon_n}\| = 1$ $(n = 1, 2, \cdots)$, that is,

$$\varphi_{\varepsilon_n}(t) = \lambda_{\varepsilon_n} \int_{a+\varepsilon_n}^{b-\varepsilon_n} \mathscr{H}(t,s)\varrho(s)\zeta_{\varepsilon_n}(s)\frac{ds}{s} = \lambda_{\varepsilon_n}\mathscr{T}_{\varepsilon_n}\zeta_{\varepsilon_n}(t),\ t \in [a,b]. \tag{3.19}$$

Take notice that

$$\|\mathscr{T}_{\varepsilon_n}\zeta_{\varepsilon_n}\| = \max_{a\le t\le b}\int_{a+\varepsilon_n}^{b-\varepsilon_n} \mathscr{H}(t,s)\varrho(s)\zeta_{\varepsilon_n}(s)\frac{ds}{s} \le \int_a^b \frac{w}{\Gamma(b-\mu)(\ln\frac{b}{a})}\varpi(s)\varrho(s)\frac{ds}{s},\ n = 1, 2, \cdots,$$

and hence $\{\mathscr{T}_{\varepsilon_n}\zeta_{\varepsilon_n}\} \subset \mathscr{E}$ is uniform boundedness. In addition, for $\forall\, n \in \mathbb{N}$ and $t_1, t_2 \in [a,b]$, one arrives

$$|\mathscr{T}_{\varepsilon_n}\zeta_{\varepsilon_n}(t_1) - \mathscr{T}_{\varepsilon_n}\zeta_{\varepsilon_n}(t_2)| \le \int_{a+\varepsilon_n}^{b-\varepsilon_n} |\mathscr{H}(t_1,s) - \mathscr{H}(t_2,s)|\varrho(s)\zeta_{\varepsilon_n}(s)\frac{ds}{s}.$$

Since $\mathscr{H}(t,s)$ has uniform continuity property on $[a,b]\times[a,b]$, we have $\{\mathscr{T}_{\varepsilon_n}\zeta_{\varepsilon_n}\} \subset \mathscr{E}$ is equicontinuous. Combing the Arzela-Ascoli theorem, Lemma 3.3 and $\lambda_{\varepsilon_n} \to \widetilde{\lambda}_1 (n \to +\infty)$, we have $\zeta_{\varepsilon_n} \to \zeta_0$ $(n \to +\infty)$. This implies $\|\zeta_0\| = 1$, then by using (3.19), we get

$$\zeta_0(t) = \widetilde{\lambda}_1 \int_a^b \mathscr{H}(t,s)\varrho(s)\zeta_0(s)\frac{ds}{s},\ t \in [a,b],$$

in short, $\zeta_0 = \widetilde{\lambda}_1\mathscr{T}\zeta_0$. □

Theorem 2. *Suppose $(A_1$-$A_2)$ hold, and*

$$\limsup_{\substack{x_i\to 0+\\ i=1,2}} \max_{t\in[a,b]} \frac{\ell(t,x_1,x_2)}{x_2} < \lambda_1, \tag{3.20}$$

$$\liminf_{x_1+x_2\to+\infty} \min_{t\in[a,b]} \frac{\ell(t,x_1,x_2)}{x_1+x_2} > \widetilde{\lambda}_1, \tag{3.21}$$

where λ_1 represents the first appeared eigenvalue of $\mathscr{T}$ and $\widetilde{\lambda}_1$ is another eigenvalue of $\mathscr{T}$, then the BVP (1.1,1.2) has at least one positive solution.

Proof. First of all, from (3.20), for $\forall\, t \in [a,b]$, there is $\iota_0 > 0$ such that

$$\ell(t,x_1,x_2) \le \lambda_1 x_2, 0 < x_1 \le \frac{(\ln b - \ln a)^\mu}{\Gamma(\mu+1)}\iota_0, 0 < x_2 \le \iota_0,\ t \in [a,b]. \tag{3.22}$$

Hence, for any $\mathscr{U} \in \partial K_{\iota_0}$, noticing ${}^H I^\mu_{a^+}\mathscr{U}(s) \le \frac{(\ln \varrho - \ln a)^\mu}{\Gamma(\mu+1)}\iota_0$, $|\mathscr{U}(s)| \le \|\mathscr{U}\| = \iota_0$. By (3.22), we get

$$\begin{aligned}(A\mathscr{U})(t) &= \int_a^b \mathscr{H}(t,s)\varrho(s)\ell(s, {}^H I^\mu_{a^+}\mathscr{U}(s), \mathscr{U}(s))\frac{ds}{s}\\ &\le \lambda_1 \int_a^b \mathscr{H}(t,s)\varrho(s)\mathscr{U}(s)\frac{ds}{s}\\ &= \lambda_1(\mathscr{T}\mathscr{U})(t),\ t \in [a,b].\end{aligned} \tag{3.23}$$

As a matter of fact, we assume there are no fixed points for on A on $\partial\mathscr{K}_{\iota_0}$. It is necessary to prove

$$A\mathscr{U} \ne d\mathscr{U},\ \text{for any } \mathscr{U} \in \partial\mathscr{K}_{\iota_0},\ d \ge 1. \tag{3.24}$$

If not, there is $\mathscr{U}_0 \in \partial\mathscr{K}_{\iota_0}$ and $d_0 \ge 1$ such that $A\mathscr{U}_0 = d_0\mathscr{U}_0$. Then $d_0 > 1$ and from (3.23), we obtain

$$d_0\mathscr{U}_0 = A\mathscr{U}_0 \le \lambda_1\mathscr{T}\mathscr{U}_0. \tag{3.25}$$

According to induction from (3.25), we achieve

$$d_0^n \mathscr{U}_0 \le \lambda_1^n \mathscr{T}^n \mathscr{U}_0,\ n = 1, 2, \cdots,$$

we easily get

$$\|\mathscr{T}^n\| \ge \frac{\|\mathscr{T}^n \mathscr{U}_0\|}{\|\mathscr{U}_0\|} \ge \frac{\mu_0^n \|\mathscr{U}_0\|}{\lambda_1^n \|\mathscr{U}_0\|} = \frac{\mu_0^n}{\lambda_1^n}.$$

Additionally, one derives the following result through the Gelfand formula

$$r(\mathscr{T}) = \lim_{n\to\infty} \sqrt[n]{\|\mathscr{T}^n\|} \ge \frac{\mu_0}{\lambda_1} > \frac{1}{\lambda_1},$$

but it contradict with $r(\mathscr{T}) = \lambda_1^{-1}$. So (3.23) holds. Applying Lemma 3.2, one demonstrates

$$i(A, \mathscr{K}_{\iota_0}, \mathscr{K}) = 1. \tag{3.26}$$

According to (3.21) and $\lambda_\varepsilon \to \widetilde{\lambda}_1 (\varepsilon \to 0^+)$, it exists a sufficiently tiny $\varepsilon \in (0, \frac{1}{2})$ and $\jmath > \iota$, we have

$$\ell(t, x_1, x_2) \ge \lambda_\varepsilon (x_1 + x_2) \ge \rho_\varepsilon \jmath,\ t \in [a, b], \tag{3.27}$$

where λ_ε represents the first appeared eigenvalue $\mathscr{T}_\varepsilon, \rho_\varepsilon = 2(\ln(a+\varepsilon))^{\flat-\mu-1}$. Assume ζ_ε be the positive eigenfunction of $\mathscr{T}_\varepsilon$ with respect to λ_ε, then $\zeta_\varepsilon = \lambda_\varepsilon \mathscr{T}_\varepsilon \zeta_\varepsilon$.

For an any $\mathscr{U} \in \partial\mathscr{K}_\jmath,\ s \in [a+\varepsilon, b-\varepsilon]$, by (2.13)–(2.15), we achieve

$$\begin{aligned} {}^H I_{a^+}^\mu \mathscr{U}(s) + \mathscr{U}(s) =& \frac{1}{\Gamma(\mu)} \int_a^t (\ln t - \ln s)^{\mu-1} \mathscr{U}(s) \frac{ds}{s} + \mathscr{U}(s) \\ &\ge \|\mathscr{U}\| (\ln t)^{\flat-\mu-1} \varrho + \|\mathscr{U}\| (\ln t)^{\flat-\mu-1} \varrho \\ &\ge 2\|\mathscr{U}\| \varrho (\ln(a+\varepsilon))^{\flat-\mu-1} = \rho_\varepsilon \jmath. \end{aligned} \tag{3.28}$$

Combing (3.27) with (3.28) , one has

$$\begin{aligned} (A\mathscr{U})(t) &= \int_a^b \mathscr{H}(t,s)\varrho(s)\ell(s, {}^H I_{a^+}^\mu \mathscr{U}(s), \mathscr{U}(s)) \frac{ds}{s} \\ &\ge \int_{a+\varepsilon}^{b-\varepsilon} \mathscr{H}(t,s)\varrho(s)\ell(s, {}^H I_{a^+}^\mu \mathscr{U}(s), \mathscr{U}(s)) \frac{ds}{s} \\ &\ge \lambda_\varepsilon \int_{a+\varepsilon}^{b-\varepsilon} \mathscr{H}(t,s)\varrho(s) \left({}^H I_{a^+}^\mu \mathscr{U}(s) + \mathscr{U}(s) \right) \frac{ds}{s} \\ &\ge \lambda_\varepsilon \int_{a+\varepsilon}^{b-\varepsilon} \mathscr{H}(t,s)\varrho(s)\mathscr{U}(s) \frac{ds}{s} \\ &= \lambda_\varepsilon (\mathscr{T}_\varepsilon \mathscr{U})(t),\ t \in [a, b]. \end{aligned}$$

By the similar method with the proof of Theorem 1, one gets

$$\mathscr{U} - A\mathscr{U} \ne d\zeta_\varepsilon,\ \mathscr{U} \in \partial\mathscr{K}_\jmath,\ d \ge 0,$$

by using Lemma 8, one arrives at

$$i(A, \mathscr{K}_\jmath, \mathscr{K}) = 0. \tag{3.29}$$

Combing (3.26) with (3.29), one obtains

$$i(A, \mathscr{K}_\jmath \setminus \overline{\mathscr{K}}_{\iota_0}, \mathscr{K}) = i(A, \mathscr{K}_\jmath, \mathscr{K}) - i(A, \mathscr{K}_{\iota_0}, \mathscr{K}) = -1.$$

Hence, A at least has one fixed point in $K_R \setminus \overline{K}_{r_0}$, which means BVP (2.1,2.2) at least has a positive solution, i.e., the BVP (1.1,1.2) has at least one positive solution. The proof is completed. □

4. Example

Example 1. *For a boundary value problem show as follows*

$$\begin{cases} -{}^H D^{\frac{5}{2}}_{a^+} v(t) = g(t)\ell\Big(t, v(t), {}^H D^{\frac{z}{z}}_{a^+} v(t)\Big), \ a < t < b, \\ {}^H D^{\frac{1}{4}}_{a^+} v(a) = {}^H D^{\frac{5}{4}}_{a^+} v(a) = 0, \\ {}^H D^{\frac{1}{4}}_{a^+} v(b) = \sum_{j=1}^{\infty} \eta_j {}^H D^{\frac{1}{4}}_{a^+} v(\xi_j), \end{cases} \tag{4.1}$$

where $\flat = \frac{5}{2}$, $\mu = \frac{1}{4}, a = 1, b = 3$, $a = 1, b = 3$, $\kappa_j = \frac{1}{16}\frac{1}{j^4}, \varsigma_j = e^{\frac{1}{j^4}}$, $\varrho(t) = \frac{1}{3(1-t)(t-3)}$, $\ell(t,x,y) = (x+y)^{-\frac{1}{3}} + |\ln y|$. *Clearly,* $\varrho(t)$ *is singular at* $t = a$ *or/and* $t = b$, $\ell(t,x,y)$ *is singular at* $x = y = 0$. *Let* $\mathscr{U}(t) = {}^H D^{\frac{1}{4}}_{a^+} \mathscr{V}(t)$, *the BVP (4.1) can be reduced to an amended boundary value problem shown as follows*

$$\begin{cases} -{}^H D^{\frac{9}{4}}_{a^+} \mathscr{U}(t) = \varrho(t)\ell(t, {}^H I^{\frac{1}{4}}_{a^+} \mathscr{U}(t), \mathscr{U}(t)), \ a < t < b, \\ \mathscr{U}(a) = \mathscr{U}'(a) = 0, \mathscr{U}(b) = \sum_{j=1}^{\infty} \eta_j \mathscr{U}(\xi_j), \end{cases} \tag{4.2}$$

by simple calculations, we have

$$\Delta = (\ln \frac{b}{a})^{\flat-\mu-1} - \sum_{j=1}^{\infty} \kappa_j (\ln \frac{\varsigma_j}{a})^{\flat-\mu-1} = 0.1554,$$

clearly, we have

$$\mathscr{G}(t,s) = \frac{1}{(\ln 3)^{\frac{5}{4}} \Gamma(\frac{9}{4})} \begin{cases} (\ln t)^{\frac{5}{4}} (\ln 3 - \ln s)^{\frac{5}{4}} - (\ln t - \ln s))^{\frac{5}{4}} (\ln 3)^{\frac{5}{4}}, \ 1 \le s \le t \le 3, \\ (\ln t)^{\frac{5}{4}} (\ln 3 - \ln s)^{\frac{5}{4}}, \ 1 \le t \le s \le 3, \end{cases} \tag{4.3}$$

$$\mathscr{H}(t,s) = \mathscr{G}(t,s) + \frac{(\ln t)^{\frac{5}{4}}}{\Delta} \sum_{j=1}^{\infty} \kappa_j \mathscr{G}(\varsigma_j, s),$$

$$w = 1 + \frac{(\ln \frac{b}{a})^{\flat-\mu-1}}{\Delta} \sum_{j=1}^{\infty} \eta_j = 1 + \frac{(\ln 3)^{\frac{5}{4}}}{0.1554} \sum_{j=1}^{\infty} \frac{1}{16} \frac{1}{j^4} \approx 1.4895,$$

then the cone

$$\mathscr{K} = \{\mathscr{U} \in C[1,e] : \mathscr{U}(t) \ge \frac{\vartheta(t)}{(\ln 3)w} \|\mathscr{U}\| \ge 0.6150\vartheta(t)\|\mathscr{U}\|\}.$$

For any $0 < r < R < +\infty$ *and* $u \in \overline{\mathscr{K}}_R \setminus \mathscr{K}_r$, *we have*

$$(\ln t)^{\frac{5}{4}} \varrho r \le \mathscr{U}(t) \le R\overline{\varrho},$$

$$(\ln t)^{\frac{5}{4}} \varrho r \le \|u\|(\ln t)^{\flat-\mu-1}\varrho \le v(t) = {}^H I^{\frac{1}{4}}_{1^+} \mathscr{U}(t) \le \frac{1}{\Gamma(\frac{5}{4})} (\ln 3)^{\frac{1}{4}} \|\mathscr{U}\| \le \overline{\varrho}\|\mathscr{U}\|, \ t \in [1,3],$$

$$\varrho = \min\left\{ \frac{1}{(\ln \frac{b}{a})^{\gamma-\mu-1} w}, \frac{(\ln a)^{\flat-\mu}(-1)^{\flat-\mu-1}}{\Gamma(\mu)\gamma(\ln \frac{b}{a})^{\flat-\mu} w}, 1 \right\} = \min\left\{ \frac{1}{(\ln 3)^{\frac{5}{4}} w}, \frac{(-1)^{\frac{5}{4}}}{\Gamma(\frac{1}{4})\gamma(\ln 3)^{\frac{9}{4}} w}, 1 \right\},$$

$$\overline{\varrho} = \max\left\{ 1, \frac{(\ln \frac{b}{a})^{\mu}}{\Gamma(\mu+1)} \right\} = \max\left\{ 1, \frac{(\ln 3)^{\frac{1}{4}}}{\Gamma(\frac{5}{4})} \right\}. \tag{4.4}$$

Since $|\ln u|$ *decreases on* $(0,1)$*, and increases on* $(1,+\infty)$*, one must have*

$$|\ln u(t)| \leq 2|\ln R| + |(\ln t)^4|, t \in [1,e],$$

$$[x(t)+y(t)]^{-\frac{1}{3}} \leq \left[(\ln t)^4 r + \frac{B(5,\frac{1}{2})}{\Gamma(\mu)} r(\ln t)^{-\frac{1}{2}}\right]^{-\frac{1}{3}}, t \in [1,e].$$

where $B(5,\frac{1}{2})$ *is a Beta function. By considering the absolute continuity of the obtained integral, one can derive that*

$$\lim_{m\to\infty}\int_{e(m)} \varpi(s)\varrho(t)\left((\ln t)^4 r + \left[(\ln t)^4 r + B(5,\frac{1}{2})r(\ln t)^{-\frac{1}{2}}\right]^{-\frac{1}{3}}\right)\frac{ds}{s} = 0.$$

Hence,

$$\begin{aligned}
&\limsup_{m\to+\infty} \sup_{\substack{x_1\in \overline{K}_{\overline{R}}\setminus K_r \\ x_2\in \overline{K}_R\setminus K_r}} \int_{e(m)} \varpi(s)\varrho(s)f(s,x_1(s),x_2(s))\frac{ds}{s} \\
\leq &\limsup_{m\to+\infty} \sup_{\substack{x_1\in \overline{K}_{\overline{R}}\setminus K_r \\ x_2\in \overline{K}_R\setminus K_r}} \int_{e(m)} (\ln\frac{s}{a})(\ln\frac{b}{s})^{\flat-\mu-1}\frac{1}{3(1-s)(s-3)}\left[(x_1+x_2)^{-\frac{1}{3}} + |\ln x_2|\right]\frac{ds}{s} \\
\leq &\limsup_{m\to+\infty} \int_{e(m)} (\ln\frac{s}{a})(\ln\frac{b}{a})^{\frac{5}{4}}\frac{1}{3(1-s)(s-3)}\left((\ln t)^4 r + \left[(\ln t)^4 r + B(5,\frac{1}{2})r(\ln t)^{-\frac{1}{2}}\right]^{-\frac{1}{3}}\right)\frac{ds}{s} = 0,
\end{aligned}$$

where $\overline{R} = \frac{1}{\Gamma(\mu+1)}(\ln\frac{b}{a})^{\mu}R = \frac{1}{\Gamma(\frac{5}{4})}(\ln 3)^{\frac{1}{4}}R$*. So* (H_1) *holds. On the other hand, it is obvious that*

$$\liminf_{\substack{x_1\to 0^+ \\ x_2\to 0^+}} \inf_{t\in[1,3]} \frac{f(t,x_1,x_2)}{x_1+x_2} = \liminf_{\substack{x_1\to 0^+ \\ x_2\to 0^+}} \frac{(x_1+x_2)^{-\frac{1}{3}} + |\ln x_2|}{x_1+x_2} = +\infty,$$

$$\limsup_{\substack{x_1+x_2\to+\infty \\ x_2\to+\infty}} \sup_{t\in[1,3]} \frac{f(t,x_1,x_2)}{x_2} = \limsup_{\substack{x_1+x_2\to+\infty \\ x_2\to+\infty}} \sup_{t\in[1,3]} \frac{((x_1+x_2)^{-\frac{1}{3}} + |\ln x_2|)}{x_2} = 0,$$

which implies that

$$\liminf_{\substack{x_1\to 0^+ \\ x_2\to 0^+}} \inf_{t\in[1,3]} \frac{f(t,x_1,x_2)}{x_1+x_2} > \lambda_1 > \limsup_{\substack{x_1+x_2\to+\infty \\ x_2\to+\infty}} \sup_{t\in[1,3]} \frac{f(t,x_1,x_2)}{x_2}.$$

So far, all the condition proposed in Theorem 1 have been meet. Consequently, Theorem 1 ensures that there must be one or more positive solutions to Equation (4.2). Equivalently, one or more positive solutions can be determined for Equation (4.2). It is worth mentioning that the calculation of fractional derivatives generally involves the beta function. In this example, because a specific explicit expression is provided, the beta function appears. However, in the general setting, if the function is abstract, then such the beta function will not appear.

5. Conclusions

The present study was concerned with the positive solutions and their existence for a nonlinear infinite-point Hadamard fractional differential equation. Using the fixed point theorem in cones, the nonlinear singularities on time and spatial variables are achieved through the spectral analysis of the relevant linear operator and Gelfand's formula. Compared with some existing literature discussing the same interval we derived herein, we overcome a significant difficulty that the consideration of solutions on this interval brings about a series of problems. Compared with some other existing literature, the nonlinear term of the equation contains the derivative term, and the infinite point is involved in the boundary conditions. These are the main contributions and innovations of the present research.

Author Contributions: C.L.: Conceptualization, Investigation, Formal analysis, Writing—original draft; L.G.: Methodology, Validation, Project administration, Resources, Writing—review and editing, Funding acquisition. All authors have read and agreed to the published version of the manuscript.

Funding: This research was supported by the National Natural Science Foundation ofChina (12272064, 12101086), Major Project of Basic Science (Natural Science) Research in Colleges andUniversities of Jiangsu Province (23KJA580001), and Young and Middle-aged Academic Leaders ofthe Blue Project in Colleges and Universities in Jiangsu Province (311202223005).

Institutional Review Board Statement: Not applicable.

Informed Consent Statement: All authors have read and agreed to the published version of the manuscript.

Data Availability Statement: No new data were created or analyzed in this study. Data sharing is not applicable to this article.

Acknowledgments: We appreciate the constructive comments and suggestions from the anonymous reviewers.

Conflicts of Interest: The authors declare no conflicts of interest.

References

1. Boulham, I.; Boubakir, A.; Labiod, S. Fuzzy L1 adaptive controller for chaos synchronization of uncertain fractional-order chaotic systems with input nonlinearities. *Tran. Inst. Meas. Cont.* **2023**, *45*, 2310–2323. [CrossRef]
2. Wu, J.; Wang, W.; Cao, Y.; Liu, S.; Zhang, Q.; Chu, W. A novel nonlinear fractional viscoelastic–viscoplastic damage creep model for rock-like geomaterials. *Compu. Geot.* **2023**, *163*, 105726. [CrossRef]
3. Ma, W.; Ma, N.; Dai, C.; Chen, Y.; Wang, X. Fractional modeling and optimal control strategies for mutated COVID-19 pandemic. *Math. Meth. Appl. Sci.* **2023**. [CrossRef]
4. Yang, B.; Zhu, T.; Zhang, X.; Wang, J.; Shu, H.; Li, S.; He, T.; Yang, L.; Yu, T. Design and implementation of Battery/SMES hybrid energy storage systems used in electric vehicles: A nonlinear robust fractional-order control approach. *Energy* **2020**, *191*, 116510. [CrossRef]
5. Ouhsaine, L.; Boukal, Y.; Ganaoui, M.; Darouach, M.; Zasadzinski, M.; Mimet, A.; Eddine Radhy, N. A General fractional-order heat transfer model for Photovoltaic/Thermal hybrid systems and its observer design. *Energy Procedia* **2017**, *139*, 49–54. [CrossRef]
6. Tomasz, P.; Jacek, G. Signal propagation in electromagnetic media described by fractional-order models. *Commun. Nonlinear Sci. Nume. Simul.* **2020**, *82*, 105029.
7. Piotrowska, E.; Rogowski, K. Analysis of fractional electrical circuit using Caputo and Conformable derivative definitions. *Non-Integer Order Calc. Appl.* **2019**, *15*, 183–194.
8. Huang, P.; Zhang, J.; Damascene, N.; Dong, C.; Wang, Z. A fractional order viscoelastic-plastic creep model for coal sample considering initial damage accumulation. *Alex. Engi. J.* **2021**, *60*, 3921–3930. [CrossRef]
9. Jena, R.; Chakraverty, S.; Rezazadeh, H.; Domiri Ganji, D. On the solution of time- fractional dynamical model of Brusselator reaction-diffusion system arising in chemical reactions. *Bull. Sci. Math.* **2020**, *43*, 3903–3913. [CrossRef]
10. Munoz-Pacheco, J.; Posadas-Castillo, C.; Zambrano-Serrano, E. The Effect of a Non-Local Fractional Operator in an Asymmetrical Glucose-Insulin Regulatory System: Analysis, Synchronization and Electronic Implementation. *Symmetry* **2020**, *12*, 1395. [CrossRef]
11. Yusuf, A.; Acay, B.; Tasiu, U.; Mustafa, I.; Baleanu, D. Mathematical modeling of pine wilt disease with Caputo fractional operator. *Chaos Solitons Fractals* **2021**, *143*, 110569. [CrossRef]
12. Guo, L.; Li, C.; Zhao, J. Existence of monotone positive solutions for Caputo-Hadamard nonlinear fractional differential equation with infinite-point boundary value conditions. *Symmetry* **2023**, *15*, 970. [CrossRef]
13. Li, C.; Zhu, C.; Lim, C.; Li, S. Nonlinear in-plane thermal buckling of rotationally restrained functionally graded carbon nanotube reinforced composite shallow arches under uniform radial loading. *Appl. Math. Mech. Engl. Ed.* **2022**, *43*, 1821–1840. [CrossRef]
14. Guo, L.; Cai, J.; Xie, Z.; Li, C. Mechanical responses of symmetric straight and curved composite microbeams. *J. Vibr. Eng. Technol.* **2024**, *12*, 1537–1549. [CrossRef]
15. Li, C.; Zhu, C.; Zhang, N.; Sui, S.; Zhao, J. Free vibration of self-powered nanoribbons subjected to thermal-mechanical-electrical fields based on a nonlocal strain gradient theory. *Appl. Math. Model.* **2022**, *110*, 583–602. [CrossRef]
16. Chabane, F.; Abbas, S.; Benbachir, M.; Benchohra, M.; N'Guérékata, G. Existence of Concave Positive Solutions for Nonlinear Fractional Differential Equation with p-Laplacian Operator. *Viet. J. Math.* **2023**, *51*, 505–543. [CrossRef]

17. Guo, L.; Zhao, J.; Liao, L.; Liu, L. Existence of multiple positive solutions for a class of infinite-point singular p-Laplacian fractional differential equation with singular source terms. *Nonli. Anal. Model. Control* **2022**, *27*, 609–629. [CrossRef]
18. Xu, J.; Liu, J.; O'Regan, D. Solvability for a Hadamard-type fractional integral boundary value problem. *Nonli. Anal. Model. Control* **2023**, *4*, 1–25. [CrossRef]
19. Arul, R.; Karthikeyan, P. Integral boundary value problems for implicit fractional differential equations involving Hadamard and Caputo-Hadamard fractional derivatives. *Krag. J. Math.* **2019**, *45*, 331–341.
20. Wang, J.; Feckan, M.; Zhou, Y. Fractional order iterative functional differential equations with parameter. *Appl. Math. Model.* **2013**, *37*, 6055–6067. [CrossRef]
21. Ardjouni, A.; Djoudi, A. Positive solutions for nonlinear Caputo-Hadamard fractional differential equations with integral boundary conditions. *Open J. Math. Anal.* **2019**, *3*, 62–69. [CrossRef]
22. Zhang, S. Positive solutions for a class of singular fractional differential equation with infinite-point boundary value conditions. *Appl. Math. Lett.* **2015**, *39*, 22–27. [CrossRef]
23. Boutiara, A.; Benbachir, M.; Guerbati, K. Boundary value problem for nonlinear Caputo-Hadamard fractional differential equation with Hadamard fractional integral and anti-periodic conditions. *Facta. Univer. Seri. Math. Infor.* **2021**, *36*, 735–748.
24. Ahmad, B.; Alsaedi, A.; Ntouyas, S.; Tariboon, J. *Hadamard-Type Fractional Differential Equations, Inclusions and Inequalities*; Springer: Cham, Switzerland, 2017.
25. Kilbas, A.; Srivastava, H.; Trujillo, J. *Theory and Applications of Fractional Differential Equations*; Elsevier Science BV: Amsterdam, The Netherlands, 2006.
26. Naseer, A. Positive solution for a Hadamard fractional singular boundary value problem of order $\mu \in (2, 3)$. *arXiv* **2021**, arXiv:2108.13127.
27. Guo, D.; Lakshmikantham, V. *Nonlinear Problems in Abstract Cones*; Academic Press: Boston, MA, USA, 1988.

fractal and fractional

MDPI

Article

Well-Posedness of the Mild Solutions for Incommensurate Systems of Delay Fractional Differential Equations

Babak Shiri

Data Recovery Key Laboratory of Sichuan Province, College of Mathematics and Information Science, Neijiang Normal University, Neijiang 641100, China; shiribabak2018@gmail.com

Abstract: Systems of incommensurate delay fractional differential equations (DFDEs) with non-vanishing constant delay of retarded type are investigated. It is shown that the mild solutions are well-posed in Hadamard sense on the space of continuous functions. The analysis is local and carried out for finite intervals. The strong results are obtained with weak conditions by using state-of-the-art new methods. No condition on the Lipschitz parameter is added for well-posedness results. Application of this theorem for the Hopfield neural network is carried out.

Keywords: fractional differential equation; well-posed in Hadamard sense; non-vanishing constant delay of retarded type; incommensurate systems; Hopfield neural networks

Academic Editors: Libo Feng, Yang Liu and Lin Liu

Received: 29 December 2024
Revised: 11 January 2025
Accepted: 19 January 2025
Published: 21 January 2025

Citation: Shiri, B. Well-Posedness of the Mild Solutions for Incommensurate Systems of Delay Fractional Differential Equations. *Fractal Fract.* **2025**, 9, 60. https://doi.org/10.3390/fractalfract9020060

1. Introduction

Fractional differential equations (FDEs) have found extensive use in the modeling of complex systems, primarily due to their ability to represent memory in systems dynamics [1,2]. Fortunately, there are various fractional derivatives, each having different memory kernels. This diversity, provide us with a rich set of operators capable of capturing the intricate dynamics of complex natural systems [3].

Similar to FDEs, delay differentiate equations (DDEs) have been widely applied in mathematical models, especially in biological phenomena [4]. However, DDEs are rarely studied and well-analyzed in connection to FDEs. The analysis of the linear delay fractional differential equation (DFDE),

$$\mathfrak{D}^{\psi}y(t) = ay(\theta(t)) + by(t) + f(t), \quad t > 0, \quad \psi \in (0,1), \quad a,b \in \mathbb{R}, \quad f \in C[0,\mathfrak{T}], \tag{1}$$

has been explored in [5]. In their paper, the lag term is defined by $\theta(t) := t - \tau$ with a constant time delay $\tau > 0$. This study covers the uniqueness of the mild solution and the dependency of the mild solution on its parameters. The stability of the problem (1) has also been studied [6].

We note that such DFDEs are known as retarded ones since the lag term does not involve fractional derivatives. If the equation involves terms like $\mathfrak{D}^{\psi}y(\theta(t))$, the equation becomes a neutral type, which we do not study here. The existence result of fractional-order neutral time-delay systems can be found in [7].

Most applied problems can not be modeled by a simple one-D equations like (1). A state of natural systems involves more than one component that are connected to each other to achieve an aim. In mathematics, a system is described by more than one equation. If the dynamics of each state depend on other states, it is called a coupled system. FDEs and DFDEs are utilized in diverse modeling. Currently, there is abundant work in biology, mechanics, electronics, and other branch of science that use systems of FDEs in their modeling.

However, they usually use a common order for all equations. But there is no need for all states to have the same memory and thus the same order of derivatives [8,9]. Therefore, the systems that may have different orders for each equation must be important for modeling. Such systems are known as incommensurate systems [10,11]. Incommensurate FDEs have been used in the structure of fractional Hopfield neural networks (FHNNs) [11–13].

A linear incommensurate system of DFDEs with multiple delays (multi-delay) can be described by

$$\begin{cases} {}^{C}\mathfrak{D}^{\psi_1} y_1(t) &= a_{11}y_1(t) + \ldots + a_{1\nu}y_\nu(t) + b_{11}y_1(\theta_1(t)) + \ldots + b_{1\nu}y_\nu(\theta_\nu(t)) + f_1(t), \\ &\vdots \\ {}^{C}\mathfrak{D}^{\psi_\nu} y_\nu(t) &= a_{\nu 1}y_1(t) + \ldots + a_{\nu\nu}y_\nu(t) + b_{\nu 1}y_1(\theta_1(t)) + \ldots + b_{\nu\nu}y_\nu(\theta_\nu(t)) + f_\nu(t), \end{cases} \tag{2}$$

where $\psi_i \in (0,1)$, $a_{ij}, b_{i,j} \in \mathbb{R}$ and $f_i : [0,\mathfrak{T}] \to \mathbb{R} \in C[0,\mathfrak{T}]$, $y_i : [0,\mathfrak{T}] \to \mathbb{R} \in C[0,\mathfrak{T}]$, for $i,j = 1,\ldots,\nu$. This system can be expressed in shorthand as

$$\mathfrak{D}^{\Psi}Y(t) = AY(t) + BY(\Theta(t)) + F(t) \tag{3}$$

where $\Psi = [\psi_1,\ldots,\psi_\nu]^T \in (0,1)^\nu$, $Y = [y_1,\ldots,y_\nu]^T \in (C[0,\mathfrak{T}])^\nu$, $F = [f_1,\ldots,f_\nu]^T \in (C[0,\mathfrak{T}])^\nu$, $\Theta = [\theta_1,\ldots,\theta_\nu]^T$, $\theta_i(t) = t - \tau_i$, $\tau_i \in \mathbb{R}$, $A = (a_{ij})$ and $B = (b_{ij})$. The meaning of such vector operators later will be recalled. Problem (2) has been studied subject to delay condition

$$y_i(t) = g_i(t), \quad t \in (-\tau_i, 0]. \tag{4}$$

If we assume that all delays are equal (i.e., $\tau_i = \tau$ where $\tau > 0$ is a positive constant), then the initial condition and prehistoric conditions (4) can be expressed as

$$Y(t) = G(t), \quad t \in (-\tau, 0]. \tag{5}$$

Such delays are also referred to as a single/constant delay. In this case, the lag term is a single function, and we simply use θ to represent it, where $\theta(t) := t - \tau$. The nonlinear DFDE with a constant delay can be described by

$$\begin{cases} {}^{C}\mathfrak{D}^{\psi_1} y_1(t) &= u_1(t, y_1(t),\ldots,y_\nu(t), y_1(\theta(t)),\ldots,y_\nu(\theta(t))), \\ &\vdots \\ {}^{C}\mathfrak{D}^{\psi_\nu} y_\nu(t) &= u_\nu(t, y_1(t),\ldots,y_\nu(t), y_1(\theta(t)),\ldots,y_\nu(\theta(t))), \end{cases} \tag{6}$$

subject to condition (5). It can be written in a compact form as

$$\mathfrak{D}^{\Psi}Y(t) = U(t, Y(t), Y(\Theta(t))). \tag{7}$$

where $\Theta(t) = [\theta,\ldots,\theta]^T$. We note that the related operations of $Y(\Theta(t))$ are explained for the algebra of the vector-valued function in the next sections, and they are not composite functions.

There are studies on systems with various delays for each state [14]. Specifically, ref. [14] addresses the stability of a class of incommensurate DFDEs with multiple delays. A numerical method for nonlinear systems of DFDE with a single/constant delay $\tau_i = \tau$ and commensurate order $\psi_i = \psi$ has been studied in [15]. The stability of a class of commensurate systems (3) with a constant delay, where A is a zero matrix, is studied in [16].

The objective of this paper is to study System (6) with constant delay. Such systems may appear in research papers, but fundamental questions regarding these systems, such as the existence of a unique solution and well-posedness are still unstudied. Existence results for systems of fractional equations typically utilize fixed-point theorems like the Banach fixed-point or Schauder fixed-point theorems. However, as discussed in [17] and other relative works, these types of theorems impose additional conditions on U or even $\mathfrak{T}$. Therefore, we apply the state-of-the-art method used in [17] to obtain strong results under weaker conditions. In the main theorems, we will find that continuity of the functions U and G and Lipschitz continuity of the function U with respect to Y is sufficient to guaranty the existence of the unique mild solution. Moreover, we show it is sufficient to ensure the continuous dependency of the mild solution to G, if we assume U is also Lipschitz continuous with respect to W.

Remark 1. *The fractional derivative and integral operators have two parameters. For example, in RL integral*

$$\mathfrak{I}^{\psi}y(t)(s)=\frac{1}{\Gamma(\psi)}\int_0^s (s-t)^{1-\psi}y(t)dt.$$

the variable t is the dummy variable, and s is the active variable. Usually, in the literature, the active variable is deleted. However, in this paper, for clarity, we may add the second variable to separate the dummy variable and active variable.

2. Vector-Valued Operational Algebra

Here, we clarify the algebra of vector-valued functions and their related operations through exact definitions. Suppose $\mathcal{Y}$ is a space of functions, and

$$F=[f_1,\dots,f_\nu]^T:[0,\mathfrak{T}]\to\mathbb{R}^\nu\in\mathcal{Y}^\nu$$

is a vector-valued function. For operators $o_i:\mathcal{Y}\to\mathcal{Y}$, we define the vector-operator

$$\mathfrak{O}=[o_1,\dots,o_\nu]^T:\mathcal{Y}^\nu\to\mathcal{Y}^\nu$$

by

$$\mathfrak{O}(F)=[o_1(f_1),\dots,o_\nu(f_\nu)]^T=\text{Diag}([o_1,\dots,o_\nu])F.$$

Such an $\mathfrak{O}$ can be ${}^CD^{\Psi}=[{}^CD^{\psi_1},\dots,{}^CD^{\psi_\nu}]^T$ or $\mathcal{I}^{\Psi}=[\mathcal{I}^{\psi_1},\dots,\mathcal{I}^{\psi_\nu}]^T$.

The elementary operations such as addition, difference, product, division, and power operations of vector-valued functions are performed element-wise. For example, if $G=[g_1,\dots,g_\nu]:[0,\mathfrak{T}]\to\mathbb{R}^\nu$ is another vector-valued function

$$F\pm G=[f_1\pm g_1,\dots,f_\nu\pm g_\nu]^T.$$

Let $\Theta=[\theta_1,\dots,\theta_\nu]^T:[0,\mathfrak{T}]\to\mathbb{R}^\nu$ be a vector-lag term. Clearly, $F(\Theta)$ is not well-defined as a composite function. However, we define it as vector-wise composite functional,

$$F(\Theta)=[f_1(\theta_1),\dots,f_\nu(\theta_\nu)]^T. \tag{8}$$

We note that, in our study $\Theta=[\theta,\dots,\theta]^T$ can be substituted by θ and serve as a composite function! Consequently, in our study, $F(\Theta)=F(\theta)$. However, in multi-delay cases, the use of element-wise composition is indispensable.

The operation of one-dimensional functions and operators with a scalar is inherited by all elements of the vector. For example, if $f:[0,\mathfrak{T}]\to\mathbb{R}$ is a scalar function, as well as $\Psi=[\psi_1,\dots,\psi_\nu]$, then $G=f^{\Psi}$ is defined by

$$G=[f^{\psi_1},\dots,f^{\psi_\nu}]^T.$$

Example 1. *For* $\Psi = [\psi_1, \ldots, \psi_v]^T$ *and* $P = [p_1, \ldots, p_v]^T$, *we have*

$$\mathfrak{I}^{\Psi} t^P = \frac{\Gamma(P+1)}{\Gamma(P+\Psi+1)} t^{\Psi+P}. \tag{9}$$

We note that this equation is exactly similar to the computation of the non-vector case of $I^{\alpha} t^{\beta}$. *However, we should note that the notation here has a different meaning. To be more clear, the right-hand side of Equation* (9) *is*

$$[\frac{\Gamma(p_1+1)}{\Gamma(p_1+\psi_1+1)} t^{p_1+\psi_1}, \ldots, \frac{\Gamma(p_v+1)}{\Gamma(p_v+\psi_v+1)} t^{p_v+\psi_v}]^T!$$

Theorem 1. *Let* $0 < \psi_i \leq 1, i = 1, \ldots, v$ *and* $Y \in (\mathcal{AC}[0, \mathfrak{T}])^v$. *Then,*

$$\mathfrak{I}^{\Psi}\, {}^C D^{\Psi} Y = Y(t) - Y(0) \tag{10}$$

and

$${}^C D^{\Psi} \mathfrak{I}^{\Psi} Y(t) = Y(t). \tag{11}$$

Theorem 2 (Generalized Gronwall inequality [18,19]). *Suppose* $q, \mathfrak{T} > 0$, $c \geq 0$ *and* $y : [0, \mathfrak{T}] \to \mathbb{R}$ *is a locally integrable and no-negative function satisfying*

$$y(t) \leq c + M \int_0^t (t-a)^{\psi-1} y(s) ds \tag{12}$$

Then, $y(t) \leq c E_{\psi}(M \Gamma(\psi) t^{\psi})$.

3. Existence of a Unique Continuous Mild Solution with Single Delay

Applying the RL integral $\mathfrak{I}^{\Psi}$ to both sides of (3), we obtain

$$Y(t) = Y(0) + \mathfrak{I}^{\Psi} A Y(t) + \mathfrak{I}^{\Psi} B Y(\Theta(t)). \tag{13}$$

It is well-known that the solution to (13) may not be differentiated, or may not be in $\mathcal{AC}$ (see, for example, [20]). So, the solution of (13) may not satisfies the original Equation (3). But, in most applied mathematics, we need a solution of the model based on (3). Therefore, we use the adjective "mild" before the word solution to emphasize that the solution of associate integral may not be the solution of the original Equation (3).

We should note that interchanging the place of the fractional integral and constant matrix in System (13) can be problematic. While this interchange is valid for commensurate systems, it is incorrect for incommensurate systems. In particular, we have

$$\mathfrak{I}^{\Psi} A Y \neq A \mathfrak{I}^{\Psi} Y. \tag{14}$$

This leads to additional complexities for incommensurate FDEs compared to commensurate FDEs.

Similarly, and in general, from (7), we can infer that the mild solution satisfies

$$Y(t) = Y(0) + \mathfrak{I}^{\Psi} U(t, Y(t), Y(\Theta(t))). \tag{15}$$

Definition 1. *We define* Y *as a mild solution of* (7) *under Condition* (5) *if it fulfills System* (15).

Remark 2. *A solution to* (15) *might not be differentiable and, therefore, might not satisfy the original Problem* (7) *under Condition* (5). *However, the solution to the original problem satisfies* (15).

The existence of a mild solution of incommensurate systems of FDEs has been established in [17]. Based on the discussion in that paper, the existence of a mild solution based on the Banach fixed-point theorem and the Schauder fixed-point theorem require stronger conditions. To obtain the weaker condition, the authors of [17] proposed a direct method using Cauchy sequences. They first established the result on the interval $[0,1]$ and then extended the global existence using the tail part of the RL integral.

Now, consider a single delay, τ. Let $t \in (0,\tau]$, which implies that $t-\tau \in (-\tau, 0]$. Then, $Y(\Theta(t)) = G(\Theta(t))$, and Equation (15) becomes equivalent to

$$Y(t) = G(0) + \mathfrak{I}^{\Psi} U(t, Y(t), G(\Theta(t))), \quad t \in [0,\tau]. \tag{16}$$

We note that $G(\Theta(0)) = G(-\tau)$ is not defined. We assume $G(\Theta(0^+)) = G(-\tau^+) < \infty$, and extend G on $[-\tau, 0]$ by $G(-\tau) = G(-\tau^+)$ to use compactness of interval. It follows immediately that $G \in \mathcal{C}[-\tau, 0]$. Also, it is clear the value of $\mathfrak{I}^{\Psi}$ at $t=0$ is zero for any continuous function.

We use the absolute value notation as a norm in $\mathbb{R}^{\nu}$, defined by

$$|Y| = \max_{i=1,\dots,\nu} |y_i|.$$

However, we use the norm notation for spaces of functions, especially for $F \in (C[0,\mathfrak{T}])^{\nu}$

$$\|F\| = \sup_{t\in[0,\mathfrak{T}]} |F(t)| = \sup_{t\in[0,\mathfrak{T}]} \max_{i=1,\dots,\nu} |f_i(t)|.$$

Now, consider the following hypotheses:

(H1) U is globally Lipschitz continuous with respect to Y, i.e., $\exists\, L_i \in \mathbb{R}$, such that

$$|u_i(t,Y,W) - u_i(t,Z,W)| \le L_i|Y-Z|$$

for all $t \in [0,\mathfrak{T}]$, and for all $Z, Y, W \in \mathbb{R}^{\nu}$.

(H2) $U : [0,\mathfrak{T}] \times \mathbb{R}^{\nu} \times \mathbb{R}^{\nu} \to \mathbb{R}^{\nu}$ is continuous with respect to its domain.

(H3) G is a continuous function on $(-\tau, 0]$ and $\lim_{t\to\tau} g_i(t) < \infty$.

Remark 3. *Letting $L = \max_{i=1,\dots,\nu} |L_i|$, it follows from (H1) that*

$$|U(t,Y,W) - U(t,Z,W)| \le L|Y-Z|, \quad \forall t \in [0,\mathfrak{T}], \forall Z, Y, W \in \mathbb{R}^{\nu}.$$

Remark 4. *More precisely, (H3) states that if we extend G to $[-\tau, 0]$ by $G(-\tau) = \lim_{t\to\tau} G(t)$, then it is continuous on $[-\tau, 0]$. Thus, the condition (H3) can be expressed as "G is continuous on $[-\tau, 0]$".*

Theorem 3. *Let Hypotheses (H1)–(H3) hold. Then, System* (7) *subject to Condition* (5) *processes a mild continuous solution on $[0,\tau]$.*

Proof. The proof is similar to Theorem 7 of [17], so we provide a proof sketch here. First, we assume $\tau \le 1$, introduce a Picard operator $\mathfrak{P}$ by

$$\mathfrak{P}Y = G(0) + \mathfrak{I}^{\Psi} U(t, Y(t), G(\Theta(t))) \tag{17}$$

and we show that the functions

$$Y_{n+1} = \mathfrak{P}Y_n, \quad Y_0 = G_0, \quad n = 0, 1, \dots$$

are Cauchy sequences in $(\mathcal{C}[0,\tau])^{\nu}$. Consequently, they have a uniform limit, say Y. By the uniform convergence theorem for fractional integrals, we have

$$\begin{aligned}\lim_{N\to\infty} Y_{n+1} &= \lim_{N\to\infty} \mathfrak{P}Y_n = G(0) + \mathfrak{I}^{\Psi}U(t, Y_N(t), G(\Theta(t))) \\ &= G(0) + \mathfrak{I}^{\Psi} \lim_{N\to\infty} U(t, Y_N(t), G(\Theta(t))).\end{aligned} \tag{18}$$

Since U is continuous, we can rewrite the above as

$$\lim_{N\to\infty} Y_{n+1} = G(0) + \mathfrak{I}^{\Psi}U(t, \lim_{N\to\infty} Y_N(t), G(\Theta(t))). \tag{19}$$

Thus, Y satisfies Equation (16). Now, assume that $1 < \tau \leq 2$. Given that a solution exists on $[0,1]$, w can decompose the $\mathfrak{I}^{\Psi}$ into two operators: the Fredholm operator $\mathfrak{F}^{\Psi}$ and the Volterra operator $\mathfrak{V}^{\Psi}$. The Fredholm operator is defined as

$$\mathfrak{F}^{\Psi}Y(t) = \frac{1}{\Gamma(\Psi)} \int_0^1 (t-z)^{\Psi-1} Y(z)dz \tag{20}$$

and the Volterra operator is defined as

$$\mathfrak{V}^{\Psi}Y(t) = \frac{1}{\Gamma(\Psi)} \int_1^t (t-z)^{\Psi-1} Y(z)dz. \tag{21}$$

We know that $\mathfrak{I}^{\Psi} = \mathfrak{F}^{\Psi} + \mathfrak{V}^{\Psi}$. Therefore, Equation (16) can be written as

$$Y(t) = G(0) + \mathfrak{F}^{\Psi}U(.,Y(.),G(\Theta(.)))(t) + \mathfrak{V}^{\Psi}U(.,Y(.),G(\Theta(.)))(t), \quad t \in [0,\tau]. \tag{22}$$

We have already established the existence of a solution of (16) on $[0,1]$. For convenience, we rename Y on $[0,1]$ by Y_1. Then, the tail of Equation (22) is known function say Z, i.e.,

$$Z(t) := G(0) + \mathfrak{F}^{\Psi}U(.,Y_1(.),G(\Theta(.)))(t).$$

Thus,

$$Y(t) = Z(t) + \mathfrak{V}^{\Psi}U(.,Y(.),G(\Theta(.)))(t), \quad t \in [1,\tau]. \tag{23}$$

Substituting $t = s+1$ gives

$$Y(s+1) = Z(s+1) + \frac{1}{\Gamma(\Psi)} \int_1^{s+1} (s+1-z)^{\Psi-1} U(z, Y(z), G(\Theta(z)))dz, \tag{24}$$

where $s \in [0, \tau-1]$. If we substitute $z = r+1$, we obtain

$$Y(s+1) = Z(s+1) + \frac{1}{\Gamma(\Psi)} \int_0^{s} (s-r)^{\Psi-1} U(r+1, Y(r+1), G(\Theta(r+1)))dr. \tag{25}$$

It is interesting to note that when considering the Volterra operator at the peak of dynamical System (25), it becomes an RL integral operate, and by renaming $Y_2(s) = Y(s+1)$, we obtain

$$Y_2(s) = Z(s+1) + \mathfrak{I}^{\Psi}U(r+1, Y_2(r), G(\Theta(r+1)))(s), \quad s \in [0, \tau-1]. \tag{26}$$

Equation (26) is of the same form as (16). Consequently, the existence of a unique $Y_2 \in (\mathcal{C}[0,\tau-1])^{\nu}$ is guaranteed in the same way. Conclusively,

$$Y(t) = \begin{cases} Y_1(t), & t \in [0,1], \\ Y_2(t-1), & t \in [1,\tau]. \end{cases} \tag{27}$$

Clearly, $Y_2(1-1) = Y_2(0) = Z(1) = Y_1(1)$ and Y is a continuous solution to (16) on $[0,\tau]$. By induction, (16) has a unique continuous solution for any $\tau > 0$. □

The uniqueness of the solution follows from the generalized Gronwall inequality.

Theorem 4. *Let Hypotheses (H1)–(H3) hold. Then, System* (7) *subject to condition* (5) *has a unique mild continuous solution on* $[0,\tau]$.

Proof. The proof is similar to Theorem 10 of [17]. First, we show the uniqueness for $\tau \leq 1$. Let X and Y be two solutions. Then,

$$|x_i(t) - y_i(t)| \leq |\mathfrak{I}^{\psi_i} u_i(t, x_i(t), g_i(\theta_i(t))) - \mathfrak{I}^{\psi_i} u_i(t, y_i(t), g_i(\theta_i(t)))|, \quad t \in [0,\tau] \subseteq [0,1].$$

Since $t \leq 1$, it follows from Hypotheses (H1) that

$$\begin{aligned} |x_i(t) - y_i(t)| &\leq \frac{L_i}{\Gamma(\alpha_i)} \int_0^t (t-z)^{\psi_i - 1} |X(z) - Y(z)| dz \\ &\leq M^* \int_0^t (t-z)^{\psi^* - 1} |X(z) - Y(z)| dz \end{aligned} \tag{28}$$

where $M* = \max_{i=1,\dots,\nu} \frac{L_i}{\Gamma(\alpha_i)}$ and $\psi^* = \min_{i=1,\dots,\nu} \psi_i$. We note that Inequality (28) is independent of i. Thus,

$$|X(t) - Y(t)| \leq M^* \int_0^t (t-z)^{\psi^* - 1} |X(z) - Y(z)| dz. \tag{29}$$

Immediately, it follows from the generalized Gronwall inequity that $|X(t) - Y(t)| = 0$, and $X(t) = Y(t)$ for all $t \in [0,1]$. Now, let $\tau \in [0,2]$. From a previous argument, we already know that $X(t) = Y(t)$ for $t \in [0,1]$. From an argument similar to the proof of Theorem (3), Y and X satisfies Equation (25) while the definition of Z only uses the information of $Y(t)$ on $[0,1]$. Since Y is unique on $[0,1]$, the tail function Z is unique for all $t \in [0,\tau]$. It follows from (25) that

$$|X(s+1) - Y(s+1)| \leq M^* \int_0^s (s-z)^{\psi^* - 1} |X(r+1) - Y(r+1)| dr, \tag{30}$$

for $s \in [0, \tau - 1] \subseteq [0,1]$. Therefore, from the generalized Gronwall inequality,

$$X(s+1) = Y(s+1), \quad s \in [0, \tau - 1]$$

or, equivalently, $X(t) = Y(t)$ on $[1,\tau]$ and, thus, on $[0,\tau]$. Similar induction can be used to prove that $X(t) = Y(t)$ for any delay $\tau > 0$. □

Up to this point, we have determined that System (1) has a unique mild solution within the interval $[0,\tau]$, and this solution is continuous. Now, we establish the existence of a unique continuous solution for any arbitrary interval $[0,\mathfrak{T}]$.

Theorem 5. *Let Hypotheses (H1)–(H3) hold. Then, System* (7) *subject to Condition* (5) *processes a unique mild continuous solution on* $[0,\mathfrak{T}]$.

Proof. If $\mathfrak{T} < \tau$, the claim follows from Theorem 4. Suppose $\tau < \mathfrak{T} \leq 2\tau$. We already know that there exists a solution on $[0,\tau]$. Denote this solution as Y_1. Then, System (15) is equivalent to

$$Y(t) = G(0) + \mathfrak{F}_\tau^\Psi U(z, Y_1(z), G(z-\tau))(t) + \mathfrak{V}_\tau^\Psi U(z, Y(z), Y_1(z-\tau))(t) \tag{31}$$

for $t \in [\tau, \mathfrak{T}]$, where

$$\mathfrak{F}_{\tau}^{\Psi} Y(t) = \frac{1}{\Gamma(\Psi)} \int_{0}^{\tau} (t-z)^{\Psi-1} Y(z) dz \tag{32}$$

is a Fredholm operator, and

$$\mathfrak{V}_{\tau}^{\Psi} Y(t) = \frac{1}{\Gamma(\Psi)} \int_{\tau}^{t} (t-z)^{\Psi-1} Y(z) dz \tag{33}$$

is a Volterra operator. For the tail of System (31) $Z(t) = G(0) + \mathfrak{F}_{\tau}^{\Psi} U(s, Y_1(s), G(s-\tau))(t)$ the Fredholm operator only depends on the values of $Y_1(s)$ and $G_1(s-\tau)$ on $s \in [0, \tau]$. Since Y_1 is known to be a unique, Z is well-defined unique function. Conclusively, by substituting $t = r + \tau$, we have (31),

$$Y(r+\tau) = Z(r+\tau) + \mathfrak{V}_{\tau}^{\Psi} U(s, Y(s), Y_1(s-\tau))(r+\tau). \tag{34}$$

Regarding the peak operator, by substituting $z = s + \tau$, we obtain

$$\begin{aligned} \mathfrak{V}_{\tau}^{\Psi} U(s, Y(s), Y_1(s-\tau))(r+\tau) =& \frac{1}{\Gamma(\Psi)} \int_{\tau}^{r+\tau} (r+\tau-z)^{\Psi-1} U(z, Y(z), Y_1(z-\tau)) dz \\ =& \frac{1}{\Gamma(\Psi)} \int_{0}^{r} (t-z)^{\Psi-1} U(s+\tau, Y(s+\tau), Y_1(s)) ds \\ =& \mathfrak{J}^{\Psi} U(s+\tau, Y(s+\tau), Y_1(s))(r). \end{aligned} \tag{35}$$

If we rename $Y_2(t) := Y(t+\tau)$ and consider Equation (35), then Equation (34) can be written as

$$Y_2(r) = Z(r+\tau) + \mathfrak{J}^{\Psi} U(s+\tau, Y_2(s), Y_1(s))(r), \quad r \in [0, \mathfrak{T} - \tau]. \tag{36}$$

Noticeably, we convert an equation on $[\tau, \mathfrak{T}]$ into a similar type of equation on the interval $[0, \mathfrak{T} - \tau] \subseteq [0, \tau]$. Therefore, we can use the same reasoning as in Theorems 3 and 4 to show that System (36) has a unique continuous solution on $[0, \mathfrak{T} - \tau]$.

We note that the only difference between Equations (16) and (36) is that $G(0)$ is replaced by $Z(r+\tau)$, which does not effect the proof. Thus, System (15) has a unique solution, given by

$$y(t) = \begin{cases} G(t), & t \in (-\tau, 0], \\ Y_1(t), & t \in (0, \tau], \\ Y_2(t-\tau), & t \in (\tau, \mathfrak{T}]. \end{cases} \tag{37}$$

It is clear that $Y(\tau^+) = Y_2(0^+) = Z(\tau^+) = Z(\tau) = Y_1(\tau)$, indicating that Y is continuous at τ. Moreover, $Y(0^+) = Y_1(0^+) = Y_1(0) = G(0) = Y(0)$, showing that Y is also continuous at 0. Conclusively, $Y \in (\mathcal{C}[0, \mathfrak{T}])^{\nu}$. This completes the proof for this case. The same recursive argument can be applied to prove the cases when $\mathfrak{T}$ belongs to $[2\tau, 3\tau]$, $[3\tau, 4\tau]$ and so on, thereby completing the proof. □

4. Transforming DFDEs into Fractional Integral Equations Without Delays

The existence analysis above presents a constructive approach to study DFDEs. Let $n = \mathrm{Ciel}\left(\frac{T}{\tau}\right)$, meaning that $n - 1 < \frac{T}{\tau} \leq n$. Let Y_j represent the solution on the interval $[(j-1)\tau, j\tau]$ for $j = 1, \ldots, n$. As stated previously, for $j < n$, we can define the Fredholem operators related to the system's memory on $[(j-1)\tau, j\tau]$ as follows:

$$\mathfrak{F}_{j\tau}^{\Psi} Y(t) = \frac{1}{\Gamma(\Psi)} \int_{(j-1)\tau}^{j\tau} (t-z)^{\Psi-1} Y_j(z) dz. \tag{38}$$

It should be noted that while this operator utilizes the information of Y at $[(j-1)\tau, j\tau]$, the variable t can take on arbitrary values. In particular, this operator is well-defined for $[0,\infty)$. Consequently, the contribution of Y within the jth interval regarded to the memory of the Fredholm operator (38) is distributed over future time.

Theorem 6. *The memory associated with the fixed interval j is fading as t increases.*

Proof. From Theorem (5), Y_j is continuous. Assume $t_2 > t_1 \geq 1 + j\tau$. Then, $t_2 - z > t_1 - z > 1$ and $(t_2 - z)^{1-\psi_i} > (t_1 - z)^{1-\psi_i} \geq 1$ for $z \in [(j-1)\tau, j\tau]$. Therefore,

$$(t_2 - z)^{\Psi-1}|Y(z)| < (t_1 - z)^{\Psi-1}|Y(z)| \leq 1$$

where the inequities are component-wise. Thus,

$$\mathfrak{F}^{\Psi}_{j\tau}|Y|(t_2) \leq \mathfrak{F}^{\Psi}_{j\tau}|Y|(t_1).$$

Therefore, $\mathfrak{F}^{\Psi}_{j\tau}|Y|$ is decreasing component-wise for $t > 1 + j\tau$. Since Y is continuous, and $(t-z)^{\Psi-1}Y(z) \to [0,\ldots,0]^T$ uniformly as $t \to \infty$, we conclude that

$$\mathfrak{F}^{\Psi}_{j\tau}Y(t) \to [0,\ldots,0]^T$$

as $t \to \infty$. This indicates that the memory of jth interval fades to zero. □

Finally, we can define the peak information of $\mathfrak{J}^{\Psi}$ by Volterra operator,

$$\mathfrak{V}^{\Psi}_{j\tau}Y(t) = \frac{1}{\Gamma(\Psi)} \int_{j\tau}^{t} (t-z)^{\Psi-1}Y(z)dz \tag{39}$$

where $t \in [j\tau, (j+1)\tau]$, for $j = 0,\ldots,n-2$, and $t \in [(n-1)\tau, \mathfrak{T}]$ when $j = n-1$. We show that the translation of the peak operator is precisely the RL operator of a translation of Y.

Theorem 7. *For $j = 0,\ldots,n-1$, we have*

$$\mathfrak{V}^{\Psi}_{j\tau}Y(t + j\tau) = \mathfrak{J}^{\Psi}Y(s + j\tau)(t). \tag{40}$$

Proof. Substitute $s = z - j\tau$ into (39). This gives

$$\mathfrak{V}^{\Psi}_{j\tau}Y(t) = \frac{1}{\Gamma(\Psi)} \int_{0}^{t-j\tau} (t - s - j\tau)^{\Psi-1}Y(s + j\tau)ds. \tag{41}$$

Now, consider the value of this operator when t is replaced by $t + j\tau$,

$$\mathfrak{V}^{\Psi}_{j\tau}Y(t + j\tau) = \frac{1}{\Gamma(\Psi)} \int_{0}^{t} (t-s)^{\Psi-1}Y(s + j\tau)ds = \mathfrak{J}^{\Psi}Y(s + j\tau)(t) \tag{42}$$

where $t \in [0,\tau]$ for $j < n-1$ and $t \in [0, \mathfrak{T} - (n-1)\tau]$ if $j = n-1$. □

For convenience, we define $Y_0(t) = G(t)$ on the interval $(-\tau, 0]$. Then, we have the following series of equations: For $t \in (0,\tau]$, Y_1 is given by

$$Y_1(t) = G(0) + \mathfrak{J}^{\Psi}U(s, Y_1(s), Y_0(\Theta(s)))(t), \tag{43}$$

for $t \in (\tau, 2\tau]$, Y_2 is expressed as

$$Y_2(t) = G(0) + \mathfrak{F}^{\Psi}_{\tau}U(s, Y_1(s), Y_0(\Theta(s)))(t) + \mathfrak{V}^{\Psi}_{\tau}U(s, Y_2(s), Y_1(\Theta(s)))(t),$$

and for $j < n$, and $t \in ((j-1)\tau, j\tau]$, Y_j is obtained by

$$\begin{aligned} Y_j(t) =& G(0) + \sum_{i=1}^{j-1} \mathfrak{F}^{\Psi}_{i\tau} U(s, Y_i(s), Y_{i-1}(\Theta(s)))(t) \\ &+ \mathfrak{V}^{\Psi}_{(j-1)\tau} U(s, Y_j(s), Y_{j-1}(\Theta(s)))(t), \end{aligned} \tag{44}$$

For final interval $t \in ((n-1)\tau, \mathfrak{T}]$, where $j = n$, we have

$$\begin{aligned} Y_n(t) =& G(0) + \sum_{i=1}^{n-1} \mathfrak{F}^{\Psi}_{i\tau} U(s, Y_i(s), Y_{i-1}(\Theta(s)))(t) \\ &+ \mathfrak{V}^{\Psi}_{(n-1)\tau} U(s, Y_n(s), Y_{n-1}(\Theta(s)))(t). \end{aligned} \tag{45}$$

We define the tail function as

$$Z_j(t) = G(0) + \sum_{i=1}^{j-1} \mathfrak{F}^{\Psi}_{i\tau} U(s, Y_i(s), Y_{i-1}(\Theta(s)))(t).$$

By using Theorem 7, we obtain integral equations

$$Y_j(t + j\tau) = Z_j(t + j\tau) + \mathfrak{J}^{\Psi} U(s + j\tau, Y_j(s + j\tau), Y_{j-1}(\Theta(s + j\tau)))(t) \tag{46}$$

for $j = 1, \ldots, n-1$, $t \in (0, \tau]$, and $j = n$, $t \in (0, \mathfrak{T} - (n-1)\tau] \subseteq (0, \tau]$. It is clear that Equation (46) is an integral equation with respect to $Y_j(t + j\tau)$.

5. Well-Posedness

The well-posedness of a mathematical problem in the Hadamard sense requires three conditions: existence, uniqueness, and stability. Until now, we have provided the first two conditions. In this section, we investigated the third condition. The stability for the well-posedness of a problem implies that the solution depends continuously to the data. In particular, for initial value problems, the data consists of initial conditions. Here, the given data are a function G which is well-defined at $(-\tau, 0]$. To ensure the stability, let emphasize the dependence of Y to G through $Y(t, G)$. Let $\tilde{Y}(t, \tilde{G})$ be a solution to (46) corresponding with $\tilde{G}$. This implies $\tilde{Y}_0(t) = \tilde{G}(t)$ for $t \in (0, \tau]$ and $\tilde{Y}_j$ satisfies the systems described by (46). To establish well-posedness we need a slightly stronger assumption than (H1). This assumption involves adding Lipschitz continuity for the third parameter of U.

(H4) U is globally Lipschitz continuous with respect to Y and W, i.e., $\exists\, L_i \in \mathbb{R}$, such that

$$|u_i(t, Y, W) - u_i(t, \hat{Y}, \hat{W})| \leq L_i \max\{|Y - \hat{Y}|, |W - \hat{W}|\}$$

for all $t \in [0, \mathfrak{T}]$ and for all $Y, W, \hat{Y}, \hat{W} \in \mathbb{R}^{\nu}$.

Theorem 8. *Let Hypotheses (H2)–(H4) hold. Then, the mild solution of System* (7) *subject to Condition* (5) *is Lipschitz continuous with respect to G on the finite interval* $[0, \mathfrak{T}]$.

Proof. Let $M = \max \frac{L_i}{\Gamma(\alpha_i)}$. Then,

$$\begin{aligned} |Y_1(t) - \tilde{Y}_1(t)| \leq& |G(0) - \tilde{G}(0)| \\ &+ M \int_0^t (t-z)^{\psi^* - 1} \max\{|Y_1(z) - \hat{Y}_1(z)|, |G(z) - \hat{G}(z)|\} \end{aligned} \tag{47}$$

for $t \leq 1$, where $\psi^* = \min \psi_i$. Therefore,

$$\begin{aligned}|Y_1(t) - \tilde{Y}_1(t)| \leq &|G(0) - \tilde{G}(0)| \\ &+ M \int_0^t (t-z)^{\psi^*-1}|Y_1(z) - \hat{Y}_1(z)|dz \\ &+ M \int_0^t (t-\tau)^{\psi^*-1}|G(z) - \hat{G}(z)|dz\end{aligned}$$

for $t \leq 1$. Thus,

$$\begin{aligned}|Y_1(t) - \tilde{Y}_1(t)| \leq &|G(0) - \tilde{G}(0)| + M \int_0^t (t-z)^{\psi^*-1}|Y_1(z) - \hat{Y}_1(z)|dz \\ &+ M \int_0^t (t-\tau)^{\psi^*-1}|G(z) - \hat{G}(z)|dz.\end{aligned}$$

It follows that

$$|Y_1(t) - \tilde{Y}_1(t)| \leq \left(1 + \frac{Mt^{\psi^*}}{\psi^*}\right)||G - \tilde{G}|| + M \int_0^t (t-z)^{\psi^*-1}|Y_1(z) - \hat{Y}_1(z)|dz \tag{48}$$

and, thus,

$$|Y_1(t) - \tilde{Y}_1(t)| \leq \left(1 + \frac{Mt^{\psi^*}}{\psi^*}\right)E_{\psi^*}(M\Gamma(\psi^*)t^{\psi^*})||G - \tilde{G}||. \tag{49}$$

Consequently,

$$|Y_1(t) - \tilde{Y}_1(t)| = C_1||G - \tilde{G}||, \quad t \in [0,1], \tag{50}$$

where $C_1 := \max_{t\in[0,1]}\left(1 + \frac{Mt^{\psi^*}}{\psi^*}\right)E_{\psi^*}(M\Gamma(\psi^*)t^{\psi^*})$. This completes the proof on $[0,\tau]$ when $\tau \leq 1$. Now, consider $\tau \in [0,2]$. For $t > 1$, we have

$$\begin{aligned}|Y_1(t) - \tilde{Y}_1(t)| \leq &|G(0) - \tilde{G}(0)| + M \int_0^t |(t-z)^{\Psi-1}||Y_1(z) - \hat{Y}_1(z)|dz \\ &+ M \int_0^t |(t-z)^{\Psi-1}||G(z) - \hat{G}(z)|dz \\ \leq &|G(0) - \tilde{G}(0)| + M \int_0^1 |(t-z)^{\Psi-1}||Y_1(z) - \hat{Y}_1(z)|dz \\ &+ M \int_1^t |(t-z)^{\Psi-1}||Y_1(z) - \hat{Y}_1(z)|dz + M \int_0^t |(t-z)^{\Psi-1}||G(z) - \hat{G}(z)|dz \\ \leq &||G - \tilde{G}|| + M \max_{z\in[0,1]} |Y_1(z) - \hat{Y}_1(z)|\Omega + M \int_1^t |(t-z)^{\Psi-1}||Y_1(z) - \hat{Y}_1(z)|dz \\ &+ M||G - \hat{G}||\Omega \\ \leq &(1 + MC_1\Omega + M\Omega)||G - \hat{G}|| + M \int_1^t |(t-z)^{\Psi-1}||Y_1(z) - \hat{Y}_1(z)|dz\end{aligned} \tag{51}$$

where

$$\Omega := \max_{t\in[0,2]} \int_0^t |(t-z)^{\Psi-1}|dz$$

and, clearly, $\Omega < \infty$. It immediately follows that

$$|Y_1(t+1) - \tilde{Y}_1(t+1)| \leq C_2||G - \hat{G}|| + M \int_0^t (t-s)^{\psi^*-1}|Y_1(s+1) - \hat{Y}_1(s+1)|ds \tag{52}$$

for $t \in [0,1]$ where $C_2 = 1 + MC_1\Omega + M\Omega$. By applying the generalized Gronwall inequality, we obtain

$$|Y_1(t+1) - \tilde{Y}_1(t+1)| \leq C_2 E_{\psi^*}(M\Gamma(\psi^*)t^{\psi^*})\|G - \hat{G}\|, \quad t \in [0,1] \tag{53}$$

and, thus,

$$|Y_1(t) - \tilde{Y}_1(t)| \leq C_2 E_{\psi^*}(M\Gamma(\psi^*)(t-1)^{\psi^*})\|G - \hat{G}\|, \quad t \in [1,2]. \tag{54}$$

Particularly, by defining $C_3 = \max\{C_1, \max_{t\in[0,1]} C_2 E_{\psi^*}(M\Gamma(\psi^*)(t-1)^{\psi^*})\}$, we have

$$|Y_1(t) - \tilde{Y}_1(t)| \leq C_3\|G - \hat{G}\|, \quad t \in [0,\tau]. \tag{55}$$

Using the same induction, Equation (55) holds for any $\tau < \infty$. Now we consider the dynamics of the system for $t > \tau$. Suppose that

$$|Y(t) - \tilde{Y}(t)| \leq C\|G - \hat{G}\|, \quad t \in [0,(j-1)\tau], j \geq 1 \tag{56}$$

for some constant C. We provide the same result on $[(j-1)\tau, j\tau]$. We note that, by definition of Z_j,

$$Z_j(t) = G(0) + \frac{1}{\Gamma(\Psi)}\sum_{i=1}^{j-1}\int_{(i-1)\tau}^{i\tau}(t-s)^{\Psi-1}U(s,Y_i(s),Y_{i-1}(\Theta(s)))ds. \tag{57}$$

Thus,

$$\begin{aligned}|Z_j(t) - \tilde{Z}_j(t)| \leq& |G(0) - \tilde{G}(0)| \\ &+ M\sum_{i=1}^{j-1}\int_{(i-1)\tau}^{i\tau}|(t-s)^{\Psi-1}|\max\{|Y_i(s) - \tilde{Y}_i|, |Y_{i-1}(\Theta(s)) - \tilde{Y}_{i-1}(\Theta(s))|\}ds\end{aligned} \tag{58}$$

For $i \leq j-1$, based on the assumption, $|Y_i(s) - \tilde{Y}_i| \leq C\|G - \hat{G}\|$ and $|Y_{i-1}(\Theta(s)) - Y_{i-1}(\Theta(s))| \leq C\|G - \hat{G}\|$. Therefore, for $t \in [(j-1)\tau, j\tau]$,

$$\begin{aligned}|Z_j(t) - \tilde{Z}_j(t)| \leq& |G(0) - \tilde{G}(0)| + MC\|G - \hat{G}\|\sum_{i=1}^{j-1}\int_{(i-1)\tau}^{i\tau}|(t-s)^{\Psi-1}|ds \\ \leq& \|G - \tilde{G}\| + MC\|G - \hat{G}\|\int_0^{(j-1)\tau}|(t-s)^{\Psi-1}|ds, \\ \leq& (1 + MC\Omega_j)\|G - \tilde{G}\|\end{aligned} \tag{59}$$

where

$$\Omega_j := \max_{t\in[(j-1)\tau, j\tau]}\int_0^t|(t-z)^{\Psi-1}|dz \leq \infty.$$

Using (46), we obtain

$$\begin{aligned}&|Y_j(j\tau+t) - \tilde{Y}_j(j\tau+t)| \leq |Z_j(j\tau+t) - \tilde{Z}_j(j\tau+t)| \\ &\quad + M\int_0^t|(t-s)^{\Psi-1}|\max\{|Y_i(j\tau+s) - \tilde{Y}_i(j\tau+s)|, |Y_{i-1}(\Theta(s)) - \tilde{Y}_{i-1}(\Theta(j\tau+s))|\}ds \\ &\leq (1 + MC\Omega_j)\|G - \tilde{G}\| \\ &\quad + M\int_0^t|(t-s)^{\Psi-1}|\max\{|Y_i(j\tau+s) - \tilde{Y}_i(j\tau+s)|, |Y_{i-1}(\Theta(j\tau+s)) - \tilde{Y}_{i-1}(\Theta(j\tau+s))|\}ds\end{aligned} \tag{60}$$

This equation is similar to (47). A similar analysis shows that $|Y_j(j\tau+t) - \tilde{Y}_j(j\tau+t)|$ is proportional to $\|G - \tilde{G}\|$. We need prove this separately for τ in $[0,1]$, $[1,2]$, and so on. We briefly carry the proof for the case $\tau \in [0,1]$, while it is also similar to previous discussions. From (60), we have

$$\begin{aligned}|Y_j(j\tau+t)-\tilde{Y}_j(j\tau+t)| \leq& (1+MC\Omega_j)\|G-\tilde{G}\| \\ &+M\int_0^t |(t-s)^{\psi^*-1}||Y_i(j\tau+s)-\tilde{Y}_i(j\tau+s)|ds \\ &+M\int_0^t |(t-s)^{\psi^*-1}||Y_{i-1}(\Theta(j\tau+s))-\tilde{Y}_{i-1}(\Theta(j\tau+s))|\}ds \\ \leq& (1+MC\Omega_j+\frac{MCt^{\psi^*}}{\psi^*})\|G-\tilde{G}\| \\ &+M\int_0^t |(t-s)^{\psi^*-1}||Y_i(j\tau+s)-\tilde{Y}_i(j\tau-s)|ds\end{aligned} \tag{61}$$

By applying the generalized Gronwall inequality, we obtain

$$|Y_j(j\tau+t)-\tilde{Y}_j(j\tau+t)| \leq (1+MC\Omega_j+\frac{MCt^{\psi^*}}{\psi^*})E_{\psi^*}(M\Gamma(\psi^*)t^{\psi^*})\|G-\tilde{G}\|. \tag{62}$$

This shows that

$$|Y_j(t)-\tilde{Y}_j(t)| \leq C_4\|G-\tilde{G}\|, \quad t\in j\tau+[0,\tau] = [j\tau,(j+1)\tau], \quad \tau\leq 1, \tag{63}$$

where C_4 is a finite constant, defined by

$$C_4 = \max_{t\in[0,1]}(1+MC\Omega_j+\frac{MCt^{\psi^*}}{\psi^*})E_{\psi^*}(M\Gamma(\psi^*)t^{\psi^*}).$$

□

6. Examples and Applications

6.1. Applications in Hopfield Neural Network

An interesting incommensurate system of DFDEs is used in [11] to describe a Hopfield neural network (HNN). HNNs, a form of recurrent artificial neural network, are particularly known for their ability to store and retrieve memory patterns efficiently, making them suitable for numerous practical uses.

In this section, we aim to study the well-posedness of a class of such HNNs. In terms of our notation, such HNNs can be described by the incommensurate System (6), with

$$\begin{aligned}u_1(t,Y,W) &= -0.1y_1+1.5\tanh(w_3)+0.8\tanh(w_4)+1.9\tanh(w_5), \\ u_2(t,Y,W) &= -0.6y_2-0.6\tanh(w_3)-0.5\tanh(w_4)+2.2\tanh(u_5), \\ u_3(t,Y,W) &= -0.1y_3-1.5\tanh(y_1)+2.5\tanh(y_2), \\ u_4(t,Y,W) &= -0.8y_4+0.5\tanh(y_1)-0.8\tanh(y_2), \\ u_5(t,Y,W) &= -1.3y_5+\tanh(y_1)-2\tanh(y_2).\end{aligned} \tag{64}$$

The stability of such systems has been studied in [11]. The hyperbolic tangent function tanh is a Lipschitz continuous function with Lipschitz constant 1. Therefore, the functions u_i for $i=1,\ldots,5$ are also Lipschitz continuous with respect to Y and W. Thus, Hypotheses (H1) and (H4) automatically hold. Particularly, $L_1 = \max\{0.1,1.5,0.8,1.9\} = 1.9$, $L_2 = \max\{0.6,0.5,2.2\} = 2.2$, $L_3 = \max\{0.1,1.5,2.5\} = 2.5$, $L_4 = \max\{0.8,0.5\} = 0.8$, and $L_5 = \max\{1.3,1,2\} = 2$. Also, each u_i is continuous with respect to its variable, thus (H2) holds. Assuming that G satisfies (H3), according to Theorem 5, the Hopfield neural network (64) has a unique mild solution. According to Theorem 8, the solution is Lipschitz continuous with respect to G, indicating that the Hopfield neural network (64) is well-posed.

6.2. A General Linear Case

Equation (3) represents a linear incommensurate systems of DFDEs with constant coefficients. However, this equation can be generalized to the form with variable coefficients as follows:

$$\mathfrak{D}^{\Psi}Y(t) = A(t)Y(t) + B(t)Y(\Theta(t)) + F(t) \tag{65}$$

where $A \in (\mathcal{C}[0,\mathfrak{T}])^{\nu\times\nu}$ and $B \in (\mathcal{C}[0,\mathfrak{T}])^{\nu\times\nu}$.

In terms of presentations (7), we have

$$U(t,Y,W) = A(t)Y + B(t)W + F(t).$$

Since a_{ij} and b_{ij} are assumed to be continuous, they are bounded on the compact interval $[0,\mathfrak{T}]$. Therefore,

$$L_i = \max_{j=1,\dots,\nu}\{\max_{t\in[a,b]}|a_{ij}(t)|, \max_{t\in[a,b]}|b_{ij}(t)|\}$$

is a well-defined real number, and Hypothesis (H4) holds with this L_i for all $i = 1,\dots,\nu$. Consequently, according to Theorems 5 and 8, it has a unique continuous mild solution, and the solution is Lipschitz continuous with respect to any continuous G. For instance, a complex system described by

$$\begin{aligned}\mathfrak{D}^{0.5}y_1(t) &= \sin(t)y_1(t) - \exp(t)y_2(t-0.5) + \cos(t),\\ \mathfrak{D}^{0.5}y_2(t) &= \cos(t)y_1(t) - t^2(t)y_2(t-0.5) + |t|,\end{aligned} \tag{66}$$

with prehistoric conditions

$$\begin{aligned}y_1(t) &= t\sin(t), \quad t \in (-0.5, 0],\\ y_2(t) &= t^2 + t + 1, \quad t \in (-0.5, 0],\end{aligned} \tag{67}$$

has a unique solution. This example demonstrates the strength of Theorems 5–8, which imposes no further requirements for well-posedness of such systems. Due to the significance of this example, we add the following corollary for reference in future research.

Corollary 1. *Assume* $A, B, F \in (\mathcal{C}[0,\mathfrak{T}])^{\nu\times\nu}$, $G \in (\mathcal{C}[-\tau,0])^{\nu\times\nu}$, $\tau,\mathfrak{T} > 0$ *and* $\Psi \in (0,1)^{\nu}$. *Then, System* (65) *has a unique mild solution, and is Lipschitz continuous with respect to G on* $[0,\mathfrak{T}]$.

Proof. The proof is a straightforward conclusion of Theorems 5–8. □

7. Conclusions and Remaining Works on This Topic

We have demonstrated that the mild solutions of incommensurate systems for DFDEs satisfy delay RL integral equations. To ensure well-posedness, we divided the interval into $[0,\tau]$, $[\tau,2\tau]$, ..., $[(n-1)\tau,\mathfrak{T}]$, and transformed studied integral equations into RL integral equations without delays as described by (46). This decomposition has been utilized throughout the analysis. Then, we proceed by assuming $\tau \in [0,1]$, and extended it for $\tau < \infty$ in any interval $[n, n+1]$, through induction. To obtain the existence result, we used Picard iterations to obtain a sequence of Cauchy continuous functions. We employed the completeness of the space of continuous functions to establish uniform convergence of such Cauchy sequences. Subsequently, we showed that the limit of the Cauchy sequence satisfies the original RL integral equation. Thus, we have established the existence of a solution. For uniqueness and stability with respect to G, we applied generalized Gronwall inequality. The results of this paper can be summarized as follows.

Theorem 9. *Assume $\psi_i \in (0,1)$, G is continuous function on $(-\tau, 0]$, $\lim_{t\to-\tau} G(t) < \infty$, and $0 \leq \tau < \infty$. Additionally, assume that U is a continuous vector-valued function and each component of U is Lipschitz continuous with respect to both the second and third variables. Then, the incommensurate system of DFDEs* (6) *with prehistoric condition* (5) *has a unique solution. Furthermore, this solution is continuously dependent on G.*

While we provided a well-posedness of the problem on the spaces of continuous functions, there are fundamental questions regarding the need to investigated in dynamic of the solution. One key question is related to the regularity of the solutions. In terms of the classical analysis, regularity speaks about the existence and behavior of derivatives of the solution. Knowledge of the differentiability of the solution is crucial for constructing efficient numerical solutions. In the context of wider analysis, regularity speaks about the existence of a solution within a specific space. Studies by Liang and Stynes [21] have investigated the regularity of a wide class of singular Volterra integral equations of the second kind in weighted space $C^{m,\alpha}$. It is an ongoing area of research related to unified theories for weakly singular integrals (including logarithm singularity and singularity with power function), mainly performed in the past three decades by Vainikko and Pedas [22–24]. Recently, in a book published by Brunner [19], the regularity of the solution of a weakly singular integral equation is studied in detail. Among the related books, this book devotes a chapter to this topic. We hope to investigate the regularity of DFDEs in future studies.

Funding: This research is supported by the Neijiang Normal University school-level science and technology project (key project, No. XJ2024008301).

Data Availability Statement: The original contributions presented in this study are included in the article. Further inquiries can be directed to the corresponding author.

Conflicts of Interest: The author declares no conflicts of interest.

Abbreviations

The following abbreviations are used in this manuscript:

FDEs	Fractional differential equations
DFDEs	delay fractional differential equations
DDEs	delay differentiate equations
HNNs	Hopfield Neural Network

References

1. Diethelm, K.; Ford, N.J. *The Analysis of Fractional Differential Equations*; Lecture Notes in Mathematics; Springer: New York, NY, USA, 2010.
2. Kilbas, A.A.; Srivastava, H.M.; Trujillo, J.J. *Theory and Applications of Fractional Differential Equations*; Elsevier: Amsterdam, The Netherlands, 2006; Volume 204, pp. 1–523.
3. Shiri, B.; Baleanu, D. All linear fractional derivatives with power functions' convolution kernel and interpolation properties. *Chaos Solit. Fractals* **2023**, *2023*, 113399. [CrossRef]
4. Smith, H.L. *An Introduction to Delay Differential Equations with Applications to the Life Sciences*; Springer: New York, NY, USA, 2011.
5. Morgado, M.L.; Ford, N.J.; Lima, P.M. Analysis and numerical methods for fractional differential equations with delay. *J. Comput. Appl. Math.* **2013**, *252*, 159–168. [CrossRef]
6. Čermák, J.; Došlá, Z.; Kisela, T. Fractional differential equations with a constant delay: Stability and asymptotics of solutions. *Appl. Math. Comput.* **2017**, *1*, 336–350. [CrossRef]

7. Huseynov, I.T.; Mahmudov, N.I. Analysis of positive fractional-order neutral time-delay systems. *J. Franklin Inst.* **2022**, *359*, 294–330. [CrossRef]
8. Shiri, B.; Baleanu, D. Numerical solution of some fractional dynamical systems in medicine involving non-singular kernel with vector order. *Res. Nonlinear Anal.* **2019**, *2*, 160–168.
9. Das, S; Gupta, P.K. A mathematical model on fractional Lotka–Volterra equations. *J. Theor. Biol.* **2011**, *277*, 1–6. [CrossRef] [PubMed]
10. Tavazoei, M; Asemani, M.H. On robust stability of incommensurate fractional-order systems. *Commun. Nonlinear Sci. Numer. Simul.* **2020**, *90*, 105–344. [CrossRef]
11. Zhang, Y.; Li L.; Huang, J.; Gorbachev, S.; Aravind, R.V. Exploration on bifurcation for an incommensurate five-neuron fractional-order BAM neural network involving multiple delays. *Phys. D Nonlinear Phenom.* **2024**, *460*, 134047. [CrossRef]
12. Kaslik, E.; Sivasundaram, S. Nonlinear dynamics and chaos in fractional-order neural networks. *Neural Netw.* **2012**, *32*, 245–256. [CrossRef] [PubMed]
13. Karoun, R.C.; Ouannas, A.; Horani, M.A.; Grassi, G. The effect of Caputo fractional variable difference operator on a discrete-time hopfield neural network with non-commensurate order. *Fractal. Fract.* **2022**, *6*, 575. [CrossRef]
14. Deng, W.; Li, C.; Lu, J. Stability analysis of linear fractional differential system with multiple time delays. *Nonlinear Dyn.* **2007**, *48*, 409–416. [CrossRef]
15. Nouri, K.; Nazari, M.; Torkzadeh, L. Numerical approximation of the system of fractional differential equations with delay and its applications. *Eur. Phys. J. Plus.* **2020**, *135*, 1–4. [CrossRef]
16. Čermák, J.; Horníček, J.; Kisela, T. Stability regions for fractional differential systems with a time delay. *Commun. Nonliner Sci. Numer. Simul.* **2016**, *31* 108–123. [CrossRef]
17. Shiri, B.; Shi, Y.G.; Baleanu, D. The Well-Posedness of Incommensurate FDEs in the space of continuous functions. *Symmetry* **2024**, *16*, 1058. [CrossRef]
18. Ye, H.; Gao, J.; Ding, Y. A generalized Gronwall inequality and its application to a fractional differential equation. *J. Math. Anal. Appl.* **2007**, *328*, 1075–1081. [CrossRef]
19. Brunner, H. *Volterra Integral Equations: An Introduction to Theory and Applications*; Cambridge University Press: Cambridge, UK, 2017.
20. Lan, K.; Webb, J.R. A new Bihari inequality and initial value problems of first order fractional differential equations. *Fract. Calc. Appl. Anal.* **2023**, *26*, 962–988. [CrossRef] [PubMed]
21. Liange, H.; Stynes, M. Regularity of the solution of a nonlinear Volterra integral equation of the second kind. *Discret. Contin. Dyn.-Syst.-Ser. B* **2023**, *28*, 2211. [CrossRef]
22. Gennadi, V. *Multidimensional Weakly Singular Integral Equations*; Lecture Notes in Mathematics; Springer: Berlin/Heidelberg, Germany, 1993; p. 1549.
23. Pedas, A.; Vainikko, G. The smoothness of solutions to nonlinear weakly singular integral equations. *Z. Anal. Anwendungen* **1994**, *13*, 463–476. [CrossRef]
24. Hermann, B.; Arvet, P.; Gennadi, V. The piecewise polynomial collocation method for nonlinear weakly singular Volterra equations. *Math. Comp.* **1999**, *68*, 1079–1095.

fractal and fractional

MDPI

Article

Certain Analytic Solutions for Time-Fractional Models Arising in Plasma Physics via a New Approach Using the Natural Transform and the Residual Power Series Methods

Asad Freihat [1], Mohammed Alabedalhadi [1], Shrideh Al-Omari [2,*], Sharifah E. Alhazmi [3], Shaher Momani [4,5] and Mohammed Al-Smadi [1,5,6]

1 Department of Applied Science, Ajloun College, Al-Balqa Applied University, Ajloun 26816, Jordan; asad.freihat@bau.edu.jo (A.F.); mohmdnh@bau.edu.jo (M.A.); malsmadi@lu.edu.qa (M.A.-S.)
2 Department of Mathematics, Faculty of Science, Al Balqa Applied University, Salt 19117, Jordan
3 Department of Mathematics, College of Education for Girls at Al-Qunfudah, Umm Al-Qura University, Mecca 11942, Saudi Arabia; sehazmi@uqu.edu.sa
4 Department of Mathematics, Faculty of Science, University of Jordan, Amman 11942, Jordan; s.momani@ju.edu.jo
5 Nonlinear Dynamics Research Center (NDRC), Ajman University, Ajman 20550, United Arab Emirates
6 College of Commerce and Business, Lusail University, Lusail 9717, Qatar
* Correspondence: shridehalomari@bau.edu.jo; Tel.: +962-772-061-029

Abstract: This paper studies three time-fractional models that arise in plasma physics: the modified Korteweg–deVries–Zakharov–Kuznetsov equation, the stochastic potential Korteweg–deVries equation, and the forced Korteweg–deVries equation. These equations are significant in plasma physics for modeling nonlinear ion acoustic waves and thus helping us to understand wave dynamics in plasmas. We introduce a new approach that relies on a new fractional expansion in the natural transform space and residual power series method to construct analytical solutions to the governing models. We investigate the theoretical analysis of the proposed method for these equations to expose this approach's applicability, efficiency, and effectiveness in constructing analytical solutions to the governing equations. Moreover, we present a comparative discussion between the solutions derived during the work and those given in the literature to confirm that the proposed approach generates analytical solutions that rapidly converge to exact solutions, which proves the effectiveness of the proposed method.

Keywords: fractional model; plasma physics; Korteweg–deVries equation; natural transform; residual power series method

Academic Editors: Libo Feng, Yang Liu and Lin Liu

Received: 26 January 2025
Revised: 21 February 2025
Accepted: 23 February 2025
Published: 28 February 2025

Citation: Freihat, A.; Alabedalhadi, M.; Al-Omari, S.; Alhazmi, S.E.; Momani, S.; Al-Smadi, M. Certain Analytic Solutions for Time-Fractional Models Arising in Plasma Physics via a New Approach Using the Natural Transform and the Residual Power Series Methods. *Fractal Fract.* **2025**, *9*, 152. https://doi.org/10.3390/fractalfract9030152

1. Introduction

Nonlinear wave phenomena are of fundamental importance in the context of physics in general and plasma dynamics in particular [1]. A comprehensive understanding of ion acoustic waves and their behavior is essential for many applications in the context of plasma physics, such as energy transfer, plasma containment, and wave–particle interactions [2,3]. In light of this, many mathematical models have been presented to describe such phenomena, including the Korteweg–de Vries (KdV) equation [4], the nonlinear Schrödinger equation [5], Burgers' equation [6], the Zakharov system [7], the sine-Gordon equation [8], the Benjamin–Bona–Mahony equation [9], the Kadomtsev–Petviashvili equation [10], and others. Since the emergence of the concept of fractional differential and integral calculus and the work carried out by researchers in studying its impact on mathematical systems in various fields of science, researchers have turned their attention to researching the importance of fractional differential and integral calculus in plasma physics [11,12]. It has been shown that fractional calculations have great importance for the effects of memory and generic properties of systems, which play a prominent role in providing an accurate description of complex plasma dynamics [13,14]. Considering the long-term interactions in plasma systems, the fractional derivative provides a more general approach than classical derivatives, which in turn helps in modeling the wave propagation, diffusion processes, and anomalous transport phenomena that may not be available in standard models to capture the complex behavior of plasma [15]. This enables scientists to gain a deeper understanding of plasma behavior in physical environments and laboratories by developing fractional mathematical models for ion acoustic waves, plasma turbulence, and other nonlinear phenomena [16]. The main interest of this work is to study three important equations in the context of plasma physics: the fractional modified Korteweg–deVries–Zakharov–Kuznetsov (mKdV-ZK) equation, the fractional stochastic potential Korteweg–deVries (spKdV) equation, and the fractional forced Korteweg–deVries (FKdV) equation. We consider the following time-fractional KdV-ZK equation [17]:

$$\mathcal{D}_t^{\alpha}\psi + \mu^2\psi\frac{\partial\psi}{\partial x} + \frac{\partial^3\psi}{\partial x^3} + \frac{\partial^3\psi}{\partial x\partial y^2} + \frac{\partial^3\psi}{\partial x\partial z^2} = 0,\ t \geq 0,\ x,y,z \in \mathbb{R}, \tag{1.1}$$

subject to the initial condition:

$$\psi(x,y,z,0) = f(x,y,z), x,y,z \in \mathbb{R}, \tag{1.2}$$

where $\mathcal{D}_t^{\alpha}\psi$ represents the Caputo derivative of the electric field potential $\psi(x,y,z,t)$ with respect to the time t of the fractional derivative order $\alpha \in (0,1]$. The variables x, y and z are the scaled space coordinates, while μ is the dispersion coefficient that stands for positive correlation with negative correlation, Boltzmann distribution, and fluid species. The Korteweg–DeVries–Zakharov–Kuznetsov equation was introduced by the authors [18] with the aim of using the reductive perturbation approach for mixing both warm adiabatic fluids and cold stationary isothermal solid background species in a magnetized plasma where they discussed acoustic electron, acoustic ion, and acoustic dust waves. In Ref. [19], the mKdV-ZK equation is presented using the reductive perturbation approach to control the diagonal diffusion in nonlinear electrostatic modes. In Ref. [20], multiple-soliton solutions for the mKdV-ZK equation were obtained. Solitary solutions were explored using the modified extended direct algebraic method for the mKdV-ZK equation in Ref. [21]. Younas et al. constructed traveling wave solutions by utilizing two different approaches to the mKdV-ZK equation [22]. The second equation considered in this work is the time-fractional spKdV equation [23]:

$$\mathcal{D}_t^{\alpha}\psi + \omega\frac{\partial\psi}{\partial x} + \beta\left(\frac{\partial\psi}{\partial x}\right)^2 + \gamma\frac{\partial^3\psi}{\partial x^3} = 0,\ t \geq 0,\ x \in \mathbb{R}, \tag{1.3}$$

subject to the initial condition:

$$\psi(x,0) = g(x), x \in \mathbb{R}, \tag{1.4}$$

where ω is the stochastic parameter, β is the nonlinearity coefficient, and γ refers to the dispersion coefficient. Equation (1.3) is a nonlinear model that appears in applications of multicomponent plasmas and electrical circuits and is used to predict the weak dispersive effects on the propagation of nonlinear optical waves and photons. The spKdV equation is presented in Ref. [23], where the authors obtained lumps, breathers, and multi-soliton solutions using the simplified Hirota method and the Cole–Hopf transformation. In Ref. [24], the authors suggested the fractional spKdV equation and they constructed closed-form solutions for the proposed model. In Ref. [25], the spKdV equation was investigated and the Lie symmetry approach was utilized to carry out the symmetry generators to obtain traveling wave solutions and conservation laws for the proposed model. The final equation that will be investigated in our paper is the following fractional FKdV equation [26]:

$$\frac{1}{c}\mathcal{D}_t^{\alpha}\psi + \left((\mathcal{F}_r - 1) - \frac{3}{2\chi}\psi\right)\frac{\partial\psi}{\partial x} - \frac{\chi^2}{6}\frac{\partial^3\psi}{\partial x^3} = \frac{1}{2}\frac{d\phi(x)}{dx},\ t \geq 0,\ x \in \mathbb{R}, \tag{1.5}$$

subject to the initial condition:

$$\psi(x,0) = h(x), x \in \mathbb{R}. \tag{1.6}$$

The term $\left((\mathcal{F}_r - 1) - \frac{3}{2\chi}\psi\right)\frac{\partial\psi}{\partial x}$ describes nonlinear interactions in a wave that are measured using the parameter χ which arises from the amplitude of the plasma wave where higher amplitudes affect the wave velocity. The parameter $\mathcal{F}_r$ represents some force or property of the plasma. The term $\frac{1}{2}\frac{d\phi(x)}{dx}$ represents an external influence on the plasma wave such as an electric or magnetic field or any other disturbance. The importance of Equation (1.5) lies in allowing the integration of external forces, memory effects, nonlinearity, and dispersion to model the propagation of nonlinear ion-acoustic waves, as these components are necessary to accurately describe the wave behavior in plasma environments. This model is of great value in many contexts such as astrophysical plasmas affected by large-scale forces, fusion plasmas affected by electric and magnetic fields, and the study of stability and turbulence in plasma systems. In Ref. [26], the authors investigated the fractional FKdV equation to construct traveling wave solutions using the fractional natural decomposition method. The distributed-order time-fractional FKdV equation is presented in Ref. [27], where the authors used the shifted Legendre operational matrix to infer approximate solutions for the proposed model. In Ref. [28], the authors studied the FKdV equation and obtained approximate solutions for the model using the q-homotopy analysis transform technique. The fractional FKdV equation was discussed with the help of the Yang homotopy perturbation method and the Yang transform decomposition method to construct approximate solutions for the model in Ref. [29].

With the wide development in scientific research regarding mathematical systems that describe and model physical phenomena and the emergence and discovery of complex systems in various fields of science, the need to develop methods to derive accurate and approximate solutions for these presented systems arose. Researchers worked to develop such techniques, including the reproducing kernel Hilbert space method [30], the multi-step reduced differential transformation method [31], the Atangana–Baleanu fractional framework for the reproducing kernel technique [32], the multistep generalized differential transform method [33], the homotopy analysis method [34], the shifted Jacobi polynomials method [35], the Chebyshev polynomials method [36], auxiliary equation method [37], the B-spline schemes method [38], the fractional variational iteration method [39–41], the Kudryashov method [42], and others. Abu Arqub et al. [43] presented a new analytical method, the residual power series (RPS) method, for obtaining solitary wave solutions of time-fractional dispersive partial differential equations based on the generalized Taylor series formula and residual error function. Researchers have employed this method to derive analytical solutions for a wide range of fractional equations as it is an effective method [44,45]. In Ref. [46], El-Ajou proposed a new technique that combines the Laplace transform and RPS method, called the Laplace residual power series (LRPS) method, where the proposed method was utilized to explore exact solitary solutions to the nonlinear time-fractional dispersive partial differential equations (PDEs). The main goal of this work is to suggest a new attractive approach, the natural residual power series (NRPS) method, that depends on the natural transformation and RPS method. Moreover, we seek to utilize the NRPS method to infer analytical solutions for the governing Equations (1.1)–(1.6). The NRPS method is based on using natural transformation to transform the time-fractional differential equation into the natural space, then using a suitable expansion to deal with the newly obtained equation together with the RPS method, which in turn helps in deducing the expansion coefficients based on the concept of limit, which reduces the calculations required to reach accurate approximate solutions to the governing equation.

We organized the paper with the introduction as the first section. Section 2 presents some preliminaries about the Caputo fractional derivative. Section 3 is devoted to introducing the definition of natural transformation and investigating its essential properties, then presenting the natural fractional expansion. In Section 4, we introduce the steps of the proposed method. Applications of the NRPS method to construct approximate solutions for the governing models are introduced in Section 5. Some conclusions are presented in Section 6.

2. Preliminaries

Since the advent of fractional calculus, many fractional operators have appeared, such as the Riemann–Liouville operator [47], the Caputo operator [48], Atangana–Baleanu operator [49], and others. Here, we present the definition of the Caputo fractional derivative and some of its essential properties that are useful in our work.

Definition 1 ([49]). *The Caputo derivative of fractional order $\alpha > 0$ of an integrable function $\psi(t)$ is given by:*

$$\mathcal{D}_t^{\alpha}\psi(t,x) = \frac{1}{\Gamma(k-\alpha)}\int_0^t \frac{\frac{\partial^k \psi(\tau,x)}{\partial \tau^k}}{(t-\tau)^{\alpha-k+1}}d\tau, \tag{2.1}$$

where $t \geq 0,\ k = [\alpha]+1$. The Riemann–Liouville fractional integral of order $\alpha,\ Re(\alpha) > 0$ is given by:

$$\mathcal{J}_t^{\alpha}\psi(t) = \frac{1}{\Gamma(\alpha)}\int_0^t (t-\tau)^{\alpha-1}\psi(\tau,x)d\tau. \tag{2.2}$$

Lemma 1 ([49]). *For $\alpha > 0,\ \eta > -1,\ p \in \mathbb{R}$ and $t \geq 0$, the Caputo fractional derivative has the property:*

(a) $\mathcal{D}_t^{\alpha} p = 0.$ (2.3)

(b) $\mathcal{D}_t^{\alpha} t^{\eta} = \dfrac{\Gamma(\eta+1)}{\Gamma(\eta+1-\alpha)} t^{\eta-\alpha}.$ (2.4)

(c) $\mathcal{D}_t^{\alpha}\mathcal{J}_t^{\alpha}\psi(t,x) = \psi(t,x).$ (2.5)

(d) $\mathcal{J}_t^{\alpha}\mathcal{D}_t^{\alpha}\psi(t,x) = \psi(t,x) - \sum\limits_{i=0}^{k-1}\dfrac{\partial^i \psi(0^+,x)}{\partial t^i}\dfrac{t^i}{i!}.$ (2.6)

3. Natural Transformation and Fractional Expansion

The natural transformation combines the features of the Laplace and the Sumudu transforms into a powerful tool for dealing with a variety of differential and integral equations. Natural transformation provides a simple and flexible approach to transforming functions from the time domain to the complex domain, making it powerful for dealing with problems involving fractional time models. Furthermore, the natural transform can handle a wide range of functions with singularities and simplifies the process of finding analytical solutions to complex problems. Therefore, it is considered a valuable tool widely used in applied mathematics, engineering, and physics [50–54]. The general integral transform of the function $\psi(t,x),\ t,x \in \mathbb{R}$ is defined as:

$$\mathcal{T}\{\psi(t,x)\} = \int_{-\infty}^{\infty} \mathcal{K}(s,t)\psi(t,x)dt, \tag{3.1}$$

where $\mathcal{K}(s,t)$ is the kernel of the transform and s represents the real (complex) number that is independent of t. The Laplace transform, Mellin transform, and Hankel transform can be obtained by letting the kernel of the transform in (3.1) be: e^{-st}, $t^{s-1}(st)$, and $tJ_n(st)$, respectively. To define the natural transformation, we define the following set:

$$\mathcal{C} = \left\{\psi(t,x) : \exists\ B,\ \sigma_1,\sigma_2,\ |\psi(t,x)| < Be^{\left(\frac{|t|}{\sigma_i}\right)}, t \in (-1)^i \times [0,\infty),\ x \in A \subseteq \mathbb{R}, i \in \mathbb{Z}^+\right\} \tag{3.2}$$

Definition 2 ([51]). *Let $\psi(t,x) \in \mathcal{C}$. The natural transformation of the function $\psi(t,x)$ is defined as:*

$$\Psi(s,w,x) = \mathcal{N}\{\psi(t,x)\} = \int_0^{\infty} e^{-st}\psi(wt,x)dt,\ s,w \in (0,\infty), \tag{3.3}$$

where $\mathcal{N}\{\psi(t,x)\}$ is the natural transformation of $\psi(t,x)$ and the variables s and w are the natural transform variables. The inverse natural transformation of the function $\Psi(s,w,x)$ is defined as follows:

$$\psi(t,x) = \mathcal{N}^{-1}\{\Psi(s,w,x)\} = \frac{1}{2\pi i}\int_{u-i\infty}^{u+i\infty} e^{-\frac{st}{w}}\Psi(s,w,x)ds,\ t \in [0,\infty), \tag{3.4}$$

where the integral is applied for a complex number $s = a + ib$ along $s = u$.

The following theorem shows the existence of natural transformations.

Theorem 1 ([51]). *Let $\psi(t,x) \in \mathcal{C}$ be piecewise continuous on $[0,\infty) \times A$ and of an exponential order $\frac{\theta}{w}$ as $t \to \infty$. Then, the natural transformation of $\psi(t,x)$ exists for all $s \in (\theta,\infty), w \in (0,\infty)$.*

Remark 1. *The Laplace and Sumudu transforms can be obtained by letting $w = 1$ and $s = 1$, respectively, in the natural transform (3.3).*

Remark 2. *Let $\Psi(s,w,x) = \mathcal{N}\{\psi(t,x)\}$ be the natural transform of the function and let*

$$\Psi_L(s,x) = \mathcal{L}\{\psi(t,x)\} = \int_0^{\infty} e^{-st}\psi(t,x)dt,$$

$$\Psi_S(w,x) = \mathcal{S}\{\psi(t,x)\} = \int_0^{\infty} e^{-t}\psi(wt,x)dt,$$

be the Laplace and Sumudu transforms of the function $\psi(t,x)$, respectively. Then, the duality relationship between the Natural-Laplace and Natural-Sumudu transforms is given by:

$$\mathcal{N}\{\psi(t,x)\} = \Psi(s,w,x) = \frac{1}{w}\Psi_L\left(\frac{s}{w},x\right), \tag{3.5}$$

$$\mathcal{N}\{\psi(t,x)\} = \Psi(s,w,x) = \frac{1}{s}\Psi_S\left(\frac{w}{s},x\right). \tag{3.6}$$

Lemma 2 ([52]). *Let $\psi_1(t,x), \psi_2(t,x) \in C$ be piecewise continuous on $[0,\infty) \times A$ and of an exponential order $\frac{\theta_1}{w}$ and $\frac{\theta_2}{w}$, respectively, with $\theta_1 < \theta_2$. Then, the natural transform satisfies the following properties:*

(i) $\mathcal{N}(1) = \frac{1}{s}$. (3.7)

(ii) $\mathcal{N}\left(t^{k\eta}\right) = \frac{\Gamma(k\eta+1)w^{k\eta}}{s^{k\eta+1}}, \ \eta > -1.$ (3.8)

(iii) $\mathcal{N}(p_1\psi_1(t,x) + p_2\psi_2(t,x)) = p_1\Psi_1(s,w,x) + p_2\Psi_2(s,w,x), \ p_1, p_2 \in \mathbb{R}.$ (3.9)

(iv) $\mathcal{N}^{-1}(p_1\Psi_1(s,w,x) + p_2\Psi_2(s,w,x)) = p_1\psi_1(t,x) + p_2\psi_1(t,x), \ p_1, p_2 \in \mathbb{R}.$ (3.10)

Lemma 3. *Let $\psi(t,x) \in C$ be piecewise continuous on $[0,\infty) \times A$ and of exponential order $\frac{\theta}{w}$. Then, the natural transform satisfies the following properties:*

(i) $\mathcal{N}\left\{\frac{\partial^k}{\partial t^k}\psi(t,x)\right\} = \frac{s^k}{w^k}\Psi(s,w,x) - \sum_{i=0}^{k-1}\frac{s^{k-i-1}}{w^{k-i}}\frac{\partial^i}{\partial t^i}\psi(0,x), k \in \mathbb{Z}.$ (3.11)

(ii) $\lim_{s\to\infty} s\Psi(s,w,x) = \psi(0,x).$ (3.12)

(iii) $\mathcal{N}\{\mathcal{D}_t^\alpha\psi(t,x)\} = \frac{s^\alpha}{w^\alpha}\Psi(s,w,x) - \frac{s^{\alpha-1}}{w^\alpha}\psi(0,x), 0 < \alpha < 1.$ (3.13)

(iv) $\mathcal{N}\left\{\mathcal{D}_t^{k\alpha}\psi(t,x)\right\} = \frac{s^{k\alpha}}{w^{k\alpha}}\Psi(s,w,x) - \sum_{i=0}^{k-1}\frac{s^{(k-i)\alpha-1}}{w^{(k-i)\alpha}}\mathcal{D}_t^{i\alpha}\psi(0,x), 0 < \alpha < 1.$ (3.14)

where $\mathcal{D}_t^{k\alpha} = \mathcal{D}_t^\alpha\mathcal{D}_t^\alpha \dots \mathcal{D}_t^\alpha$ (k-times).

Proof. The properties (i) can be found in Ref. [48]. To prove (ii), we use the property (i) at $k = 1$, we obtain:

$$\mathcal{N}\left\{\frac{\partial}{\partial t}\psi(t,x)\right\} = \frac{s}{w}\Psi(s,w,x) - \frac{1}{w}\psi(0,x). \tag{3.15}$$

Which gives:

$$s\Psi(s,w,x) = w\int_{-\infty}^{\infty} e^{-st}\frac{\partial}{\partial t}\psi(wt,x)dt + \psi(0,x). \tag{3.16}$$

Taking the limit for both sides of (3.16) as $s \to \infty$ to obtain the property (ii). To prove (iii), we use the following fact for the Sumudu transform for the Caputo fractional derivative:

$$\Psi_S^*(w,x) = \mathcal{S}\{\mathcal{D}_t^\alpha\psi(t,x)\} = \frac{1}{w^\alpha}(\Psi_S(w,x) - \psi(0,x)). \tag{3.17}$$

Substitute $w = \frac{w}{s}$ in (3.17) to obtain:

$$\Psi_S^*\left(\frac{w}{s},x\right) = \frac{s^\alpha}{w^\alpha}\left(\Psi_S\left(\frac{w}{s},x\right) - \psi(0,x)\right). \tag{3.18}$$

Multiply (3.18) by $\frac{1}{s}$, we have:

$$\frac{1}{s}\Psi_S^*\left(\frac{w}{s},x\right) = \frac{s^\alpha}{w^\alpha}\frac{1}{s}\Psi_S\left(\frac{w}{s},x\right) - \frac{s^{\alpha-1}}{w^\alpha}\psi(0,x). \tag{3.19}$$

Therefore, using the relationship (3.6) to obtain the property (iii):

$$\mathcal{N}\{\mathcal{D}_t^\alpha\psi(t,x)\} = \frac{s^\alpha}{w^\alpha}\Psi(s,w,x) - \frac{s^{\alpha-1}}{w^\alpha}\psi(0,x). \tag{3.20}$$

We use mathematical induction to verify the property (iv). For $k = 1$, the property is proved in (iii). We assume the formula is satisfied at $k = n$, then we have:

$$\mathcal{N}\{\mathcal{D}_t^{n\alpha}\psi(t,x)\} = \frac{s^{n\alpha}}{w^{n\alpha}}\Psi(s,w,x) - \sum_{i=0}^{n-1}\frac{s^{(n-i)\alpha-1}}{w^{(n-i)\alpha}}\mathcal{D}_t^{i\alpha}\psi(0,x). \tag{3.21}$$

We seek to verify it at $k = n+1$. Now,

$$\mathcal{N}\left\{\mathcal{D}_t^{(n+1)\alpha}\psi(t,x)\right\} = \mathcal{N}\{\mathcal{D}_t^{\alpha}(\mathcal{D}_t^{n\alpha}\psi(t,x))\} = \mathcal{N}\{\mathcal{D}_t^{\alpha}\phi(t,x)\}, \tag{3.22}$$

where $\phi(t,x) = \mathcal{D}_t^{n\alpha}\psi(t,x)$. Using (iii), we obtain:

$$\mathcal{N}\{\mathcal{D}_t^{\alpha}\phi(t,x)\} = \frac{s^{\alpha}}{w^{\alpha}}\mathcal{N}\{\mathcal{D}_t^{n\alpha}\psi(t,x)\} - \frac{s^{\alpha-1}}{w^{\alpha}}\mathcal{D}_t^{n\alpha}\psi(0,x). \tag{3.23}$$

Using (3.21) into (3.23) to infer:

$$\begin{aligned}\mathcal{N}\{\mathcal{D}_t^{\alpha}\phi(t,x)\} = {} & \frac{s^{\alpha}}{w^{\alpha}}\left(\frac{s^{n\alpha}}{w^{n\alpha}}\Psi(s,w,x) - \sum_{i=0}^{n-1}\frac{s^{(n-i)\alpha-1}}{w^{(n-i)\alpha}}\mathcal{D}_t^{i\alpha}\psi(0,x)\right) - \frac{s^{\alpha-1}}{w^{\alpha}}\mathcal{D}_t^{n\alpha}\psi(0,x)\\ = {} & \frac{s^{(n+1)\alpha}}{w^{(n+1)\alpha}}\Psi(s,w,x) - \sum_{i=0}^{n-1}\frac{s^{(n+1-i)\alpha-1}}{w^{(n+1-i)\alpha}}\mathcal{D}_t^{i\alpha}\psi(0,x) - \frac{s^{\alpha-1}}{w^{\alpha}}\mathcal{D}_t^{n\alpha}\psi(0,x)\\ = {} & \frac{s^{(n+1)\alpha}}{w^{(n+1)\alpha}}\Psi(s,w,x) - \sum_{i=0}^{n}\frac{s^{(n+1-i)\alpha-1}}{w^{(n+1-i)\alpha}}\mathcal{D}_t^{i\alpha}\psi(0,x).\end{aligned} \tag{3.24}$$

From (3.22) and (3.24), we obtain:

$$\mathcal{N}\left\{\mathcal{D}_t^{(n+1)\alpha}\psi(t,x)\right\} = \frac{s^{(n+1)\alpha}}{w^{(n+1)\alpha}}\Psi(s,w,x) - \sum_{i=0}^{n}\frac{s^{(n+1-i)\alpha-1}}{w^{(n+1-i)\alpha}}\mathcal{D}_t^{i\alpha}\psi(0,x). \tag{3.25}$$

Therefore, the poof is complete. □

The next theorem presents a new fractional expansion which is essential for our proposed method, the NRPS method.

Theorem 2. *Let $\psi(t,x) \in C$ be piecewise continuous on $[0,\infty) \times A$ and of an exponential order $\frac{\theta}{w}$, and let $\Psi(s,w,x) = \mathcal{N}\{\psi(t,x)\}$ have the following fractional expansion:*

$$\Psi(s,w,x) = \sum_{k=0}^{\infty}\frac{\Pi_k(x)w^{k\alpha}}{s^{1+k\alpha}},\ \alpha \in (0,1], x \in A,\ s \in (\theta,\infty), w \in (0,\infty). \tag{3.26}$$

Then, the coefficients $\Pi_k(x) = \mathcal{D}_t^{k\alpha}\psi(0,x), k = 0,1,2,\ldots$. Consequently, $\psi(t,x)$ can be written in the following fractional Taylor's formula:

$$\psi(t,x) = \sum_{k=0}^{\infty}\frac{\mathcal{D}_t^{k\alpha}\psi(0,x)t^{k\alpha}}{\Gamma(1+k\alpha)}, \alpha \in (0,1], x \in \mathbb{R}, t \in (0,\infty). \tag{3.27}$$

Proof. Assume that $\Psi(s,w,x)$ has the fractional expansion in (3.26). Multiply both sides of (3.26) by s, then take the limit as $s \to \infty$ to obtain:

$$\lim_{s\to\infty} s\Psi(s,w,x) = \lim_{s\to\infty} s\sum_{k=0}^{\infty}\frac{\Pi_k(x)w^{k\alpha}}{s^{1+k\alpha}} = \lim_{s\to\infty}\left(\Pi_0(x) + \sum_{k=1}^{\infty}\frac{\Pi_k(x)w^{k\alpha}}{s^{k\alpha}}\right) = \Pi_0(x).$$

Using (3.12), it is possible to obtain $\Pi_0(x) = \psi(0,x)$. Hence, we can write the fractional expansion (3.26) in the form:

$$\Psi(s,w,x) = \frac{\psi(0,x)}{s} + \sum_{k=1}^{\infty} \frac{\Pi_k(x)w^{k\alpha}}{s^{1+k\alpha}},\ \alpha \in (0,1], x \in \mathbb{R}, s \in (\theta,\infty), w \in (0,\infty). \tag{3.28}$$

Now, multiply both sides of (3.28) by $s^{\alpha+1}$ to obtain:

$$s^{\alpha+1}\Psi(s,w,x) = s^{\alpha}\psi(0,x) + \Pi_1(x)w^{\alpha} + \sum_{k=2}^{\infty} \frac{\Pi_k(x)w^{k\alpha}}{s^{(k-1)\alpha}}. \tag{3.29}$$

Consequently, with the aid of (3.13), we have:

$$\begin{aligned} \Pi_1(x) = \frac{s^{\alpha+1}}{w^{\alpha}}\Psi(s,w,x) - \frac{s^{\alpha}}{w^{\alpha}}\psi(0,x) - \sum_{k=2}^{\infty} \frac{\Pi_k(x)w^{(k-1)\alpha}}{s^{(k-1)\alpha}} \\ = s\left(\frac{s^{\alpha}}{w^{\alpha}}\Psi(s,w,x) - \frac{s^{\alpha-1}}{w^{\alpha}}\psi(0,x)\right) - \sum_{k=2}^{\infty} \frac{\Pi_k(x)w^{(k-1)\alpha}}{s^{(k-1)\alpha}} \\ = s\mathcal{N}\{\mathcal{D}_t^{\alpha}\psi(t,x)\} - \sum_{k=2}^{\infty} \frac{\Pi_k(x)w^{(k-1)\alpha}}{s^{(k-1)\alpha}}. \end{aligned} \tag{3.30}$$

Take the limit as $s \to \infty$ for both sides of (3.30) with the aid of (3.12) to infer:

$$\Pi_1(x) = \lim_{s\to\infty} s\mathcal{N}\{\mathcal{D}_t^{\alpha}\psi(t,x)\} = \mathcal{D}_t^{\alpha}\psi(0,x). \tag{3.31}$$

To find $\Pi_2(x)$, multiply both sides of (3.28) by $s^{2\alpha+1}$ to obtain:

$$s^{2\alpha+1}\Psi(s,w,x) = s^{2\alpha}\psi(0,x) + s^{\alpha}\Pi_1(x)w^{\alpha} + \Pi_2(x)w^{2\alpha} + \sum_{k=3}^{\infty} \frac{\Pi_k(x)w^{k\alpha}}{s^{(k-2)\alpha}}. \tag{3.32}$$

Thus, using (3.31) and (3.14), we infer:

$$\begin{aligned} \Pi_2(x) = \frac{s^{2\alpha+1}}{w^{2\alpha}}\Psi(s,w,x) - \frac{s^{2\alpha}}{w^{2\alpha}}\psi(0,x) - \frac{s^{\alpha}\Pi_1(x)}{w^{\alpha}} - \sum_{k=3}^{\infty} \frac{\Pi_k(x)w^{(k-2)\alpha}}{s^{(k-2)\alpha}} \\ = s\left(\frac{s^{2\alpha}}{w^{2\alpha}}\Psi(s,w,x) - \frac{s^{2\alpha-1}}{w^{2\alpha}}\psi(0,x) - \frac{s^{\alpha-1}\mathcal{D}_t^{\alpha}\psi(0,x)}{w^{\alpha}}\right) - \sum_{k=3}^{\infty} \frac{\Pi_k(x)w^{(k-2)\alpha}}{s^{(k-2)\alpha}} \\ = s\mathcal{N}\{\mathcal{D}_t^{2\alpha}\psi(t,x)\} - \sum_{k=3}^{\infty} \frac{\Pi_k(x)w^{(k-2)\alpha}}{s^{(k-2)\alpha}}. \end{aligned} \tag{3.33}$$

Take the limit as $s \to \infty$ for both sides of (3.33) with the aid of (3.12) to obtain:

$$\Pi_2(x) = \lim_{s\to\infty} s\mathcal{N}\left\{\mathcal{D}_t^{2\alpha}\psi(t,x)\right\} = \mathcal{D}_t^{2\alpha}\psi(0,x). \tag{3.34}$$

Now, we use the induction to complete the proof. Then, assume that $\Pi_k(x) = \mathcal{D}_t^{k\alpha}\psi(0,x), k = 0,1,2,\ldots,n$. Then, the fractional expansion (3.28) can be written as:

$$\Psi(s,w,x) = \frac{\psi(0,x)}{s} + \sum_{k=1}^{n} \frac{\mathcal{D}_t^{k\alpha}\psi(0,x)w^{k\alpha}}{s^{1+k\alpha}} + \frac{\Pi_{n+1}(x)w^{(n+1)\alpha}}{s^{1+(n+1)\alpha}} + \sum_{k=n+2}^{\infty} \frac{\Pi_k(x)w^{k\alpha}}{s^{1+k\alpha}}. \tag{3.35}$$

To find the coefficient $\Pi_{n+1}(x)$, we multiply both sides of (3.35) by $s^{(n+1)\alpha+1}$, one has:

$$\begin{aligned} &s^{(n+1)\alpha+1}\Psi(s,w,x) \\ &\quad = s^{(n+1)\alpha}\psi(0,x) + \sum_{k=1}^{n} \frac{\mathcal{D}_t^{k\alpha}\psi(0,x)w^{k\alpha}}{s^{(k-n-1)\alpha}} + \Pi_{n+1}(x)w^{(n+1)\alpha} \\ &\quad + \sum_{k=n+2}^{\infty} \frac{\Pi_k(x)w^{k\alpha}}{s^{(k-n-1)\alpha}}. \end{aligned} \tag{3.36}$$

Consequently, with the aid of (3.14), we obtain:

$$\begin{aligned} \Pi_{n+1}(x) &= \frac{s^{(n+1)\alpha+1}}{w^{(n+1)\alpha}}\Psi(s,w,x) - \frac{s^{(n+1)\alpha}}{w^{(n+1)\alpha}}\psi(0,x) - \sum_{k=1}^{n} \frac{\mathcal{D}_t^{k\alpha}\psi(0,x)s^{(n+1-k)\alpha}}{w^{(n+1-k)\alpha}} - \sum_{k=n+2}^{\infty} \frac{\Pi_k(x)w^{(k-n-1)\alpha}}{s^{(k-n-1)\alpha}} \\ &= s\left(\frac{s^{(n+1)\alpha}}{w^{(n+1)\alpha}}\Psi(s,w,x) - \frac{s^{(n+1)\alpha-1}}{w^{(n+1)\alpha}}\psi(0,x) - \sum_{k=1}^{n} \frac{\mathcal{D}_t^{k\alpha}\psi(0,x)s^{(n+1-k)\alpha-1}}{w^{(n+1-k)\alpha}} \right) \\ &- \sum_{k=n+2}^{\infty} \frac{\Pi_k(x)w^{(k-n-1)\alpha}}{s^{(k-n-1)\alpha}} = s\mathcal{N}\left\{\mathcal{D}_t^{(n+1)\alpha}\psi(t,x)\right\} - \sum_{k=n+2}^{\infty} \frac{\Pi_k(x)w^{(k-n-1)\alpha}}{s^{(k-n-1)\alpha}}. \end{aligned} \tag{3.37}$$

Take the limit as $s \to \infty$ for both sides of (3.37) with the aid of (3.12) to obtain:

$$\Pi_{n+1}(x) = \lim_{s\to\infty} s\mathcal{N}\left\{\mathcal{D}_t^{(n+1)\alpha}\psi(t,x)\right\} = \mathcal{D}_t^{(n+1)\alpha}\psi(0,x). \tag{3.38}$$

Which prove that $\Pi_k(x) = \mathcal{D}_t^{k\alpha}\psi(0,x), k = 0,\ 1,\ 2,\ldots$.. Hence, the function $\Psi(s,w,x)$ can be written as:

$$\Psi(s,w,x) = \sum_{k=0}^{\infty} \frac{\mathcal{D}_t^{k\alpha}\psi(0,x)w^{k\alpha}}{s^{1+k\alpha}},\ \alpha \in (0,1], x \in A,\ s \in (\theta,\infty), w \in (0,\infty). \tag{3.39}$$

Take the inverse natural transform of (3.39) with the aid of (3.8) to obtain:

$$\psi(t,x) = \sum_{k=0}^{\infty} \frac{\mathcal{D}_t^{k\alpha}\psi(0,x)t^{k\alpha}}{\Gamma(1+k\alpha)}, \alpha \in (0,1], x \in \mathbb{R}, t \in (0,\infty). \tag{3.40}$$

Which completes the proof. □

We are now going to prove that the fractional expansion for $\Psi(s,w,x)$ in (3.26) converges under certain conditions.

Theorem 3. *Let $\psi(t,x) \in C$ be piecewise continuous on $[0,\infty) \times A$ and of an exponential order $\frac{\theta}{w}$, and let $\Psi(s,w,x) = \mathcal{N}\{\psi(t,x)\}$ have the fractional expansion (3.26). Then, the reminder $\mathfrak{R}_{\mathcal{K}}(s,w,x)$ of (3.26) is satisfied:*

$$|\mathfrak{R}_{\mathcal{K}}(s,w,x)| \le \frac{w^{(\mathcal{K}+1)\alpha}}{s^{(\mathcal{K}+1)\alpha+1}}\mathfrak{M}(x),\ \alpha \in (0,1], x \in A,\ s \in (\theta,\theta^*], w \in (0,\infty), \tag{3.41}$$

provided that:

$$\left|s\mathcal{N}\left\{\mathcal{D}_t^{(\mathcal{K}+1)\alpha}\psi(t,x)\right\}\right| \le \mathfrak{M}(x),\ \alpha \in (0,1], x \in A,\ s \in (\theta,\theta^*], w \in (0,\infty). \tag{3.42}$$

Proof. The reminder $\mathfrak{R}_k(s,w,x)$ of the fractional expansion (3.26) can be written as:

$$\mathfrak{R}_{\mathcal{K}}(s,w,x) = \Psi(s,w,x) - \sum_{k=0}^{\mathcal{K}} \frac{\mathcal{D}_t^{k\alpha}\psi(0,x)w^{k\alpha}}{s^{1+k\alpha}}. \tag{3.43}$$

Multiply both sides of (3.43) by $\frac{s^{(\mathcal{K}+1)\alpha+1}}{w^{(\mathcal{K}+1)\alpha}}$; with the aid of (3.14) we obtain:

$$\begin{aligned}\frac{s^{(\mathcal{K}+1)\alpha+1}}{w^{(\mathcal{K}+1)\alpha}}\mathfrak{R}_{\mathcal{K}}(s,w,x) &= \frac{s^{(\mathcal{K}+1)\alpha+1}}{w^{(\mathcal{K}+1)\alpha}}\Psi(s,w,x) - \sum_{k=0}^{\mathcal{K}}\frac{\mathcal{D}_t^{k\alpha}\psi(0,x)s^{(\mathcal{K}+1-k)\alpha}}{w^{(\mathcal{K}+1-k)\alpha}} \\ &= s\left(\frac{s^{(\mathcal{K}+1)\alpha}}{w^{(\mathcal{K}+1)\alpha}}\Psi(s,w,x) - \sum_{k=0}^{\mathcal{K}}\frac{\mathcal{D}_t^{k\alpha}\psi(0,x)s^{(\mathcal{K}+1-k)\alpha-1}}{w^{(\mathcal{K}+1-k)\alpha}}\right) \\ &= s\mathcal{N}\left\{\mathcal{D}_t^{(\mathcal{K}+1)\alpha}\psi(t,x)\right\}.\end{aligned} \tag{3.44}$$

Using (3.42) and (3.44), we obtain the result in (3.41). □

4. A Description of the Proposed NRPS Method

This section is devoted to presenting the NRPS method to establish approximate solutions to the nonlinear fractional PDE:

$$\mathcal{D}_t^{\alpha}\psi(t,x) = \mathcal{G}_x[\psi(t,x)],\ \alpha \in (0,1], x \in A, t \geq 0, \tag{4.1}$$

with the initial condition:

$$\psi(0,x) = g(x), x \in A. \tag{4.2}$$

The symbol $\mathcal{G}_x$ denotes a nonlinear operator relative to the x of degree $r, x \in A, t \geq 0$, $\mathcal{D}_t^{\alpha}$ refers to the α-th Caputo fractional derivative for $\alpha \in (0,1]$, and $\psi(t,x)$ is an unknown function to be determined. To obtain the desired analytical solution for (4.1)–(4.2) using the NRPS method, we can perform the following steps:

Step 1. Apply the natural transformation for both sides of (4.1) by taking advantage of the facts in Lemmas 2 and 3, we obtain:

$$\Psi(s,w,x) - \frac{1}{s}g(x) - \frac{w^{\alpha}}{s^{\alpha}}\mathcal{N}\left\{\mathcal{G}_x\left[\mathcal{N}^{-1}\{\Psi(s,w,x)\}\right]\right\} = 0,\ \alpha \in (0,1], x \in A,\ s \in (\theta,\infty), w \in (0,\infty), \tag{4.3}$$

where $\Psi(s,w,x) = \mathcal{N}\{\psi(t,x), x \in A, s \in (\theta,\infty), w \in (0,\infty)$.

Step 2. Suppose that the solution of (4.3) can be written in the following expansion:

$$\Psi(s,w,x) = \sum_{k=0}^{\infty}\frac{\Pi_k(x)w^{k\alpha}}{s^{1+k\alpha}},\ \alpha \in (0,1], x \in A,\ s \in (\theta,\infty), w \in (0,\infty). \tag{4.4}$$

Define the $\mathcal{K}$th truncated series of $\Psi(s,w,x)$ as:

$$\Psi_{\mathcal{K}}(s,w,x) = \sum_{k=0}^{\mathcal{K}}\frac{\Pi_k(x)w^{k\alpha}}{s^{1+k\alpha}},\ \alpha \in (0,1], x \in A,\ s \in (\theta,\infty), w \in (0,\infty). \tag{4.5}$$

Using the results in Theorem 2, we deduce that the initial guess $\Pi_0(x) = g(x)$. Upon this fact, we write the $\mathcal{K}$th truncated series of $\Psi(s,w,x)$ as:

$$\Psi_{\mathcal{K}}(s,w,x) = \frac{g(x)}{s} + \sum_{k=1}^{\mathcal{K}}\frac{\Pi_k(x)w^{k\alpha}}{s^{1+k\alpha}},\ \alpha \in (0,1], x \in A,\ s \in (\theta,\infty), w \in (0,\infty). \tag{4.6}$$

Step 3. We define the so-called natural residual function (NRF) as:

$$\begin{aligned}\mathcal{N}Res(s,w,x) = \Psi(s,w,x) - \tfrac{1}{s}g(x) - \tfrac{w^{\alpha}}{s^{\alpha}}\mathcal{N}\{\mathcal{G}_x[\mathcal{N}^{-1}\{\Psi(s,w,x)\}]\}, \alpha \\ \in (0,1], x \in A,\ s \in (\theta,\infty), w \in (0,\infty),\end{aligned} \tag{4.7}$$

and the $\mathcal{K}$th truncated NRF as:

$$\mathcal{N}Res_{\mathcal{K}}(s,w,x) = \Psi_{\mathcal{K}}(s,w,x) - \frac{1}{s}g(x) - \frac{w^{\alpha}}{s^{\alpha}}\mathcal{N}\{\mathcal{G}_x[\mathcal{N}^{-1}\{\Psi_{\mathcal{K}}(s,w,x)\}]\},\ \alpha \in (0,1], x \in A,\ s \in (\theta,\infty), w \in (0,\infty), \tag{4.8}$$

Now, we introduce some essential facts to obtain the desired coefficients $\Pi_k(x)$ in the expansion (4.6).

Remark 3. *The NRF and the $\mathcal{K}$th truncated NRF in (4.7) and (4.8), respectively, satisfy the following properties:*

(i) $\mathcal{N}Res(s,w,x) = 0, x \in A, s \in (\theta,\infty), w \in (0,\infty).$ (4.9)

(ii) $\lim\limits_{k\to\infty} \mathcal{N}Res_{\mathcal{K}}(s,w,x) = \mathcal{N}Res(s,w,x), x \in A,\ s \in (\theta,\infty), w \in (0,\infty).$ (4.10)

(iii) $\lim\limits_{s\to\infty} s^{1+\mathcal{K}\alpha}\mathcal{N}Res_{\mathcal{K}}(s,w,x) = 0,\ \mathcal{K} = 1,2,\ldots, x \in A,\ s \in (\theta,\infty), w \in (0,\infty).$ (4.11)

Proof. The properties (i) and (ii) are clear. To prove (iii), we represent the NRF (4.7) in fractional expansion as:

$$\mathcal{N}Res(s,w,x) = \sum_{k=1}^{\infty} \frac{\left(\Pi_k(x) - \mathcal{G}_x^{(k)}[\Pi_i(x)]\right)w^{k\alpha}}{s^{1+k\alpha}},\ \alpha \in (0,1], x \in A,\ s \in (\theta,\infty), w \in (0,\infty),\ i \in \{0,1,2,\ldots,k-1\}, \tag{4.12}$$

where $\mathcal{G}_x^{(k)}, k = 1,2,3,\ldots$ are differential operators related to the x of degree r. Using (4.9), we obtain $\Pi_k(x) - \mathcal{G}_x^{(k)}[\Pi_i(x)] = 0$ for $k = 1,2,3,\ldots$ and $i \in \{0,1,2,\ldots,k-1\}$. Substitute the $\mathcal{K}$th truncated series $\Psi_{\mathcal{K}}(s,w,x)$ into (4.12) that can be written in the following expression:

$$\mathcal{N}Res_{\mathcal{K}}(s,w,x) = \sum_{k=1}^{\mathcal{K}} \frac{\left(\Pi_k(x) - \mathcal{G}_x^{(k)}[\Pi_i(x)]\right)w^{k\alpha}}{s^{1+k\alpha}} + \sum_{k=\mathcal{K}+1}^{\mathcal{K}r+1} \frac{\left(\Pi_k(x) - \mathcal{G}^{*(\mathcal{K}k)}_x[\Pi_j(x)]\right)w^{k\alpha}}{s^{1+k\alpha}}, \tag{4.13}$$

where $i \in \{0,1,2,\ldots,k-1\}, j \in \{0,1,2,\ldots,\mathcal{K}\}$, and $\mathcal{G}^{*(\mathcal{K}k)}_x, k = \mathcal{K}+1, \mathcal{K}-2,\ldots,\mathcal{K}r+1$ are differential operators related to x and $\Pi_k(x) - \mathcal{G}^{*(\mathcal{K}k)}_x[\Pi_j(x)] \neq 0$. By multiplying both sides of (4.13) by $s^{1+k\alpha}$ and taking the limit as $s \to \infty$, we obtain:

$$\lim_{s\to\infty} s^{1+k\alpha}\mathcal{N}Res_k(s,w,x) = \left(\Pi_k(x) - \mathcal{G}_x^{(k)}[\Pi_i(x)]\right)w^{k\alpha} = 0,\ k = 1,2,\ldots, i \in \{0,1,2,\ldots,k-1\}. \tag{4.14}$$

Step 4. Substitute the $\mathcal{K}$th truncated series $\Psi_{\mathcal{K}}(s,w,x)$ in (4.6) into the $\mathcal{K}$th truncated NRF (4.8).

Step 5. To determine the coefficients $\Pi_k(x)$, $k = 1,2,\ldots,\mathcal{K}$, we solve the system that was obtained using $\lim\limits_{s\to\infty} s^{1+k\alpha}\mathcal{N}Res_k(s,w,x) = 0, k = 1,2,\ldots,\mathcal{K}$ and collect the results into the $\mathcal{K}$th truncated series $\Psi_{\mathcal{K}}(s,w,x)$ in (4.6).

Step 6. Apply the inverse natural operator for both sides of the inferred $\mathcal{K}$th truncated series $\Psi_{\mathcal{K}}(s,w,x)$ to obtain the $\mathcal{K}$th approximate solution for the nonlinear fractional PDE (4.1). □

5. Applications

In this section, we investigate the time-fractional KdV-ZK Equations (1.1) and (1.2), the time-fractional spKdV Equations (1.3) and (1.4), and the fractional FKdV E quations (1.5) and (1.6) using the proposed NRPS method. We used the mathematical software package Mathematica 13 to implement all symbolic and numerical computations.

Application 1. *Consider the time-fractional KdV-ZK equation [17]:*

$$\mathcal{D}_t^{\alpha}\psi+36\psi\frac{\partial\psi}{\partial x}+\frac{\partial^3\psi}{\partial x^3}+\frac{\partial^3\psi}{\partial x\partial y^2}+\frac{\partial^3\psi}{\partial x\partial z^2}=0,\ t\geq 0,\ x,y,z\in\mathbb{R}, \tag{5.1}$$

subject to the initial condition:

$$\psi(0,x,y,z)=\frac{1}{\sqrt{2}}\text{sech}(x+y+z),x,y,z\in\mathbb{R}. \tag{5.2}$$

Apply the natural transformation for both sides of (5.1) with the aid of (3.13) to obtain:

$$\begin{aligned}\tfrac{s^{\alpha}}{w^{\alpha}}\Psi(s,w,x,y,z)&-\tfrac{s^{\alpha-1}}{w^{\alpha}}\psi(0,x,y,z)+36\mathcal{N}\left\{\mathcal{N}^{-1}\{\Psi(s,w,x,y,z)\}\tfrac{\partial}{\partial x}\mathcal{N}^{-1}\{\Psi(s,w,x,y,z)\}\right\}\\&+\mathcal{N}\left\{\tfrac{\partial^3}{\partial x^3}\mathcal{N}^{-1}\{\Psi(s,w,x,y,z)\}\right\}+\mathcal{N}\left\{\tfrac{\partial^3}{\partial x\partial y^2}\mathcal{N}^{-1}\{\Psi(s,w,x,y,z)\}\right\}\\&+\mathcal{N}\left\{\tfrac{\partial^3}{\partial x\partial z^2}\mathcal{N}^{-1}\{\Psi(s,w,x,y,z)\}\right\}=0,\end{aligned} \tag{5.3}$$

where $\Psi(s,w,x,y,z)=\mathcal{N}\{\psi(t,x,y,z)\}$. *Using (5.2) in (5.3), we obtain:*

$$\begin{aligned}\Psi(s,w,x,y,z)-&\ \tfrac{1}{s\sqrt{2}}\text{sech}(x+y+z)\\&+\tfrac{w^{\alpha}}{s^{\alpha}}\Big(36\mathcal{N}\left\{\mathcal{N}^{-1}\{\Psi(s,w,x,y,z)\}\tfrac{\partial}{\partial x}\mathcal{N}^{-1}\{\Psi(s,w,x,y,z)\}\right\}\\&+\mathcal{N}\left\{\tfrac{\partial^3}{\partial x^3}\mathcal{N}^{-1}\{\Psi(s,w,x,y,z)\}\right\}\\&+\mathcal{N}\left\{\tfrac{\partial^3}{\partial x\partial y^2}\mathcal{N}^{-1}\{\Psi(s,w,x,y,z)\}\right\}\\&+\mathcal{N}\left\{\tfrac{\partial^3}{\partial x\partial z^2}\mathcal{N}^{-1}\{\Psi(s,w,x,y,z)\}\right\}\Big)=0,\end{aligned} \tag{5.4}$$

Suppose that the solution of (5.4) can be written in the following expansion:

$$\Psi(s,w,x,y,z)=\sum_{k=0}^{\infty}\frac{\Pi_k(x,y,z)w^{k\alpha}}{s^{1+k\alpha}},\ \alpha\in(0,1],x,y,z\in A,\ s\in(\theta,\infty),w\in(0,\infty). \tag{5.5}$$

Define the $\mathcal{K}$*th truncated series of* $\Psi(s,w,x,y,z)$ *as:*

$$\Psi_{\mathcal{K}}(s,w,x,y,z)=\sum_{k=0}^{\mathcal{K}}\frac{\Pi_k(x,y,z)w^{k\alpha}}{s^{1+k\alpha}},\ \alpha\in(0,1],x,y,z\in A,\ s\in(\theta,\infty),w\in(0,\infty). \tag{5.6}$$

Using the results in Theorem 2, we deduce that the initial guess $\Pi_0(x,y,z)=\sqrt{\frac{1}{2}}sech((x+y+z))$. *Upon this fact, we write the* $\mathcal{K}$*th truncated series of* $\Psi(s,w,x,y,z)$ *as:*

$$\Psi_{\mathcal{K}}(s,w,x,y,z)=\frac{1}{s\sqrt{2}}\text{sech}(x+y+z)+\sum_{k=1}^{\mathcal{K}}\frac{\Pi_k(x,y,z)w^{k\alpha}}{s^{1+k\alpha}}. \tag{5.7}$$

We define the NRF as:

$$\begin{aligned}
\mathcal{N}Res(s,w,x,y,z) \\
&= \Psi(s,w,x,y,z) - \tfrac{1}{s\sqrt{2}}\text{sech}(x+y+z) \\
&+ \tfrac{w^\alpha}{s^\alpha}\Big(36\mathcal{N}\Big\{\mathcal{N}^{-1}\{\Psi(s,w,x,y,z)\}\tfrac{\partial}{\partial x}\mathcal{N}^{-1}\{\Psi(s,w,x,y,z)\}\Big\} \\
&+ \mathcal{N}\Big\{\tfrac{\partial^3}{\partial x^3}\mathcal{N}^{-1}\{\Psi(s,w,x,y,z)\}\Big\} \\
&+ \mathcal{N}\Big\{\tfrac{\partial^3}{\partial x\partial y^2}\mathcal{N}^{-1}\{\Psi(s,w,x,y,z)\}\Big\} \\
&+ \mathcal{N}\Big\{\tfrac{\partial^3}{\partial x\partial z^2}\mathcal{N}^{-1}\{\Psi(s,w,x,y,z)\}\Big\}\Big),
\end{aligned} \tag{5.8}$$

and the $\mathcal{K}$th truncated NRF as:

$$\begin{aligned}
\mathcal{N}Res_{\mathcal{K}}(s,w,x,y,z) \\
&= \Psi_{\mathcal{K}}(s,w,x,y,z) - \tfrac{1}{s\sqrt{2}}\text{sech}(x+y+z) \\
&+ \tfrac{w^\alpha}{s^\alpha}\Big(36\mathcal{N}\Big\{\mathcal{N}^{-1}\{\Psi_{\mathcal{K}}(s,w,x,y,z)\}\tfrac{\partial}{\partial x}\mathcal{N}^{-1}\{\Psi_{\mathcal{K}}(s,w,x,y,z)\}\Big\} \\
&+ \mathcal{N}\Big\{\tfrac{\partial^3}{\partial x^3}\mathcal{N}^{-1}\{\Psi_{\mathcal{K}}(s,w,x,y,z)\}\Big\} \\
&+ \mathcal{N}\Big\{\tfrac{\partial^3}{\partial x\partial y^2}\mathcal{N}^{-1}\{\Psi_{\mathcal{K}}(s,w,x,y,z)\}\Big\} \\
&+ \mathcal{N}\Big\{\tfrac{\partial^3}{\partial x\partial z^2}\mathcal{N}^{-1}\{\Psi_{\mathcal{K}}(s,w,x,y,z)\}\Big\}\Big).
\end{aligned} \tag{5.9}$$

For $\mathcal{K}=1$, we obtain the 1st NRF with using (5.7) as follows:

$$\begin{aligned}
\mathcal{N}Res_1(s,w,x,y,z) \\
&= \tfrac{\Pi_1(x,y,z)w^\alpha}{s^{1+\alpha}} \\
&+ \tfrac{w^\alpha}{s^\alpha}\Big(36\mathcal{N}\Big\{\mathcal{N}^{-1}\Big\{\tfrac{1}{s\sqrt{2}}\text{sech}(x+y+z) + \tfrac{\Pi_1(x,y,z)w^\alpha}{s^{1+\alpha}}\Big\}\tfrac{\partial}{\partial x}\mathcal{N}^{-1}\Big\{\tfrac{1}{s\sqrt{2}}\text{sech}(x+y+z) \\
&+ \tfrac{\Pi_1(x,y,z)w^\alpha}{s^{1+\alpha}}\Big\}\Big\} + \mathcal{N}\Big\{\tfrac{\partial^3}{\partial x^3}\mathcal{N}^{-1}\Big\{\tfrac{1}{s\sqrt{2}}\text{sech}(x+y+z) + \tfrac{\Pi_1(x,y,z)w^\alpha}{s^{1+\alpha}}\Big\}\Big\} \\
&+ \mathcal{N}\Big\{\tfrac{\partial^3}{\partial x\partial y^2}\mathcal{N}^{-1}\Big\{\tfrac{1}{s\sqrt{2}}\text{sech}(x+y+z) + \tfrac{\Pi_1(x,y,z)w^\alpha}{s^{1+\alpha}}\Big\}\Big\} \\
&+ \mathcal{N}\Big\{\tfrac{\partial^3}{\partial x\partial z^2}\mathcal{N}^{-1}\Big\{\tfrac{1}{s\sqrt{2}}\text{sech}(x+y+z) + \tfrac{\Pi_1(x,y,z)w^\alpha}{s^{1+\alpha}}\Big\}\Big\}\Big).
\end{aligned} \tag{5.10}$$

By multiplying both sides of (5.10) by $s^{1+\alpha}$ and using $\lim_{s\to\infty} s^{1+\alpha}\mathcal{N}Res_1(s,w,x) = 0$, we obtain:

$$\Pi_1(x,y,z) = 3\sqrt{3}\text{sech}(x+y+z)\tanh(x+y+z). \tag{5.11}$$

Similarly, we can obtain the following:

$$\Pi_2(x,y,z) = \frac{9}{2}\sqrt{3}(-3+\cosh(2(x+y+z)))\text{sech}(x+y+z)^3. \tag{5.12}$$

$$\Pi_3(x,y,z) = \frac{27}{4}\sqrt{3}\text{sech}(x+y+z)^4(-23\sinh(x+y+z)+\sinh(3(x+y+z))). \tag{5.13}$$

$$\Pi_4(x,y,z) = \frac{81\sqrt{3}}{8}(115-76\cosh(2(x+y+z))+\cosh(4(x+y+z)))\text{sech}(x+y+z)^5. \tag{5.14}$$

The CPU time that is needed to obtain these results (5.11)–(5.14) was 35.390625 s. Substitute (5.11)–(5.14) into (5.7) to obtain the $\mathcal{K}$th truncated series of $\Psi(s,w,x)$ as:

$$\begin{aligned}
\Psi_{\mathcal{K}}(s,w,x,y,z) = \tfrac{1}{s\sqrt{2}}\text{sech}(x+y+z) + \tfrac{\Pi_1(x,y,z)w^\alpha}{s^{1+\alpha}} + \tfrac{\Pi_2(x,y,z)w^{2\alpha}}{s^{1+2\alpha}} \\
+ \tfrac{\Pi_3(x,y,z)w^{3\alpha}}{s^{1+3\alpha}} + \tfrac{\Pi_4(x,y,z)w^{4\alpha}}{s^{1+4\alpha}} + \ldots + \tfrac{\Pi_{\mathcal{K}}(x,y,z)w^{\mathcal{K}\alpha}}{s^{1+\mathcal{K}\alpha}}.
\end{aligned} \tag{5.15}$$

Apply the inverse natural operator of (5.15) to obtain the $\mathcal{K}$th NRPS solution for the time-fractional KdV-ZK equation as:

$$\psi_{\mathcal{K}}(t,x,y,z) = \frac{1}{\sqrt{2}}\text{sech}(x+y+z) + \frac{\Pi_1(x,y,z)t^{\alpha}}{\Gamma(1+\alpha)} + \frac{\Pi_2(x,y,z)t^{2\alpha}}{\Gamma(1+2\alpha)} + \frac{\Pi_3(x,y,z)t^{3\alpha}}{\Gamma(1+3\alpha)} + \frac{\Pi_4(x,y,z)t^{4\alpha}}{\Gamma(1+4\alpha)} + \ldots + \frac{\Pi_{\mathcal{K}}(x,y,z)t^{\mathcal{K}\alpha}}{\Gamma(1+\mathcal{K}\alpha)}. \tag{5.16}$$

We present a numerical simulation of the results obtained. We assume that $y = z = 0$. Table 1 compares the exact solution for the time-fractional KdV-ZK Equation (5.1) and the inferred fourth NRPS solution (5.16) where the absolute and relative errors are presented. Figure 1 presents the absolute error between the exact solution and the explored approximate fourth NRPS solution (5.16) for the time-fractional KdV-ZK Equation (5.1). To our knowledge, no works in the literature have been presented to deduce approximate solutions to Equation (5.1). Still, all the works presented are concerned with deriving exact solutions to the equation. From Table 1 and Figure 1, we can see the convergence of the numerical solutions derived through the proposed method with the exact solutions, which provides insight into the high accuracy of the proposed method. Figure 2 shows the surfaces of the exact and the fourth NRPS solutions (5.16) at $\alpha = 1$. Figure 3 presents the fourth NRPS solution (5.16) surfaces at different fractional orders $\alpha = 0.95$ and $\alpha = 0.9$ to demonstrate the effect of the fractional derivative on the established surfaces. For more explanation, we depicted, in Figures 4–6, the NRPS solutions explored in 2D plots at different fractional orders α where the time t is considered at various values.

Table 1. A comparison between the exact solution and the NRPS solution for the time-fractional KdV-ZK equation.

x	t	**Exact Solution**	**NRPS Solution**	**Absolute Error**	**Relative Error**	**CPU Time**
−1	0.0	1.122462928047994	1.122462928047995	2.220446×10^{-16}	1.978190×10^{-16}	0.015625
	0.1	0.878805774963547	0.878736016940593	6.975802×10^{-5}	7.937820×10^{-5}	
	0.2	0.671997937091220	0.669464247695219	2.533689×10^{-3}	3.770382×10^{-3}	
	0.3	0.506783751732631	0.486544390701797	2.023936×10^{-2}	3.993687×10^{-2}	
0	0.0	1.732050807568877	1.732050807568877	2.220446×10^{-16}	1.281975×10^{-16}	0.015625
	0.1	1.656928147349753	1.657031356966050	1.032096×10^{-5}	6.228973×10^{-5}	
	0.2	1.461072649720631	1.467047034010839	5.974384×10^{-3}	4.089039×10^{-3}	
	0.3	1.208615771635065	1.267319925263052	5.870415×10^{-2}	4.857139×10^{-2}	
1	0.0	1.122462928047994	1.122462928047995	2.220446×10^{-16}	1.978190×10^{-16}	0.015625
	0.1	1.379934335375192	1.379959423861089	2.508848×10^{-5}	1.818092×10^{-5}	
	0.2	1.602159904080592	1.601751160457130	4.087436×10^{-4}	2.551203×10^{-4}	
	0.3	1.723426491772224	1.709575007146963	1.385148×10^{-2}	8.037177×10^{-3}	

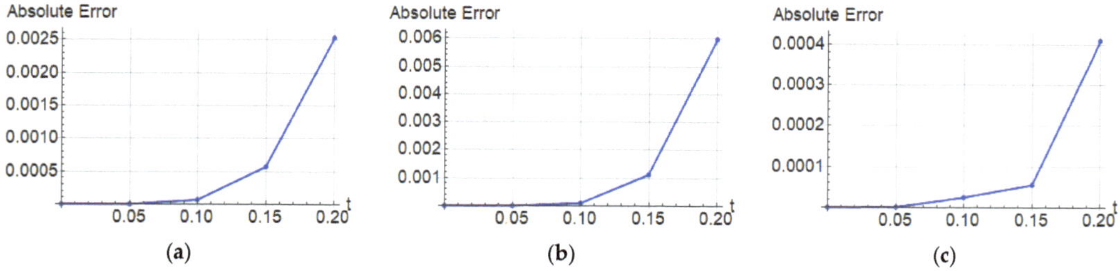

Figure 1. The absolute error between the approximate and exact solutions for the time-fractional KdV-ZK Equation (5.1) at $\alpha = 1$: (**a**) $x = -1$, (**b**) $x = 0$, and (**c**) $x = 1$.

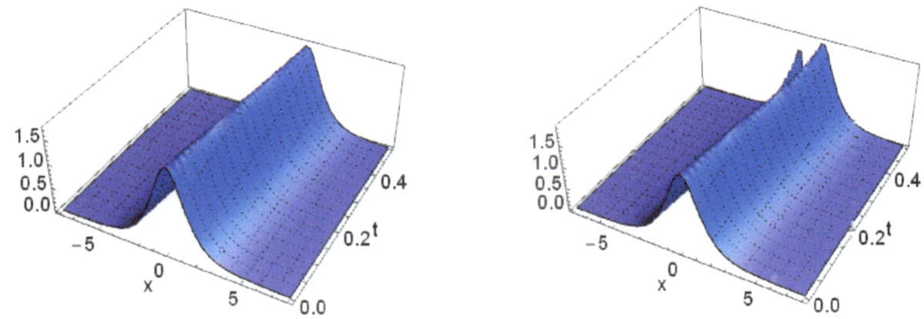

(a) (b)

Figure 2. The solution of the time-fractional KdV-ZK Equation (5.1) at $\alpha = 1$: (**a**) exact solution, (**b**) 4th NRPS solution (5.16).

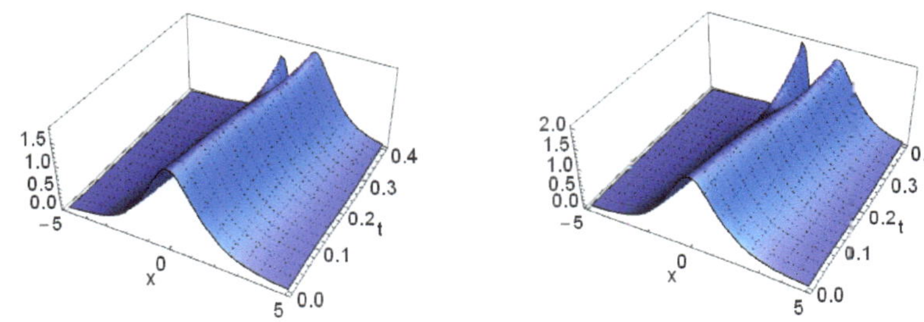

(a) (b)

Figure 3. The 4th NRPS solution (5.16) of the time-fractional KdV-ZK Equation (5.1): (**a**) $\alpha = 0.95$; (**b**) $\alpha = 0.9$.

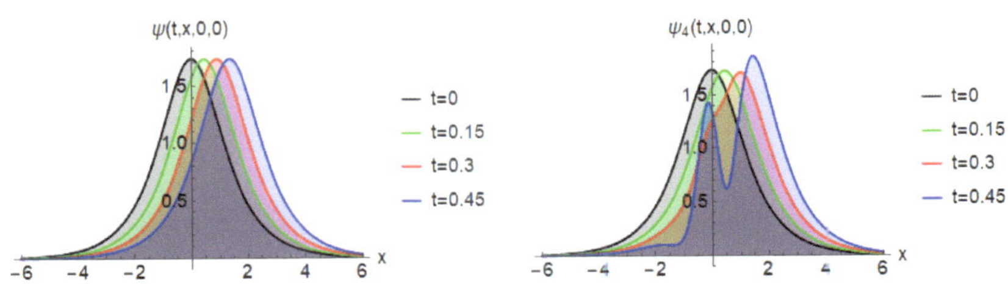

(a) (b)

Figure 4. Solution of the time-fractional KdV-ZK Equation (5.1) at $\alpha = 1, t \in \{0, 0.15, 0.3, 0.45\}$: (**a**) exact solution; (**b**) 4th NRPS solution (5.16).

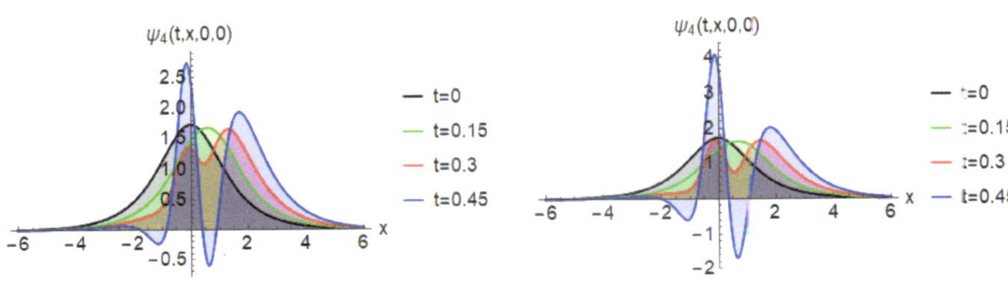

(a) (b)

Figure 5. The 4th NRPS solution (5.16) of the time-fractional KdV-ZK Equation (5.1) at $t \in \{0,\ 0.15,\ 0.3,\ 0.45\}$: (**a**) $\alpha = 0.9$; (**b**) $\alpha = 0.85$.

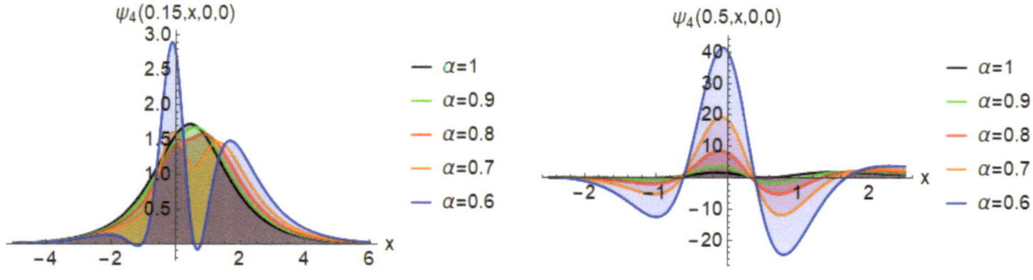

(a) (b)

Figure 6. The 4th NRPS solution (5.16) of the time-fractional KdV-ZK Equation (5.1) at different fractional orders $\alpha \in \{1,\ 0.9,\ 0.8,\ 0.7,\ 0.6\}$: (**a**) $t = 0.15$; (**b**) $t = 0.5$.

Applications 2. *Consider the following time-fractional spKdV equation:*

$$\mathcal{D}_t^{\alpha}\psi - 8\frac{\partial\psi}{\partial x} + \left(\frac{\partial\psi}{\partial x}\right)^2 + 2\frac{\partial^3\psi}{\partial x^3} = 0,\ t \geq 0,\ x \in \mathbb{R}, \tag{5.17}$$

subject to the initial condition:

$$\psi(x,0) = g(x) = 1 + 12\tanh(x), x \in \mathbb{R}. \tag{5.18}$$

Apply the natural transformation for both sides of (5.17) with the aid of (3.13) to obtain:

$$\begin{aligned}\tfrac{s^\alpha}{w^\alpha}\Psi(s,w,x) - \tfrac{s^{\alpha-1}}{w^\alpha}\psi(0,x) - 8\mathcal{N}\left\{\tfrac{\partial}{\partial x}\mathcal{N}^{-1}\{\Psi(s,w,x)\}\right\} + \mathcal{N}\left\{\left(\tfrac{\partial}{\partial x}\mathcal{N}^{-1}\{\Psi(s,w,x)\}\right)^2\right\} \\ +2\mathcal{N}\left\{\tfrac{\partial^3}{\partial x^3}\mathcal{N}^{-1}\{\Psi(s,w,x)\}\right\} = 0,\end{aligned} \tag{5.19}$$

where $\Psi(s,w,x) = \mathcal{N}\{\psi(t,x)\}$. *Using (5.18) in (5.19), we obtain:*

$$\begin{aligned}\Psi(s,w,x) - \tfrac{1}{s}(1+12\tanh(x)) \\ +\tfrac{w^\alpha}{s^\alpha}\left(-8\mathcal{N}\left\{\tfrac{\partial}{\partial x}\mathcal{N}^{-1}\{\Psi(s,w,x)\}\right\} + \mathcal{N}\left\{\left(\tfrac{\partial}{\partial x}\mathcal{N}^{-1}\{\Psi(s,w,x)\}\right)^2\right\}\right. \\ \left.+2\mathcal{N}\left\{\tfrac{\partial^3}{\partial x^3}\mathcal{N}^{-1}\{\Psi(s,w,x)\}\right\}\right) = 0,\end{aligned} \tag{5.20}$$

Suppose that the solution of (5.20) can be written in the following expansion:

$$\Psi(s,w,x) = \sum_{k=0}^{\infty}\frac{\Pi_k(x)w^{k\alpha}}{s^{1+k\alpha}},\ \alpha \in (0,1], x \in A,\ s \in (\theta,\infty), w \in (0,\infty). \tag{5.21}$$

Define the $\mathcal{K}$*th truncated series of* $\Psi(s,w,x)$ *as:*

$$\Psi_{\mathcal{K}}(s,w,x) = \sum_{k=0}^{\mathcal{K}}\frac{\Pi_k(x)w^{k\alpha}}{s^{1+k\alpha}},\ \alpha \in (0,1], x \in A,\ s \in (\theta,\infty), w \in (0,\infty). \tag{5.22}$$

Using the results in Theorem 2, we deduce that the initial guess $\Pi_0(x) = 1 + 12tanh(x)$. *Based on this fact, we write the* $\mathcal{K}$*th truncated series of* $\Psi(s,w,x)$ *as:*

$$\Psi_{\mathcal{K}}(s,w,x) = \frac{1}{s}(1 + 12\tanh(x)) + \sum_{k=1}^{\mathcal{K}}\frac{\Pi_k(x)w^{k\alpha}}{s^{1+k\alpha}}. \tag{5.23}$$

We define the NRF as:

$$\begin{aligned}\mathcal{N}Res(s,w,x) &= \Psi(s,w,x) - \tfrac{1}{s}(1+12\tanh(x)) \\ &+ \tfrac{w^{\alpha}}{s^{\alpha}}\Big(-8\mathcal{N}\Big\{\tfrac{\partial}{\partial x}\mathcal{N}^{-1}\{\Psi(s,w,x)\}\Big\} + \mathcal{N}\Big\{\Big(\tfrac{\partial}{\partial x}\mathcal{N}^{-1}\{\Psi(s,w,x)\}\Big)^{2}\Big\} \\ &+ 2\mathcal{N}\Big\{\tfrac{\partial^3}{\partial x^3}\mathcal{N}^{-1}\{\Psi(s,w,x)\}\Big\},\end{aligned} \tag{5.24}$$

and the $\mathcal{K}$th truncated NRF as:

$$\begin{aligned}\mathcal{N}Res_{\mathcal{K}}(s,w,x) &= \Psi_{\mathcal{K}}(s,w,x) - \tfrac{1}{s}(1+12\tanh(x)) \\ &+ \tfrac{w^{\alpha}}{s^{\alpha}}\Big(-8\mathcal{N}\Big\{\tfrac{\partial}{\partial x}\mathcal{N}^{-1}\{\Psi_{\mathcal{K}}(s,w,x)\}\Big\} + \mathcal{N}\Big\{\Big(\tfrac{\partial}{\partial x}\mathcal{N}^{-1}\{\Psi_{\mathcal{K}}(s,w,x)\}\Big)^{2}\Big\} \\ &+ 2\mathcal{N}\Big\{\tfrac{\partial^3}{\partial x^3}\mathcal{N}^{-1}\{\Psi_{\mathcal{K}}(s,w,x)\}\Big\},\end{aligned} \tag{5.25}$$

Using the fact that $\lim_{s\to\infty} s^{1+k\alpha}\mathcal{N}Res_k(s,w,x) = 0,\ k = 1,2,\ldots,$ *we obtain.*

$$\Pi_1(x) = 8g'(x) - (g'(x))^2 - 2g'''(x). \tag{5.26}$$

$$\Pi_2(x) = 8\Pi_1'(x) - 2g'(x)\Pi_1'(x) - 2\Pi_1'''(x). \tag{5.27}$$

$$\Pi_3(x) = \frac{-\Gamma(1+2\alpha)(\Pi_1'(x))^2 - (\Gamma(1+\alpha))^2(-8\Pi_2'(x) + 2g'(x)\Pi_2'(x) + 2\Pi_2'''(x))}{(\Gamma(1+\alpha))^2}. \tag{5.28}$$

$$\Pi_4(x) = \frac{-2\Gamma(1+3\alpha)\Pi_1'(x)\Pi_2'(x) + 8\Gamma(1+\alpha)\Gamma(1+2\alpha)(-8\Pi_3'(x) + 2g'(x)\Pi_3'(x) + 2\Pi_3'''(x))}{\Gamma(1+\alpha)\Gamma(1+2\alpha)}. \tag{5.29}$$

The CPU time needed to obtain these results (5.26)–(5.29) was 8.53125 s. Substitute (5.26)–(5.29) into (5.23) to obtain the $\mathcal{K}$th truncated series of $\Psi(s,w,x)$ as:

$$\Psi_{\mathcal{K}}(s,w,x) = \frac{1}{s}(1+12\tanh(x)) + \frac{\Pi_1(x)w^{\alpha}}{s^{1+\alpha}} + \frac{\Pi_2(x)w^{2\alpha}}{s^{1+2\alpha}} + \frac{\Pi_3(x)w^{3\alpha}}{s^{1+3\alpha}} + \frac{\Pi_4(x)w^{4\alpha}}{s^{1+4\alpha}} + \ldots + \frac{\Pi_{\mathcal{K}}(x)w^{\mathcal{K}\alpha}}{s^{1+\mathcal{K}\alpha}}. \tag{5.30}$$

Apply the inverse natural operator of (5.30) to obtain the $\mathcal{K}$th NRPS solution for the time-fractional spKdV equation as:

$$\psi_{\mathcal{K}}(t,x,y,z) = 1 + 12\tanh(x) + \frac{\Pi_1(x)t^{\alpha}}{\Gamma(1+\alpha)} + \frac{\Pi_2(x)t^{2\alpha}}{\Gamma(1+2\alpha)} + \frac{\Pi_3(x)t^{3\alpha}}{\Gamma(1+3\alpha)} + \frac{\Pi_4(x)t^{4\alpha}}{\Gamma(1+4\alpha)} + \ldots + \frac{\Pi_{\mathcal{K}}(x)t^{\mathcal{K}\alpha}}{\Gamma(1+\mathcal{K}\alpha)}. \tag{5.31}$$

Table 2 shows the absolute and relative errors for the constructed numerical results to demonstrate the accuracy of the NRPS method. Figure 7 presents the absolute error between the exact solution and the explored approximate 4th NRPS solution (5.31) for the time-fractional spKdV Equation (5.17). When comparing the solution explored using the proposed method and the exact solution and calculating the absolute error and relative error, it becomes clear that the proposed method achieves an accuracy within 10^{-4} in dealing with Equation (5.17), which is a suitable level for numerical modeling applications that require accurate and stable solutions. It is worth noting here that based on the literature review, it was found that there are no works that focused on extracting approximate solutions to Equation (5.17), which prompted us to compare the approximate solutions derived here with the exact solutions to the equation. This may be a strong motivation to work on presenting other works that focus on deducing approximate solutions to the equation using new techniques and comparing them with the numerical solutions presented in this work. Figure 8 depicts the exact and 4th NRPS solution surfaces in (5.31) for the time-fractional spKdV Equation (5.17) where we can note the harmony between them, which ensures the efficiency of the proposed method. In Figure 9, we present the 2D plots for the exact and 4th NRPS solutions for the time-fractional spKdV Equation (5.17) at $t = 0.1$ and $t = 1$. Note that the harmony between the depicted solutions reduces with time. This harmony can be increased for a longer time by increasing the terms of the NRPS solution (5.17). To illustrate the effect of the fractional order α on the behavior of the constructed approximate solution, we present Figure 10 which depicts the 4th NRPS solution (5.17) at different fractional orders $\alpha = 0.95, 0.75, 0.5, 0.25$.

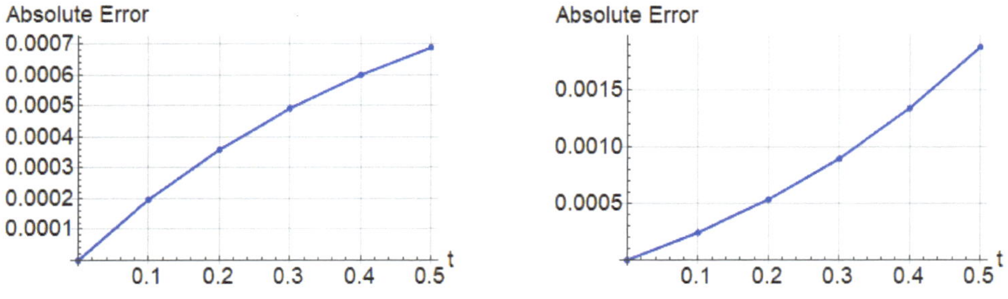

(**a**) (**b**)

Figure 7. Absolute error between the approximate and exact solutions for the time-fractional spKdV Equation (5.17) at $\alpha = 1$ where (**a**) $x = -5$; (**b**) $x = 5$.

Table 2. Comparison between the exact solution and NRPS solution for the time-fractional sp-KdV equation.

x	t	Exact Solution	NRPS Solution	Absolute Error	Relative Error	CPU Time
-5	0.0	-10.99891045115114	-10.998910451151142	0.0	0.0	2.328125
	0.1	-10.99866923540665	-10.998910451151142	2.412157×10^{-4}	2.193135×10^{-5}	
	0.2	-10.99837462040912	-10.998910451151142	5.358307×10^{-4}	4.871908×10^{-5}	
	0.3	-10.99801478665156	-10.998910451151142	8.956644×10^{-4}	8.143874×10^{-5}	
	0.4	-10.99757529934621	-10.998910451151142	1.335151×10^{-3}	1.214041×10^{-4}	
	0.5	-10.99703853017633	-10.998910451151142	1.871920×10^{-3}	1.702204×10^{-4}	
5	0.0	12.99891045115114	12.998910451151142	0.0	0.0	1.703125
	0.1	12.99910794550953	12.998910451151142	1.974943×10^{-4}	1.519291×10^{-5}	
	0.2	12.99926964263438	12.998910451151142	3.591914×10^{-4}	2.763166×10^{-5}	
	0.3	12.99940203066545	12.998910451151142	4.915795×10^{-4}	3.781554×10^{-5}	
	0.4	12.99951042190528	12.998910451151142	5.999707×10^{-4}	4.615333×10^{-5}	
	0.5	12.99959916587564	12.998910451151142	6.887147×10^{-4}	5.297968×10^{-5}	

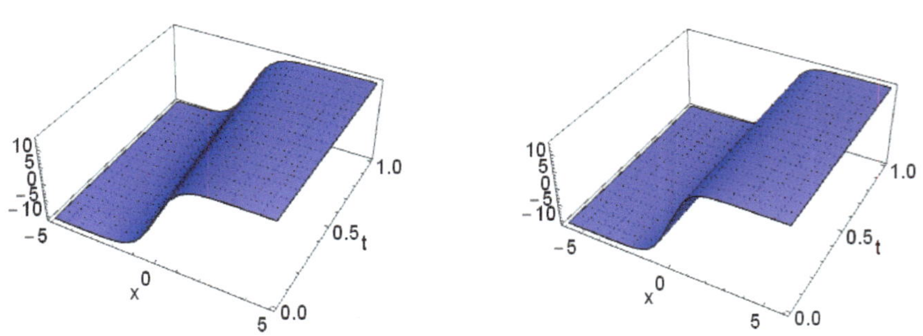

(**a**) (**b**)

Figure 8. Solution of the time-fractional spKdV Equation (5.17) at $\alpha = 1$ where (**a**) exact solution; (**b**) 4th NRPS solution (5.31).

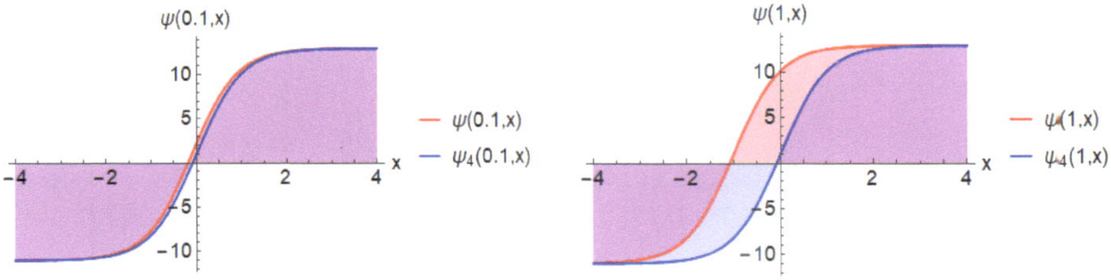

(**a**) (**b**)

Figure 9. Solution of the time-fractional spKdV Equation (5.17) at $\alpha = 1$ where (**a**) $t = 0.1$; (**b**) $t = 1$.

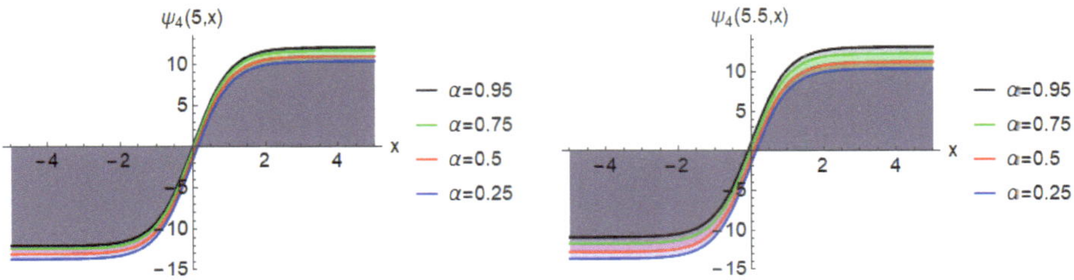

(**a**) (**b**)

Figure 10. The 4th NRPS solution (5.31) of the time-fractional spKdV Equation (5.17) at different fractional orders $\alpha \in \{0.95,\ 0.75,\ 0.5,\ 0.25\}$ where (**a**) $t = 5$; (**b**) $t = 5.5$.

Application 3. *Consider the time-fractional FKdV equation:*

$$\frac{1}{\sqrt{9.8}}\mathcal{D}_t^{\alpha}\psi + \left(-2 - \frac{3}{2}\psi\right)\frac{\partial\psi}{\partial x} - \frac{1}{6}\frac{\partial^3\psi}{\partial x^3} = \frac{-x}{40}e^{\frac{-x^2}{4}},\ t \geq 0,\ x \in \mathbb{R}, \tag{5.32}$$

where $\phi(x) = 1 + 0.1e^{\frac{-x^2}{4}}$. *We assume the initial condition as follows:*

$$\psi(x,0) = h(x) = \frac{-2e^x}{(1+e^x)^2}, x \in \mathbb{R}. \tag{5.33}$$

By applying the same argument above for the NRPS method, we obtain the following results:

$$\Pi_1(x) = \frac{\sqrt{9.8}}{6}\left(3\phi'(x) + 6h'(x) + 6h'(x) + 9h(x)h'(x) + h'''(x)\right). \tag{5.34}$$

$$\Pi_2(x) = \frac{\sqrt{9.8}}{6}\left(6\Pi_1'(x) + 6\Pi_1'(x) + 9h(x)\Pi_1'(x) + 9\Pi_1(x)h'(x) + \Pi_1'''(x)\right). \tag{5.35}$$

$$\begin{aligned}\Pi_3(x) = \tfrac{\sqrt{9.8}}{6(\Gamma(1+\alpha))^2}\Big(&9\Gamma(1+2\alpha)\Pi_1(x)\Pi_1'(x)\\ &+(\Gamma(1+\alpha))^2\left(12\Pi_2'(x) + 9h(x)\Pi_2'(x) + 9\Pi_2(x)h'(x) + \Pi_2'''(x)\right)\Big).\end{aligned} \tag{5.36}$$

$$\begin{aligned}\Pi_4(x) = \tfrac{\sqrt{9.8}}{6\Gamma(1+\alpha)\Gamma(1+2\alpha)}\Big(&9\Gamma(1+3\alpha)\left(\Pi_2(x)\Pi_1'(x) + \Pi_1(x)\Pi_2'(x)\right)\\ &+\Gamma(1+\alpha)\Gamma(1+2\alpha)\left(12\Pi_3'(x) + 9h(x)\Pi_3'(x) + 9h'(x)\Pi_3(x) + \Pi_3'''(x)\right)\Big).\end{aligned} \tag{5.37}$$

The CPU time needed to obtain these results (5.34)–(5.37) was 16.59375 *s. Using these results, we obtain the Kth NRPS solution for the time-fractional FKdV equation as:*

$$\psi_{\mathcal{K}}(t,x,y,z)=\frac{-2e^{x}}{(1+e^{x})^{2}}+\frac{\Pi_1(x)t^{\alpha}}{\Gamma(1+\alpha)}+\frac{\Pi_2(x)t^{2\alpha}}{\Gamma(1+2\alpha)}+\frac{\Pi_3(x)t^{3\alpha}}{\Gamma(1+3\alpha)}+\frac{\Pi_4(x)t^{4\alpha}}{\Gamma(1+4\alpha)}+\ldots+\frac{\Pi_{\mathcal{K}}(x)t^{\mathcal{K}\alpha}}{\Gamma(1+\mathcal{K}\alpha)}. \quad (5.38)$$

Table 3 shows the numerical results at $x \in \{-0.05, 0, 0.05\}$ and $t \in \{0, 0.01, 0.02, 0.03, 0.04, 0.05\}$ and the absolute and relative errors correspond to these numerical results which show convergences between the approximate and exact solutions for the time-fractional FKdV Equation (5.38). Figure 11 presents the absolute error between the exact solution and the explored approximate fourth NRPS solution (5.31) for the time-fractional FKdV Equation (5.32). It is clear from Table 3 that the absolute and relative errors were calculated at different values of the variable x and the variable t, and the accuracy of the proposed method was within 10^{-3}, which is a suitable level to confirm that the proposed method in this work is effective and suitable for dealing with such issues in the context of numerical modeling. Moreover, the approximate solutions presented here are in agreement with those presented in Ref. [28]. It is worth noting here that the work in Ref. [28] did not present the absolute or relative error of the derived solutions and did not provide a reading for comparing the solutions with other approximate solutions of Equation (5.32). Figures 12 and 13 present the surface of the fourth NRPS approximate solution and the exact solution to be able to observe the harmony between them. In Figure 9, we consider $\alpha = 1$, while we consider different fractional orders $\alpha \in \{0.9, 0.8\}$ in Figure 13. Figure 14 compares the exact solution and the fourth NRPS solution (5.38) at $t = 0.05$ and $t = 0.15$. The harmony between the depicted solutions is a little larger at $t = 0.05$ than at $t = 0.15$. To further illustrate the effect of the fractional derivative on the behavior of the solution derived using the proposed method for the time-fractional FKdV Equation (5.32), we present Figures 15–17. Figure 15 was performed by considering $t \in \{0, 0.1, 0.2\}$ at $\alpha = 0.9$ and $\alpha = 0.75$, while we performed Figure 16 by considering $t \in \{0, 0.15, 0.3, 0.5\}$ where $\alpha = 0.95$ and $\alpha = 0.75$. Figure 17 depicts the fourth NRPS solution at $t \in \{0.05, 0.15, 0.25, 0.5\}$ where $\alpha \in \{0.95, 0.8, 0.7, 0.6\}$.

Table 3. A comparison between the exact solution and NRPS solution for the time-fractional FKdV equation.

x	t	Exact Solution	NRPS Solution	Absolute Error	Relative Error	CPU Time
−0.05	0.0	−0.49968763016223	−0.49968763016223	0.0	0.0	41.265625
	0.01	−0.49968763016223	−0.49987856920723	1.909390×10^{-4}	3.821168×10^{-4}	
	0.02	−0.49968763016223	−0.49981332801340	1.256978×10^{-4}	2.515528×10^{-4}	
	0.03	−0.49968763016223	−0.49949449146807	1.931386×10^{-4}	3.865188×10^{-4}	
	0.04	−0.49968763016223	−0.49892579802621	7.618321×10^{-4}	1.524616×10^{-3}	
	0.05	−0.49968763016223	−0.49811213971041	1.575490×10^{-3}	3.152950×10^{-3}	
0	0.0	−0.5	−0.5	0.0	0.0	58.0625
	0.01	−0.5	−0.49987108520902	1.289147×10^{-4}	2.578295×10^{-4}	
	0.02	−0.5	−0.49948494667779	5.150533×10^{-4}	1.030106×10^{-3}	
	0.03	−0.5	−0.49884340193134	1.156598×10^{-3}	2.313196×10^{-3}	
	0.04	−0.5	−0.49794948017807	2.050519×10^{-3}	4.101039×10^{-3}	
	0.05	−0.5	−0.49680742230976	3.192577×10^{-3}	6.385155×10^{-3}	
0.05	0.0	−0.49968763016223	−0.49968763016223	0.0	0.0	39.9375
	0.01	−0.49968763016223	−0.49923907955869	4.485506×10^{-4}	8.976620×10^{-4}	
	0.02	−0.49968763016223	−0.49853263964456	1.154990×10^{-3}	2.311425×10^{-3}	
	0.03	−0.49968763016223	−0.49756918623542	2.118443×10^{-3}	4.239536×10^{-3}	
	0.04	−0.49968763016223	−0.49635074871447	3.336881×10^{-3}	6.677934×10^{-3}	
	0.05	−0.49968763016223	−0.49488051003255	4.807120×10^{-3}	9.620250×10^{-3}	

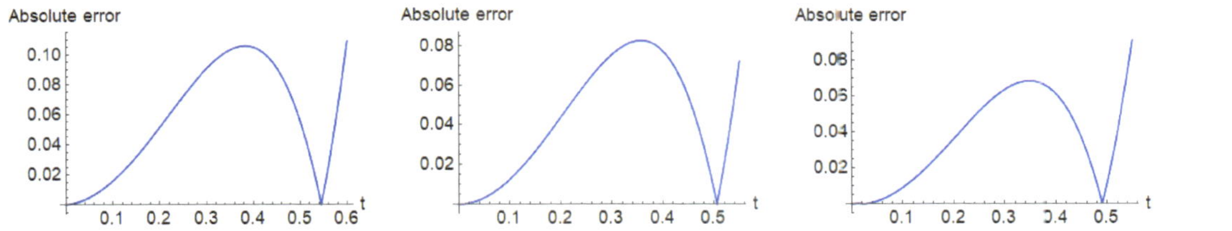

(a) (b) (c)

Figure 11. The absolute error between the approximate and exact solutions for the time-fractional FKdV Equation (5.32) at $\alpha = 1$: (**a**) $x = -0.05$, (**b**) $x = 0$, and (**c**) $x = 0.05$.

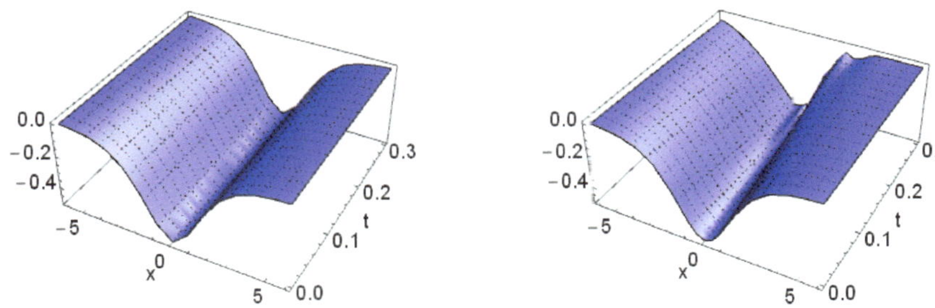

(a) (b)

Figure 12. The solution of the time-fractional FKdV Equation (5.32) at $\alpha = 1$: (**a**) exact solution, (**b**) 4th NRPS solution (5.38).

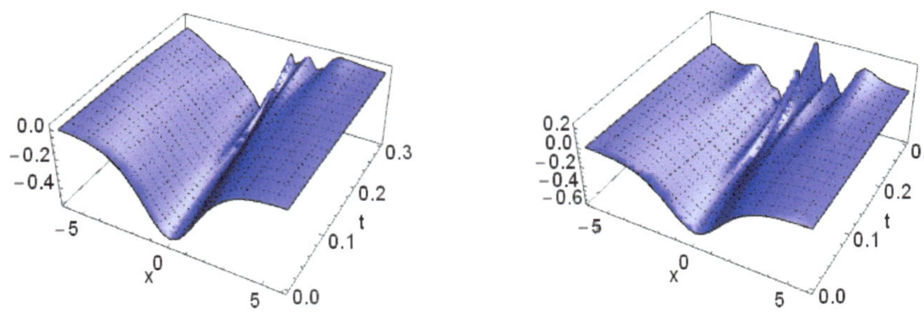

(a) (b)

Figure 13. The 4th NRPS solution (5.38) of the time-fractional FKdV Equation (5.32): (**a**) $\alpha = 0.9$; (**b**) $\alpha = 0.8$.

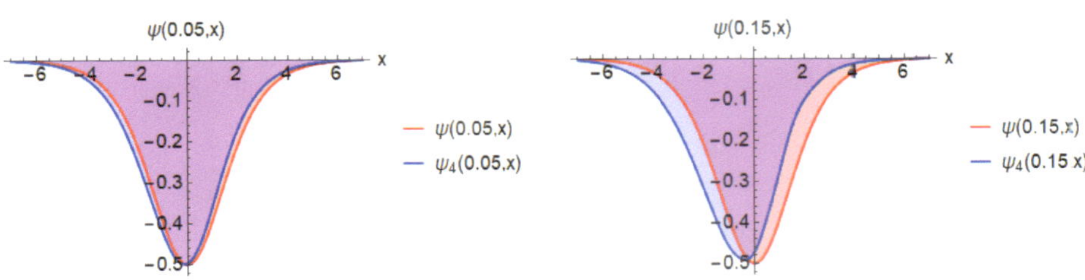

(a) (b)

Figure 14. The solution of the time-fractional FKdV Equation (5.32) at $\alpha = 1$: (**a**) $t = 0.05$, (**b**) $t = 0.15$ on (5.38).

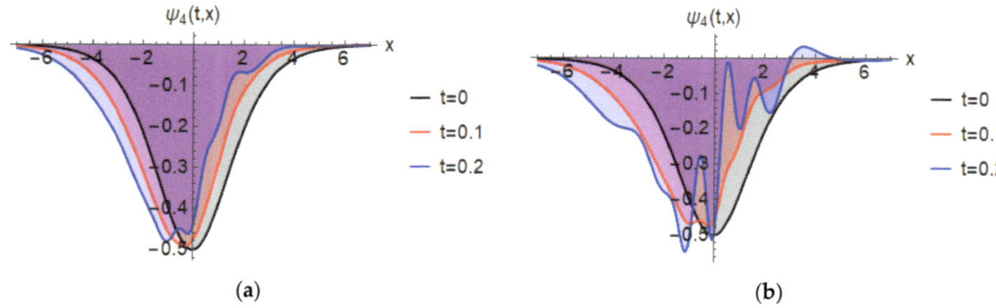

Figure 15. The 4th NRPS solution (5.38) of the time-fractional FKdV Equation (5.32) at $t \in \{0, 0.1, 0.2\}$: (**a**) $\alpha = 0.9$, (**b**) $\alpha = 0.75$.

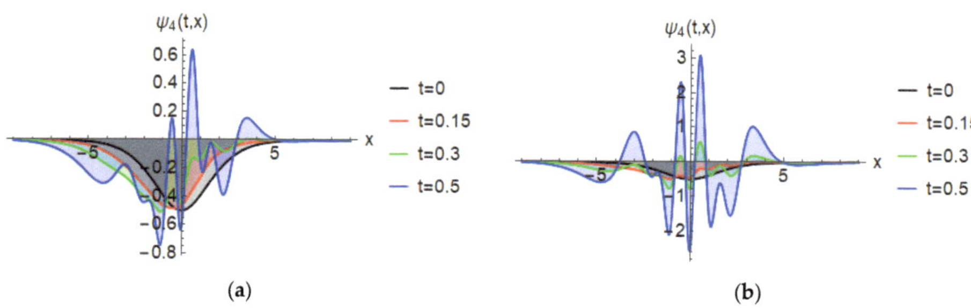

Figure 16. The 4th NRPS solution (5.38) of the time-fractional FKdV Equation (5.32) at $t \in \{0,\ 0.15,\ 0.3,\ 0.5\}$: (**a**) $\alpha = 0.95$, (**b**) $\alpha = 0.75$.

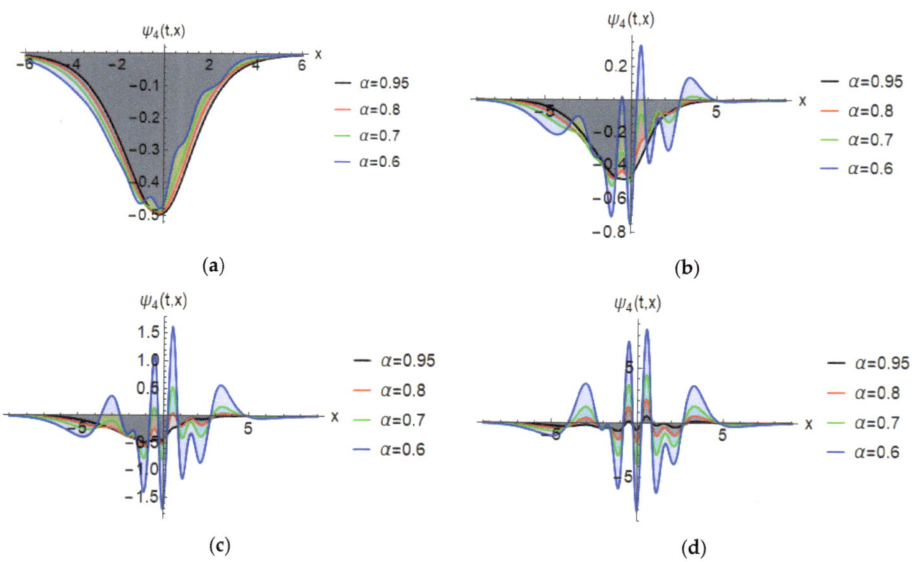

Figure 17. The 4th NRPS solution (5.38) of the time-fractional FKdV Equation (5.32) at different fractional orders $\alpha \in \{0.95,\ 0.8,\ 0.7,\ 0.6\}$: (**a**) $t = 0.05$, (**b**) $t = 0.15$, (**c**) $t = 0.25$, and (**d**) $t = 0.5$.

From the introduced numerical and graphical results, we can deduce that the proposed NRPS method is efficient in dealing with the governing models by noticing the harmony and convergences between the approximate and exact solutions. We utilized four terms of the NRPS solutions to show their efficiency and we can achieve higher efficiency by obtaining further terms of the approximate solutions.

6. Conclusions

This work investigated three time-fractional models that arise in plasma physics: the mKdV-ZK equation, the spKdV equation, and the FKdV equation. The governing equations in this work are of considerable importance in plasma physics by modeling nonlinear ion acoustic waves which play a fundamental role in understanding wave dynamics in plasma. We considered the fractional derivative in the Caputo sense. In this work, we investigate a new work based on the new fractional expansion in the natural transformation space and the RPS method to construct analytical solutions for the governing models. The RPS method is one of the important methods in deriving approximate solutions for fractional models, but it may require complex calculations to deal with more complex fractional models. To address this point, the Laplace transform was integrated with the RPS method in the literature. In this work, we seek to compose the natural transformation with the RPS method, as the natural transformation is more general than the Laplace transform, which enables us to deal with a wider range of complex fractional systems. To achieve our goal, we investigated the theoretical analysis of the proposed method to reveal the applicability, efficiency, and effectiveness of this approach to dealing with the governing models. Clear steps were set for the proposed method to obtain analytical solutions that are consistent and close to the exact solutions. These steps were implemented for the three governing equations, and an analytical study of the derived solutions for each equation was presented to confirm that the proposed approach generates analytical solutions that converge quickly to exact solutions, which proves the effectiveness of the proposed method.

Author Contributions: Conceptualization, S.M. and S.A.-O.; methodology, M.A.; software, S.E.A.; validation, M.A. and S.M.; formal analysis, S.A.-O.; investigation, M.A.-S.; writing—original draft preparation, S.A.-O. and M.A.-S.; writing—review and editing, A.F.; visualization, A.F.; supervision and project administration, S.E.A.; funding acquisition, S.A.-O. All authors have read and agreed to the published version of the manuscript.

Funding: This research work was funded by Umm Al-Qura University, Saudi Arabia under grant number: 25UQU4282396GSSR01.

Data Availability Statement: Data are contained within the article.

Acknowledgments: The authors extend their appreciation to Umm Al-Qura University, Saudi Arabia for funding this research work through grant number: (25UQU4282396GSSR01).

Conflicts of Interest: The authors declare no conflict of interest.

References

1. Saha, A.; Banerjee, S. *Dynamical Systems and Nonlinear Waves in Plasmas*; CRC Press: Boca Raton, FL, USA, 2021.
2. Varghese, S.S.; Singh, K.; Kourakis, I. Electrostatic Supersolitary Waves: A Challenging Paradigm in Nonlinear Plasma Science and Beyond—State of the Art and Overview of Recent Results. *Fundam. Plasma Phys.* **2024**, *10*, 100048. [CrossRef]
3. Altuijri, R.; Abdel-Aty, A.H.; Nisar, K.S.; Khater, M. Exploring novel wave characteristics in a nonlinear model with complexity arising in plasma physics. *Opt. Quantum Electron.* **2024**, *56*, 1128. [CrossRef]
4. Johnson, R.S. Water waves and Korteweg–de Vries equations. *J. Fluid Mech.* **1980**, *97*, 701–719. [CrossRef]
5. Seadawy, A.R.; Alsaedi, B. Contraction of variational principle and optical soliton solutions for two models of nonlinear Schrodinger equation with polynomial law nonlinearity. *AIMS Math.* **2024**, *9*, 6336–6367.
6. Mubaraki, A.M.; Nuruddeen, R.I.; Ali, K.K.; Gómez-Aguilar, J.F. Additional solitonic and other analytical solutions for the higher-order Boussinesq-Burgers equation. *Opt. Quantum Electron.* **2024**, *56*, 165. [CrossRef]
7. Gulay, T.A.; Zakharov, V.V.; Kron, R.V.; Popova, S.V.; Shibaev, V.P. Algorithmic support of the integrated aircraft control system using the systems embedding technology. *AIP Conf. Proc.* **2024**, *2969*, 060009.

8. Halcrow, C.; John, R.R.; Anusree, N. Modified sine-Gordon theory with static multikinks. *Phys. Rev. D* **2024**, *110*, 065010. [CrossRef]
9. Sagib, M.; Hossain, M.A.; Saha, B.K.; Khan, K. On traveling wave solutions with stability and phase plane analysis for the modified Benjamin-Bona-Mahony equation. *PLoS ONE* **2024**, *19*, e0306196.
10. Hosseini, K.; Hincal, E.; Sadri, K.; Rabiei, F.; Ilie, M.; Akgül, A.; Osman, M.S. The positive multi-complexiton solution to a generalized Kadomtsev–Petviashvili equation. *Partial Differ. Equ. Appl. Math.* **2024**, *9*, 100647.
11. Faridi, W.A.; Iqbal, M.; Riaz, M.B.; AlQahtani, S.A.; Wazwaz, A.M. The fractional soliton solutions of dynamical system arising in plasma physics: The comparative analysis. *Alex. Eng. J.* **2024**, *95*, 247–261. [CrossRef]
12. Alam, M.M.; Alam, M.S. Study of Time-Fractional Dust Ion Acoustic Waves Propagation in Collisionless Unmagnetized Dusty Plasmas. *Braz. J. Phys.* **2024**, *54*, 192.
13. Phoosree, S.; Khongnual, N.; Sanjun, J.; Kammanee, A.; Thadee, W. Riccati sub-equation method for solving fractional flood wave equation and fractional plasma physics equation. *Partial Differ. Equ. Appl. Math.* **2024**, *10*, 100672. [CrossRef]
14. Abdelwahed, H.G.; El-Shewy, E.K.; Alghanim, S.; Abdelrahman, M.A. On the physical fractional modulations on Langmuir plasma structures. *Fractal Fract.* **2022**, *6*, 430. [CrossRef]
15. Poolcharuansin, P.; Bradley, J.W. Short-and long-term plasma phenomena in a HiPIMS discharge. *Plasma Sources Sci. Technol.* **2010**, *19*, 025010. [CrossRef]
16. Alsatami, K.A. Exploring Nonlinear Fractional (2+1)-Dimensional KP–Burgers Model: Derivation and Ion Acoustic Solitary Wave Solution. *Int. J. Anal. Appl.* **2024**, *22*, 147. [CrossRef]
17. Alabedalhadi, M.; Al-Omari, S.; Al-Smadi, M.; Alhazmi, S. Traveling wave solutions for time-fractional mKdV-ZK equation of weakly nonlinear ion-acoustic waves in magnetized electron–positron plasma. *Symmetry* **2023**, *15*, 361. [CrossRef]
18. Lazarus, I.J.; Bharuthram, R.; Hellberg, M.A. Modified Korteweg–de Vries–Zakharov–Kuznetsov solitons in symmetric two-temperature electron–positron plasmas. *J. Plasma Phys.* **2008**, *74*, 519–529. [CrossRef]
19. Verheest, F.; Mace, R.L.; Pillay, S.R.; Hellberg, M.A. Unified derivation of Korteweg-de Vries-Zakharov-Kuznetsov equations in multispecies plasmas. *J. Phys. A Math. Gen.* **2002**, *35*, 795.
20. Zhou, T.Y.; Tian, B.; Zhang, C.R.; Liu, S.H. Auto-Bäcklund transformations, bilinear forms, multiple-soliton, quasi-soliton and hybrid solutions of a (3+1)-dimensional modified Korteweg-de Vries-Zakharov-Kuznetsov equation in an electron-positron plasma. *Eur. Phys. J. Plus* **2022**, *137*, 912. [CrossRef]
21. Rehman, H.U.; Seadawy, A.R.; Younis, M.; Rizvi ST, R.; Anwar, I.; Baber, M.Z.; Althobaiti, A. Weakly nonlinear electron-acoustic waves in the fluid ions propagated via a (3+1)-dimensional generalized Korteweg–de-Vries–Zakharov–Kuznetsov equation in plasma physics. *Results Phys.* **2022**, *33*, 105069.
22. Younas, U.; Ren, J.; Baber, M.Z.; Yasin, M.W.; Shahzad, T. Ion-acoustic wave structures in the fluid ions modeled by higher dimensional generalized Korteweg-de Vries–Zakharov–Kuznetsov equation. *J. Ocean Eng. Sci.* **2023**, *8*, 623–635. [CrossRef]
23. Alhami, R.; Alquran, M. Extracted different types of optical lumps and breathers to the new generalized stochastic potential-KdV equation via using the Cole-Hopf transformation and Hirota bilinear method. *Opt. Quantum Electron.* **2022**, *54*, 553. [CrossRef]
24. Alquran MA RW, A.N. Investigating the revisited generalized stochastic potential-KdV equation: Fractional time-derivative against proportional time-delay. *Rom. J. Phys.* **2023**, *68*, 106.
25. Abbas, N.; Hussain, A.; Riaz, M.B.; Ibrahim, T.F.; Birkea, F.O.; Tahir, R.A. A discussion on the Lie symmetry analysis, travelling wave solutions and conservation laws of new generalized stochastic potential-KdV equation. *Results Phys.* **2024**, *56*, 107302. [CrossRef]
26. Wu TY, T. Generation of upstream advancing solitons by moving disturbances. *J. Fluid Mech.* **1987**, *184*, 75–99.
27. Derakhshan, M.H.; Aminataei, A. A numerical method for finding solution of the distributed-order time-fractional forced Korteweg–de Vries equation including the Caputo fractional derivative. *Math. Methods Appl. Sci.* **2022**, *45*, 3144–3165. [CrossRef]
28. Veeresha, P.; Yavuz, M.; Baishya, C. A computational approach for shallow water forced Korteweg–De Vries equation on critical flow over a hole with three fractional operators. *Int. J. Optim. Control Theor. Appl. (IJOCTA)* **2021**, *11*, 52–67. [CrossRef]
29. Alhejaili, W.; Az-Zo'bi, E.; Shah, R.; El-Tantawy, S.A. On the analytical soliton approximations to fractional forced Korteweg-de Vries equation arising in fluids and Plasmas using two novel techniques. *Commun. Theor. Phys.* **2024**, *76*, 085001. [CrossRef]
30. Abu Arqub, O.; Al-Smadi, M.; Momani, S.; Hayat, T. Numerical solutions of fuzzy differential equations using reproducing kernel Hilbert space method. *Soft Comput.* **2016**, *20*, 3283–3302. [CrossRef]
31. Al-Smadi, M.; Freihat, A.; Khalil, H.; Momani, S.; Ali Khan, R. Numerical multistep approach for solving fractional partial differential equations. *Int. J. Comput. Methods* **2017**, *14*, 1750029. [CrossRef]
32. Hasan, S.; El-Ajou, A.; Hadid, S.; Al-Smadi, M.; Momani, S. Atangana-Baleanu fractional framework of reproducing kernel technique in solving fractional population dynamics system. *Chaos Solitons Fractals* **2020**, *133*, 109624.
33. Momani, S.; Freihat, A.; Al-Smadi, M. Analytical Study of Fractional-Order Multiple Chaotic FitzHugh-Nagumo Neurons Model Using Multistep Generalized Differential Transform Method. *Abstr. Appl. Anal.* **2014**, *2014*, 276279. [CrossRef]

34. Al-Smadi, M.; Gumah, G. On the homotopy analysis method for fractional SEIR epidemic model. *Res. J. Appl. Sci. Eng. Technol.* **2014**, *7*, 3809–3820.
35. Khalil, H.; Khan, R.A.; HAl-Smadi, M.; Freihat, A.A. Approximation of solution of time fractional order three-dimensional heat conduction problems with Jacobi Polynomials. *Punjab Univ. J. Math.* **2020**, *47*, 35–56.
36. Heydari, M.H.; Zhagharian, S.; Razzaghi, M. Discrete Chebyshev polynomials for the numerical solution of stochastic fractional two-dimensional Sobolev equation. *Commun. Nonlinear Sci. Numer. Simul.* **2024**, *130*, 107742. [CrossRef]
37. Sivasankar, K.; Nadhaprasadh, M.; Sathish, K.; Al-Omari, S.; Udhayakuma, R. New study on Cauchy problems of fractional stochastic evolution systems on an infinite interval. *Math. Methods Appl. Sci.* **2024**, *48*, 890–904.
38. Roul, P.; Rohil, V. A high-accuracy computational technique based on L 2-1 σ and B-spline schemes for solving the nonlinear time-fractional Burgers' equation. *Soft Comput.* **2024**, *28*, 6153–6169.
39. Nawaz, R.; Fewster-Young, N.; Katbar, N.M.; Ali, N.; Zada, L.; Ibrahim, R.W.; Alqahtani, H. Numerical inspection of (3 + 1)-perturbed Zakharov–Kuznetsov equation via fractional variational iteration method with Caputo fractional derivative. *Numer. Heat Transf. Part B Fundam.* **2024**, *85*, 1162–1177.
40. Momani, S.; Odibat, Z.; Alawneh, A. Variational iteration method for solving the space-and time-fractional KdV equation. *Numer. Methods Partial Differ. Equ. Int. J.* **2008**, *24*, 262–271.
41. El-Wakil, S.A.; Abulwafa, E.M.; Zahran, M.A.; Mahmoud, A.A. Time-fractional KdV equation: Formulation and solution using variational methods. *Nonlinear Dyn.* **2011**, *65*, 55–63.
42. Bulut, H.; Pandir, Y.; Demiray, S.T. Exact solutions of time-fractional KdV equations by using generalized Kudryashov method. *Int. J. Model. Optim.* **2014**, *4*, 315.
43. Arqub, O.A.; El-Ajou, A.; Momani, S. Constructing and predicting solitary pattern solutions for nonlinear time-fractional dispersive partial differential equations. *J. Comput. Phys.* **2015**, *293*, 385–399.
44. Abu Arqub, O. Application of residual power series method for the solution of time-fractional Schrödinger equations in one-dimensional space. *Fundam. Inform.* **2019**, *166*, 87–110. [CrossRef]
45. Tchier, F.; Inc, M.; Korpinar, Z.S.; Baleanu, D. Solutions of the time fractional reaction–diffusion equations with residual power series method. *Adv. Mech. Eng.* **2016**, *8*, 1687814016670867.
46. El-Ajou, A. Adapting the Laplace transform to create solitary solutions for the nonlinear time-fractional dispersive PDEs via a new approach. *Eur. Phys. J. Plus* **2021**, *136*, 229.
47. Džrbašjan, M.M. A generalized Riemann-Liouville operator and some of its applications. *Math. USSR-Izv.* **1968**, *2*, 1027.
48. Farman, M.; Alfiniyah, C. A constant proportional caputo operator for modeling childhood disease epidemics. *Decis. Anal. J.* **2024**, *10*, 100393.
49. Trujillo, J.J.; Scalas, E.; Diethelm, K.; Baleanu, D. *Fractional Calculus: Models and Numerical Methods*; World Scientific: Singapore, 2016; Volume 5.
50. Sivasankar, S.; Udhayakumar, R.; Muthukumaran, V.; Gokul, G.; Al-Omari, S. Existence of Hilfer fractional neutral stochastic differential systemswith infinite delay, Bulletin of the Karaganda University. *Math. Ser.* **2024**, *1*, 119–174.
51. Belgacem FB, M.; Silambarasan, R.; Sivasundaram, S. Advances in the natural transform. *AIP Conf. Proc.-Am. Inst. Phys.* **2012**, *1493*, 106.
52. Rawashdeh, M.S.; Al-Jammal, H. New approximate solutions to fractional nonlinear systems of partial differential equations using the FNDM. *Adv. Differ. Equ.* **2016**, *2016*, 235.
53. Maitama, S.; Zhao, W. Beyond Sumudu transform and natural transform: J-transform properties and applications. *J. Appl. Anal. Comput.* **2020**, *10*, 1223–1241.
54. Rawashdeh, M.S.; Maitama, S. Solving nonlinear ordinary differential equations using the NDM. *J. Appl. Anal. Comput.* **2015**, *5*, 77–88.

fractal and fractional

MDPI

Article

A Mixed Finite Element Method for the Multi-Term Time-Fractional Reaction–Diffusion Equations

Jie Zhao [1], Shubin Dong [2] and Zhichao Fang [2,*]

1 School of Statistics and Mathematics, Inner Mongolia University of Finance and Economics, Hohhot 010070, China; zhaoj@imufe.edu.cn
2 School of Mathematical Sciences, Inner Mongolia University, Hohhot 010021, China; madongshubin@126.com
* Correspondence: zcfang@imu.edu.cn

Abstract: In this work, a fully discrete mixed finite element (MFE) scheme is designed to solve the multi-term time-fractional reaction–diffusion equations with variable coefficients by using the well-known $L1$ formula and the Raviart–Thomas MFE space. The existence and uniqueness of the discrete solution is proved by using the matrix theory, and the unconditional stability is also discussed in detail. By introducing the mixed elliptic projection, the error estimates for the unknown variable u in the discrete $L^\infty(L^2(\Omega))$ norm and for the auxiliary variable $\boldsymbol{\lambda}$ in the discrete $L^\infty\left(\left(L^2(\Omega)\right)^2\right)$ and $L^\infty(\boldsymbol{H}(\text{div}, \Omega))$ norms are obtained. Finally, three numerical examples are given to demonstrate the theoretical results.

Keywords: multi-term time-fractional reaction–diffusion equations; mixed finite element method; $L1$ formula; unconditional stability; error estimate

1. Introduction

Fractional calculus and fractional partial differential equations (FPDEs) have been confirmed to be very important tools in describing some anomalous phenomena and processes with memory and nonlocal properties [1–6]. Moreover, some underlying and complex processes can be described more appropriately by multi-term FPDEs [7–9], as they contains multiple fractional derivative or calculus terms. In recent years, many numerical methods have been increasingly used by scholars to solve multi-term FPDEs. Liu et al. [10] constructed some finite difference (FD) schemes to solve the multi-term time-fractional wave-diffusion equations by using two fractional predictor–corrector methods. Dehghan et al. [11] devised two high-order numerical schemes to solve the multi-term time-fractional diffusion-wave equations by using the compact FD method and Galerkin spectral technique. Ren and Sun [12] established an efficient compact FD scheme for the multi-term time-fractional diffusion-wave equation by using the $L1$ formula. Zheng et al. [13] proposed a high-order space–time spectral method for the multi-term time-fractional diffusion equations by using the Legendre polynomials in the temporal direction and the Fourier-like basis functions in the spatial direction. Du and Sun [14] constructed an FD scheme for multi-term time-fractional mixed diffusion and wave equations by using the $L2-1_\sigma$ formula. Hendy and Zaky [15] proposed a spectral method for a coupled system of nonlinear multi-term time–space fractional diffusion equations by using the $L1$ formula on a time-graded mesh. Liu et al. [16] developed an ADI Legendre spectral method for solving a multi-term time-fractional Oldroyd-B fluid-type diffusion equation. Wei and Wang [17] constructed a higher-order numerical scheme for the multi-term variable-order time-fractional diffusion equation by using the local discontinuous Galerkin method. She et al. [18] considered a spectral method for solving the multi-term time-fractional diffusion problem by using a modified $L1$ formula.

Meanwhile, many scholars selected the finite element (FE) method for solving the multi-term FPDEs and have achieved excellent results. Jin et al. [19] developed an FE

Citation: Zhao, J.; Dong, S.; Fang, Z. A Mixed Finite Element Method for the Multi-Term Time-Fractional Reaction–Diffusion Equations. *Fractal Fract.* **2024**, *8*, 51. https://doi.org/10.3390/fractalfract8010051

Academic Editor: Jordan Hristov

Received: 5 December 2023
Revised: 4 January 2024
Accepted: 8 January 2024
Published: 12 January 2024

method for a multi-term time-fractional diffusion equation and considered the case of smooth and nonsmooth initial data. Li et al. [20] proposed an FE method to solve a higher-dimensional multi-term fractional diffusion equation on nonuniform time meshes. Zhou et al. [21] developed a weak Galerkin FE method for solving multi-term time-fractional diffusion equations by using a convolution quadrature formula. Bu et al. [22] proposed a space–time FE method for solving the multi-term time–space fractional diffusion equation based on the suitable graded time mesh. Feng et al. [23] proposed an FE method for a multi-term time-fractional mixed subdiffusion and diffusion-wave equation on the convex domain by using mixed L-type schemes. Meng and Stynes [24] considered an $L1$ FE method for a multi-term time-fractional initial-boundary value problem on the temporal graded mesh. Yin et al. [25] constructed a class of efficient time-stepping FE schemes for multi-term time-fractional reaction–diffusion-wave equations by using the shifted convolution quadrature method. Huang et al. [26] proposed an α-robust FE method for a multi-term time-fractional diffusion problem on a graded mesh by using the $L1$ formula. Liu et al. [27] proposed an FE method for solving a multi-term variable-order time-fractional diffusion equation and developed an efficient parallel-in-time algorithm to reduce the computational costs.

In this work, we will construct a fully discrete mixed finite element (MFE) scheme for the following multi-term time-fractional reaction–diffusion (TFRD) equations with variable coefficients:

$$\begin{cases} P(D_t)u(\boldsymbol{x},t) - \operatorname{div}(\mathcal{A}(\boldsymbol{x})\nabla u(\boldsymbol{x},t)) + p(\boldsymbol{x})u(\boldsymbol{x},t) = f(\boldsymbol{x},t), & (\boldsymbol{x},t) \in \Omega \times J, \\ u(\boldsymbol{x},t) = 0, & (\boldsymbol{x},t) \in \partial\Omega \times \overline{J}, \\ u(\boldsymbol{x},0) = u_0(\boldsymbol{x}), & \boldsymbol{x} \in \overline{\Omega}, \end{cases} \tag{1}$$

where $\Omega \subset R^2$ is a convex and bounded polygon region with boundary $\partial\Omega$, $J = (0,T]$ with $0 < T < \infty$. We assume that the source function $f(\boldsymbol{x},t)$, initial data $u_0(\boldsymbol{x})$, and non-negative coefficient $p(\boldsymbol{x})$ are smooth enough. Specifically, for the symmetric diffusion coefficient matrix $\mathcal{A}(\boldsymbol{x})$, we should assume that there exist two constants A_0, $A_1 > 0$ such that

$$A_0 \boldsymbol{z}^T \boldsymbol{z} \le \boldsymbol{z}^T \mathcal{A}(\boldsymbol{x}) \boldsymbol{z} \le A_1 \boldsymbol{z}^T \boldsymbol{z}, \forall \boldsymbol{z} \in R^2, \forall \boldsymbol{x} \in \overline{\Omega}.$$

Moreover, the multi-term time-fractional derivative $P(D_t)u(\boldsymbol{x},t)$ is defined by

$$P(D_t)u(\boldsymbol{x},t) = \sum_{i=1}^{m} b_i D_t^{\alpha_i} u(\boldsymbol{x},t), 0 < \alpha_m < \alpha_{m-1} < \cdots < \alpha_1 < 1,$$

where $b_i \quad (i = 1, 2, \cdots, m)$ are the positive real numbers and $D_t^{\alpha_i} u$ is the Caputo time-fractional derivative as follows:

$$D_t^{\alpha_i} u(\boldsymbol{x},t) = \frac{1}{\Gamma(1-\alpha_i)} \int_0^t \frac{\partial u(\boldsymbol{x},s)}{\partial s} \frac{1}{(t-s)^{\alpha_i}} ds,$$

where $\Gamma(\cdot)$ denotes the Γ-function.

It should be noted that the MFE method, as an important numerical calculation method, has been widely used to solve FPDEs [28–32], and some scholars have also used this method to solve the multi-term FPDEs [33–35]. In [33], Shi et al. proposed an H^1-Galerkin mixed finite element (MFE) method for the multi-term time-fractional diffusion equations and gave a superconvergence result. In [34], Li et al. proposed an MFE method for the multi-term time-fractional diffusion and diffusion-wave equations by using an MFE space contained in $\left(L^2(\Omega)\right)^d \times H_0^1(\Omega)$, where $d = 2, 3$. In [35], Cao et al constructed a nonconforming MFE scheme for the multi-term time-fractional mixed diffusion and diffusion-wave equations. Motivated by the above excellent works, we will construct a fully discrete MFE scheme for the multi-term TFRD equation (1) by using the Raviart–Thomas MFE space and the $L1$ formula, analyze the existence, uniqueness, and unconditional stability in detail, and give error estimates for u (in discrete $L^\infty(L^2(\Omega))$ norm) and auxiliary variable λ (in discrete $L^\infty\left((L^2(\Omega))^2\right)$ and discrete $L^\infty(\boldsymbol{H}(\text{div},\Omega))$ norms). Finally, we give numerical experiments to demonstrate the efficiency of the proposed method.

The remainder of this paper is arranged as follows. In Section 2, we construct a fully discrete MFE scheme for the multi-term TFRD equations by using the Raviart–Thomas MFE space and the $L1$ -formula. In Section 3, we give a fractional Grönwall inequality and analyze the existence and uniqueness of the discrete solution. We derive the unconditional stability results and a priori error estimates in detail in Sections 4 and 5, respectively. Finally, three numerical examples are given to verify the theoretical results.

2. Mixed Finite Element Method

We introduce the flux $\boldsymbol{\lambda}(\boldsymbol{x},t) = -\mathcal{A}(\boldsymbol{x})\nabla u(\boldsymbol{x},t)$ as the auxiliary variable. Then, the original problem (1) can be rewritten as follows:

$$\begin{cases} (a)P(D_t)u(\boldsymbol{x},t) + \text{div}\boldsymbol{\lambda}(\boldsymbol{x},t) + p(\boldsymbol{x})u(\boldsymbol{x},t) = f(\boldsymbol{x},t), & (\boldsymbol{x},t) \in \Omega \times J, \\ (b)\mathcal{A}^{-1}(\boldsymbol{x})\boldsymbol{\lambda}(\boldsymbol{x},t) + \nabla u(\boldsymbol{x},t) = 0, & (\boldsymbol{x},t) \in \Omega \times J, \\ (c)u(\boldsymbol{x},t) = 0, & (\boldsymbol{x},t) \in \partial\Omega \times \overline{J}, \\ (d)u(\boldsymbol{x},0) = u_0(\boldsymbol{x}), & \boldsymbol{x} \in \overline{\Omega}. \end{cases} \tag{2}$$

Let $V = L^2(\Omega)$ and $\boldsymbol{W} = \boldsymbol{H}(\text{div},\Omega) = \left\{\boldsymbol{w} \in (L^2(\Omega))^2 : \text{div}\boldsymbol{w} \in L^2(\Omega)\right\}$. Then, we obtain the mixed variational formulation of (2): find $(u,\boldsymbol{\lambda}) \in V \times \boldsymbol{W}$ such that

$$\begin{cases} (a)(P(D_t)u,v) + (\text{div}\boldsymbol{\lambda},v) + (pu,v) = (f,v), & \forall v \in V, \\ (b)(\mathcal{A}^{-1}\boldsymbol{\lambda},\boldsymbol{w}) - (u,\text{div}\boldsymbol{w}) = 0, & \forall \boldsymbol{w} \in \boldsymbol{W}, \\ (c)u(\boldsymbol{x},0) = u_0(\boldsymbol{x}), & \forall \boldsymbol{x} \in \overline{\Omega}. \end{cases} \tag{3}$$

Let K_h be a quasi-uniform triangulation of the domain Ω, h_T be the diameter of the triangle $T \in K_h$ and denote $h = \max\{h_T\}$. We select the Raviart–Thomas MFE space $V_h \times \boldsymbol{W}_h \subset V \times \boldsymbol{W}$, that is,

$$\begin{aligned} V_h(K) &= \{v_h \in V : v_h|_T \in P_r(T), \forall T \in K_h\}, \\ \boldsymbol{W}_h(K) &= \left\{\boldsymbol{w}_h \in \boldsymbol{W} : \boldsymbol{w}_h\Big|_T \in (P_r(T))^2 \oplus (\boldsymbol{x}P_r(T)), \forall T \in K_h\right\}, \end{aligned}$$

where the notation $\oplus$ indicates a direct sum, $\boldsymbol{x}P_r(T) = (x_1P_r(T), x_2P_r(T)), \boldsymbol{x} = (x_1,x_2)$ and $r \geq 0$ is a given integer.

Let $\tau = T/N$ and $t_n = n\tau$ for $n = 0,1,2,\cdots,N$, where N is a positive integer. For the parameters α_i and $i = 1,2,\cdots,m$, the Caputo time-fraction derivative $D_t^{\alpha_i}u(\boldsymbol{x},t)$ at $t = t_n$ is approximated by using the well-known $L1$ formula [36,37] as follows:

$$\begin{aligned} D_t^{\alpha_i}u(\boldsymbol{x},t_n) =& \frac{1}{\Gamma(1-\alpha_i)}\int_0^{t_n}\frac{\partial u(x,s)}{\partial s}\frac{1}{(t_n-s)^{\alpha_i}}ds \\ =& \frac{1}{\Gamma(2-\alpha_i)}\sum_{k=0}^{n-1}\frac{u^{k+1}-u^k}{\tau}\left[(t_n-t_k)^{1-\alpha_i} - (t_n-t_{k+1})^{1-\alpha_i}\right] + Q_{\alpha_i}^n(x) \\ =& \frac{1}{\Gamma(2-\alpha_i)}\left[d_{\alpha_i,1}^n u^n + \sum_{k=1}^{n-1}\left(d_{\alpha_i,k+1}^n - d_{\alpha_i,k}^n\right)u^{n-k} - d_{\alpha_i,n}^n u^0\right] + Q_{\alpha_i}^n(x) \\ =& \frac{1}{\Gamma(2-\alpha_i)}\sum_{k=0}^{n}\widetilde{d}_{\alpha_i,k}^n u^k + Q_{\alpha_i}^n(x), \end{aligned} \tag{4}$$

where $d^n_{\alpha_i,k} = \frac{(t_n - t_{n-k})^{1-\alpha_i} - (t_n - t_{n-k+1})^{1-\alpha_i}}{\tau}$, $\widetilde{d}^n_{\alpha_i,0} = -d^n_{\alpha_i,n}$, $\widetilde{d}^n_{\alpha_i,n} = -d^n_{\alpha_i,1}$, and $\widetilde{d}^n_{\alpha_i,k} = d^n_{\alpha_i,n-k+1} - d^n_{\alpha_i,n-k}(0 < k \leq n-1)$. Setting $D^{\alpha_i}_N u^n = \frac{1}{\Gamma(2-\alpha_i)} \sum\limits_{k=0}^{n} \widetilde{d}^n_{\alpha_i,k} u^k$, we have $D^{\alpha_i}_t u(x, t_n) = D^{\alpha_i}_N u^n + Q^n_{\alpha_i}(x)$, where $Q^n_{\alpha_i}(x)$ is the truncation error.

Based on the above definitions, and setting u^n_h and $\boldsymbol{\lambda}^n_h$ to be the discrete solutions of u and λ at $t = t_n$, respectively, then we can design a fully discrete MFE scheme for the original problem (1): find $(u^n_h, \lambda^n_h) \in V_h \times \mathbf{W}_h$ such that

$$\begin{cases} (a)\left(\sum\limits_{i=1}^{m} b_i D^{\alpha_i}_N u^n_h, v_h\right) + (\mathrm{div}\boldsymbol{\lambda}^n_h, v_h) + (pu^n_h, v_h) = (f^n, v_h), & \forall v_h \in V_h, \\ (b)\,(\mathcal{A}^{-1}\boldsymbol{\lambda}^n_h, \boldsymbol{w}_h) - (u^n_h, \mathrm{div}\boldsymbol{w}_h) = 0, & \forall \boldsymbol{w}_h \in \mathbf{W}_h, \end{cases} \tag{5}$$

where $(u^0_h, \boldsymbol{\lambda}^0_h) \in V_h \times \mathbf{W}_h$ satisfies

$$\begin{cases} (a)\,(\mathcal{A}^{-1}\boldsymbol{\lambda}^0_h, \boldsymbol{w}_h) - (u^0_h, \mathrm{div}\boldsymbol{w}_h) = 0, & \forall \boldsymbol{w}_h \in \mathbf{W}_h, \\ (b)\,(\mathrm{div}\boldsymbol{\lambda}^0_h, v_h) = (\mathrm{div}\boldsymbol{\lambda}_0, v_h), & \forall v_h \in V_h, \end{cases} \tag{6}$$

where $\boldsymbol{\lambda}_0(\boldsymbol{x}) = -\mathcal{A}(\boldsymbol{x})\nabla u_0(\boldsymbol{x})$.

Remark 1. *(I) In the MFE scheme (5)–(6), we particularly emphasize the calculation of initial values $(u^0_h, \boldsymbol{\lambda}^0_h)$, as this calculation will be used in stability and convergence analyses. Moreover, from the mixed elliptic projection R_h defined in Section 5, we can see that $(u^0_h, \boldsymbol{\lambda}^0_h) = (R_h u_0, R_h \boldsymbol{\lambda}_0)$.*

(II) Compared with the standard FE methods, it is well known that the MFE method can not only reduce the smoothness requirement of the finite element space, but also simultaneously calculate multiple physical quantities. These advantages are very important and popular in practical applications.

3. Existence and Uniqueness

In this section, we shall prove the existence and uniqueness for the MFE scheme (5)–(6). We first give some lemmas, which are important in subsequent theoretical analysis.

Lemma 1 ([38]). *There exist two positive constants μ_0 and μ_1 such that*

$$\mu_0 \| \boldsymbol{w} \|^2 \leq \| \boldsymbol{w} \|^2_{\mathcal{A}^{-1}} \leq \mu_1 \| \boldsymbol{w} \|^2, where \| \boldsymbol{w} \|^2_{\mathcal{A}^{-1}} = (\mathcal{A}^{-1}\boldsymbol{w}, \boldsymbol{w}), \forall \boldsymbol{w} \in \mathbf{W}.$$

Lemma 2 ([28]). *Let $\{z^n\}_{n=0}^{\infty}$ be a sequence on $\mathbf{W}_h$. Then, the following identity holds:*

$$\begin{aligned} \sum_{k=0}^{n} \widetilde{d}^n_{\alpha_i,k}\left(\mathcal{A}^{-1}z^k, z^n\right) = & \frac{1}{2}\Big[\widetilde{d}^n_{\alpha_i,n}\left(\mathcal{A}^{-1}z^n, z^n\right) - \sum_{k=0}^{n-1} \widetilde{d}^n_{\alpha_i,k}\left(\mathcal{A}^{-1}z^k, z^k\right) \\ & + \sum_{k=0}^{n-1} \widetilde{d}^n_{\alpha_i,k}\left(\mathcal{A}^{-1}\left(z^n - z^k\right), z^n - z^k\right)\Big]. \end{aligned}$$

Lemma 3. *Let $\left\{\varphi^k : 0 \leq k \leq N\right\}$ be a non-negative sequence, $\left\{\xi^k : 0 \leq k \leq N\right\}$ be a nondecreasing positive sequence, and $C_0 \geq 1$ be a constant, which satisfy*

$$\sum_{i=1}^{m} \frac{b_i}{\Gamma(2-\alpha_i)} \widetilde{d}^n_{\alpha_i,n} \varphi^n \leq -C_0 \sum_{i=1}^{m} \frac{b_i}{\Gamma(2-\alpha_i)} \sum_{k=0}^{n-1} \widetilde{d}^n_{\alpha_i,k} \varphi^k + \xi^n, 1 \leq n \leq N. \tag{7}$$

Then, we have

$$\varphi^n \leq C^n_0\Big(\varphi^0 + \frac{1}{\sum\limits_{i=1}^{m} \frac{b_i}{\Gamma(2-\alpha_i)} d^n_{\alpha_i,n}} \xi^n\Big), 1 \leq n \leq N. \tag{8}$$

Further, we can further write (8) as

$$\varphi^n \leq C_0^n(\varphi^0 + \sum_{i=1}^{m} \frac{\Gamma(1-\alpha_i)t_n^{\alpha_i}}{b_i}\xi^n), 1 \leq n \leq N. \tag{9}$$

Proof. When $n = 1$ in (7), we have

$$\sum_{i=1}^{m} \frac{b_i}{\Gamma(2-\alpha_i)}\widetilde{d}_{\alpha_i,1}^{1}\varphi^1 \leq -C_0 \sum_{i=1}^{m} \frac{b_i}{\Gamma(2-\alpha_i)}\widetilde{d}_{\alpha_i,0}^{1}\varphi^0 + \xi^1. \tag{10}$$

Noting that $\widetilde{d}_{\alpha_i,0}^{n} = -d_{\alpha_i,n}^{n}, \widetilde{d}_{\alpha_i,n}^{n} = d_{\alpha_i,1}^{n}$, we have

$$\sum_{i=1}^{m} \frac{b_i}{\Gamma(2-\alpha_i)}d_{\alpha_i,1}^{1}\varphi^1 \leq C_0 \sum_{i=1}^{m} \frac{b_i}{\Gamma(2-\alpha_i)}d_{\alpha_i,1}^{1}\varphi^0 + \xi^1. \tag{11}$$

Then, we can obtain

$$\varphi^1 \leq C_0(\varphi^0 + \frac{1}{\sum\limits_{i=1}^{m} \frac{b_i}{\Gamma(2-\alpha_i)}d_{\alpha_i,1}^{1}}\xi^1). \tag{12}$$

It means that the conclusion (8) is valid for the case of $n = 1$. Assume that (8) is valid for $n = 1, 2, \cdots, r$. We now need to prove that it also holds for $n = r + 1$. Selecting $n = r + 1$ in (7), we obtain

$$\begin{aligned}
&\sum_{i=1}^{m} \frac{b_i}{\Gamma(2-\alpha_i)}\widetilde{d}_{\alpha_i,j+1}^{j+1}\varphi^{j+1} \\
\leq& -C_0 \sum_{i=1}^{m} \frac{b_i}{\Gamma(2-\alpha_i)} \sum_{k=0}^{j} \widetilde{d}_{\alpha_i,k}^{j+1}\varphi^k + \xi^{j+1} \\
=& C_0 \sum_{i=1}^{m} \frac{b_i}{\Gamma(2-\alpha_i)} \sum_{k=1}^{j} \left(d_{\alpha_i,j-k+1}^{j+1} - d_{\alpha_i,j-k+2}^{j+1}\right)\varphi^k + C_0 \sum_{i=1}^{m} b_i d_{\alpha_i,j+1}^{j+1}\varphi^0 + \xi^{j+1} \\
=& C_0 \sum_{i=1}^{m} \frac{b_i}{\Gamma(2-\alpha_i)} \sum_{k=0}^{j-1} \left(d_{\alpha_i,k+1}^{j+1} - d_{\alpha_i,k+2}^{j+1}\right)\varphi^{j-k} + C_0 \sum_{i=1}^{m} \frac{b_i}{\Gamma(2-\alpha_i)}d_{\alpha_i,j+1}^{j+1}\varphi^0 + \xi^{j+1}.
\end{aligned} \tag{13}$$

Noting that $0 < d_{\alpha_i,k+1}^{k+1} < d_{\alpha_i,k}^{k}$ and $0 < d_{\alpha_i,k+1}^{n} < d_{\alpha_i,k}^{n}, (k = 0, 1 \cdots j)$, we have

$$\begin{aligned}
&\sum_{i=1}^{m} \frac{b_i}{\Gamma(2-\alpha_i)}\widetilde{d}_{\alpha_i,j+1}^{j+1}\varphi^{j+1} \\
\leq& C_0 \sum_{i=1}^{m} \frac{b_i}{\Gamma(2-\alpha_i)} \sum_{k=0}^{j-1} \left(d_{\alpha_i,k+1}^{j+1} - d_{\alpha_i,k+2}^{j+1}\right)[C_0^{j-k}(\varphi^0 + \frac{1}{\sum\limits_{i=1}^{m} \frac{b_i}{\Gamma(2-\alpha_i)}d_{\alpha_i,j-k}^{j-k}}\xi^{j-k})] \\
&+ C_0 \sum_{i=1}^{m} \frac{b_i}{\Gamma(2-\alpha_i)}\mathrm{d}_{\alpha_i,j+1}^{j+1}(\varphi^0 + \frac{1}{\sum\limits_{i=1}^{m} \frac{b_i}{\Gamma(2-\alpha_i)}d_{\alpha_i,j+1}^{j+1}}\xi^{j+1}) \\
\leq& C_0^{j+1} \sum_{i=1}^{m} \frac{b_i}{\Gamma(2-\alpha_i)}[\sum_{k=0}^{j-1}\left(d_{\alpha_i,k+1}^{j+1} - d_{\alpha_i,k+2}^{j+1}\right) + d_{\alpha_i,j+1}^{j+1}](\varphi^0 + \frac{1}{\sum\limits_{i=1}^{m} \frac{b_i}{\Gamma(2-\alpha_i)}d_{\alpha_i,j+1}^{j+1}}\xi^{j+1}) \\
\leq& C_0^{j+1} \sum_{i=1}^{m} \frac{b_i}{\Gamma(2-\alpha_i)}d_{\alpha_i,1}^{j+1}(\varphi^0 + \frac{1}{\sum\limits_{i=1}^{m} \frac{b_i}{\Gamma(2-\alpha_i)}d_{\alpha_i,j+1}^{j+1}}\xi^{j+1}).
\end{aligned} \tag{14}$$

Therefore, using the mathematical induction method, we can complete the proof of (8). From [37], we know $n^{\alpha_i}\left(n^{1-\alpha_i}-(n-1)^{1-\alpha_i}\right) \geq 1-\alpha_i$, and then

$$d^n_{\alpha_i,n} = \frac{(n\tau)^{1-\alpha_i}-(n\tau-\tau)^{1-\alpha_i}}{\tau} = \frac{\left(n^{1-\alpha_i}-(n-1)^{1-\alpha_i}\right)}{\tau^{\alpha_i}} \geq \frac{(1-\alpha_i)}{\tau^{\alpha_i}n^{\alpha_i}}. \tag{15}$$

Thus, making use of (8) and (15), we can complete the proof of (9). □

Next, we give the existence and uniqueness results for the MFE scheme (5).

Theorem 1. *The MFE scheme (5) has a unique solution.*

Proof. Let $V_h = \mathrm{span}\{\phi_1, \phi_2, \cdots, \phi_{M_1}\}$ and $\mathbf{W}_h = \mathrm{span}\{\boldsymbol{\psi}_1, \boldsymbol{\psi}_2, \cdots, \boldsymbol{\psi}_{M_2}\}$. Then, u^n_h and $\boldsymbol{\lambda}^n_h$ can be written as

$$u^n_h = \sum_{i=1}^{M_1} \widetilde{u}^n_i \phi_i, \boldsymbol{\lambda}^n_h = \sum_{j=1}^{M_2} \widetilde{s}^n_j \boldsymbol{\psi}_j. \tag{16}$$

Substituting 16 into (5) and selecting $v_h = \phi_i \ (i = 1, 2, \cdots, M_1)$ and $\mathbf{w}_h = \boldsymbol{\psi}_j \ (j = 1, 2, \cdots, M_2)$, we have

$$\begin{bmatrix} \sum_{i=1}^{m} \frac{b_i}{\Gamma(2-\alpha_i)} \widetilde{d}^n_{\alpha_i,n} B_1 + B_3 & D^T \\ -D & B_2 \end{bmatrix} \begin{bmatrix} U^n \\ L^n \end{bmatrix} = \begin{bmatrix} F^n - \sum_{i=1}^{m} \frac{b_i}{\Gamma(2-\alpha_i)} \sum_{k=0}^{n-1} \widetilde{d}^n_{\alpha_i,k} B_1 U^k \\ 0 \end{bmatrix}, \tag{17}$$

where

$$\begin{aligned} &U^n = \left(\widetilde{u}^n_1, \widetilde{u}^n_2, \cdots, \widetilde{u}^n_{M_1}\right)^T, && L^n = \left(\widetilde{s}^n_1, \widetilde{s}^n_2, \cdots, \widetilde{s}^n_{M_2}\right)^T, \\ &B_1 = ((\phi_i, \phi_j))_{M_1\times M_1}, && B_2 = ((\mathcal{A}^{-1}\boldsymbol{\psi}_i, \boldsymbol{\psi}_j))_{M_2\times M_2}, \\ &B_3 = ((p\phi_i, \phi_j))_{M_1\times M_1}, && D = ((\mathrm{div}\boldsymbol{\psi}_i, \phi_j))_{M_2\times M_1}, \\ &F^n = ((f^n, \phi_i))_{M_1\times 1}, \end{aligned}$$

Noting that B_1 and B_2 are symmetric positive definite matrices and B_3 is a symmetric semi-positive matrix, we have

$$\begin{bmatrix} E & -D^T B_2^{-1} \\ 0 & E \end{bmatrix} \begin{bmatrix} \sum_{i=1}^{m} \frac{b_i}{\Gamma(2-\alpha_i)} \widetilde{d}^n_{\alpha_i,n} B_1 + B_3 & D^T \\ -D & B_2 \end{bmatrix} = \begin{bmatrix} G & 0 \\ -D & B_2 \end{bmatrix}. \tag{18}$$

where $G = \sum_{i=1}^{m} \frac{b_i}{\Gamma(2-\alpha_i)} \widetilde{d}^n_{\alpha_i,n} B_1 + B_3 + D^T B_2^{-1} D$. It is easy to see that G is invertible, so the coefficient matrix of linear Equation (17) is invertible. This means that the MFE scheme (5) has a unique solution. □

Remark 2. *For Lemma 3, when $C_0 = 1$, a similar conclusion can be seen from the proof of Theorem 3.1 in [20]. When $C_0 > 1$, some special applications can be seen from [39]. It should be noted that this lemma can be considered a fractional Grönwall inequality without any other conditions for its existence, which will play a crucial role in the subsequent proof process of stability and convergence analyses.*

4. Stability Analysis

In this section, we will discuss the unconditional stability for the MFE scheme (5)–(6).

Theorem 2. *Let $\left(u_h^n, \lambda_h^n\right)_{n=1}^N$ be the solutions of the MFE scheme (5). Then, there exists a constant $C > 0$ independent of h and N such that*

$$\| u_h^n \| \le \| u_h^0 \| + \sum_{i=1}^m \frac{\Gamma(1-\alpha_i)t_n^{\alpha_i}}{b_i} \sup_{t\in[0,T]} \| f(t) \| \triangleq U_h^\diamond,$$

$$\| \lambda_h^n \| \le C\left(\| \lambda_h^0 \| + \left(\sum_{i=1}^m \frac{\Gamma(1-\alpha_i)t_n^{\alpha_i}}{b_i}\right)^{1/2} \left(\sup_{t\in[0,T]} \| f(t) \| + \| p \|_\infty U_h^\diamond\right)\right).$$

Proof. Taking $v_h = u_h^n$ and $w_h = \lambda_h^n$ in (5), we have

$$\left(\sum_{i=1}^m b_i D_N^{\alpha_i} u_h^n, u_h^n\right) + \left(\mathcal{A}^{-1}\lambda_h^n, \lambda_h^n\right) + (pu_h^n, u_h^n) = (f^n, u_h^n). \tag{19}$$

Using Lemma 1 and the definition of $D_N^{\alpha_i} u_h^n$, we have

$$\sum_{i=1}^m \frac{b_i}{\Gamma(2-\alpha_i)} \widetilde{d}_{\alpha_i,n}^n (u_h^n, u_h^n) + \mu_0 \| \lambda_h^n \|^2 \le \sum_{i=1}^m \frac{b_i}{\Gamma(2-\alpha_i)} \sum_{k=0}^{n-1} \widetilde{d}_{\alpha_i,k}^n \left(u_h^k, u_h^n\right) + (f^n, u_h^n). \tag{20}$$

Applying the Cauchy–Schwarz inequality yields

$$\sum_{i=1}^m \frac{b_i}{\Gamma(2-\alpha_i)} \widetilde{d}_{\alpha_i,n}^n \| u_h^n \| \le \sum_{i=1}^m \frac{b_i}{\Gamma(2-\alpha_i)} \sum_{k=0}^{n-1} \widetilde{d}_{\alpha_i,k}^n \| u_h^k \| + \| f^n \|. \tag{21}$$

Using Lemma 3, we obtain

$$\| u_h^n \| \le \| u_h^0 \| + \sum_{i=1}^m \frac{\Gamma(1-\alpha_i)t_n^{\alpha_i}}{b_i} \sup_{t\in[0,T]} \| f(t) \| \triangleq U_h^\diamond. \tag{22}$$

Next, using (5) (b) and (6) (a), we have

$$\left(A^{-1}\sum_{i=1}^m b_i D_N^{\alpha_i}\lambda_h^n, \boldsymbol{w}_h\right) - \left(\sum_{i=1}^m b_i D_N^{\alpha_i} u_h^n, \mathrm{div}\boldsymbol{w}_h\right) = 0, \forall \boldsymbol{w}_h \in \mathbf{W}_h. \tag{23}$$

Choosing $v_h = \sum_{i=1}^m b_i D_N^{\alpha_i} u_h^n$ and $\boldsymbol{w}_h = \lambda_h^n$ in (5) (a) and (23), respectively, we obtain

$$\| \sum_{i=1}^m b_i D_N^{\alpha_i} u_h^n \|^2 + \left(A^{-1}\sum_{i=1}^m b_i D_N^{\alpha_i}\lambda_h^n, \lambda_h^n\right) + \left(pu_h^n, \sum_{i=1}^m b_i D_N^{\alpha_i} u_h^n\right) = \left(f^n, \sum_{i=1}^m b_i D_N^{\alpha_i} u_h^n\right). \tag{24}$$

Using Lemma 2 in (24) yields

$$\begin{aligned} &\| \sum_{i=1}^m b_i D_N^{\alpha_i} u_h^n \|^2 + \frac{1}{2}\sum_{i=0}^m \frac{b_i}{\Gamma(2-\alpha_i)}\Bigg[\widetilde{d}_{\alpha_i,n}^n\left(A^{-1}\lambda_h^n, \lambda_h^n\right) + \sum_{k=0}^{n-1}\widetilde{d}_{\alpha_i,k}^n\left(A^{-1}\lambda_h^k, \lambda_h^k\right) \\ &- \sum_{k=0}^{n-1}\widetilde{d}_{\alpha_i,k}^n\left(A^{-1}\left(\lambda_h^n - \lambda_h^k\right), \lambda_h^n - \lambda_h^k\right)\Bigg] = \left(f^n, \sum_{i=1}^m b_i D_N^{\alpha_i} u_h^n\right) - \left(pu_h^n, \sum_{i=1}^m b_i D_N^{\alpha_i} u_h^n\right). \end{aligned} \tag{25}$$

Because of $\widetilde{d}^n_{\alpha_i,k} < 0, 0 < k \le n-1$, we have

$$\begin{aligned}&\| \sum_{i=1}^{m} b_i D_N^{\alpha_i} u_h^n \|^2 + \frac{1}{2}\sum_{i=0}^{m} \frac{b_i}{\Gamma(2-\alpha_i)} \widetilde{d}^n_{\alpha_i,n}\left(A^{-1}\lambda_h^n, \lambda_h^n\right) \\ \le& -\frac{1}{2}\sum_{i=0}^{m} \frac{b_i}{\Gamma(2-\alpha_i)} \sum_{k=0}^{n-1} \widetilde{d}^n_{\alpha_i,k}\left(A^{-1}\lambda_h^k, \lambda_h^k\right) \\ &+ \left(f^n, \sum_{i=1}^{m} b_i D_N^{\alpha_i} u_h^n\right) - \left(p u_h^n, \sum_{i=1}^{m} b_i D_N^{\alpha_i} u_h^n\right).\end{aligned} \tag{26}$$

Apply the Cauchy–Schwarz inequality and the Young inequality in (26) to obtain

$$\begin{aligned}&\| \sum_{i=1}^{m} b_i D_N^{\alpha_i} u_h^n \|^2 + \frac{1}{2}\sum_{i=0}^{m} \frac{b_i}{\Gamma(2-\alpha_i)} \widetilde{d}^n_{\alpha_i,n}\left(A^{-1}\lambda_h^n, \lambda_h^n\right) \\ \le& -\frac{1}{2}\sum_{i=0}^{m} \frac{b_i}{\Gamma(2-\alpha_i)} \sum_{k=0}^{n-1} \widetilde{d}^n_{\alpha_i,k}\left(A^{-1}\lambda_h^k, \lambda_h^k\right) \\ &+ \frac{1}{2}\| \sum_{i=1}^{m} b_i D_N^{\alpha_i} u_h^n \|^2 + \| f^n \|^2 + \| v \|_\infty^2 \| u_h^n \|^2.\end{aligned} \tag{27}$$

Using Lemma 3 in (27), we obtain

$$\| \lambda_h^n \| \le C\left(\| \lambda_h^0 \| + \left(\sum_{i=1}^{m} \frac{\Gamma(1-\alpha_i)t_n^{\alpha_i}}{b_i}\right)^{1/2} \left(\sup_{t\in[0,T]} \| f(t) \| + \| p \|_\infty U_h^\diamond\right)\right). \tag{28}$$

Thus, we complete the proof. □

5. Convergence Analysis

In this section, we will present the convergence results. For this purpose, we first introduce the mixed elliptic projection $(R_h u, R_h \lambda) \in V_h \times \mathbf{W}_h$ defined by

$$\begin{cases} (a)\left(\mathcal{A}^{-1}(\lambda - R_h\lambda), w_h\right) - (u - R_h u, \mathrm{div} w_h) = 0, & \forall w_h \in \mathbf{W}_h, \\ (b)(\mathrm{div}(\lambda - R_h\lambda), v_h) = 0, & \forall v_h \in V_h. \end{cases} \tag{29}$$

Then, the above projection satisfies the classical estimates as follows.

Lemma 4 ([40,41])**.** *There exists a constant $C > 0$ independent of h and N such that*

$$\begin{aligned}&\| \lambda - R_h\lambda \| \le C h^{r+1} \| \lambda \|_{r+1}, for \lambda \in \left(H^{r+1}(\Omega)\right)^2, \\ &\| \mathrm{div}(\lambda - R_h\lambda) \| \le C h^{r+1} \| \mathrm{div}\lambda \|_{r+1}, for \mathrm{div}\lambda \in H^{r+1}(\Omega), \\ &\| u - R_h u \| \le C h^{r+1}\left(\| u \|_{r+1} + \| \lambda \|_{r+1}\right), for u \in H^{r+1}(\Omega), \lambda \in \left(H^{r+1}(\Omega)\right)^2.\end{aligned}$$

For the truncation error $Q^n_{\alpha_i}$ $(i = 1, 2, \cdots, m)$ of the $L1$ formula, from [36,37], we give the following estimates.

Lemma 5. *Let $u \in C^2\left(\overline{J}, L^2(\Omega)\right)$. Then, we have*

$$\begin{aligned}&\| Q^n_{\alpha_i} \| \le C N^{-(2-\alpha_i)}, i = 1, 2, \cdots, m, \\ &\| \sum_{i=1}^{m} b_i Q^n_{\alpha_i} \| \le C N^{-(2-\alpha_1)},\end{aligned}$$

where $C > 0$ is a constant independent of h and N.

Now, we write the errors $u(t_n) - u_h^n = u(t_n) - R_h u(t_n) + R_h u(t_n) - u_h^n = \rho^n + \theta^n$ and $\boldsymbol{\lambda}(t_n) - \boldsymbol{\lambda}_h^n = \boldsymbol{\lambda}(t_n) - R_h\boldsymbol{\lambda}(t_n) + R_h\boldsymbol{\lambda}(t_n) - \boldsymbol{\lambda}_h^n = \boldsymbol{\xi}^n + \boldsymbol{\eta}^n$. From (3) and (5), making use of the mixed elliptic projection R_h, we have the following error equations:

$$\begin{cases} (a)\left(\sum_{i=1}^{m} b_i D_N^{\alpha_i}(\theta^n + \rho^n), v_h\right) + (\mathrm{div}\boldsymbol{\eta}^n, v_h) + (p(\theta^n + \rho^n), v_h) \\ = -\left(\sum_{i=1}^{m} b_i Q_{\alpha_i}^n, v_h\right), & \forall v_h \in V_h, \\ (b)(\mathcal{A}^{-1}\boldsymbol{\eta}^n, \boldsymbol{w}_h) - (\theta^n, \mathrm{div}\boldsymbol{w}_h) = 0, & \forall \boldsymbol{w}_h \in \boldsymbol{W}_h. \end{cases} \tag{30}$$

Noting that $(u_h^0, \boldsymbol{\lambda}_h^0) = (R_h u_0, R_h \boldsymbol{\lambda}_0)$, we have $\theta^0 = 0$ and $\boldsymbol{\eta}^0 = \mathbf{0}$. We next give the convergence results for the MFE scheme (5)–(6).

Theorem 3. *Let $(u^n, \boldsymbol{\lambda}^n) \in V \times \boldsymbol{W}$ and $(u_h^n, \boldsymbol{\lambda}_h^n) \in V_h \times \boldsymbol{W}_h$ be the solutions of (3) and (5), respectively. Assume that $u, div\boldsymbol{\lambda} \in C^2(\bar{J}, H^{r+1}(\Omega))$, $\boldsymbol{\lambda} \in C^2\left(\bar{J}, (H^{r+1}(\Omega))^2\right)$. Then, we have*

$$\max_{1\le n\le N} \| u(t_n) - u_h^n \| + \max_{1\le n\le N} \| \boldsymbol{\lambda}(t_n) - \boldsymbol{\lambda}_h^n \| \le C\left(h^{r+1} + N^{-(2-\alpha_1)}\right),$$

$$\max_{1\le n\le N} \| \boldsymbol{\lambda}(t_n) - \boldsymbol{\lambda}_h^n \|_{\boldsymbol{H}(\mathrm{div},\Omega)} \le C\left(1 + N^{\frac{\alpha_m}{2}}\right)\left(h^{r+1} + N^{-(2-\alpha_1)}\right),$$

where $C > 0$ is a constant independent of h and N.

Proof. Taking $v_h = \theta^n$ and $\boldsymbol{w}_h = \boldsymbol{\eta}^n$ in (30), we can obtain

$$\begin{aligned} &\left(\sum_{i=1}^{m} b_i D_N^{\alpha_i}\theta^n, \theta^n\right) + \left(\mathcal{A}^{-1}\boldsymbol{\eta}^n, \boldsymbol{\eta}^n\right) + (p\theta^n, \theta^n) \\ =&-\left(\sum_{i=1}^{m} b_i Q_{\alpha_i}^n, \theta^n\right) - (p\rho^n, \theta^n) - \left(\sum_{i=1}^{m} b_i D_N^{\alpha_i}\rho^n, \theta^n\right). \end{aligned} \tag{31}$$

Noting that $p(\boldsymbol{x}) \ge 0$, using the Lemma 1 and the definition of $D_N^{\alpha_i} u_h^n$, we have

$$\begin{aligned} &\sum_{i=1}^{m} \frac{b_i}{\Gamma(2-\alpha_i)} \tilde{d}_{\alpha_i,n}^n (\theta^n, \theta^n) + \mu_0 \| \boldsymbol{\eta}^n \|^2 \\ \le& -\sum_{i=1}^{m} \frac{b_i}{\Gamma(2-\alpha_i)} \sum_{k=0}^{n-1} \tilde{d}_{\alpha_i,k}^n \left(\theta^k, \theta^n\right) - \left(\sum_{i=1}^{m} b_i Q_{\alpha_i}^n, \theta^n\right) - (p\rho^n, \theta^n) - \left(\sum_{i=1}^{m} b_i D_N^{\alpha_i}\rho^n, \theta^n\right). \end{aligned} \tag{32}$$

Applying the Cauchy–Schwarz inequality, we obtain

$$\begin{aligned} &\sum_{i=1}^{m} \frac{b_i}{\Gamma(2-\alpha_i)} \tilde{d}_{\alpha_i,n}^n \| \theta^n \|^2 + \mu_0 \| \boldsymbol{\eta}^n \|^2 \\ \le& -\sum_{i=1}^{m} \frac{b_i}{\Gamma(2-\alpha_i)} \sum_{k=0}^{n-1} \tilde{d}_{\alpha_i,k}^n \| \theta^k \| \| \theta^n \| \\ &+(\| \sum_{i=1}^{m} b_i Q_{\alpha_i}^n \| + \| p \|_\infty \| \rho \| + \| \sum_{i=1}^{m} b_i D_N^{\alpha_i}\rho^n \|) \| \theta^n \|, \end{aligned} \tag{33}$$

and then

$$\begin{aligned}&\sum_{i=1}^{m}\frac{b_i}{\Gamma(2-\alpha_i)}\widetilde{d}^n_{\alpha_i,n}\|\theta^n\|^2\\&\leq-\sum_{i=1}^{m}\frac{b_i}{\Gamma(2-\alpha_i)}\sum_{k=0}^{n-1}\widetilde{d}^n_{\alpha_i,k}\|\theta^k\|+(\|\sum_{i=1}^{m}b_iQ^n_{\alpha_i}\|+\|p\|_\infty\|\rho\|+\|\sum_{i=1}^{m}b_iD_N^{\alpha_i}\rho^n\|).\end{aligned}\tag{34}$$

Using Lemmas 4 and 5, we obtain

$$\sum_{i=1}^{m}\frac{b_i}{\Gamma(2-\alpha_i)}\widetilde{d}^n_{\alpha_i,n}\|\theta^n\|\leq-\sum_{i=1}^{m}\frac{b_i}{\Gamma(2-\alpha_i)}\sum_{k=0}^{n-1}\widetilde{d}^n_{\alpha_i,k}\|\vartheta^k\|+C\Big(N^{-(2-\alpha_1)}+h^{r+1}\Big).\tag{35}$$

Noting that $\theta^0=0$ and using Lemma 3, we obtain

$$\|\theta^n\|\leq C\Big(h^{r+1}+N^{-(2-\alpha_1)}\Big).\tag{36}$$

Now, from (30) (b), we obtain

$$\left(\mathcal{A}^{-1}\sum_{i=1}^{m}b_iD_N^{\alpha_i}\boldsymbol{\eta}^n,\boldsymbol{w}_h\right)-\left(\sum_{i=1}^{m}b_iD_N^{\alpha_i}\theta^n,\mathrm{div}\boldsymbol{w}_h\right)=0,\forall\boldsymbol{w}_h\in\mathbf{W}_h.\tag{37}$$

Choosing $v_h=\sum\limits_{i=1}^{m}b_iD_N^{\alpha_i}\theta^n$ and $\boldsymbol{w}_h=\boldsymbol{\eta}^n$ in (30) (a) and (37), respectively, we can obtain

$$\begin{aligned}&\|\sum_{i=1}^{m}b_iD_N^{\alpha_i}\theta^n\|^2+\left(\mathcal{A}^{-1}\sum_{i=1}^{m}b_iD_N^{\alpha_i}\boldsymbol{\eta}^n,\boldsymbol{\eta}^n\right)\\&=-\left(\sum_{i=1}^{m}b_iD_N^{\alpha_i}\rho^n,\sum_{i=1}^{m}b_iD_N^{\alpha_i}\theta^n\right)-\left(p(\rho^n+\theta^n),\sum_{i=1}^{m}b_iD_N^{\alpha_i}\theta^n\right)-\left(\sum_{i=1}^{m}b_iQ^n_{\alpha_i},\sum_{i=1}^{m}b_iD_N^{\alpha_i}\theta^n\right).\end{aligned}\tag{38}$$

Using Lemma 2, we have

$$\begin{aligned}&\|\sum_{i=1}^{m}b_iD_N^{\alpha_i}\theta^n\|^2+\frac{1}{2}\sum_{i=1}^{m}\frac{b_i}{\Gamma(2-\alpha_i)}[\widetilde{d}^n_{\alpha_i,n}\Big(\mathcal{A}^{-1}\boldsymbol{\eta}^n,\boldsymbol{\eta}^n\Big)-\sum_{k=0}^{n-1}\widetilde{d}^n_{\alpha_i,k}\Big(\mathcal{A}^{-1}\Big(\boldsymbol{\eta}^n-\boldsymbol{\eta}^k,\boldsymbol{\eta}^n-\boldsymbol{\eta}^k\Big)]\\&=-\frac{1}{2}\sum_{i=1}^{m}\frac{b_i}{\Gamma(2-\alpha_i)}\sum_{k=0}^{n-1}\widetilde{d}^n_{\alpha_i,k}\Big(\mathcal{A}^{-1}\boldsymbol{\eta}^k,\boldsymbol{\eta}^k\Big)-\left(\sum_{i=1}^{m}b_iD_N^{\alpha_i}\rho^n,\sum_{i=1}^{m}b_iD_N^{\alpha_i}\theta^n\right)\\&-\left(\sum_{i=1}^{m}b_iQ^n_{\alpha_i},\sum_{i=1}^{m}b_iD_N^{\alpha_i}\theta^n\right)-\left(p(\rho^n+\theta^n),\sum_{i=1}^{m}b_iD_N^{\alpha_i}\theta^n\right).\end{aligned}\tag{39}$$

Noting that $\widetilde{d}^n_{\alpha_i,k}<0,0<k\leq n-1$ and using Lemma 1, we obtain

$$\begin{aligned}&\|\sum_{i=1}^{m}b_iD_N^{\alpha_i}\theta^n\|^2+\frac{1}{2}\sum_{i=1}^{m}\frac{b_i}{\Gamma(2-\alpha_i)}\widetilde{d}^n_{\alpha_i,n}\Big(\mathcal{A}^{-1}\boldsymbol{\eta}^n,\boldsymbol{\eta}^n\Big)\\&\leq-\frac{1}{2}\sum_{i=1}^{m}\frac{b_i}{\Gamma(2-\alpha_i)}\sum_{k=0}^{n-1}\widetilde{d}^n_{\alpha_i,k}\Big(\mathcal{A}^{-1}\boldsymbol{\eta}^k,\boldsymbol{\eta}^k\Big)-\left(\sum_{i=1}^{m}b_iD_N^{\alpha_i}\rho^n,\sum_{i=1}^{m}b_iD_N^{\alpha_i}\theta^n\right)\\&-\left(\sum_{i=1}^{m}b_iQ^n_{\alpha_i},\sum_{i=1}^{m}b_iD_N^{\alpha_i}\theta^n\right)-\left(p(\rho^n+\theta^n),\sum_{i=1}^{m}b_iD_N^{\alpha_i}\theta^n\right).\end{aligned}\tag{40}$$

Applying the Cauchy–Schwarz and the Young inequality in (40) yields

$$\begin{aligned} &\| \sum_{i=1}^{m} b_i D_N^{\alpha_i}\theta^n \|^2 + \frac{1}{2}\sum_{i=1}^{m} \frac{b_i}{\Gamma(2-\alpha_i)} \widetilde{d}_{\alpha_i,n}^n \left(\mathcal{A}^{-1}\boldsymbol{\eta}^n, \boldsymbol{\eta}^n\right) \\ \leq &-\frac{1}{2}\sum_{i=1}^{m} \frac{b_i}{\Gamma(2-\alpha_i)} \sum_{k=0}^{n-1} \widetilde{d}_{\alpha_i,k}^n \left(\mathcal{A}^{-1}\boldsymbol{\eta}^k, \boldsymbol{\eta}^k\right) + \frac{1}{2}\| \sum_{i=1}^{m} b_i D_N^{\alpha}\theta^n \|^2 \\ &+2\left(\| \sum_{i=1}^{m} b_i D_N^{\alpha_i}\rho^n \| + \| \sum_{i=1}^{m} b_i Q_{\alpha_i}^n \|^2 + \| p \|_\infty^2 (\| \rho^n \|^2 + \| \theta^n \|^2) \right). \end{aligned} \tag{41}$$

Using Lemmas 4 and 5, we obtain

$$\begin{aligned} &\sum_{i=1}^{m} \frac{b_i}{\Gamma(2-\alpha_i)} \widetilde{d}_{\alpha_i,n}^n \left(\mathcal{A}^{-1}\boldsymbol{\eta}^n, \boldsymbol{\eta}^n\right) \\ \leq &-\sum_{i=1}^{m} \frac{b_i}{\Gamma(2-\alpha_i)} \sum_{k=0}^{n-1} \widetilde{d}_{\alpha_i,k}^n \left(\mathcal{A}^{-1}\boldsymbol{\eta}^k, \boldsymbol{\eta}^k\right) + C\left(N^{-2(2-\alpha_1)} + h^{2r+2}\right). \end{aligned} \tag{42}$$

Noting that $\boldsymbol{\eta}^0 = \mathbf{0}$ and using Lemma 3, we obtain

$$\| \boldsymbol{\eta}^n \| \leq C\left(h^{r+1} + N^{-(2-\alpha_1)}\right). \tag{43}$$

We now estimate $\| \boldsymbol{\lambda}^n - \boldsymbol{\lambda}_h^n \|_{\boldsymbol{H}(\text{div},\Omega)}$. Taking $v_h = \sum_{i=1}^{m} b_i D_N^{\alpha_i}\theta^n$ and $\boldsymbol{w}_h = \boldsymbol{\eta}^n$ in (30) (a) and (37), respectively, we have

$$\begin{aligned} \| \text{div}\boldsymbol{\eta}^n \|^2 = &-\left(\mathcal{A}^{-1}\sum_{i=1}^{m} b_i D_N^{\alpha_i}\boldsymbol{\eta}^n, \boldsymbol{\eta}^n\right) - \left(\sum_{i=1}^{m} b_i D_N^{\alpha_i}\rho^n, \text{div}\boldsymbol{\eta}^n\right) \\ &-(p(\rho^n + \theta^n), \text{div}\boldsymbol{\eta}^n) - \left(\sum_{i=1}^{m} b_i Q_{\alpha_i}^n, \text{div}\boldsymbol{\eta}^n\right). \end{aligned} \tag{44}$$

For the term $-\left(\mathcal{A}^{-1}\sum_{i=1}^{m} b_i D_N^{\alpha_i}\boldsymbol{\eta}^n, \boldsymbol{\eta}^n\right)$, noting that $\sum_{k=0}^{n-1}\left(-\widetilde{d}_{\alpha_i,k}^n\right) = T^{-\alpha_i}N^{\alpha_i}$, we obtain

$$\begin{aligned} -\left(\mathcal{A}^{-1}\sum_{i=1}^{m} b_i D_N^{\alpha_i}\boldsymbol{\eta}^n, \boldsymbol{\eta}^n\right) = &-\sum_{i=1}^{m} \frac{b_i}{\Gamma(2-\alpha_i)} \left(\sum_{k=0}^{n-1} \widetilde{d}_{\alpha_i,k}^n \left(\mathcal{A}^{-1}\boldsymbol{\eta}^k, \boldsymbol{\eta}^n\right) + \widetilde{d}_{\alpha_i,n}^n \left(\mathcal{A}^{-1}\boldsymbol{\eta}^n, \boldsymbol{\eta}^n\right)\right) \\ \leq &\mu_1 \sum_{i=1}^{m} \frac{b_i}{\Gamma(2-\alpha_i)} \sum_{k=0}^{n-1}\left(-\widetilde{d}_{\alpha_i,k}^n\right) \| \boldsymbol{\eta}^k \| \| \boldsymbol{\eta}^n \| \\ \leq &C\sum_{i=1}^{m} \frac{b_i}{\Gamma(2-\alpha_i)} T^{-\alpha_i}N^{\alpha_i}\left(N^{-(2-\alpha_1)} + h^{r+1}\right)^2. \end{aligned} \tag{45}$$

Then, it holds from (45) that

$$\begin{aligned} \| \text{div}\boldsymbol{\eta}^n \|^2 = &2\left(\| \sum_{i=1}^{m} b_i D_N^{\alpha_i}\rho^n \| + \| \sum_{i=1}^{m} b_i Q_{\alpha_i}^n \|^2 + \| p \|_\infty^2 (\| \rho^n \|^2 + \| \theta^n \|^2) \right) \\ &+C\sum_{i=1}^{m} \frac{b_i}{\Gamma(2-\alpha_i)} T^{-\alpha_i}N^{\alpha_i}\left(N^{-(2-\alpha_1)} + h^{r+1}\right)^2 + \frac{1}{2}\| \text{div}\boldsymbol{\eta}^n \|^2. \end{aligned} \tag{46}$$

Using Lemmas 4 and 5, we have

$$\| \text{div}\boldsymbol{\eta}^n \| \leq C\left(1 + N^{\frac{\alpha_m}{2}}\right)\left(h^{r+1} + N^{-(2-\alpha_1)}\right). \tag{47}$$

Then, we finish the proof. □

Remark 3. *(I) For variables u and $\boldsymbol{\lambda}$, we define the discrete norms of the errors as follows:*

$$\begin{aligned}
\| u-u_h \|_{\hat{L}^\infty(L^2(\Omega))} &= \max_{1\le n\le N} \| u(t_n) - u_h^n \|, \\
\| \boldsymbol{\lambda}-\boldsymbol{\lambda}_h \|_{\hat{L}^\infty((L^2(\Omega))^2)} &= \max_{1\le n\le N} \| \boldsymbol{\lambda}(t_n) - \boldsymbol{\lambda}_h^n \|, \\
\| \boldsymbol{\lambda}-\boldsymbol{\lambda}_h \|_{\hat{L}^\infty(\boldsymbol{H}(\mathrm{div},\Omega))} &= \max_{1\le n\le N} \| \boldsymbol{\lambda}(t_n) - \boldsymbol{\lambda}_h^n \|_{\boldsymbol{H}(\mathrm{div},\Omega)}.
\end{aligned}$$

From Theorem 3, we obtain the optimal a priori error estimate results for u in the discrete $L^\infty(L^2(\Omega))$ norm and $\boldsymbol{\lambda}$ in the discrete $L^\infty\left(\left(L^2(\Omega)\right)^2\right)$ norm and obtain the suboptimal error estimate for $\boldsymbol{\lambda}$ in the discrete $L^\infty(\boldsymbol{H}(div,\Omega))$ norm. In the actual calculation in the next section, we achieve the optimal convergence rates for variables u and $\boldsymbol{\lambda}$ based on the above discrete norms.

(II) It should be pointed out that the solutions of many FPDEs have an initial layer at $t = 0$ (see [42,43]). To overcome this difficulty, some scholars have adopted nonuniform mesh methods and achieved excellent results [24,26,42,44–46]. Moreover, it is noted that $\left\{\xi^k : 0 \le k \le N\right\}$ in Lemma 3 is required to be a nondecreasing positive sequence, so the error estimates for the MFE scheme (5)–(6) with the temporal nonuniform method should adopt some other techniques. It is gratifying that the numerical results in Example 3 show that the MFE scheme (5)–(6) with the temporal graded mesh is feasible and effective.

6. Numerical Examples

In this section, we given three test examples to verify the effectiveness and convergence accuracy of the proposed MFE scheme (5)–(6) and adopt the lowest-order Raviart–Thomas MFE space for variables u and $\boldsymbol{\lambda}$ in the numerical experiments.

Example 1. *Consider the following two-term TFRD equation:*

$$\begin{cases} D_t^{\alpha_1} u(\boldsymbol{x},t) + D_t^{\alpha_2} u(\boldsymbol{x},t) - \Delta u(\boldsymbol{x},t) + p(\boldsymbol{x})u(\boldsymbol{x},t) = f(\boldsymbol{x},t), & (\boldsymbol{x},t) \in \Omega \times J, \\ u(\boldsymbol{x},t) = 0, & (\boldsymbol{x},t) \in \partial\Omega \times \overline{J}, \\ u(\boldsymbol{x},0) = u_0(\boldsymbol{x}), & \boldsymbol{x} \in \overline{\Omega}, \end{cases} \quad (48)$$

where $J = (0,1]$, $\Omega = (0,1)^2$, $p(\boldsymbol{x}) = 1 + x_1^2 + x_2^2$, $\boldsymbol{x} = (x_1, x_2) \in \Omega$, $u(\boldsymbol{x},0) = 0$, and the source function f is taken by

$$\begin{aligned} f(\boldsymbol{x},t) = &\left(\frac{\Gamma(3+\alpha_1+\alpha_2)}{\Gamma(3+\alpha_2)} t^{2+\alpha_2} + \frac{\Gamma(3+\alpha_1+\alpha_2)}{\Gamma(3+\alpha_1)} t^{2+\alpha_1} + \left(2\pi^2 + p(\boldsymbol{x})\right) t^{2+\alpha_1+\alpha_2} \right) \\ &\times \sin(\pi x_1)\sin(\pi x_2). \end{aligned}$$

And we can find the analytical solutions for variables u and $\boldsymbol{\lambda}$ as follows:

$$\begin{aligned} u(\boldsymbol{x},t) &= t^{2+\alpha_1+\alpha_2}\sin(\pi x_1)\sin(\pi x_2), \\ \boldsymbol{\lambda}(\boldsymbol{x},t) &= -\pi t^{2+\alpha_1+\alpha_2}(\cos(\pi x_1)\sin(\pi x_2), \sin(\pi x_1)\cos(\pi x_2)). \end{aligned}$$

In the numerical simulation, we select fractional parameters $\alpha_1 = 0.9, 0.7, 0.5$ and $\alpha_2 = 0.1, 0.4$ in Equation (48) and know that among these different fractional parameters, the convergence rates are only related to the largest fractional parameter α_1 from Theorem 3. By taking $N = 5, 8, 10, 16$ and the corresponding $h = \sqrt{2}/N^{2-\alpha_1}$, we give the error results and convergence rates in Tables 1–3 for the MFE scheme (5)–(6), which show that the convergence rates in the temporal direction for u (in the discrete $L^\infty(L^2(\Omega))$ norm) and $\boldsymbol{\lambda}$ (in the discrete $L^\infty\left(\left(L^2(\Omega)\right)^2\right)$ and $L^\infty(\boldsymbol{H}(\mathrm{div},\Omega))$ norms) are close to $2-\alpha_1$. Moreover, in order to test convergence rates in the spatial direction, by fixing $N = 100$ and taking $h = \sqrt{2}/4, \sqrt{2}/8, \sqrt{2}/16, \sqrt{2}/32$, we give the error results and convergence rates in Tables 4–6, which show that the convergence rates in the spatial direction for u (in the discrete $L^\infty(L^2(\Omega))$ norm) and λ (in the discrete $L^\infty\left(\left(L^2(\Omega)\right)^2\right)$ and $L^\infty(\boldsymbol{H}(\mathrm{div},\Omega))$ norms) are close to 1.

Table 1. Numerical results with $h \approx \sqrt{2}/N^{2-\alpha_1}$ and $\alpha_1 = 0.9$ in Example 1.

α_2	N	u-$\hat{L}^\infty(L^2)$	Rates	λ-$\hat{L}^\infty((L^2)^2)$	Rates	λ-$\hat{L}^\infty(H(\mathrm{div}))$	Rates
0.1	5	8.6930×10^{-2}	–	3.3630×10^{-1}	–	$1.7545 \times 10^{+0}$	–
	8	5.2424×10^{-2}	1.0760	2.0229×10^{-1}	1.0814	$1.0569 \times 10^{+0}$	1.0784
	10	4.0389×10^{-2}	1.1687	1.5578×10^{-1}	1.1709	8.1390×10^{-1}	1.1708
	16	2.3908×10^{-2}	1.1157	9.2174×10^{-2}	1.1165	4.8141×10^{-1}	1.1172
0.4	5	8.7218×10^{-2}	–	3.3776×10^{-1}	–	$1.7631 \times 10^{+0}$	–
	8	5.2624×10^{-2}	1.0750	2.0332×10^{-1}	1.0799	$1.0619 \times 10^{+0}$	1.0786
	10	4.0555×10^{-2}	1.1674	1.5663×10^{-1}	1.1692	8.1786×10^{-1}	1.1704
	16	2.4011×10^{-2}	1.1153	9.2701×10^{-2}	1.1160	4.8371×10^{-1}	1.1175

Table 2. Numerical results with $h \approx \sqrt{2}/N^{2-\alpha_1}$ and $\alpha_1 = 0.7$ in Example 1.

α_2	N	u-$\hat{L}^\infty(L^2)$	Rates	λ-$\hat{L}^\infty((L^2)^2)$	Rates	λ-$\hat{L}^\infty(H(\mathrm{div}))$	Rates
0.1	5	6.5262×10^{-2}	–	2.5184×10^{-1}	–	$1.3146 \times 10^{+0}$	–
	8	3.4916×10^{-2}	1.3308	1.3451×10^{-1}	1.3344	7.0291×10^{-1}	1.3321
	10	2.6204×10^{-2}	1.2863	1.0092×10^{-1}	1.2875	5.2740×10^{-1}	1.2873
	16	1.4175×10^{-2}	1.3073	5.4581×10^{-2}	1.3077	2.8520×10^{-1}	1.3080
0.4	5	6.5379×10^{-2}	–	2.5244×10^{-1}	–	$1.3184 \times 10^{+0}$	–
	8	3.4998×10^{-2}	1.3296	1.3493×10^{-1}	1.3328	7.0499×10^{-1}	1.3318
	10	2.6269×10^{-2}	1.2858	1.0125×10^{-1}	1.2869	5.2895×10^{-1}	1.2875
	16	1.4212×10^{-2}	1.3070	5.4771×10^{-2}	1.3073	2.8602×10^{-1}	1.3081

Table 3. Numerical results with $h \approx \sqrt{2}/N^{2-\alpha_1}$ and $\alpha_1 = 0.5$ in Example 1.

α_2	N	u-$\hat{L}^\infty(L^2)$	Rates	λ-$\hat{L}^\infty((L^2)^2)$	Rates	λ-$\hat{L}^\infty(H(\mathrm{div}))$	Rates
0.1	5	4.7518×10^{-2}	–	1.8311×10^{-1}	–	9.5625×10^{-1}	–
	8	2.2765×10^{-2}	1.5657	8.7633×10^{-2}	1.5679	4.5800×10^{-1}	1.5663
	10	1.6367×10^{-2}	1.4786	6.2996×10^{-2}	1.4792	3.2925×10^{-1}	1.4791
	16	8.1859×10^{-3}	1.4742	3.1504×10^{-2}	1.4743	1.6465×10^{-1}	1.4744
0.4	5	4.7568×10^{-2}	–	1.8337×10^{-1}	–	9.5799×10^{-1}	–
	8	2.2802×10^{-2}	1.5645	8.7824×10^{-2}	1.5662	4.5893×10^{-1}	1.5658
	10	1.6396×10^{-2}	1.4781	6.3145×10^{-2}	1.4785	3.2993×10^{-1}	1.4790
	16	8.2016×10^{-3}	1.4738	3.1585×10^{-2}	1.4739	1.6499×10^{-1}	1.4744

Table 4. Numerical results with $\tau = T/N = 1/100$ and $\alpha_1 = 0.9$ in Example 1.

α_2	h	u-$\hat{L}^\infty(L^2)$	Rates	λ-$\hat{L}^\infty((L^2)^2)$	Rates	λ-$\hat{L}^\infty(H(\mathrm{div}))$	Rates
0.1	$\sqrt{2}/4$	1.2927×10^{-1}	–	5.0254×10^{-1}	–	$2.5970 \times 10^{+0}$	–
	$\sqrt{2}/8$	6.5234×10^{-2}	0.9867	2.5171×10^{-1}	0.9975	$1.3118 \times 10^{+0}$	0.9853
	$\sqrt{2}/16$	3.2696×10^{-2}	0.9965	1.2589×10^{-1}	0.9996	6.5762×10^{-1}	0.9963
	$\sqrt{2}/32$	1.6361×10^{-2}	0.9989	6.2964×10^{-2}	0.9996	3.2908×10^{-1}	0.9988
0.4	$\sqrt{2}/4$	1.2926×10^{-1}	–	5.0244×10^{-1}	–	$2.5971 \times 10^{+0}$	–
	$\sqrt{2}/8$	6.5231×10^{-2}	0.9866	2.5169×10^{-1}	0.9973	$1.3119 \times 10^{+0}$	0.9853
	$\sqrt{2}/16$	3.2696×10^{-2}	0.9964	1.2589×10^{-1}	0.9995	6.5766×10^{-1}	0.9962
	$\sqrt{2}/32$	1.6363×10^{-2}	0.9987	6.2973×10^{-2}	0.9994	3.2913×10^{-1}	0.9987

Table 5. Numerical results with $\tau = T/N = 1/100$ and $\alpha_1 = 0.7$ in Example 1.

α_2	h	u-$\hat{L}^\infty(L^2)$	Rates	λ-$\hat{L}^\infty((L^2)^2)$	Rates	λ-$\hat{L}^\infty(H(\text{div}))$	Rates
0.1	$\sqrt{2}/4$	1.2929×10^{-1}	–	5.0266×10^{-1}	–	$2.5969 \times 10^{+0}$	–
	$\sqrt{2}/8$	6.5241×10^{-2}	0.9867	2.5174×10^{-1}	0.9976	$1.3118 \times 10^{+0}$	0.9853
	$\sqrt{2}/16$	3.2698×10^{-2}	0.9966	1.2590×10^{-1}	0.9997	6.5757×10^{-1}	0.9963
	$\sqrt{2}/32$	1.6359×10^{-2}	0.9991	6.2954×10^{-2}	0.9999	3.2900×10^{-1}	0.9991
0.4	$\sqrt{2}/4$	1.2928×10^{-1}	–	5.0258×10^{-1}	–	$2.5970 \times 10^{+0}$	–
	$\sqrt{2}/8$	6.5239×10^{-2}	0.9867	2.5173×10^{-1}	0.9975	$1.3118 \times 10^{+0}$	0.9853
	$\sqrt{2}/16$	3.2697×10^{-2}	0.9966	1.2590×10^{-1}	0.9996	6.5758×10^{-1}	0.9963
	$\sqrt{2}/32$	1.6359×10^{-2}	0.9991	6.2954×10^{-2}	0.9999	3.2900×10^{-1}	0.9990

Table 6. Numerical results with $\tau = T/N = 1/100$ and $\alpha_1 = 0.5$ in Example 1.

α_2	h	u-$\hat{L}^\infty(L^2)$	Rates	λ-$\hat{L}^\infty((L^2)^2)$	Rates	λ-$\hat{L}^\infty(H(\text{div}))$	Rates
0.1	$\sqrt{2}/4$	1.2930×10^{-1}	–	5.0273×10^{-1}	–	$2.5969 \times 10^{+0}$	–
	$\sqrt{2}/8$	6.5243×10^{-2}	0.9868	2.5176×10^{-1}	0.9977	$1.3118 \times 10^{+0}$	0.9852
	$\sqrt{2}/16$	3.2699×10^{-2}	0.9966	1.2591×10^{-1}	0.9997	6.5756×10^{-1}	0.9963
	$\sqrt{2}/32$	1.6359×10^{-2}	0.9991	6.2955×10^{-2}	0.9999	3.2899×10^{-1}	0.9991
0.4	$\sqrt{2}/4$	1.2929×10^{-1}	–	5.0266×10^{-1}	–	$2.5969 \times 10^{+0}$	–
	$\sqrt{2}/8$	6.5242×10^{-2}	0.9867	2.5175×10^{-1}	0.9976	$1.3118 \times 10^{+0}$	0.9853
	$\sqrt{2}/16$	3.2698×10^{-2}	0.9966	1.2590×10^{-1}	0.9997	6.5757×10^{-1}	0.9963
	$\sqrt{2}/32$	1.6359×10^{-2}	0.9991	6.2955×10^{-2}	0.9999	3.2899×10^{-1}	0.9991

Example 2. *Consider the following three-term TFRD equation:*

$$\begin{cases} D_t^{\alpha_1} u(\mathbf{x},t) + D_t^{\alpha_2} u(\mathbf{x},t) + D_t^{\alpha_3} u(\mathbf{x},t) - \Delta u(\mathbf{x},t) + p(\mathbf{x})u(\mathbf{x},t) = f(\mathbf{x},t), & (\mathbf{x},t) \in \Omega \times J, \\ u\,(\mathbf{x},t) = 0, & (\mathbf{x},t) \in \partial\Omega \times \bar{J}, \\ u(\mathbf{x},0) = u_0(\mathbf{x}), & \mathbf{x} \in \bar{\Omega}, \end{cases} \tag{49}$$

where the spatial domain Ω, temporal domain J, coefficient $p(\mathbf{x})$, and initial data $u(\mathbf{x},0)$ are as in Example 1 and the source function f is taken by

$$\begin{aligned} f(\mathbf{x},t) = &\bigg(\frac{\Gamma(3+\alpha_1+\alpha_2+\alpha_3)}{\Gamma(3+\alpha_2+\alpha_3)} t^{2+\alpha_2+\alpha_3} + \frac{\Gamma(3+\alpha_1+\alpha_2+\alpha_3)}{\Gamma(3+\alpha_1+\alpha_3)} t^{2+\alpha_1+\alpha_3} \\ &+ \frac{\Gamma(3+\alpha_1+\alpha_2+\alpha_3)}{\Gamma(3+\alpha_1+\alpha_2)} t^{2+\alpha_1+\alpha_2} + \Big(2\pi^2 + p(\mathbf{x})\Big) t^{2+\alpha_1+\alpha_2+\alpha_3} \bigg) \sin(\pi x_1)\sin(\pi x_2). \end{aligned}$$

And we can also find the analytical solutions for variables u and $\boldsymbol{\lambda}$ as follows:

$$\begin{aligned} u(\mathbf{x},t) &= t^{2+\alpha_1+\alpha_2+\alpha_3} \sin(\pi x_1)\sin(\pi x_2), \\ \boldsymbol{\lambda}(\mathbf{x},t) &= -\pi t^{2+\alpha_1+\alpha_2+\alpha_3} (\cos(\pi x_1)\sin(\pi x_2), \sin(\pi x_1)\cos(\pi x_2)). \end{aligned}$$

In this example, since the Equation (49) contains three Caputo time-fractional derivative terms, we specifically take the fractional parameters $\alpha_1 = 0.9, 0.7, 0.5$ and $(\alpha_2, \alpha_3) = (0.4, 0.2), (0.3, 0.1)$. From Theorem 3, we also point out that the convergence rates are only related to the maximum fractional parameter α_1. In Tables 7–9, for different $N = 5, 8, 10, 16$, we give the error results and convergence rates for the MFE scheme (5)–(6), where the spatial grid sizes are also taken as $h = \sqrt{2}/N^{2-\alpha_1}$. We can also see that the convergence rates in the temporal direction for u (in the discrete $L^\infty(L^2(\Omega))$ norm) and $\boldsymbol{\lambda}$ (in the discrete $L^\infty\Big(\big(L^2(\Omega)\big)^2\Big)$ and $L^\infty(\mathbf{H}(\text{div},\Omega))$ norms) are close to $2-\alpha_1$. Furthermore, in Tables 10–12, we also fix $N = 100$ and take $h = \sqrt{2}/4, \sqrt{2}/8, \sqrt{2}/16, \sqrt{2}/32$, give the error results and convergence rates, and see that the convergence rates in the spatial direction for u and λ in the above corresponding discrete norms are also close to 1.

Table 7. Numerical results with $h \approx \sqrt{2}/N^{2-\alpha_1}$ and $\alpha_1 = 0.9$ in Example 2.

α_2	α_3	N	u-$\hat{L}^\infty(L^2)$	Rates	λ-$\hat{L}^\infty((L^2)^2)$	Rates	λ-$\hat{L}^\infty(H(\text{div}))$	Rates
0.4	0.2	5	8.7400×10^{-2}	–	3.3868×10^{-1}	–	$1.7681 \times 10^{+0}$	–
		8	5.2741×10^{-2}	1.0747	2.0392×10^{-1}	1.0795	$1.0648 \times 10^{+0}$	1.0790
		10	4.0650×10^{-2}	1.1670	1.5711×10^{-1}	1.1686	8.2004×10^{-1}	1.1704
		16	2.4067×10^{-2}	1.1152	9.2988×10^{-2}	1.1159	4.8494×10^{-1}	1.1177
0.3	0.1	5	8.7138×10^{-2}	–	3.3735×10^{-1}	–	$1.7608 \times 10^{+0}$	–
		8	5.2567×10^{-2}	1.0753	2.0303×10^{-1}	1.0804	$1.0605 \times 10^{+0}$	1.0788
		10	4.0507×10^{-2}	1.1679	1.5638×10^{-1}	1.1698	8.1673×10^{-1}	1.1706
		16	2.3980×10^{-2}	1.1154	9.2546×10^{-2}	1.1162	4.8304×10^{-1}	1.1175

Table 8. Numerical results with $h \approx \sqrt{2}/N^{2-\alpha_1}$ and $\alpha_1 = 0.7$ in Example 2.

α_2	α_3	N	u-$\hat{L}^\infty(L^2)$	Rates	λ-$\hat{L}^\infty((L^2)^2)$	Rates	λ-$\hat{L}^\infty(H(\text{div}))$	Rates
0.4	0.2	5	6.5465×10^{-2}	–	2.5288×10^{-1}	–	$1.3208 \times 10^{+0}$	–
		8	3.5052×10^{-2}	1.3291	1.3521×10^{-1}	1.3321	7.0629×10^{-1}	1.3319
		10	2.6310×10^{-2}	1.2857	1.0146×10^{-1}	1.2868	5.2989×10^{-1}	1.2877
		16	1.4234×10^{-2}	1.3070	5.4886×10^{-2}	1.3073	2.8651×10^{-1}	1.3083
0.3	0.1	5	6.5343×10^{-2}	–	2.5225×10^{-1}	–	$1.3173 \times 10^{+0}$	–
		8	3.4972×10^{-2}	1.3300	1.3479×10^{-1}	1.3334	7.0434×10^{-1}	1.3321
		10	2.6247×10^{-2}	1.2860	1.0114×10^{-1}	1.2872	5.2845×10^{-1}	1.2876
		16	1.4200×10^{-2}	1.3071	5.4707×10^{-2}	1.3075	2.8575×10^{-1}	1.3082

Table 9. Numerical results with $h \approx \sqrt{2}/N^{2-\alpha_1}$ and $\alpha_1 = 0.5$ in Example 2.

α_2	α_3	N	u-$\hat{L}^\infty(L^2)$	Rates	λ-$\hat{L}^\infty((L^2)^2)$	Rates	λ-$\hat{L}^\infty(H(\text{div}))$	Rates
0.4	0.2	5	4.7614×10^{-2}	–	1.8360×10^{-1}	–	9.5930×10^{-1}	–
		8	2.2831×10^{-2}	1.5638	8.7972×10^{-2}	1.5654	4.5961×10^{-1}	1.5656
		10	1.6417×10^{-2}	1.4779	6.3255×10^{-2}	1.4782	3.3041×10^{-1}	1.4790
		16	8.2126×10^{-3}	1.4738	3.1641×10^{-2}	1.4738	1.6523×10^{-1}	1.4745
0.3	0.1	5	4.7550×10^{-2}	–	1.8327×10^{-1}	–	9.5743×10^{-1}	–
		8	2.2788×10^{-2}	1.5650	8.7752×10^{-2}	1.5669	4.5859×10^{-1}	1.5661
		10	1.6385×10^{-2}	1.4783	6.3087×10^{-2}	1.4788	3.2967×10^{-1}	1.4791
		16	8.1952×10^{-3}	1.4740	3.1552×10^{-2}	1.4742	1.6486×10^{-1}	1.4745

Table 10. Numerical results with $\tau = T/N = 1/100$ and $\alpha_1 = 0.9$ in Example 2.

α_2	α_3	h	u-$\hat{L}^\infty(L^2)$	Rates	λ-$\hat{L}^\infty((L^2)^2)$	Rates	λ-$\hat{L}^\infty(H(\text{div}))$	Rates
0.4	0.2	$\sqrt{2}/4$	1.2924×10^{-1}	–	5.0230×10^{-1}	–	$2.5973 \times 10^{+0}$	–
		$\sqrt{2}/8$	6.5229×10^{-2}	0.9865	2.5168×10^{-1}	0.9970	$1.3119 \times 10^{+0}$	0.9853
		$\sqrt{2}/16$	3.2696×10^{-2}	0.9964	1.2589×10^{-1}	0.9994	6.5768×10^{-1}	0.9962
		$\sqrt{2}/32$	1.6364×10^{-2}	0.9986	6.2979×10^{-2}	0.9992	3.2916×10^{-1}	0.9986
0.3	0.1	$\sqrt{2}/4$	1.2925×10^{-1}	–	5.0236×10^{-1}	–	$2.5972 \times 10^{+0}$	–
		$\sqrt{2}/8$	6.5231×10^{-2}	0.9865	2.5169×10^{-1}	0.9971	$1.3119 \times 10^{+0}$	0.9853
		$\sqrt{2}/16$	3.2696×10^{-2}	0.9964	1.2589×10^{-1}	0.9995	6.5765×10^{-1}	0.9962
		$\sqrt{2}/32$	1.6362×10^{-2}	0.9988	6.2971×10^{-2}	0.9994	3.2912×10^{-1}	0.9987

Table 11. Numerical results with $\tau = T/N = 1/100$ and $\alpha_1 = 0.7$ in Example 2.

α_2	α_3	h	u–$\hat{L}^\infty(L^2)$	Rates	λ–$\hat{L}^\infty((L^2)^2)$	Rates	λ–$\hat{L}^\infty(H(\text{div}))$	Rates
0.4	0.2	$\sqrt{2}/4$	1.2926×10^{-1}	–	5.0244×10^{-1}	–	$2.5971 \times 10^{+0}$	–
		$\sqrt{2}/8$	6.5237×10^{-2}	0.9866	2.5172×10^{-1}	0.9972	$1.3118 \times 10^{+0}$	0.9853
		$\sqrt{2}/16$	3.2697×10^{-2}	0.9965	1.2590×10^{-1}	0.9996	6.5758×10^{-1}	0.9963
		$\sqrt{2}/32$	1.6359×10^{-2}	0.9991	6.2954×10^{-2}	0.9999	3.2901×10^{-1}	0.9990
0.3	0.1	$\sqrt{2}/4$	1.2927×10^{-1}	–	5.0249×10^{-1}	–	$2.5970 \times 10^{+0}$	–
		$\sqrt{2}/8$	6.5238×10^{-2}	0.9866	2.5172×10^{-1}	0.9973	$1.3118 \times 10^{+0}$	0.9853
		$\sqrt{2}/16$	3.2697×10^{-2}	0.9965	1.2590×10^{-1}	0.9996	6.5758×10^{-1}	0.9963
		$\sqrt{2}/32$	1.6359×10^{-2}	0.9991	6.2954×10^{-2}	0.9999	3.2900×10^{-1}	0.9991

Table 12. Numerical results with $\tau = T/N = 1/100$ and $\alpha_1 = 0.5$ in Example 2.

α_2	α_3	h	u–$\hat{L}^\infty(L^2)$	Rates	λ–$\hat{L}^\infty((L^2)^2)$	Rates	λ–$\hat{L}^\infty(H(\text{div}))$	Rates
0.4	0.2	$\sqrt{2}/4$	1.2927×10^{-1}	–	5.0252×10^{-1}	–	$2.5970 \times 10^{+0}$	–
		$\sqrt{2}/8$	6.5240×10^{-2}	0.9866	2.5173×10^{-1}	0.9973	$1.3118 \times 10^{+0}$	0.9853
		$\sqrt{2}/16$	3.2698×10^{-2}	0.9965	1.2590×10^{-1}	0.9996	6.5757×10^{-1}	0.9963
		$\sqrt{2}/32$	1.6359×10^{-2}	0.9991	6.2955×10^{-2}	0.9999	3.2899×10^{-1}	0.9991
0.3	0.1	$\sqrt{2}/4$	1.2928×10^{-1}	–	5.0257×10^{-1}	–	$2.5970 \times 10^{+0}$	–
		$\sqrt{2}/8$	6.5241×10^{-2}	0.9866	2.5174×10^{-1}	0.9974	$1.3118 \times 10^{+0}$	0.9853
		$\sqrt{2}/16$	3.2698×10^{-2}	0.9966	1.2590×10^{-1}	0.9996	6.5757×10^{-1}	0.9963
		$\sqrt{2}/32$	1.6359×10^{-2}	0.9991	6.2955×10^{-2}	0.9999	3.2899×10^{-1}	0.9991

Based on the numerical results in Tables 1–12 obtained from the above two test examples, we can see that the convergence rates in the spatial and temporal directions for u (in the discrete $L^\infty(L^2(\Omega))$ norm) and $\boldsymbol{\lambda}$ (in the discrete $L^\infty\left((L^2(\Omega))^2\right)$ norm) are in agreement with the theoretical results in Theorem 3, and those for $\boldsymbol{\lambda}$ (in the discrete $L^\infty(\boldsymbol{H}(\text{div},\Omega))$ norm) are higher than the theoretical result. These results fully demonstrate that the proposed MFE method for the multi-term TFRD equations is effective.

Example 3. *Consider the two-term TFRD equation in Example 1 with weak regularity solutions near the initial time $t = 0$, where the source function f is taken by*

$$
\begin{aligned}
f(\boldsymbol{x},t) =& \Bigg(\frac{2}{\Gamma(3-\alpha_1)} t^{2-\alpha_1} + \Gamma(1+\alpha_1) + \frac{\Gamma(2+\alpha_2)}{\Gamma(2+\alpha_2-\alpha_1)} t^{1+\alpha_2-\alpha_1} \\
&+ \frac{2}{\Gamma(3-\alpha_2)} t^{2-\alpha_2} + \frac{\Gamma(1+\alpha_1)}{\Gamma(1+\alpha_1-\alpha_2)} t^{\alpha_1-\alpha_2} + \Gamma(2+\alpha_2) t \\
&+ \left(2\pi^2 + p(\boldsymbol{x})\right)\left(t^2 + t^{\alpha_1} + t^{1+\alpha_2}\right) \Bigg) \sin(\pi x_1)\sin(\pi x_2).
\end{aligned}
$$

And we can also find the analytical solutions for variables u and $\boldsymbol{\lambda}$ as follows:

$$
\begin{aligned}
u(\boldsymbol{x},t) &= \left(t^2 + t^{\alpha_1} + t^{1+\alpha_2}\right)\sin(\pi x_1)\sin(\pi x_2), \\
\boldsymbol{\lambda}(\boldsymbol{x},t) &= -\pi\left(t^2 + t^{\alpha_1} + t^{1+\alpha_2}\right)(\cos(\pi x_1)\sin(\pi x_2), \sin(\pi x_1)\cos(\pi x_2)).
\end{aligned}
$$

In this example, we will select the graded mesh to discretize the interval $[0,T]$ and set $t_n = T(n/N)^\gamma$, for $n = 0,1,2,\cdots,N$, where constant $\gamma \geq 1$ is the temporal graded mesh parameter. The ideal optimal error estimates for u (in the discrete $L^\infty(L^2(\Omega))$ norm) and $\boldsymbol{\lambda}$ (in the discrete $L^\infty\left((L^2(\Omega))^2\right)$ and $L^\infty(\boldsymbol{H}(\text{div},\Omega))$ norms) should be $O\left(N^{-\min\{\gamma\alpha_1, 2-\alpha_1\}} + h\right)$. Here, we will mainly test the convergence rates in the temporal direction with the graded mesh parameter $\gamma = 1$ and $(2-\alpha_1)/\alpha_1$. We first conduct numerical experiments with

$\gamma = 1$. Then, the optimal convergence rate in the temporal direction is α_1. For fractional parameters $\alpha_1 = 0.9, 0.7, 0.5$ and $\alpha_2 = 0.1, 0.4$, we take the time mesh parameter $N = 20, 40, 80, 160$ and special spatial grid parameters: (i) when $\alpha_1 = 0.9$, take $h \approx 2\sqrt{2}/N^{\alpha_1}$; (ii) when $\alpha_1 = 0.7$, take $h \approx \sqrt{2}/N^{\alpha_1}$; (iii) when $\alpha_1 = 0.5$, take $h \approx \sqrt{2}/(2N^{\alpha_1})$. Then, we give the numerical results in Tables 13–15, which show that the convergence rates in the temporal direction for u (in the discrete $L^\infty(L^2(\Omega))$ norm) and λ (in the discrete $L^\infty\left((L^2(\Omega))^2\right)$ and $L^\infty(\boldsymbol{H}(\mathrm{div}, \Omega))$ norms) are close to α_1.

Table 13. Numerical results with $\alpha_1 = 0.9$ and graded mesh parameter $\gamma = 1$ in Example 3.

α_2	N	u-$\hat{L}^\infty(L^2)$	Rates	λ-$\hat{L}^\infty((L^2)^2)$	Rates	λ-$\hat{L}^\infty(H(\mathrm{div}))$	Rates
0.1	20	1.9572×10^{-1}	–	7.5520×10^{-1}	–	$3.9354 \times 10^{+0}$	–
	40	1.1208×10^{-1}	0.8042	4.3165×10^{-1}	0.8070	$2.2539 \times 10^{+0}$	0.8041
	80	6.0396×10^{-2}	0.8920	2.3245×10^{-1}	0.8930	$1.2146 \times 10^{+0}$	0.8920
	160	3.2722×10^{-2}	0.8842	1.2591×10^{-1}	0.8845	6.5806×10^{-1}	0.8842
0.4	20	1.9571×10^{-1}	–	7.5516×10^{-1}	–	$3.9354 \times 10^{+0}$	–
	40	1.1208×10^{-1}	0.8042	4.3164×10^{-1}	0.8069	$2.2540 \times 10^{+0}$	0.8041
	80	6.0396×10^{-2}	0.8920	2.3244×10^{-1}	0.8929	$1.2146 \times 10^{+0}$	0.8920
	160	3.2722×10^{-2}	0.8842	1.2591×10^{-1}	0.8845	6.5806×10^{-1}	0.8842

Table 14. Numerical results with $\alpha_1 = 0.7$ and graded mesh parameter $\gamma = 1$ in Example 3.

α_2	N	u-$\hat{L}^\infty(L^2)$	Rates	λ-$\hat{L}^\infty((L^2)^2)$	Rates	λ-$\hat{L}^\infty(H(\mathrm{div}))$	Rates
0.1	20	1.9573×10^{-1}	–	7.5526×10^{-1}	–	$3.9353 \times 10^{+0}$	–
	40	1.2068×10^{-1}	0.6976	4.6487×10^{-1}	0.7002	$2.4269 \times 10^{+0}$	0.6974
	80	7.4765×10^{-2}	0.6908	2.8779×10^{-1}	0.6918	$1.5035 \times 10^{+0}$	0.6907
	160	4.4872×10^{-2}	0.7365	1.7268×10^{-1}	0.7369	9.0241×10^{-1}	0.7365
0.4	20	1.9572×10^{-1}	–	7.5523×10^{-1}	–	$3.9354 \times 10^{+0}$	–
	40	1.2068×10^{-1}	0.6976	4.6486×10^{-1}	0.7001	$2.4269 \times 10^{+0}$	0.6974
	80	7.4765×10^{-2}	0.6908	2.8779×10^{-1}	0.6918	$1.5035 \times 10^{+0}$	0.6907
	160	4.4872×10^{-2}	0.7365	1.7268×10^{-1}	0.7369	9.0241×10^{-1}	0.7365

Table 15. Numerical results with $\alpha_1 = 0.5$ and graded mesh parameter $\gamma = 1$ in Example 3.

α_2	N	u-$\hat{L}^\infty(L^2)$	Rates	λ-$\hat{L}^\infty((L^2)^2)$	Rates	λ-$\hat{L}^\infty(H(\mathrm{div}))$	Rates
0.1	20	1.7410×10^{-1}	–	6.7141×10^{-1}	–	$3.5006 \times 10^{+0}$	–
	40	1.2069×10^{-1}	0.5286	4.6487×10^{-1}	0.5303	$2.4269 \times 10^{+0}$	0.5285
	80	8.7212×10^{-2}	0.4687	3.3576×10^{-1}	0.4694	$1.7538 \times 10^{+0}$	0.4686
	160	6.2812×10^{-2}	0.4735	2.4175×10^{-1}	0.4739	$1.2632 \times 10^{+0}$	0.4735
0.4	20	1.7410×10^{-1}	–	6.7139×10^{-1}	–	$3.5006 \times 10^{+0}$	–
	40	1.2069×10^{-1}	0.5286	4.6487×10^{-1}	0.5303	$2.4269 \times 10^{+0}$	0.5285
	80	8.7212×10^{-2}	0.4687	3.3576×10^{-1}	0.4694	$1.7538 \times 10^{+0}$	0.4686
	160	6.2812×10^{-2}	0.4735	2.4175×10^{-1}	0.4739	$1.2632 \times 10^{+0}$	0.4735

Next, we conduct numerical experiments with $\gamma = (2 - \alpha_1)/\alpha_1$. Then, the optimal convergence rate is $2 - \alpha_1$. We take the time mesh parameter $N = 5, 8, 10, 16$ and the spatial grid parameter $h = \sqrt{2}/N^{2-\alpha_1}$. Then, we give the numerical results in Tables 16–18 and find that the convergence rates in the temporal direction for u (in the discrete $L^\infty(L^2(\Omega))$ norm) and λ (in the discrete $L^\infty\left((L^2(\Omega))^2\right)$ and $L^\infty(\boldsymbol{H}(\mathrm{div}, \Omega))$ norms) are close to $2 - \alpha_1$. Based on the above discussion, we know that the MFE scheme (5)–(6) with the temporal graded mesh for solving the multi-term TFRD equations with the initial layer is also feasible and effective.

Table 16. Numerical results with $\alpha_1 = 0.9$ and graded mesh parameter $\gamma = \frac{2-\alpha_1}{\alpha_1}$ in Example 3.

α_2	N	u-$\hat{L}^\infty(L^2)$	Rates	λ-$\hat{L}^\infty((L^2)^2)$	Rates	λ-$\hat{L}^\infty(H(\text{div}))$	Rates
0.1	5	2.6024×10^{-1}	–	$1.0062 \times 10^{+0}$	–	$5.2342 \times 10^{+0}$	–
	8	1.5673×10^{-1}	1.0789	6.0412×10^{-1}	1.0855	$3.1526 \times 10^{+0}$	1.0787
	10	1.2067×10^{-1}	1.1718	4.6479×10^{-1}	1.1750	$2.4273 \times 10^{+0}$	1.1718
	16	7.1368×10^{-2}	1.1175	2.7470×10^{-1}	1.1189	$1.4355 \times 10^{+0}$	1.1176
0.4	5	2.6021×10^{-1}	–	$1.0060 \times 10^{+0}$	–	$5.2351 \times 10^{+0}$	–
	8	1.5673×10^{-1}	1.0787	6.0410×10^{-1}	1.0852	$3.1531 \times 10^{+0}$	1.0787
	10	1.2067×10^{-1}	1.1716	4.6479×10^{-1}	1.1748	$2.4276 \times 10^{+0}$	1.1718
	16	7.1372×10^{-2}	1.1174	2.7472×10^{-1}	1.1188	$1.4357 \times 10^{+0}$	1.1176

Table 17. Numerical results with $\alpha_1 = 0.7$ and graded mesh parameter $\gamma = \frac{2-\alpha_1}{\alpha_1}$ in Example 3.

α_2	N	u-$\hat{L}^\infty(L^2)$	Rates	λ-$\hat{L}^\infty((L^2)^2)$	Rates	λ-$\hat{L}^\infty(H(\text{div}))$	Rates
0.1	5	1.9568×10^{-1}	–	7.5499×10^{-1}	–	$3.9360 \times 10^{+0}$	–
	8	1.0462×10^{-1}	1.3323	4.0284×10^{-1}	1.3365	$2.1043 \times 10^{+0}$	1.3323
	10	7.8497×10^{-2}	1.2872	3.0217×10^{-1}	1.2887	$1.5789 \times 10^{+0}$	1.2873
	16	4.2450×10^{-2}	1.3079	1.6336×10^{-1}	1.3086	8.5380×10^{-1}	1.3081
0.4	5	1.9566×10^{-1}	–	7.5492×10^{-1}	–	$3.9370 \times 10^{+0}$	–
	8	1.0462×10^{-1}	1.3320	4.0289×10^{-1}	1.3361	$2.1049 \times 10^{+0}$	1.3323
	10	7.8506×10^{-2}	1.2870	3.0221×10^{-1}	1.2885	$1.5793 \times 10^{+0}$	1.2874
	16	4.2457×10^{-2}	1.3078	1.6339×10^{-1}	1.3084	8.5400×10^{-1}	1.3081

Table 18. Numerical results with $\alpha_1 = 0.5$ and graded mesh parameter $\gamma = \frac{2-\alpha_1}{\alpha_1}$ in Example 3.

α_2	N	u-$\hat{L}^\infty(L^2)$	Rates	λ-$\hat{L}^\infty((L^2)^2)$	Rates	λ-$\hat{L}^\infty(H(\text{div}))$	Rates
0.1	5	1.4253×10^{-1}	–	5.4923×10^{-1}	–	$2.8673 \times 10^{+0}$	–
	8	6.8272×10^{-2}	1.5661	2.6278×10^{-1}	1.5685	$1.3733 \times 10^{+0}$	1.5663
	10	4.9083×10^{-2}	1.4788	1.8890×10^{-1}	1.4794	9.8727×10^{-1}	1.4790
	16	2.4548×10^{-2}	1.4742	9.4464×10^{-2}	1.4744	4.9373×10^{-1}	1.4744
0.4	5	1.4255×10^{-1}	–	5.4933×10^{-1}	–	$2.8688 \times 10^{+0}$	–
	8	6.8306×10^{-2}	1.5654	2.6296×10^{-1}	1.5675	$1.3743 \times 10^{+0}$	1.5659
	10	4.9113×10^{-2}	1.4783	1.8905×10^{-1}	1.4788	9.8804×10^{-1}	1.4787
	16	2.4567×10^{-2}	1.4739	9.4561×10^{-2}	1.4740	4.9416×10^{-1}	1.4742

7. Conclusions

This work presents a Raviart–Thomas MFE method for solving the multi-term TFRD equations with variable coefficients by using the well-known $L1$ formula. The existence, uniqueness, and unconditional stability of the discrete solution are discussed, and the optimal a priori error estimates for u (in the discrete $L^\infty(L^2(\Omega))$ norm) and λ (in the discrete $L^\infty\left(\left(L^2(\Omega)\right)^2\right)$ norm) and the suboptimal a priori error estimate for λ (in the discrete $L^\infty(\boldsymbol{H}(\text{div},\Omega))$ norm) are obtained in this work. In addition, some numerical results are given to demonstrate the effectiveness of the proposed MFE method. In future research, we will try to give theoretical analysis for the MFE method with the temporal graded mesh to solve some FPDEs with the initial layer at $t = 0$ and apply the MFE method to solve more FPDEs in scientific and engineering fields.

Author Contributions: Conceptualization, J.Z. and S.D.; methodology, Z.F.; software, Z.F.; validation, J.Z., S.D. and Z.F.; formal analysis, J.Z. and S.D.; writing—original draft preparation, J.Z. and S.D.; writing—review and editing, J.Z. and Z.F.; funding acquisition, J.Z. and Z.F. All authors have read and agreed to the published version of the manuscript.

Funding: This work was supported by the National Natural Science Foundation of China (11701299), the Scientific Research Projection of Higher Schools of the Inner Mongolia Autonomous Region

(NJZY23055), and the Central Government Guided Local Science and Technology Development Fund Project of China (2022ZY0175).

Data Availability Statement: Data are contained within the article.

Acknowledgments: The authors are very grateful to the editors and reviewers for their helpful comments and suggestions on improving this work.

Conflicts of Interest: The authors declares no conflicts of interest.

References

1. Ross, B. *Fractional Calculus and Its Applications*; Lecture Notes in Mathematics; Springer: Berlin/Heidelberg, Germany, 1975.
2. Hilfer, R. *Applications of Fractional Calculus in Physics*; World Scientific: Singapore, 2000.
3. Diethelm, K. *The Analysis of Fractional Differential Equations*; Springer: Berlin/Heidelberg, Germany, 2010.
4. Hendy, M.H.; Amin, M.M.; Ezzat, M.A. Two-dimensional problem for thermoviscoelastic materials with fractional order heat transfer. *J. Therm. Stress.* **2019**, *42*, 1298–1315. [CrossRef]
5. Zhou, J.C.; Salahshour, S.; Ahmadian, A.; Senu, N. Modeling the dynamics of COVID-19 using fractal-fractional operator with a case study. *Results Phys.* **2022**, *33*, 105103. [CrossRef] [PubMed]
6. Shymanskyi, V.; Sokolovskyy, I.; Sokolovskyy, Y.; Bubnyak, T. Variational Method for Solving the Time-Fractal Heat Conduction Problem in the Claydite-Block Construction. In *Advances in Computer Science for Engineering and Education. ICCSEEA 2022*; Hu, Z., Dychka, I., Petoukhov, S., He, M., Eds.; Lecture Notes on Data Engineering and Communications Technologies; Springer: Cham, Switzerland, 2022; Volume 134.
7. Bagley, R.L.; Torvik, R.J. On the appearance of the fractional derivative in the behaviour of real materials. *Appl. Mech.* **1984**, *51*, 294–298.
8. Ming, C.Y.; Liu, F.W.; Zheng, L.C.; Turner, I.; Anh, V. Analytical solutions of multi-term time fractional differential equations and application to unsteady flows of generalized viscoelastic fluid. *Comput. Math. Appl.* **2006**, *72*, 2084–2097. [CrossRef]
9. Fetecau, C.; Athar, M.; Fetecau, C. Unsteady flow of a generalized Maxwell fluid with fractional derivative due to a constantly accelerating plate. *Comput. Math. Appl.* **2009**, *57*, 596–603. [CrossRef]
10. Liu, F.W.; Meerschaert, M.M.; McGough, R.J.; Zhuang, P.H.; Liu, Q.X. Numerical methods for solving the multi-term time-fractional wave-diffusion equation. *Fract. Calc. Appl. Anal.* **2013**, *16*, 9–25. [CrossRef]
11. Dehghan, M.; Safarpoor, M.; Abbaszadeh, M. Two high-order numerical algorithms for solving the multi-term time fractional diffusion-wave equations. *J. Comput. Appl. Math.* **2015**, *290*, 174–195. [CrossRef]
12. Ren, J.C.; Sun, Z.Z. Efficient Numerical Solution of the Multi-Term Time Fractional Diffusion-Wave Equation. *East Asian J. Appl. Math.* **2015**, *5*, 1–28. [CrossRef]
13. Zheng, M.; Liu, F.W.; Anh, V.; Turner, I. A high-order spectral method for the multi-term time-fractional diffusion equations. *Appl. Math. Model.* **2016**, *40*, 4970–4985. [CrossRef]
14. Du, R.L.; Sun, Z.Z. Temporal second-order difference methods for solving multi-term time fractional mixed diffusion and wave equations. *Numer. Algorithms* **2021**, *88*, 191–226. [CrossRef]
15. Hendy, A.S.; Zaky, M.A. Graded mesh discretization for coupled system of nonlinear multi-term time-space fractional diffusion equations. *Eng. Comput.* **2022**, *38*, 1351–1363. [CrossRef]
16. Liu, Y.Q.; Yin, X.L.; Liu, F.W.; Xin, X.Y.; Shen, Y.F.; Feng, L.B. An alternating direction implicit Legendre spectral method for simulating a 2D multi-term time-fractional Oldroyd-B fluid type diffusion equation. *Comput. Math. Appl.* **2022**, *113*, 160–173. [CrossRef]
17. Wei, L.L.; Wang, H.H. Local discontinuous Galerkin method for multi-term variable-order time fractional diffusion equation. *Math. Comput. Simulat.* **2023**, *203*, 685–698. [CrossRef]
18. She, M.F.; Li, D.F.; Sun, H.W. A transformed $L1$ method for solving the multi-term time-fractional diffusion problem. *Math. Comput. Simulat.* **2022**, *193*, 584–606. [CrossRef]
19. Jin, B.T.; Lazarov, R.; Liu, Y.K.; Zhou, Z. The Galerkin finite element method for a multi-term time-fractional diffusion equation. *J. Comput. Phys.* **2015**, *281*, 825–843. [CrossRef]
20. Li, M.; Huang, C.M.; Jiang, F.Z. Galerkin finite element method for higher dimensional multi-term fractional diffusion equation on non-uniform meshes. *Appl. Anal.* **2017**, *96*, 1269–1284. [CrossRef]
21. Zhou, J.; Xu, D.; Chen, H.B. A Weak Galerkin Finite Element Method for Multi-Term Time-Fractional Diffusion Equations. *East Asian J. Appl. Math.* **2018**, *8*, 181–193. [CrossRef]
22. Bu, W.P.; Shu, S.; Xue, X.Q.; Xiao, A.G.; Zeng, W. Space-time finite element method for the multi-term time-space fractional diffusion equation on a two-dimensional domain. *Comput. Math. Appl.* **2019**, *78*, 1367–1379. [CrossRef]
23. Feng, L.B.; Liu, F.W.; Turner, I. Finite difference/finite element method for a novel 2D multi-term time-fractional mixed sub-diffusion and diffusion-wave equation on convex domains. *Commun. Nonlinear Sci. Numer. Simulat.* **2019**, *70*, 354–371. [CrossRef]
24. Meng, X.Y.; Stynes, M. Barrier function local and global analysis of an $L1$ finite element method for a multiterm time-fractional initial-boundary value problem. *J. Sci. Comput.* **2020**, *84*, 5. [CrossRef]

25. Yin, B.L.; Liu, Y.; Li, H.; Zeng, F.H. A class of efficient time-stepping methods for multi-term time-fractional reaction-diffusion-wave equations. *Appl. Numer. Math.* **2021**, *165*, 56–82. [CrossRef]
26. Huang, C.B.; Stynes, M.; Chen, H. An α-robust finite element method for a multi-term time-fractional diffusion problem. *J. Comput. Appl. Math.* **2021**, *389*, 113334. [CrossRef]
27. Liu, H.; Zheng, X.C.; Fu, H.F. Analysis of a multi-term variable-order time-fractional diffusion equation and its Galerkin finite element approximation. *J. Comput. Math.* **2022**, *40*, 814–834. [CrossRef]
28. Zhao, Y.M.; Chen, P.; Bu, W.P.; Liu, X.T.; Tang, Y.F. Two mixed finite element methods for time-fractional diffusion equations. *J. Sci. Comput.* **2017**, *70*, 407–428. [CrossRef]
29. Liu, Y.; Fang, Z.C.; Li, H.; He, S. A mixed finite element method for a time-fractional fourth-order partial differential equation. *Appl. Math. Comput.* **2014**, *243*, 703–717. [CrossRef]
30. Li, X.C.; Yang, X.Y.; Zhang, Y.H. Error estimates of mixed finite element methods for time-fractional Navier-Stokes equations. *J. Sci. Comput.* **2017**, *70*, 500–515. [CrossRef]
31. Liu, Y.; Du, Y.W.; Li, H.; Li, J.C.; He, S. A two-grid mixed finite element method for a nonlinear fourth-order reaction-diffusion problem with time-fractional derivative. *Comput. Math. Appl.* **2015**, *70*, 2474–2492. [CrossRef]
32. Li, X.W.; Tang, Y.L. Interpolated coefficient mixed finite elements for semilinear time fractional diffusion equations. *Fractal Fract.* **2023**, *7*, 482. [CrossRef]
33. Shi, Z.G.; Zhao, Y.M.; Tang, Y.F.; Wang, F.L.; Shi, Y.H. Superconvergence analysis of an H^1-Galerkin mixed finite element method for two-dimensional multi-term time fractional diffusion equations. *Int. J. Comput. Math.* **2018**, *95*, 1845–1857 [CrossRef]
34. Li, M.; Huang, C.M.; Ming, W.Y. Mixed finite-element method for multi-term time-fractional diffusion and diffusion-wave equations. *Comp. Appl. Math.* **2018**, *37*, 2309–2334. [CrossRef]
35. Cao, F.F.; Zhao, Y.M.; Wang, F.L.; Shi, Y.H.; Yao, C.H. Nonconforming mixed FEM analysis for multi-term time-fractional mixed sub-diffusion and diffusion-wave equation with time-space coupled derivative. *Adv. Appl. Math. Mech.* **2023**, *15*, 322–358. [CrossRef]
36. Sun, Z.Z.; Wu, X.N. A fully discrete scheme for a diffusion-wave system. *Appl. Numer. Math.* **2006**, *56*, 193–209. [CrossRef]
37. Lin, Y.M.; Xu, C.J. Finite difference/spectral approximations for the time-fractional diffusion equation. *J. Comput. Phys.* **2007**, *225*, 1533–1552. [CrossRef]
38. Sinha, R.K.; Ewing, R.E.; Lazarov, R.D. Mixed finite element approximations of parabolic integro-differential equations with nonsmooth initial data. *SIAM J. Numer. Anal.* **2009**, *47*, 3269–3292. [CrossRef]
39. Fang, Z.C.; Zhao, J.; Li, H.; Liu, Y. Finite volume element methods for two-dimensional time fractional reaction-diffusion equations on triangular grids. *Appl. Anal.* **2023**, *102*, 2248–2270. [CrossRef]
40. Douglas, J.; Roberts, J.E. Global Estimates for Mixed Methods for Second Order Elliptic Equations. *Math. Comput.* **1985**, *44*, 39–52. [CrossRef]
41. Durán, R.G. *Mixed Finite Element Methods*; Lecture Notes in Mathematics; Springer: Berlin/Heidelberg, Germany, 2008; Volume 1939.
42. Stynes, M.; O'Riordan, E.; Gracia, J.L. Error analysis of a finite difference method on graded meshes for a time-fractional diffusion equation. *SIAM J. Numer. Anal.* **2017**, *55*, 1057–1079. [CrossRef]
43. Jin, B.T.; Li, B.Y.; Zhou, Z. Numerical analysis of nonlinear subdiffusion equations. *SIAM J. Numer. Anal.* **2018**, *55*, 1–23. [CrossRef]
44. Liao, H.L.; Li, D.F.; Zhang, J.W. Sharp error estimate of the nonuniform $L1$ formula for linear reaction-subdiffusion equations. *SIAM J. Numer. Anal.* **2018**, *56*, 1112–1133. [CrossRef]
45. Ren, J.C.; Liao, H.L.; Zhang, Z.M. Superconvergence error estimate of a finite element method on nonuniform time meshes for reaction-subdiffusion equations. *J. Sci. Comput.* **2020**, *84*, 38. [CrossRef]
46. Huang, C.B.; Stynes, M. Optimal H^1 spatial convergence of a fully discrete finite element method for the time-fractional Allen-Cahn equation. *Adv. Comput. Math.* **2020**, *46*, 63. [CrossRef]

fractal and fractional

MDPI

Article

Fractional Second-Grade Fluid Flow over a Semi-Infinite Plate by Constructing the Absorbing Boundary Condition

Jingyu Yang [1], Lin Liu [1,*], Siyu Chen [1], Libo Feng [2] and Chiyu Xie [3]

[1] School of Mathematics and Physics, University of Science and Technology Beijing, Beijing 100083, China; yjy20030104@163.com (J.Y.); csy15110422531@163.com (S.C.)
[2] School of Mathematical Sciences, Queensland University of Technology, GPO Box 2434, Brisbane, QLD 4001, Australia; l6.feng@qut.edu.au
[3] School of Astronautics, Beihang University, Beijing 100191, China; chiyuxie@ustb.edu.cn
* Correspondence: liulin1020@126.com

Abstract: The modified second-grade fluid flow across a plate of semi-infinite extent, which is initiated by the plate's movement, is considered herein. The relaxation parameters and fractional parameters are introduced to express the generalized constitutive relation. A convolution-based absorbing boundary condition (ABC) is developed based on the artificial boundary method (ABM), addressing issues related to the semi-infinite boundary. We adopt the finite difference method (FDM) for deriving the numerical solution by employing the L1 scheme to approximate the fractional derivative. To confirm the precision of this method, a source term is added to establish an exact solution for verification purposes. A comparative evaluation of the ABC versus the direct truncated boundary condition (DTBC) is conducted, with their effectiveness and soundness being visually scrutinized and assessed. This study investigates the impact of the motion of plates at different fluid flow velocities, focusing on the effects of dynamic elements influencing flow mechanisms and velocity. This research's primary conclusion is that a higher fractional parameter correlates with the fluid flow. As relaxation parameters decrease, the delay effect intensifies and the fluid velocity decreases.

Keywords: absorbing boundary condition; second-grade fluid; fractional derivative; finite difference method

Citation: Yang, J.; Liu, L.; Chen, S.; Feng, L.; Xie, C. Fractional Second-Grade Fluid Flow over a Semi-Infinite Plate by Constructing the Absorbing Boundary Condition. *Fractal Fract.* **2024**, *8*, 309. https://doi.org/10.3390/fractalfract8060309

Academic Editor: Riccardo Caponetto

Received: 28 March 2024
Revised: 16 May 2024
Accepted: 21 May 2024
Published: 23 May 2024

1. Introduction

Viscoelastic fluid is widely used in petroleum exploitation, medicine, biology, and other fields [1–3]. Viscoelastic fluid combines the viscosity of the fluid and the elasticity of the elastic solid, which shows the complex behavior between them and makes it a typical non-Newtonian fluid. Viscoelastic fluid has a time-dependent stress–strain response, showing shear rate-dependent viscosity and elastic recovery after unloading. Second-grade fluids are a specialized subclass of non-Newtonian fluids [4,5] which exhibit unique viscoelastic characteristics. A feature of these fluids is that their velocity field includes two derivatives within the correlation linking stress to strain, whereas the velocity field of Newtonian fluid contains only a first-order derivative. To better understand the flow characteristics of second-grade fluids, scholars often choose a simple model for analysis, which is helpful to understand the motion nature of second-grade fluid more deeply and provide theoretical support for related industrial applications. Ho et al. [6] investigated the migration of rigid spheres in a flowing second-grade fluid by simplifying the fluid flow problem between parallel plane walls. Khan et al. [7] gave an exact analytical solution for the fluid flow of a generalized second-grade fluid between two walls perpendicular to a flat plate. Tassaddiq [8] analyzed the flow of a second-grade fluid by simplifying the problem to a non-constant flow of the incompressible fluid over an inclined plate with an inclined magnetic field. Further studies are cited in references [9,10]. This paper focuses on the behavior of generalized second-grade fluids over a semi-infinite plate which contains broad applications in various fields, as noted in references [11–14], including the flow of biological fluids, liquid metals and alloys, plasma, and blood.

The traditional second-grade fluid constitutive relation is grounded in integer differentials [15,16]. By considering the historical memory during their flow process, meaning that their current state is not solely determined by the present stress state but is also influenced by the stress history of the past, the fractional second-grade fluid is developed. The fractional differential equations [17–20] surpass the capabilities of integer-based models and offer a more precise capture of the historical dependency. Furthermore, the fractional derivatives possess a nonlocal nature; the derivative value at any given instant is intimately connected to the entire historical trajectory. This attribute equips fractional differential equations with the ability to describe physical processes exhibiting long-range or spatially extensive correlations [21]. The fractional operator addresses the significant inconsistencies between classical integer differential models and experimental outcomes, which compensates for their serious shortcomings. Consequently, it can achieve superior fitting results with experimental data by utilizing a reduced number of parameters. Sene et al. [22] used the double integration method to solve the fractional differential equation in the two-level fluid model. Chan et al. [23] carried out research into the behavior of deformable droplets when suspended within a second-grade fluid environment. Jiang et al. [24] made contributions to the field by examining the unsteady magnetohydrodynamic flow of a modified second-grade fluid by taking into account the Hall effect, as it pertains to movement through porous media. For more research on fractional derivatives in the field of viscoelastic fluids, the readers are encouraged to refer to [25–27]. In this work, we integrate the fundamental equation governing fluid dynamics with a model representing a modified second-grade fluid which includes the application of Riemann–Liouville time fractional derivative (RLTFD) operators.

Numerical simulations, when analyzing flow mechanisms, can demonstrate the flow properties of complex fluid models, eliminating the necessity of experimental procedures. Within this study, the formulation of the equation is achieved through the application of the FDM, as introduced in references [28,29]. The FDM is a numerical technique that divides a continuous domain into discrete grids and approximates the equation at these grid points. This method is adept at transforming a partial differential equation into a collection of either linear or nonlinear algebraic equation. This conversion facilitates straightforward numerical solutions that can be efficiently computed with the aid of computer systems. The FDM is not only intuitive in concept and easy to implement in programming, but also can flexibly handle problems with various boundary conditions and complex geometric shapes. Mei et al. [30] studied the flow around a sphere with free flow velocity oscillation and unsteady resistance under a finite Reynolds number by using FDM. Dennis et al. [31] proposed a new FDM for calculating steady flow in curved pipes. Kim et al. [32] developed a computational approach specifically tailored for the second-grade gradient theory of incompressible fluid dynamics. For more related research, one can refer to [33–35]. In addition, in the numerical simulation of the fractional derivatives, a variety of interpolation approximation methods has been reported, such as Grünwald–Letnikov definition [36], the L1 scheme [37], the $L2-1_\sigma$ scheme [38] and the fast algorithm [39].

The critical challenge in numerical methods pertains to the effective reduction of the effects that the unbounded regions exert on computational outcomes. Historically, the prevalent approach to dealing with the unbounded regions is through the employment of a direct truncation method (DTM), as documented in reference [40]. This approach involves selecting an exceedingly large value at the boundary, rather than attempting to define an actual infinite boundary condition. The DTM is less favorable for performing computer-based numerical simulations, notably for those that are extended over a long duration. In this paper, we employ an alternative strategy referred to as the artificial boundary method to establish the ABC. The core concept of this approach entails partitioning the unbounded domain into two distinct regions: a confined computational zone, which is the internal area, and an unbounded zone, representing the external area. The chosen boundary points need to ensure that the source term is compactly supported within a finite truncation area. Subsequently, the relation at the truncation point is determined through the use of the Laplace transform in the external region. The use of this approach retains the effect exerted by the external area on the truncated boundary, which ensures the restricted area has a more reasonable boundary condition. The mentioned approach has been utilized extensively in various studies. For instance, Barucq et al. [41] analyzed the Helmholtz equation numerically by using ABC. They conducted an in-depth analysis of the impact of various parameters on these numerical results. Li et al. [42] solved the effective second-grade Schrödinger equation with the FDM by constructing ABCs. Muhr et al. [43] applied the ABC to address the Westervelt wave equation related to sound velocity potential, and the effectiveness and efficiency were confirmed. Further information on ABC is available in [44–46].

This paper studies the impulsive motion of a modified second-grade fluid on a plate of semi-infinite extent by introducing an ABC. In Section 2, we propose the constitutive relation and formulate the governing equation. Section 3 details the development and construction of the ABC. In Section 4, the numerical framework for the governing equation, along with its initial and boundary conditions (IBCs), is presented. In Section 5, four numerical instances are provided, detailing numerical methods' efficacy, the dynamic parameters' impacts on velocity, the ABC, and a comparison of this study with other methods. Section 6 offers a summary of these findings.

2. The Construction of the Governing Equation

We are now focusing on an infinitely extended flat plate with a fluid situated on a single side of the plate. In the initial state, the plate and the fluid above it remain stationary. Suddenly, the flat plate initiates movement with a velocity denoted as $u_w(t)$. As Figure 1 shows, we use a coordinate system where the flow progresses in the x direction, while the y direction is perpendicular to the plate. The positions of the plate are denoted as $y = 0$. The fluid's velocity on the surface of the plate is equal to the plate's velocity, which considers the no-slip condition. The fluid is regarded as a generalized second-grade fluid with stresses that depend not only on the current strain rate, but also possibly on the historical one. We neglect sidewall effects because the plate is considered to be infinitely long, such that the flow problem is mainly controlled by the motion of the plate and the viscoelastic properties of the fluid. In addition, the fluid is considered incompressible. The flow is supposed to be laminar, and no turbulence occurs. Finally, the deformation of the fluid is small and can be analyzed using linear theory. Collectively, these stipulations and suppositions constitute the foundational framework upon which the problem's analysis and resolution are predicated.

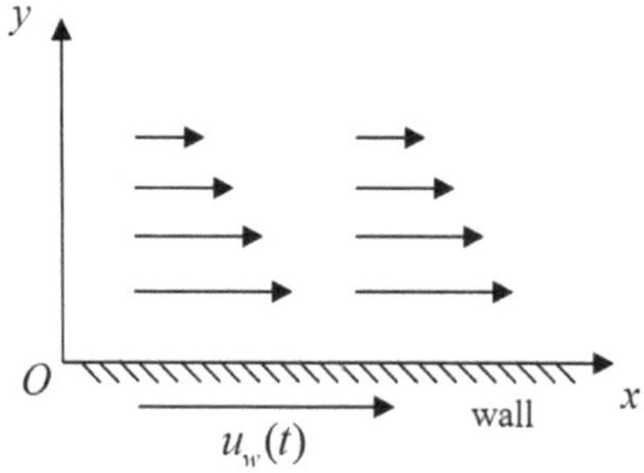

Figure 1. Schematic diagram of fractional second-grade fluid flow problem on a plate of semi-infinite extent.

The constitutive relation for a second-grade fluid satisfies [45]:

$$\mathbf{T} = -p\mathbf{I} + \mu\mathbf{A_1} + \alpha_1\mathbf{A_2} + \alpha_2\mathbf{A_1}^2 \tag{1}$$

in which the identity tensor is expressed as $\mathbf{I}$, the hydrostatic pressure is expressed as p, the stress tensor is expressed as $\mathbf{T}$, μ represents the coefficients of viscosity, α_1 and α_2 denote as the normal stress moduli, and the symbols $\mathbf{A_1}$ and $\mathbf{A_2}$ refer to the kinematic tensors [47] separately described by the following equation.

$$\mathbf{A}_1 = \nabla\mathbf{V} + (\nabla\mathbf{V})^{\mathrm{T}} \tag{2}$$

$$\mathbf{A}_2 = \frac{\partial \mathbf{A}_1}{\partial t} + \mathbf{A}_1(\nabla\mathbf{V}) + (\nabla\mathbf{V})^{\mathrm{T}}\mathbf{A}_1 \tag{3}$$

in which $\frac{\partial}{\partial t}$ is indicative of the material time partial derivative, $\mathbf{V}$ expresses the velocity vector field, and ∇ is the operator of the gradient. Suppose that the fluid described by Equation (1) conforms to thermodynamics, then all motions of the fluid satisfy the Clausius–Duhem inequality. Assume that the specific Helmholtz free energy of the fluid is minimized when the fluid is locally stationary [48]; then,

$$\mu \geqslant 0,\ \alpha_1 \geqslant 0,\ \alpha_1 + \alpha_2 = 0 \tag{4}$$

The fractional derivatives are able to describe the nonlocality and memory effects [17,21] of the generalized second-grade fluid. Since the fractional derivatives are nonlocal, this allows fractional differential equations to describe physical processes with long-range correlations or spatial-extent correlations. In general, the constitutive relation for the fractional second-grade fluid [49,50] also assumes the form (1), yet $\mathbf{A}_2$ satisfies

$$\mathbf{A}_2 = {}_0^{RL}\mathcal{D}_t^{\beta}\mathbf{A}_1 + \mathbf{A}_1(\nabla\mathbf{V}) + (\nabla\mathbf{V})^{\mathrm{T}}\mathbf{A}_1 \tag{5}$$

in which ${}_0^{RL}\mathcal{D}_t^{\beta}$ denotes the RLTFD operator [51].

The RLTFD operator ${}_0^{RL}\mathcal{D}_t^{\beta}$ satisfies

$${}_0^{RL}\mathcal{D}_t^{\beta}u(y,t) = \frac{\partial}{\partial t}\left(\frac{1}{\Gamma(1-\beta)}\int_0^t \frac{u(y,\tau)d\tau}{(t-\tau)^{\beta}}\right),\ 0<\beta<1 \tag{6}$$

where $\Gamma(\cdot)$ denotes the Gamma function. When $\beta = 1$, using the RLTFD's property, one obtains ${}_0^{RL}\mathcal{D}_t^{\beta}f(t) = {}_0^{RL}\mathcal{D}_t^{1}({}_0^{RL}\mathcal{D}_t^{0}f(t)) = \frac{\mathrm{d}f(t)}{\mathrm{d}t}$, namely, Equation (5) simplifies to Equation (3). When $\alpha_1 = 0$ and $\alpha_2 = 0$, the classical viscous Newtonian fluid is recovered [52,53]. It should be indicated that we have kept the same notation for the constant α_1 in (1) for the sake of simplicity, but it refers to a new material constant with the dimensions $kg \cdot m^{-1} \cdot s^{\beta-1}$. It reduces to $kg \cdot m^{-1}$ for $\beta \to 1$. For the other relations between (1) and (3), they remain formally unchanged under the above dimensional understanding [54].

Neglecting the external forces, the equation of motion satisfies

$$\rho\frac{\mathrm{D}\mathbf{V}}{\mathrm{D}t} = \nabla\cdot\mathbf{T} \tag{7}$$

in which the fluid's density is expressed as ρ, and the material derivative is represented as $\frac{\mathrm{D}}{\mathrm{D}t}$.

The continuity equation for velocity is expressed as

$$\nabla\cdot\mathbf{V} = 0 \tag{8}$$

Consider a modified second-grade fluid flowing close to a plate, moving with the velocity $u_w(t)$ within its own plane suddenly. The x-axis is defined following the motion adjacent to the wall, and the y-axis and the wall are perpendicular. Assuming the effects of the side walls can be disregarded and implying the wall is infinitely long, we aim to find solutions in the form of the velocity field as

$$\mathbf{V} = u(y,t)\mathbf{i} \tag{9}$$

in which the x-direction unit vector is represented as $\mathbf{i}$, and the x-direction velocity is expressed as u.

The stresses exerted on the plate initiate the movement of the fluid, which is given by (1), (2), (4), and (5). Substituting (9) into (1)–(6), we obtain

$$\mathbf{T}_{xy} = \mu\frac{\partial u}{\partial y} + \alpha_1 {}_0^{RL}\mathcal{D}_t^{\beta}\frac{\partial u}{\partial y} \tag{10}$$

where $\mathbf{T}_{xy} = \mathbf{T}_{yx}$ and $\mathbf{T}_{xx} = \mathbf{T}_{yy} = \mathbf{T}_{zz} = \mathbf{T}_{xz} = \mathbf{T}_{yz} = 0$. Note that we have kept the same notation for the constant α_1 in (10), but it refers to a new material constant with the dimensions $kg \cdot m^{-1} \cdot s^{\beta-1}$.

By introducing (9) and (10) into (7), it yields

$$\rho\frac{\partial u}{\partial t} = \mu\frac{\partial^2 u}{\partial y^2} + \alpha_1 {}_0^{RL}\mathcal{D}_t^{\beta}\frac{\partial^2 u}{\partial y^2} \tag{11}$$

The IBCs satisfy

$$u(y,0) = 0,\ y > 0 \tag{12}$$

$$u(0,t) = u_w(t),\ u(+\infty,t) = 0,\ t \in [0,T] \tag{13}$$

where $u_w(t)$ is the velocity of the plate.

Dimensionless variables are introduced to facilitate the analysis

$$\widehat{u} = \frac{u}{U},\ \widehat{y} = \frac{yU\rho}{\mu},\ \widehat{t} = \frac{tU^2\rho}{\mu},\ \widehat{u}_w(\widehat{t}) = \frac{u_w(t)}{U} \tag{14}$$

where U represents the characteristic velocity.

For generalization purposes, the form of $f(y,t)$ is presented. Subsequently, the dimensionless governing equation along with its IBCs are presented as follows (the superscript "$\frown$" is omitted)

$$\frac{\partial u(y,t)}{\partial t} = \frac{\partial^2 u(y,t)}{\partial y^2} + \eta {}_0^{RL}\mathcal{D}_t^{\beta}\frac{\partial^2 u(y,t)}{\partial y^2} + f(y,t) \tag{15}$$

$$u(y,0) = 0,\ y > 0 \tag{16}$$

$$u(0,t) = u_w(t),\ u(+\infty,t) = 0,\ t \in [0,T] \tag{17}$$

where $\eta = \frac{\alpha_1 U^{2\beta}\rho^{\beta}}{\mu^{\beta+1}}$.

3. The Establishment of the ABC

By using the ABM, we formulate the exact ABC. First, the unbounded region $\Omega := \{y|0 \le y < +\infty\}$ is truncated by the point y_r. In such a case, we divide the unbounded region as the unbounded area $\Omega_r := \{y|y_r \le y < +\infty\}$ located to the right, and the bounded area $\Omega_c := \{y|0 \le y \le y_r\}$ is designated for calculation. The selection of y_r demands that the initial conditions and the source term be compactly supported in region Ω_c.

In the unbounded region Ω_r, we have

$$\frac{\partial u(y,t)}{\partial t} = \frac{\partial^2 u(y,t)}{\partial y^2} + \eta {}_0^{RL}\mathcal{D}_t^{\beta}\frac{\partial^2 u(y,t)}{\partial y^2} + f(y,t) \tag{18}$$

$$u(y,0) = 0 \text{ for } y > 0 \tag{19}$$

$$u(+\infty,t) = 0 \text{ when } t \in [0,T] \tag{20}$$

Since β is between zero and one, the Laplace transform for ${}_0^{RL}\mathcal{D}_t^{\beta}$ [55] has the form $L\left\{{}_0^{RL}\mathcal{D}_t^{\beta}f(t),s\right\} = s^{\beta}L\{f(t),s\} - \left[{}_0^{RL}\mathcal{D}_t^{\beta-1}f(t)\right]_{t=0}$. By referring to the principle of the Laplace transform for the RLTFD, we obtain

$$\frac{\partial^2 \hat{u}(y,s)}{\partial y^2} - \frac{s}{1+\eta s^{\beta}}\hat{u}(y,s) = 0 \tag{21}$$

in which the Laplace transform of $u(y,t)$ is represented as $\hat{u}(y,s)$.

Solving the above equation yields

$$\hat{u}(y,s) = e^{-\sqrt{\frac{s}{1+\eta s^{\beta}}}y} \tag{22}$$

For Equation (22), we take the derivative for y

$$\frac{\partial \hat{u}(y,s)}{\partial y} = -\frac{1}{\left(1+\eta s^{\beta}\right)^{\frac{1}{2}}}\hat{u}(y,s) \tag{23}$$

Before performing the inverse Laplace transform, it is crucial to define the generalized Mittag–Leffler function [55]:

$$E_{k,l}^{\gamma}(z) = \sum_{n=0}^{\infty}\frac{(\gamma)_n}{n!\Gamma(nk+l)}z^n,\ \mathrm{Re}k > 0,\ l \in C,\ \gamma > 0 \tag{24}$$

where $(\gamma)_n = \gamma(\gamma+1)\cdots(\gamma+n-1) = \Gamma(\gamma+n)/\Gamma(\gamma)$.

The Mittag–Leffler function [55] has the following property:

$$\mathcal{L}\left\{t^{l-1}E_{k,l}^{\gamma}\left(-\lambda t^{k}\right),s\right\} = \frac{s^{k\gamma-l}}{\left(\lambda+s^{k}\right)^{\gamma}} \tag{25}$$

where $\mathrm{Re}k > 0$ and $\mathrm{Re}l > 0$.

By conducting the inverse Laplace transformation for (23), we infer the ABC at $y = y_r$

$$\frac{\partial u(y_r,t)}{\partial y} = -\sqrt{\frac{1}{\eta}}\left[t^{\frac{1}{2}\beta-1}E_{\beta,\frac{1}{2}\beta}^{\frac{1}{2}}\left(-\frac{1}{\eta}t^{\beta}\right)\right] * {}_0^{RL}\mathcal{D}_t^{\frac{1}{2}}u(y_r,t) \tag{26}$$

where $*$ stands for the convolution.

To facilitate explanation, we introduce the notation as

$$K_1(t) = \sqrt{\frac{1}{\eta}}\left[t^{\frac{1}{2}\beta-1}E_{\beta,\frac{1}{2}\beta}^{\frac{1}{2}}\left(-\frac{1}{\eta}t^{\beta}\right)\right] \tag{27}$$

After introducing the symbol $K_1(t)$, the Equation (26) changes:

$$\frac{\partial u(y_r,t)}{\partial y} = \int_0^t -K_1(t-\tau) * {}_0^{RL}\mathcal{D}_{\tau}^{\frac{1}{2}}u(y_r,\tau)d\tau \tag{28}$$

Taking Equation (28) as the exact ABC, we formulate the governing equation based on the IBCs:

$$\frac{\partial u(y,t)}{\partial t} = \frac{\partial^2 u(y,t)}{\partial y^2} + \eta\,{}_0^{RL}\mathcal{D}_t^{\beta}\frac{\partial^2 u(y,t)}{\partial y^2} + f(y,t) \tag{29}$$

$$u(y,0) = 0 \tag{30}$$

$$u(0,t) = 0 \tag{31}$$

$$\frac{\partial u(y_r,t)}{\partial y} = \int_0^t -K_1(t-\tau) * {}_0^{RL}\mathcal{D}_\tau^{\frac{1}{2}} u(y_r,\tau)d\tau \tag{32}$$

Now, we discuss the well-posedness of the considered problem. Since the initial condition of our question is zero, ${}_0^{RL}\mathcal{D}_t^\beta u(t)$ and ${}_0^C\mathcal{D}_t^\beta u(t)$ are equivalent. Then, the property for the Caputo fractional derivative is also applicable to the RLTFD. Initially, we introduce the following lemma.

Lemma 1. *Assume $u(t)$ to be an absolutely continuous function on $[0,T]$. Considering $0 < \beta < 1$ and $u(0) = 0$, it yields [56]*

$$u(t){}_0^{RL}\mathcal{D}_t^\beta u(t) \geq \frac{1}{2}{}_0^{RL}\mathcal{D}_t^\beta u^2(t) \tag{33}$$

Theorem 1. *The problem (29)–(32) is L^2 -stable and adheres to the subsequent estimation:*

$$\left\|\frac{\partial u(y,t)}{\partial y}\right\|_{L^2[0,y_r]}^2 \leq \|f(y,t)\|_{L^2[0,y_r]}^2 \tag{34}$$

where $\|u(y,t)\|_{L^2[a,b]}^2(t) = \int_a^b |u(y,t)|^2 \mathrm{d}y$.

Proof. In the outer region $(y_r,+\infty]$, the source term becomes zero due to the compactly supported assumption in $[0,y_r]$. Multiplying both sides of Equation (29) simultaneously by $\frac{\partial^2 u(y,t)}{\partial y^2}$ and integrating over y from y_r to $+\infty$ and t from 0 to T, it yields:

$$\begin{aligned}\int_0^T \int_{y_r}^{+\infty} \frac{\partial u(y,t)}{\partial t}\frac{\partial^2 w(y,t)}{\partial y^2}\mathrm{d}y\mathrm{d}t = \int_0^T \int_{y_r}^{+\infty} \frac{\partial^2 u(y,t)}{\partial y^2}\frac{\partial^2 u(y,t)}{\partial y^2}\mathrm{d}y\mathrm{d}t \\ +\eta\int_0^T \int_{y_r}^{+\infty} \frac{\partial^2 u(y,t)}{\partial y^2}{}_0^{RL}\mathcal{D}_t^\beta\frac{\partial^2 u(y,t)}{\partial y^2}\mathrm{d}y\mathrm{d}t.\end{aligned} \tag{35}$$

Employing the method of integration by parts and Lemma 1 yields:

$$\int_0^T \int_{y_r}^{+\infty} \frac{\partial u(y,t)}{\partial t}\frac{\partial^2 u(y,t)}{\partial y^2}\mathrm{d}y\mathrm{d}t = -\int_0^T \frac{\partial u(y_r,t)}{\partial t}\frac{\partial u(y_r,t)}{\partial y}\mathrm{d}t - \frac{1}{2}\int_0^T \frac{\partial}{\partial t}\left\|\frac{\partial u(y,t)}{\partial y}\right\|_{L^2(y_r,+\infty]}^2 \mathrm{d}t \tag{36}$$

Considering (30) and (31), we have:

$$\begin{aligned}&-\int_0^T \frac{\partial u(y_r,t)}{\partial t}\frac{\partial u(y_r,t)}{\partial y}\mathrm{d}t - \frac{1}{2}\int_0^T \frac{\partial}{\partial t}\left\|\frac{\partial u(y,t)}{\partial y}\right\|_{L^2(y_r,+\infty]}^2 dt \\ &= -\int_0^T \frac{\partial u(y_r,t)}{\partial t}\frac{\partial u(y_r,t)}{\partial y}\mathrm{d}t - \frac{1}{2}\left\|\frac{\partial u(y,T)}{\partial y}\right\|_{L^2(y_r,+\infty]}^2 \\ &\leq -\int_0^T \frac{\partial u(y_r,t)}{\partial t}\frac{\partial u(y_r,t)}{\partial y}\mathrm{d}t.\end{aligned} \tag{37}$$

The second term satisfies that:

$$\int_0^T \int_{y_r}^{+\infty} \frac{\partial^2 u(y,t)}{\partial y^2}\frac{\partial^2 u(y,t)}{\partial y^2}\mathrm{d}y\mathrm{d}t \geq 0 \tag{38}$$

By using Lemma 1, we obtain:

$$\eta\int_0^T\int_{y_r}^{+\infty}\frac{\partial^2u(y,t)}{\partial y^2}{}_0^{RL}\mathcal{D}_t^{\beta}\frac{\partial^2u(y,t)}{\partial y^2}\mathrm{d}y\mathrm{d}t\geq\frac{\eta}{2}\int_0^T{}_0^{RL}\mathcal{D}_t^{\beta}\left\|\frac{\partial^2u(y,t)}{\partial y^2}\right\|^2_{L^2(y_r,+\infty]}\mathrm{d}t$$
$$=\frac{\eta}{2}\int_0^T\frac{\mathrm{d}}{\mathrm{d}t}\left(\frac{1}{\Gamma(1-\beta)}\int_0^t\left\|\frac{\partial^2u(y,\tau)}{\partial y^2}\right\|^2_{L^2(y_r,+\infty]}\frac{1}{(t-\tau)^{\beta}}\mathrm{d}\tau\right)\mathrm{d}t \tag{39}$$
$$=\frac{\eta}{2}\left(\frac{1}{\Gamma(1-\beta)}\int_0^T\left\|\frac{\partial^2u(y,\tau)}{\partial y^2}\right\|^2_{L^2(y_r,+\infty]}\frac{1}{(T-\tau)^{\beta}}\mathrm{d}\tau\right)\geq 0.$$

Combining (35)–(39), we can deduce:

$$\int_0^T\frac{\partial u(y_r,t)}{\partial t}\frac{\partial u(y_r,t)}{\partial y}\mathrm{d}t\leq 0 \tag{40}$$

Consider the inner region $[0,y_r]$. Both sides of Equation (29) are simultaneously multiplied by $\frac{\partial^2u(y,t)}{\partial y^2}$ and integrated over y in region $[0,y_r]$ and t in region $[0,T]$. Then, we have

$$\int_0^T\int_0^{y_r}\frac{\partial u(y,t)}{\partial t}\frac{\partial^2u(y,t)}{\partial y^2}\mathrm{d}y\mathrm{d}t=\int_0^T\int_0^{y_r}\frac{\partial^2u(y,t)}{\partial y^2}\frac{\partial^2u(y,t)}{\partial y^2}\mathrm{d}y\mathrm{d}t$$
$$+\eta\int_0^T\int_0^{y_r}{}_0^{RL}\mathcal{D}_t^{\beta}\frac{\partial^2u(y,t)}{\partial y^2}\frac{\partial^2u(y,t)}{\partial y^2}\mathrm{d}y\mathrm{d}t+\int_0^T\int_0^{y_r}f(y,t)\frac{\partial^2u(y,t)}{\partial y^2}\mathrm{d}y\mathrm{d}t. \tag{41}$$

For the first term, considering the boundary condition, we can derive:

$$\int_0^T\int_0^{y_r}\frac{\partial w(y,t)}{\partial t}\frac{\partial^2w(y,t)}{\partial y^2}\mathrm{d}y\mathrm{d}t=\int_0^T\int_0^{y_r}\frac{\partial w(y,t)}{\partial t}\mathrm{d}\frac{\partial w(y,t)}{\partial y}\mathrm{d}t$$
$$=\int_0^T\frac{\partial w(y_r,t)}{\partial t}\frac{\partial w(y_r,t)}{\partial y}\mathrm{d}t-\int_0^T\int_0^{y_r}\frac{\partial w(y,t)}{\partial y}\frac{\partial}{\partial t}\frac{\partial w(y,t)}{\partial y}\mathrm{d}y\mathrm{d}t. \tag{42}$$

Considering the initial condition, it yields:

$$-\int_0^T\int_0^{y_r}\frac{\partial w(y,t)}{\partial y}\frac{\partial}{\partial t}\frac{\partial w(y,t)}{\partial y}\mathrm{d}y\mathrm{d}t\leq-\int_0^{y_r}\left(\frac{\partial w(y,T)}{\partial y}\right)^2\mathrm{d}y+\int_0^{y_r}\left[\int_0^T\frac{\partial w(y,t)}{\partial y}\frac{\partial}{\partial t}\frac{\partial w(y,t)}{\partial y}\mathrm{d}t\right]\mathrm{d}y \tag{43}$$

then one can obtain

$$-\int_0^T\int_0^{y_r}\frac{\partial w(y,t)}{\partial y}\frac{\partial}{\partial t}\frac{\partial w(y,t)}{\partial y}\mathrm{d}y\mathrm{d}t\leq-\frac{1}{2}\int_0^{y_r}\left(\frac{\partial w(y,T)}{\partial y}\right)^2\mathrm{d}y\leq 0 \tag{44}$$

Combining (40), (42) and (44), the first term changes:

$$\int_0^T\int_0^{y_r}\frac{\partial w(y,t)}{\partial t}\frac{\partial^2w(y,t)}{\partial y^2}\mathrm{d}y\mathrm{d}t\leq 0 \tag{45}$$

The second term satisfies that:

$$\int_0^T\int_0^{y_r}\frac{\partial^2w(y,t)}{\partial y^2}\frac{\partial^2w(y,t)}{\partial y^2}\mathrm{d}y\mathrm{d}t\geq 0 \tag{46}$$

Using Lemma 1, the third term satisfies

$$\eta\int_0^T\int_0^{y_r}\frac{\partial^2w(y,t)}{\partial y^2}{}_0^{RL}\mathcal{D}_t^{\beta}\frac{\partial^2w(y,t)}{\partial y^2}\mathrm{d}y\mathrm{d}t\geq\frac{\eta}{2}\int_0^T\int_0^{y_r}{}_0^{RL}\mathcal{D}_t^{\beta}\left(\frac{\partial^2w(y,t)}{\partial y^2}\right)^2\mathrm{d}y\mathrm{d}t \tag{47}$$

Considering the definition (6), we obtain

$$\begin{aligned}&\frac{\eta}{2}\int_0^T\int_0^{y_r}{}_0^{RL}\mathcal{D}_t^{\beta}\left(\frac{\partial^2 w(y,t)}{\partial y^2}\right)^2 \mathrm{d}y\mathrm{d}t\\&=\frac{\eta}{2}\int_0^T\int_0^{y_r}\frac{\mathrm{d}}{\mathrm{d}t}\left(\frac{1}{\Gamma(1-\beta)}\int_0^t\frac{1}{(t-\tau_1)^{\beta}}\left(\frac{\partial^2 w(y,\tau_1)}{\partial y^2}\right)^2\mathrm{d}\tau_1\right)\mathrm{d}y\mathrm{d}t\\&=\frac{\eta}{2}\int_0^{y_r}\left(\frac{1}{\Gamma(1-\beta)}\int_0^t\frac{1}{(T-\tau_1)^{\beta}}\left(\frac{\partial^2 w(y,\tau_1)}{\partial y^2}\right)^2\mathrm{d}\tau_1\right)\mathrm{d}y\geq 0.\end{aligned} \tag{48}$$

According to the Cauchy–Schwartz inequality, one can estimate the fourth term as

$$-\int_0^T\int_0^{y_r} f(y,t)\frac{\partial^2 u(y,t)}{\partial y^2}\mathrm{d}y\mathrm{d}t\leq\frac{1}{2}\int_0^T\int_0^{y_r}(f(y,t))^2\mathrm{d}y\mathrm{d}t+\frac{1}{2}\int_0^T\int_0^{y_r}\left(\frac{\partial^2 u(y,t)}{\partial y^2}\right)^2\mathrm{d}y\mathrm{d}t \tag{49}$$

Then, Equation (41) changes:

$$\int_0^T\int_0^{y_r}\frac{\partial^2 u(y,t)}{\partial y^2}\frac{\partial^2 u(y,t)}{\partial y^2}\mathrm{d}y\mathrm{d}t\leq\int_0^T\int_0^{y_r}(f(y,t))^2\mathrm{d}y\mathrm{d}t \tag{50}$$

Therefore, the proof is complete and Equation (34) has been proven. □

Theorem 2. *Given that $u(y,t)$ and $v(y,t)$ are solutions to (29)–(32), it is concluded that $u(y,t)$ and $v(y,t)$ are identical, ensuring the uniqueness of the solution.*

Proof. Denote $w(y,t)=u(y,t)-v(y,t)$. $w(y,t)$ satisfies the following equation:

$$\frac{\partial w(y,t)}{\partial t}=\frac{\partial^2 w(y,t)}{\partial y^2}+\eta{}_0^{RL}\mathcal{D}_t^{\beta}\frac{\partial^2 w(y,t)}{\partial y^2} \tag{51}$$

which is subjected to IBCs:

$$w(y,0)=0 \tag{52}$$

$$w(0,t)=0 \tag{53}$$

$$\frac{\partial w(y_r,t)}{\partial y}=\int_0^t -K_1(t-\tau)*{}_0^{RL}\mathcal{D}_t^{\frac{1}{2}}w(y_r,\tau)\mathrm{d}\tau \tag{54}$$

According to Theorem 1, it follows that

$$\int_0^t\left\|\frac{\partial^2 w(\cdot,\tau)}{\partial y^2}\right\|^2\mathrm{d}\tau\leq 0 \tag{55}$$

where $\|w(\cdot,t)\|^2$ is defined as $\|w(\cdot,t)\|^2=\int_0^{y_r}w^2(y,t)\mathrm{d}y$.

Since $\left\|\frac{\partial^2 w(\cdot,\tau)}{\partial y^2}\right\|^2$ is no less than zero, we obtain

$$\frac{\partial^2 w(\cdot,\tau)}{\partial y^2}=0 \tag{56}$$

According to (56), we assume that $w(y,t)$ has the form

$$w(y,t)=C_1(t)y+C_2(t) \tag{57}$$

Then, one can obtain

$$\frac{\partial w(y,t)}{\partial y}=C_1(t) \tag{58}$$

Considering the boundary condition (53), we have:

$$C_2(t) = 0 \tag{59}$$

Combining Equations (54)–(58), we can deduce:

$$C_1(t) = -y_r \int_0^t K_1(t-\tau) * {}_0^{RL}\mathcal{D}_\tau^{\frac{1}{2}} C_1(\tau) d\tau. \tag{60}$$

Since the parameter y_r is arbitrary, the following condition must hold for (60):

$$C_1(t) = 0 \tag{61}$$

which means $w(y,t) = 0$. The uniqueness of the solution has been established. The proof of Theorem 2 has been finished. □

4. Construction of the FDM

This section is dedicated to formulating a numerical scheme based on Equation (29) to Equation (32), with the aim of solving the problem on a bounded domain Ω_c. In the domain $\Omega_c \times [0,T]$, we define the temporal step size as $\tau = T/N$ and the spatial step size as $h = \frac{y_r}{M}$, where the integers N and M correspond to the count of grid points in the temporal and spatial dimensions, respectively. The grid points are defined as $t_n := n\tau$ for $0 \le n \le N$ and $y_j = jh_y$ for $0 \le j \le M$. Furthermore, $u(y_j,t_n)$ and u_j^n represent the exact solutions and numerical solutions of the velocity at grid points (y_j,t_n) for $0 \le n \le N$ and $0 \le j \le M$, respectively. For clarity, we introduce some notations below [28,29]

$$\delta_t u_j^n = \frac{u_j^n - u_j^{n-1}}{\tau}, \ \delta_y^2 u_j^n = \frac{u_{j+1}^n - 2u_j^n + u_{j-1}^n}{h^2} \tag{62}$$

At $t = t_n$, the time-fractional derivative of order α in the RLTFD is approximated by the L1 scheme [49] by considering the equivalent relationship between the RLTFD and the Caputo fractional derivative:

$${}_0^{RL}\mathcal{D}_t^\alpha u^n = {}_0^C\widetilde{D}_t^\alpha u^n + R_1 = \frac{\tau^{-\alpha}}{\Gamma(2-\alpha)}\left[a_0^{(\alpha)}u^n - \sum_{k=1}^{n-1}\left(a_{n-k-1}^{(\alpha)} - a_{n-k}^{(\alpha)}\right)u^k - a_{n-1}^{(\alpha)}u^0 + \frac{1-\alpha}{n^\alpha}u^0\right] + R_1 \tag{63}$$

where $a_0^{(\alpha)} = 1$ and $a_k^{(\alpha)} = (k+1)^{1-\alpha} - k^{1-\alpha}$, $k = 1,2,\ldots,n-1$. The symbol R_1 refers to the error and satisfies $|R_1| \le C\tau^{2-\alpha}$.

Apart from that, the central difference formula is utilized to approximate $\frac{\partial^2 u}{\partial y^2}$, and the error terms are disregarded. Subsequently, the discretized scheme of the equation is obtained as follows:

$$\delta_t u_j^n = \frac{\tau^{-\beta}}{\Gamma(2-\beta)}\left[a_0{}^{(\beta)}\delta_y^2 u_j^n - \sum_{k=1}^{n-1}\left(a_{n-k-1}^{(\beta)} - a_{n-k}^{(\beta)}\right)\delta_y^2 u_j^k - a_{n-1}^{(\beta)}\delta_y^2 u_j^0 + \frac{1-\beta}{n^\beta}u_j^0\right] + \delta_y^2 u_j^n \tag{64}$$

Furthermore, using the definitions given by Equation (62), the aforementioned equation can be rearranged as:

$$\begin{aligned}
&-\left(\frac{\eta\tau^{-\beta}}{\Gamma(2-\beta)}a_0{}^{(\beta)}+1\right)\frac{u_{j+1}^n}{h^2}+\left(\frac{\eta\tau^{-\beta}}{\Gamma(2-\beta)}a_0{}^{(\beta)}\frac{2}{h^2}+\frac{2}{h^2}+\frac{1}{\tau}\right)u_j^n-\left(\frac{\eta\tau^{-\beta}}{\Gamma(2-\beta)}a_0{}^{(\beta)}+1\right)\frac{u_{j-1}^n}{h^2}\\
&=-\frac{\eta\tau^{-\beta}}{\Gamma(2-\beta)}\left[\sum_{k=1}^{n-1}\left(a_{n-k-1}^{(\beta)}-a_{n-k}^{(\beta)}\right)\delta_y^2u_j^k+a_{n-1}^{(\beta)}\delta_y^2u_j^0-\frac{1-\beta}{n^\beta}u_j^0\right]+\frac{u_j^{n-1}}{\tau}.
\end{aligned} \tag{65}$$

The integral term (32) under the ABC is approximated using the following form:

$$\begin{aligned}
&\int_{t_{k-1}}^{t_k}K_1(t-\tau)*{}_0^{RL}\mathcal{D}_t^{1/2}ud\tau\\
&\approx\tau K_1(t_n-t_k)\frac{\tau^{-1/2}}{\Gamma(3/2)}\left[a_0^{(1/2)}u^k-\sum_{j=1}^{k-1}\left(a_{k-j-1}^{(1/2)}-a_{k-j}^{(1/2)}\right)u^j-a_{k-1}^{(1/2)}u^0-\frac{1}{2n^{1/2}}u_j^0\right].
\end{aligned} \tag{66}$$

By using the backward difference scheme with first-order spatial derivatives, we derive the final difference scheme for the ABC:

$$\frac{u_j^n-u_{j-1}^n}{h}=-\sum_{k=1}^{n}K_1((n-k)\tau)\frac{\tau^{1/2}}{\Gamma(3/2)}\left[a_0^{(1/2)}u^k-\sum_{j=1}^{k-1}\left(a_{k-j-1}^{(1/2)}-a_{k-j}^{(1/2)}\right)u^j-a_{k-1}^{(1/2)}u^0+\frac{1}{2n^{1/2}}u_j^0\right] \tag{67}$$

Subsequently, the remaining conditions can be discretized as follows:

$$u_j^0=0,\ j=0,1,2,\ldots,M,\ u_0^n=u_w(t_n),\ n=0,1,2,\ldots.$$

Remark 1. *Since the governing equation in this work constitutes a particular instance of the issue in [57] with new boundary condition, the stability and convergence of the proposed numerical scheme can be proved through a comparable approach. This paper primarily focuses on the formulation of the ABC and an evaluation of its advantages. As such, a detailed theoretical analysis of the numerical scheme has been excluded.*

5. Numerical Examples

5.1. The Verification of the Solution and the ABC

The first step is to validate the numerical solution by contrasting it with the exact solution. According to (29) to (32), the governing equation with a source term $f(y,t)$ is as follows:

$$\frac{\partial u(y,t)}{\partial t}=\frac{\partial^2u(y,t)}{\partial y^2}+\eta{}_0^{RL}\mathcal{D}_t^\beta\frac{\partial^2u(y,t)}{\partial y^2}+f(y,t) \tag{68}$$

subsequently, the IBCs are presented by

$$u(y,0)=0\text{ for }y>0 \tag{69}$$

$$u(0,t)=Wt^2\text{and }\frac{\partial u(y_r,t)}{\partial y}+\int_0^tK_1(t-\tau)*{}_0^{RL}\mathcal{D}_\tau^{1/2}u(y_r,\tau)d\tau=0,\ \text{for }t>0 \tag{70}$$

where y_r is selected to be $y_r=1$.

We define the exact solution as

$$u(y,t)=\begin{cases}W(1-y)^6(1+y)^6t^6, & \text{for}\quad 0\le y\le 1,\\ 0, & \text{elsewhere}.\end{cases} \tag{71}$$

Bringing the above exact solution (71) into Equation (68) yields the expression for the source term:

$$f(y,t) = \begin{cases} 6Wt^5(1-y)^6(y+1)^6 - Wt^6(30(y+1)^6(y-1)^4 \\ +72(y+1)^5(y-1)^5 + 30(y+1)^4(y-1)^6)\left[\frac{\Gamma(7)t^{-\beta}}{\Gamma(7-\beta)} + 1\right], & \text{for} \quad 0 \leq y \leq 1, \\ 0, & \text{elsewhere.} \end{cases}$$

It should be indicated that the source term complies with the condition of the ABC that it is compactly supported in the inner region $[0,1]$.

In conjunction with the matrix form of the difference scheme, which can be deduced by (65), we use MATLAB to solve the governing equation. This aims at generating the numerical solutions on successive time levels. The computational domain is specified as $[0,1]$, with a termination time denoted by $T = 1$. The calculation parameters are set to $\beta = 0.5$ and $\eta = 1$, while the temporal and spatial steps are $\tau = 1/2^{10}$ and $h = 1/2^{10}$, respectively. As apparently revealed in Figure 2, the comparative velocity profiles of the numerical and analytical solutions at $t = 1$ exhibit similar distribution patterns, which is sufficient to prove that the numerical scheme we developed is precise.

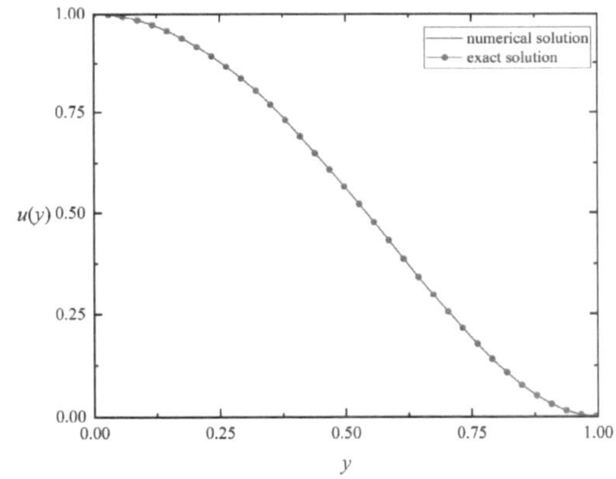

Figure 2. The comparison curve of error distribution for $T = 1$ between the numerical solution and the exact solution when $\beta = 0.5$, $\eta = 1$ and $W = 1$.

Figure 3 provides a clear visualization of the disparities in velocity distribution when comparing the ABC against the DTBC, facilitated by the choice of $f(y,t) = 0$. Notably, the distribution curves exhibit significant divergence at the right boundary. This difference stems from the fact that the DTBC specifies zero velocity at the selected right boundary, whereas the velocity at the right boundary follows a specific functional relationship when subject to the ABC, which is determined through meticulous derivation. In numerous practical situations, the velocity at the boundary on the right does not maintain a constant zero over time, which potentially results in imprecise numerical solutions based on the DTBC. However, the ABC averts the artificial error at the truncation point and is consistent with the zero boundary condition at infinite locations. This proves that the ABC can effectively deal with problems related to extension to infinite domains.

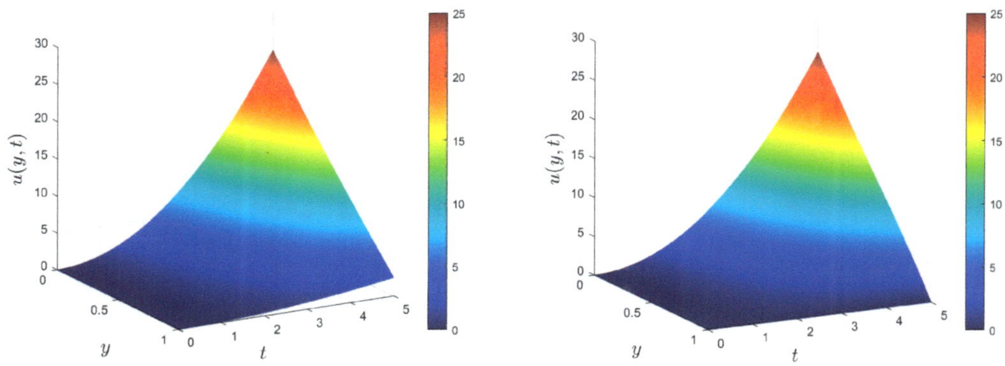

Figure 3. The velocity distribution between the ABC (**left**) and the DTBC (**right**) when $f(y,t) = 0$, $T = 5$, $\beta = 0.5$, $\eta = 1$, and $W = 1$.

5.2. Impact of Parameter Variations on the Velocity Field Induced by a Moving Plate

We assume that the fluid is set in motion by a plate moving at the velocity $u_w(t) = Wt^2$. With this assumption, the governing equation with its associated IBCs reduces to:

$$\frac{\partial u(y,t)}{\partial t} = \frac{\partial^2 u(y,t)}{\partial y^2} + \eta {}_0^{RL}\mathcal{D}_t^{\beta} \frac{\partial^2 u(y,t)}{\partial y^2} \tag{72}$$

$$u(y,0) = 0 \text{ for } y > 0 \tag{73}$$

$$u(0,t) = Wt^2 \text{ and } \frac{\partial u(y_r,t)}{\partial y} + \int_0^t K_1(t-\tau) * {}_0^{RL}\mathcal{D}_\tau^{1/2} u(y_r,\tau) d\tau = 0 \tag{74}$$

The examination of parameter influence on the velocity field is conducted. In this example, we select the cut-off point as $y_r = 8$. Figure 4 illustrates the impact of the varying time parameter on the velocity distribution. Comparing multiple curves at different time points, it can be found that the fluid velocity at the same location increases with time. Analyzing a single curve, it can be observed that the closer the fluid is to the plate, the greater the velocity will be. The figure describes the time evolution of the velocity well.

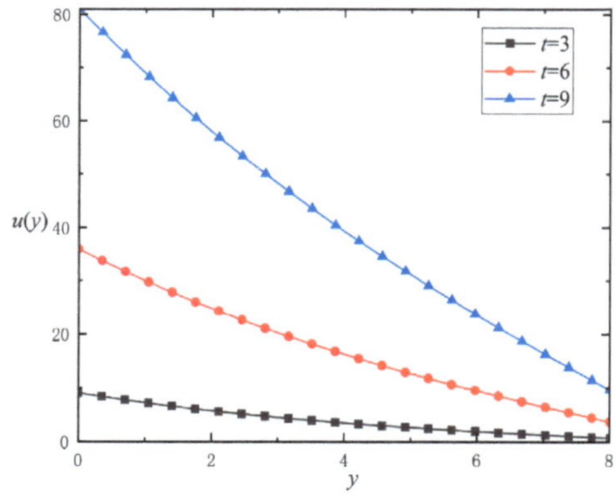

Figure 4. The impacts exerted by the time parameters on the velocity distribution for $\beta = 0.5$, $\eta = 1$, and $W = 1$.

As shown in Figure 5 for y (left) and t (right), the variation in fractional orders significantly influences the velocity distribution. This plate's motion changes the shear force, causing the fluid near the plate to begin to flow. Therefore, it can be observed that in the actual image, the velocity at the starting point is very high, whereas the velocity at locations further away is relatively low. A trend is observable in these images, where the flow velocity diminishes with an increase in y. And at a fixed position, the velocity rises with the augmentation of t. As shown in the figure on the left of Figure 5, at a fixed time, the increase in the fractional parameter results in a smaller value of the distribution curve. From this, it can be observed that as the fractional parameter diminishes, the fluid's memory properties become more pronounced, and the velocity transmission occurs more rapidly. Regarding the relationship between the distribution curve and time, when the spatial position is fixed, the larger fractional parameter results in the decreasing velocity of the fluid.

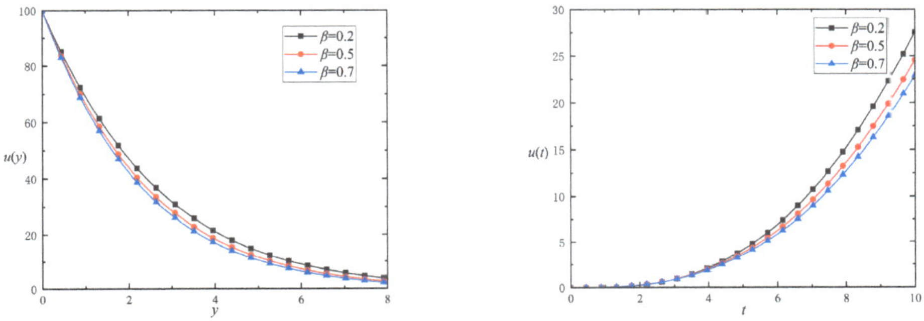

Figure 5. The impacts exerted of the fractional parameters on the velocity distribution versus y ($t = 1$) and t ($y = 1$) for $\eta = 1$ and $W = 1$.

In Figure 6, the effect of relaxation parameters on the velocity distribution is shown. The relaxation time parameter exerts a delayed impact on the fluid's flow. As illustrated on the left side of Figure 6, at a fixed time, an increase in the relaxation parameter results in a higher value of the distribution curve. From this observation, a conclusion can be drawn that there is a correlation between the larger relaxation parameter and the delay effect. For the relationship between the distribution curve and time, when the space y is fixed, the velocity increases as the relaxation parameter becomes larger.

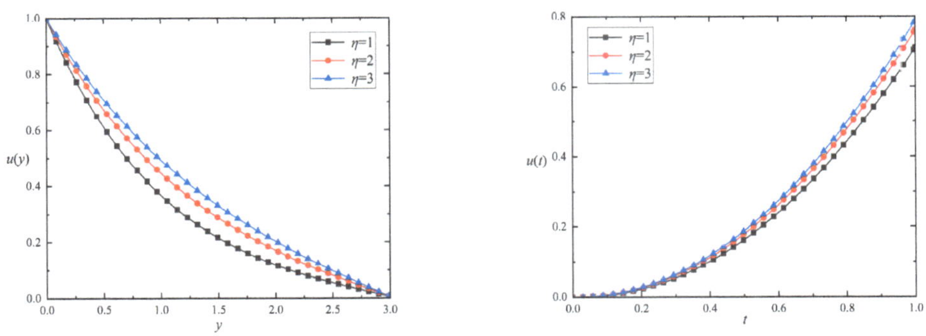

Figure 6. The impacts of the relaxation parameters on the velocity distribution versus y ($t = 1$) and t ($y = 1$) for $\beta = 0.5$ and $W = 1$.

Figure 7 demonstrates the distribution versus β with the influence of various relaxation time parameters, highlighting the relaxation time parameter's delayed influence on fluid flow. As depicted in Figure 7, for a constant fractional parameter β, an increase in the relaxation parameter leads to a higher value of the distribution curve. Concurrently, with the fractional parameter increasing, the fluid velocity diminishes progressively. It is inferred that a larger relaxation parameter results in a diminished delay effect and an increased velocity.

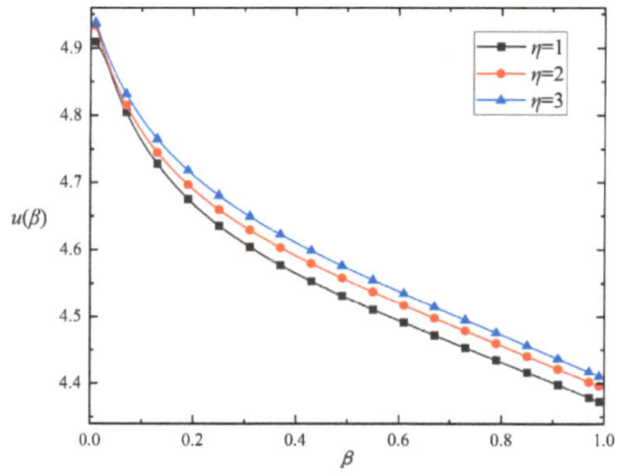

Figure 7. The impacts of the relaxation parameters on the velocity distribution versus β for $W = 1$.

5.3. Various Velocities Induced by Vibrating Plates with Different Physical Parameters

In this example, the emphasis is placed on the impact of dynamic parameters on the flow mechanism of fluid, propelled by a plate with oscillating velocity. The governing equation for the IBCs is presented below

$$\frac{\partial u(y,t)}{\partial t} = \frac{\partial^2 u(y,t)}{\partial y^2} + \eta\,{}_0^{RL}D_t^{\beta}\frac{\partial^2 u(y,t)}{\partial y^2} \tag{75}$$

$$u(y,0) = 0 \text{ for } y > 0 \tag{76}$$

$$u(0,t) = U\sin\left(\omega t^2\right) \text{ and } \frac{\partial u(y_r,t)}{\partial y} + \int_0^t K_1(t-\tau) * {}_0^{RL}\mathcal{D}_\tau^{1/2} u(y_r,\tau)\mathrm{d}\tau = 0 \tag{77}$$

Figure 8 shows the velocity distribution of y and t. Contrary to previous numerical instances, the observations from the figure reveal that the oscillating distribution results directly from a moving plate exhibiting oscillatory velocity. The effects of frequency ω and amplitude U on the velocity distribution on y and t are shown in Figures 9 and 10. It should be noted that we took the space calculation region as $[0,5]$, and the termination time of the analysis was $T = 5$.

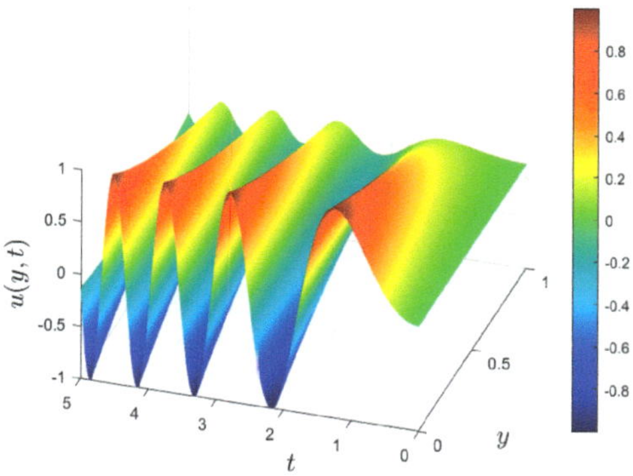

Figure 8. The velocity distributions with y and t for $\beta = 0.5$, $\eta = 1$, $\omega = 1$, and $U = 1$.

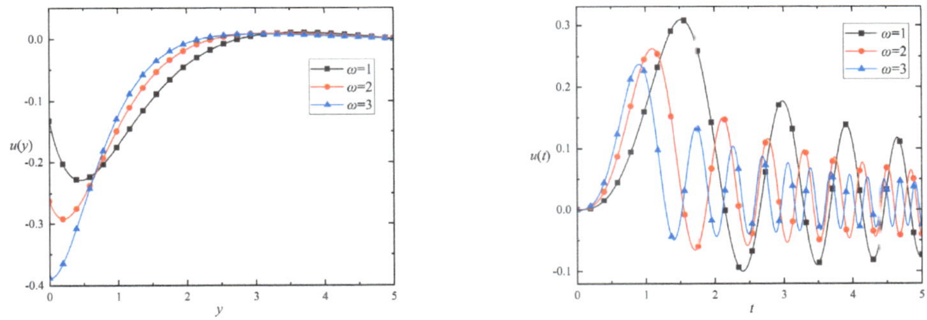

Figure 9. The velocity distributions with different frequency ω versus y ($t = 5$) and t ($y = 1$) for $\beta = 0.5$, $U = 1$, and $\eta = 1$.

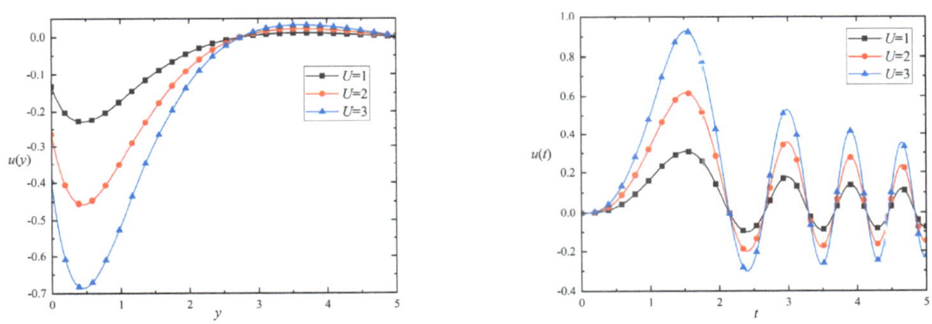

Figure 10. The velocity distributions with different amplitudes U versus y ($t = 5$) and t ($y = 1$) for $\beta = 0.5$, $\omega = 1$, and $\eta = 1$.

Figure 9 illustrates the impact of the oscillation frequency ω on the velocity distribution. In the left figure, at a fixed time, the fluid near the moving plate has a larger negative velocity in the initial position. The fluid's velocity increases with the moving plate's vibration frequency ω. Due to the vibration of the moving plate, the peak value of the velocity distribution curve occurs within a certain distance from the plate when $\omega = 1$ and $\omega = 2$, while the curve does not oscillate when $\omega = 3$. With the increase in y, the influence of oscillations on the fluid weakens and the fluid velocity decreases in the negative direction. By comparing the velocity distribution curves at different frequencies, it can be observed that the velocity distribution curve of the fluid changes more rapidly as the frequency ω of the moving plate increases. Finally, when the distance from the plate is far enough, the vibration of the plate has no effect on the fluid. The velocity distribution curve of the fluid becomes smoother, and the velocity distribution curves of different vibration frequencies tend to be close. In the figure on the right, the time change of the fluid near the plate, which is affected by the moving plate, is described. For the same vibration frequency ω, the amplitude of the velocity distribution curve and the vibration period gradually decrease with the passage of time, because the moving plate whose velocity oscillates in the form of a sinusoidal function causes the fluctuation of the velocity distribution. The peak value of the velocity distribution curve is earlier for a larger vibration frequency ω. The results show that, as the frequency ω increases, both the period and the amplitude of the distribution curve decrease.

Figure 10 depicts the impact of the amplitude on the velocity distribution. Similar to Figure 9, the negative velocity of the fluid close to the moving plate first increases to the peak and then decreases. At a fixed position, the oscillation period and amplitude of the fluid decrease gradually with the passage of time. By comparing the velocity distribution curves of different amplitudes, the larger the amplitude U is, the greater the peak value of the velocity distribution curve at a fixed time is, and the faster the curve changes. In a fixed position, the velocity distribution curve's amplitude is directly proportional to the amplitude U.

5.4. Comparison of This Study with Other Methods in Ref. [58]

To further verify the accuracy of the numerical solutions presented in this paper, other studies on generalized second-order fluid flow on a semi-infinite plate are selected for comparison. Shahed [58] studied the impulsive flow of a fractional second-grade fluid on a flat plate. The IBCs were $u(y,0) = 0, y > 0, u(0,t) = 1, u(+\infty,t) = 0$, and $t \in [0,T]$.

Since the initial condition is zero, the equation containing the Caputo fractional derivative in reference [58] is equivalent to the equation with RLTFD in this paper. The ABC (32) is applied. By selecting the same parameters and comparing them with Figure 2 in [58], the comparison curve with the numerical solution in our study is given in Figure 11. The red dotted line is the curve derived from the method of this study, and the black solid line is the curve given by Shahed [58]. The comparison shows that the curves derived from the two methods match very well, which indicates the accuracy of our numerical method.

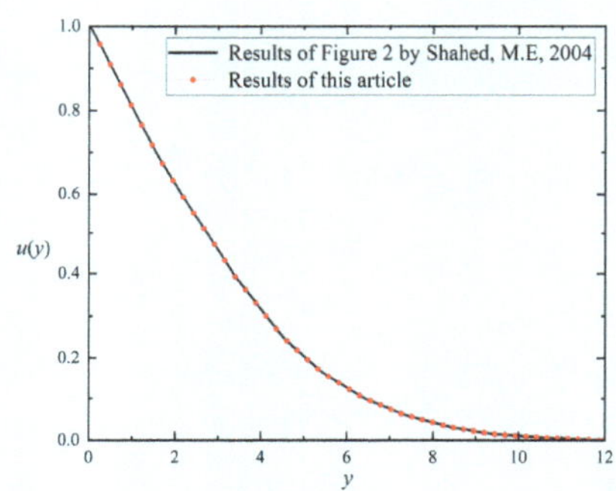

Figure 11. Comparison between the numerical solution in our study and other methods in Ref. [58] by selecting parameters $\eta = 1$, $t = 2$, and $\beta = 0.5$.

6. Conclusions

In this study, the fractional governing equation that describes generalized second-grade fluid's flow over semi-infinite plates, induced by varying plate velocities, was rigorously derived. For semi-bounded domains, the study employed the ABM. This resulted in the formulation of ABC as convolutions within a finite region. The use of the artificial boundary method made the numerical simulation of this fractional governing equation highly applicable. Subsequently, the governing equation was numerically discretized by incorporating the initial condition and the ABC by utilizing the L1 scheme. The paper then presented four numerical examples: the first one validated the difference scheme's effectiveness through introducing the source term and demonstrated the superiority of the ABC compared to the DTBC, and the second one examined how dynamic parameters within the governing equation affect velocity distribution. In the third example, a numerical example was given to analyze the flow mechanism arising from an oscillating plate, and the effects of the frequency and amplitude on velocity were described, respectively. Finally, the comparison of the numerical example with the ABC and the results in Ref. [58] was discussed. Some of the major findings are summarized below.

(i) A higher fractional parameter leads to a slower rate of fluid flow;

(ii) The delay effect increases with the decrease in relaxation parameters, and the fluid velocity rises as the relaxation parameter increases;
(iii) The higher the oscillation frequency is, the faster the velocity will change, and the velocity distribution curve will exhibit a decreased period and amplitude;
(iv) With a greater amplitude, the fluid velocity changes more rapidly and the amplitude of the velocity distribution curve also increases.

The important conclusions drawn from this study not only advance the theory of non-Newtonian fluid dynamics, but also extend to its engineering applications in a broad spectrum of fields, including chemical, petroleum, aerospace, and biomedical engineering. In addition, this research contributes to the design and optimization of new materials in materials science, the prediction of pollutant dispersion in environmental science, and the diagnosis and treatment of cardiovascular diseases in the medical field. It also promotes the advancement of numerical simulation technology to solve the problem of infinite boundaries using artificial boundary techniques and provides a theoretical basis and computational methods for the simulation of complex fluid systems. We intend to explore more comprehensive non-Newtonian fluid models with ABC is our future work.

Author Contributions: Conceptualization, J.Y.; Methodology, L.L., S.C. and C.X.; Validation, J.Y.; Formal analysis, J.Y. and S.C.; Writing—original draft, J.Y.; Writing—review & editing, J.Y., L.L., S.C., L.F. and C.X.; Visualization, J.Y. and C.X.; Supervision, L.L. and L.F.; Project administration, L.L. All authors have read and agreed to the published version of the manuscript.

Funding: The work is supported by the Project funded by the National Natural Science Foundation of China (Nos. 11801029, 12302326), Fundamental Research Funds for the Central Universities (Nos. QNXM20220048, FRF-EYIT-23-07), and Beijing Natural Science Foundation (No. I523027), 2023 Fund for Fostering Young Scholars of the School of Mathematics and Physics, USTB (No. FRF-BR-23-01B).

Data Availability Statement: The data that support the findings of this study are available from the corresponding author upon reasonable request.

Conflicts of Interest: The authors declare no conflicts of interest.

References

1. Mohamed, A.A.; Khishvand, M.; Piri, M. Entrapment and mobilization dynamics during the flow of viscoelastic fluids in natural porous media: A micro-scale experimental investigation. *Phys. Fluids* **2023**, *35*, 047119. [CrossRef]
2. Steinhaus, B.; Shen, A.Q.; Sureshkumar, R. Dynamics of viscoelastic fluid filaments in microfluidic devices. *Phys. Fluids* **2007**, *19*, 073103. [CrossRef]
3. Lauga, E. Propulsion in a viscoelastic fluid. *Phys. Fluids* **2007**, *19*, 083104. [CrossRef]
4. VeeraKrishna, M.; Chamkha, A.J. Hall effects on unsteady MHD flow of second grade fluid through porous medium with ramped wall temperature and ramped surface concentration. *Phys. Fluids* **2018**, *30*, 053101. [CrossRef]
5. Wang, S.; Tai, C.W.; Narsimhan, V. Dynamics of spheroids in an unbound quadratic flow of a general second-grade fluid. *Phys. Fluids* **2020**, *32*, 113106. [CrossRef]
6. Ho, B.P.; Leal, L.G. Migration of rigid spheres in a two-dimensional unidirectional shear flow of a second-grade fluid. *J. Fluid Mech.* **1976**, *76*, 783–799. [CrossRef]
7. Khan, M.; Wang, S. Flow of a generalized second-grade fluid between two side walls perpendicular to a plate with a fractional derivative model. *Nonlinear Anal. Real World Appl.* **2009**, *10*, 203–208. [CrossRef]
8. Tassaddiq, A. MHD flow of a fractional second grade fluid over an inclined heated plate. *Chaos Soliton. Fract.* **2019**, *123*, 341–346. [CrossRef]
9. Metzne, A.B.; Park, M.G. Turbulent flow characteristics of viscoelastic fluids. *J. Fluid Mech.* **1964**, *20*, 291–303. [CrossRef]
10. Li, G.; McKinley, G.H.; Ardekani, A.M. Dynamics of particle migration in channel flow of viscoelastic fluids. *J. Fluid Mech.* **2015**, *785*, 486–505. [CrossRef]
11. Billingham, J.; King, A.C. The interaction of a moving fluid/fluid interface with a flat plate. *J. Fluid Mech.* **1995**, *296*, 325–351. [CrossRef]
12. Traugott, S.C. Impulsive Motion of an Infinite Plate in a Compressible Fluid with Non-Uniform External Flow. *J. Fluid Mech.* **1962**, *13*, 400–416. [CrossRef]
13. Jaworski, J.W.; Peake, N. Aerodynamic noise from a poroelastic edge with implications for the silent flight of owls. *J. Fluid Mech.* **2013**, *723*, 456–479. [CrossRef]
14. Sahoo, T.; Yip, T.L.; Chwang, A.T. Scattering of surface waves by a semi-infinite floating elastic plate. *Phys. Fluids* **2001**, *13*, 3215–3222. [CrossRef]

15. Baranovskii, E.S. Exact solutions for non-isothermal flows of second grade fluid between parallel plates. *Nanomaterials* **2023**, *13*, 1409. [CrossRef] [PubMed]
16. Baranovskii, E.S. Analytical Solutions to the Unsteady Poiseuille Flow of a Second Grade Fluid with Slip Boundary Conditions. *Polymers* **2024**, *16*, 179. [CrossRef] [PubMed]
17. Du, M.; Wang, Z.; Hu, H. Measuring memory with the grade of fractional derivative. *Sci. Rep.* **2013**, *3*, 3431. [CrossRef] [PubMed]
18. Carpinteri, A.; Francesco, M. *Fractals and Fractional Calculus in Continuum Mechanics*; Springer: Berlin/Heidelberg, Germany, 2014; Volume 378.
19. Uchaikin, V.; Renat, S. *Fractional Kinetics in Space: Anomalous Transport Models*; World Scientific: Singapore, 2018.
20. Tarasov, V.E. *Fractional Dynamics: Applications of Fractional Calculus to Dynamics of Particles, Fields and Media*; Springer Science & Business Media: Berlin/Heidelberg, Germany, 2011.
21. Shen, M.; Chen, L.; Zhang, M.; Liu, F. A renovated Buongiorno's model for unsteady Sisko nanofluid with fractional Cattaneo heat flux. *Int. J. Heat Mass Transf.* **2018**, *126*, 277–286. [CrossRef]
22. Sene, N. Second-grade fluid model with Caputo-Liouville generalized fractional derivative. *Chaos Soliton. Fract.* **2020**, *133*, 109631. [CrossRef]
23. Chan, P.C.H.; Leal, L.G. The motion of a deformable drop in a second-grade fluid. *J. Fluid Mech.* **1979**, *92*, 131–170. [CrossRef]
24. Jiang, X.; Zhang, H.; Wang, S. Unsteady magnetohydrodynamic flow of generalized second grade fluid through porous medium with Hall effects on heat and mass transfer. *Phys. Fluids* **2020**, *32*, 113105. [CrossRef]
25. Deswal, S.C.R.; Kalkal, K.K. Fractional grade heat conduction law in micropolar thermo-viscoelasticity with two temperatures. *Int. J. Heat Mass Transf.* **2013**, *66*, 451–460. [CrossRef]
26. Wang, X.; Xu, H.; Qi, H. Transient magnetohydrodynamic flow and heat transfer of fractional Oldroyd-B fluids in a microchannel with slip boundary condition. *Phys. Fluids* **2020**, *32*, 103104. [CrossRef]
27. Awad, E. Dual-phase-lag in the balance Sufficiency bounds for the class of Jeffreys' equations to furnish physical solutions. *Int. J. Heat Mass Transf.* **2020**, *158*, 119742. [CrossRef]
28. Coppola, G.; Veldman, A.E.P. Global and local conservation of mass, momentum and kinetic energy in the simulation of compressible flow. *J. Comput. Phys.* **2023**, *475*, 111879. [CrossRef]
29. Tamim, S.; Bostwick, J.B. Spreading of a thin droplet on a soft substrate. *J. Fluid Mech.* **2023**, *971*, A32. [CrossRef]
30. Mei, R.; Adrian, R.J. Flow past a sphere with an oscillation in the free-stream velocity and unsteady drag at finite Reynolds number. *J. Fluid Mech.* **1992**, *237*, 323–341. [CrossRef]
31. Dennis, S.C.R. Calculation of the steady flow through a curved tube using a new finite-difference method. *J. Fluid Mech.* **1980**, *99*, 449–467. [CrossRef]
32. Kim, T.Y.; Dolbow, J.; Fried, E. A numerical method for a second-gradient theory of incompressible fluid flow. *J. Comput. Phys.* **2007**, *223*, 551–570. [CrossRef]
33. Gilmanov, A.; Sotiropoulos, F. A hybrid Cartesian/immersed boundary method for simulating flows with 3D, geometrically complex, moving bodies. *J. Comput. Phys.* **2005**, *207*, 457–492. [CrossRef]
34. Li, B.; Zhang, J.; Zheng, C. Stability and error analysis for a second-grade fast approximation of the one-dimensional Schrödinger equation under absorbing boundary conditions. *SIAM J. Sci. Comput.* **2018**, *40*, A4083–A4104. [CrossRef]
35. Uhlmann, M. An immersed boundary method with direct forcing for the simulation of particulate flows. *J. Comput. Phys.* **2005**, *209*, 448–476. [CrossRef]
36. Özdemir, N.; Karadeniz, D.; İskender, B.B. Fractional optimal control problem of a distributed system in cylindrical coordinates. *Phys. Lett. A* **2009**, *373*, 221–226. [CrossRef]
37. Yan, Y.; Khan, M.; Ford, N.J. An analysis of the modified L1 scheme for time-fractional partial differential equations with nonsmooth data. *SIAM J. Numer. Anal.* **2018**, *56*, 210–227. [CrossRef]
38. Lyu, P.; Vong, S. A nonuniform L2 formula of Caputo derivative and its application to a fractional Benjamin-Bona-Mahony-type equation with nonsmooth solutions. *Numer. Meth. Part Differ. Equ.* **2020**, *36*, 579–600. [CrossRef]
39. Xing, Z.; Wen, L. Numerical Analysis of the Nonuniform Fast L1 Formula for Nonlinear Time-Space Fractional Parabolic Equations. *J. Sci. Comput.* **2023**, *95*, 58. [CrossRef]
40. Chen, Z.; Zheng, W. Convergence of the uniaxial perfectly matched layer method for time-harmonic scattering problems in two-layered media. *SIAM J. Numer. Anal.* **2010**, *48*, 2158–2185. [CrossRef]
41. Barucq, H.; Rouxelin, N.; Tordeux, S. Low-grade Prandtl-Glauert-Lorentz based Absorbing Boundary Conditions for solving the convected Helmholtz equation with Discontinuous Galerkin methods. *J. Comput. Phys.* **2022**, *468*, 111450. [CrossRef]
42. Li, B.; Zhang, J.; Zheng, C. An efficient second-grade finite difference method for the one-dimensional Schrödinger equation with absorbing boundary conditions. *SIAM J. Numer. Anal.* **2018**, *56*, 766–791. [CrossRef]
43. Muhr, M.; Nikolić, V.; Wohlmuth, B. Self-adaptive absorbing boundary conditions for quasilinear acoustic wave propagation. *J. Comput. Phys.* **2019**, *388*, 279–299. [CrossRef]
44. Fu, W.; Wang, W.; Li, C.; Huang, S. An investigation of natural convection in parallel square plates with a heated bottom surface by an absorbing boundary condition. *Int. J. Heat Mass Transf.* **2013**, *56*, 35–44. [CrossRef]
45. Baffet, D.; Grote, M.J. On wave splitting, source separation and echo removal with absorbing boundary conditions. *J. Comput. Phys.* **2019**, *387*, 589–596. [CrossRef]

46. Hwang, J.; Jang, J.; Jung, J. The Fokker-Planck equation with absorbing boundary conditions in bounded domains. *SIAM J. Math. Anal.* **2018**, *50*, 2194–2232. [CrossRef]
47. Tan, W.; Masuoka, T. Stokes' first problem for a second grade fluid in a porous half-space with heated boundary. *Int. J. Non-Linear Mech.* **2005**, *40*, 515–522. [CrossRef]
48. Rivlin, R.S.; Ericksen, J.L.V. *Stress-Deformation Relations for Isotropic Materials*; Collected Papers of RS Rivlin Volume I and II; Springer: Berlin/Heidelberg, Germany, 1997; pp. 911–1013.
49. Tan, W.; Xu, M. The impulsive motion of flat plate in a generalized second grade fluid. *Mech. Res. Commun.* **2002**, *29*, 3–9. [CrossRef]
50. Khan, M.; Nadeem, S.; Hayat, T.; Siddiqui, A.M. Unsteady motions of a generalized second-grade fluid. *Math. Comput.* **2005**, *41*, 629–637. [CrossRef]
51. Hilfer, R. *Applications of Fractional Calculus in Physics*; World Scientific: Singapore, 2000.
52. Hristov, J. Linear viscoelastic responses and constitutive equations in terms of fractional operators with non-singular kernels-pragmatic approach, memory kernel correspondence requirement and analyses. *Eur. Phys. J. Plus* **2019**, *134*, 283. [CrossRef]
53. Li, C.; Qian, D.; Chen, Y. On Riemann-Liouville and Caputo derivatives. *Discrete. Dyn. Nat. Soc.* **2011**, *2011*, 562494. [CrossRef]
54. Mahmood, A.; Fetecau, C.; Khan, N.A.; Jamil, M. Some exact solutions of the oscillatory motion of a generalized second grade fluid in an annular region of two cylinders. *Acta Mech. Sin.* **2010**, *26*, 541–550. [CrossRef]
55. Podlubny, I. *Fractional Differential Equation*; Academic Press: San Diego, LA, USA, 1999.
56. Alikhanov, A.A. A priori estimates for solutions of boundary value problems for fractional-grade equations. *Differ. Equ.* **2010**, *46*, 660–666. [CrossRef]
57. Feng, L.; Liu, F.; Turner, I.; Zheng, L. Novel numerical analysis of multi-term time fractional viscoelastic non-Newtonian fluid models for simulating unsteady MHD Couette flow of a generalized Oldroyd-B fluid. *Fract. Calc Appl. Anal.* **2018**, *21*, 1073–1103. [CrossRef]
58. Shahed, M.E. On the impulsive motion of flat plate in a generalized second grade fluid. *Z. Naturfors. Sect. A-J. Phys. Sci.* **2004**, *59*, 829–837.

fractal and fractional

MDPI

Article

Dynamical Analysis of Two-Dimensional Fractional-Order-in-Time Biological Population Model Using Chebyshev Spectral Method

Ishtiaq Ali

Department of Mathematics and Statistics, College of Science, King Faisal University, P.O. Box 400, Al-Ahsa 31982, Saudi Arabia; iamirzada@kfu.edu.sa

Abstract: In this study, we investigate the application of fractional calculus to the mathematical modeling of biological systems, focusing on fractional-order-in-time partial differential equations (FTPDEs). Fractional derivatives, especially those defined in the Caputo sense, provide a useful tool for modeling memory and hereditary characteristics, which are problems that are frequently faced with integer-order models. We use the Chebyshev spectral approach for spatial derivatives, which is known for its faster convergence rate, in conjunction with the $L1$ scheme for time-fractional derivatives because of its high accuracy and robustness in handling nonlocal effects. A detailed theoretical analysis, followed by a number of numerical experiments, is performed to confirmed the theoretical justification. Our simulation results show that our numerical technique significantly improves the convergence rates, effectively tackles computing difficulties, and provides a realistic simulation of biological population dynamics.

Keywords: fractional-order-in-time biological population model; Chebyshev spectral method; error analysis; numerical examples

Citation: Ali, I. Dynamical Analysis of Two-Dimensional Fractional-Order-in-Time Biological Population Model Using Chebyshev Spectral Method. *Fractal Fract.* **2024**, *8*, 325. https://doi.org/10.3390/fractalfract8060325

Academic Editors: Libo Feng, Lin Liu and Yang Liu

Received: 28 April 2024
Revised: 23 May 2024
Accepted: 27 May 2024
Published: 29 May 2024

1. Introduction

Fractional calculus is an extension of conventional calculus that explores the possibility of taking real-numbered powers of the differentiation and integration operators. Unlike classical calculus, which only works with integer orders, fractional calculus accepts integrals and derivatives of any fractional order, positive or negative. The fractional integral of order α is a generalization of the n-fold integral applied n times, where α can be any real number. It is typically defined using a convolution with a power function, leading to an integral operator known as the Riemann–Liouville integral. Fractional calculus has a wide range of applications in disciplines including biology, engineering, physics, economics, and more. It is employed in physics to simulate systems that have memory and inherited characteristics, such as viscoelastic materials or processes in which the system's future state is dependent on both its past behavior and its current state. Fractional calculus is helpful to engineers in control theory, especially for designing controllers that provide reliable performance under a variety of circumstances [1–3]. Fractional derivatives are incorporated into standard partial differential equations in order to generate fractional partial differential equations (FPDEs), which enable the modeling of memory and hereditary features in a variety of materials and systems. Time-fractional differential equations, in which the order of the time derivative is a fraction instead of an integer, are the subject of this research. When modeling systems where the rate of change is not constant but rather depends on the process's whole history, these equations are especially helpful. Most recently, the application of fractional-order equation-based vegetation–water in an arid flat environment has been studied in [4].

The most common form of a time-fractional partial differential equation involves the Caputo or Riemann–Liouville definition of fractional derivatives. Time-fractional PDEs are extensively used to model physical phenomena exhibiting anomalous diffusion, which is diffusion where the flux at any point in space depends on the historical states of the system, rather than being directly proportional to the gradient of concentration as in classical diffusion. These models have been used to simulate subdiffusion or superdiffusion processes in complex systems where particle trajectories diverge from those of traditional Brownian motion in domains like physics. When molecular movement is impeded by barriers or binding effects that lead to anomalous diffusion, biology uses it to describe the transport dynamics within cells. In finance, on the other hand, it is employed to capture memory and hereditary effects in financial markets, enhancing asset price and risk management modeling [5–8].

The goal of the latest developments in fractional calculus has been to enhance the numerical techniques for solving fractional differential equations, which are crucial for real-world applications across a range of industries. Several novel techniques have been presented by researchers in an effort to more effectively handle the nonlocal characteristics included in fractional derivatives. Adaptive methods are among them; they improve accuracy and save computing costs by modifying the time steps based on the behavior of the answer. The ability to numerically solve differential equations with fractional derivatives, which are more difficult by nature because of their nonlocal nature, has been made possible by advances in computational mathematics. Modern numerical techniques and algorithms are still developing, which makes fractional calculus more useful in both industry and scientific study [9–14]. As fractional calculus advances, it will offer a more comprehensive mathematical framework that goes beyond the constraints of traditional differential and integral calculus, providing up new possibilities for modeling and understanding the natural world.

Biological population models are crucial and have contributed significantly to our understanding of ecology, public health, and the behavior of the environment. Because they aid in forecasting and controlling species interactions, population growth, and the impact of changing environmental circumstances on ecosystems, these models are crucial to ecology. They serve as crucial tools in conservation biology as well, helping scientists identify a species' vulnerability and develop effective conservation strategies. In order to evaluate programs, predict the beginning of epidemics, and comprehend how infectious diseases propagate, biological population models are essential to public health research [15,16]. A more comprehensive study on the role of fractional calculus in modeling a biological phenomena can be found in [17]. Fractional derivatives are used in biology to simulate anomalous diffusion, which deviates from the standard equations of diffusion in cells and tissues due to its ability to capture memory and hereditary properties inherent in such systems as compared to the integer-order differential equation. Due to the fact that these mathematical models, which consist of fractional-order derivatives, have a nonlocal nature, involve complicated boundary conditions and memory effects, and therefore lack closed-form solutions, the analytical solution of fractional differential equations is more complicated, and sometimes it is even not possible to find it. These models are especially helpful in ecosystems where organisms have lengthy lifespans or where environmental changes occur slowly, which reduces the efficacy of standard models.

A sophisticated method of applying fractional calculus to study biological population dynamics is introduced by the idea of a two-dimensional fractional-order-in-time biological population model. By utilizing fractional derivatives, which are expansions of ordinary derivatives, this mathematical framework expands upon the traditional population models by including memory and hereditary features. This model takes into account both the history and present population sizes when determining each population's growth rate. This is accomplished through the use of fractional differential equations, in which the degree to which the past influences the dynamics of the present is indicated by the derivative's order, typically between 0 and 1. These models are better able to capture the intricacies of biological processes that, as is common in many ecological systems, display long-term memory or power-law waiting times. Predators and prey are two examples of interacting animals that could be included in a two-dimensional model. Fractional derivatives are useful in modeling scenarios in which the dynamics of both populations in the future are influenced throughout time by the residual effects of past interactions (e.g., resource competition or predation rates). When contrasted to traditional models, this may result in oscillations and stability conditions that are more realistic. In general, biologists and ecologists can benefit greatly from the use of two-dimensional fractional-order models in biological populations since they offer a greater understanding of the dynamics of intricate ecosystems where past conditions have a substantial influence on future states. Analytically solving population models is often infeasible for complex boundary and initial conditions; thus, numerical solutions such as finite difference methods, finite element methods, and spectral methods are employed. These methods involve discretizing the time and space variables and approximating the fractional derivatives using techniques adapted for their nonlocal nature. The integration of fractional calculus into partial differential equations offers a powerful tool for exploring and describing dynamic systems with memory and hereditary characteristics, which are not adequately modeled by classical PDEs for these models. The development of robust numerical methods continues to be a vital area of research, enabling the practical application of FPDEs across various scientific disciplines [18,19].

Fractional-order diffusion equations, which are generalizations of classical diffusion equations and address super-diffusive flow processes, are among the most basic instances of the former. A large portion of published work on FPDEs has been focused on fractional diffusion equations. In [20], Agrawal uses a finite sine transform method to find the general solution of fractional-order diffusion equations. A theoretical framework using the least-squares finite element technique has been investigated in [21], while a spectral collocation scheme for two-dimensional nonlinear fractional diffusion equations and a radial basis function approximation method is used in [22,23], respectively. A mathematical model based on nonlinear fractional-order equations for the description of the behavior of viscoplastic materials was developed in [24], while a fractional advection–dispersion equation has been numerically investigated in [25,26]. In order to find the scale-invariant solution of the TFDE in terms of the Wright function, Gorenflo et al. employed the similarity approach and the Laplace transform method [27,28]. Numerous authors have examined these models in analytical and numerical frameworks. A few of these researchers have attempted to find analytical solutions for differential equations using temporal fractions. For instance, time-fractional diffusion-wave equations were taken into consideration by Schneider and Wyss and Wyss [29,30]. The temporal fractional PDE using finite difference in fractional time and a higher numerical scheme was investigated in detail in [31]. Capturing memory and hereditary properties is crucial in biological systems due to their inherent dependence on past states. Fractional differential equations are particularly adept at modeling these characteristics because they incorporate nonlocal properties, meaning the rate of change at any point in time depends on all previous states. However, these equations are complex and often lack closed-form solutions, making analytical solutions challenging and necessitating robust numerical methods. The complexity of fractional derivatives, especially their nonlocal nature, poses significant computational challenges that need to be addressed with efficient numerical techniques.

In this work, we consider solving numerically a fractional-order-in-time biological population model in two dimensions. We use the $L1$ scheme for fractional-order-in-time derivatives. This scheme is very useful in handling the nonlocal properties of fractional-order-in-time derivatives. For spatial derivatives, we use an efficient numerical scheme based on the Chebyshev spectral method. Spectral methods are a class of techniques used in numerical analysis to solve differential equations whose solutions are essentially represented by the sum of globally defined basis functions. These functions are typically orthogonal or trigonometric polynomials, depending on the boundary conditions and nature of the problem. When the solution has smooth properties, spectral methods are well known for their high accuracy and exponential rates of convergence, which greatly exceed the performance provided by conventional finite difference or finite element methods. The most notable method among these is the Chebyshev spectral method due to its particular effectiveness in handling problems with complex geometries or boundary conditions. This method utilizes Chebyshev polynomials, which are a set of orthogonal polynomials defined on the interval $[-1, 1]$. The main advantage of Chebyshev polynomials is their near-minimal property for polynomial approximation. This property significantly minimizes the maximum error between the numerical solution and the true solution. The efficiency of this approach is further increased by the employment of Chebyshev nodes, or the roots of these polynomials, which assist prevent the Runge phenomenon and guarantee stability and dependability in numerical approximations [32–40]. Several studies have applied fractional calculus to model biological systems. For instance, fractional derivatives have been used to simulate anomalous diffusion in cells and tissues, capturing more realistic dynamics than traditional models. Hattaf and Yousfi explored the global stability of fractional diffusion equations in biological systems, highlighting the advantages of fractional models in capturing complex biological behaviors [41]. However, many of these studies focus on one-dimensional models or rely on analytical solutions, which are not feasible for complex systems. This paper addresses this gap by developing a numerical scheme for a two-dimensional fractional-order-in-time biological population model [42,43].

The rest of the paper is organized as follows: In Section 2, we present some preliminaries and some basic definitions, which play a key role in the analysis of the scheme, followed by the mathematical description of the model and the discretization scheme. Section 4 consists of a detailed error analysis, followed by numerical examples in Section 5. Section 6 consists of concluding remarks.

2. Preliminaries and Some Basic Definitions

In this section, we present some basic definitions related to fractional calculus theory and orthogonal polynomials that are useful in the error analysis of our proposed numerical scheme [44–47].

Definition 1. *A real function $g(\tau)$, $\tau > 0$ is defined to belong to the space C_ν, $\nu \in \mathbb{R}$, if there exists a real number $q > \nu$, such that $g(\tau) = \tau^q g_1(\tau)$, where $g_1(\tau) \in C(0,\infty)$, and g is in the space $C_{n\nu}$ if and only if $g^{(n)} \in C_\nu$, $n \in \mathbb{N}$.*

Definition 2. *The Riemann–Liouville fractional integral operator I^β ($\beta \geq 0$) of a function $g \in C_\nu$, $\nu \geq -1$ is given by*

$$I^\beta g(\tau) = \frac{1}{\Gamma(\beta)} \int_0^\tau (\tau - s)^{\beta-1} g(s)\, ds, \quad (\beta \geq 0) \tag{1}$$

where $\Gamma(\cdot)$ is the gamma function. The properties of the operator I^β include

$$I^\beta I^\gamma g(\tau) = I^{\beta+\gamma} g(\tau), \quad (\beta \geq 0, \gamma \geq 0) \tag{2}$$

$$I^\beta \tau^\delta = \frac{\Gamma(1+\delta)}{\Gamma(1+\delta+\beta)} \tau^{\beta+\delta}, \quad (\delta \geq -1) \tag{3}$$

Definition 3. *The Caputo fractional derivative C^α of a function $g(\tau)$ is defined as*

$${}^C D_\tau^\alpha g(\tau) = \frac{1}{\Gamma(n-\alpha)} \int_0^\tau g^{(n)}(s) \frac{ds}{(\tau - s)^{\alpha+1-n}}, \quad (n-1 < Re(\alpha) \leq n, n \in \mathbb{N}) \tag{4}$$

The properties of the Caputo fractional derivative include

$${}^C D_\tau^\alpha \tau^\beta = \frac{\Gamma(1+\beta)}{\Gamma(1+\beta-\alpha)} \tau^{\beta-\alpha}, \tag{5}$$

$$(I^{\alpha C}D^{\alpha})g(\tau) = g(\tau) - \sum_{k=0}^{n-1} \frac{g^{(k)}(0+)}{k!}\tau^k, \tag{6}$$

The Caputo fractional derivative is chosen for its ability to incorporate conventional initial and boundary conditions in problem formulations.

Definition 4. *The Mittag–Leffler function $E_{\alpha,\beta}(z)$ for complex numbers z and parameters α and β is defined by the series*

$$E_{\alpha,\beta}(z) = \sum_{k=0}^{\infty} \frac{z^k}{\Gamma(\alpha k + \beta)}, \tag{7}$$

where $\alpha > 0$ and β are real numbers, and $\Gamma(\cdot)$ denotes the Gamma function. This function generalizes the exponential function, which is recovered as a special case when $\alpha = 1$ and $\beta = 1$.

$E_{\alpha,\beta}(z)$ represents a two-parameter family of functions, where α and β allow for various forms of convergence and divergence characteristics depending on their values. When $\alpha = 1$ and $\beta = 1$, the Mittag–Leffler function simplifies to the exponential function e^z, that is $E_{1,1}(z) = e^z$.

Definition 5. *Chebyshev polynomials, represented as $C_n(x)$, are established over the domain $[-1,1]$ and can be formulated by the expression*

$$C_n(x) = \cos(n \arccos(x)), \quad n = 0,1,2,\ldots \tag{8}$$

Definition 6. *The collection of Chebyshev polynomials $\{C_n(x)\}$ set within the interval $[-1,1]$ adhere to this orthogonality condition with respect to the weighted scalar product*

$$\langle C_i, C_j\rangle_\omega := \int_{-1}^{1} C_i(x)C_j(x)\omega(x)\,dx = \begin{cases} 0, & \text{if } i \neq j, \\ \pi, & \text{if } i = j = 0, \\ \frac{\pi}{2}, & \text{if } i = j \neq 0. \end{cases} \tag{9}$$

where the weighting function $\omega(x)$ is defined by

$$\omega(x) = \frac{1}{\sqrt{1-x^2}}. \tag{10}$$

Definition 7. *Let δ_n denote the scaling coefficients for orthogonality:*

$$\delta_n = \begin{cases} 2, & \text{if } n = 0, \\ 1, & \text{if } n > 0. \end{cases} \tag{11}$$

Considering practical approaches in mathematical representations, when examining polynomials of a degree at most N, the matrix representation of the weighted scalar product can be succinctly illustrated as

$$H = \operatorname{diag}\{h_{ii}\}, \quad h_{ii} := \frac{\sigma}{2}\delta_n \tag{12}$$

Here, the matrix H is diagonal with size $(N+1) \times (N+1)$, and each diagonal element h_{ii} is influenced by the coefficient δ_n from the previous equation.

Definition 8. *A smooth, continuous function $\sigma(\theta)$, defined on $[-1,1]$, can be approximated using Chebyshev polynomials $C_n(\theta)$ as follows:*

$$\sigma(\theta) \approx \sum_{n=0}^{N} \widetilde{\sigma}_n C_n(\theta), \tag{13}$$

where N represents the truncation level in the spectral method, and $\widetilde{\sigma}_n$ is the Chebyshev expansion coefficient for σ. This coefficient is calculated by

$$\widetilde{\sigma}_n = \frac{1}{\delta_n}\int_{-1}^{1} \sigma(\theta)C_n(\theta)\omega(\theta)\,d\theta, \quad \delta_n = \int_{-1}^{1} C_n^2(\theta)\sqrt{1-\theta^2}\,d\theta = \begin{cases} \pi, & \text{if } n = 0, \\ \frac{\pi}{2}, & \text{if } n > 0. \end{cases} \tag{14}$$

Definition 9. *In the computational process known as the Chebyshev forward transformation, the Gauss–Lobatto integration method is typically employed for computing the weighted integral:*

$$\widetilde{\sigma}_n \approx \frac{1}{\delta_n}\sum_{i=0}^{N} \sigma(\theta_i)C_n(\theta_i)w_i \tag{15}$$

where N represents the number of integration nodes, matching the truncation level. These nodes and weights are specified as

$$\theta_0 = 1, \theta_N = -1, \theta_i = \cos\left(\frac{\pi i}{N}\right), w_0 = w_N = \frac{\pi}{2N}, w_i = \frac{\pi}{N} \tag{16}$$

Definition 10. *For any continuous and smooth function $\sigma(\theta)$ defined over the interval $[-1,1]$, if the derivative $\sigma'(\theta)$ maintains smoothness, it can be expanded using $C_n(\theta)$ as*

$$\sigma'(\theta) \approx \sum_{n=0}^{N} \widetilde{\sigma}_n' C_n(\theta), \tag{17}$$

It is established that the expansion coefficient $\widetilde{\sigma}_n'$ for the derivative $\sigma'(\theta)$ and the coefficient $\widetilde{\sigma}_n$ for the original function satisfy the relation

$$\widetilde{\sigma}_n' \approx \frac{2}{\delta_n}\sum_{q=n+1,\,q+n\,odd}^{N} q\widetilde{\sigma}_q, \quad n \geq 0 \tag{18}$$

This relation can then be reformulated in matrix terminology:

$$\widetilde{\sigma}' = D\widetilde{\sigma}, \tag{19}$$

where $\widetilde{\sigma}'$ and $\widetilde{\sigma}$ are arrays with dimensions $1 \times (N+1)$, defined as $\widetilde{\sigma}' = [\widetilde{\sigma}_0', \widetilde{\sigma}_1', \ldots, \widetilde{\sigma}_N']$ and $\widetilde{\sigma} = [\widetilde{\sigma}_0, \widetilde{\sigma}_1, \ldots, \widetilde{\sigma}_N]$. The matrix D is an upper triangular square matrix, sized $(N-1) \times (N+1)$.

Definition 11. *The natural Sobolev norms, appropriate for gauging approximation errors within the Chebyshev framework, integrate the Chebyshev weight into the quadratic means of the error and its derivatives over the span $(-1,1)$. Thus, we establish the weighted Sobolev norm as*

$$\|g\|_{\mathcal{H}_m^w(-1,1)} = \left(\sum_{n=0}^{m} \|g^{(n)}\|^2_{L^2_w(-1,1)}\right)^{1/2}. \tag{20}$$

The associated Hilbert space is denoted $\mathcal{H}_m^w(-1,1)$, where:

- *$\|g^{(n)}\|_{L^2_w(-1,1)}$ signifies the L^2-norm of the n-th derivative of g, weighted over the interval $(-1,1)$.*
- *$\mathcal{H}_m^w(-1,1)$ is the weighted Sobolev space capturing the behavior of functions and their derivatives up to order m under the weighted norm.*

3. Fractional-Order-in-Time Dispersal in Population Dynamics

According to biologists, migration or dispersal has a significant impact on the regulation of species populations. The diffusion of a biological species in a given region C is described by three functions of position $\mathbf{x} = (\eta, \xi)$ and time t: the population supply $s(\mathbf{x}, t)$, the diffusion velocity $\mathbf{u}(\mathbf{x}, t)$, and the population density $\rho(\mathbf{x}, t)$. The population density $\rho(\mathbf{x}, t)$ quantifies the number of individuals per unit volume at position $\mathbf{x}$ and at time t. By integrating ρ over any subregion D of C, the total population in D at time t is obtained. The term $s(\mathbf{x}, t)$ indicates the rate at which individuals are added or removed per unit volume at position $\mathbf{x}$ due to births and deaths. The diffusion velocity $\mathbf{u}(\mathbf{x}, t)$ represents the mean velocity of individuals at position $\mathbf{x}$ at time t, facilitating the description of population movement from one location to another.

The following equation governs the dynamics:

$$\frac{\partial^\alpha}{\partial t^\alpha} \int_D \rho \, dV + \int_{\partial D} \rho \mathbf{u} \cdot \mathbf{n} \, dA = \int_D s \, dV, \tag{21}$$

where $\mathbf{n}$ is the outward unit normal on the boundary ∂D. The Caputo fractional derivative is used to interpret the derivative. This basic equation means that the rate at which new individuals are delivered directly to D must equal the rate at which the population within D changes plus the rate at which people leave D over its boundary. With $s = s(\rho)$ and $\mathbf{u} = \lambda(\rho)\nabla\rho$ as assumptions, one can derive the following two-dimensional nonlinear degenerate parabolic partial differential equation for ρ, where $\lambda(\rho) > 0$ for $\rho > 0$ and ∇ is the gradient operator:

$$\frac{\partial^\alpha \rho}{\partial t^\alpha} = \Delta(s(\rho)) + s(\rho), \quad t > 0, \quad (\eta, \xi) \in \mathbb{R}^2, \tag{22}$$

where the order of the fractional derivative with respect to time t is denoted by α. The temporal fractional-order biological population model (TFBPM) is the name given to this equation. The analysis presents $\phi(\rho) = \rho^2$ as a specific instance for modeling animal populations, as explored by Gurney and Nisbet [48]. The migrations usually happen as a result of individuals traveling down the population density gradient, which moves more quickly at higher densities, in search of less congested areas to breed in a model that took into account an animal walking through a rectangular grid was created to mimic this behavior. With each step, the animal might either remain in the same spot or migrate toward the area with the lowest density. The size of the population density gradient at the relevant grid boundary determines the probability distribution for these movements. This model leads to

$$\frac{\partial^\alpha \rho}{\partial t^\alpha} = \Delta(\rho^2) + s(\rho), \quad t > 0, \quad (\eta, \xi) \in \mathbb{R}^2, \tag{23}$$

with the initial condition $\rho(\eta, \xi, 0)$ provided. When $\alpha = 1$, this equation simplifies to the normal biological population model (NBPM):

$$\frac{\partial^\alpha \rho}{\partial t^\alpha} = \Delta(\rho^2) + s(\rho), \quad t > 0, \quad (\eta, \xi) \in \mathbb{R}^2. \tag{24}$$

Additionally, various properties such as Hölder estimates and solutions of this model have been explored.

Constitutive equations for $s(\rho)$ may include the following:

- Malthusian Law: $s(\rho) = c\rho$, where c is a constant.
- Verhulst Law: $s(\rho) = c_1\rho - c_2\rho^2$, where c_1, c_2 are positive constants.
- Porous Media: $s(\rho) = c\rho^q$, where $c > 0$ and $0 < q < 1$.

For a generalized form, consider

$$\frac{\partial^\alpha \rho}{\partial t^\alpha} = \frac{\partial^2 \rho^2}{\partial \eta^2} + \frac{\partial^2 \rho^2}{\partial \xi^2} + h\rho^a(1 - \ell\rho^b), \quad t > 0, \quad (\eta, \xi) \in \mathbb{R}^2, \tag{25}$$

subject to some appropriate initial conditions and where h, a, ℓ, b are real numbers. Under some specific parameter conditions, both the Verhulst and the Malthusian laws are covered by Equation (25).

Discretization Methodology

We consider the time-fractional partial differential equation given by Equation (25):

The $L1$ scheme is used to discretize the fractional time derivative $\frac{\partial^\alpha \rho}{\partial t^\alpha}$. This scheme is particularly suitable for fractional derivatives due to its capability to handle the memory effect inherent in such derivatives. The $L1$ approximation at a time t_n is defined as: The $L1$ scheme approximates the fractional time derivative as follows:

$$\frac{\partial^\alpha \rho_{n+1}}{\partial t^\alpha} = \frac{(\delta t)^{-\alpha}}{\Gamma(2-\alpha)} \sum_{k=0}^{n} (\rho_{n+1-k} - \rho_{n-k}) \left((k+1)^{1-\alpha} - k^{1-\alpha} \right) + O(\delta t^{2-\alpha}), \tag{26}$$

where δt is the time step, and α is the fractional order of the derivative.

The correction term B_n, defined as

$$B_n = \frac{(\delta t)^{-\alpha}}{\Gamma(2-\alpha)} \sum_{k=1}^{n} \left((k+1)^{1-\alpha} - k^{1-\alpha} \right) (\rho_{n+1-k} - \rho_{n-k}), \tag{27}$$

becomes zero when $n = 0$. Therefore, the discretization formula simplifies as

$$\frac{\partial^\alpha \rho_{n+1}}{\partial t^\alpha} = \begin{cases} \frac{(\delta t)^{-\alpha}}{\Gamma(2-\alpha)} (\rho_{n+1} - \rho_n) + B_n + O(\delta t^{2-\alpha}), & \alpha \in (0,1), \\ \frac{\rho_{n+1} - \rho_n}{\delta t} + O(\delta t), & \alpha = 1. \end{cases} \tag{28}$$

The Chebyshev spectral method involves representing the solution ρ as a series expansion in terms of Chebyshev polynomials. These polynomials are particularly effective for approximating functions on bounded intervals due to their excellent approximation properties and the clustering of nodes at the endpoints, which can help in resolving boundary layer effects. The approximation of ρ at a fixed time t_n is given by

$$\rho(\eta, \xi, t_n) \approx \sum_{i=0}^{N} \sum_{j=0}^{N} c_{ij}^n T_i(\eta) T_j(\xi), \tag{29}$$

where T_i are Chebyshev polynomials and c_{ij}^n are the spectral coefficients.

The derivatives in the spatial domain are computed using the derivative properties of Chebyshev polynomials:

$$\frac{\partial^2 \rho^2}{\partial \eta^2} \approx \sum_{i=0}^{N} \sum_{j=0}^{N} \left(c_{ij}^n \right)^2 \frac{d^2}{d\eta^2} T_i(\eta) T_j(\xi), \tag{30}$$

$$\frac{\partial^2 \rho^2}{\partial \xi^2} \approx \sum_{i=0}^{N} \sum_{j=0}^{N} \left(c_{ij}^n \right)^2 T_i(\eta) \frac{d^2}{d\xi^2} T_j(\xi). \tag{31}$$

Combining the discretized forms of the time and space derivatives, we obtain the fully discretized version of the the model Equation (25), given by

$$\begin{aligned} &\frac{1}{\Gamma(2-\alpha)} \sum_{k=0}^{n} \frac{\rho^{k+1} - \rho^k}{\Delta t^\alpha} ((k+1)\Delta t)^{1-\alpha} - (k\Delta t)^{1-\alpha} \\ &\quad = \sum_{i=0}^{N} \sum_{j=0}^{N} \left[(c_{ij}^n)^2 \left(\frac{d^2}{d\eta^2} T_i(\eta) + \frac{d^2}{d\xi^2} T_j(\xi) \right) \right] \\ &\quad + h \left(\sum_{i=0}^{N} \sum_{j=0}^{N} c_{ij}^n T_i(\eta) T_j(\xi) \right)^a \left(1 - \ell \left(\sum_{i=0}^{N} \sum_{j=0}^{N} c_{ij}^n T_i(\eta) T_i(\xi) \right)^b \right). \end{aligned} \tag{32}$$

4. Error Analysis

This section deals with the error analysis of our proposed numerical scheme. Before the main results, we state some useful results in the form of lemmas [49].

Lemma 1 (Estimate for the Truncation Error). *Consider the truncation error of a function u when approximated by its truncated Chebyshev series, denoted as $P_N u$, where*

$$P_N u = \sum_{k=0}^{N} \hat{u}_k T_k,$$

and $\hat{u}_k$ represents the Chebyshev-series coefficients of u. The truncation error $u - P_N u$, measured in the weighted L^2 norm over the interval $(-1,1)$, satisfies the inequality

$$\|u - P_N u\|_{L^2_w(-1,1)} \leq CN^{-m} \|u\|_{\mathcal{H}^w_{m;N}(-1,1)}, \tag{33}$$

for all functions u belonging to the weighted Sobolev space $\mathcal{H}^w_m(-1,1)$ with $m \geq 0$. Here, C is a constant that depends on N and m, and $\mathcal{H}^w_{m;N}(-1,1)$ represents the Sobolev space characterized by integrating the function and its derivatives up to order m, each weighted by the Chebyshev weight.

Lemma 2 (Interpolation Error Estimate). *The interpolation error associated with approximating a function u by its interpolant $I_N u$, which is defined at Chebyshev Gauss points across three different families. For this interpolant, which belongs to the polynomial space P_N, an important error estimates. The error between the function u and its interpolant, measured in the weighted L^2 norm over the interval $[-1,1]$, adheres to the following bound:*

$$\|u - I_N u\|_{L^2_w(-1,1)} \leq CN^{-m} \|u\|_{H^w_{m;N}(-1,1)}, \tag{34}$$

$$\|u - I_N u\|_{L^\infty_w(-1,1)} \leq CN^{\frac{1}{2}-m} \|u\|_{H^w_{m;N}(-1,1)}, \tag{35}$$

where C represents a constant and $m \geq 1$ reflects the smoothness level of the function. This estimate is valid under the condition that u is an element of the weighted Sobolev space $H^w_m(-1,1)$. This space is characterized by considering functions that maintain their derivatives up to the m-th order, each weighted appropriately over the interval.

Lemma 3 (Integration Error Estimate). *Given a function u from the weighted Sobolev space $H^w_m(-1,1)$ with $m \geq 1$ and a polynomial φ from the space P_N, the error produced by applying a Gauss-type quadrature formula for integration relative to the Chebyshev weight can be estimated as follows. Consider the integral of the product of u and φ, weighted by the Chebyshev weight $w(x)$, and its approximation by the quadrature formula:*

$$\left| \int_{-1}^{1} u(x)\varphi(x)w(x)\,dx - (u,\varphi)_N \right| \leq CN^{-m} \|u\|_{H^w_{m;N}(-1,1)} \|\varphi\|_{L^2_w(-1,1)}, \tag{36}$$

where

- *$(u,\varphi)_N$ denotes the approximation of the integral by the Gauss-type quadrature.*
- *C is a constant dependent on N and m, indicating the rate at which the error diminishes as the polynomial degree N or the smoothness m of the function u increases.*
- *$\|u\|_{H^w_{m;N}(-1,1)}$ and $\|\varphi\|_{L^2_w(-1,1)}$ are the norms measuring the magnitude of u and φ in their respective function spaces.*

One can state and prove the following error estimates for the fractional-order time discretization.

Theorem 1 ([31]). *Let ρ be the exact solution of the model Equation (25), and let $\{\rho_k\}_{k=0}^K$ be the time-discrete solution of with the initial condition $\rho_0(x) = \rho(x,0)$; then, the following error estimates hold:*

1. *For $0 \leq a < 1$,*

$$\|\rho(t_k) - \rho_k\|_1 \leq c_{u,a} T^a \Delta t^{2-a}, \quad k = 1,2,\ldots,K, \tag{37}$$

where $c_{u,a} := \frac{c_u}{1-a}$ with c_u a constant defined in (3.3).

2. *As $a \to 1$,*

$$\|\rho(t_k) - \rho_k\|_1 \leq c_u T \Delta t, \quad k = 1,2,\ldots,K. \tag{38}$$

Similarly for the space discretization, using Lemmas 1–3, one can state and prove the following results of exponential order of convergence.

Theorem 2. *Let $\{\rho_k^N\}_{k=0}^K$ be the solution of the problem given in Equation (25) with the initial condition ρ_0^N taken to be $I_N \rho(0)$, and $\{\rho_k\}_{k=0}^K$ be the numerical solution. Suppose $\rho_k \in H^m(K) \cap H_1^0(K)$, $m > 1$; then,*

- *For $0 \leq \alpha < 1$,*

$$\|\rho_k - \rho_k^N\|_{L_2} \leq c_\alpha \Delta t^\alpha N^{-m} \max_{0 \leq j \leq k} \|\rho_j\|_m, \quad k = 1,2,\ldots,K, \tag{39}$$

where $c_\alpha = \frac{c}{1-\alpha}$ with c a constant.

- *As $\alpha \to 1$,*

$$\|\rho_k - \rho_k^N\|_{L_2} \leq c \Delta t^{-1} N^{-m} \max_{0 \leq j \leq k} \|\rho_j\|_m, \quad k = 1,2,\ldots,K, \tag{40}$$

where c only depends on T.

Proof. Assume $r^N = \rho_{k+1}^N - w^N$ where w^N is any function in $P_0^N(K)$. By direct calculation,

$$A^N(r^N, r^N) = A(\rho_{k+1} - w^N, r^N) + A(w^N, r^N) - A^N(w^N, r^N) + F^N(r^N) - F(r^N), \tag{41}$$

which leads to

$$\begin{aligned} \|r^N\|_{1,N}^2 \leq \|\rho_{k+1} - w^N\|_1 \|r^N\|_1 + \left|A(w^N, r^N) - A^N(w^N, r^N)\right| \\ + \left|F(r^N) - F^N(r^N)\right| \quad \forall w^N \in P_0^{N-1}(K). \end{aligned} \tag{42}$$

$$\begin{aligned} \left|F(r^N) - F^N(r^N)\right| = \Big|(1-b_1)\left[(\rho_k, r^N) - (\rho_k^N, r^N)^N\right] \\ + \sum_{j=1}^{k-1} (b_j - b_{j+1})\left[(\rho_{k-j}, r^N) - (\rho_{k-j}^N, r^N)^N\right] \\ + b_k\left[(\rho_0, r^N) - (\rho_0^N, r^N)^N\right]\Big|. \end{aligned} \tag{43}$$

$$(g, r^N) - (g^N, r^N)^N \leq (cN^{-m}\|g\|_m + \|g - g^N\|_{L_2})\|r^N\|_{L_2}. \tag{44}$$

Applying this to $g = \rho_j$; $g^N = \rho_j^N$ for all $j = 1,2,\ldots,k$, we derive

$$\begin{aligned} \left|F(r^N) - F^N(r^N)\right| \leq (1-b_1)\|e_k^N\|_{L_2} + \sum_{j=1}^{k-1}(b_j - b_{j+1})\|e_{k-j}^N\|_{L_2} \\ + b_k\|e_0^N\|_{L_2} + cN^{-m} \max_{0 \leq j \leq k} \|\rho_j\|_m \|r^N\|_{L_2}. \end{aligned} \tag{45}$$

This shows that the order of convergence of our combined numerical scheme is $O(\delta t^{2-\alpha} + N^{-m})$ in the L_2 norm in the case of spatial coordinates. Similarly, if we use the L_∞ norm in the space coordinate, then the error estimate is of $O(\delta t^{2-\alpha} + N^{\frac{1}{2}-m})$. □

5. Numerical Results

We present in this section a number of numerical examples to confirm our theoretical justification, as given in Theorems 1 and 2. In our simulations, we use a fixed time $t = 0.1$ in comparison between the exact solution and the solution obtained by using the proposed scheme. In addition to this, in all of our simulations, we use a PC with processor 12th Gen Intel(R) Core(TM) i7-1255U and 16 GB RAM. The average CPU time in all these simulations is 0.03 s for collocation points from 15–50. The algorithm for these simulations is given in Appendix A.

Example 1. *Consider the model Equation (25) with $a = 0, b = 1$, that is, we have the following linear fractional-order-in-time biological population model of the form*

$$\frac{\partial^\alpha \rho}{\partial t^\alpha} = \frac{\partial^2 \rho^2}{\partial \eta^2} + \frac{\partial^2 \rho^2}{\partial \xi^2} + h\rho, \tag{46}$$

subject to the initial condition:

$$\rho_0 = \sqrt{\eta\xi}. \tag{47}$$

Equation (46), subject to the initial condition given in Equation (47), corresponds to Malthusian law. The exact solution is given by

$$\rho(\eta, \xi, t) = \sqrt{\eta\xi} e^{ht} \mathcal{E}_\alpha h t^\alpha, \tag{48}$$

where $\mathcal{E}_\alpha h t^\alpha$ is the Mittag–Leffler function. When $\alpha \to 1$, then the exact solution becomes

$$\rho(\eta, \xi, t) = \sqrt{\eta\xi} \sum_{n=0}^{\infty} \frac{(ht)^n}{\Gamma(1+n)} = \sqrt{\eta\xi} e^{ht}. \tag{49}$$

In our simulations, we use the fractional-order $\alpha = 0.5$. We compare our results for different collocation points with the exact solution as shown in Figures 1–3. The error behavior between the exact and numerical solution is shown in Figure 4. One can see that the error between the exact and approximate solution decays exponentially by increasing the number of collocation points. These results emphasize the reliability of our numerical scheme, which is crucial for accurately predicting population dynamics in more detailed biological studies, making it practical for real-world biological population models where computational resources may be limited.

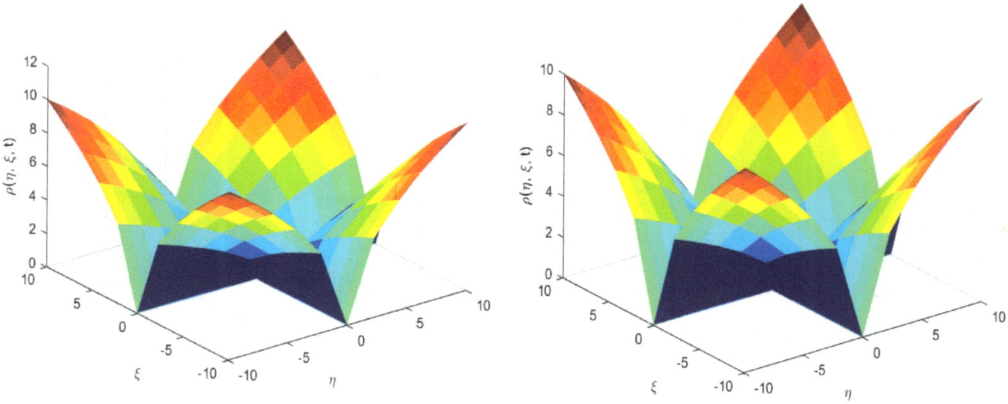

Figure 1. Example 1: Exact (**left**) vs. numerical solution (**right**) at $N = 15$ collocation points.

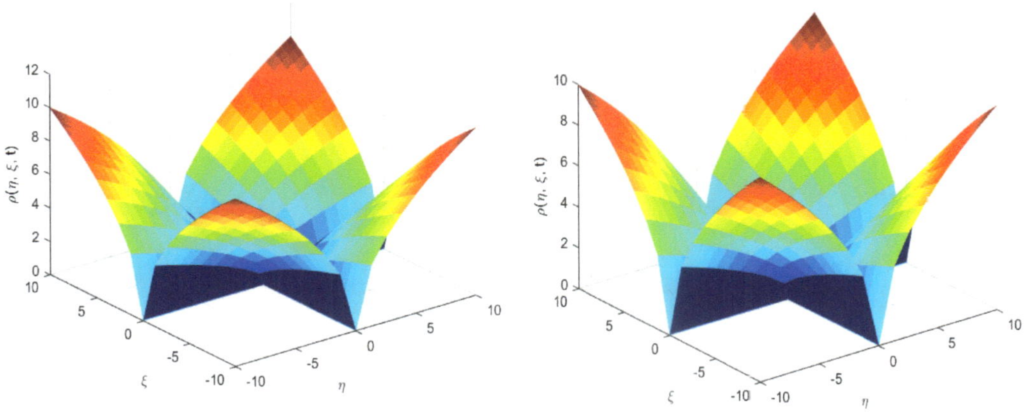

Figure 2. Example 1: Exact (**left**) vs. numerical solution (**right**) at $N = 25$ collocation points.

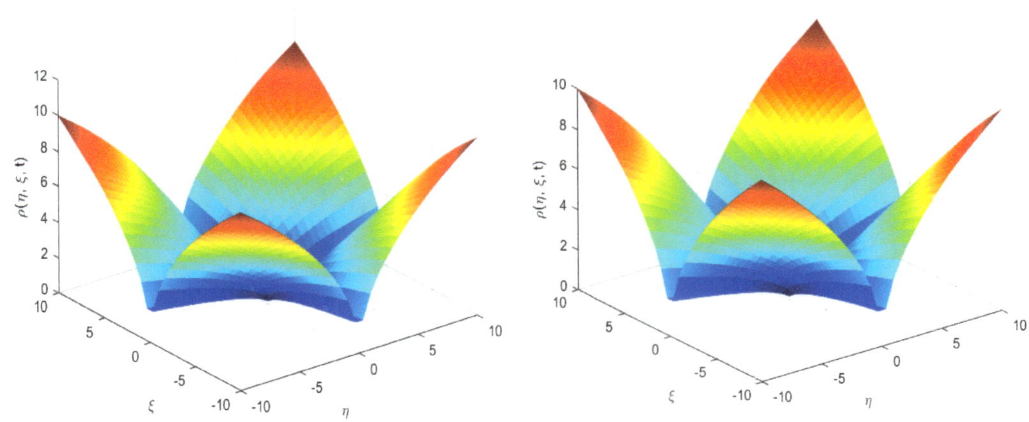

Figure 3. Example 1: Exact (**left**) vs. numerical solution (**right**) at $N = 50$ collocation points.

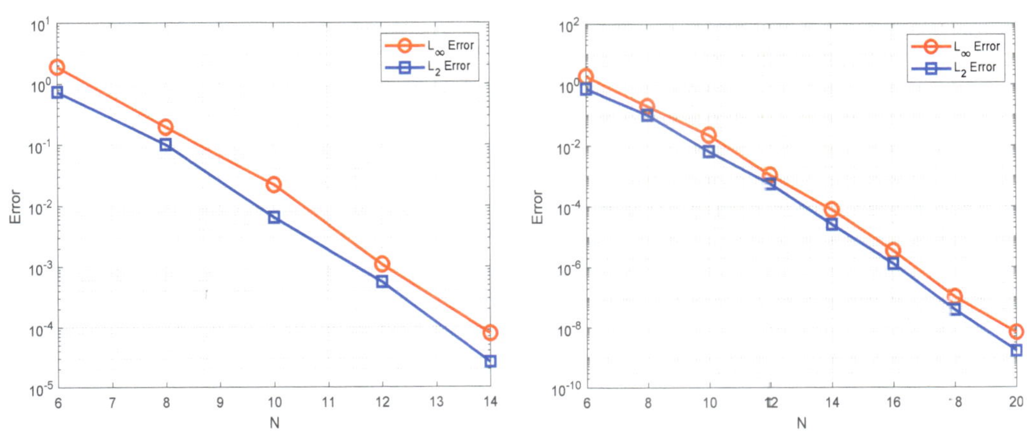

Figure 4. Example 1: Error behavior of exact solution vs. numerical solution at $N = 15$ (**left**) and at $N = 20$ (**right**).

Example 2. *Consider the model Equation (25) with $a = 1, b = 1$. This gives us Verhulst law, and the model Equation (25) becomes*

$$\frac{\partial^\alpha \rho}{\partial t^\alpha} = \frac{\partial^2 \rho^2}{\partial \eta^2} + \frac{\partial^2 \rho^2}{\partial \xi^2} + h\rho(1 - \ell\rho), \tag{50}$$

subject to the initial condition

$$\rho_0 = e^{\sqrt{\frac{h\ell}{8}}(\eta+\xi)}, \tag{51}$$

The exact solution is given by

$$\rho_0 = e^{\sqrt{\frac{h\ell}{8}}(\eta+\xi)} \mathcal{E}_\alpha h t^\alpha \tag{52}$$

For $\alpha \to 1$, the exact solution has the form

$$\rho_0 = e^{\sqrt{\frac{h\ell}{8}}(\eta+\xi)+ht}. \tag{53}$$

The numerical experiments are performed between exact and approximate solution for $\alpha = 0.5$ and different collocations points, as shown in Figures 5–7. The error behavior between the exact and numerical solution is given in Table 1. Again, one can see how fast the error is decreasing while increasing the number of collocation points. The numerical solution accurately captures the behavior of the exact solution, which represents the population growth with a carrying capacity, highlighting how the population stabilizes over time, which is a key aspect in ecological studies. The high-level accuracy is essential for making reliable predictions, especially in scenarios involving critical decisions about species conservation or resource management.

Table 1. Example 2: Error behavior of exact vs. numerical solution.

N	L_∞	L_2	N	L_∞	L_2
6	9.541×10^{-1}	4.086×10^{-1}	8	4.842×10^{-2}	2.485×10^{-2}
10	8.145×10^{-3}	3.139×10^{-3}	12	2.670×10^{-4}	1.210×10^{-4}
14	1.650×10^{-5}	7.114×10^{-6}	16	4.706×10^{-7}	2.099×10^{-7}
18	2.132×10^{-8}	6.028×10^{-9}	20	5.213×10^{-10}	1.906×10^{-10}

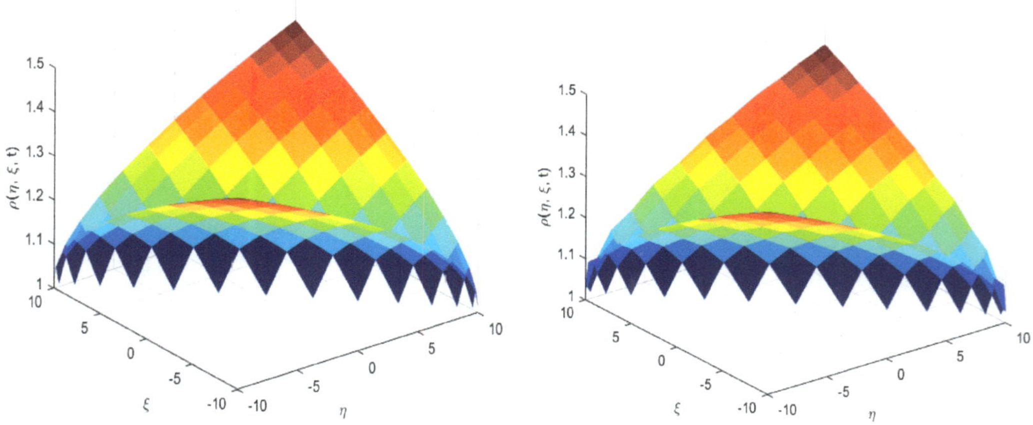

Figure 5. Example 2: Exact (**left**) vs. numerical solution (**right**) at $N = 15$ collocation points.

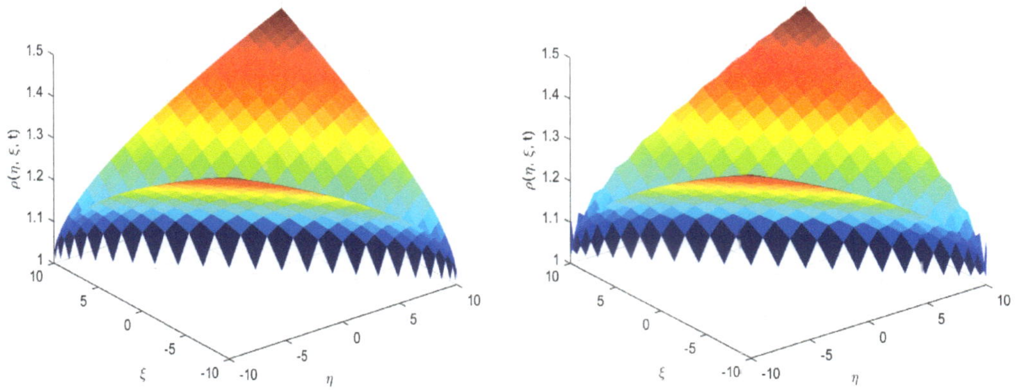

Figure 6. Example 2: Exact (**left**) vs. numerical solution (**right**) at $N = 25$ collocation points.

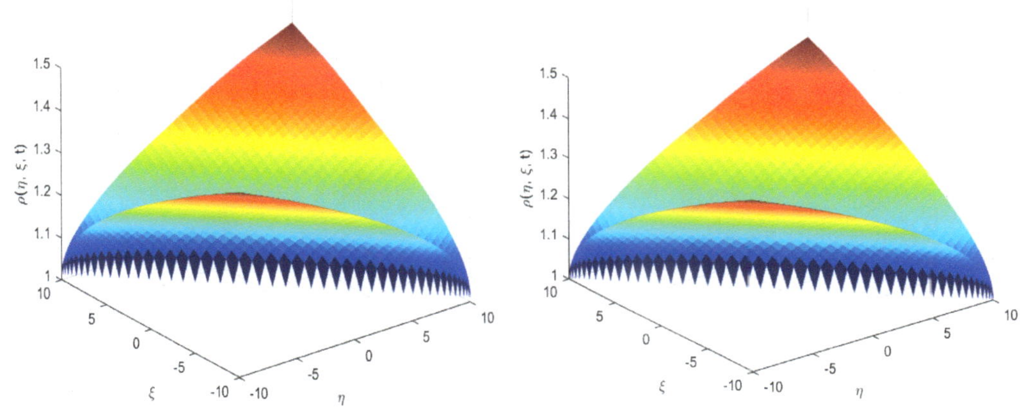

Figure 7. Example 2: Exact (**left**) vs. numerical solution (**right**) at $N = 50$ collocation points.

Example 3. *In this example, we choose $a = -1, b = 1$. This gives us the model Equation (25) of the form*

$$\frac{\partial^\alpha \rho}{\partial t^\alpha} = \frac{\partial^2 \rho^2}{\partial \eta^2} + \frac{\partial^2 \rho^2}{\partial \xi^2} + h(\rho^{-1} - \ell), \tag{54}$$

subject to the initial condition

$$\rho_0 = \sqrt{\frac{h\ell}{4}\eta^2 + \frac{h\ell}{4}\xi^2 + \xi - 5}, \tag{55}$$

The exact solution is given by

$$\rho(\eta, \xi, t) = \rho_0 + \sum_{n=0}^{\infty} \frac{(ht^\alpha)\rho_0}{\Gamma(1 + (n+1)\alpha)} + \left(\frac{ht^\alpha}{\rho_0^2}\right)^n \tag{56}$$

For $\alpha \to 1$, the transform exact solution is

$$\rho(\eta, \xi, t) = \rho_0 + \frac{ht}{\rho_0} e^{\frac{ht}{\rho_0^2}}. \tag{57}$$

Figures 8–10 show the comparison between the actual and numerical solution for $\alpha = 0.5$ with different collocations points. Our simulations confirmed that our numerical method performed very well against the exact solution. Further validation of exponential error decays is shown in Table 2. These results illustrate the model's capability to handle nonlinear dynamics, which are common in biological systems where interactions between species or within populations can lead to complex behaviors.

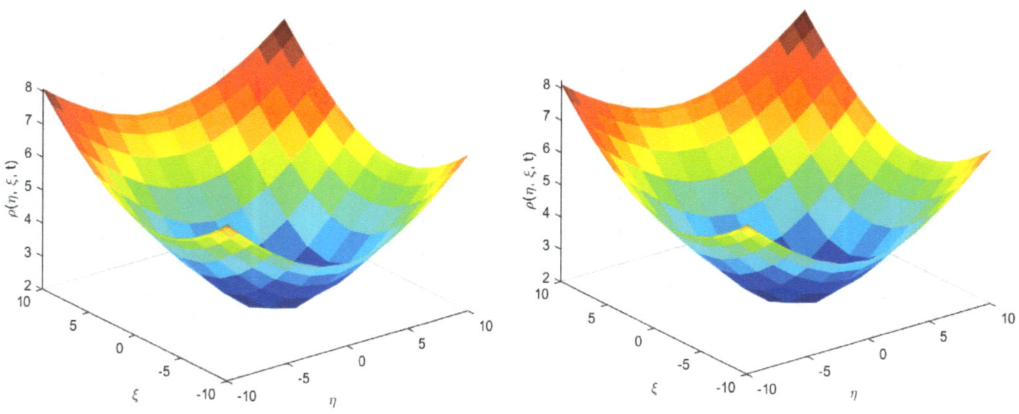

Figure 8. Example 3: Exact (**left**) vs. numerical solution (**right**) at $N = 15$ collocation points.

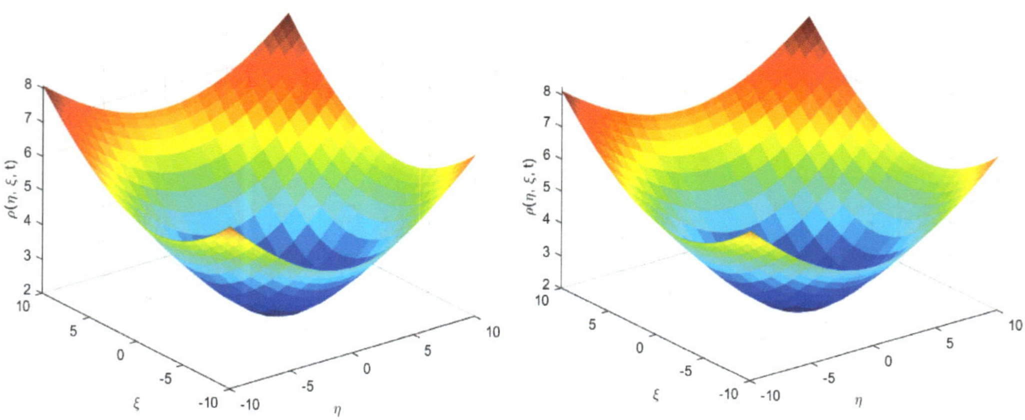

Figure 9. Example 3: Exact (**left**) vs. numerical solution (**right**) at $N = 25$ collocation points.

Table 2. Example 3: Error behavior of exact vs. numerical solution.

N	L_∞	L_2	N	L_∞	L_2
6	1.901×10^{0}	7.400×10^{-1}	8	1.951×10^{-1}	1.009×10^{-1}
10	2.218×10^{-2}	6.456×10^{-3}	12	1.093×10^{-3}	5.585×10^{-4}
14	7.824×10^{-5}	2.584×10^{-5}	16	3.480×10^{-6}	1.295×10^{-6}
18	1.039×10^{-7}	3.863×10^{-8}	20	6.763×10^{-9}	1.602×10^{-9}

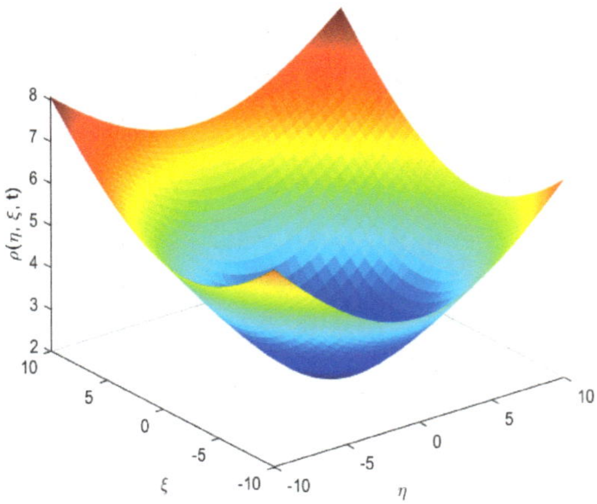

Figure 10. Example 3: Exact (**left**) vs. numerical solution (**right**) at $N = 50$ collocation points.

6. Conclusions

This study has successfully shown how fractional calculus may be used to describe fractional-order-in-time partial differential equations, which are used to simulate the dynamic behavior of biological populations. It has been shown that using the $L1$ scheme for time-fractional derivatives is a reliable and accurate way to deal with the nonlocal properties included in these kinds of equations. This method is especially crucial for capturing the genetic and memory components that are essential to biological systems. Additionally, it has been demonstrated that the Chebyshev spectral method for spatial derivatives is a highly effective method that offers an exponential rate of convergence, making it perfect for managing intricate boundary conditions and geometries that are common in biological modeling. Our algorithms provide a powerful tool for approximation solutions in situations where analytical methods are not sufficient, as confirmed by our theoretical and numerical investigation. Our approach is accurate and reliable, as evidenced by the well-aligned numerical experiments with the theoretical predictions.

Funding: This research received no external funding.

Data Availability Statement: The original contributions presented in the study are included in the article, further inquiries can be directed to the corresponding author.

Conflicts of Interest: The author declares no conflicts of interest.

Appendix A

Algorithm A1 Algorithm for Solving the Fractional PDE (25):

1. **Parameter Initialization:**
   - Define h, a, ℓ, b, and fractional order α.
   - Define time step size Δt and total simulation time T.
2. **Spatial Domain Discretization:**
   - Define Chebyshev nodes $\eta_j = \cos\left(\frac{j\pi}{N_\eta}\right)$ and $\xi_i = \cos\left(\frac{i\pi}{N_\xi}\right)$ for $j, i = 0, 1, \ldots, N$.
   - Construct Chebyshev differentiation matrices D_η and D_ξ.
3. **Initial Condition:**

$$\rho(\eta, \xi, 0) = \text{given initial condition.} \tag{A1}$$

Algorithm A1 *Cont.*

4. **Time Discretization Using L1 Scheme:**

$$\frac{\partial^\alpha \rho}{\partial t^\alpha} \approx \frac{1}{\Gamma(2-\alpha)} \sum_{k=0}^{n} w_k(\rho^{n-k} - \rho^{n-k-1}), \tag{A2}$$

where $w_k = (\Delta t)^\alpha (k^{1-\alpha} - (k-1)^{1-\alpha})$.

5. **Assembling the System:**

$$\frac{1}{\Gamma(2-\alpha)} \sum_{k=0}^{n} w_k \rho^{n-k} = D_\eta^2 (\rho^n)^2 + D_\xi^2 (\rho^n)^2 + h(\rho^n)^a (1 - \ell(\rho^n)^b), \tag{A3}$$

6. **Matrix Formulation:**
   - Represent the problem in matrix form to prepare for numerical solving:

$$A\rho^{n+1} = b^n, \tag{A4}$$

   where A is derived from the $L1$ and Chebyshev methods, and b^n contains the known terms from time step n.

7. **Iterative Solving:**
   - Solve the matrix equation using an appropriate iterative method:

$$\rho^{n+1} = A^{-1} b^n. \tag{A5}$$

8. **Advance Time Step:**
   - Increment n and repeat from step 4 until $t = T$.
9. **Postprocessing:**
   - Plot $\rho(\eta, \xi, t)$.

References

1. Singh, H.; Srivastava, H.M.; Nieto, J.J. *Handbook of Fractional Calculus for Engineering and Science*; Chapman and Hall/CRC: Boca Raton, FL, USA, 2022.
2. Inoue, T. *The Advanced Theory of Fractional Calculus*, 3rd ed.; Academic Research Publication: New York, NY, USA, 2021.
3. Hilfer, R. *Applications of Fractional Calculus in Physics*; World Scientific: Singapore, 2000.
4. Gao, X.-L.; Zhang, H.-L.; Wang, Y.-L.; Li, Z.-Y. Research on Pattern Dynamics Behavior of a Fractional Vegetation-Water Model in Arid Flat Environment. *Fractal Fract.* **2024**, *8*, 264. [CrossRef]
5. Magin, R.L. Fractional Calculus Models of Complex Dynamics in Biological Tissues. *Comput. Math. Appl.* **2010**, *59*, 1586–1593. [CrossRef]
6. Hattaf, K. On the Stability and Numerical Scheme of Fractional Differential Equations with Application to Biology. *Computation* **2022**, *10*, 97. [CrossRef]
7. Hattaf, K. A New Mixed Fractional Derivative with Applications in Computational Biology. *Computation* **2024**, *12*, 7. [CrossRef]
8. Nguyen Tien, D. Fractional Stochastic Differential Equations with Applications to Finance. *J. Math. Anal. Appl.* **2013**, *397*, 334–348. [CrossRef]
9. Wu, Z.; Zhang, X.; Wang, J.; Zeng, X. Applications of Fractional Differentiation Matrices in Solving Caputo Fractional Differential Equations. *Fractal Fract.* **2023**, *7*, 374. [CrossRef]
10. Podlubny, I. *Fractional Differential Equations: An Introduction to Fractional Derivatives, Fractional Differential Equations, to Methods of Their Solution and Some of Their Applications*; Elsevier: Amsterdam, The Netherlands, 1998.
11. Bin Jebreen, H. The Müntz–Legendre Wavelet Collocation Method for Solving Weakly Singular Integro-Differential Equations with Fractional Derivatives. *Fractal Fract.* **2023**, *7*, 763. [CrossRef]
12. Xu, Y.; Liu, Y.; Yin, X.; Feng, L.; Wang, Z. A Compact Scheme Combining the Fast Time Stepping Method for Solving 2D Fractional Subdiffusion Equations. *Fractal Fract.* **2023**, *7*, 186. [CrossRef]
13. Sytnyk, D.; Wohlmuth, B. Exponentially Convergent Numerical Method for Abstract Cauchy Problem with Fractional Derivative of Caputo Type. *Mathematics* **2023**, *11*, 2312. [CrossRef]
14. Caputo, M. Linear Models of Dissipation Whose Q Is Almost Frequency Independent II. *Geophys. J. Int.* **1967**, *13*, 529–539. [CrossRef]

15. Djilali, S. Threshold Asymptotic Dynamics for a Spatial Age-Dependent Cell-to-Cell Transmission Model with Nonlocal Disperse. *Discret. Contin. Dyn. Syst. Ser. B* **2023**, *28*, 4108–4143. [CrossRef]
16. Hathout, F.Z.; Touaoula, T.M.; Djilali, S. Efficiency of Protection in the Presence of Immigration Process for an Age-Structured Epidemiological Model. *Acta Appl. Math.* **2023**, *185*, 3. [CrossRef]
17. Ionescu, C.; Lopes, A.; Copot, D.; Machado, J.A.T.; Bates, J.H.T. The role of fractional calculus in modeling biological phenomena: A review. *Commun. Nonlinear Sci. Numer. Simul.* **2017**, *51*, 141–159. [CrossRef]
18. Caputo, M.; Fabrizio, M. A New Definition of Fractional Derivative without Singular Kernel. *Prog. Fract. Differ. Appl.* **2015**, *1*, 1–13.
19. Chen, Y.-G.; Yang, F.; Tian, F. The Landweber Iterative Regularization Method for Identifying the Unknown Source of Caputo-Fabrizio Time Fractional Diffusion Equation on Spherically Symmetric Domain. *Symmetry* **2023**, *15*, 1468. [CrossRef]
20. Agrawal, O.P. Solution for a Fractional Diffusion-Wave Equation Defined in a Bounded Domain. *J. Nonlinear Dynam.* **2002**, *29*, 145–155. [CrossRef]
21. Fix, G.J.; Roop, J.P. Least Squares Finite Element Solution of a Fractional Order Two-Point Boundary Value Problem. *Comput. Math. Appl.* **2004**, *48*, 1017–1033. [CrossRef]
22. Ali, I.; Haq, S.; Hussain, M.; Nisar, K.S.; Ul Arifeen, S. On the Analysis and Application of a Spectral Collocation Scheme for the Nonlinear Two-Dimensional Fractional Diffusion Equation. *Results Phys.* **2024**, *56*, 107222. [CrossRef]
23. Alam, M.; Haq, S.; Ali, I.; Ebadi, M.J.; Salahshour, S. Radial Basis Functions Approximation Method for Time-Fractional FitzHugh–Nagumo Equation. *Fractal Fract.* **2023**, *7*, 882. [CrossRef]
24. Diethelm, K.; Freed, A.D. On the Solution of Nonlinear Fractional-Order Differential Equations Used in the Modeling of Viscoplasticity. In *Proceedings of the Second Conference on Scientific Computing in Chemical Engineering*; Springer: Berlin/Heidelberg, Germany, 1999; pp. 217–224.
25. Deng, Z.; Singh, V.P.; Bengtsson, L. Numerical Solution of Fractional Advection–Dispersion Equation. *J. Hydraul. Eng.* **2004**, *130*, 422–431. [CrossRef]
26. Meerschaert, M.; Tadjeran, C. Finite Difference Approximations for Fractional Advection-Dispersion Flow Equations. *J. Comp. Appl. Math.* **2004**, *172*, 65–77. [CrossRef]
27. Gorenflo, R.; Mainardi, F.; Moretti, D.; Paradisi, P. Time Fractional Diffusion: A Discrete Random Walk Approach. *Nonlinear Dyn.* **2002**, *29*, 129–143. [CrossRef]
28. Gorenflo, R.; Luchko, Y.; Mainardi, F. Wright Function as Scale-Invariant Solutions of the Diffusion-Wave Equation. *J. Comp. Appl. Math.* **2000**, *118*, 175–191. [CrossRef]
29. Schneider, W.R.; Wyss, W. Fractional Diffusion and Wave Equations. *J. Math. Phys.* **1989**, *30*, 134–144. [CrossRef]
30. Wyss, W. The Fractional Diffusion Equation. *J. Math. Phys.* **1986**, *27*, 2782–2785. [CrossRef]
31. Lin, Y.; Xu, C. Finite Difference/Spectral Approximations for the Time-Fractional Diffusion Equation. *J. Comput. Phys.* **2007**, *225*, 1533–1552. [CrossRef]
32. Gottlieb, D.; Hussaini, M.Y.; Orszag, S.A. Theory and application of spectral methods. In *Spectral Methods for Partial Differential Equations*; Voigt, R.G., Gotlieb, D., Hussaini, M.Y., Eds.; SIAM: Philadelphia, PA, USA, 1984; pp. 1–55.
33. Gottlieb, D.; Orszag, S.A. *Numerical Analysis of Spectral Methods: Theory and Applications*; SIAM: Philadelphia, PA, USA, 1977.
34. Trefethen, L.N. *Spectral Methods in MATLAB*; SIAM: Philadelphia, PA, USA, 2000.
35. Mason, J.C.; Handscomb, D.C. *Chebyshev Polynomials*; CRC Press LLC: New York, NY, USA, 2003.
36. Ali, I.; Saleem, M.T. Applications of Orthogonal Polynomials in Simulations of Mass Transfer Diffusion Equation Arising in Food Engineering. *Symmetry* **2023**, *15*, 527. [CrossRef]
37. Ali, I.; Khan, S.U. Asymptotic Behavior of Three Connected Stochastic Delay Neoclassical Growth Systems Using Spectral Technique. *Mathematics* **2022**, *10*, 3639. [CrossRef]
38. Ali, I.; Khan, S.U. Threshold of Stochastic SIRS Epidemic Model from Infectious to Susceptible Class with Saturated Incidence Rate Using Spectral Method. *Symmetry* **2022**, *14*, 1838. [CrossRef]
39. Ali, I.; Khan, S.U. Dynamics and Simulations of Stochastic COVID-19 Epidemic Model Using Legendre Spectral Collocation Method. *AIMS Math.* **2023**, *8*, 4220–4236. [CrossRef]
40. Khan, S.U.; Ali, I. Convergence and Error Analysis of a Spectral Collocation Method for Solving System of Nonlinear Fredholm Integral Equations of Second Kind. *Comp. Appl. Math.* **2019**, *38*, 125. [CrossRef]
41. Hattaf, K.; Yousfi, N. Global Stability for Fractional Diffusion Equations in Biological Systems. *Complexity* **2020**, *2020*, 5476842. [CrossRef]
42. Rida, S.Z.; Arafa, A.A.M. Exact Solutions of Fractional-Order Biological Population Model. *Commun. Theor. Phys.* **2009**, *52*, 992–996.
43. Shakeri, E.; Dehghan, M. Numerical Solutions of a Biological Population Model Using He's Variational Iteration Method. *Comput. Math. Appl.* **2007**, *54*, 1197–1207. [CrossRef]
44. Liu, Y.; Li, Z.; Zhang, Y. Homotopy Perturbation Method to Fractional Biological Population Equation. *Fract. Differ. Calc.* **2011**, *1*, 117–124. [CrossRef]
45. Ma, X.; Wang, Y.; Zhu, X.; Liu, W.; Lan, Q.; Xiao, W. A Spectral Method for Two-Dimensional Ocean Acoustic Propagation. *J. Mar. Sci. Eng.* **2021**, *9*, 892. [CrossRef]
46. Shen, J.; Tang, T. *Spectral and High-Order Methods with Applications*; Science Press: Beijing, China, 2006.
47. Boyd, J.P. *Chebyshev and Fourier Spectral Methods*, 2nd ed.; Dover: New York, NY, USA, 2001.

48. Gurney, W.S.C.; Nisbet, R.M. *Ecological Dynamics*; Oxford University Press: Oxford, UK, 1998.
49. Canuto, C.; Hussaini, M.Y.; Quarteroni, A.; Zang, T.A. *Spectral Methods: Fundamentals in Single Domains*; Springer Series in Scientific Computation; Springer: New York, NY, USA, 2006.

fractal and fractional

MDPI

Article

An Efficient Numerical Solution of a Multi-Dimensional Two-Term Fractional Order PDE via a Hybrid Methodology: The Caputo–Lucas–Fibonacci Approach with Strang Splitting

Imtiaz Ahmad [1,*], Abdulrahman Obaid Alshammari [2], Rashid Jan [3], Normy Norfiza Abdul Razak [3] and Sahar Ahmed Idris [4]

1 Institute of Informatics and Computing in Energy (IICE), Universiti Tenaga Nasional (UNITEN), Kajang 43000, Selangor, Malaysia
2 Department of Mathematics, College of Science, Jouf University, Sakaka 72388, Saudi Arabia; aoshammari@ju.edu.sa
3 Institute of Energy Infrastructure (IEI), Department of Civil Engineering, College of Engineering, Universiti Tenaga Nasional (UNITEN), Putrajaya Campus, Jalan IKRAM-UNITEN, Kajang 43000, Selangor, Malaysia; rashid.jan@uniten.edu.my (R.J.); normy@uniten.edu.my (N.N.A.R.)
4 Faculty of Engineering, Department of Industrial Engineering, King Khalid University, Abha 61421, Saudi Arabia; smohammedali@kku.edu.sa
* Correspondence: imtiazkakakhil@gmail.com

Abstract: The utilization of time-fractional PDEs in diverse fields within science and technology has attracted significant interest from researchers. This paper presents a relatively new numerical approach aimed at solving two-term time-fractional PDE models in two and three dimensions. We combined the Liouville–Caputo fractional derivative scheme with the Strang splitting algorithm for the temporal component and employed a meshless technique for spatial derivatives utilizing Lucas and Fibonacci polynomials. The rising demand for meshless methods stems from their inherent mesh-free nature and suitability for higher dimensions. Moreover, this approach demonstrates the effective approximation of solutions across both regular and irregular domains. Error norms were used to assess the accuracy of the methodology across both regular and irregular domains. A comparative analysis was conducted between the exact solution and alternative numerical methods found in the contemporary literature. The findings demonstrate that our proposed approach exhibited better performance while demanding fewer computational resources.

Keywords: Strang splitting algorithm; Caputo derivative; Fibonacci polynomials; Lucas polynomials; two-term fractional-order PDE

Citation: Ahmad, I.; Alshammari, A.O.; Jan, R.; Razak, N.N.A.; Idris, S.A. An Efficient Numerical Solution of a Multi-Dimensional Two-Term Fractional Order PDE via a Hybrid Methodology: The Caputo–Lucas–Fibonacci Approach with Strang Splitting. *Fractal Fract.* **2024**, *8*, 364. https://doi.org/10.3390/fractalfract8060364

Academic Editors: Ricardo Almeida, Libo Feng, Lin Liu and Yang Liu

Received: 29 May 2024
Revised: 13 June 2024
Accepted: 17 June 2024
Published: 20 June 2024

1. Introduction

Over the last decade, there has been considerable interest in fractional partial differential equations (PDEs), making it a vibrant research field for scientists and engineers. PDEs possess the capability to describe numerous complex phenomena across diverse fields, including biology, fluid mechanics, plasma physics, optics, acoustics, financial mathematics, climate modeling, materials science, and electromagnetics [1,2]. Fractional calculus, which deals with derivatives and integrals of non-integer order, finds broad applications in science and technology. In physics, it models complex systems with non-local behavior, such as anomalous diffusion in fluid dynamics and memory effects in viscoelastic materials [3]. Engineering benefits from fractional calculus are found in signal processing, control theory, and optimization, offering robustness and efficiency in dynamic systems. Telecommunications utilize it for modeling wireless communication channels and improving channel estimation algorithms. Fractional calculus enables advancements in understanding and optimizing real-world processes and systems across various domains [4]. The fractional spatio-temporal PDE models are crucial in describing complex phenomena [5], whereas the two-scale fractal PDE models are essential for simulating intricate structures in materials science and physics, offering insights into multiscale behaviors and enhancing our understanding of fractal geometries in natural and engineered systems [6]. In particular, the Sobolev model equation is integral across various scientific fields, providing a robust framework for investigating phenomena with spatial variations. Its relevance extends across disciplines, encompassing physics, engineering, biology, and finance. In physics, Sobolev PDEs are commonly utilized to model phenomena such as heat conduction, fluid dynamics, and quantum mechanics [7,8]. Engineers employ them to analyze structural mechanics, electromagnetism, and signal-processing problems [7]. Furthermore, Sobolev PDEs find applications in image processing, medical imaging, and geophysical exploration. Their adaptability lies in their capacity to handle irregular domains and boundary conditions while providing solutions that demonstrate smoothness properties.

Numerous studies explored the applications and theoretical foundations of Sobolev PDEs, underscoring their significance in both theoretical and applied contexts. However, numerous researchers faced challenges in deriving and formulating various complex phenomena within nonlinear PDEs with integer orders [9]. In response, fractional calculus is regarded as a viable solution to this issue, as it incorporates a nonlocal property that is absent in nonlinear PDEs with integer orders [10]. Observations indicate that multi-term time-fractional PDEs were proposed to enhance the modeling accuracy in depicting anomalous diffusion processes, capturing various types of viscoelastic damping, accurately representing power-law frequency dependence, and simulating the flow of a fractional Maxwell fluid [11]. This study focused on the two-term time-fractional Sobolev equation in two and three dimensions, defined as follows:

$$\begin{aligned}&\frac{\partial^{\xi_1}\mathcal{U}}{\partial t^{\xi_1}}+\frac{\partial^{\xi_2}\mathcal{U}}{\partial t^{\xi_2}}-\beta\frac{\partial}{\partial t}\left(\frac{\partial^2\mathcal{U}}{\partial r^2}+\frac{\partial^2\mathcal{U}}{\partial s^2}\right)-\delta\left(\frac{\partial^2\mathcal{U}}{\partial r^2}+\frac{\partial^2\mathcal{U}}{\partial s^2}\right)+\eta\left(\mathcal{U}\left(\frac{\partial^2\mathcal{U}}{\partial r^2}+\frac{\partial^2\mathcal{U}}{\partial s^2}\right)\right)\\&+\eta\left(\frac{\partial\mathcal{U}}{\partial r}+\frac{\partial\mathcal{U}}{\partial s}\right)^2+\gamma\mathcal{U}=F(r,s,t),\ \ r,s\in\Omega\subset\mathbb{R}^2,\ \ 0<\xi_2\leq\xi_1\leq 1,\ \ t>0,\end{aligned}\tag{1}$$

and

$$\begin{aligned}&\frac{\partial^{\xi_1}\mathcal{U}}{\partial t^{\xi_1}}+\frac{\partial^{\xi_2}\mathcal{U}}{\partial t^{\xi_2}}-\beta\frac{\partial}{\partial t}\left(\frac{\partial^2\mathcal{U}}{\partial r^2}+\frac{\partial^2\mathcal{U}}{\partial s^2}+\frac{\partial^2\mathcal{U}}{\partial z^2}\right)-\delta\left(\frac{\partial^2\mathcal{U}}{\partial r^2}+\frac{\partial^2\mathcal{U}}{\partial s^2}+\frac{\partial^2\mathcal{U}}{\partial z^2}\right)\\&+\eta\left(\mathcal{U}\left(\frac{\partial^2\mathcal{U}}{\partial r^2}+\frac{\partial^2\mathcal{U}}{\partial s^2}+\frac{\partial^2\mathcal{U}}{\partial z^2}\right)\right)+\eta\left(\frac{\partial\mathcal{U}}{\partial r}+\frac{\partial\mathcal{U}}{\partial s}+\frac{\partial\mathcal{U}}{\partial z}\right)^2+\gamma\mathcal{U}\\&=F(r,s,z,t),\ \ r,s,z\in\Omega\subset\mathbb{R}^3,\ \ 0<\xi_2\leq\xi_1\leq 1,\ \ t>0,\end{aligned}\tag{2}$$

with the following initial and boundary conditions:

$$\mathcal{U}(\bar{\boldsymbol{r}},0)=\mathcal{U}_0(\bar{\boldsymbol{r}}),\tag{3}$$

$$\mathcal{U}(\bar{\boldsymbol{r}},t)=f_1(\bar{\boldsymbol{r}},t),\ \ \bar{\boldsymbol{r}}\in\partial\Omega\subset\mathbb{R}^d,\ \ d=2,3.\tag{4}$$

where the constants β, δ, η, and γ have known values. Moreover, $\frac{\partial^{\xi_1}}{\partial t^{\xi_1}}$ and $\frac{\partial^{\xi_2}}{\partial t^{\xi_2}}$ denote the Caputo derivatives in accordance with Caputo [12] and applied to $\mathcal{U}(\bar{\boldsymbol{r}},t)$, where $0<\xi_2\leq\xi_1\leq 1$.

Several works delved into the existence and uniqueness of solutions for Sobolev equations, as documented in [13]. Ewing [14] applied the finite difference method to tackle 2D Sobolev equations. The study [15] used a computationally appealing and precise local meshless technique to provide numerical solutions for three-dimensional two- and three-term time-fractional PDE models. Various test problems were employed to assess the accuracy and reliability of the proposed approach. The paper by Luo et al. [16] introduces a reduced-order extrapolated finite difference iterative scheme for 2D Sobolev equations, employing proper orthogonal decomposition to construct a reduced-order solution space model, thus reducing the computational costs. Numerical experiments validated the scheme's efficiency and accuracy, indicating its potential for efficient Sobolev equation solving. An accurate and efficient local meshless technique is used in the article [17] to explore numerical solutions for two-term time-fractional Sobolev models. The method approximates the solution on a uniform or scattered set of nodes, yielding sparse and well-conditioned coefficient matrices. Another paper by Luo et al. [18] introduces a novel method that employs reduced-order techniques and extrapolation within a Crank–Nicolson finite volume framework for 2D Sobolev equations. It aims to improve the computational efficiency and accuracy, showcasing promising outcomes via numerical simulations. The paper by Li et al. [19] introduces an extended mixed finite element approach for solving 2D Sobolev equations, integrating a novel formulation with stable discretization for accurate solutions and computational efficiency. Numerical experiments highlighted its effectiveness and superiority over existing methods, especially in maintaining stability and convergence properties. Heydari et al. [20] introduced a novel method that employs orthonormal Bernoulli polynomials to solve distributed-order time-fractional 2D Sobolev equations, which demonstrated accuracy through four test problems. Gao et al. [21] used the local discontinuous Galerkin finite element method to solve a particular class of 2D Sobolev equations, while another work [22] applied the weak Galerkin finite element method, providing an error estimate. Abu et al. [23] presented a method for solving time- and space-fractional Sobolev equations with Caputo fractional derivatives in n-dimensional space, utilizing the reproducing kernel Hilbert space method, which is particularly effective for Caputo class derivatives. A meshless RBFs technique for solving 2D time-fractional Sobolev equations was presented by Hussain et al. [24]. This method uses a finite difference formula for time-fractional derivative approximation and RBFs for spatial operator approximation. Using a Liouville–Caputo derivative technique for the time derivative, Ahmad et al. [25] suggested an efficient meshless method for estimating the numerical solution of 3D time-fractional Sobolev equations. Furthermore, Zhang et al. [26] used a characteristic splitting mixed finite element approach to solve convection-dominated Sobolev equations.

In response to its extensive applications, researchers explored a multitude of numerical and analytical methodologies [27,28] for addressing complex PDEs. These methods encompass various techniques, such as the alternating direction implicit method [29], the homotopy perturbation method [30], the Laplace transform technique [31], the variational approach [32,33], finite element methods (FEMs) [34], finite difference methods (FDMs) [16,35], gradient descent iterative method [36], spectral methods [37], the exp-function method [38], the alternating direction method [39], and meshless methods [40]. Notably, the straightforward nature of the FDM, FEM, and meshless techniques is noteworthy. Recently, hybrid methodologies have emerged to enhance the efficiency and accuracy of numerical solutions, including meshless methods, the method of lines utilizing Fibonacci polynomials, and combinations of the FDM and FEM.

This investigation focused on computing numerical solutions for the suggested model using a hybrid methodology based on the Caputo derivative. This approach integrates Fibonacci polynomials with the established Caputo derivative concept, exploiting the relationship between Fibonacci and Lucas polynomials. This integration provides a significant advantage in the straightforward implementation of higher-order derivatives. Moreover, the proposed approach reduces computational costs by enhancing the accuracy, even with a limited number of collocation sites. Remarkably, these polynomials find diverse real-world applications in the realm of differential equations.

Addressing boundary value problems accurately involves exploring the interplay between Chebyshev and Lucas polynomials, as demonstrated in previous studies [41,42]. For instance, the Lucas sequence was utilized to approximate integro-differential equations [43], while Lucas polynomials were applied to solve higher-order differential equations [44]. Additionally, the efficacy of a Fibonacci polynomial methodology in resolving Volterra–Fredholm integral differential equations was demonstrated [45]. Furthermore, a hybrid Taylor–Lucas polynomial approach was introduced for addressing delay difference equations [46]. Notably, novel methodologies for solving time-dependent PDEs were proposed by combining hybrid Fibonacci and Lucas polynomial schemes [47,48]. Moreover, researchers employed finite differences and Lucas polynomials to achieve effective numerical solutions for various PDE models [49,50].

Motivation

The primary objective of this research was to introduce a relatively new numerical approach designed to solve two-term time-fractional PDE models in both two and three dimensions. This method combines the Liouville–Caputo fractional derivative scheme with the Strang splitting algorithm for the temporal component and utilizes a meshless technique for spatial derivatives, incorporating Lucas and Fibonacci polynomials. Below are some of the highlighted key features of the proposed study:

- The complex characteristics of fractional nonlinear PDEs make calculating analytical solutions challenging, driving ongoing research efforts to develop accurate and efficient numerical methodologies, with the two-term fractional order Sobolev model equation in both two and three dimensions holding significant importance across multiple scientific domains.
- This study aimed to introduce an efficient numerical framework specifically designed for solving PDEs with temporal fractions.
- The proposed methodology adopts a hybrid approach, integrating Fibonacci and Lucas polynomials with finite difference techniques, while also addressing the temporal direction through the utilization of the Liouville–Caputo fractional derivative in conjunction with a splitting mechanism.
- Lucas and Fibonacci polynomials, unlike orthogonal counterparts, like Chebyshev polynomials, are non-orthogonal, eliminating the need for interval transformations. Additionally, they facilitate the straightforward approximation of higher-order derivatives for unknown functions.
- Furthermore, the approach is characterized by its simplicity and ability to enhance the accuracy, even in scenarios involving fewer nodal points, with the aim to provide a robust and effective numerical solution to the intricate challenges posed by nonlinear PDEs.

This paper is organized as follows: Section 2 provides an overview of the fundamental terms and concepts. The proposed technique for the underlying model equations is discussed in Section 3, while theoretical results regarding stability and error analysis are presented in Section 4. Section 5 utilizes numerical experiments to validate the method's efficacy, and finally, Section 6 summarizes the outcomes and presents concluding remarks to finalize the work.

2. Basic Concepts in Fractional Calculus and Polynomial Theory

In fractional calculus, fractional derivatives are crucial. Fractional calculus finds applications in various fields, including physics, engineering, biology, and finance, providing a powerful framework for modeling complex phenomena characterized by memory, nonlocality, and fractal behavior. The fundamental definition of fractional calculus involves the generalization of differentiation and integration to non-integer orders. Here are some basic definitions of frequently used fractional derivatives.

Definition 1. *The Riemann–Liouville derivative [51,52] is a mathematical concept used to generalize the notion of a derivative to functions that are not necessarily differentiable in the traditional sense. It is commonly applied in the field of fractional calculus to describe the behavior of functions with fractional orders of differentiation.*

$$\frac{\partial^{\xi}\mathcal{U}(\bar{\mathbf{r}},t)}{\partial t^{\xi}} = \frac{1}{\Gamma(1-\xi)}\frac{d}{dt}\int_{t}^{T}\frac{(\mathcal{U}(\bar{\mathbf{r}},\vartheta)-\mathcal{U}(\bar{\mathbf{r}},T))}{(\vartheta-t)^{\xi}}d\vartheta,\ 0<\xi<1. \tag{5}$$

Definition 2. *Caputo's fractional derivative [12] is a widely used mathematical tool in fractional calculus, providing a powerful approach for describing and analyzing non-local and memory-dependent phenomena in various scientific and engineering fields.*

$$\frac{\partial^{\xi}\mathcal{U}(\bar{\mathbf{r}},t)}{\partial t^{\xi}} = \frac{1}{\Gamma(1-\xi)}\int_{0}^{t}\frac{\partial\mathcal{U}(\bar{\mathbf{r}},\zeta)}{\partial\zeta}(t-\zeta)^{-\xi}d\zeta,\ 0<\xi<1 \tag{6}$$

Definition 3. *The Atangana and Baleanu fractional derivative [53], proposed by Atangana and Baleanu, is a non-local operator that extends the concept of fractional calculus. It provides a novel approach for modeling complex systems that exhibit memory and non-local behavior.*

$${}^{ABC}_{a}\frac{\partial^{\xi}\mathcal{U}(\bar{\mathbf{r}},t)}{\partial t^{\xi}} = \frac{B(\xi)}{1-\xi}\int_{a}^{t}\mathcal{U}'(\bar{\mathbf{r}})E_{\xi}\left(-\frac{\xi(t-\bar{\mathbf{r}})^{\xi}}{1-\xi}\right)d\bar{\mathbf{r}},\ 0<\xi<1. \tag{7}$$

Definition 4. *He's fractional derivative [54]:*

$$\frac{\partial^{\xi}\mathcal{U}(\bar{\mathbf{r}},t)}{\partial t^{\xi}} = \frac{1}{\Gamma(1-\xi)}\frac{d}{d\bar{\mathbf{r}}}\int_{t_0}^{t}(t-\zeta)^{-\xi}[\mathcal{U}_0(\zeta)-\mathcal{U}(\zeta)],\ 0<\xi<1. \tag{8}$$

2.1. Fibonacci and Lucas Polynomial Theory

Below are explanations outlining the definitions, practical implications, and utilization of Lucas and Fibonacci polynomials, encompassing tasks such as approximating unfamiliar functions and their derivatives.

Fibonacci polynomials [49]:

The Fibonacci polynomial, which is an extension of Fibonacci numbers, is defined through a three-term recurrence relation:

$$F_k(r) = kF_{k-1}(r) + F_{k-2}(r),\ k\ \geq 2, \tag{9}$$

where the sequence begins with $F_0(r) = 0$ and $F_1(r) = 1$. When $r = 1$, Equation (9) generates the familiar sequence of Fibonacci numbers.

Lucas polynomials [49]:

One can define the Lucas polynomials using the three-term recurrence relation:

$$L_k(r) = kL_{k-1}(r) + L_{k-2}(r),\ k\ \geq 2, \tag{10}$$

with $L_0(r) = 2$ and $L_1(r) = r$ as initial values. This allows Equation (10) to produce a series of Lucas numbers with $r = 1$.

Lemma 1 ([49]). *The kth Fibonacci polynomial $F_k(r)$ can be used to define the mth-order derivative of the kth Lucas polynomial $L_k(r)$:*

$$L_k^{(m)}(r) = kF_k(r)D^{m-1},\quad D^{m-1} = \underbrace{D\times D\times D\cdots D}_{(m-1)time} \tag{11}$$

where an $(M+1)\times(M+1)$ matrix is represented by D, which can be described as

$$D = \begin{bmatrix} 0 & 0 & \dots & 0 \\ 0 & & & \\ \vdots & & d & \\ 0 & & & \end{bmatrix},$$

where d is computed according to [49]:

$$d_{ij} = \begin{cases} i \sin \frac{(j-i)\pi}{2}, & \text{if } j > i, \\ 0, & \text{otherwise.} \end{cases}$$

2.2. Function Approximation

Suppose $\mathscr{U} \in L^2(\mathbb{R})$, and let $\mathscr{U}(\bar{\mathbf{r}})$ be continuous. Then, through the linear combination of the kth Lucas polynomials, it is possible to represent $\mathscr{U}$ as

$$\mathscr{U}(\bar{\mathbf{r}}) = \sum_{k=0}^{\infty} \Lambda_k L_k(\bar{\mathbf{r}}), \tag{12}$$

where the Lucas polynomials are denoted as $L_k(\bar{\mathbf{r}})$, with the corresponding unknown coefficients represented by Λ_k.

Likewise, $\mathscr{U}(\bar{\mathbf{r}})$ can also be expanded under the same conditions by using a linear combination of the kth Fibonacci polynomials, as demonstrated below:

$$\mathscr{U}(\bar{\mathbf{r}}) = \sum_{k=0}^{\infty} \Lambda_k F_k(\bar{\mathbf{r}}).$$

In this context, $F_k(\bar{\mathbf{r}})$ denotes the Fibonacci polynomials, while Λ_k stands for the coefficients yet to be determined.

The Lucas polynomial series can be used to expand the function $\mathscr{U}(\bar{\mathbf{r}})$ and find the first-order derivative:

$$\mathscr{U}'(\bar{\mathbf{r}}) = \sum_{k=0}^{\infty} \Lambda_k L'_k(\bar{\mathbf{r}}), \tag{13}$$

and the function $\mathscr{U}(\bar{\mathbf{r}})$'s associated mth-order derivative is given as

$$\mathscr{U}^m(\bar{\mathbf{r}}) = \sum_{k=0}^{\infty} \Lambda_k L_k^{(m)}(\bar{\mathbf{r}}), \tag{14}$$

where

$$\mathscr{U}^m(\bar{\mathbf{r}}) = \frac{\mathrm{d}^m \mathscr{U}(\bar{\mathbf{r}})}{\mathrm{d}\bar{\mathbf{r}}^m}, \quad L_k^{(m)}(\bar{\mathbf{r}}) = \frac{\mathrm{d}^m L_k(\bar{\mathbf{r}})}{\mathrm{d}\bar{\mathbf{r}}^m}.$$

Equations (13) and (14) can be written by incorporating the relation (11):

$$\mathscr{U}'(\bar{\mathbf{r}}) = \sum_{k=0}^{\infty} \Lambda_k k F_k(\bar{\mathbf{r}}), \tag{15}$$

and similarly, the mth derivative can be computed as

$$\mathscr{U}^{(m)}(\bar{\mathbf{r}}) = \sum_{k=0}^{\infty} \Lambda_k k F_k(\bar{\mathbf{r}}) D^{m-1}, \tag{16}$$

in which D and D^{m-1} are defined above.

Remark 1. *In numerical computations, the expansion of $\mathscr{U}(\bar{\mathbf{r}})$ and its mth derivative often involves utilizing truncated Fibonacci and Lucas polynomial series. In other words, we take*

$$\mathscr{U}(\bar{\mathbf{r}}) \simeq \sum_{k=0}^{M} \Lambda_k L_k(\bar{\mathbf{r}}), \quad M \in \mathbb{N},$$

and

$$\mathscr{U}^{(m)}(\bar{\mathbf{r}}) \simeq \sum_{k=0}^{M} \Lambda_k L_k^{(m)}(\bar{\mathbf{r}}) = \sum_{k=0}^{M} \Lambda_k k F_k(\bar{\mathbf{r}}) D^{m-1}, \quad M \in \mathbb{N}, \tag{17}$$

3. Suggested Methodology

This section is devoted to formulating the suggested meshless method for approximating the 2D time-fractional Sobolev model equations given in Equations (20) and (21). To simplify the notation throughout the discussion in this section, we introduce the following:

$$\mathscr{U}^{n+1}(\bar{r}) = \mathscr{U}(r,s,t^{n+1}), \quad \mathscr{U}_{ij}^{n+1} = \mathscr{U}(r_i,s_j,t^{n+1}),$$

where r_i and s_j are the collocation points specified as follows, and $t^n = n \times \delta t$, with δt representing the time step size:

$$r_i = a + ih_r, \quad s_j = c + jh_s, \quad (i,j = 1,2,\cdots,M,\ M \in \mathbb{N}),$$

where $h_r = (b-a)/M$ and $h_s = (d-c)/M$ are the mesh step sizes in the r and s spatial directions throughout the spatial domain $\Xi = [a,b] \times [c,d] \subset \mathbb{R}^2$.

3.1. Time Discretization

The widely recognized $L1$-formula, with $\mathcal{O}(\delta t^{2-\xi_1})$ as the approximation order error, where $0 < \xi_1 \leq 1$, is utilized to derive the discrete representation of the $(n+1)$th time level, as outlined in [55,56].

$$\begin{aligned}\frac{\partial^{\xi_1}\mathscr{U}(\bar{r},t^{n+1})}{\partial t^{\xi_1}} &= \frac{1}{\Gamma(1-\xi_1)}\int_0^{t^{n+1}} \frac{\partial \mathscr{U}(\bar{r},\zeta)}{\partial \zeta}(t^{n+1}-\zeta)^{-\xi_1}\mathrm{d}\zeta,\\ &= \frac{1}{\Gamma(1-\xi_1)}\sum_{j=0}^{n}\int_{j\times\delta t}^{(j+1)\times\delta t} \frac{\partial \mathscr{U}(\bar{r},\zeta)}{\partial \zeta}(t^{n+1}-\zeta)^{-\xi_1}\mathrm{d}\zeta,\\ &= \frac{1}{\Gamma(1-\xi_1)}\sum_{j=0}^{n}\left[\frac{\mathscr{U}^{j+1}(\bar{r})-\mathscr{U}^{j}(\bar{r})}{\delta t} + O(\delta t)\right]\int_{j\times\delta t}^{(j+1)\times\delta t}((j+1)\delta t-\zeta)^{-\xi_1}\mathrm{d}\zeta.\end{aligned}$$

Following integration, it gives

$$\frac{\partial^{\xi_1}\mathscr{U}(s,r,t^{n+1})}{\partial t^{\xi_1}} = \begin{cases} A_{\xi_1}\sum_{j=0}^{n} K_{\xi_1}(j)\left[\mathscr{U}^{n-j+1}(\bar{r}) - \mathscr{U}^{n-j}(\bar{r})\right] + O(\delta t^{2-\xi_1}), & 0<\xi_1<1,\\ \frac{\mathscr{U}^{n+1}(\bar{r})-\mathscr{U}^{n}(\bar{r})}{\delta t} + O(\delta t), & \xi_1 = 1, \end{cases} \tag{18}$$

where $A_{\xi_1} = \frac{\delta t^{-\xi_1}}{\Gamma(2-\xi_1)}$ and $K_{\xi_1}(j) = (j+1)^{1-\xi_1} - (j)^{1-\xi_1}$. Therefore, we may write the following for $0 < \xi_1 < 1$ after ignoring the error term:

$$\frac{\partial^{\xi_1}\mathscr{U}(\bar{r},t^{n+1})}{\partial t^{\xi_1}} = A_{\xi_1}\left[\mathscr{U}^{n+1}(\bar{r}) - \mathscr{U}^{n}(\bar{r})\right] + A_{\xi_1}\sum_{j=1}^{n} K_{\xi_1}(j)\left[\mathscr{U}^{n-j+1}(\bar{r}) - \mathscr{U}^{n-j}(\bar{r})\right], \tag{19}$$

with $K_{\xi_1}(j) = 1$ and $j = 0$. The fractional derivative of order ξ_2 can be found above.

Before applying the θ-weighted rule to the suggested Equations (1)–(4), let us rewrite them in compact form as follows:

$$\frac{\partial^{\xi_1}\mathcal{U}(\bar{\boldsymbol{r}},t)}{\partial t^{\xi_1}}+\frac{\partial^{\xi_2}\mathcal{U}(\bar{\boldsymbol{r}},t)}{\partial t^{\xi_2}}-\beta\frac{\partial\nabla^2\mathcal{U}(\bar{\boldsymbol{r}},t)}{\partial t}-\delta\nabla^2\mathcal{U}(\bar{\boldsymbol{r}},t)-\eta\nabla(\mathcal{U}(\bar{\boldsymbol{r}},t)\nabla\mathcal{U}(\bar{\boldsymbol{r}},t))+\gamma\mathcal{U}(\bar{\boldsymbol{r}},t)=f(\bar{\boldsymbol{r}},t),$$
$$\bar{\boldsymbol{r}}\in\Omega\subset\mathbb{R}^d,\ d=2,3,\ 0<\xi_2\leq\xi_1\leq1,\ t>0, \tag{20}$$

with the following conditions:

$$\mathcal{U}(\bar{\boldsymbol{r}},0)=\mathcal{U}_0(\bar{\boldsymbol{r}}),\ \mathcal{U}(\bar{\boldsymbol{r}},t)=f_1(\bar{\boldsymbol{r}},t),\ \bar{\boldsymbol{r}}\in\partial\Omega, \tag{21}$$

where the Laplacian and gradient operators are represented by the symbols ∇^2 and ∇, respectively.

The θ-weighted rule is now applied, and Equation (20) becomes

$$\begin{aligned}&A_{\xi_1}\mathcal{U}^{n+1}(\bar{\boldsymbol{r}})+A_{\xi_2}\mathcal{U}^{n+1}(\bar{\boldsymbol{r}})-\beta\left(\frac{\nabla^2\mathcal{U}^{n+1}(\bar{\boldsymbol{r}})-\nabla^2\mathcal{U}^{n}(\bar{\boldsymbol{r}})}{\delta t}\right)-\theta\delta\nabla^2\mathcal{U}^{n+1}(\bar{\boldsymbol{r}})+\theta\gamma\mathcal{U}^{n+1}(\bar{\boldsymbol{r}})\\&-\theta\nabla\left(\mathcal{U}^{n+1}(\bar{\boldsymbol{r}})\nabla\mathcal{U}^{n+1}(\bar{\boldsymbol{r}})\right)=A_{\xi_1}\mathcal{U}^{n}(\bar{\boldsymbol{r}})+A_{\xi_2}\mathcal{U}^{n}(\bar{\boldsymbol{r}})+(1-\theta)\delta\nabla^2\mathcal{U}^{n}(\bar{\boldsymbol{r}})+(1-\theta)\eta\nabla(\mathcal{U}^{n}(\bar{\boldsymbol{r}})\nabla\mathcal{U}^{n}(\bar{\boldsymbol{r}}))\\&-(1-\theta)\gamma\mathcal{U}^{n}(\bar{\boldsymbol{r}})+f_\theta^{n+1}(\bar{\boldsymbol{r}})-Q_\xi^n(\bar{\boldsymbol{r}}),\end{aligned} \tag{22}$$

where

$$f_\theta^{n+1}(\bar{\boldsymbol{r}})=\theta f^{n+1}(\bar{\boldsymbol{r}})+(1-\theta)f^n(\bar{\boldsymbol{r}}),$$

and

$$Q_\xi^n(\bar{\boldsymbol{r}})=A_{\xi_1}\sum_{j=1}^{n}K_{\xi_1}(j)\left[\mathcal{U}^{n-j+1}(r,s)-\mathcal{U}^{n-j}(r,s)\right]+A_{\xi_2}\sum_{j=1}^{n}K_{\xi_2}(j)\left[\mathcal{U}^{n-j+1}(r,s)-\mathcal{U}^{n-j}(r,s)\right].$$

Since

$$\nabla\left(\mathcal{U}^{n+1}(\bar{\boldsymbol{r}})\nabla\mathcal{U}^{n+1}(\bar{\boldsymbol{r}})\right)=\mathcal{U}^{n+1}(\bar{\boldsymbol{r}})\nabla^2\mathcal{U}^{n+1}(\bar{\boldsymbol{r}})+\left(\mathcal{U}_x^{n+1}(\bar{\boldsymbol{r}})\right)^2+\left(\mathcal{U}_y^{n+1}(\bar{\boldsymbol{r}})\right)^2,$$

the nonlinear terms are linearized as follows [24]:

$$\mathcal{U}^{n+1}(\bar{\boldsymbol{r}})\nabla^2\mathcal{U}^{n+1}(\bar{\boldsymbol{r}})\equiv\mathcal{U}^{n+1}(\bar{\boldsymbol{r}})\nabla^2\mathcal{U}^{n}(\bar{\boldsymbol{r}})+\mathcal{U}^{n}(\bar{\boldsymbol{r}})\nabla^2\mathcal{U}^{n+1}(\bar{\boldsymbol{r}})-2\mathcal{U}^{n}(\bar{\boldsymbol{r}})\nabla^2\mathcal{U}^{n}(\bar{\boldsymbol{r}}), \tag{23}$$

$$\left(\mathcal{U}_r^{n+1}(\bar{\boldsymbol{r}})\right)^2\equiv2\mathcal{U}_r^{n}(\bar{\boldsymbol{r}})\mathcal{U}_r^{n+1}(\bar{\boldsymbol{r}})-(\mathcal{U}_r^{n}(\bar{\boldsymbol{r}}))^2, \tag{24}$$

$$\left(\mathcal{U}_s^{n+1}(\bar{\boldsymbol{r}})\right)^2\equiv2\mathcal{U}_s^{n}(\bar{\boldsymbol{r}})\mathcal{U}_s^{n+1}(\bar{\boldsymbol{r}})-(\mathcal{U}_s^{n}(\bar{\boldsymbol{r}}))^2. \tag{25}$$

The insertion of Equations (23)–(25) and simple rearrangement reduces Equation (22) to

$$\left(A_{\xi_1}+A_{\xi_2}+\theta\gamma-\theta\nabla^2\mathcal{U}^n(\bar{\boldsymbol{r}})\right)\mathcal{U}^{n+1}(\bar{\boldsymbol{r}})-\left(\frac{\beta}{\delta t}+\theta\delta-\theta\eta\mathcal{U}^n(\bar{\boldsymbol{r}})\right)\nabla^2\mathcal{U}^{n+1}(\bar{\boldsymbol{r}}) \tag{26}$$

$$=(A_{\xi_1}+A_{\xi_2}-(1-\theta)\gamma)\mathcal{U}^n(\bar{\boldsymbol{r}})-\left(\frac{\beta}{\delta t}-(1-\theta)\delta\right)\nabla^2\mathcal{U}^n(\bar{\boldsymbol{r}})+H_\theta^{n+1}(\bar{\boldsymbol{r}})-G_\theta^n(\bar{\boldsymbol{r}}), \tag{27}$$

where

$$H_\theta^{n+1}(\bar{\boldsymbol{r}})=\begin{cases}f_\theta^{n+1}-Q_\xi^n\mathcal{U}^n(\bar{\boldsymbol{r}}), & n\geq1,\\ f_\theta^1, & n=1,\end{cases} \tag{28}$$

and

$$G_\theta^n(\bar{\boldsymbol{r}})=\eta\theta\left(2\mathcal{U}^n(\bar{\boldsymbol{r}})\nabla^2\mathcal{U}^n(\bar{\boldsymbol{r}})+(\mathcal{U}_x^n(\bar{\boldsymbol{r}}))^2+\left(\mathcal{U}_y^n(\bar{\boldsymbol{r}})\right)^2\right)+(1-\theta)\eta\nabla(\mathcal{U}^n(\bar{\boldsymbol{r}})\nabla\mathcal{U}^n(\bar{\boldsymbol{r}})).$$

Let us label

$$p_\theta^n(\bar{\boldsymbol{r}}) = A_{\xi_1} + A_{\xi_2} + \theta\gamma - \theta\eta\nabla^2\mathscr{U}^n(\bar{\boldsymbol{r}}),\ q_\theta^n = \frac{\beta}{\delta t} + \delta\theta - \theta\eta\mathscr{U}^n(\bar{\boldsymbol{r}}),$$
$$r_\theta = A_{\xi_1} + A_{\xi_2} - (1-\theta)\gamma,\ s_\theta = \frac{\beta}{\delta t} - (1-\theta)\delta,$$

which, when substituted into Equation (26), give us

$$p_\theta^n\mathscr{U}^{n+1}(\bar{\boldsymbol{r}}) - q_\theta^n\nabla^2\mathscr{U}^{n+1}(\bar{\boldsymbol{r}}) = r_\theta\mathscr{U}^n(\bar{\boldsymbol{r}}) + s_\theta\nabla^2\mathscr{U}^n(\bar{\boldsymbol{r}}) + H_\theta^{n+1}(\bar{\boldsymbol{r}}) - G_\theta^n(\bar{\boldsymbol{r}}). \tag{29}$$

We have approximated in the time direction at this point using Equation (29). Furthermore, the time direction is coupled with the splitting technique known as the Strang splitting algorithm, which is given as follows [4,57–59]:

$$\frac{\partial\mathscr{U}_1}{\partial t} = \mathscr{L}_1\mathscr{U}_1(t) \quad \text{with } t \in [t^n, t^{n+1/2}] \text{ and } \mathscr{U}_1(t^n) = \mathscr{U}_{sp}^n \tag{30}$$

$$\frac{\partial\mathscr{U}_2}{\partial t} = \mathscr{L}_2\mathscr{U}_2(t) \quad \text{with } t \in [t^n, t^{n+1}] \text{ and } \mathscr{U}_2(t^n) = \mathscr{U}_1(t^{n+1/2}) \tag{31}$$

$$\frac{\partial\mathscr{U}_3}{\partial t} = \mathscr{L}_1\mathscr{U}_3(t) \quad \text{with } t \in [t^{n+1/2}, t^{n+1}] \text{ and } \mathscr{U}_3(t^{n+1/2}) = \mathscr{U}_2(t^{n+1}) \tag{32}$$

where $\mathcal{L}_1\mathscr{U}(t) = \delta\nabla^2\mathscr{U}(t) - \eta(\mathscr{U} * \nabla^2\mathscr{U}(t))$, $\mathcal{L}_2\mathscr{U}(t) = -\eta(\nabla\mathscr{U}(t))^2 - \gamma\mathscr{U}(t)$, and $\mathscr{U}_{sp}^n = \mathscr{U}_0$ is the initial solution defined for Equation (20). The Strang splitting method achieves second-order accuracy. Further information can be found in reference [4].

3.2. Spatial Discretization

We can expand the function $\mathscr{U}(\bar{\boldsymbol{r}})$, which is a two-dimensional function denoted as $\mathscr{U}(r,s)$, into the following truncated series of Lucas polynomials, expressed as a product:

$$\mathscr{U}(r,s) \simeq \sum_{k=0}^{M}\sum_{m=0}^{M}\Lambda_{km}L_k(r)L_m(s) = \sum_{k=0}^{M}\sum_{m=0}^{M}\Lambda_{km}\mathcal{L}_{km}(r,s), \tag{33}$$

where $\mathcal{L}_{km}(r,s) = L_k(r)L_m(s)$. Equation (33) can be expressed in compact form as

$$\mathscr{U}(r,s) \simeq \mathbf{L}^{\mathrm{T}}(r,s)\mathbf{\Lambda}, \tag{34}$$

with $\mathbf{\Lambda} = [\Lambda_{00}, \Lambda_{01}, \Lambda_{10}, \cdots, \Lambda_{MM}]^{\mathrm{T}}$ and $\mathbf{L}^{\mathrm{T}}(r,s) = [\mathcal{L}_{00}(r,s), \mathcal{L}_{01}(r,s), \mathcal{L}_{10}(r,s), \cdots, \mathcal{L}_{MM}(r,s)]$.

One can compute and briefly represent the consolidated partial derivatives of $\mathscr{U}(r,s)$ as follows:

$$\frac{\partial\mathscr{U}(r,s)}{\partial r} \simeq \sum_{k=0}^{M}\sum_{m=0}^{M}\Lambda_{km}\frac{\mathrm{d}L_k(r)}{\mathrm{d}r}L_m(s) = \sum_{k=0}^{M}\sum_{m=0}^{M}\Lambda_{km}k(F_k(r))L_m(s) = \mathrm{L}_r^{\mathrm{T}}(r,s)\mathbf{\Lambda},$$

and

$$\frac{\partial^2\mathscr{U}(r,s)}{\partial r^2} \simeq \sum_{k=0}^{M}\sum_{m=0}^{M}\Lambda_{km}\frac{\mathrm{d}^2L_k(r)}{\mathrm{d}r^2}L_m(s) = \sum_{k=0}^{M}\sum_{m=0}^{M}\Lambda_{km}k(F_k(r)D)L_m(s) = \mathbf{L}_{rr}^{\mathrm{T}}(r,s)\mathbf{\Lambda},$$

where

$$\mathbf{L}_r^{\mathrm{T}}(r,s) = \{\frac{\partial}{\partial r}\mathcal{L}_{km}(r,s)\}_{k,m=0}^{M} = \{\frac{\mathrm{d}L_k(r)}{\mathrm{d}r}L_m(s)\}_{k,m=0}^{M} = \{k(F_k(r))L_m(s)\}_{k,m=0}^{M},$$

$$\mathbf{L}_{rr}^{\mathrm{T}}(r,s) = \{\frac{\partial^2}{\partial r^2}\mathcal{L}_{km}(r,s)\}_{k,m=0}^{M} = \{\frac{\mathrm{d}^2L_k(r)}{\mathrm{d}r^2}L_m(s)\}_{k,m=0}^{M} = \{k(F_k(r)D)L_m(s)\}_{k,m=0}^{M}.$$

Similarly

$$\frac{\partial \mathcal{U}(r,s)}{\partial s} \simeq \sum_{k=0}^{M}\sum_{m=0}^{M} \Lambda_{km} L_k(r) \frac{\mathrm{d}L_m(s)}{\mathrm{d}r} = \sum_{k=0}^{M}\sum_{m=0}^{M} \Lambda_{km} m L_k(r)(F_m(s)) = \mathbf{L}_s^{\mathrm{T}}(r,s)\mathbf{\Lambda},$$

and

$$\frac{\partial^2 \mathcal{U}(r,s)}{\partial s^2} \simeq \sum_{k=0}^{M}\sum_{m=0}^{M} \Lambda_{km} L_k(r) \frac{\mathrm{d}^2 L_m(s)}{\mathrm{d}s^2} = \sum_{k=0}^{M}\sum_{m=0}^{M} \Lambda_{km} m L_k(r)(F_m(s)D) = \mathbf{L}_{ss}^{\mathrm{T}}(r,s)\mathbf{\Lambda},$$

with

$$\mathbf{L}_s^{\mathrm{T}}(r,s) = \{\frac{\partial}{\partial s}\mathcal{L}_{km}(r,s)\}_{k,m=0}^{M} = \{L_k(r)\frac{\mathrm{d}^2 L_m(s)}{\mathrm{d}s^2}\}_{k,m=0}^{M} = \{mL_k(r)(F_m(s))\}_{k,m=0}^{M}.$$

$$\mathbf{L}_{ss}^{\mathrm{T}}(r,s) = \{\frac{\partial^2}{\partial s^2}\mathcal{L}_{km}(r,s)\}_{k,m=0}^{M} = \{L_k(r)\frac{\mathrm{d}^2 L_m(s)}{\mathrm{d}s^2}\}_{k,m=0}^{M} = \{mL_k(r)(F_m(s)D)\}_{k,m=0}^{M}.$$

These results can be further refined as

$$\nabla \mathcal{U}(r,s) = \frac{\partial \mathcal{U}}{\partial r} + \frac{\partial \mathcal{U}}{\partial s} \simeq \left(\mathbf{L}_r^{\mathrm{T}}(r,s) + \mathbf{L}_s^{\mathrm{T}}(r,s)\right)\mathbf{\Lambda}. \tag{35}$$

$$\nabla^2 \mathcal{U}(r,s) = \frac{\partial^2 \mathcal{U}}{\partial r^2} + \frac{\partial^2 \mathcal{U}}{\partial s^2} \simeq \left(\mathbf{L}_{rr}^{\mathrm{T}}(r,s) + \mathbf{L}_{ss}^{\mathrm{T}}(r,s)\right)\mathbf{\Lambda}. \tag{36}$$

It is consequently possible to express the time-dependent function $\mathcal{U}(r,s,t)$ and its compact form partial derivatives as follows:

$$\mathcal{U}(r,s,t) \approx \mathbf{L}^{\mathrm{T}}(r,s)\mathbf{\Lambda}(t),\ \frac{\partial^2 \mathcal{U}(r,s,t)}{\partial r^2} \approx \mathbf{L}_{rr}^{\mathrm{T}}(r,s)\mathbf{\Lambda}(t),\ \frac{\partial^2 \mathcal{U}(r,s,t)}{\partial s^2} \approx \mathbf{L}_{ss}^{\mathrm{T}}(r,s)\mathbf{\Lambda}(t), \tag{37}$$

with

$$\nabla \mathcal{U}(r,s,t) \approx \left(\mathbf{L}_r^{\mathrm{T}}(r,s) + \mathbf{L}_s^{\mathrm{T}}(r,s)\right)\mathbf{\Lambda}(t),\ \nabla^2 \mathcal{U}(r,s,t) \approx \left(\mathbf{L}_{rr}^{\mathrm{T}}(r,s) + \mathbf{L}_{ss}^{\mathrm{T}}(r,s)\right)\mathbf{\Lambda}(t), \tag{38}$$

where $\mathbf{\Lambda}(t)$ represents the vector of the time-varying unknown coefficients. Once the information obtained from Equations (37) and (38) is combined with Equation (29), we obtain

$$\begin{aligned} &p_\theta^n \mathbf{L}(r,s)\mathbf{\Lambda}^{n+1} - q_\theta^n\left(\mathbf{L}_{rr}^{\mathrm{T}}(r,s) + \mathbf{L}_{ss}^{\mathrm{T}}(r,s)\right)\mathbf{\Lambda}^{n+1} \\ &= r_\theta \mathbf{L}(r,s)\mathbf{\Lambda}^n + r_\theta\left(\mathbf{L}_{rr}^{\mathrm{T}}(r,s) + \mathbf{L}_{ss}^{\mathrm{T}}(r,s)\right)\mathbf{\Lambda}^n + \mathbf{H}_\theta^{n+1}(r,s) - \mathbf{G}_\theta^n(r,s), \quad \forall (r,s) \in \Omega. \end{aligned} \tag{39}$$

The boundary conditions are stated as

$$\mathbb{B}\mathbf{L}(r,s)\mathbf{\Lambda}^{n+1} \approx f_1(r,s,t^{n+1}) = f_1^{n+1}(r,s), \quad \forall (r,s) \in \partial\Omega, \tag{40}$$

where $\mathbf{\Lambda}^n = \mathbf{\Lambda}(t^n)$.

3.3. Full Discretization

We collocate Equations (39) and (40) at discrete mesh points r_i and s_j at their respective time levels to derive the discrete form of the time-fractional differential equation as expressed in Equation (20), resulting in the subsequent set of equations:

$$\begin{aligned} &p_\theta^n \mathbf{L}(r_i,s_j)\mathbf{\Lambda}^{n+1} - q_\theta^n\left(\mathbf{L}_{rr}^{\mathrm{T}}(r_i,s_j) + \mathbf{L}_{ss}^{\mathrm{T}}(r_i,s_j)\right)\mathbf{\Lambda}^{n+1} \\ &= s_\theta \mathbf{L}(r_i,s_j)\mathbf{\Lambda}^n + r_\theta\left(\mathbf{L}_{rr}^{\mathrm{T}}(r_i,s_j) + \mathbf{L}_{ss}^{\mathrm{T}}(r_i,s_j)\right)\mathbf{\Lambda}^n + \mathbf{H}_\theta^{n+1}(r_i,s_j) - \mathbf{G}_\theta^n(r_i,s_j), \quad \forall (r_i,s_j) \in \Omega, \end{aligned}$$

and

$$\mathbb{B}\mathbf{L}(r_i, s_j)\mathbf{\Lambda}^{n+1} \approx f_1(r_i, s_j, t^{n+1}) = f_1^{n+1}(r_i, s_j), \quad \forall (r_i, s_j) \in \partial\Omega,$$

where $i, j = 0, 1, 2, \cdots, M$. This implies the following matrix–vector form:

$$\mathcal{G}\mathbf{\Lambda}^{n+1} = \mathcal{H}\mathbf{\Lambda}^n + \mathcal{Q}^{n+1}. \tag{41}$$

In this case, the matrix entries are provided by

$$(\mathcal{G})_{ij} = \begin{cases} p_\theta^n \mathbf{L}(r_i, s_j) - q_\theta^n \Big(\mathbf{L}_{rr}^{\mathrm{T}}(r_i, s_j) + \mathbf{L}_{ss}^{\mathrm{T}}(r_i, s_j)\Big), & (r_i, s_j) \in \Omega, \\ \mathbb{B}\mathbf{L}(r_i, s_j), & (r_i, s_j) \in \partial\Omega, \end{cases} \tag{42}$$

$$(\mathcal{H})_{ij} = \begin{cases} s_\theta \mathbf{L}(r_i, s_j) + r_\theta \Big(\mathbf{L}_{rr}^{\mathrm{T}}(r_i, s_j) + \mathbf{L}_{ss}^{\mathrm{T}}(r_i, s_j)\Big), & (r_i, s_j) \in \Omega, \\ 0, & (r_i, s_j) \in \partial\Omega, \end{cases} \tag{43}$$

and

$$(\mathcal{Q})_{ij} = \begin{cases} \mathbf{H}_\theta^{n+1}(r_i, s_j) - \mathbf{G}_\theta^n(r_i, s_j), & (r_i, s_j) \in \Omega, \\ f_1(r_i, s_j, t^{n+1}), & (r_i, s_j) \in \partial\Omega. \end{cases} \tag{44}$$

The construction of the scheme for the two-dimensional model equation is finalized. Simplifying it to a one-dimensional form involves eliminating the variable s so that $\overline{r} = r$. The linear system represented by Equation (41) benefits from the assurances provided by [47], ensuring the existence of a solution. Solving Equation (41) allows for the determination of the vector containing unknown coefficients at every time step. The initial condition allows for initialization in the following manner:

$$\mathbf{L}^{\mathrm{T}}(r_i, s_j)\mathbf{\Lambda}^0 = \mathcal{U}(r_i, s_j, 0) = \mathcal{U}_0 \Longrightarrow \mathbf{A}\mathbf{\Lambda}^0 = \mathcal{U}_0.$$

When solved, it yields $\mathbf{\Lambda}^0$. Utilizing Equation (41), the iteration is subsequently repeated up to the relevant time level. Upon obtaining the coefficients vector, the solution at the corresponding time level can be determined by

$$\mathcal{U}(r_i, s_j, t^n) = \mathbf{L}^{\mathrm{T}}(r_i, s_j)\mathbf{\Lambda}^n, \quad \forall (r_i, s_j) \in \Omega, \ (n \geq 0).$$

A similar procedure can be used in the case of a three-dimensional problem.

4. Analyzing Errors

To conduct the error analysis, we referred to the following theorem.

Theorem 1. *The error can be defined as follows when denoting the exact solution as* $\widehat{\mathbb{U}}$ *and the computed solution obtained by the suggested meshless method as* $\mathcal{U}$ *for the underlying problems.*

The error term $\mathcal{E}$ *is constrained within the following expression:*

$$|\mathcal{E}| \leq \frac{4\exp(2\kappa)\cosh^2(2\mathcal{P})\kappa^{2(N+1)}}{((N+1)!)^2}.$$

Proof. Consider the absolute error between the exact solution and the approximate numerical solution using the proposed hybrid meshless method:

$$|\mathcal{E}| = |\widehat{\mathbb{U}} - \mathcal{U}|,$$

where $\widehat{\mathbb{U}} = \sum_{k=0}^{\infty} \sum_{m=0}^{\infty} \Lambda_k \Lambda_m L_k(r) L_m(s)$ and $\mathcal{U} = \sum_{k=0}^{M} \sum_{m=0}^{M} \Lambda_k \Lambda_m L_k(r) L_m(s)$. Next, the truncated term is as follows:

$$|\mathcal{E}| = \sum_{k=N+1}^{\infty} \sum_{m=N+1}^{\infty} \Lambda_k \Lambda_m L_k(r) L_m(s) \tag{45}$$

In [60], it is demonstrated that

$$L_m(r) \leq 2\vartheta^m,$$

$$|\Lambda_m| \leq \frac{\mathcal{P}^m \cosh(2\mathcal{P})}{m!},$$

where the famous golden ratio is denoted by ϑ. Consequently, Equation (45) suggests that

$$|\mathcal{E}| = 4\cosh^2(2\mathcal{P}) \sum_{k=N+1}^{\infty} \sum_{m=N+1}^{\infty} \frac{\kappa^{k+m}}{k!m!},$$

where $\kappa = \mathcal{P}\vartheta$. This inequality can alternatively be expressed as

$$|\mathcal{E}| = 4\exp(2\kappa)\cosh^2(2\mathcal{P}) \left[1 - \frac{\Gamma(N+1,\kappa)}{\Gamma(N+1)}\right]^2. \tag{46}$$

The incomplete gamma function in this case is $\Gamma(N+1,\kappa)$, and the complete gamma function is $\Gamma(N+1)$ [61]. Equation (46) can be expressed in integral form as follows:

$$|\mathcal{E}| \leq \frac{4\exp(2\kappa)\cosh^2(2\mathcal{P})}{(N!)^2} \left[\int_0^{\kappa} t^N \exp(-t)dt\right]^2.$$

Since $\exp(-t) < 1, \forall t > 0$, we obtain:

$$|\mathcal{E}| \leq \frac{4\exp(2\kappa)\cosh^2(2\mathcal{P})\kappa^{2(N+1)}}{((N+1)!)^2}.$$

□

Stability Analysis

The stability of the meshless method is crucial for ensuring the accuracy and convergence of numerical solutions over time. By mitigating numerical instabilities, the meshless method maintains the solution quality and reliability across diverse computational domains. The stability analysis was carried out in this work by utilizing a matrix technique. The methodology employed was consistent with that described in previous studies [24,62].

Theorem 2. *Consider w as the computed approximate solution to the underlying problem. In this context, the amplification matrix $\digamma$ is defined as $\digamma = L\mathcal{G}^{-1}\mathcal{H}L^{-1}$. It is essential to confirm that the absolute maximum eigenvalue of $\digamma$, denoted as $\rho(\digamma)$, stays less than or equal to 1 to ensure stability.*

Proof. We aim to show that $\digamma = L\mathcal{G}^{-1}\mathcal{H}L^{-1}$. From Equation (17), we write

$$\Lambda^n = L^{-1}\mathcal{U}^n. \tag{47}$$

The relationship can be described utilizing Equation (41):

$$\mathcal{U}^{n+1} = L\mathcal{G}^{-1}\mathcal{H}L^{-1}\mathcal{U}^n + L\mathcal{G}^{-1}\mathcal{Q}^{n+1}. \tag{48}$$

Moreover, the error vector

$$\mathcal{E} = \widehat{\mathbb{U}} - \mathcal{U}, \tag{49}$$

satisfies the following condition:

$$\mathcal{E}^{n+1} = \digamma \mathcal{E}^{n}.$$

Thus, the amplification matrix is defined as $\digamma = L\mathcal{G}^{-1}\mathcal{H}L^{-1}$. Adherence to the Lax–Richtmyer stability criterion [63], which requires that

$$||\digamma|| \leq 1,$$

ensures the stability of the method. An illustration demonstrating that the spectral radius $\rho(\digamma)$ must not exceed 1 to fulfill this condition is provided in the subsequent section. □

5. Discussion and Numerical Results

The proposed hybrid meshless method (HMM) was assessed for its effectiveness and precision in solving model Equations (1) and (2). The evaluation includes various scenarios with both uniform and scattered nodes in non-rectangular and rectangular domains. The CPU time was measured in seconds for all cases. All computations were performed on an HP PC laptop equipped with an Intel(R) Core(TM) i5-7200U CPU @ 2.50 GHz 2.71 GHz and 8 GB of RAM, using MATLAB (R2012a) software.

The accuracy was measured using error norms, including the following:

$$e_{\infty} = \max(|\widehat{\mathbb{U}} - \mathcal{U}|), \; e_{rms} = \sqrt{\frac{\sum_{i=1}^{N}\left(\widehat{\mathbb{U}}_i - \mathcal{U}_i\right)^2}{N}}, \; e_{L2} = \sqrt{\delta h \sum_{h=1}^{N}\left(\widehat{\mathbb{U}}_h - \mathcal{U}_h\right)^2}, \tag{50}$$

where the exact solution is $\widehat{\mathbb{U}}$, while the approximate solution is $\mathcal{U}$.

Problem 1. *The exact solution of model* (1) *with* $\delta = 1$ *and* $\eta = \gamma = 0$ *is*

$$\mathcal{U}(\bar{\mathbf{r}}, t) = e^{-t}\sin(\pi r)\sin(\pi s), \quad \bar{\mathbf{r}} = (r, s) \in \Omega. \tag{51}$$

The numerical results of the proposed hybrid meshless method were used to compute the numerical solution of Problem 1, as shown in Table 1. The ϵ_{∞} *norm was obtained for the final times* $t = 1$ *and* $t = 5$*. The obtained numerical results were compared with the method reported in [17]. The results using explicit, implicit, and Crank–Nicolson schemes for the proposed method were computed for different time step sizes, and one could conclude that as the time step size* δt *decreased, the accuracy of the proposed method increased. The comparison shows that the HMM was more accurate than the method given in [17]. In Table 2, the numerical results are presented, where they were computed by the proposed method using different fractional orders* ξ*, various nodal points* N*,* $\delta t = 0.001$*, and* $t = 1$*. The results were computed as* e_{∞} *and* e_{rms} *norms. It can be observed that increasing the number of nodes improved the accuracy to some extent, whereas changing the fractional order did not significantly affect the accuracy. In this case, the HMM yielded good and accurate results. Additionally, the CPU time is provided in this table, demonstrating the efficiency of the suggested method.*

Comparative results of the HMM and [17] are presented in Figure 1 for various nodes in the form of different error norms. It can be observed that the results produced by the HMM were superior to those of [17]. Furthermore, Figure 2 depicts the comparison of the numerically computed solution with the exact solution, along with the absolute error.

Table 1. Problem 1 numerical results using HMM and [17].

		e_∞			
Method	δt	$t = 1$		$t = 5$	
		MQ **[17]**	**HMM**	MQ **[17]**	**HMM**
Explicit	0.1	1.8985×10^{-2}	8.5002×10^{-3}	1.5971×10^{-3}	8.3765×10^{-4}
	0.05	9.2780×10^{-3}	5.3215×10^{-3}	8.2371×10^{-4}	5.1759×10^{-4}
	0.025	4.5880×10^{-3}	1.0717×10^{-3}	4.1804×10^{-4}	1.0352×10^{-4}
	0.0125	2.2815×10^{-3}	9.0701×10^{-3}	2.1057×10^{-4}	9.0047×10^{-5}
	0.00625	1.1377×10^{-3}	8.0520×10^{-4}	1.0567×10^{-4}	8.6471×10^{-5}
Implicit	0.1	1.7500×10^{-2}	8.3842×10^{-3}	1.7904×10^{-3}	8.5478×10^{-4}
	0.05	8.9133×10^{-3}	5.2547×10^{-3}	8.7217×10^{-4}	5.5662×10^{-4}
	0.025	4.4982×10^{-3}	1.1406×10^{-3}	4.3024×10^{-4}	1.0639×10^{-4}
	0.0125	2.2594×10^{-3}	9.9320×10^{-4}	2.1364×10^{-4}	9.9331×10^{-5}
	0.00625	1.1323×10^{-3}	8.1481×10^{-4}	1.0645×10^{-4}	8.1101×10^{-5}
Crank–Nicolson	0.1	3.0659×10^{-4}	9.0152×10^{-5}	3.0957×10^{-5}	8.3810×10^{-6}
	0.05	8.4249×10^{-5}	1.2486×10^{-5}	8.4718×10^{-6}	1.1754×10^{-6}
	0.025	2.4213×10^{-5}	6.2692×10^{-6}	2.3901×10^{-6}	3.1455×10^{-7}
	0.0125	7.1414×10^{-6}	8.6427×10^{-7}	6.8938×10^{-7}	8.0157×10^{-8}
	0.00625	2.1392×10^{-6}	1.1503×10^{-7}	2.0203×10^{-7}	5.8023×10^{-8}

Table 2. Problem 1 numerical results using HMM.

N	$\xi_1 = \xi_2 = 0.1$		$\xi_1 = \xi_2 = 0.5$		$\xi_1 = \xi_2 = 0.9$		**CPU Time**
	e_∞	e_{ems}	e_∞	e_{ems}	e_∞	e_{ems}	
10	3.1204×10^{-9}	2.7813×10^{-9}	4.2101×10^{-9}	3.7823×10^{-9}	5.9105×10^{-9}	3.7930×10^{-9}	15.64
20	2.6534×10^{-9}	2.6591×10^{-9}	3.6587×10^{-9}	3.6901×10^{-9}	3.8257×10^{-9}	3.7133×10^{-9}	18.89
25	2.0673×10^{-9}	2.2873×10^{-9}	3.2510×10^{-9}	3.1246×10^{-9}	3.2780×10^{-9}	3.3854×10^{-9}	19.05

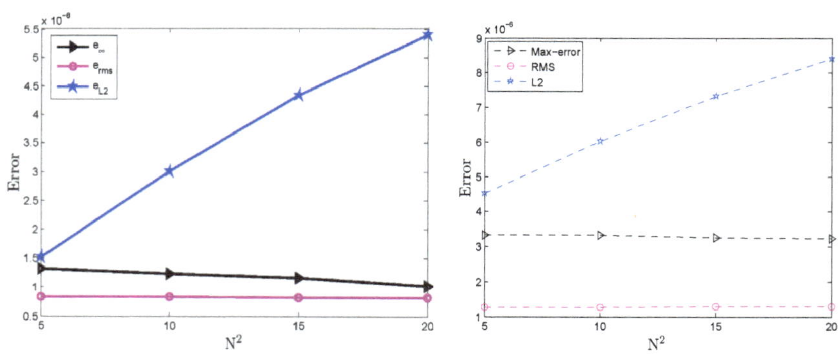

Figure 1. Problem 1 error norms using HMM (**left**) and [17] (**right**).

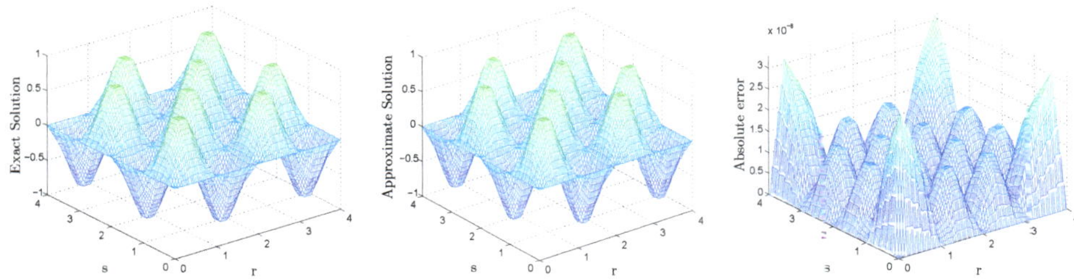

Figure 2. Problem 1 exact and numerical solutions using HMM with absolute error.

Problem 2. *Consider the model Equation* (1) *with* $\beta = 1$ *and* $\gamma = \delta = 0$ *having the exact solution*

$$\mathcal{U}(\bar{\mathbf{r}}, t) = e^{r-s-t} \sin(\pi r) \sin(\pi s), \quad \bar{\mathbf{r}} = (r, s) \in \Omega. \tag{52}$$

Meshless methods offer the advantage of computing solutions without being constrained by the computational geometry. These methods naturally adjust to the geometry of the domain and offer greater flexibility in handling boundary conditions. Additionally, meshless methods often exhibit excellent scalability and parallelizability, making them suitable for large-scale simulations. To assess the true effectiveness of the proposed meshless method, various irregular geometries were examined. To evaluate the performance of the proposed HMM in non-rectangular domains, as shown in Figure 3 for Problem 2, Table 3 presents the numerical results obtained using $\xi_1 = \xi_2 = 0.5$ *and* $\delta t = 0.0001$, *along with various time values, in comparison with the method outlined in [17]. The table illustrates that the HMM achieved greater accuracy compared with [17]. Additionally, in Table 4, the numerical outcomes derived from the proposed HMM were utilized to compute the* e_{L_2} *norm for* δt, *which was then compared with the results obtained using the method reported in [17]. Employing explicit, implicit, and Crank–Nicolson schemes, the proposed method's results were calculated for different time step sizes, demonstrating that decreasing* δt *enhanced the accuracy of the proposed approach. This comparison highlights the superior performance of the HMM over the method described in [17].*

Figure 4 presents a comparison of the HMM results with those in [17] using various error norms for different fractional orders ξ. *Once again, the HMM demonstrated superior performance. Figure 5 illustrates the close agreement between the exact and numerical solutions of Problem 2 with the parameters* $\xi = 0.5$, $N = 10^2$, $\delta t = 0.001$, *and time* $t = 0.5$ *in terms of the absolute error.*

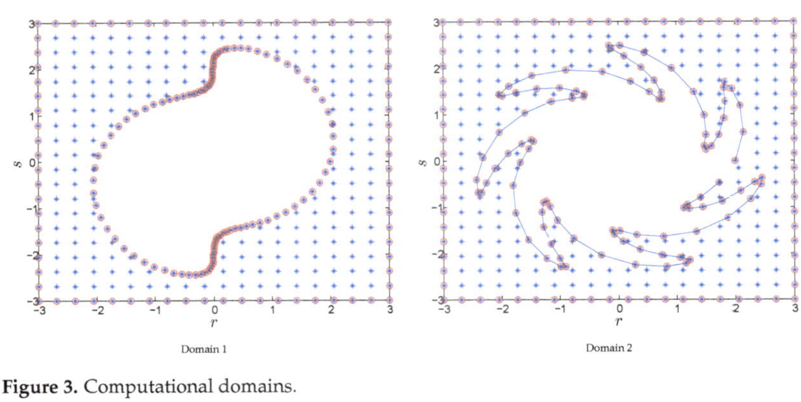

Figure 3. Computational domains.

Table 3. Problem 2 numerical results of HMM using domains given in Figure 3.

Time	Error Norm	Domain 1 HMM	Domain 2 HMM	Domain 2 IQ [17]	Domain 2 MQ [17]	Domain 2 IMQ [17]
t = 1	e_∞	7.0321×10^{-5}	8.1482×10^{-4}	2.8503×10^{-3}	1.0762×10^{-4}	1.6486×10^{-4}
	e_{rms}	6.2456×10^{-6}	8.2453×10^{-5}	5.3969×10^{-4}	1.2498×10^{-5}	1.6465×10^{-5}
	e_{L2}	6.2710×10^{-5}	9.3608×10^{-4}	5.8179×10^{-3}	1.3473×10^{-4}	1.7749×10^{-4}
t = 2	e_∞	3.1438×10^{-5}	4.1930×10^{-5}	6.9177×10^{-3}	7.4827×10^{-5}	1.2021×10^{-4}
	e_{rms}	4.4711×10^{-6}	5.591×10^{-6}	1.1019×10^{-3}	8.9038×10^{-6}	1.2001×10^{-5}
	e_{L2}	5.5379×10^{-5}	5.6283×10^{-5}	1.1879×10^{-2}	9.5983×10^{-5}	1.2937×10^{-4}
t = 3	e_∞	9.3510×10^{-6}	1.8646×10^{-5}	1.9053×10^{-2}	3.9556×10^{-5}	6.5942×10^{-5}
	e_{rms}	1.0392×10^{-6}	2.8366×10^{-6}	2.6854×10^{-3}	4.8116×10^{-6}	6.5818×10^{-6}
	e_{L2}	2.0406×10^{-5}	2.9815×10^{-5}	2.8949×10^{-2}	5.1869×10^{-5}	7.0953×10^{-5}

Table 4. Problem 2 results of HMM and [17] for $\xi_1 = \xi_2 = 0.6$.

		e_{L2}			
		t = 1	**t = 1**	**t = 2**	**t = 2**
Method	δt	**HMM**	**[17]**	**HMM**	**[17]**
Explicit	0.1	2.0148×10^{-2}	5.1872×10^{-2}	1.3867×10^{-2}	3.7523×10^{-2}
	0.05	9.5461×10^{-3}	2.5444×10^{-2}	8.4984×10^{-3}	1.8637×10^{-2}
	0.025	8.3610×10^{-3}	1.2612×10^{-2}	5.5718×10^{-3}	9.2941×10^{-3}
	0.125	3.4513×10^{-3}	6.2822×10^{-3}	1.9845×10^{-3}	4.6433×10^{-3}
	0.00625	9.3719×10^{-4}	3.1364×10^{-3}	8.3805×10^{-4}	2.3216×10^{-3}
Implicit	0.1	1.3584×10^{-2}	4.8894×10^{-2}	9.7926×10^{-3}	3.7120×10^{-2}
	0.05	9.4241×10^{-3}	2.4862×10^{-2}	7.3200×10^{-3}	1.8657×10^{-2}
	0.025	7.3721×10^{-3}	1.2528×10^{-2}	5.7313×10^{-3}	9.3447×10^{-3}
	0.0125	2.6219×10^{-3}	6.2845×10^{-3}	9.9378×10^{-4}	4.6732×10^{-3}
	0.00125	8.9866×10^{-4}	3.1458×10^{-3}	7.5566×10^{-4}	2.3357×10^{-3}
Crank–Nicolson	0.1	4.3021×10^{-4}	3.5095×10^{-4}	1.5380×10^{-4}	2.5348×10^{-4}
	0.05	7.6320×10^{-5}	9.2813×10^{-5}	7.3817×10^{-5}	7.1868×10^{-5}
	0.025	9.8431×10^{-6}	4.9749×10^{-5}	8.0116×10^{-6}	3.8280×10^{-5}
	0.0125	4.9669×10^{-6}	2.4637×10^{-5}	2.0167×10^{-6}	1.8688×10^{-5}
	0.00625	7.3499×10^{-7}	1.0941×10^{-5}	7.9173×10^{-7}	8.2362×10^{-6}

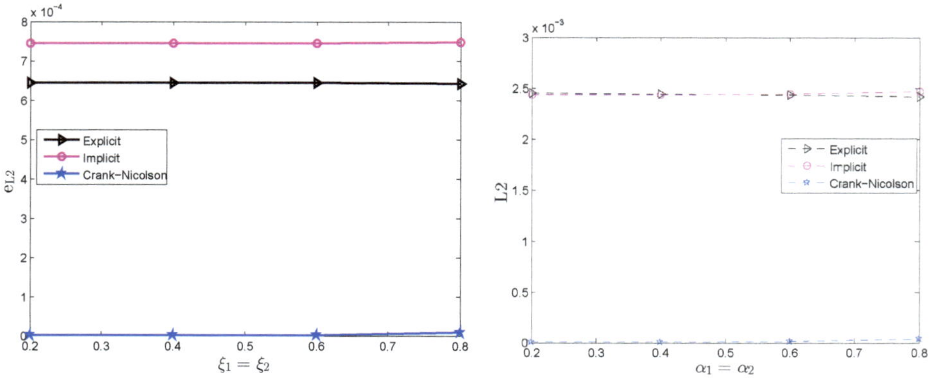

Figure 4. Problem 2 error norms of HMM and [17].

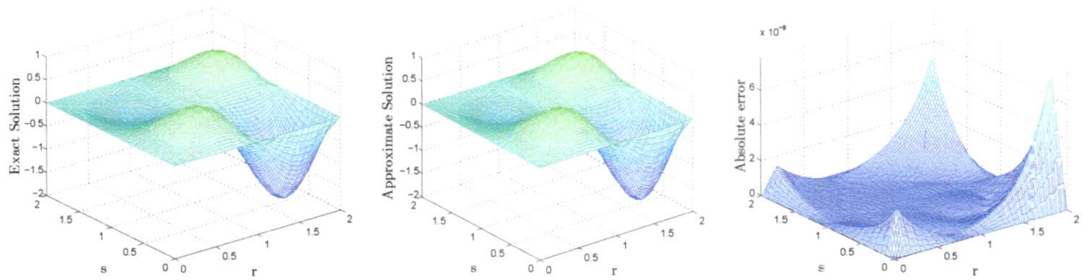

Figure 5. Problem 1 exact and numerical solutions of HMM with absolute error.

Problem 3. *The exact solution of model* (2) *with* $\delta = 1$ *and* $\eta = \gamma = 0$ *is as follows:*

$$\mathcal{U}(\bar{\mathbf{r}}, t) = e^{-t} \sin(\pi r) \sin(\pi s) \sin(\pi z), \quad \bar{\mathbf{r}} = (r, s, z) \in \Omega. \tag{53}$$

The results for Problem 3 were compared with the exact solution and the approach provided in [15] for different values of δt, $N = 10$, *and using* $t = 1$ *and* $\xi_1 = \xi_2 = 0.3$. *This served to illustrate the accurateness and effectiveness of the suggested strategy. These findings are presented in Table 5, along with the CPU time. The table shows that after a few iterations, the suggested meshless technique yielded better results. The accuracy also improved with an increase in the number of iterations, with both error norms reaching as low as* 10^{-10}.

The findings of Problem 3 were computed using the proposed HMM in Table 6, with varying values of fractional order ξ *for* $N = 10$, $t = 1$, *and* $t = 2$. *We evaluated the outcomes in terms of* e_{rms} *and* e_∞. *Accurate results were found for several time-fractional orders in this problem, and these results were also compared with the findings published in [15].*

Table 5. Problem 3 numerical results of HMM and [15].

δt	e_∞	e_{rms}	e_∞ [15]	e_{rms} [15]	CPU Time of HMM
0.1	7.7488×10^{-5}	3.7354×10^{-5}	2.9285×10^{-4}	9.2558×10^{-5}	1.39
0.05	2.6377×10^{-5}	8.5621×10^{-6}	7.3125×10^{-5}	2.3112×10^{-5}	1.54
0.01	6.7313×10^{-7}	3.5478×10^{-7}	2.9195×10^{-6}	9.2271×10^{-7}	2.60
0.005	2.4307×10^{-7}	8.5271×10^{-8}	7.2896×10^{-7}	2.3039×10^{-7}	3.96
0.001	8.1268×10^{-9}	3.2773×10^{-9}	2.9007×10^{-8}	9.1677×10^{-9}	17.73
0.0005	2.4501×10^{-10}	9.9388×10^{-10}	7.2236×10^{-9}	2.2830×10^{-9}	42.14

Table 6. Problem 3 numerical results of HMM and [15].

ξ	$t = 1$				$t = 2$			
	e_∞	e_{rms}	e_∞ [15]	e_{rms} [15]	e_∞	e_{rms}	e_∞ [15]	e_{rms} [15]
0.2	8.69430×10^{-8}	5.3992×10^{-8}	7.3170×10^{-7}	2.3126×10^{-7}	7.9456×10^{-8}	3.9267×10^{-8}	5.3832×10^{-7}	1.7014×10^{-7}
0.4	9.65010×10^{-8}	6.5001×10^{-8}	7.3053×10^{-7}	2.3089×10^{-7}	8.1655×10^{-8}	4.9126×10^{-8}	5.3746×10^{-7}	1.6987×10^{-7}
0.6	9.27650×10^{-8}	8.4952×10^{-8}	7.2584×10^{-7}	2.2940×10^{-7}	7.8037×10^{-8}	5.7601×10^{-8}	5.3402×10^{-7}	1.6878×10^{-7}
0.8	8.72000×10^{-8}	7.3528×10^{-8}	7.0765×10^{-7}	2.2364×10^{-7}	7.6584×10^{-8}	3.9773×10^{-8}	5.2065×10^{-7}	1.6455×10^{-7}

For Problem 3, we considered some non-rectangular domains shown in Figure 6 and in Figure 7 (left) to evaluate the performance of the suggested HMM. In domain 3, scattered nodes were taken into account, as shown in Figure 6 (left). The results obtained in this case are shown in Figure 8, whereas the results for domain 5 are shown in Figure 7 (right) for final times up to $t = 4$. *These results provide evidence of the accurate performance of the HMM in non-rectangular domains.*

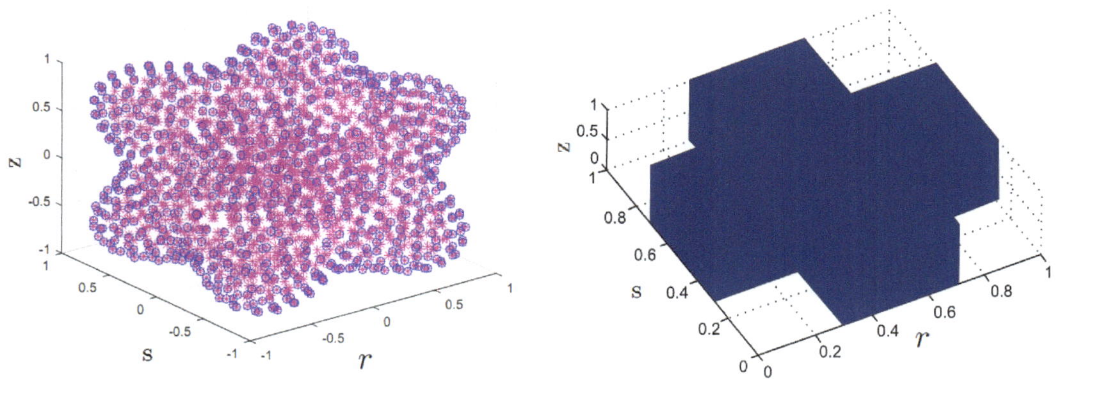

Domain 3 Domain 4

Figure 6. Computational domains.

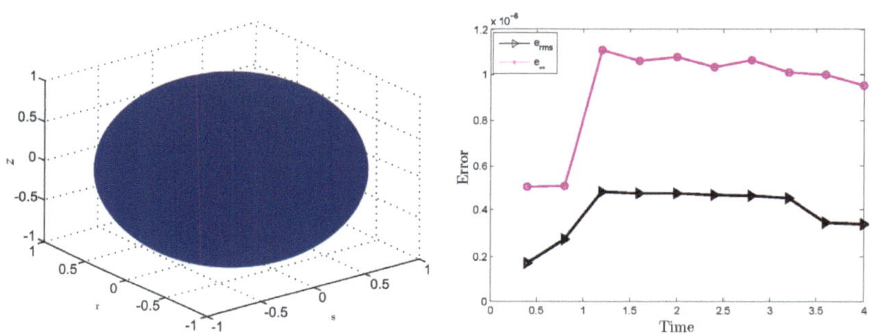

Domain 5

Figure 7. Problem 3 domain 5 (**left**) and error (**right**).

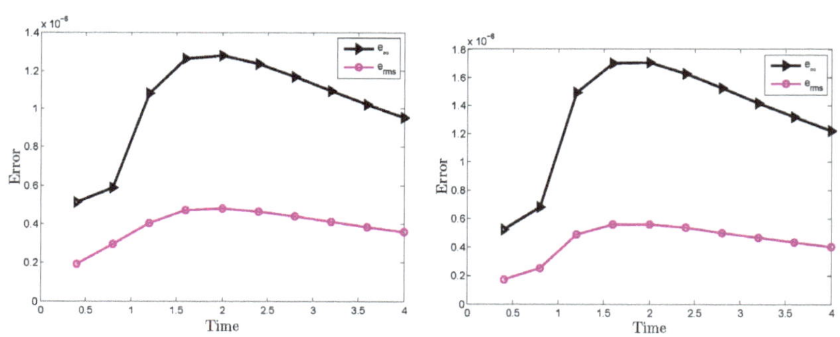

Figure 8. Problem 3 error norms using domain 3 (**left**) and domain 4 (**right**).

Problem 4. *The exact solution for Equation* (2) *with* $\beta = 1$, $\gamma = 1$, *and* $\delta = \pi^2$ *is*

$$\mathcal{U}(\bar{\mathbf{r}}, t) = e^t \sin(\pi r) \sin(\pi s) \sin(\pi z), \quad \bar{\mathbf{r}} = (r, s, z) \in \Omega. \tag{54}$$

For Problem 4, the behavior of the numerical and exact solutions using $N = 21$ *and* $t = 0.01$ *are visualized in Figure 9 at* $z = 0.5$ *and* $\xi_1 = \xi_2 = 0.5$, *along with the absolute error. One can see from these figures that the approximate solution was very compatible with the exact solution. Also, Figure 10 shows the contour graph of the problem at* $\xi_1 = \xi_2 = 0.3$ *and* $\xi_1 = \xi_2 = 0.9$.

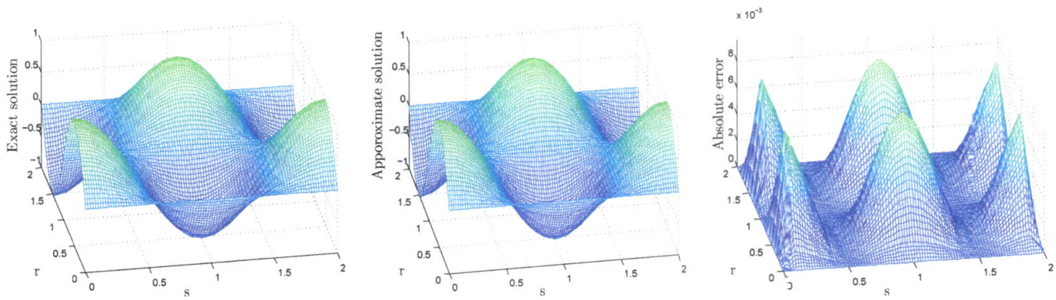

Figure 9. Problem 4 exact and numerical solutions of HMM with absolute error at $z = 0.5$.

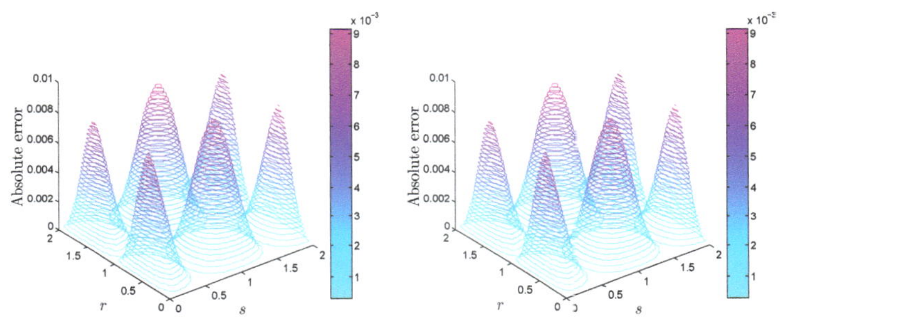

Figure 10. Problem 4 absolute error at $\xi_1 = \xi_2 = 0.3$ (**left**) and $\xi_1 = \xi_2 = 0.9$ (**right**), whereas $z = 0.5$.

6. Conclusions

This paper introduces a relatively new numerical methodology for addressing two-term time-fractional PDE models in two and three dimensions. We combined the Louiville–Caputo fractional derivative scheme with the Strang splitting algorithm for the temporal component and employed a meshless technique for spatial derivatives utilizing Lucas and Fibonacci polynomials. Error norms were utilized to evaluate the accuracy of the methodology across regular and irregular domains. A comparative study was conducted between the exact solution and alternative numerical methods outlined in the contemporary literature. This inquiry illustrated that the proposed approach yielded superior performance while requiring reduced computational resources. Moreover, the study illustrated how minimizing the errors could be accomplished by refining the spatial and temporal stages. With minor modifications, the proposed approach could be applied to various intricate fractional partial differential equations that arise in fields such as fluid dynamics, quantum mechanics, and financial mathematics, where they model phenomena with significant spatial and temporal dependencies.

Author Contributions: Conceptualization, I.A.; Methodology, I.A. and A.O.A.; Software, A.O.A.; Validation, R.J.; Formal analysis, N.N.A.R.; Investigation, I.A. and R.J.; Resources, R.J., N.N.A.R. and S.A.I.; Data curation, N.N.A.R. and S.A.I.; Writing—original draft, I.A. and R.J.; Writing—review & editing, A.O.A., N.N.A.R. and S.A.I.; Visualization, R.J.; Supervision, N.N.A.R. and S.A.I.; Project administration, N.N.A.R.; Funding acquisition, I.A. and S.A.I. All authors read and agreed to the published version of this manuscript.

Funding: The authors extend their appreciation to the Deanship of Scientific Research at King Khalid University for funding this work through Large Groups (project under grant number 2/478/44).

Data Availability Statement: Data are contained within the article and further inquiries can be directed to the corresponding author.

Conflicts of Interest: The authors declare no conflicts of interest.

References

1. Jan, A.; Srivastava, H.M.; Khan, A.; Mohammed, P.O.; Jan, R.; Hamed, Y. In vivo HIV dynamics, modeling the interaction of HIV and immune system via non-integer derivatives. *Fractal Fract.* **2023**, *7*, 361. [CrossRef]
2. Oderinu, R.A.; Owolabi, J.A.; Taiwo, M. Approximate solutions of linear time-fractional differential equations. *J. Math. Comput. Sci.* **2023**, *29*, 60–72. [CrossRef]
3. Ahmad, I.; Ali, I.; Jan, R.; Idris, S.A.; Mousa, M. Solutions of a three-dimensional multi-term fractional anomalous solute transport model for contamination in groundwater. *PLoS ONE* **2023**, *18*, e0294348. [CrossRef] [PubMed]
4. Siraj-ul-Islam; Ahmad, I. A comparative analysis of local meshless formulation for multi-asset option models. *Eng. Anal. Bound. Elem.* **2016**, *65*, 159–176. [CrossRef]
5. Wang, K.L.; He, C.H. A remark on Wang's fractal variational principle. *Fractals* **2019**, *27*, 1950134. [CrossRef]
6. Anjum, N.; He, C.H.; He, J.H. Two-scale fractal theory for the population dynamics. *Fractals* **2021**, *29*, 2150182. [CrossRef]
7. Zheltov, I.; Kochina, I.; Barenblatt, G. Basic concepts in the theory of seepage of homogeneous liquids in fissured rocks (strata). *J. Appl. Math. Mech.* **1960**, *24*, 852–864.
8. Ting, T.W. A cooling process according to two-temperature theory of heat conduction. *J. Math. Anal. Appl.* **1974**, *45*, 23–31. [CrossRef]
9. Rizvi, S.T.R.; Afzal, I.; Ali, K. Chirped optical solitons for Triki–Biswas equation. *Mod. Phys. Lett. B* **2019**, *33*, 1950264. [CrossRef]
10. Attia, R.A.; Lu, D.; MA Khater, M. Chaos and relativistic energy-momentum of the nonlinear time fractional Duffing equation. *Math. Comput. Appl.* **2019**, *24*, 10. [CrossRef]
11. Kelly, J.F.; McGough, R.J.; Meerschaert, M.M. Analytical time-domain Green's functions for power-law media. *J. Acoust. Soc. Am.* **2008**, *124*, 2861–2872. [CrossRef] [PubMed]
12. Caputo, M. Linear models of dissipation whose Q is almost frequency independent—II. *Geophys. J. Int.* **1967**, *13*, 529–539. [CrossRef]
13. Showalter, R. Existence and representation theorems for a semilinear Sobolev equation in Banach space. *SIAM J. Math. Anal.* **1972**, *3*, 527–543. [CrossRef]
14. Ewing, R.E. Numerical solution of Sobolev partial differential equations. *SIAM J. Numer. Anal.* **1975**, *12*, 345–363. [CrossRef]
15. Wang, F.; Ahmad, I.; Ahmad, H.; Alsulami, M.; Alimgeer, K.; Cesarano, C.; Nofal, T.A. Meshless method based on RBFs for solving three-dimensional multi-term time fractional PDEs arising in engineering phenomenons. *J. King Saud Univ.-Sci.* **2021**, *33*, 101604. [CrossRef]
16. Luo, Z.; Teng, F. A reduced-order extrapolated finite difference iterative scheme based on POD method for 2D Sobolev equation. *Appl. Math. Comput.* **2018**, *329*, 374–383. [CrossRef]
17. Li, J.F.; Ahmad, I.; Ahmad, H.; Shah, D.; Chu, Y.M.; Thounthong, P.; Ayaz, M. Numerical solution of two-term time-fractional PDE models arising in mathematical physics using local meshless method. *Open Phys.* **2020**, *18*, 1063–1072. [CrossRef]
18. Luo, Z.; Teng, F.; Chen, J. A POD-based reduced-order Crank–Nicolson finite volume element extrapolating algorithm for 2D Sobolev equations. *Math. Comput. Simul.* **2018**, *146*, 118–133. [CrossRef]
19. Li, N.; Lin, P.; Gao, F. An expanded mixed finite element method for two-dimensional Sobolev equations. *J. Comput. Appl. Math.* **2019**, *348*, 342–355. [CrossRef]
20. Heydari, M.; Rashid, S.; Jarad, F. A numerical method for distributed-order time fractional 2D Sobolev equation. *Results Phys.* **2023**, *45*, 106211. [CrossRef]
21. Gao, F.; Qiu, J.; Zhang, Q. Local discontinuous Galerkin finite element method and error estimates for one class of Sobolev equation. *J. Sci. Comput.* **2009**, *41*, 436–460. [CrossRef]
22. Gao, F.; Cui, J.; Zhao, G. Weak Galerkin finite element methods for Sobolev equation. *J. Comput. Appl. Math.* **2017**, *317*, 188–202. [CrossRef]
23. Abu Arqub, O.; Alsulami, H.; Alhodaly, M. Numerical Hilbert space solution of fractional Sobolev equation in 1+1-dimensional space. *Math. Sci.* **2022**, *18*, 217–228. [CrossRef]
24. Hussain, M.; Haq, S.; Ghafoor, A. Meshless RBFs method for numerical solutions of two-dimensional high order fractional Sobolev equations. *Comput. Math. Appl.* **2020**, *79*, 802–816. [CrossRef]
25. Ahmad, I.; Ahsan, M.; Elamin, A.E.A.; Abdel-Khalek, S.; Inc, M. Numerical simulation of 3-D Sobolev equation via local meshless method. *Therm. Sci.* **2022**, *26*, 457–462. [CrossRef]
26. Zhang, J.; Zhang, Y.; Guo, H.; Fu, H. A mass-conservative characteristic splitting mixed finite element method for convection-dominated Sobolev equation. *Math. Comput. Simul.* **2019**, *160*, 180–191. [CrossRef]
27. Wang, K.L.; Yao, S.W.; Liu, Y.P.; Zhang, L.N. A fractal variational principle for the telegraph equation with fractal derivatives. *Fractals* **2020**, *28*, 2050058. [CrossRef]
28. Wang, K.L.; Wang, K.J.; He, C.H. Physical insight of local fractional calculus and its application to fractional Kdv–Burgers–Kuramoto equation. *Fractals* **2019**, *27*, 1950122. [CrossRef]
29. Dong, G.; Guo, Z.; Yao, W. Numerical methods for time-fractional convection-diffusion problems with high-order accuracy. *Open Math.* **2021**, *19*, 782–802. [CrossRef]
30. He, J.H. Homotopy perturbation method: A new nonlinear analytical technique. *Appl. Math. Comput.* **2003**, *135*, 73–79. [CrossRef]

31. Salama, F.M.; Ali, N.H.M.; Abd Hamid, N.N. Fast O(N) hybrid Laplace transform-finite difference method in solving 2D time fractional diffusion equation. *J. Math. Comput. Sci.* **2021**, *23*, 110–123. [CrossRef]
32. Wang, K.L. A novel variational approach to fractal Swift–Hohenberg model arising in fluid dynamics. *Fractals* **2022**, *30*, 2250156. [CrossRef]
33. Wang, K. New variational theory for coupled nonlinear fractal Schrödinger system. *Int. J. Numer. Methods Heat Fluid Flow* **2022**, *32*, 589–597. [CrossRef]
34. Deng, W. Finite element method for the space and time fractional Fokker–Planck equation. *SIAM J. Numer. Anal.* **2009**, *47*, 204–226. [CrossRef]
35. Li, L.; Jiang, Z.; Yin, Z. Compact finite-difference method for 2D time-fractional convection–diffusion equation of groundwater pollution problems. *Comput. Appl. Math.* **2020**, *39*, 142. [CrossRef]
36. Tansri, K.; Kittisopaporn, A.; Chansangiam, P. Numerical solutions of the space-time fractional diffusion equation via a gradient-descent iterative procedure. *J. Math. Comput. Sci.* **2023**, *31*, 353–366. [CrossRef]
37. Dehghan, M.; Abbaszadeh, M.; Mohebbi, A. Legendre spectral element method for solving time fractional modified anomalous sub-diffusion equation. *Appl. Math. Model.* **2016**, *40*, 3635–3654. [CrossRef]
38. He, J.H.; Wu, X.H. Exp-function method for nonlinear wave equations. *Chaos Solitons Fractals* **2006**, *30*, 700–708. [CrossRef]
39. Guo, T.; Nikan, O.; Avazzadeh, Z.; Qiu, W. Efficient alternating direction implicit numerical approaches for multi-dimensional distributed-order fractional integro differential problems. *Comput. Appl. Math.* **2022**, *41*, 236. [CrossRef]
40. Qiao, H.; Cheng, A. A fast finite difference/RBF meshless approach for time fractional convection-diffusion equation with non-smooth solution. *Eng. Anal. Bound. Elem.* **2021**, *125*, 280–289. [CrossRef]
41. Abd-Elhameed, W.M.; Youssri, Y.H.; El-Sissi, N.; Sadek, M. New hypergeometric connection formulae between Fibonacci and Chebyshev polynomials. *Ramanujan J.* **2017**, *42*, 347–361. [CrossRef]
42. Abd-Elhameed, W.; Youssri, Y. Connection formulae between generalized Lucas polynomials and some Jacobi polynomials: Application to certain types of fourth-order BVPs. *Int. J. Appl. Comput. Math.* **2020**, *6*, 45. [CrossRef]
43. Nadir, M. Lucas polynomials for solving linear integral equations. *J. Theor. Appl. Comput. Sci.* **2017**, *11*, 13–19.
44. Çetin, M.; Sezer, M.; Güler, C. Lucas polynomial approach for system of high-order linear differential equations and residual error estimation. *Math. Probl. Eng.* **2015**, *2015*, 625984. [CrossRef]
45. Mirzaee, F.; Hoseini, S.F. Application of Fibonacci collocation method for solving Volterra–Fredholm integral equations. *Appl. Math. Comput.* **2016**, *273*, 637–644. [CrossRef]
46. Baykuş-Savaşaneril, N.; Sezer, M. Hybrid Taylor-Lucas collocation method for numerical solution of high-order Pantograph type delay differential equations with variables delays. *Appl. Math. Inf. Sci.* **2017**, *11*, 1795–1801.
47. Oruç, Ö. A new algorithm based on Lucas polynomials for approximate solution of 1D and 2D nonlinear generalized Benjamin–Bona–Mahony–Burgers equation. *Comput. Math. Appl.* **2017**, *74*, 3042–3057. [CrossRef]
48. Oruç, Ö. A new numerical treatment based on Lucas polynomials for 1D and 2D sinh-Gordon equation. *Commun. Nonlinear Sci. Numer. Simul.* **2018**, *57*, 14–25. [CrossRef]
49. Ali, I.; Haq, S.; Nisar, K.S.; Baleanu, D. An efficient numerical scheme based on Lucas polynomials for the study of multidimensional Burgers-type equations. *Adv. Differ. Equ.* **2021**, *2021*, 43. [CrossRef]
50. Ahmad, I.; Bakar, A.A.; Ali, I.; Haq, S.; Yussof, S.; Ali, A.H. Computational analysis of time-fractional models in energy infrastructure applications. *Alex. Eng. J.* **2023**, *82*, 426–436. [CrossRef]
51. Jumarie, G. Stock exchange fractional dynamics defined as fractional exponential growth driven by (usual) Gaussian white noise. Application to fractional Black-Scholes equations. *Insur. Math. Econ.* **2008**, *42*, 271–287. [CrossRef]
52. Jumarie, G. Derivation and solutions of some fractional Black-Scholes equations in coarse-grained space and time. Application to Merton's optimal portfolio. *Comput. Math. Appl.* **2010**, *59*, 1142–1164. [CrossRef]
53. Atangana, A.; Baleanu, D. New fractional derivatives with non-local and nonsingular kernel theory and application to heat transfer model. *Therm. Sci.* **2016**, *20*, 763. [CrossRef]
54. He, J.H. A new fractal derivation. *Therm. Sci.* **2011**, *15*, 145–147. [CrossRef]
55. Ali, I.; Haq, S.; Aldosary, S.F.; Nisar, K.S.; Ahmad, F. Numerical solution of one-and two-dimensional time-fractional Burgers' equation via Lucas polynomials coupled with Finite difference method. *Alex. Eng. J.* **2022**, *61*, 6077–6087. [CrossRef]
56. Hussain, M.; Haq, S.; Ghafoor, A.; Ali, I. Numerical solutions of time-fractional coupled viscous Burgers' equations using meshfree spectral method. *Comput. Appl. Math.* **2020**, *39*, 6. [CrossRef]
57. Marchuk, G.I. Some applicatons of splitting-up methods to the solution of problems in mathematical physics. *Aplikace Matematiky* **1968**, *1*, 103–132.
58. Strang, G. On the construction and comparision of difference schemes. *SIAM J. Numer. Anal.* **1968**, *5*, 506–517. [CrossRef]
59. Geiser, J.; Tanoglu, G.; Gucuyenenb, N. Higher order operator splitting methods via Zassenhaus product formula: Theory and applications. *Comput. Math. Appl.* **2011**, *62*, 1994–2015. [CrossRef]
60. Abd-Elhameed, W.; Youssri, Y. Spectral solutions for fractional differential equations via a novel Lucas operational matrix of fractional derivatives. *Rom. J. Phys* **2016**, *61*, 795–813.
61. Ali, I.; Haq, S.; Nisar, K.S.; Arifeen, S.U. Numerical study of 1D and 2D advection-diffusion-reaction equations using Lucas and Fibonacci polynomials. *Arab. J. Math.* **2021**, *10*, 513–526. [CrossRef]

62. Garmanjani, G.; Cavoretto, R.; Esmaeilbeigi, M. A RBF partition of unity collocation method based on finite difference for initial–boundary value problems. *Comput. Math. Appl.* **2018**, *75*, 4066–4090. [CrossRef]
63. Lax, P.D. Weak solutions of nonlinear hyperbolic equations and their numerical computation. *Commun. Pure Appl. Math.* **1954**, *7*, 159–193. [CrossRef]

fractal and fractional

MDPI

Article

A Fast High-Order Compact Difference Scheme for Time-Fractional KS Equation with the Generalized Burgers' Type Nonlinearity

Huifa Jiang [1,*] and Da Xu [2,*]

1 School of Science, Hunan University of Technology, Zhuzhou 412007, China
2 School of Mathematics and Statistics, Hunan Normal University, Changsha 410081, China
* Correspondence: huifajiang98@163.com (H.J.); daxu@hunnu.edu.cn (D.X.)

Abstract: This work integrates the fast Alikhanov method with a compact scheme to solve the time-fractional Kuramoto–Sivashinsky (KS) equation with the generalized Burgers' type nonlinearity. Initially, the Alikhanov algorithm, designed to handle the Caputo fractional derivative on non-uniform time grids, effectively avoids the initial singularity. Additionally, the combination of the Alikhanov method with the sum-of-exponentials (SOE) technique significantly reduces both computational cost and memory requirements. By discretizing the spatial direction using a compact finite difference method, a fully discrete scheme is developed, achieving fourth-order convergence in the spatial domain. Stability and convergence are analyzed through energy methods. Several numerical examples are provided to validate the theoretical framework, demonstrating that the proposed algorithm is accurate, stable, and efficient.

Keywords: time-fractional Kuramoto–Sivashinsky equation; fast Alikhanov method; sum-of-exponentials technique; non-uniform time steps; compact difference method

Academic Editor: Luis Manuel Sánchez Ruiz

Received: 17 December 2024
Revised: 22 March 2025
Accepted: 28 March 2025
Published: 30 March 2025

Citation: Jiang, H.; Xu, D. A Fast High-Order Compact Difference Scheme for Time-Fractional KS Equation with the Generalized Burgers' Type Nonlinearity. *Fractal Fract.* **2025**, *9*, 218. https://doi.org/10.3390/fractalfract9040218

1. Introduction

The Kuramoto–Sivashinsky (KS) equation belongs to an important class of chaotic partial differential equations that serve as a bridge between the partial differential equation and dynamical system frameworks. Kuramoto [1] originally proposed the KS equation in his study of phase turbulence in reaction–diffusion systems, while Sivashinsky [2] introduced it in his analysis of flame combustion propagation models. The KS equation has a wide range of applications in simulation science and engineering, such as the evolution of the flow of fluid films on inclined planes [3], flame front instability [4], interfacial turbulence [5], etc.

In this paper, we study the time-fractional KS equation with the generalized Burgers' type nonlinearity:

$$
\begin{aligned}
{}_0^C D_t^{\alpha} u(x,t) + u^p(x,t)u_x(x,t) - \mu u_{xx}(x,t) + \lambda u_{xxxx}(x,t) = f(x,t),& \\
0 < x < L, \quad 0 < t \le T,&
\end{aligned} \tag{1}
$$

with the initial condition (IC)

$$
u(x,0) = u_0(x), \qquad 0 < x < L, \tag{2}
$$

and the boundary conditions (BCs)

$$\begin{aligned} u(0,t) &= u(L,t) = 0, \quad 0 \le t \le T, \\ u_{xx}(0,t) &= u_{xx}(L,t) = 0, \quad 0 \le t \le T, \end{aligned} \tag{3}$$

where $\mu > 0$ is the viscosity coefficient, $p \geq 1$ is a positive integer, perturbation parameters $0 < \lambda \leq 1$, $L > 0, T > 0$ are fixed values, and $u_0(x)$ and $f(x,t)$ are prescribed real functions. The equations are complex, and their combination of fractional-order derivative KS equations and Burgers' nonlinear terms provides them with greater flexibility to capture the intricate dynamical behaviors observed in various physical systems.

- When the fourth-order derivative u_{xxxx} and the source term are not included in the equation, Equation (1) transforms into a time-fractional generalized Burgers equation [6] as follows:

$$ {}_0^C D_t^\alpha u(x,t) + u^p(x,t)u_x(x,t) - \mu u_{xx}(x,t) = 0, \quad 0 < x < L, \quad 0 < t \leq T, \tag{4}$$

  where $u^p u_x$ is the generalized Burgers' type nonlinearity.
- When $\alpha = 1$, $p = 1$, Equation (1) is the usual KS equation as follows:

$$u_t(x,t) + u(x,t)u_x(x,t) - \mu u_{xx}(x,t) + u_{xxxx}(x,t) = 0, \quad 0 < x < L, \quad 0 < t \leq T, \tag{5}$$

  and many pioneering studies have been carried out [7–9].
- When $p = 1$, Equation (1) is the generalized fractional KS equation [10] as follows:

$$ {}_0^C D_t^\alpha u(x,t) + u(x,t)u_x(x,t) - \mu u_{xx}(x,t) + u_{xxxx}(x,t) = 0, \quad 0 < x < L, \quad 0 < t \leq T, \tag{6}$$

In recent years, numerical computational algorithms for solving the time-fractional KS equation have been developed and become imperative. Many numerical methods are already available for studying KS equations, including those of the form presented in Equation (5). Mittal [11] proposed a quintic B-spline collocation method based on the Crank–Nicolson scheme for temporal discretization, aimed at approximating the solution to the KS equation. In [12], a locally discontinuous Galerkin method is implemented to obtain the numerical solution to the KS equation. The implicit–explicit BDF method was developed by Akrivis and Smyrlis [13] for simulating the KS equation.

However, there are few research studies on effective numerical algorithms for solving fractional KS equations. For the fractional KS equation of the type Equation (6), the weak singularity of the solution generated by the fractional derivative presents a difficult problem to solve. At present, the effective method to solve this problem is to use the variable step numerical method, which concentrates more grid points around the weak singularity and employs a sparse grid where the solution changes slowly. Indeed, the $L1$ formula for the construction of piecewise linear interpolation was proposed by Langlands [14], which leads to a temporal convergence order of $1 + \gamma$ for $0 < \gamma < 1$. Since then, the $L1$ formula has been widely used to deal with Caputo fractional derivatives [15–20]. Furthermore, for the numerical solution to pure sub-diffusion equations, there are several high-order discrete convolution forms for the approximation of the Caputo fractional-order time derivatives, such as the $L1$-2 scheme [21–23] and the $L2$-1_σ formula, which is based on linear interpolation on the last subinterval $[t_{k-1}, t_k]$ and quadratic polynomial interpolation on the rest of the subintervals $[t_{k-1}, t_k]$ (for details, please refer to [24]). These formulas can achieve second-order temporal accuracy for sufficiently smooth solutions.

Due to the advantages of the $L2$-1_σ formula, we will study the approximation of the Caputo fractional derivative by the variable time-step Alikhanov formula. Influenced by [25,26], we also give the regularity assumptions on the exact solution u as follows:

$$\left\| \frac{\partial^k u}{\partial t^k} \right\|_{H^2(0,L)} \le C\left(1 + t^{\sigma - k}\right), \qquad k = 0,1,2,3, \tag{7}$$

$$\left\| {}_0^C D_t^{\alpha} u \right\|_{H^6(0,L)} + \|u\|_{H^8(0,L)} \le C, \tag{8}$$

where $0 < t \le T$, and $\sigma \in (0,1) \cup (1,2)$ is a regularity parameter. In this paper, we only consider the case of $\lambda = 1$.

In order to obtain a more efficient algorithm for solving the problem and avoid the issue of large computational costs caused by the historical dependence of the fractional derivative, one of the most effective methods is the SOE technique [27–29], which is applied in approximating the convolution kernel $t^{-\alpha}$ with a uniform absolute error ϵ. There have been many research studies on the nonlinear KS equation with a spatial fourth derivative, employing various methods such as the quintic B-spline collocation method [30], cubic Hermite collocation [31], Chebyshev collocation method [10], and Sinc–Galerkin method [32]. The compact difference scheme has good characteristics and high accuracy and is widely used to solve a variety of equations [33–37]. The application of a compact difference scheme to the spatial discretization of fractional KS equations is exactly the starting point of our present work. The theoretical part of the proof makes use of the classical energy method [38–40]. Moreover, the main contributions of this work can be summarized as follows:

- For the complex KS equation, we first apply the non-uniform Alikhanov formula to approximate the Caputo fractional derivative, and then treat the generalized Burgers' nonlinear term. Then, we combine the second-order time scheme and the SOE technique to establish the fast Alikhanov second-order scheme. This approach improves numerical accuracy and reduces computational costs.
- The high-order compact difference scheme is proposed for the first time to be applied to the KS equation with the generalized Burgers' nonlinear term, which can improve the spatial convergence rate to the fourth order.
- We demonstrate in detail the convergence and stability of the fully discrete scheme by the energy argument. In addition, we provide numerical examples to validate the theoretical analysis and demonstrate the efficiency of the algorithm by comparing fast and non-fast Alikhanov schemes.

The outline of this paper is as follows: In Section 2, we introduce some useful notations and lemmas. Additionally, some properties of compact difference formulas and fast Alikhanov formulas are analyzed. In Section 3, the high-order fast Alikhanov compact difference scheme is constructed. The stability and convergence of a full-discrete scheme are investigated. Numerical examples verify the stability and efficiency of the algorithm and the correctness of the theory in Section 4. The conclusion is summarized in Section 5.

2. Preliminaries

In this section, we first recall some basic formulations and useful lemmas. For ease of exposition and proof, we define some notations. To make our analysis extendable, the conditions and properties of the discrete convolution kernel in the Alikhanov formula are also introduced in this section.

2.1. Properties of Compact Difference Scheme

Let the spatial step size $h := L/M$, where M is a given positive integer. We denote $x_j = jh$ $(0 \le j \le M)$ and $\overline{\mathcal{U}}_h := \{u | u = (u_0, u_1, \dots, u_M), \ 0 \le j \le M\}$. For any $u, \ v \in \overline{\mathcal{U}}_h$, we denote the following notations:

$$\nabla^+ u_{j+1} = \frac{1}{h}(u_{j+1} - u_j), \quad \nabla^- u_{j-1} = \frac{1}{h}(u_j - u_{j-1}),$$
$$\nabla_h u_j = \frac{1}{2h}(u_{j+1} - u_{j-1}), \quad \Delta_x u_j = \frac{u_{j+1} - 2u_j + u_{j-1}}{h^2}.$$

We also need to introduce the subspace of $\overline{\mathcal{U}}_h$ as $\mathcal{U}_h := \{u \in \overline{\mathcal{U}}_h | u_0 = u_M = 0\}$, and define the inner product and norm as follows:

$$\langle u, v\rangle = h\sum_{j=1}^{M-1} u_j v_j, \quad (u,v) = h\sum_{j=1}^{M}\left(\nabla^+ u_j\right)\left(\nabla^+ v_j\right),$$
$$\|u\| = \sqrt{\langle u,u\rangle}, \quad |u|_1 = \sqrt{(u,u)}, \quad \|u\|_\infty = \max_{0\le j\le M}|u_j|.$$

Furthermore, we denote the following:

$$\mathcal{A}_1 u_j := (1 + \frac{h^2}{6}\Delta_x)u_j, \qquad \mathcal{A}_2 u_j := (1 + \frac{h^2}{12}\Delta_x)u_j.$$

Thus, for the discretization of the second and fourth derivatives of the function $u = u(x,t)$, we have the following formulas:

$$u_x(x_j) = \mathcal{A}_1^{-1}\nabla_h U_j + \mathcal{O}(h^4), \qquad u_{xx}(x_j) = \mathcal{A}_2^{-1}\Delta_x U_j + \mathcal{O}(h^4). \tag{9}$$

The following transformation can be advanced for the nonlinear term:

$$u^p u_x = v u_x = \frac{1}{p+2}[v u_x + (vu)_x], \tag{10}$$

where $v := u^p$, and

$$v_j\frac{\partial u_j}{\partial x} + \frac{\partial u_j v_j}{\partial x} = v_j\mathcal{A}_1^{-1}\nabla_h u_j + \mathcal{A}_1^{-1}\nabla_h(v_j u_j) + \mathcal{O}(h^4) = \varphi(u_j, v_j) + \mathcal{O}(h^4). \tag{11}$$

Lemma 1 ([41]). *For any grid functions $u,\ v \in \mathcal{U}_h$, then the following holds:*

$$\langle \nabla_h u, v\rangle = -\langle u, \nabla_h v\rangle, \qquad \langle \Delta_x u, v\rangle = -(u,v).$$

Lemma 2 ([41]). *For any grid functions $u \in \mathcal{U}_h$, then the following holds:*

$$\|u\|_\infty \le \frac{\sqrt{L}}{2}|u|_1, \quad \|u\| \le \frac{L}{\sqrt{6}}|u|_1, \quad |u|_1 \le \frac{2}{h}\|u\|.$$

Lemma 3 ([42]). *For real symmetric positive-definite matrices $\mathcal{A}_1^{-1}$, $\mathcal{A}_2^{-1}$, and $\omega, u \in \mathcal{U}_h$, we obtain the following:*

$$\left(\mathcal{A}_1^{-1}\omega, u\right) = (\mathcal{V}_1\omega, \mathcal{V}_1 u) = \left(\omega, \mathcal{A}_1^{-1}u\right), \left(\mathcal{A}_2^{-1}\omega, u\right) = (\mathcal{V}_2\omega, \mathcal{V}_2 u) = \left(\omega, \mathcal{A}_2^{-1}u\right),$$

where $\mathcal{V}_1$ and $\mathcal{V}_2$ are upper triangular matrices, which can be obtained by the Cholesky factorization of $\mathcal{A}_1^{-1}$ and $\mathcal{A}_2^{-1}$, namely, $\mathcal{A}_1^{-1} = (\mathcal{V}_1)^T\mathcal{V}_1$, $\mathcal{A}_2^{-1} = (\mathcal{V}_2)^T\mathcal{V}_2$.

Proof. The proof of the theorem can be referred to in ([42], Lemma 2.1c). □

Lemma 4 ([43]). *Let $\omega, u \in \mathcal{U}_h$, then we can obtain the following:*

$$\left(\mathcal{A}_1^{-1}\nabla_h\omega, u\right) = \left(\nabla_h\mathcal{A}_1^{-1}\omega, u\right) = -\left(\mathcal{A}_1^{-1}\omega, \nabla_h u\right) = -\left(\omega, \mathcal{A}_1^{-1}\nabla_h u\right),$$
$$\left(\mathcal{A}_2^{-1}\Delta_x\omega, u\right) = \left(\Delta_x\mathcal{A}_2^{-1}\omega, u\right) = -\left(\nabla^+\mathcal{A}_2^{-1}\omega, \nabla^+ u\right) = -\left(\mathcal{A}_2^{-1}\nabla^+\omega, \nabla^+ u\right).$$

Lemma 5. *Let $u,\ v \in \mathcal{U}_h$, we arrive at*

$$\Big\langle \varphi(u,v)\ ,\ u \Big\rangle = 0.$$

Proof. For $u,\ v \in \mathcal{U}_h$, using Lemmas 1 and 4, we can obtain the following:

$$\begin{aligned}
\Big\langle \varphi(u,v)\ ,\ u \Big\rangle &= \Big\langle v_j \mathcal{A}_1^{-1} \nabla_h u_j + \mathcal{A}_1^{-1} \nabla_h (v_j u_j)\ ,\ u_j \Big\rangle \\
&= \Big\langle v_j \mathcal{A}_1^{-1} \nabla_h u_j\ ,\ u_j \Big\rangle + \Big\langle \mathcal{A}_1^{-1} \nabla_h (v_j u_j)\ ,\ u_j \Big\rangle, \\
&= \Big\langle \mathcal{A}_1^{-1} \nabla_h u_j\ ,\ v_j u_j \Big\rangle - \Big\langle v_j u_j\ ,\ \mathcal{A}_1^{-1} \nabla_h u_j \Big\rangle, \\
&= 0.
\end{aligned}$$

□

Lemma 6 ([43]). *For real symmetric positive-definite matrices $\mathcal{A}_1^{-1}$, $\mathcal{A}_2^{-1}$, and $\omega \in \mathcal{U}_h$, we obtain the following:*

$$\|\omega\|^2 \le \Big(\mathcal{A}_1^{-1}\omega, \omega\Big) = \|\mathcal{V}_1 \omega\|^2 \le 3\|\omega\|^2,\ \|\omega\|^2 \le \Big(\mathcal{A}_2^{-1}\omega, \omega\Big) = \|\mathcal{V}_2 \omega\ |^2 \le \tfrac{3}{2}\|\omega\|^2.$$

2.2. Properties of the Fast Alikhanov Scheme

In order to construct the fast Alikhanov scheme, we first consider the time levels $0 = t_0 < t_1 < t_2 < \cdots < t_N = T$, where N is a given positive integer. We define the nth step size $\tau_n := t_n - t_{n-1}$ for $1 \le n \le N$, the maximum step size $\tau := \max_{0\le n\le N} \tau_n$. We set the time level $t_{n-\theta} := \theta t_{n-1} + (1-\theta) t_n$ with a fixed parameter $\theta = \alpha/2$. It is necessary to set the local step-size ratios:

$$\rho_n := \frac{\tau_n}{\tau_{n+1}} \text{ for } 1 \le n \le N-1 \text{ and } \rho := \max_{1\le n\le N-1} \rho_n.$$

For any time sequence $(v^n)_{n=0}^N$, we define $\nabla_\tau v^n := v^n - v^{n-1}$ and the interpolated value $v^{n-\theta} := \theta v^{n-1} + (1-\theta) v^n$. Thus, it is advisable to denote $\Pi_{1,k} u$ as the linear interpolant function of a function, u, with respect to t_{k-1}, t_k, and to denote the quadratic interpolant of u with respect to t_{k-1}, t_k and t_{k+1} as $\Pi_{2,k} u$, respectively. Recalling the definition of $\nabla_\tau u^k$, it follows that for $k \ge 1$, we have the following:

$$\begin{aligned}
(\Pi_{1,k} u)'(t) &= \frac{\nabla_\tau u^k}{\tau_k}, \\
(\Pi_{2,k} u)'(t) &= \frac{\nabla_\tau u^k}{\tau_k} + \frac{2(t - t_{k-1/2})}{\tau_k(\tau_k + \tau_{k+1})} \Big(\rho_k \nabla_\tau u^{k+1} - \nabla_\tau u^k\Big).
\end{aligned}$$

Next, the Caputo fractional derivative can be approximated by the defined interpolant polynomial at $t_{n-\theta}$ to obtain the Alikhanov approximation:

$$\begin{aligned}
(D_\tau^\alpha u)^{n-\theta} &:= \int_{t_{n-1}}^{t_{n-\theta}} \omega_{1-\alpha}(t_{n-\theta} - s)(\Pi_{1,n} u)'(s) \mathrm{d}s + \sum_{k=1}^{n-1} \int_{t_{k-1}}^{t_k} \omega_{1-\alpha}(t_{n-\theta} - s)(\Pi_{2,k} u)'(s) \mathrm{d}s \\
&= a_0^{(n)} \nabla_\tau u^n + \sum_{k=1}^{n-1} \Big(a_{n-k}^{(n)} \nabla_\tau u^k + \rho_k b_{n-k}^{(n)} \nabla_\tau u^{k+1} - b_{n-k}^{(n)} \nabla_\tau u^k\Big), \\
&= \sum_{k=1}^{n} A_{n-k}^{(\alpha,n)} \nabla_\tau v^k,
\end{aligned} \tag{12}$$

we have $A_0^{(1)} := a_0^{(1)}$ if $n = 1$, and for $n \ge 2$,

$$A_{n-k}^{(n)} := \begin{cases} a_0^{(n)} + \rho_{n-1} b_1^{(n)}, & k = n, \\ a_{n-k}^{(n)} + \rho_{k-1} b_{n-k+1}^{(n)} - b_{n-k}^{(n)}, & 2 \le k \le n-1, \\ a_{n-1}^{(n)} - b_{n-1}^{(n)}, & k = 1. \end{cases} \quad (13)$$

Here, $A_{n-k}^{(n)}$ is the discrete convolution kernel of the Alikhanov formula; next, the SOE technique is applied to derive the discrete convolution kernel of the fast Alikhanov formula.

Lemma 7 ([27]). *(SOE) Let $\alpha \in (0,1)$, $\epsilon \ll 1$, and $\Delta t > 0$. Assume there exists a positive integer N_q, and positive quadrature nodes s^l and $\varpi^l > 0$ $(1 \le l \le N_q)$, such that we have the following:*

$$\left| \omega_{1-\alpha}(t) - \sum_{l=1}^{N_q} \varpi^l e^{-s^l t} \right| \le \epsilon, \quad t \in [\Delta t, T],$$

and the quadrature node number N_q satisfies

$$N_q = O\left(\log \frac{1}{\epsilon} \left(\log\log \frac{1}{\epsilon} + \log \frac{T}{\Delta t} \right) + \log \frac{1}{\Delta t} \left(\log\log \frac{1}{\epsilon} + \log \frac{1}{\Delta t} \right) \right).$$

Now, the fractional derivative can be approximated in combination with the SOE technique in Lemma 7 at the time point $t_{n-\theta}$ for $1 \le n \le N$

$$\begin{aligned} {}_0^C D_f^\alpha u(t_{n-\theta}) &\approx \int_{t_{n-1}}^{t_{n-\theta}} \omega_{1-\alpha}(t_{n-\theta} - s)(\Pi_{1,n} u)'(s) ds + \sum_{k=1}^{n-1} \int_{t_{k-1}}^{t_k} \sum_{l=1}^{N_q} \varpi^l e^{-s^l (t_{n-\theta} - s)} u'(s) ds \\ &= \hat{\mathbf{a}}_0^{(n)} \nabla_\tau u^n + \sum_{k=1}^{n-1} \left(\hat{\mathbf{a}}_{n-k}^{(n)} \nabla_\tau u^k + \rho_k \hat{\mathbf{b}}_{n-k}^{(n)} \nabla_\tau u^{k+1} - \hat{\mathbf{b}}_{n-k}^{(n)} \nabla_\tau u^k \right) \\ &= \sum_{k=1}^{n} \hat{\mathbf{A}}_{n-k}^{(n)} \nabla_\tau u^k, \end{aligned} \quad (14)$$

where coefficients $\hat{\mathbf{a}}_{n-k}^{(n)}$ and $\hat{\mathbf{b}}_{n-k}^{(n)}$ are defined by the following:

$$\hat{\mathbf{a}}_0^{(n)} := a_0^{(n)}, \qquad \hat{\mathbf{a}}_{n-k}^{(n)} := \frac{1}{\tau_k} \int_{t_{k-1}}^{t_k} \sum_{l=1}^{N_q} \varpi^l e^{-s^l (t_{n-\theta} - s)} ds, \quad 1 \le k \le n-1,$$

$$\hat{\mathbf{b}}_{n-k}^{(n)} := \frac{2}{\tau_k(\tau_k + \tau_{k+1})} \int_{t_{k-1}}^{t_k} \sum_{l=1}^{N_q} \varpi^l e^{-s^l (t_{n-\theta} - s)} e^{-s^l (t_{k+1-\theta} - s)} (s - t_{k-\frac{1}{2}}) ds, \quad 1 \le k \le n-1.$$

We reformulate the discrete convolution kernel $\hat{\mathbf{A}}_{n-k}^{(n)}$ as follows: $\hat{\mathbf{A}}_0^{(1)} := a_0^{(1)}$ if $n = 1$ and for $n \ge 2$

$$\hat{\mathbf{A}}_{n-k}^{(n)} := \begin{cases} \hat{\mathbf{a}}_0^{(n)} + \rho_{n-1} \hat{\mathbf{b}}_1^{(n)}, & \text{for } k = n, \\ \hat{\mathbf{a}}_{n-k}^{(n)} + \rho_{k-1} \hat{\mathbf{b}}_{n-k+1}^{(n)} - \hat{\mathbf{b}}_{n-k}^{(n)}, & \text{for } 2 \le k \le n-1, \\ \hat{\mathbf{a}}_{n-1}^{(n)} - \hat{\mathbf{b}}_{n-1}^{(n)}, & \text{for } k = 1. \end{cases} \quad (15)$$

In order to effectively assess some necessary properties of the discrete convolution kernel $\hat{\mathbf{A}}_{n-k}^{(n)}$, the time step needs to satisfy the following two conditions:

M1. The maximum time-step ratio is $\rho = 7/4$.

M2. There exists a positive constant C_γ, such that $\tau_k \le C_\gamma \tau \min\{1, t_k^{1-1/\gamma}\}$ for $1 \le k \le N$, with $t_k \le C_\gamma t_{k-1}$ and $\tau_k / t_k \le C_\gamma \tau_{k-1} / t_{k-1}$ for $2 \le k \le N$.

Lemma 8 ([27]). *Assume that the tolerance error ϵ of the SOE approximation satisfies*

$$\epsilon \le \min\left\{ \frac{\alpha_p}{2(1-\alpha_p)} \omega_{1-\alpha_p}(T), \frac{1}{26}\omega_{1-\alpha_p}(T) \right\},$$

then the discrete kernel $\hat{\mathbf{A}}^{(n)}_{n-k}$ satisfies the following:

(I) $\hat{\mathbf{A}}^{(n)}_{n-k}$ *are bounded,* $\hat{\mathbf{A}}^{(n)}_{n-k} \le \frac{2}{\tau_n}\omega_{2-\alpha_p}(\tau_n)$ *and*

$$\hat{\mathbf{A}}^{(n)}_{n-k} \ge \frac{1}{2\tau_k}\int_{t_{k-1}}^{t_k} \omega_{1-\alpha}(t_n - s)\mathrm{d}s, \quad 1 \le k \le n;$$

(II) $\hat{\mathbf{A}}^{(n)}_{n-k}$ *are monotone, and*

$$\hat{\mathbf{A}}^{(n)}_{n-k-1} - \hat{\mathbf{A}}^{(n)}_{n-k} \ge (1+\rho_k)\left(\frac{1}{5\rho_k} + 1\right) b^{(n)}_{n-k}, \quad 1 \le k \le n-1;$$

(III) $\hat{\mathbf{A}}^{(n)}_0 - \hat{\mathbf{A}}^{(n)}_1 > \theta\left(2\hat{\mathbf{A}}^{(n)}_0 - \hat{\mathbf{A}}^{(n)}_1\right)$ *for* $n \ge 2$.

In this case, for the later analysis of global consistency, we introduce the complementary discrete convolution kernels $P^{(n)}_{n-k}$ of $\hat{\mathbf{A}}^{(n)}_{n-k}$ that satisfy

$$\sum_{i=k}^{n} P^{(n)}_{n-i}\hat{\mathbf{A}}^{(i)}_{i-k} \equiv 1. \tag{16}$$

which yields the following recursive formulas:

$$P^{(n)}_0 := \frac{1}{\hat{\mathbf{A}}^{(n)}_0}, \quad P^{(n)}_i := \frac{1}{\hat{\mathbf{A}}^{(i)}_0}\sum_{k=i+1}^{n}\left(\hat{\mathbf{A}}^{(k)}_{k-i-1} - \hat{\mathbf{A}}^{(k)}_{k-i}\right)P^{(n)}_k, \quad 1 \le i \le n-1.$$

In fact, according to Lemma 8, the complementary discrete kernels are non-negative and satisfy the following:

$$\sum_{i=1}^{n} P^{(n)}_{n-i}\omega_{1+(m-1)\alpha}(t_i) \le \pi_A \omega_{1+m\alpha}(t_n), \quad m = 0, 1, \quad 1 \le n \le N. \tag{17}$$

Lemma 9 ([44]). *Let λ_s be nonnegative constants with $0 < \sum_{s=0}^{n}\lambda_s \le \Lambda$ for $n \ge 1$, where Λ is some positive constant independent of n. Suppose that the nonnegative sequences ξ_n and η_n are bounded, and the nonnegative grid function $\{v_n \mid n \ge 0\}$ satisfies the following:*

$$\sum_{k=1}^{n} A^{(n)}_{n-k}\nabla_\tau(v^k)^2 \le \sum_{s=1}^{n}\lambda_{n-s}(v^{s,\theta_1})^2 + \xi^n v^{n,\theta_2} + (\eta^n)^2, \quad n \ge 1, \tag{18}$$

where for $i = 1, 2$ and each n, we set $v^{n,\theta_1} := \theta_i v^{n-1} + (1-\theta_i)v^n$ for some constant $\theta_i \in [0,1]$. If the nonuniform grid satisfies the maximum time-step criterion $\tau_N \le [\frac{11}{2}\Gamma(2-\alpha)\Lambda]^{-1/\alpha}$, then we have the following:

$$v^n \le 2E_\alpha(\frac{11}{2}\Lambda t_n^\alpha)\left[v^0 + \max_{1\le k\le n}\sum_{j=1}^{k} P^{(k)}_{k-j}(\xi^j + \eta^j) + \max_{1\le j\le n}\eta^j\right], \quad 1 \le n \le N,$$

where $E_\alpha(z) := \sum_{k=0}^{\infty} z^k/\Gamma(1+k\alpha)$, and $E_\alpha(\cdot)$ represents the Mittag–Leffler function.

Lemma 10 ([45]). *Under the condition* **M1** *and Lemma 8, the fast Alikhanov Formula* (14) *satisfies the following:*

$$\left\langle (\mathcal{D}_{\tau}^{\alpha} v)^{n-\theta}, v^{n-\theta} \right\rangle \geq \frac{1}{2}\sum_{k=1}^{n} A_{n-k}^{(n)} \nabla_{\tau}(\|v^k\|^2), \quad 1 \leq n \leq N.$$

3. Establishment and Analysis of Compact Scheme

In this section, we shall derive the second-order fast compact difference scheme for problems (1)–(3). After that, the convergence and stability of the fast compact difference algorithm are rigorously proved.

3.1. Numerical Scheme

First, combined with (11), the temporal direction of Equation (1) at the grid points $(x_j, t_{n-\theta})$ is discretized by the fast nonuniform Alikhanov Formula (14) and the spatial direction is discretized by the compact difference Formula (9), where we obtain the following:

$$\begin{cases} \left(\mathcal{D}_f^{\alpha} U_j\right)^{n-\theta} + \dfrac{1}{p+2}\varphi(V_j^{n-\theta}, U_j^{n-\theta}) - \mu \mathcal{A}_2^{-1}\Delta_x U_j^{n-\theta} + \mathcal{A}_2^{-1}\Delta_x W_j^{n-\theta} = f_j^{n-\theta} + \Lambda_j^{n-\theta}, \\ \quad 1 \leq j \leq M-1, \quad 1 \leq n \leq N, & (19) \\ V_j^n = \left(U_j^n\right)^p, \quad 0 \leq j \leq M, \quad 0 \leq n \leq N, & (20) \\ W_j^{n-\theta} = \Delta_x U_j^{n-\theta} + R_3^{n-\theta}, \quad 0 \leq j \leq M, \quad 0 \leq n \leq N, & (21) \end{cases}$$

where $\Lambda_j^{n-\theta}$ and $R_3^{n-\theta}$ denote the truncation errors.

Omitting the small terms $\Lambda_j^{n-\theta}$ and $R_3^{n-\theta}$ in (19), we replace the functions U_j^n, W_j^n, and V_j^n with their numerical approximations u_j^n, ω_j^n, and v_j^n, respectively. Meanwhile, combining IC (2) and BC (3), we can obtain the fully discrete difference scheme:

$$\begin{cases} \left(\mathcal{D}_f^{\alpha} u_j\right)^{n-\theta} + \dfrac{1}{p+2}\varphi(v_j^{n-\theta}, u_j^{n-\theta}) - \mu \mathcal{A}_2^{-1}\Delta_x u_j^{n-\theta} + \mathcal{A}_2^{-1}\Delta_x \omega_j^{n-\theta} = f_j^{n-\theta}, & (22) \\ 1 \leq j \leq M-1, \quad 1 \leq n \leq N, \\ v_j^n = \left(u_j^n\right)^p, \quad 0 \leq j \leq M, \quad 0 \leq n \leq N, & (23) \\ \omega_j^{n-\theta} = \Delta_x u_j^{n-\theta}, \quad 0 \leq j \leq M, \quad 0 \leq n \leq N, & (24) \\ u_j^0 = u_0(x_j), \quad 1 \leq j \leq M-1, & (25) \\ u_0^n = u_M^n = \omega_0^n = \omega_M^n = 0, \quad 0 \leq n \leq N. & (26) \end{cases}$$

In what follows, the stability and convergence analysis of the above fully discrete schemes (22)–(26) are carried out by the energy method.

3.2. Stability

Here, we present the stability result as follows:

Theorem 1. *Suppose that* $\{u^n\}_{n=0}^{N}$ *is the solution to* (22)–(26)*, and under the conditions of Lemma 8 and* **M2***, then we have the following:*

$$\|u^n\| \leq \|u_0\| + \frac{11}{2}\Gamma(1-\alpha)\max_{1\leq k\leq n} t_k^{\alpha}\|f^{k-\theta}\|.$$

Proof. Taking an inner product of (22) with $u^{n-\theta}$, we obtain the following:

$$\left\langle \left(\mathcal{D}_f^\alpha u\right)^{n-\theta}, u^{n-\theta} \right\rangle + \frac{1}{p+2}\left\langle \varphi(v^{n-\theta}, u^{n-\theta}), u^{n-\theta} \right\rangle - \mu\left\langle \mathcal{A}_2^{-1}\Delta_x u^{n-\theta}, u^{n-\theta} \right\rangle + \left\langle \mathcal{A}_2^{-1}\Delta_x \omega^{n-\theta}, u^{n-\theta} \right\rangle$$
$$= \left\langle f^{n-\theta}, u^{n-\theta} \right\rangle. \tag{27}$$

By Lemmas 5 and 10, we have the following:

$$\left\langle \left(\mathcal{D}_f^\alpha u\right)^{n-\theta}, u^{n-\theta} \right\rangle \geq \frac{1}{2}\sum_{k=1}^{n} A_{n-k}^{(n)} \nabla_\tau(\|u^k\|^2), \tag{28}$$

$$\frac{1}{p+2}\left\langle \varphi(v^{n-\theta}, u^{n-\theta}), u^{n-\theta} \right\rangle = 0, \tag{29}$$

and

$$-\mu\left\langle \mathcal{A}_2^{-1}\Delta_x u^{n-\theta}, u^{n-\theta} \right\rangle = \mu\left\langle \mathcal{A}_2^{-1}\nabla^+ u^{n-\theta}, \nabla^+ u^{n-\theta} \right\rangle \geq \mu \mid u^{n-\theta} \mid_1^2, \tag{30}$$

$$\left\langle \mathcal{A}_2^{-1}\Delta_x \omega^{n-\theta}, u^{n-\theta} \right\rangle = \left\langle \mathcal{A}_2^{-1}\omega^{n-\theta}, \Delta_x u^{n-\theta} \right\rangle = \left\langle \mathcal{A}_2^{-1}\omega^{n-\theta}, \omega^{n-\theta} \right\rangle \geq 0. \tag{31}$$

Under the relations (28)–(31), we have the following:

$$\sum_{k=1}^{n} A_{n-k}^{(n)} \nabla_\tau(\|u^k\|^2) \leq 2\|f^{n-\theta}\| \|u^{n-\theta}\|.$$

Taking $\lambda_s = 0$ for $s \geq 0$, $\eta^n = 0$ and $\xi^n = 2\|f^{n-\theta}\|$ for $n \geq 1$ in Lemma 9 and $\pi_A = \frac{11}{4}$ in Lemma 8, it is easy to obtain

$$\|u^n\| \leq \|u_0\| + \max_{1\leq k\leq n} \sum_{i=1}^{k} 2P_{k-i}^{(k)} \|f^{k-\theta}\| \leq \|u_0\| - 2\pi_A\Gamma(1-\alpha) \max_{1\leq k\leq n} t_k^\alpha \|f^{k-\theta}\|.$$

Consequently, we infer the following:

$$\|u^n\| \leq \|u_0\| + \frac{11}{2}\Gamma(1-\alpha) \max_{1\leq k\leq n} t_k^\alpha \|f^{k-\theta}\|.$$

□

3.3. Convergence

Before analyzing the convergence of the fast Alikhanov schemes (22)–(26), we first analyze the truncation errors $\Lambda_j^{n-\theta}$ and $(R_3)_j^{n-\theta}$. In fact, $\Lambda_j^{n-\theta}$ is composed of the following four parts:

$$\Lambda_j^{n-\theta} := (R_{t1})_j^{n-\theta} + (R_{t2})_j^{n-\theta} + (R_3)_j^{n-\theta} + (R_s)_j^{n-\theta}, \qquad 1 \leq j \leq M-1, \quad 1 \leq n \leq N,$$

where $(R_s)_j^{n-\theta}$ means that the error is caused by approximating space derivatives; it can easily be seen that $(R_s)_j^{n-\theta} = \mathcal{O}(h^4)$ from (9) and (11). Moreover, we have the following:

$$(R_{t1})_j^{n-\theta} := \left(\mathcal{D}_f^\alpha u_j\right)^{n-\theta} - {}_0^C D_t^\alpha u(x_j, t_{n-\theta}),$$
$$(R_{t2})_j^{n-\theta} := -\frac{1}{p+2}\varphi((R_v)_j^{n-\theta}, (R_u)_j^{n-\theta}) + \mu \mathcal{A}_2^{-1}\Delta_x (R_u)_j^{n-\theta} - \mathcal{A}_2^{-1}\Delta_x (R_\omega)_j^{n-\theta},$$

where $(R_u)_j^{n-\theta}$, $(R_v)_j^{n-\theta}$ and $(R_\omega)_j^{n-\theta}$ denote the errors of the weighted time approximation at $t_{n-\theta}$, namely, we have the following:

$$(R_u)_j^{n-\theta} := u(x_j, t_{n-\theta}) - [\theta U_j^{n-1} + (1-\theta)U_j^n],$$
$$(R_v)_j^{n-\theta} := v(x_j, t_{n-\theta}) - [\theta V_j^{n-1} + (1-\theta)V_j^n],$$
$$(R_\omega)_j^{n-\theta} := \omega(x_j, t_{n-\theta}) - [\theta W_j^{n-1} + (1-\theta)W_j^n].$$

Based on the mesh condition **M1** and the regularity (7), and consulting Lemma 3.3 and Lemma 3.4 in [25,27], we arrive at the following:

$$\sum_{k=1}^{n} P_{n-k}^{(n)}\left|(R_{t1})_j^{k-\theta}\right| \le C\left(\frac{\tau_1^\sigma}{\sigma} + \frac{\epsilon}{\sigma} t_n^\alpha \hat{t}_{n-1}^2 + \frac{1}{1-\alpha}\max_{2\le k\le n} t_k^\alpha t_{k-1}^{\sigma-3}\tau_k^3/\tau_{k-1}^\alpha\right), \quad 1 \le n \le N, \tag{32}$$
$$\sum_{k=1}^{n} P_{n-k}^{(n)}\left|(R_u)_j^{k-\theta}\right| \le C\left(\frac{\tau_1^{\sigma+\alpha}}{\sigma} + t_n^\alpha \max_{2\le k\le n} t_{k-1}^{\sigma-2}\tau_k^2\right), \quad 1 \le n \le N, \tag{33}$$

where $\hat{t}_n = \max\{1, t_n\}$.

Combining (17) and $(R_s)_j^{n-\theta} = \mathcal{O}(h^4)$, we can also obtain the following:

$$\max_{1\le k\le n}\sum_{i=1}^{k} P_{k-i}^{(k)}|(R_s)_j^{i-\theta}| \le Ch^4, \quad 1 \le n \le N. \tag{34}$$

Therefore, combining (20)–(22), the following holds:

$$\max_{1\le k\le n}\sum_{i=1}^{k} P_{k-i}^{(k)}\|\Lambda^{i-\theta}\| \le C\left(\frac{\tau_1^\sigma}{\sigma} + \frac{\epsilon}{\sigma} t_n^\alpha \hat{t}_{n-1}^2 + \max_{2\le i\le k} t_i^\alpha t_{i-1}^{\sigma-3}\tau_i^3/\tau_{i-1}^\alpha + t_k^\alpha \max_{2\le i\le k} t_{i-1}^{\sigma-2}\tau_i^2 + h^4\right), \quad 1 \le n \le N. \tag{35}$$

To obtain a pointwise error estimate of (22)–(26), we need to apply an important method, namely, the truncated function method. Based on the regularity assumption (8), we denote

$Q := \max_{(x,t)\in[0,L]\times[0,T]} |u(x,t)|$. We can find a second-order smooth function as follows:

$$g(u) := \begin{cases} u^p, & |u| \le Q+1, \\ 0, & |u| \ge Q+2. \end{cases}$$

Furthermore, let $c_1 := \max_{u\in R}|g(u)|$, $c_2 := \max_{u\in R}|g'(u)|$. It is not difficult to verify that the constants c_1 and c_2 depend only on Q and p by the Hermite polynomial interpolant. Clearly, (20) and (23) are equivalent to the following:

$$\begin{cases} V_j^n = g(U_j^n), & 0 \le j \le M, \quad 0 \le n \le N, & (36) \\ v_j^n = g(u_j^n), & 0 \le j \le M, \quad 0 \le n \le N. & (37) \end{cases}$$

The convergence of the full discrete schemes (22)–(26) is studied below. Subtracting (19)–(21) from (22)–(24), respectively, the error equation can be written as follows:

$$\begin{cases} (\mathcal{D}_t^\alpha e_j)^{n-\theta} + \frac{1}{p+2}\left[\varphi(V_j^{n-\theta}, U_j^{n-\theta}) - \varphi(v_j^{n-\theta}, u_j^{n-\theta})\right] - \mu \mathcal{A}_2^{-1}\Delta_x e_j^{n-\theta} + \mathcal{A}_2^{-1}\Delta_x \eta_j^{n-\theta} = \Lambda_j^{n-\theta}, \\ \qquad 1 \le j \le M-1, \quad 1 \le n \le N, & (38) \\ \rho_j^{n-\theta} = g(U_j^{n-\theta}) - g(u_j^{n-\theta}), \quad 0 \le j \le M, \quad 0 \le n \le N, & (39) \\ \eta_j^{n-\theta} = \mathcal{A}_2^{-1}\Delta_x e_j^{n-\theta} + (R_3)_j^{n-\theta}, \quad 0 \le j \le M, \quad 0 \le n \le N, & (40) \\ e_j^0 = 0, \quad 1 \le j \le M-1, & (41) \\ e_0^n = e_M^n = \eta_0^n = \eta_M^n = 0, \quad 0 \le n \le N, & (42) \end{cases}$$

where $e_j^n := U_j^n - u_j^n$, $\varrho_j^n := V_j^n - v_j^n$, $\eta_j^n := W_j^n - \omega_j^n$.

Theorem 2. *Suppose that the exact solution, u, satisfies the regularity property in* (7) *and* (8) *with the parameter $\sigma \in (0,1) \cup (1,2)$. Let **M1**, **M2** hold, and $\hat{t}_n = \max\{1, t_n\}$ for $1 \le n \le N$. If $\tau \le ((11/2)\Gamma(2-\alpha)c_3)^{-1/\alpha}$, the following estimate holds:*

$$\|e^n\| \le C(\tau^{\min\{\gamma\sigma,2\}} + h^4 + \epsilon), \quad 0 \le n \le N. \tag{43}$$

Proof. Firstly, according to the differential mean value theorem, from (39), it is clear that

$$|\varrho_j^n| \le c_2 |e_j^n|, \quad 0 \le j \le M, \quad 0 \le n \le N,$$

moreover, we obtain the following:

$$\|\varrho^n\| \le c_2 \|e^n\|, \quad 0 \le n \le N. \tag{44}$$

The next thing to do in the proof is to take an inner product of (38) with $e^{n-\theta}$; we obtain the following:

$$\begin{aligned} &\left\langle (\mathcal{D}_t^\alpha e)^{n-\theta}, e^{n-\theta}\right\rangle + \frac{1}{p+2}\left\langle \varphi(V^{n-\theta}, U^{n-\theta}) - \varphi(v^{n-\theta}, u^{n-\theta}), e^{n-\theta}\right\rangle \\ &- \mu\left\langle \mathcal{A}_2^{-1}\Delta_x e^{n-\theta}, e^{n-\theta}\right\rangle + \left\langle \mathcal{A}_2^{-1}\Delta_x \eta^{n-\theta}, e^{n-\theta}\right\rangle = \left\langle \Lambda^{n-\theta}, e^{n-\theta}\right\rangle. \end{aligned} \tag{45}$$

Furthermore, taking an inner product of (40) with $\eta^{n-\theta}$, we obtain the following:

$$\left\langle \eta^{n-\theta}, \eta^{n-\theta}\right\rangle = \left\langle \mathcal{A}_2^{-1}\Delta_x e^{n-\theta}, \eta^{n-\theta}\right\rangle + \left\langle R_3^{n-\theta}, \eta^{n-\theta}\right\rangle. \tag{46}$$

Another step is to analyze the above equation term by term. Using Lemma 10, we have the following:

$$\left\langle (\mathcal{D}_t^\alpha e)^{n-\theta}, e^{n-\theta}\right\rangle \ge \frac{1}{2}\sum_{k=1}^{n} A_{n-k}^{(n)} \nabla_\tau (\|e^k\|^2). \tag{47}$$

By the regularity assumption (7), let $c_0 := \max_{(x,t)\in[0,L]\times[0,T]}\{|u(x,t)|,\ |u_x(x,t)|\}$. Applying Lemmas 2–6 and the Cauchy–Schwarz inequality, and noting that $v^{n-\theta} = V^{n-\theta} - \varrho^{n-\theta}$ and $u^{n-\theta} = U^{n-\theta} - e^{n-\theta}$, we have the following:

$$\begin{aligned}
&\left\langle \varphi(V^{n-\theta}, U^{n-\theta}) - \varphi(v^{n-\theta}, u^{n-\theta}), e^{n-\theta} \right\rangle \\
=&\left\langle \varphi(\varrho^{n-\theta}, U^{n-\theta}) + \varphi(V^{n-\theta}, e^{n-\theta}) - \varphi(\varrho^{n-\theta}, e^{n-\theta}), e^{n-\theta} \right\rangle \\
=&\left\langle \varphi(\varrho^{n-\theta}, U^{n-\theta}), e^{n-\theta} \right\rangle \\
=&\left\langle \varrho^{n-\theta}\mathcal{A}_1^{-1}\nabla_h U^{n-\theta} + \mathcal{A}_1^{-1}\nabla_h(\varrho^{n-\theta}U^{n-\theta}), e^{n-\theta} \right\rangle \\
=&\left(\varrho^{n-\theta}\mathcal{A}_1^{-1}\nabla_h U^{n-\theta}, e^{n-\theta} \right) + \left(\mathcal{A}_1^{-1}\nabla_h(\varrho^{n-\theta}U^{n-\theta}), e^{n-\theta} \right) \\
=&\left(\nabla_h U^{n-\theta}, \mathcal{A}_1^{-1}\left(\varrho^{n-\theta}e^{n-\theta}\right) \right) - \left(\varrho^{n-\theta}U^{n-\theta}, \mathcal{A}_1^{-1}\nabla_h e^{n-\theta} \right) \\
\leq&\|\mathcal{V}_1\nabla_h U^{n-\theta}\|\|\mathcal{V}_1\left(\varrho^{n-\theta}e^{n-\theta}\right)\| + \|\mathcal{V}_1\left(\varrho^{n-\theta}U^{n-\theta}\right)\|\|\mathcal{V}_1\nabla_h e^{n-\theta}\| \\
\leq&3Lc_0\|\varrho^{n-\theta}e^{n-\theta}\| + 3c_0|e^{n-\theta}|_1\|\varrho^{n-\theta}\| \\
\leq&3Lc_0\|\varrho^{n-\theta}\|\|e^{n-\theta}\|_\infty + 3c_0c_2|e^{n-\theta}|_1\|e^{n-\theta}\| \\
\leq&3Lc_0c_2\|e^{n-\theta}\|\frac{\sqrt{L}}{2}|e^{n-\theta}|_1 + 3c_0c_2|e^{n-\theta}|_1\|e^{n-\theta}\| \\
\leq&3c_0c_2(L^{3/2}+1)\|e^{n-\theta}\||e^{n-\theta}|_1 \\
\leq&\frac{9c_0^2c_1^2(L^{3/2}+1)^2}{4\mu}\|e^{n-\theta}\|^2 + \mu|e^{n-\theta}|_1^2,
\end{aligned} \tag{48}$$

as well as

$$-\mu\left\langle \mathcal{A}_2^{-1}\Delta_x e^{n-\theta}, e^{n-\theta} \right\rangle = \mu\left\langle \mathcal{A}_2^{-1}\nabla^+ e^{n-\theta}, \nabla^+ e^{n-\theta} \right\rangle \geq \mu|e^{n-\theta}|_1^2, \tag{49}$$

$$\left\langle \Lambda^{n-\theta}, e^{n-\theta} \right\rangle \leq \|\Lambda^{n-\theta}\|\|e^{n-\theta}\|. \tag{50}$$

Substituting the results of (47)–(49) into (34), and adding (46), we have the following:

$$\frac{1}{2}\sum_{k=1}^{n} A_{n-k}^{(n)}\nabla_\tau(\|e^k\|^2) + \left\langle \eta^{n-\theta}, \eta^{n-\theta} \right\rangle \leq \frac{9c_0^2c_1^2(L^{3/2}+1)^2}{4\mu}\|e^{n-\theta}\|^2 + \|\Lambda^{n-\theta}\|\|e^{n-\theta}\| + \left\langle R_3^{n-\theta}, \eta^{n-\theta} \right\rangle.$$

$$\sum_{k=1}^{n} A_{n-k}^{(n)}\nabla_\tau(\|e^k\|^2) \leq \frac{9c_0^2c_1^2(L^{3/2}+1)^2}{2\mu}\|e^{n-\theta}\|^2 + 2\|\Lambda^{n-\theta}\|\|e^{n-\theta}\| + \frac{1}{2}\|R_3^{n-\theta}\|^2.$$

According to Lemma 9, $\lambda_0 = \frac{9c_0^2c_1^2(L^{3/2}+1)^2}{2\mu} < c_3, \xi^n = 2\|\Lambda^{n-\theta}\|, \eta^n = \frac{\sqrt{2}}{2}\|R_3^{n-\theta}\|$, we can obtain the following:

$$\|e^n\| \leq 2E_\alpha(\frac{11}{2}c_3t_n^\alpha)\left[\|e^0\| + \max_{1\leq k\leq n}\sum_{j=1}^{k} P_{k-j}^{(k)}(2\|\Lambda^{j-\theta}\| + \frac{\sqrt{2}}{2}\|R_3^{j-\theta}\|) + \frac{\sqrt{2}}{2}\max_{1\leq j\leq n} R_3^j\right], \quad 1 \leq n \leq N. \tag{51}$$

Due to $\|e^0\| = 0, 2E_\alpha(\frac{11}{2}c_3t_n^\alpha) \leq C$, and combing (35), we have the following:

$$\|e^n\| \leq C\left(\frac{\tau_1^\sigma}{\sigma} + \frac{\epsilon}{\sigma}t_n^\alpha\hat{t}_{n-1}^2 + \max_{2\leq i\leq k} t_i^\alpha t_{i-1}^{\sigma-3}\tau_i^3/\tau_{i-1}^\alpha + t_k^\alpha\max_{2\leq i\leq k} t_{i-1}^{\sigma-2}\tau_i^2 + h^4\right). \tag{52}$$

On the other hand, if **M1** and **M2** hold, $\beta = \min\{2, \gamma\sigma\}$, we have the following:

$$\begin{aligned} t_k^\alpha t_{k-1}^{\sigma-3} \tau_k^3 \tau_{k-1}^{-\alpha} &\le C_\gamma t_k^{\alpha+\sigma-3} \tau_k^{3-\alpha} \\ &\le C_\gamma t_k^{\alpha+\sigma-3} \tau_k^{3-\alpha-\beta} \left(\tau \min\left\{1, t_k^{1-1/\gamma}\right\}\right)^\beta \\ &\le C_\gamma t_k^{\sigma-\beta/\gamma} (\tau_k/t_k)^{3-\alpha-\beta} \tau^\beta \\ &\le C_\gamma t_k^{\max\{0, \sigma-(3-\alpha)/\gamma\}} \tau^\beta \quad \text{for } 2 \le k \le n, \end{aligned} \tag{53}$$

$$\begin{aligned} t_{k-1}^{\sigma-2} \tau_k^2 &\le C_\gamma t_k^{\sigma-2} \tau_k^{2-\beta} \left(\tau \min\left\{1, t_k^{1-1/\gamma}\right\}\right)^\beta \\ &\le C_\gamma t_k^{\sigma-\beta/\gamma} (\tau_k/t_k)^{2-\beta} \tau^\beta \\ &\le C_\gamma t_k^{\max\{0, \sigma-2/\gamma\}} \tau^\beta, \quad \text{for } 2 \le k \le n. \end{aligned} \tag{54}$$

As a result, the discrete solution $u_j^{n-\theta}$ is convergent in the sense that

$$\|e^n\| \le C(\tau^{\min\{\gamma\sigma, 2\}} + h^4 + \epsilon), \quad 0 \le n \le N. \tag{55}$$

This finishes the proof. □

4. Numerical Experiment

In this section, we test the stability and convergence of the fast compact schemes (22)–(26) for three different experiments on a nonuniform grid $\{t_n = T(n/N)^\gamma, n = 1, \ldots, N\}$. All experiments were performed on a Windows server with an AMD Ryzen 5 4600H processor of 16GB RAM and 3.00 GHz CPU. In the following experiments, the tolerance accuracy of the SOE algorithm is limited to $\epsilon = 10^{-12}$ in order to balance accuracy and efficiency [6]. The following suitable formulas are given to calculate the error and convergence order of the following numerical solution:

$$E^{SOE}(M, N) = \left\|U^N - u^N\right\|_2 = \sqrt{h \sum_{j=1}^{M-1} (U_j^N - u_j^N)^2},$$

$$Rate_h^{SOE} = \log_2\left(\frac{E^{SOE}(M, N)}{E^{SOE}(2M, N)}\right), \quad Rate_\tau^{SOE} = \log_2\left(\frac{E^{SOE}(M, N)}{E^{SOE}(M, 2N)}\right),$$

where U^N and u^N represent the exact and numerical solutions at the Nth time level, respectively.

In addition, we define the following notation to illustrate the numerical implementation of the proposed scheme:

$$U^* = \max_{1 \le n \le N} \|U^n\|.$$

Before proceeding further, let us remark on the numerical scheme setting of the fast Alikhanov approach and Alikhanov approach.

- Alikhanov scheme: The Caputo fractional derivative discretization scheme is (12).
- Fast Alikhanov scheme: The Caputo fractional derivative discretization scheme is (14).

Example 1. *In this example, we consider the following problem:*

$$\begin{cases} {}_0^C D_t^\alpha u + u^p u_x - \mu u_{xx} + u_{xxxx} = f(x, t), & 0 < x < 1, \quad 0 \le t \le 1, \quad (56) \\ u(x, 0) = u_0(x), & 0 < x < 1, \quad (57) \\ u_{xx}(0, t) = u_{xx}(1, t), & 0 \le t \le 1, \quad (58) \end{cases}$$

and the exact solution is set as $u(x, t) = \left(1 + \frac{t^\alpha}{\Gamma(1+\alpha)}\right) \sin(\pi x)$. *Accordingly, the initial condition* $u_0 = \sin(\pi x)$ *and the source term* $f(x, t)$ *are as follows:*

$$f(x,t) = \sin(\pi x)\left(1 + \mu\pi^2 + \mu\pi^2 \frac{t^\alpha}{\Gamma(1+\alpha)} + \pi^4 + \pi^4 \frac{t^\alpha}{\Gamma(1+\alpha)}\right) + \pi\left(1 + \frac{t^\alpha}{\Gamma(1+\alpha)}\right)^{p+1} (\sin(\pi x))^{p+1} \cos(\pi x).$$

For Example 1, in Table 1, we fix $M = 256$, $\mu = 1$, and $\gamma = 2/\alpha$ to explore the performances of the fast Alikhanov and Alikhanov schemes in terms of errors in the L^2-norm and the convergence order for different α and p cases. It can be found that the errors in the L^2-norm of the Alikhanov and fast Alikhanov schemes are very close and both of them can reach second-order convergence in the temporal direction with $\alpha \in \{0.3, 0.5, 0.7\}$ and $p \in \{2, 3, 4\}$. When $p \in \{2, 3, 4\}$, the value of α keeps on changing, the errors in Alikhanov and fast Alikhanov schemes have the same effect and also achieve second-order convergence. Table 2 shows the errors and convergence orders in the spatial direction for different μ and p when fixing $N = 512$, $\alpha = 0.5$, and $\gamma = 4$. It can be observed that regardless of whether μ changes or p changes, the spatial errors in the L^2-norm of the two schemes are similar and both can achieve fourth-order convergence in agreement with the theoretical prediction.

Table 3 considers the cases where $N = 512$, $p = 2$, $\mu = 1$, and $\gamma = 2$ are fixed, with α changing, which indicates that the fast Alikhanov and Alikhanov schemes can achieve similar spatial errors and fourth-order convergence. The CPU times of the two algorithms are given; it is obvious that the fast Alikhanov scheme is much faster than the Alikhanov scheme. In order to further confirm whether the temporal and spatial convergence orders of the fast Alikhanov scheme remain stable as p increases, the temporal convergence and spatial convergence orders are shown in Figures 1 and 2. Table 4 shows that with the gradual increase in N, the value of U^* tends to stabilize and shows no increasing trend. These results confirm the numerical stability of the proposed scheme in the time direction.

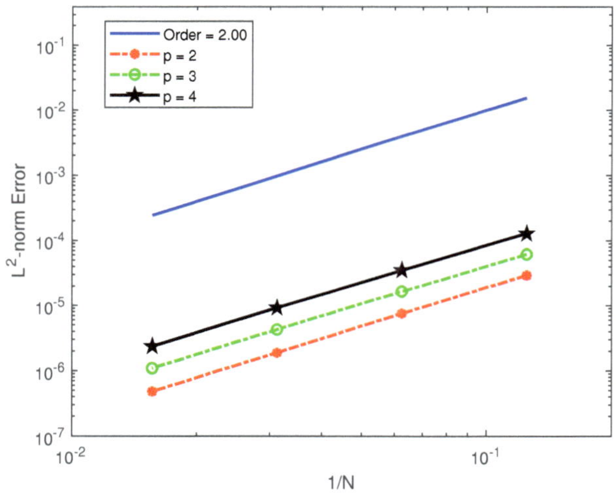

Figure 1. The temporal convergence orders of the fast compact difference scheme when $\alpha = 0.5$, $\mu = 1$, $\gamma = 4$ and $N = 256$.

Table 1. Temporal error and convergence order of Example 1 for different α and p with $M = 256$, $\mu = 1$, $\gamma = 2/\alpha$.

	N	$E^{SOE}(M,N)$	$Rate_\tau^{SOE}$	$E^{SOE}(M,N)$	$Rate_\tau^{SOE}$	$E^{SOE}(M,N)$	$Rate_\tau^{SOE}$
	Alikhanov	$p=2$		$p=3$		$p=4$	
	8	2.6080×10^{-5}	-	4.2294×10^{-5}	-	8.0635×10^{-5}	-
	16	6.2591×10^{-6}	2.0589	1.1197×10^{-5}	1.9173	2.2588×10^{-5}	1.8358
	32	1.4056×10^{-6}	2.1548	2.8265×10^{-6}	1.9861	5.9572×10^{-6}	1.9228
	64	3.2266×10^{-7}	2.1230	7.0951×10^{-7}	1.9941	1.5306×10^{-6}	1.9605
$\alpha = 0.3$	Fast Alikhanov	$p=2$		$p=3$		$p=4$	
	8	2.6081×10^{-5}	-	4.2298×10^{-5}	-	8.0632×10^{-5}	-
	16	6.2657×10^{-6}	2.0575	1.1195×10^{-5}	1.9177	2.2588×10^{-5}	1.8358
	32	1.4042×10^{-6}	2.1578	2.8276×10^{-6}	1.9852	5.9571×10^{-6}	1.9229
	64	3.2485×10^{-7}	2.1119	7.0925×10^{-7}	1.9952	1.5302×10^{-6}	1.9609
	Alikhanov	$p=2$		$p=3$		$p=4$	
	8	2.9692×10^{-5}	-	6.1797×10^{-5}	-	1.2884×10^{-4}	-
	16	7.5590×10^{-6}	1.9738	1.6557×10^{-5}	1.9001	3.5529×10^{-5}	1.8585
	32	1.8727×10^{-6}	2.0131	4.2675×10^{-6}	1.9560	9.2949×10^{-6}	1.9345
	64	4.8342×10^{-7}	1.9537	1.0864×10^{-6}	1.9739	2.3738×10^{-6}	1.9693
$\alpha = 0.5$	Fast Alikhanov	$p=2$		$p=3$		$p=4$	
	8	2.9697×10^{-5}	-	6.1799×10^{-5}	-	1.2884×10^{-4}	-
	16	7.5559×10^{-6}	1.9746	1.6555×10^{-5}	1.9003	3.5529×10^{-5}	1.8585
	32	1.8753×10^{-6}	2.0105	4.2675×10^{-6}	1.9558	9.2949×10^{-6}	1.9345
	64	4.8013×10^{-7}	1.9657	1.0853×10^{-6}	1.9752	2.3737×10^{-6}	1.9693
	Alikhanov	$p=2$		$p=3$		$p=4$	
	8	3.2325×10^{-5}	-	7.1925×10^{-5}	-	1.5150×10^{-4}	-
	16	8.3560×10^{-6}	1.9517	1.8960×10^{-5}	1.9235	4.0683×10^{-5}	1.8968
	32	2.1056×10^{-6}	1.9886	4.8485×10^{-6}	1.9674	1.0492×10^{-5}	1.9552
	64	5.3111×10^{-7}	1.9872	1.2273×10^{-6}	1.9820	2.6629×10^{-6}	1.9782
$\alpha = 0.7$	Fast Alikhanov	$p=2$		$p=3$		$p=4$	
	8	3.2332×10^{-5}	-	7.1927×10^{-5}	-	1.5149×10^{-4}	-
	16	8.3605×10^{-6}	1.9513	1.8961×10^{-5}	1.9235	4.0681×10^{-5}	1.8968
	32	2.1158×10^{-6}	1.9824	4.8494×10^{-6}	1.9672	1.0485×10^{-5}	1.9560
	64	5.2915×10^{-7}	1.9995	1.2265×10^{-6}	1.9833	2.6618×10^{-6}	1.9779

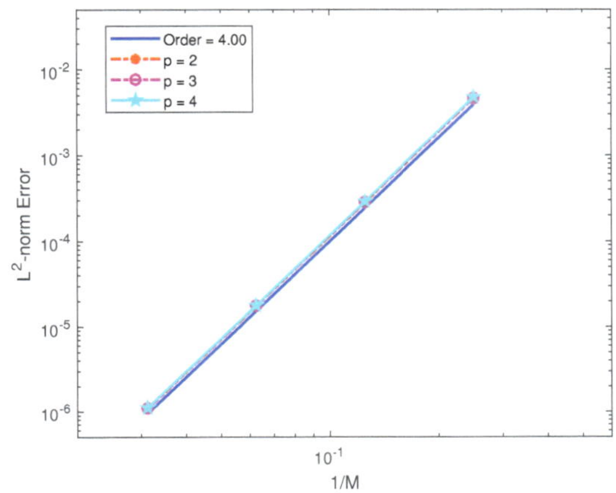

Figure 2. The spatial convergence orders of the fast compact difference scheme when $\alpha = 0.5$, $\mu = 1$, $\gamma = 4$ and $N = 512$.

Table 2. Space error and convergence order of Example 1 for different μ and p with $N = 512$, $\alpha = 0.5$, $\gamma = 4$.

	M	$E^{SOE}(M,N)$	$Rate_h^{SOE}$	$E^{SOE}(M,N)$	$Rate_h^{SOE}$	$E^{SOE}(M,N)$	$Rate_h^{SOE}$
$\mu = 0.1$	Alikhanov	$p = 2$		$p = 3$		$p = 4$	
	4	4.8317×10^{-3}	-	4.8272×10^{-3}	-	4.9532×10^{-3}	-
	8	2.9702×10^{-4}	4.0239	2.9925×10^{-4}	4.0118	3.0394×10^{-4}	4.0265
	16	1.8486×10^{-5}	4.0060	1.8543×10^{-5}	4.0124	1.8659×10^{-5}	4.0258
	32	1.1542×10^{-6}	4.0015	1.1572×10^{-6}	4.0021	1.1689×10^{-6}	3.9966
	Fast Alikhanov	$p = 2$		$p = 3$		$p = 4$	
	4	4.8317×10^{-3}	-	4.8272×10^{-3}	-	4.9532×10^{-3}	-
	8	2.9702×10^{-4}	4.0239	2.9925×10^{-4}	4.0118	3.0394×10^{-4}	4.0265
	16	1.8486×10^{-5}	4.0060	1.8543×10^{-5}	4.0124	1.8659×10^{-5}	4.0258
	32	1.1542×10^{-6}	4.0015	1.1572×10^{-6}	4.0021	1.1689×10^{-6}	3.9966
$\mu = 1$	Alikhanov	$p = 2$		$p = 3$		$p = 4$	
	4	4.6355×10^{-3}	-	4.6325×10^{-3}	-	4.7695×10^{-3}	-
	8	2.8495×10^{-4}	4.0239	2.8705×10^{-4}	4.0124	2.9150×10^{-4}	4.0323
	16	1.7735×10^{-5}	4.0060	1.7788×10^{-5}	4.0123	1.7899×10^{-5}	4.0256
	32	1.1073×10^{-6}	4.0015	1.1102×10^{-6}	4.0021	1.1212×10^{-6}	3.9967
	Fast Alikhanov	$p = 2$		$p = 3$		$p = 4$	
	4	4.6355×10^{-3}	-	4.6325×10^{-3}	-	4.7695×10^{-3}	-
	8	2.8495×10^{-4}	4.0239	2.8705×10^{-4}	4.0124	2.9150×10^{-4}	4.0323
	16	1.7735×10^{-5}	4.0060	1.7788×10^{-5}	4.0123	1.7899×10^{-5}	4.0256
	32	1.1073×10^{-6}	4.0015	1.1102×10^{-6}	4.0021	1.1212×10^{-6}	3.9967
$\mu = 10$	Alikhanov	$p = 2$		$p = 3$		$p = 4$	
	4	3.6473×10^{-3}	-	3.6500×10^{-3}	-	3.8049×10^{-3}	-
	8	2.2412×10^{-4}	4.0245	2.2550×10^{-4}	4.0167	2.2849×10^{-4}	4.0576
	16	1.3948×10^{-5}	4.0062	1.3981×10^{-5}	4.0116	1.4058×10^{-5}	4.0227
	32	8.7088×10^{-7}	4.0014	8.7269×10^{-7}	4.0019	8.8006×10^{-7}	3.9976
	Fast Alikhanov	$p = 2$		$p = 3$		$p = 4$	
	4	3.6473×10^{-3}	-	3.6500×10^{-3}	-	3.8049×10^{-3}	-
	8	2.2412×10^{-4}	4.0245	2.2550×10^{-4}	4.0167	2.2849×10^{-4}	4.0576
	16	1.3948×10^{-5}	4.0062	1.3981×10^{-5}	4.0116	1.4058×10^{-5}	4.0227
	32	8.7088×10^{-7}	4.0014	8.7269×10^{-7}	4.0019	8.8006×10^{-7}	3.9976

Table 3. Errors, convergence orders in space, and CPU times for the Alikhanov scheme and fast Alikhanov scheme by fixing $N = 512$, $p = 2$, $\mu = 1$, and $\gamma = 2$ for Example 1.

		$\alpha = 0.3$		$\alpha = 0.5$		$\alpha = 0.7$	
Alikhanov	M	$E^{SOE}(M,N)$	$Rate_h^{SOE}$	$E^{SOE}(M,N)$	$Rate_h^{SOE}$	$E^{SOE}(M,N)$	$Rate_h^{SOE}$
	4	4.6004×10^{-3}	-	4.6355×10^{-3}	-	4.5792×10^{-3}	-
	8	2.8278×10^{-4}	4.0240	2.8495×10^{-4}	4.0239	2.8149×10^{-4}	4.0240
	16	1.7600×10^{-5}	4.0061	1.7735×10^{-5}	4.0060	1.7555×10^{-5}	4.0031
	32	1.0988×10^{-6}	4.0016	1.1073×10^{-6}	4.0015	1.1327×10^{-6}	3.9541
	CPU(s)	82.777		75.941		82.308	
		$\alpha = 0.3$		$\alpha = 0.5$		$\alpha = 0.7$	
Fast Alikhanov	M	$E^{SOE}(M,N)$	$Rate_h^{SOE}$	$E^{SOE}(M,N)$	$Rate_h^{SOE}$	$E^{SOE}(M,N)$	$Rate_h^{SOE}$
	4	4.6004×10^{-3}	-	4.6355×10^{-3}	-	4.5792×10^{-3}	-
	8	2.8278×10^{-4}	4.0240	2.8495×10^{-4}	4.0239	2.8145×10^{-4}	4.0241
	16	1.7600×10^{-5}	4.0060	1.7735×10^{-5}	4.0060	1.7517×10^{-5}	4.0061
	32	1.0997×10^{-6}	4.0005	1.1073×10^{-6}	4.0015	1.0939×10^{-6}	4.0011
	CPU(s)	3.9295		3.8685		3.9980	

Table 4. Numerical solutions when $M = 256$, $p = 2$, $\mu = 1$, and $\gamma = 2/\alpha$ for Example 1.

	$\alpha = 0.3$	$\alpha = 0.5$	$\alpha = 0.7$
N	U^*	U^*	U^*
4	1.4949	1.5049	1.4853
8	1.4950	1.5050	1.4853
16	1.4950	1.5050	1.4853
32	1.4950	1.5050	1.4853
64	1.4950	1.5050	1.4853
128	1.4950	1.5050	1.4853

Example 2. *In order to verify the wide range of applications of the numerical scheme, the case where the exact solution is unknown is tested in this numerical example. We consider problems* (1)–(3) *in* $\Omega = (0, \pi)$ *to verify the effectiveness and high accuracy of the numerical scheme. Let* $T = 1$, *the initial condition is* $u_0(x) = \sin(x)$, *and the source term is* $f(x,t) = 0$.

In Table 5, the errors in the L^2-norm, spatial convergence orders, and CPU times of the Alikhanov scheme and the fast Alikhanov numerical scheme are presented for fixed values of $p = 2$, $\mu = 1$, $\gamma = 2$, and $N = 2048$, with $\alpha \in \{0.3, 0.5, 0.7\}$. It is worth mentioning that both numerical schemes can achieve the same approximation effect, but the CPU time of the fast scheme is much shorter. Table 6 demonstrates that for fixed values of $M = 512$, $\mu = 1$, $\alpha = 0.8$, and $\gamma = 2/\alpha$, the L^2-norm error decreases as the number of temporal subintervals increases. It is expected that the convergence order of the fast approximate scheme is $\mathcal{O}(\tau^2)$. Meanwhile, the error and convergence order remain as expected for changing values of p, suggesting that the approximation scheme is stable for different p. Table 7 shows the errors, temporal convergence orders, and CPU times for different μ, where we take $M = 512$, $p = 2$, and $\gamma = 2/\alpha$. In each row of the table, the errors and convergence orders change very little as μ varies, so the method is stable for the coefficient μ, as predicted by our theoretical analysis. In addition, in order to demonstrate more intuitively the speedup effect of the SOE approximation, taking $\alpha = 0.5$, $p = 2$, $\mu = 1$, $\gamma = 2$, and $M = 4$, we plot the computation time curves of the two schemes for the N increase, as shown in Figure 3. It is clear that the fast Alikhanov scheme saves a lot of computational costs compared to the Alikhanov scheme.

Table 5. Errors, convergence orders in space, and CPU times for the Alikhanov scheme and fast Alikhanov scheme by fixing $p = 2$, $\mu = 1$, $\gamma = 2$, and $N = 2048$ for Example 2.

		$\alpha = 0.3$		$\alpha = 0.5$		$\alpha = 0.7$	
Alikhanov	M	$E^{SOE}(M,N)$	$Rate_h^{SOE}$	$E^{SOE}(M,N)$	$Rate_h^{SOE}$	$E^{SOE}(M,N)$	$Rate_h^{SOE}$
	4	4.3605×10^{-4}	-	4.3568×10^{-4}	-	4.5247×10^{-4}	-
	8	2.6588×10^{-5}	4.0356	2.6570×10^{-5}	4.0354	2.7511×10^{-5}	4.0398
	16	1.6529×10^{-6}	4.0077	1.6517×10^{-6}	4.0078	1.7094×10^{-6}	4.0084
	32	1.0316×10^{-7}	4.0020	1.0308×10^{-7}	4.0021	1.0667×10^{-7}	4.0022
	CPU(s)	1253.8		1197.3		1279.9	
		$\alpha = 0.3$		$\alpha = 0.5$		$\alpha = 0.7$	
Fast Alikhanov	M	$E^{SOE}(M,N)$	$Rate_h^{SOE}$	$E^{SOE}(M,N)$	$Rate_h^{SOE}$	$E^{SOE}(M,N)$	$Rate_h^{SOE}$
	4	4.3605×10^{-4}	-	4.3568×10^{-4}	-	4.5247×10^{-4}	-
	8	2.6588×10^{-5}	4.0356	2.6570×10^{-5}	4.0354	2.7511×10^{-5}	4.0398
	16	1.6529×10^{-6}	4.0077	1.6517×10^{-6}	4.0078	1.7094×10^{-6}	4.0084
	32	1.0316×10^{-7}	4.0020	1.0308×10^{-7}	4.0021	1.0667×10^{-7}	4.0022
	CPU(s)	44.704		45.165		45.322	

Table 6. The E^{SOE}-norm errors and temporal convergence orders with $M = 512$, $\mu = 1$, $\alpha = 0.8$, and $\gamma = 2/\alpha$.

	$p=1$		$p=2$		$p=3$		$p=4$		$p=5$	
N	$E^{SOE}(M,N)$	$Rate_\tau^{SOE}$	$E^{SOE}(M,N)$	$Rate_\tau^{SOE}$	$E^{SOE}(M,N)$	$Rate_\tau^{SOE}$	$E^{SOE}(M,N)$	$Rate_\tau^{SOE}$	$E^{SOE}(M,N)$	$Rate_\tau^{SOE}$
10	1.3260×10^{-3}	-	1.3265×10^{-3}	-	1.3266×10^{-3}	-	1.3266×10^{-3}	-	1.3266×10^{-3}	-
20	3.4271×10^{-4}	1.9520	3.4279×10^{-4}	1.9522	3.4281×10^{-4}	1.9522	3.4281×10^{-4}	1.9522	3.4281×10^{-4}	1.9523
40	9.0337×10^{-5}	1.9236	9.0353×10^{-5}	1.9237	9.0356×10^{-5}	1.9237	9.0354×10^{-5}	1.9238	9.0351×10^{-5}	1.9238
80	2.3914×10^{-5}	1.9174	2.3917×10^{-5}	1.9176	2.3917×10^{-5}	1.9176	2.3914×10^{-5}	1.9177	2.3914×10^{-5}	1.9177
160	6.3120×10^{-6}	1.9217	6.3132×10^{-6}	1.9216	6.3131×10^{-6}	1.9216	6.3145×10^{-6}	1.9211	6.3158×10^{-6}	1.9208

Table 7. The E^{SOE}-norm errors, temporal convergence orders, and CPU time (seconds) with $M = 512$, $p = 2$, and $\gamma = 2/\alpha$.

	$\mu = 0.05$				$\mu = 0.1$		
N	$E^{SOE}(M,N)$	$Rate_\tau^{SOE}$	CPU(s)	N	$E^{SOE}(M,N)$	$Rate_\tau^{SOE}$	CPU(s)
4	2.8796×10^{-3}	-	0.4605	4	3.2656×10^{-3}	-	0.4406
8	6.8144×10^{-4}	2.0792	0.7595	8	7.7751×10^{-4}	2.0704	0.8029
16	1.7604×10^{-4}	1.9527	1.5174	16	2.0089×10^{-4}	1.9525	1.5480
32	4.8379×10^{-5}	1.8634	2.5469	32	5.4828×10^{-5}	1.8734	2.7098
64	1.3632×10^{-5}	1.8273	5.0930	64	1.5262×10^{-5}	1.8450	5.3880
	$\mu = 1$				$\mu = 2$		
N	$E^{SOE}(M,N)$	$Rate_\tau^{SOE}$	CPU(s)	N	$E^{SOE}(M,N)$	$Rate_\tau^{SOE}$	CPU(s)
4	8.7476×10^{-3}	-	0.4646	4	9.7976×10^{-3}	-	0.5341
8	2.0697×10^{-3}	2.0794	0.7903	8	2.3022×10^{-3}	2.0894	0.7326
16	5.2813×10^{-4}	1.9705	1.4359	16	5.8600×10^{-4}	1.9741	1.3253
32	1.3868×10^{-4}	1.9291	2.6138	32	1.5258×10^{-4}	1.9413	2.5373
64	3.6684×10^{-5}	1.9185	5.0402	64	3.9897×10^{-5}	1.9353	4.7893

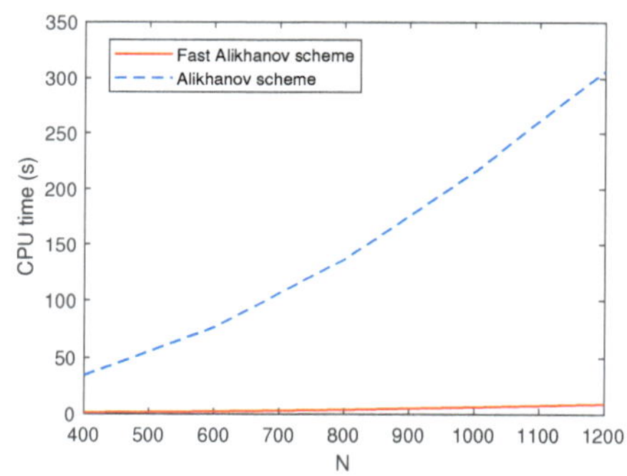

Figure 3. Comparison of the CPU time between the fast Alikhanov and Alikhanov schemes with $\alpha = 0.5$, $p = 2$, $\mu = 1$, $\gamma = 2$, and $M = 4$.

Example 3. *In this example, we consider problem (1) over domains $\Omega = (0,1)$ and $T = 1$. The analytic solution is $u(x,t) = \left(\frac{t^\alpha}{\Gamma(1+\alpha)}\right)\sin(\pi x)\cos(\pi x)$, such that the source term $f(x,t)$ is as follows:*

$$f(x,t) = \sin(\pi x)\cos(\pi x) + 2\pi^2\mu\frac{t^\alpha}{\Gamma(1+\alpha)}\sin(2\pi x) + 8\pi^4\frac{t^\alpha}{\Gamma(1+\alpha)}\sin(2\pi x) + \pi\left(\frac{t^\alpha}{\Gamma(1+\alpha)}\right)^{p+1}(\sin(\pi x))^p(\cos(\pi x))^p\cos(2\pi x).$$

Tables 8 and 9 display the temporal and spatial errors, spatial–temporal convergence orders, and CPU times for various α values. Furthermore, the results show that the proposed scheme can reach the second order of convergence in time and the fourth order of convergence in space. Next, we test the stability by fixing $M = 256$, $\mu = 1$, $p = 2$, and

$\gamma = 2/\alpha$ and selecting different T values, which means the time-step sizes are different. The numerical results are listed in Table 10, which shows that the fast Alikhanov scheme is stable and has a relaxed stability restriction. In Table 11, we tested the errors for $\alpha = 0.95, \alpha = 0.99, \alpha = 0.999,$ *and* $\alpha = 0.9999$ by making $M = 256$, $\mu = 1$, $p = 2$, and $\gamma = 2/\alpha$, which also shows that the numerical scheme is stable when $\alpha \to 1^-$.

Table 8. Errors, temporal convergence orders, and CPU times by fixing $p = 2$, $\mu = 1$, $\gamma = 4$, and $M = 256$ for Example 3.

	$\alpha = 0.3$			$\alpha = 0.5$			$\alpha = 0.7$		
N	$E^{SOE}(M,N)$	$Rate_\tau^{SOE}$	CPU	$E^{SOE}(M,N)$	$Rate_\tau^{SOE}$	CPU	$E^{SOE}(M,N)$	$Rate_\tau^{SOE}$	CPU
16	9.5519×10^{-7}	-	0.2818	3.9181×10^{-6}	-	0.2457	8.5622×10^{-6}	-	0.2060
32	2.4356×10^{-7}	1.9715	0.4170	1.0148×10^{-6}	1.9490	0.3583	2.2256×10^{-6}	1.9438	0.4410
64	5.8573×10^{-8}	2.0560	0.7595	2.5816×10^{-7}	1.9748	0.7026	5.6804×10^{-7}	1.9701	0.7078
128	1.6584×10^{-8}	1.8205	1.8818	6.6561×10^{-8}	1.9555	1.4596	1.4255×10^{-7}	1.9946	1.5336

Table 9. Errors, convergence orders in space, and CPU times by fixing $p = 2$, $\mu = 1$, $\gamma = 4$, and $N = 1000$ for Example 3.

	$\alpha = 0.3$			$\alpha = 0.5$			$\alpha = 0.7$		
M	$E^{SOE}(M,N)$	$Rate_h^{SOE}$	CPU	$E^{SOE}(M,N)$	$Rate_h^{SOE}$	CPU	$E^{SOE}(M,N)$	$Rate_h^{SOE}$	CPU
16	7.7546×10^{-5}	-	13.587	7.8530×10^{-5}	-	13.533	7.5592×10^{-5}	-	13.994
32	4.8243×10^{-6}	4.0067	15.856	4.8854×10^{-6}	4.0067	15.493	4.7648×10^{-6}	4.0067	15.609
64	3.0131×10^{-7}	4.0010	19.515	3.0497×10^{-7}	4.0017	19.445	2.9744×10^{-7}	4.0017	19.734
128	1.9031×10^{-8}	3.9848	28.411	1.9111×10^{-8}	3.9962	28.621	1.8807×10^{-8}	3.9833	28.510

Table 10. The E_2-norm errors and temporal convergence orders with $M = 256$, $\mu = 1$, $p = 2$, and $\gamma = 2/\alpha$.

	$T = 0.5$		$T = 1$		$T = 4$		$T = 10$	
N	$E^{SOE}(M,N)$	$Rate_\tau^{SOE}$	$E^{SOE}(M,N)$	$Rate_\tau^{SOE}$	$E^{SOE}(M,N)$	$Rate_\tau^{SOE}$	$E^{SOE}(M,N)$	$Rate_\tau^{SOE}$
16	1.3810×10^{-6}	-	3.9181×10^{-6}	-	3.1501×10^{-5}	-	1.2475×10^{-4}	-
32	3.5931×10^{-7}	1.9424	1.0148×10^{-6}	1.9490	8.1468×10^{-6}	1.9511	3.2269×10^{-5}	1.9507
64	9.0207×10^{-8}	1.9939	2.5816×10^{-7}	1.9748	2.0664×10^{-6}	1.9791	8.2114×10^{-6}	1.9745
128	2.3438×10^{-8}	1.9444	6.6561×10^{-8}	1.9555	5.2247×10^{-7}	1.9837	2.0765×10^{-6}	1.9835
256	6.0047×10^{-9}	1.9647	1.5317×10^{-8}	2.1196	1.3334×10^{-7}	1.9703	5.2527×10^{-7}	1.9830

Table 11. The E_2-norm errors and temporal convergence orders with $M = 256$, $\mu = 1$, $p = 2$, $\alpha = 0.5$, and $\gamma = 2/\alpha$.

	$\alpha = 0.95$		$\alpha = 0.99$		$\alpha = 0.999$		$\alpha = 0.9999$	
N	$E^{SOE}(M,N)$	$Rate_\tau^{SOE}$	$E^{SOE}(M,N)$	$Rate_\tau^{SOE}$	$E^{SOE}(M,N)$	$Rate_\tau^{SOE}$	$E^{SOE}(M,N)$	$Rate_\tau^{SOE}$
16	4.0608×10^{-6}	-	3.8864×10^{-6}	-	3.8417×10^{-6}	-	3.8366×10^{-6}	-
32	1.0171×10^{-6}	1.9973	9.7447×10^{-6}	1.9957	9.6418×10^{-7}	1.9944	9.6071×10^{-7}	1.9977
64	2.5443×10^{-7}	1.9992	2.4199×10^{-7}	2.0097	2.4196×10^{-7}	1.9946	2.3883×10^{-7}	2.0081
128	6.5139×10^{-8}	1.9657	6.0890×10^{-8}	1.9907	6.0593×10^{-8}	1.9975	5.8879×10^{-8}	2.0201
256	1.6078×10^{-8}	2.0184	1.5477×10^{-8}	1.9761	1.7048×10^{-8}	1.8295	1.4500×10^{-8}	2.0217

5. Conclusions

In this work, we propose a fast high-order compact scheme for solving the fractional KS equation with the generalized Burgers' type nonlinearity. To deal with the weak singularity at the initial time, the Alikhanov scheme is implemented to approximate the Caputo fractional derivative in the time direction. With the aid of the SOE technique, the computational efficiency is improved and the storage requirement is reduced. In the space direction, the developed compact difference formulas have been successfully applied for approximating the spatial derivatives. As a result, a fully discrete scheme is constructed.

Meanwhile, the stability and convergence with $\tau^2 + h^4$ of the proposed scheme are proved. Finally, the results of the numerical experiments demonstrate that the convergence orders for temporal and spatial convergence achieve second-order and fourth-order accuracies, respectively. This is highly consistent with the theoretical predictions and effectively verifies the theoretical analysis.

Author Contributions: Conceptualization, D.X.; methodology, H.J.; software, H.J.; validation, H.J.; formal analysis, H.J.; investigation, H.J.; re-sources, H.J.; data curation, H.J.; writing—original draft preparation, H.J.; writing—review and editing, D.X.; visu-alization, D.X.; supervision, D.X. ; project administration, D.X. All authors have read and agreed to the published version of the manuscript.

Funding: The work was supported by Scientific Research Fund of Hunan Provincial Education Department (grant number No. 24C0265).

Data Availability Statement: The manuscript has no associated data.

Conflicts of Interest: The authors declare that they have no conflicts of interest.

References

1. Kuramoto, Y.; Tsuzuki, T. On the Formation of Dissipative Structures in Reaction-Diffusion Systems. *Prog. Theor. Phys.* **1975**, *54*, 687–699.
2. Sivashinsky, G.I. Nonlinear Analysis of Hydrodynamic Instability in Laminar Flames-I. Derivation of Basic Equations. *Acta Astronaut.* **1977**, *4*, 1177–1206.
3. Sivashinsky, G.I.; Michelson, D.M. On Irregular Wavy Flow of a Liquid Film Down a Vertical Plane. *Prog. Theor. Phys.* **1980**, *63*, 2112–2114.
4. Sivashinsky, G.I. Instabilities, Pattern-Formation, and Turbulence in Flames. *Ann. Rev. Fluid Mech.* **1983**, *15*, 179–199.
5. Hyman, J.M.; Nicolaenko, B.; Zaleski, S. Order and Complexity in the Kuramoto-Sivashinsky Model of Weakly Turbulent Interfaces. *Phys. D* **1986**, *23*, 265–292.
6. Peng, X.; Qiu, W.; Hendy, A.; Zaky, M. Temporal Second-Order Fast Finite Difference/Compact Difference Schemes for Time-Fractional Generalized Burgers' Equations. *J. Sci. Comput.* **2024**, *99*, 52.
7. Tadmor, E. The Well-Posedness of the Kuramoto–Sivashinsky Equation. *SIAM J. Math. Anal.* **1986**, *17*, 884–893.
8. Zgliczynski, P. Attracting Fixed Points for the Kuramoto–Sivashinsky Equation: A Computer Assisted Proof. *SIAM J. Appl. Dyn. Syst.* **2002**, *1*, 215–235.
9. Sakthivel, R.; Ito, H. Non-Linear Robust Boundary Control of the Kuramoto–Sivashinsky Equation. *IMA J. Math. Control Inf.* **2007**, *24*, 47–55.
10. Khater, A.H.; Temsah, R.S. Numerical Solutions of the Generalized Kuramoto–Sivashinsky Equation by Chebyshev Spectral Collocation Methods. *Comput. Math. Appl.* **2008**, *56*, 1465–1472.
11. Mittal, R.C.; Arora, G. Quintic B-Spline Collocation Method for Numerical Solution of the Kuramoto–Sivashinsky Equation. *Commun. Nonlinear Sci.* **2010**, *15*, 2798–2808.
12. Xu, Y.; Shu, C.W. Local Discontinuous Galerkin Methods for the Kuramoto–Sivashinsky Equations and the Ito-Type Coupled KdV Equations. *Comput. Methods Appl. Mech. Eng.* **2006**, *195*, 3430–3447.
13. Akrivis, G.; Smyrlis, Y. Implicit—Explicit BDF Methods for the Kuramoto–Sivashinsky Equation. *Appl. Numer. Math.* **2004**, *51*, 151–169.
14. Langlands, T.A.M.; Henry, B.I. The Accuracy and Stability of an Implicit Solution Method for the Fractional Diffusion Equation. *J. Comput. Phys.* **2005**, *205*, 719–736.
15. Sun, Z.Z.; Wu, X.N. A Fully Discrete Difference Scheme for a Diffusion-Wave System. *Appl. Numer. Math.* **2006**, *56*, 193–209.
16. Li, C.; Deng, W.H.; Wu, Y.J. Numerical Analysis and Physical Simulations for the Time Fractional Radial Diffusion Equation. *Comput. Math. Appl.* **2011**, *62*, 1024–1037.
17. Li, L.M.; Xu, D.; Luo, M. Alternating Direction Implicit Galerkin Finite Element Method for the Two-Dimensional Fractional Diffusion-Wave Equation. *J. Comput. Phys.* **2013**, *255*, 471–485.
18. Lin, Y.M.; Xu, C.J. Finite Difference/Spectral Approximations for the Time-Fractional Diffusion Equation. *J. Comput. Phys.* **2007**, *225*, 1533–1552.
19. Qiu, W.L.; Chen, H.B.; Xuan, X. An Implicit Difference Scheme and Algorithm Implementation for the One-Dimensional Time-Fractional Burgers Equations. *Math. Comput. Simulat.* **2019**, *166*, 298–314.
20. Liu, T.; Liu, H.; Ma, Y. Approximate Solution of a Kind of Time-Fractional Evolution Equations Based on Fast L1 Formula and Barycentric Lagrange Interpolation. *Fractal Fract.* **2024**, *8*, 675. [CrossRef]

21. Gao, G.H.; Sun, Z.Z.; Zhang, H.W. A New Fractional Differentiation Formula to Approximate the Caputo Fractional Derivative and Its Applications. *J. Comput. Phys.* **2014**, *259*, 33–50. [CrossRef]
22. Liao, H.L.; Lyu, P.; Vong, S.; Zhao, Y. Stability of Fully Discrete Schemes with Interpolation-Type Fractional Formulas for Distributed-Order Subdiffusion Equations. *Numer. Algor.* **2017**, *75*, 845–878. [CrossRef]
23. Lv, C.W.; Xu, C.J. Error Analysis of a High Order Method for Time-Fractional Diffusion Equations. *SIAM J. Sci. Comput.* **2016**, *38*, 2699–2724. [CrossRef]
24. Alikhanov, A.A. A New Difference Scheme for the Time Fractional Diffusion Equation. *J. Comput. Phys.* **2015**, *280*, 424–438.
25. Liao, H.L.; McLean, W.; Zhang, J.W. A Second-Order Scheme with Nonuniform Time Steps for a Linear Reaction-Subdiffusion Problem. *Commun. Comput. Phys.* **2021**, *30*, 567–601.
26. Zhang, H.X.; Yang, X.H.; Xu, D. An Efficient Spline Collocation Method for a Nonlinear Fourth-Order Reaction Subdiffusion Equation. *J. Sci. Comput.* **2020**, *85*, 7. [CrossRef]
27. Li, X.; Liao, H.L.; Zhang, L.M. A Second-Order Fast Compact Scheme with Unequal Time-Steps for Subdiffusion Problems. *Numer. Algorithms* **2021**, *86*, 1011–1039.
28. Jiang, S.D.; Zhang, J.W.; Zhang, Q.; Zhang, Z.M. Fast Evaluation of the Caputo Fractional Derivative and Its Applications to Fractional Diffusion Equations. *Commun. Comput. Phys.* **2017**, *21*, 650–678. [CrossRef]
29. Lee, S.; Kim, H.; Jang, B. A Novel Numerical Method for Solving Nonlinear Fractional-Order Differential Equations and Its Applications. *Fractal Fract.* **2024**, *8*, 65. [CrossRef]
30. Yang, S.; Yang, X.H. A time two-grid difference method for nonlinear generalized viscous Burgers' equation.*J. Math. Chem.* **2024**, *62*, 1323–1356.
31. Ganaie, I.A.; Arora, S.; Kukreja, V.K. Cubic Hermite Collocation Solution of the Kuramoto–Sivashinsky Equation. *Int. J. Comput. Math.* **2016**, *93*, 223–235. [CrossRef]
32. Al-Khaled, K.; Abu-Irwaq, I. Computational Sinc-Scheme for Extracting Analytical Solution for the Model Kuramoto–Sivashinsky Equation. *Int. J. Electr. Comput. Eng.* **2019**, *9*, 3720–3731.
33. Wang, Y.M.; Zhang, H.B. Higher-Order Compact Finite Difference Method for Systems of Reaction-Diffusion Equations. *J. Comput. Appl. Math.* **2009**, *233*, 502–518.
34. Sari, M.; Gürarslan, G. A Sixth-Order Compact Finite Difference Scheme to the Numerical Solutions of Burgers' Equation. *Appl. Math. Comput.* **2009**, *208*, 475–483.
35. Sutmann, G. Compact Finite Difference Schemes of Sixth Order for the Helmholtz Equation. *J. Comput. Appl. Math.* **2007**, *203*, 15–31.
36. Tian, Z.F.; Liang, X.; Yu, P.X. A Higher Order Compact Finite Difference Algorithm for Solving the Incompressible Navier-Stokes Equations. *Int. J. Numer. Meth. Eng.* **2011**, *88*, 511–532.
37. Kuzenov, V.V.; Ryzhkov, S.V.; Varaksin, A.Y. Development of a Method for Solving Elliptic Differential Equations Based on a Nonlinear Compact-Polynomial Scheme. *J. Comput. Appl. Math.* **2024**, *451*, 116098.
38. Qiu, W.L.; Nikan, O.; Avazzadeh, Z. Numerical Investigation of Generalized Tempered-Type Integrodifferential Equations with Respect to Another Function. *Fract. Calc. Appl. Anal.* **2023**, *60*, 41.
39. Xu, D.; Guo, J.; Qiu, W.L. Time Two-Grid Algorithm Based on Finite Difference Method for Two-Dimensional Nonlinear Fractional Evolution Equations. *Appl. Numer. Math.* **2020**, *152*, 169–184.
40. Qiu, W.L.; Fairweather, G.; Yang, X.H.; Zhang, H.X. ADI Finite Element Galerkin Methods for Two-Dimensional Tempered Fractional Integro-Differential Equations. *Calcolo* **2023**, *60*, 41.
41. Sun, Z.Z. *Numerical Methods of Partial Differential Equations*; Science Press: Beijing, China, 2012.
42. Wang, T.C. Optimal Point-Wise Error Estimate of a Compact Difference Scheme for the Klein–Gordon–Schrödinger Equation. *J. Math. Anal. Appl.* **2014**, *412*, 155–167.
43. Dimitrienko, Y.I.; Li, S.G.; Niu, Y. Study on the Dynamics of a Nonlinear Dispersion Model in Both 1D and 2D Based on the Fourth-Order Compact Conservative Difference Scheme. *Math. Comput. Simulat.* **2020**, *182*. 661–689.
44. Huang, C.B.; Stynes, M. A Sharp α-Robust $L^\infty(H^1)$ Error Bound for a Time-Fractional Allen-Cahn Problem Discretised by the Alikhanov $L2-1_\sigma$ Scheme and a Standard FEM. *J. Sci. Comput.* **2022**, *91*, 43.
45. Liao, H.L.; McLean, W.; Zhang, J.W. A Discrete Grönwall Inequality with Applications to Numerical Schemes for Subdiffusion Problems. *SIAM J. Numer. Anal.* **2019**, *57*, 218–237.

fractal and fractional

MDPI

Article

Numerical Analysis of a Fractional Cauchy Problem for the Laplace Equation in an Annular Circular Region

José Julio Conde Mones [1], Julio Andrés Acevedo Vázquez [1], Eduardo Hernández Montero [1], María Monserrat Morín Castillo [2], Carlos Arturo Hernández Gracidas [3,*] and José Jacobo Oliveros Oliveros [1,*]

[1] Facultad de Ciencias Físico Matemáticas, Benemérita Universidad Autónoma de Puebla, Boulevard 18 Sur y Avenida San Claudio, Colonia San Manuel, Ciudad Universitaria, Puebla 72570, Mexico; jose.conde@correo.buap.mx (J.J.C.M.); julio.acevedova@alumno.buap.mx (J.A.A.V.); eduardo.hernandezmontero@correo.buap.mx (E.H.M.)

[2] Facultad de Ciencias de la Electrónica, Benemérita Universidad Autónoma de Puebla, Boulevard 18 Sur y Avenida San Claudio, Colonia San Manuel, Ciudad Universitaria, Puebla 72570, Mexico; maria.morin@correo.buap.mx

[3] CONAHCYT-BUAP, Facultad de Ciencias Físico Matemáticas, Benemérita Universidad Autónoma de Puebla, Boulevard 18 Sur y Avenida San Claudio, Colonia San Manuel, Ciudad Universitaria, Puebla 72570, Mexico

* Correspondence: cahernandezgr@conahcyt.mx (C.A.H.G.); jose.oliveros@correo.buap.mx (J.J.O.O.)

Abstract: The Cauchy problem for the Laplace equation in an annular bounded region consists of finding a harmonic function from the Dirichlet and Neumann data known on the exterior boundary. This work considers a fractional boundary condition instead of the Dirichlet condition in a circular annular region. We found the solution to the fractional boundary problem using circular harmonics. Then, the Tikhonov regularization is used to handle the numerical instability of the fractional Cauchy problem. The regularization parameter was chosen using the L-curve method, Morozov's discrepancy principle, and the Tikhonov criterion. From numerical tests, we found that the series expansion of the solution to the Cauchy problem can be truncated in $N = 20$, $N = 25$, or $N = 30$ for smooth functions. For other functions, such as absolute value and the jump function, we have to choose other values of N. Thus, we found a stable method for finding the solution to the problem studied. To illustrate the proposed method, we elaborate on synthetic examples and MATLAB 2021 programs to implement it. The numerical results show the feasibility of the proposed stable algorithm. In almost all cases, the L-curve method gives better results than the Tikhonov Criterion and Morozov's discrepancy principle. In all cases, the regularization using the L-curve method gives better results than without regularization.

Keywords: Cauchy problem; inverse problem; ill-posed problem; Laplace equation; fractional boundary condition; Tikhonov regularization

Academic Editors: Carlo Cattani, Libo Feng, Lin Liu and Yang Liu

Received: 5 February 2025
Revised: 8 April 2025
Accepted: 15 April 2025
Published: 27 April 2025

Citation: Conde Mones, J.J.; Acevedo Vázquez, J.A.; Hernández Montero, E.; Morín Castillo, M.M.; Hernández Gracidas, C.A.; Oliveros Oliveros, J.J. Numerical Analysis of a Fractional Cauchy Problem for the Laplace Equation in an Annular Circular Region. *Fractal Fract.* **2025**, *9*, 284. https://doi.org/10.3390/fractalfract9050284

1. Introduction

The challenge of determining a harmonic function within a bounded annular region based on partial boundary measurements (Cauchy data) is known as the Cauchy problem for the Laplace equation [1]. This problem is notoriously ill-posed in the sense of Hadamard, meaning that small perturbations in the Cauchy data can lead to significant changes in the solution, resulting in numerical instability. Consequently, regularization techniques are essential for solving this problem. To ensure a solution to the Cauchy problem, certain smoothness conditions must be imposed on the Cauchy data (refer to Theorem 1 in [2]).

Various methods have been developed to analyze the Cauchy problem. For instance, ref. [3] utilized singular value decomposition to find the solution in a circular annular region, employing the spectral cut-off of the pseudo-inverse method to manage the numerical instability. In Refs. [4,5], a novel regularization method was introduced using the method of fundamental solutions to address the Cauchy problem in both annular and multi-connected domains. To effectively solve the discrete ill-posed problem arising from a boundary collocation scheme, Tikhonov regularization and L-curve methods were applied in [5]. Additionally, the technique of layer potentials was used in [3,6,7] to derive an equivalent system of integral equations. The moment problem approach, based on Green's formula, was employed in [8] to solve the Cauchy problem in more complex annular regions. A similar technique was proposed for the three-dimensional Cauchy problem in [9], where the solution was expressed using spherical harmonics and Tikhonov regularization. In [10], a variational formulation was introduced, minimizing the cost functional through conjugate gradient iterations combined with boundary element discretization. In [1], the potential on the interior boundary of the annular region was treated as a control function to match the Cauchy input data on the exterior boundary, incorporating a penalized term in the cost function. This approach allowed for the determination of the optimal solution using an iterative conjugate gradient algorithm, with the computational cost involving the solution of two elliptic problems per iteration, solved by the finite element method. Similar techniques have been applied to other control problems, as seen in [10–14].

The Cauchy problem holds significant importance due to its numerous applications, such as estimating pipeline deterioration, calculating solutions or potentials in inaccessible regions or boundaries, and studying cracks in plates [15,16]. Furthermore, the Cauchy problem is utilized in inverse electrocardiography problems [17–19] and in solving inverse problems in electroencephalography (EEG) [14]. EEG signals are known to exhibit fractal characteristics [20–22]. Additionally, fractional derivatives can model voltage propagation in axons using a fractional cable geometry to study human neural networks [23]. In the context of EEG, these fractal characteristics may be related to the sources generating the signal. One potential approach to relate these fractal characteristics to EEG signals is through the use of fractional operators, which warrants further investigation in future studies.

In this work, we consider one variant of the Cauchy problem. More precisely, we consider that we know the action of a fractional operator on the potential on the exterior boundary instead of the potential itself. We apply the Tikhonov regularization to handle the numerical instability that presents this variant, which we call the fractional Cauchy problem. Since we consider circular geometry, we use the Fourier series method to solve the normal equations. The adjoint operator was found using its definition. From this, we found a stable algorithm for some of the parameters defining the fractional operator. To illustrate the results presented in this work, we elaborate synthetic examples and programs in MATLAB.

Regarding the fractional Cauchy problem, we found no work on it. Therefore, as a validation of the results from our proposal, we include results for the classical case that considers a Dirichlet condition, which has been extensively studied, as we attempted to demonstrate in our literature review, which included a substantial number of articles. We obtain the same results as in the classical case using our method.

The paper is organized as follows: In Section 2, the definition and some results of the classical Cauchy problem, as well as the Sturm–Liouville operator, are presented. Section 2 also finalizes the definition of the fractional Cauchy problem. Section 3 applies the Tikhonov regularization to find an algorithm to recover the potential on the interior boundary. Section 4 presents numerical examples to illustrate the algorithm presented in this work. In Section 5, we discuss the stability of the proposed algorithm. In Section 6, we give the conclusions.

2. Problem Formulation

2.1. The Cauchy Problem

Let Ω be a bounded annular region in $\mathbb{R}^2$ with sufficiently smooth interior boundary S_1 and exterior boundary S_2, as shown in Figure 1.

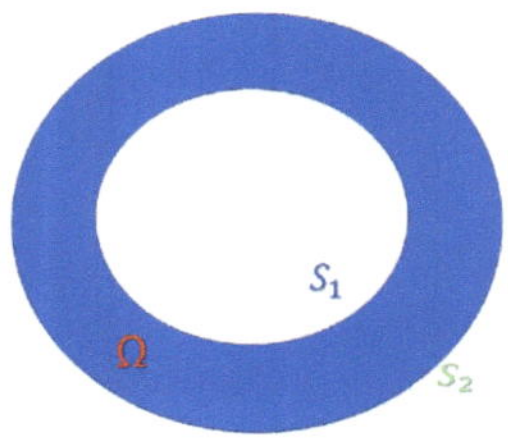

Figure 1. Bi-dimensional circular annular region Ω.

We consider the following boundary value problem: Find ω, such that

$$\begin{aligned} \Delta w &= 0, \text{ in } \Omega, \\ w &= \Phi, \text{ on } S_2, \\ \sigma\frac{\partial w}{\partial n} &= \Psi, \text{ on } S_2, \end{aligned} \tag{1}$$

where $\Phi \in \mathrm{H}^{1/2}(S_2)$, $\sigma\frac{\partial w}{\partial n}\Big|_{S_2} = \Psi \in \mathrm{H}^{-1/2}(S_2)$, n is the outward unitary vector defined on $\partial\Omega$, and $\partial w/\partial n$ denotes the outward normal derivative of w on S_2. For simplicity, we consider (1) with $\Psi \equiv 0$ by the change in variable $u = w - w_1$, where w_1 is the unique harmonic function satisfying $\sigma\, \partial w_1/\partial n|_{S_2} = \Psi$ on S_2, and $w_1|_{S_1} \equiv 0$. Then,

$$\begin{aligned} \sigma\Delta u &= 0, \text{ in } \Omega, \\ u &= V, \text{ on } S_2, \\ \sigma\frac{\partial u}{\partial n} &= 0, \text{ on } S_2, \end{aligned} \tag{2}$$

where $V = \Phi - w_1|_{S_2} \in \mathrm{H}^{1/2}(S_2)$. For the analysis of the Cauchy problem (2), the following problem is employed (see [15]):

Given a function φ defined onS_1, find u such that

$$\begin{aligned} \sigma\Delta u &= 0, \text{ in } \Omega, \\ u &= \varphi, \text{ on } S_1, \\ \sigma\frac{\partial u}{\partial n} &= 0, \text{ on } S_2. \end{aligned} \tag{3}$$

This problem is well-posed, and we will call it the *auxiliary problem.*

The inverse problem associated with the Cauchy problem can be formulated in the following way:

Recover the potential $u = \varphi$ on S_1 from the measurements $u = V$ on S_2, where u is the solution to the auxiliary problem (3).

Definition 1. *A function $u \in V_\varphi$ is a weak solution to the auxiliary problem (3) if*

$$\int_\Omega \sigma\nabla u\cdot\nabla v \, d\Omega = 0, \quad \textit{for all} \quad v \in V_0, \tag{4}$$

where

$$V_\varphi = \left\{v \in H^1(\Omega) : v = \varphi \; on \; S_1\right\}, \tag{5}$$

$$V_0 = \left\{v \in H^1(\Omega) : v = 0 \; on \; S_1\right\}. \tag{6}$$

Theorem 1 given in [1] guarantees the existence and uniqueness of the weak solution and allows us to define the lineal, injective, and compact operator $K : H^{m-1/2}(S_1) \to L_2(S_2)$ that associates to each $\varphi \in H^{m-1/2}(S_1)$ the trace over S_2 of the weak solution u to the auxiliary problem (3). Operator K is compact because it is the composition of the continuous operator $T : H^{m-1/2}(S_1) \to H^1(\Omega)$, which associates to each $\varphi \in H^{m-1/2}(S_1)$ the weak solution to the auxiliary problem (3), with the trace operator Tr from $H^m(\Omega)$ into $L_2(\partial\Omega)$, which is compact. The relationship between problem (2) and auxiliary problem (3) can be described by the operator K as follows:

A solution to the auxiliary problem (3) is also a solution to the problem (2) if we choose ϕ on S_1, such that

$$K\varphi := u(\varphi)|_{S_2} = V, \tag{7}$$

where $u(\varphi)$ denotes the solution to the auxiliary problem (3), and V is the known measurement in problem (2), so we have $\varphi = K^{-1}(V)$.

The following result is very important for the statement of the minimization problem presented in Section 3, and its demonstration can be found in [10].

Theorem 1. *$Im(K)$ is dense in $L_2(S_2)$.*

Equation (7) does not have a solution for all $V \in L_2(S_2)$. However, if we impose some smoothness conditions on V, we can find global conditions of the existence of the solution, as in [10]. As K is an injective and well-defined [15] operator, it ensures uniqueness when a solution is available. Since the operator K is linear, injective, and compact, its inverse K^{-1} is not continuous. Therefore, the inverse problem is ill-posed due to its numerical instability.

2.2. Fractional Boundary Operator

The following material has been obtained from [24]. Let $\Omega_1 = \{x : |x| < 1\}$ be a unit ball, $2 \le n$. The $\partial\Omega$ corresponds with the unit sphere; $r = |x|$, $\overline{x} = \frac{x}{r}$, let $\delta = r\frac{d}{dr}$ be a Dirac operator, where $r\frac{d}{dr} = \sum_{j=1}^{n} x_j \frac{\partial}{\partial x_j}$. Let $u(x)$ be a smooth function on the domain $\overline{\Omega}$. For any $\alpha > 0$, the following expression

$$J^\beta[u](x) = \frac{1}{\Gamma(\beta)} \int_0^r \left(ln\frac{r}{s}\right)^{\beta-1} \frac{u(s\overline{x})}{s} ds, \quad x \in \Omega_1 \tag{8}$$

is called an operator of integration of the order β in the Hadamard sense. Furthermore, we will assume that $J^0[u](x) = u(x), x \in \Omega$.

We consider the following modification of the Hadamard operator:

$$D_m^\beta[u](x) = J^{m-\beta}[\delta^m u](x) = \frac{1}{\Gamma(m-\beta)} \int_0^r \left(ln\frac{r}{s}\right)^{m-1-\beta} \left(s\frac{d}{ds}\right)^m \frac{u(s\overline{x})}{s} ds, \tag{9}$$

where m is a positive integer.

Properties and applications of the operators J^β y D^β have been studied in [24]. In that paper, the authors studied a certain generalization of the classical Neumann problem with the fractional order of boundary operators. Let $0 < \beta_n < \ldots < \beta_1 < \beta \le 1$,

$P_N(D) = D_m^{\beta} + \sum_{j=1}^{N} a_j D_m^{\beta j}$. In the domain Ω_1, the authors consider the following problem:

$$\Delta u(x) = 0, \quad x \in \Omega_1, \tag{10}$$

$$P_N(D)u(x) = f(x), \quad x \in \partial\Omega_1. \tag{11}$$

As a solution to the last problem, the authors consider a function $u \in C^2(\Omega_1) \cap C(\overline{\Omega_1})$ satisfying Equation (10) and the boundary condition (11) in a classical sense. Since $J^0[u](x) = u(x)$, then $D^1[u](x) \mid_{\partial\Omega_1} = r\frac{du}{dr} \mid_{\partial\Omega_1} = \frac{du}{dn} \mid_{\partial\Omega_1}$, where n is a normal vector to the boundary of the domain Ω. Therefore, in the case $\beta = 1$ and $a_j = 0$, $j = 1, \ldots, n$, we obtain the classical Neumann problem.

2.3. Fractional Cauchy Problem

We consider the following fractional Cauchy problem

$$\begin{aligned} \Delta u &= 0, \quad \text{in} \quad \Omega, \\ D_m^{\beta} u\Big|_{S_2} &= V, \ \text{on} \quad S_2, \\ \frac{\partial u}{\partial n} &= 0, \ \text{on} \quad S_2, \end{aligned} \tag{12}$$

where the operator D_m^{β} is given in (9).

In this case, the operator $D_m^{\beta} u(x) = \frac{1}{\Gamma(m-\beta)} \int_{R_1}^{r} (ln\frac{r}{s})^{m-1-\beta} \left(s\frac{d}{ds}\right)^m \frac{u(s\overline{x})}{s} ds$. We note that the operator is linear and continuous, and it has no singularities since $R_1 \leq r, s \leq R_2$. For the analysis of the fractional Cauchy problem (12), we also consider the auxiliary problem (3). We define the operator $K_m^{\beta}(\varphi) = \left(Tr \circ \left(D_m^{\beta} T\right)\right)(\varphi) = D_m^{\beta} u|_{S_2}$, which is a compact operator. We have the following two definitions to study the problem that concerns us.

Definition 2. *The Forward Problem (FP) related to the fractional Cauchy problem consists of finding the potential $V = K_m^{\beta}(\varphi)$ when ϕ is known.*

We can consider other fractional Cauchy problems by changing the boundary operator. We can consider different kernels for the integral operator. For example, we can take the kernel of the Riemann–Liouville and Caputo fractional derivative, which can be found in [25].

Definition 3. *Given $V \in L_2(S_2)$, the Inverse Problem (IP) related to the fractional Cauchy problem consists of finding $\varphi \in L_2(S_1)$ such that $K_m^{\beta}(\varphi) = V$.*

3. Methods

3.1. Tikhonov Regularization of the Fractional Cauchy Problem

To find an approximate solution $\varphi \in L_2(S_1)$ of Equation (7) for $K = K_m^{\beta}$ when we have a measurement with error V_δ, the minimization of the following Tikhonov functional is proposed in [26]:

$$J_\alpha(\varphi) := \frac{1}{2} \| K_m^{\beta} \varphi - V_\delta \|_{L_2(S_2)}^2 + \frac{\alpha}{2} \| \varphi \|^2, \quad \forall \varphi \in L_2(S_1), \tag{13}$$

where α is the Tikhonov regularization parameter, which will be chosen by the L-curve method, Morozov's discrepancy principle, and numerical tests. The first and second Fréchet derivatives are given by (see Appendix A):

$$D J_\alpha(\varphi) = \left(\left(K_m^\beta\right)^* K_m^\beta + \alpha I \right)\varphi - \left(K_m^\beta\right)^* V_\delta,$$
$$D^2 J_\alpha(\varphi) = \left(K_m^\beta\right)^* K_m^\beta + \alpha I$$

This least squares procedure is equivalent to solving the normal equation

$$\left(\left(K_m^\beta\right)^* K_m^\beta + \alpha I \right) \varphi = \left(K_m^\beta\right)^* V, \tag{14}$$

where $\left(K_m^\beta\right)^*$ is the adjoint operator.

According to Theorem 2.12 given in [26], the operator $\left(K_m^\beta\right)^* K_m^\beta + \alpha I$ is boundedly invertible. Given $\varphi \in L_2(S_1)$, the exact solution to the auxiliary problem (3) in a circular annular region $R_1 \leq r \leq R_2$, in polar coordinates, is given by

$$\begin{aligned} u(r,\theta) = \varphi_0 + \sum_{k=1}^{\infty} \frac{(R_2/R_1)^{2k}}{(R_2/R_1)^{2k}+1} \Bigg\{ & \left[\left(\frac{R_1}{R_2}\right)^k \left(\frac{r}{R_2}\right)^k + \left(\frac{R_1}{r}\right)^k \right] \varphi_k^1 \cos(k\theta) \\ & + \left[\left(\frac{R_1}{R_2}\right)^k \left(\frac{r}{R_2}\right)^k + \left(\frac{R_1}{r}\right)^k \right] \varphi_k^2 \sin(k\theta) \Bigg\}, \end{aligned} \tag{15}$$

where $0 \leq \theta < 2\pi$. The values φ_0, φ_k^1, φ_k^2, $k = 1, 2, \ldots$, are the Fourier coefficients of φ. The solution to the FP, called *measurement*, is given by $V = K_m^\beta \varphi$, which is obtained by applying the operator D_m^β, the identities

$$\left(s\tfrac{d}{ds}\right)^m \left(s^k \cos(k\theta)\right) = k^m s^k \cos(k\theta), \left(s\tfrac{d}{ds}\right)^m \left(s^{-k} \cos(k\theta)\right) = (-1)^m k^m s^{-k} \cos(k\theta),$$
$$\left(s\tfrac{d}{ds}\right)^m \left(s^k \sin(k\theta)\right) = k^m s^k \sin(k\theta), \left(s\tfrac{d}{ds}\right)^m \left(s^{-k} \sin(k\theta)\right) = (-1)^m k^m s^{-k} \sin(k\theta),$$

and then evaluating in $r = R_2$, i.e.,

$$\begin{aligned} V(\theta) = K_m^\beta \varphi(\theta) &= \left(Tr \circ \left(D_m^\beta T\right)\right)\varphi(\theta) = D_m^\beta u|_{S_2}(\theta) \\ &= \sum_{k=1}^{\infty} G_{m,k}^\beta \left(\varphi_k^1 \cos(k\theta) + \varphi_k^2 \sin(k\theta)\right), \end{aligned} \tag{16}$$

where the Fourier coefficients V_k^i of exact measurement V are given by $V_k^i = G_{m,k}^\beta \varphi_k^i$, for $i = 1, 2$, in which

$$G_{m,k}^\beta = \frac{k^m}{\Gamma(m-\beta)} \int_{R_1}^{R_2} \left(ln\frac{R_2}{s}\right)^{m-1-\beta} \left[\left(\frac{R_1}{R_2}\right)^k \frac{s^{k-1}}{R_2^k} + (-1)^m R_1^k s^{-k} \right] ds. \tag{17}$$

In the numerical examples, the integrals are calculated using the function *quadl* of MATLAB.

The '*exact solution*' u and the '*exact measurement*' $V = v|_{S_2}$ are generated taking $2N+1$ terms of the Fourier series (15) and (16), with $N = 15$, which is obtained from numerical tests. To find the solution to the IP, we must solve the normal equations. To do this, we calculate the adjoint operator using its definition:

$$\left\langle \left(K_m^\beta\right)^* W, \varphi \right\rangle_{L_2(S_1)} = \left\langle V, K_m^\beta \varphi \right\rangle_{L_2(S_2)}. \tag{18}$$

Without loss of generality, we consider functions in which the constant term of their series expansion is null. Using (16) and (18), we found

$$
\begin{aligned}
&\left\langle W, K_m^\beta \varphi \right\rangle_{L_2(S_2)} = \\
&\left\langle \sum_{j=1}^{\infty} W_j^1 \cos(j\theta) + W_j^2 \sin(j\theta), \sum_{k=1}^{\infty} G_{m,k}^\beta \varphi_k^1 \cos(k\theta) + G_{m,k}^\beta \varphi_k^2 \sin(k\theta) \right\rangle_{L_2(S_2)} \\
&= \left\langle \frac{R_2}{R_1} \sum_{j=1}^{\infty} G_{m,k}^\beta W_j^1 \cos(j\theta) + G_{m,k}^\beta W_j^2 \sin(j\theta), \sum_{k=1}^{\infty} \varphi_k^1 \cos(k\theta) + \varphi_k^2 \sin(k\theta) \right\rangle_{L_2(S_1)} \\
&= \left\langle \left(K_m^\beta\right)^* W, \varphi \right\rangle_{L_2(S_1)}.
\end{aligned} \tag{19}
$$

Thus, the adjoint operator is defined by $\left(K_m^\beta\right)^* : L_2(S_2) \to L_2(S_1)$,

$$
\left(K_m^\beta\right)^*(W) = \frac{R_2}{R_1} \sum_{k=1}^{\infty} G_{m,k}^\beta \Big(W_k^1 \cos(k\theta) + W_k^2 \sin(k\theta)\Big). \tag{20}
$$

After some calculations, the regularized solution $\varphi_{\alpha(\delta)}$ that minimizes the functional (13) or that solves normal Equation (14) is given by

$$
\varphi_{\alpha(\delta)}(\theta) = \sum_{k=1}^{\infty} \Big(\varphi_{k,\alpha(\delta)}^1 \cos(k\theta) + \varphi_{k,\alpha(\delta)}^2 \sin(k\theta)\Big), \text{ on } S_1, \tag{21}
$$

where

$$
\varphi_{k,\alpha(\delta)}^i = \frac{R_2 G_{m,k}^\beta}{\left(G_{m,k}^\beta\right)^2 R_2 + \alpha R_1} V_{k,\delta}^i, \text{ for } i = 1,2, \text{ and } k = 1,2,\ldots,N,
$$

and $V_{k,\delta}^i$ are the Fourier coefficients of measurement with error V_δ.

3.2. *Tikhonov Regularization for the Classical Cauchy Problem*

Given $\varphi \in L_2(S_1)$, the exact solution $u(r,\theta)$ to the auxiliary problem (3) in a circular annular region $R_1 \le r \le R_2$ is given, in polar coordinates, by (15). Therefore, the measurement $V = u|_{S_2}$ is obtained with $r = R_2$ in (15):

$$
V(\theta) = u(r,\theta)|_{r=R_2} = \varphi_0 + 2 \sum_{k=1}^{\infty} \frac{(R_2/R_1)^k}{(R_2/R_1)^{2k}+1} \Big\{\varphi_k^1 \cos(k\theta) + \varphi_k^2 \sin(k\theta)\Big\}, \tag{22}
$$

which is the solution to the FP. The Fourier coefficients of V are given by

$$
V_0 = \varphi_0 \text{ and } = V_k^i = \frac{2(R_2/R_1)^k}{(R_2/R_1)^{2k}+1} \varphi_k^i, \text{ for } i = 1,2.
$$

Therefore, the solution to the IP from the measurement with error

$$
V_\delta(\theta) = V_{0,\delta} + \sum_{k=1}^{\infty} \Big(V_{k,\delta}^1 \cos(k\theta) + V_{k,\delta}^2 \sin(k\theta)\Big) \text{ on } S_1, \tag{23}
$$

is given by the regularized solution

$$
\varphi_{\alpha(\delta)}(\theta) = \varphi_{0,\alpha(\delta)} + \sum_{k=1}^{\infty} \Big(\varphi_{k,\alpha(\delta)}^1 \cos(k\theta) + \varphi_{k,\alpha(\delta)}^2 \sin(k\theta)\Big) \text{ on } S_1, \tag{24}
$$

where

$$\varphi^i_{k,\alpha(\delta)} = \frac{\left(\frac{R_2}{R_1}\right)^k\left[1+\left(\frac{R_2}{R_1}\right)^{2k}\right]}{2\left(\frac{R_2}{R_1}\right)^{2k}+\frac{\alpha R_1}{2R_2}\left[1+\left(\frac{R_2}{R_1}\right)^{2k}\right]^2} V^i_{k,\delta},\ i = 1,2,\ k = 1,2,3,\ldots \tag{25}$$

where $V^i_{k,\delta}$ are the Fourier coefficients of V_δ and α is the Tikhonov regularization parameter. Thus, the solution to the IP (of the classical Cauchy problem) applying the Tikhonov regularization method (TRM) is given by (15), replacing the coefficients φ^i_k by the coefficients $\varphi^i_{k,\alpha(\delta)}$ given by (25).

4. Numerical Results

In this section, we illustrate the method proposed in this work using synthetic examples. We know the exact φ defined on S_1 in this case. Then, we calculated the measurement with and without noise by solving the FP for the classical and fractional Cauchy problem.

The exact measurement is calculated by solving the FP. To generate the measurements with error V_δ, we added to the exact measurement a Gaussian error using the function *random* of MATLAB. The exact measurement was calculated by solving the FP. Therefore, we define

$$V_\delta = V + E, \tag{26}$$

where $E = random(\text{'Normal'}, \mu_0, \sigma_0, 1, m)$ is a vector of random numbers of length m (numbers of nodes on S_2) with a normal distribution. The corresponding numerical solutions are denoted by $\varphi_{\alpha(\delta)}$.

In this section, we obtain the relative error between the exact source φ and the recovered source $\varphi_{\alpha(\delta)}$ shown in tables and denoted by $RE\left(\varphi, \varphi_{\alpha(\delta)}\right)$. The relative error is given by

$$RE\left(\varphi, \varphi_{\alpha(\delta)}\right) = \| \varphi - \varphi_{\alpha(\delta)} \|_{L_2(S_1)} / \| \varphi \|_{L_2(S_1)},$$

and the relative error between the exact measurement V and the measurement with error V_δ is denoted by $RE(V, V_\delta)$, which is given by

$$RE(V, V_\delta) = \| V - V_\delta \|_{L_2(S_2)} / | V \|_{L_2(S_2)},$$

where $\| \cdot \|_{L_2(S_i)}$ is the norm of the space $L_2(S_i)$, $i = 1, 2$.

4.1. Solution to the IP Related to the Classical Cauchy Problem

In the following two examples, we consider a circular annular region $R_1 \leq r \leq R_2$ with $R_1 = 1$ and $R_2 = 1.2$; then S_1 and S_2 are two circumferences of radii $R_1 = 1$ and $R_2 = 1.2$ (see Figure 1), respectively.

Example 1. *We take the 'exact potential' $\varphi(x,y) = x^2 - y^2$. $(x,y) \in S_1$, that in polar coordinates is $\varphi(\theta) = \cos(2\theta)$. In this case, $V_0 = \varphi_0 = 0$, and the solution to the forward problem, that is, the solution to the auxiliary problem (3), is given by*

$$V(\theta) = K\varphi(\theta) = u(r,\theta)|_{r = R_2} = \left[\frac{2\left(\frac{R_2}{R_1}\right)^2}{1+\left(\frac{R_2}{R_1}\right)^4}\right]\varphi^1_2 \cos(2\theta),\ \theta \in [0, 2\pi],$$

where $\varphi^1_2 = 1$. Then, the 'exact solution' V and the 'measurement with error' V_δ are generated with the first N terms of the Fourier series (22) and (23), respectively. In this case, we take values of N = 20, 25, and 30 terms. For smooth functions in the Cauchy data, these values of N are obtained by combining numerical tests and the following ideas:

$K(\varphi) = V$ is approximated by a truncation $K_N(\varphi) = V_N$ choosing N such that we can guarantee that $\| V - V_N \|^2_{L_2(S_2)} = \| K(\varphi) - K_N(\varphi) \|^2_{L_2(S_2)} < \frac{\delta^2}{4}$, where the approximation $K_N(g) = \sum_{k=1}^{N} \left[V_k^1 \cos k\theta + V_k^2 \sin k\theta \right]$. From the Parseval equality and (22), where we found that the Fourier coefficients of V decay at least as $1/k^2$, we infer

$$\| V - V_N \|^2_{L_2(S_2)} < 2\pi/N < \delta^2/4. \tag{27}$$

From this, we can take $N > \frac{8\pi}{\delta^2}$ for obtaining the inequality. With this, we obtain an error regarding the truncation of the series expansion.

Now, the measurement error is simulated by adding a random error to each Fourier coefficient $V_k^1, V_k^2, k = 1,2,\ldots,N$, such that $\| V_N - V_{\delta,N} \|^2_{L_2(S_2)} < \frac{\delta^2}{4}$, which guarantees that $\| V - V_{\delta,N} \|_{L_2(S_2)} \leq \delta$.

For other functions, such as absolute value and the jump function, we have to choose other values of N. This is shown in Examples 3 and 4, which are included in Section 4.2.6 .

Thus, we consider synthetic examples; that is, we examine the fundamental elements of the problems studied, such as how real data are generated and the inherent error within them. In this way, we attempt to emulate the characteristics of real-world problems so that our proposal is closely aligned with providing a solution to them.

Therefore, the measurement with error V_δ is given by the series

$$V_\delta(\theta) = \sum_{k=1}^{N} \left(V_{k,\delta}^1 \cos(k\theta) + V_{k,\delta}^2 \sin(k\theta) \right) \text{ on } S_1, \tag{28}$$

where $V_{k,\delta}^i$ are the Fourier coefficients of V_δ. The regularized solution $\varphi_{\alpha(\delta)}$ to the inverse problem is given by the series (24) truncated to N terms. The solution without regularization φ_δ to the IP is given by

$$\varphi_\delta(\theta) = \sum_{k=1}^{N} \left(\varphi_{k,\delta}^1 \cos(k\theta) + \varphi_{k,\delta}^2 \sin(k\theta) \right) \text{ on } S_1, \tag{29}$$

where the coefficients $\varphi_{k,\delta}^i$ are given by

$$\varphi_{k,\delta}^i = \frac{\left[1 + \left(\frac{R_2}{R_1} \right)^{2k} \right]}{2 \left(\frac{R_2}{R_1} \right)^k} V_{k,\delta}^i, \; i = 1,2. \tag{30}$$

Remark 1. *In all tables associated with the classical case, if $\alpha(\delta) = 0$, then the solution $\varphi_{\alpha(\delta)}$ is the solution without regularization φ_δ given by (29), where the coefficients $\varphi_{k,\delta}^i$ are given by (30).*

Table 1 shows the numerical results for data with and without error, applying TRM to solve the IP of the classical Cauchy problem (2). In this case, we observe that the solutions with regularization $\varphi_{\alpha(\delta)}$ have a percentage of relative errors around 10%, equal to the percentage of error included in the data with error V_δ for $\delta = 0.1$. The regularization parameter was chosen as $\alpha(\delta) = \delta$ for $N = 20$, 25, and 30. Also, we can see that the $RE\left(\varphi, \varphi_{\alpha(\delta)}\right)$ decreases when the error δ tends to zero, while the $RE(\varphi, \varphi_\delta)$ increases for each value of N. In particular, the $RE(\varphi, \varphi_\delta)$ increases faster when $N = 30$, for $\delta = 0.1$, 0.01, and y 0.001. In this case, the regularization parameter $\alpha(\delta)$ depends on V_δ.

Table 1. Numerical results applying TRM to solve the IP related to the classical Cauchy problem (2), for $\varphi(x,y) = x^2 - y^2, (x,y) \in S_1$, and different values of δ and N.

δ	N	$\alpha(\delta)$	$RE(V,V_\delta)$	$RE(\varphi,\varphi_{\alpha(\delta)})$	$RE(\varphi,\varphi_\delta)$
0	20	0	0	0	0
0.1	20	0.1	0.1069	0.1465	0.8224
0.1	25	0.1	0.1034	0.1471	1.6048
0.1	30	0.1	0.1125	0.1435	3.0063
0.01	20	0.01	0.0102	0.0342	0.0663
0.01	25	0.01	0.0112	0.0360	0.2137
0.01	30	0.01	0.0114	0.0362	0.4820
0.001	20	0.001	9.0770×10^{-4}	0.0050	0.0057
0.001	25	0.001	8.8517×10^{-4}	0.0068	0.0122
0.001	30	0.001	9.5479×10^{-4}	0.0083	0.0354

Figure 2a,b show the graphs of the exact measurement V and with error V_δ, the graphs of the exact potential φ and its approximations $\varphi_{\alpha(\delta)}$ (with regularization) and φ_δ (without regularization) taking $\alpha(\delta) = 0.1$ and $N = 30$, corresponding to Example 1, for $\delta = 0.1$ (see Table 1). In Figure 2b, we can see the ill-posedness of the inverse problem if we do not apply regularization, where $RE\left(\varphi, \varphi_{\alpha(\delta)}\right) = 0.1435$ and $RE(\varphi, \varphi_\delta) = 3.0063$.

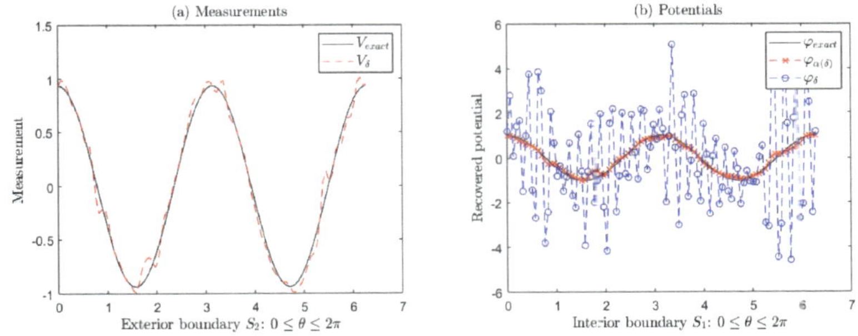

Figure 2. (**a**) Exact measurement V (black line) and with error V_δ (red line). (**b**) Exact potential φ and its approximations $\varphi_{\alpha(\delta)}$ and φ_δ applying regularization and without regularization, corresponding to Example 1 for $\delta = 0.1$ (see Table 1). We take $\alpha(\delta) = 0.1$ and $N = 30$ in this case.

Example 2. *We consider the 'exact potential' $\varphi(x,y) = e^x \sin(y)$, for $(x,y) \in S_1$. Similar to the first example, the 'exact measurement' V and the 'measurement with error' V_δ are generated with the first N terms of the Fourier series (22) and (23), respectively, such that $\| \varphi_N - \varphi \|_{L_2(S_1)} \le \epsilon_F$, with $0 < \epsilon_F < 10^{-14}$. In this case, $V_0 = \varphi_0 = 0$ and the Fourier coefficients φ_k, $1 \le k \le N$, are obtained numerically using the intrinsic function quadl of MATLAB. Here, we take values of $N = 18$, 25, and 30 terms.*

Table 2 shows the numerical results for data with and without error, applying TRM to solve the IP of the classical Cauchy problem (2). Analogous to Example 1, we can observe that the solutions with regularization $\varphi_{\alpha(\delta)}$ have a percentage of relative errors around 10%, equal to the percentage of error included in the data with error V_δ for $\delta = 0.1$. Also, we can see that the $RE\left(\varphi, \varphi_{\alpha(\delta)}\right)$ decreases when the error δ tends to zero, while the $RE(\varphi, \varphi_\delta)$ increases for each value of N. In particular, the $RE(\varphi, \varphi_\delta)$ increases when $N = 30$ for each

$\delta = 0.1, 0.01$, and 0.001. As in the previous example, the regularization parameter $\alpha(\delta)$ depends on V_δ, and we take $\alpha(\delta) = \delta$ for each value of $N = 18, 25$, and 30.

Table 2. Numerical results applying TRM to solve the IP related to the classical Cauchy problem (2), for $\varphi(x,y) = e^x \sin(y)$, $(x,y) \in S_1$ and different values of δ and N.

δ	N	$\alpha(\delta)$	$RE(V,V_\delta)$	$RE(\varphi,\varphi_{\alpha(\delta)})$	$RE(\varphi,\varphi_\delta)$
0	18	0	0	2.5338×10^{-17}	2.4537×10^{-17}
0.1	18	0.1	0.1161	0.1590	0.8005
0.1	25	0.1	0.1125	0.1571	1.6366
0.1	30	0.1	0.1340	0.1512	5.7825
0.01	18	0.01	0.0117	0.0357	0.0509
0.01	25	0.01	0.0105	0.0392	0.1485
0.01	30	0.01	0.0112	0.0364	0.3478
0.001	18	0.001	8.8274×10^{-4}	0.0046	0.0048
0.001	25	0.001	9.9199×10^{-4}	0.0095	0.0162
0.001	30	0.001	0.0011	0.0099	0.0423

Figure 3a,b show the graphs of the exact measurement V and with error V_δ, the graphs of the exact potential φ and its approximations $\varphi_{\alpha(\delta)}$ (with regularization) and φ_δ (without regularization) taking $\alpha(\delta) = 0.1$ and $N = 30$, corresponding to Example 2, for $\delta = 0.1$ (see Table 2). In Figure 3b, we can see the ill-posedness of the inverse problem if we do not apply regularization. In this case, $RE\left(\varphi, \varphi_{\alpha(\delta)}\right) = 0.1512$ and $RE(\varphi, \varphi_\delta) = 5.7825$.

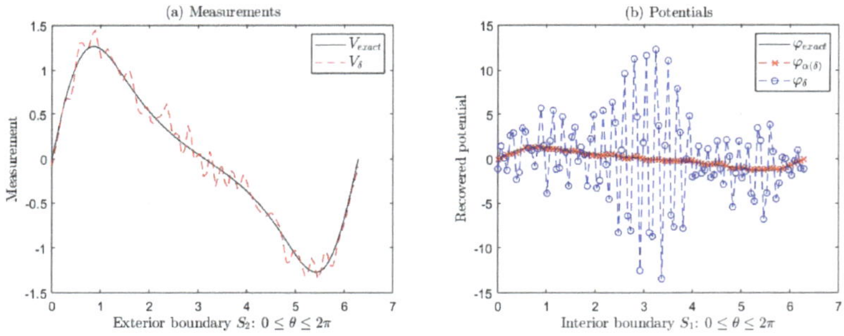

Figure 3. (**a**) Exact measurement V (black line) and with error V_δ (red line). (**b**) Exact potential φ and its approximations $\varphi_{\alpha(\delta)}$ and φ_δ applying regularization and without regularization, corresponding to Example 2 for $\delta = 0.1$ (see Table 2). In this case, we take $\alpha(\delta) = 0.1$ and $N = 30$.

4.2. Solution to the IP Related to the Fractional Cauchy Problem

In this section, we look into the performance of the TRM to solve the IP of the fractional Cauchy problem (12) in a circular annular region $R_1 \le r \le R_2$ with $R_1 = 1$ and $R_2 = 1.2$. Then, S_1 and S_2 are two circumferences of radii $R_1 = 1$ and $R_2 = 1.2$ (see Figure 1), respectively. In this case, we consider as '*exact potentials*' the two functions from the previous subsection: $\varphi(x,y) = x^2 - y^2$ and $\varphi(x,y) = e^x \sin(y)$, for $(x,y) \in S_1$.

Similar to the previous subsection, the '*exact solution*' V and the '*measurement with error*' V_δ are obtained by truncating the series (16) and (23) up to N terms, respectively; furthermore, the Fourier coefficients φ_k^1, φ_k^2, and $G_{m,k}^\beta$ (given by (17)) are obtained numerically using the function *quadl* of *MATLAB*.

In this case, we take values of $N = 18, 20, 25$, and 30 terms. Therefore, the measurement with error V_δ is given by the series (23) truncated to N terms. The regularized solution

$\varphi_{\alpha(\delta)}$ to the IP is given by the series (21) truncated to N terms. Also, the solution without regularization φ_δ to the IP is given by (29), where

$$\varphi^i_{k,\delta} = \frac{1}{G^\beta_{m,k}} V^i_{k,\delta}, \; i = 1,2, \; k = 1,2,\ldots,N. \tag{31}$$

Remark 2. *In all tables from the fractional case, if $\alpha(\delta) = 0$, the solution $\varphi_{\alpha(\delta)}$ is the solution without regularization φ_δ given by (29), where the coefficients $\varphi^i_{k,\delta}$ are given by (31).*

4.2.1. Case 1: $\beta = 0.5$ and $m = 1$, When δ Tends to Zero

In this section, we consider the case when $\beta = 0.5, m = 1$, and for different values of δ close to zero. Tables 3 and 4 show the relative errors of the approximations $\varphi_{\alpha(\delta)}$ and φ_δ, when δ tends to zero, for the two exact functions φ considered in Section 4.1. In both cases, we observe that the $RE\left(\varphi, \varphi_{\alpha(\delta)}\right)$ of the solutions with regularization $\varphi_{\alpha(\delta)}$ is less than the $RE(V, V_\delta)$ for each value of δ and N given in these tables. Additionally, the $RE\left(\varphi, \varphi_{\alpha(\delta)}\right)$ and $RE(\varphi, \varphi_\delta)$ are of the same order, i.e., the solutions without regularization φ_δ are close to regularized solutions $\varphi_{\alpha(\delta)}$ for $\beta = 0.5$ and $m = 1$. In both cases, the measurements with errors V_δ do not have much impact on recovered solution φ_δ, and they are close to $\varphi_{\alpha(\delta)}$. We observe from the relative errors that regularized approximations $\varphi_{\alpha(\delta)}$ are better than those without regularization. In this case, the regularization parameter $\alpha(\delta)$ depends on V_δ, N, m, and β.

Table 3. Numerical results applying TRM to solve IP related to the fractional Cauchy problem (12), for $\varphi(x,y) = x^2 - y^2$, $(x,y) \in S_1$, and different values of δ and N.

δ	N	$\alpha(\delta)$	β	m	$RE(V,V_\delta)$	$RE(\varphi,\varphi_{\alpha(\delta)})$	$RE(\varphi,\varphi_\delta)$
0	20	0	0.5	1	0	1.1102×10^{-16}	0
0.1	20	1×10^{-2}	0.5	1	0.0970	0.0617	0.0880
0.1	25	1×10^{-2}	0.5	1	0.0979	0.0719	0.0990
0.1	30	1×10^{-2}	0.5	1	0.1038	0.0775	0.1217
0.01	20	1×10^{-4}	0.5	1	0.0093	0.0037	0.0039
0.01	25	1×10^{-4}	0.5	1	0.0091	0.0070	0.0070
0.01	30	1×10^{-4}	0.5	1	0.0127	0.0072	0.0073
0.001	20	1×10^{-6}	0.5	1	9.6950×10^{-4}	8.9805×10^{-4}	8.9826×10^{-4}
0.001	25	1×10^{-6}	0.5	1	8.3006×10^{-4}	3.4557×10^{-4}	3.4620×10^{-4}
0.001	30	1×10^{-6}	0.5	1	9.8327×10^{-4}	5.4898×10^{-4}	5.4925×10^{-4}

Considering $\delta = 0.1, \beta = 0.5$, and $m = 1$, we show the graphs for the following potentials $\varphi(x,y) = x^2 - y^2$ (Figure 4) and $\varphi(x,y) = e^x \sin(y)$ (Figure 5) for $(x,y) \in S_1$ where the following is true:

(a) The exact measurement V and the measurement with error V_δ.

(b) The exact potential φ and its approximations $\varphi_{\alpha(\delta)}$ (with regularization) and φ_δ (without regularization) taking $\alpha(\delta) = 10^{-2}$ and $N = 30$.

Table 4. Numerical results applying TRM to solve IP related to the fractional Cauchy problem (12), for $\varphi(x,y) = e^x \sin(y)$, $(x,y) \in S_1$, and different values of δ and N.

δ	N	$\alpha(\delta)$	β	m	$RE(V,V_\delta)$	$RE(\varphi,\varphi_{\alpha(\delta)})$	$RE(\varphi,\varphi_\delta)$
0	18	0	0.5	1	0	2.4537×10^{-17}	2.4537×10^{-17}
0.1	18	1×10^{-4}	0.5	1	0.1292	0.1196	0.1212
0.1	25	1×10^{-4}	0.5	1	0.1732	0.1238	0.1296
0.1	30	1×10^{-4}	0.5	1	0.1921	0.0624	0.0672
0.01	18	1×10^{-5}	0.5	1	0.0147	0.0126	0.0129
0.01	25	1×10^{-5}	0.5	1	0.0166	0.0124	0.0129
0.01	30	1×10^{-5}	0.5	1	0.0177	0.0069	0.0071
0.001	18	1×10^{-6}	0.5	1	0.0012	8.6623×10^{-4}	9.1811×10^{-4}
0.001	25	1×10^{-6}	0.5	1	0.0017	8.6801×10^{-4}	9.1904×10^{-4}
0.001	25	1×10^{-6}	0.5	1	0.0018	7.3211×10^{-4}	7.7961×10^{-4}

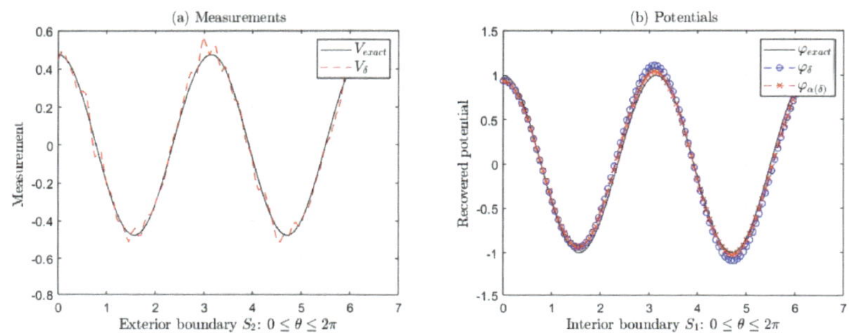

Figure 4. (**a**) Exact measurement V (black line) and with error V_δ (red line). (**b**) Exact potential φ and its approximations $\varphi_{\alpha(\delta)}$ and φ_δ, corresponding to Example 1 for $\delta = 0.1$ (see Table 3). In this case, we take $\alpha(\delta) = 10^{-2}$ and $N = 30$.

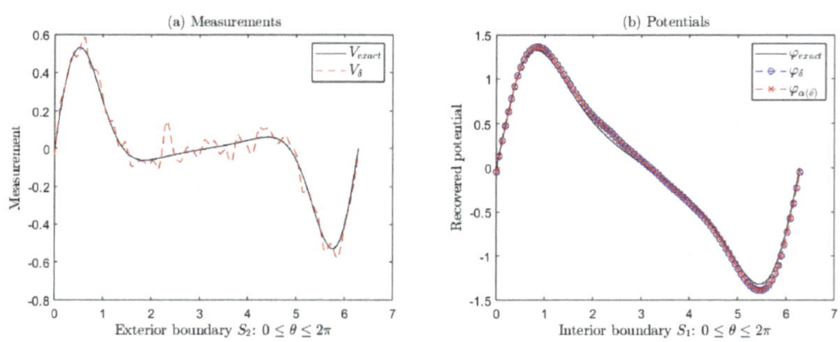

Figure 5. (**a**) Exact measurement V (black line) and with error V_δ (red line). (**b**) Exact potential φ and its approximations $\varphi_{\alpha(\delta)}$ and φ_δ, corresponding to Example 2 for $\delta = 0.1$ (see Table 4). In this case, we take $\alpha(\delta) = 10^{-4}$ and $N = 30$.

See Tables 3 and 4 for other values of the parameters N and δ.

In Figure 4b, the $RE\left(\varphi, \varphi_{\alpha(\delta)}\right) = 0.0775$ is less than $RE(\varphi, \varphi_\delta) = 0.1217$ for $\delta = 0.1$ (see Table 3). In Figure 5b, the $RE\left(\varphi, \varphi_{\alpha(\delta)}\right) = 0.0624$ is less than $RE(\varphi, \varphi_\delta) = 0.0672$ for $\delta = 0.1$ (see Table 4).

4.2.2. Case 2: $\beta \in (m-1, m)$, for $m = 2, \ldots, 12$ and $\delta = 0.1$

Tables 5 and 6 show the relative errors of the approximations $\varphi_{\alpha(\delta)}$ and φ_δ when $\delta = 0.1$ for the two exact functions φ considered in Section 4.1.

Table 5. Case 2: Numerical results applying TRM to solve IP related to the fractional Cauchy problem (12) for $\delta = 0.1$ and different values of β and m, where $\varphi(x,y) = x^2 - y^2$, $(x,y) \in S_1$.

N	$\alpha(\delta)$	β	m	$RE(V,V_\delta)$	$RE(\varphi,\varphi_{\alpha(\delta)})$	$RE(\varphi,\varphi_\delta)$
20	1×10^{-1}	1.5	2	0.1027	0.0360	0.0392
25	8×10^{-1}	1.5	2	0.1017	0.0518	0.0819
30	5×10^{-1}	1.5	2	0.1136	0.0597	0.0718
20	2×10^{-1}	2.5	3	0.0936	0.0508	0.4857
20	1×10^{1}	3.5	4	0.0969	0.0474	0.3523
20	1×10^{0}	4.5	5	0.0915	0.0421	1.5924
20	1×10^{2}	5.5	6	0.1112	0.0248	1.2606
20	1×10^{1}	6.5	7	0.1078	0.0366	5.8758
20	1×10^{2}	7.5	8	0.1045	0.0351	2.2946
20	1×10^{2}	8.5	9	0.0916	0.0197	16.6149
20	1×10^{3}	9.5	10	0.0996	0.0274	21.0258
20	1×10^{3}	10.5	11	0.0939	0.0176	46.6720
20	1×10^{3}	11.5	12	0.0861	0.0407	33.8060

Table 6. Case 2: Numerical results applying TRM to solve IP related to the fractional Cauchy problem (12) for $\delta = 0.1$ and different values of β and m, where $\varphi(x,y) = e^x \sin(y)$, $(x,y) \in S_1$.

N	$\alpha(\delta)$	β	m	$RE(V,V_\delta)$	$RE(\varphi,\varphi_{\alpha(\delta)})$	$RE(\varphi,\varphi_\delta)$
18	1×10^{-2}	1.5	2	0.1279	0.0342	0.0411
25	1×10^{-2}	1.5	2	0.1780	0.0258	0.0264
30	1×10^{-2}	1.5	2	0.1732	0.0439	0.0509
18	9×10^{-4}	2.5	3	0.1629	0.1437	0.1504
18	4×10^{-1}	3.5	4	0.1461	0.0931	0.3543
18	1.4×10^{-1}	4.5	5	0.1466	0.1385	7.4132
18	1×10^{1}	5.5	6	0.1757	0.3306	9.1616
18	1×10^{1}	6.5	7	0.1512	0.3654	478.6290
18	4.2×10^{2}	7.5	8	0.1643	0.3822	310.6255
18	3×10^{5}	8.5	9	0.1732	0.8982	633.1135
18	2×10^{7}	9.5	10	0.2062	0.9546	4.0208×10^{3}
18	2×10^{7}	10.5	11	0.1712	0.8948	1.9672×10^{5}
18	8×10^{8}	11.5	12	0.1636	0.9326	2.2607×10^{5}

In Table 5, we observe that $RE\left(\varphi, \varphi_{\alpha(\delta)}\right) < RE(V, V_\delta)$ for each value of N, β, and m given in the mentioned table. Also, the $RE\left(\varphi, \varphi_{\alpha(\delta)}\right)$ and $RE(\varphi, \varphi_\delta)$ are of the same order, i.e., the solutions without regularization φ_δ are close to regularized solutions $\varphi_{\alpha(\delta)}$, for $\delta = 0.1$, $N = 20, 25, 30$, $\beta = 1.5$, and $m = 2$. We can see similar results in Table 6 for $\delta = 0.1$, $N = 18, 25, 30$, $\beta = 1.5, 2.5$, $m = 2$, and 3; however the regularized approximates $\varphi_{\alpha(\delta)}$ are better than the solutions without regularization. Furthermore, $RE(\varphi, \varphi_\delta) < RE(V, V_\delta)$ and these increase suddenly, starting at $m = 3$ and $m = 4$ (see Tables 5 and 6) for the functions $\varphi(x,y) = x^2 - y^2$ and $\varphi(x,y) = e^x \sin(y)$ for $(x,y) \in S_1$, respectively. As in the previous case, the regularization parameter $\alpha(\delta)$ changes depending on V_δ, N, m, and β.

We show the graphs for the following functions:

- $\varphi(x,y) = x^2 - y^2$ (Figure 6 with parameters $\beta = 1.5, m = 2, \delta = 0.1, \alpha(\delta) = 5 \times 10^{-1}$, and $N = 30$; and Figure 7 with parameters $\beta = 2.5$, $m = 3$, $\delta = 0.1$, $\alpha(\delta) = 2 \times 10^{-1}$, and $N = 20$).
- $\varphi(x,y) = e^x \sin(y)$ (Figure 8 with parameters $\beta = 7.5$, $m = 8$, $\delta = 0.1$, $\alpha(\delta) = 4.2 \times 10^2$, and $N = 18$).

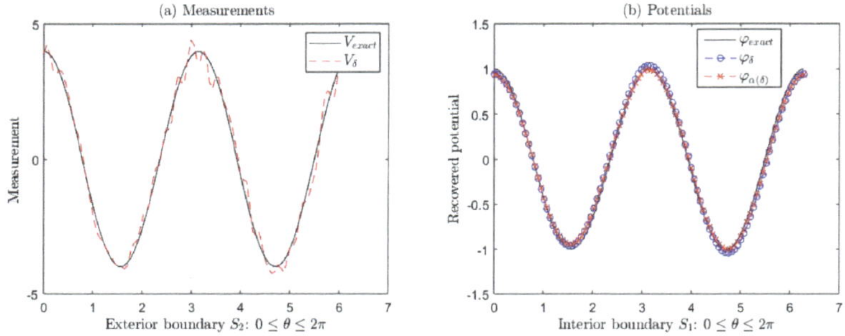

Figure 6. (**a**) Exact measurement V (black line) and with error V_δ (red line). (**b**) Exact potential φ and its approximations $\varphi_{\alpha(\delta)}$ and φ_δ, corresponding to Example 1 for $\beta = 1.5, m = 2$, and $\delta = 0.1$ (see Table 5). In this case, we take $\alpha(\delta) = 5 \times 10^{-1}$ and $N = 30$.

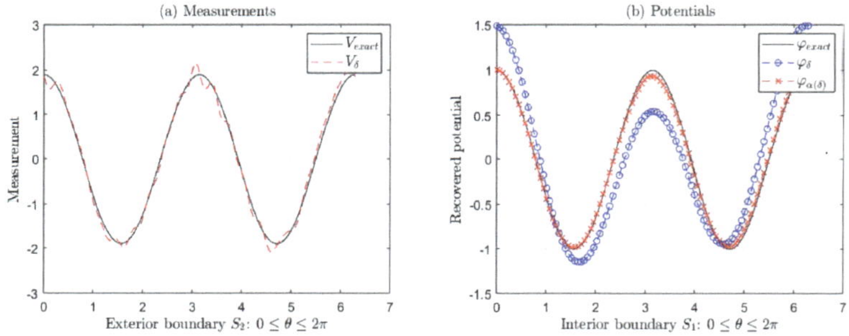

Figure 7. (**a**) Exact measurement V (black line) and with error V_δ (red line). (**b**) Exact potential φ and its approximations $\varphi_{\alpha(\delta)}$ and φ_δ, corresponding to Example 1 for $\beta = 2.5, m = 3$, and $\delta = 0.1$ (see Table 5). In this case, we take $\alpha(\delta) = 2 \times 10^{-1}$ and $N = 20$.

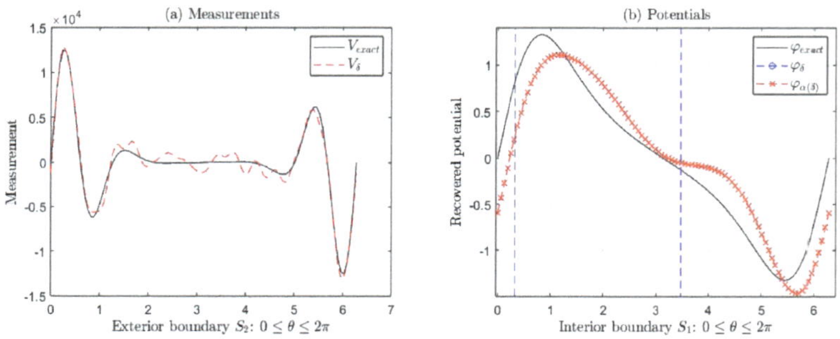

Figure 8. (**a**) Exact measurement V (black line) and with error V_δ (red line). (**b**) Exact potential φ and its approximations $\varphi_{\alpha(\delta)}$ and φ_δ, corresponding to Example 2 for $\beta = 7.5, m = 8$, and $\delta = 0.1$ (see Table 6). In this case, we take $\alpha(\delta) = 4.2 \times 10^2$ and $N = 18$.

For $(x,y) \in S_1$. These figures show the following:

(a) The exact measurement V and the measurement with error V_δ.

(b) The exact potential φ and its approximations $\varphi_{\alpha(\delta)}$ (with regularization) and φ_δ (without regularization).

In both cases, as mentioned in the previous paragraph, the errors increase suddenly, starting at $m = 3$ for the first function and $m = 4$ for the second one, as can be seen in Figures 7b and 8b, where we can see the ill-posedness of the IP if we do not apply regularization. For example, for the second function, the $RE\left(\varphi, \varphi_{\alpha(\delta)}\right) = 310.6255$ is much greater than $RE(\varphi, \varphi_\delta) = 0.3822$ for $\beta = 7.5$, $m = 8$, and $\delta = 0.1$ (see Table 6). However, in this same example, for $m = 9, 10, 11$, and 12, the $RE\left(\varphi, \varphi_{\alpha(\delta)}\right)$ increases around 90%. Nevertheless, $RE(\varphi, \varphi_\delta)$ is bigger than $RE\left(\varphi, \varphi_{\alpha(\delta)}\right)$. In this case, we could use the regularized solution as an initial point of an iterative method to recover a better solution to the IP.

4.2.3. Case 3: $\beta \in (m-1, m)$, When β Is Next to $m-1$ or m and $\delta = 0.1$

This section considers the case when $m - 1 < \beta < m$ and β is next to $m - 1$ or m. Tables 7 and 8 show the relative errors of the approximations $\varphi_{\alpha(\delta)}$ and φ_δ when $\delta = 0.1$ for the same two exact functions φ considered in Section 4.1.

Table 7. Case 3: Numerical results applying TRM to solve the IP related to the fractional Cauchy problem (12) for $\delta = 0.1$ and different values of β and m, where $\varphi(x, y) = x^2 - y^2$, $(x, y) \in S_1$.

N	$\alpha(\delta)$	β	m	$RE(V,V_\delta)$	$RE(\varphi,\varphi_{\alpha(\delta)})$	$RE(\varphi,\varphi_\delta)$
20	1×10^{-3}	0.1	1	0.1093	0.0565	0.0765
20	1×10^{-5}	0.0001	1	0.1032	0.0699	0.0705
20	1×10^{-12}	0.0000001	1	0.1040	0.0394	0.0394
20	1×10^{-3}	0.9	1	0.0922	0.0809	0.0817
20	1×10^{-8}	0.9999	1	0.0988	0.0581	0.0583
20	1×10^{-12}	0.9999999	1	0.0967	0.0921	0.1277
20	1×10^{-1}	1.1	2	0.1081	0.0563	0.0647
20	5×10^{-2}	1.0001	2	0.0931	0.0791	0.0984
20	1×10^{-2}	1.0000001	2	0.0979	0.0901	0.0953
20	1×10^{-1}	1.9	2	0.0980	0.0726	0.0746
20	1×10^{-5}	1.9999	2	0.0971	0.0449	0.0488
20	1×10^{-11}	1.9999999	2	0.1033	0.0997	0.1105
20	1×10^{-2}	2.1	3	0.0998	0.0392	0.2108
20	1×10^{-2}	2.0001	3	0.0905	0.0479	0.3250
20	1×10^{-2}	2.0000001	3	0.0894	0.0264	0.1021
20	1×10^{-1}	2.9	3	0.1110	0.0860	0.1731
20	1×10^{-5}	2.9999	3	0.0999	0.0804	0.4756
20	1×10^{-11}	2.9999999	3	0.1027	0.0672	0.3546
20	1×10^{0}	3.1	4	0.1061	0.0328	0.1583
20	1×10^{2}	3.9	4	0.1061	0.0738	0.5624
20	1×10^{-1}	4.1	5	0.0863	0.0421	1.8856
20	1×10^{0}	4.9	5	0.1030	0.0401	0.2528
20	1×10^{1}	5.1	6	0.0751	0.0368	1.5197
20	1×10^{2}	5.9	6	0.0957	0.0438	1.0323
20	1×10^{0}	6.1	7	0.0917	0.0267	5.3155
20	1×10^{0}	6.9	7	0.1020	0.0600	5.5627
20	1×10^{2}	7.1	8	0.0935	0.0199	3.3483
20	1×10^{2}	7.0000001	8	0.0827	0.0206	1.7258
20	1×10^{3}	7.9	8	0.0870	0.0212	4.5019
20	1×10^{-8}	7.9999999	8	0.0936	0.0454	4.2454

Table 7. *Cont.*

N	$\alpha(\delta)$	β	m	$RE(V,V_\delta)$	$RE(\varphi,\varphi_{\alpha(\delta)})$	$RE(\varphi,\varphi_\delta)$
20	1×10^{1}	8.1	9	0.1155	0.0276	14.3075
20	1×10^{2}	8.9	9	0.0953	0.0342	34.0155
20	1×10^{2}	9.1	10	0.1013	0.0508	20.6047
20	1×10^{4}	9.9	10	0.0972	0.0095	24.1306
20	1×10^{2}	10.1	11	0.0941	0.0271	35.0730
20	1×10^{3}	10.9	11	0.0956	0.0090	90.0697
20	1×10^{3}	11.1	12	0.0951	0.0226	65.9229
20	1×10^{4}	11.9	12	0.0967	0.0475	101.7022

Table 8. Case 3: Numerical results applying TRM to solve IP related to the fractional Cauchy problem (12) for $\delta = 0.1$ and different values of β and m, where $\varphi(x,y) = e^x\sin(y)$, $(x,y)\in S_1$.

N	$\alpha(\delta)$	β	m	$RE(V,V_\delta)$	$\delta RE(\varphi,\varphi_{\alpha(\delta)})$	$RE(\varphi,\varphi_\delta)$
30	1×10^{-5}	0.1	1	0.1429	0.0210	0.0230
30	1×10^{-6}	0.0001	1	0.1849	0.0396	0.0401
30	1×10^{-13}	0.0000001	1	0.1762	0.0314	0.0314
30	1×10^{-4}	0.9	1	0.1674	0.0448	0.0452
30	1×10^{-8}	0.9999	1	0.1299	0.0544	0.0575
30	1×10^{-15}	0.9999999	1	0.1748	0.0271	0.0272
30	1×10^{-5}	1.1	2	0.1613	0.0454	0.0454
30	1×10^{-5}	1.0001	2	0.1829	0.0492	0.0493
30	1×10^{-5}	1.0000001	2	0.1842	0.0584	0.0585
30	1×10^{-1}	1.9	2	0.1739	0.0353	0.0431
30	1×10^{-6}	1.9999	2	0.1870	0.0380	0.0413
30	1×10^{-12}	1.9999999	2	0.1690	0.0609	0.0674
30	1×10^{-4}	2.1	3	0.2071	0.0760	0.1016
30	1×10^{-8}	2.0001	3	0.2114	0.1688	0.1689
30	1×10^{-9}	2.0000001	3	0.1877	0.1176	0.1176
30	3×10^{-2}	2.9	3	0.2256	0.1156	0.2868
30	1×10^{-6}	2.9999	3	0.1676	0.0761	0.4059
30	5×10^{-13}	2.9999999	3	0.2155	0.1581	0.1751
30	5×10^{-2}	3.1	4	0.2057	0.1258	0.1717
30	1×10^{0}	3.9	4	0.2001	0.0881	0.3453
30	5×10^{-2}	4.1	5	0.2545	0.3090	16.5294
30	1×10^{0}	4.9	5	0.2011	0.2408	6.8619
30	4×10^{0}	5.1	6	0.1983	0.4226	12.4482
30	5×10^{1}	5.9	6	0.2049	0.3878	7.9340
30	7×10^{-1}	6.1	7	0.2124	0.4678	243.7232
30	6×10^{1}	6.9	7	0.2553	0.5497	363.0725
30	4×10^{1}	7.1	8	0.2463	0.4505	123.0513
30	5×10^{1}	7.0001	8	0.2452	0.2910	240.1527
30	2×10^{1}	7.0000001	8	0.2335	0.1839	119.2328
30	2×10^{3}	7.9	8	0.2184	0.1710	402.7594
30	1.3×10^{-2}	7.9999	8	0.2308	0.3440	181.8997
30	2×10^{-8}	7.9999999	8	0.1833	0.5806	164.4764

Table 8. *Cont.*

N	$\alpha(\delta)$	β	m	$RE(V,V_\delta)$	$\delta RE(\varphi,\varphi_{\alpha(\delta)})$	$RE(\varphi,\varphi_\delta)$
30	3×10^4	8.1	9	0.2548	0.8971	3.4981×10^3
30	2×10^6	8.9	9	0.2054	0.9059	2.8978×10^3
30	2×10^6	9.1	10	0.2260	0.8857	6.8133×10^3
30	7×10^7	9.9	10	0.2241	0.8896	6.2431×10^3
30	4×10^6	10.1	11	0.2233	0.9011	2.5172×10^5
30	2×10^8	10.9	11	0.2069	0.8934	1.9773×10^5
30	5×10^8	11.1	12	0.2479	0.9212	1.1346×10^5
30	5×10^9	11.9	12	0.2372	0.8996	3.4904×10^5

In Table 7, we observe that the $RE\left(\varphi, \varphi_{\alpha(\delta)}\right)$ are less than the $RE(V, V_\delta)$ for each value of N, β, and m given in this table. We can see that $RE\left(\varphi, \varphi_{\alpha(\delta)}\right)$ and $RE(\varphi, \varphi_\delta)$ are of the same order for $m = 1, 2$, i.e., the solutions without regularization φ_δ are close to regularized solutions $\varphi_{\alpha(\delta)}$. However, the regularized approximates $\varphi_{\alpha(\delta)}$ are better than the solutions without regularization for $m = 1, 2$. Nonetheless, $RE\left(\varphi, \varphi_{\alpha(\delta)}\right)$ increases more than $RE(V, V_\delta)$ starting at $m = 3$. Furthermore, we can observe similar results in Table 8, where the $RE\left(\varphi, \varphi_{\alpha(\delta)}\right)$ are less than the $RE(V, V_\delta)$ for $m = 1, 2, 3, 4$, with $\delta = 0.1$, $N = 30$, and the different values of β are close to m or $m-1$ given in this table. For the values of $m = 5, 6, 7$, and 8, the $RE\left(\varphi, \varphi_{\alpha(\delta)}\right)$ are around the percentage of the $RE(V, V_\delta)$. For the other values of $m = 9, 10, 11$, and 12, given in Table 8, the corresponding $RE\left(\varphi, \varphi_{\alpha(\delta)}\right)$ increases around 90%, but no more than $RE(\varphi, \varphi_\delta)$, i.e., the TRM does not provide a good approximate solution to the IP. In this case, we could use the regularized solution $\varphi_{\alpha(\delta)}$ as an initial point of an iterative method to recover a better solution to the IP. Furthermore, the relative errors of the recovered solutions φ_δ without applying regularization increase suddenly, starting at $m = 7$ and $m = 5$ (see Tables 7 and 8) for the functions $\varphi(x, y) = x^2 - y^2$ and $\varphi(x, y) = e^x \sin(y)$ for $(x, y) \in S_1$, respectively. As in the previous cases, the regularization parameter $\alpha(\delta)$ changes depending on V_δ, N, m, and β.

We show the graphs for the following functions:

- $\varphi(x, y) = x^2 - y^2$ (Figure 9 with parameters $\beta = 7.1$, $m = 8$, $\delta = 0.1$, $\alpha(\delta) = 10^2$, and $N = 20$).
- $\varphi(x, y) = e^x \sin(y)$ (Figure 10 with parameters $\beta = 7.9999999$, $m = 8$, $\delta = 0.1$, $\alpha(\delta) = 2 \times 10^{-8}$, and $N = 30$).

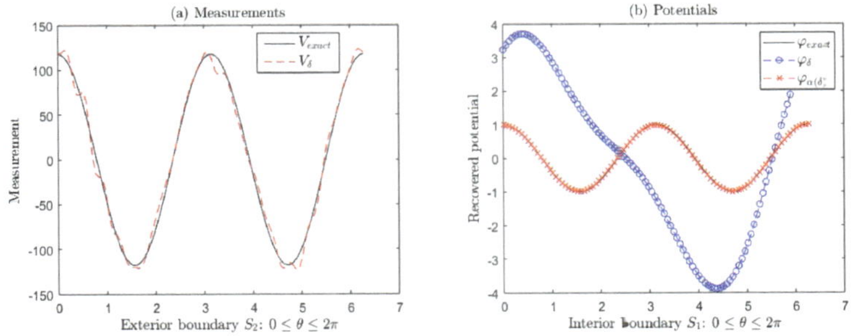

Figure 9. (**a**) Exact measurement V (black line) and with error V_δ (red line). (**b**) Exact potential φ and its approximations $\varphi_{\alpha(\delta)}$ and φ_δ, corresponding to Example 1 for $\beta = 7.1$, $m = 8$, and $\delta = 0.1$ (see Table 7). In this case, we take $\alpha(\delta) = 10^2$ and $N = 20$.

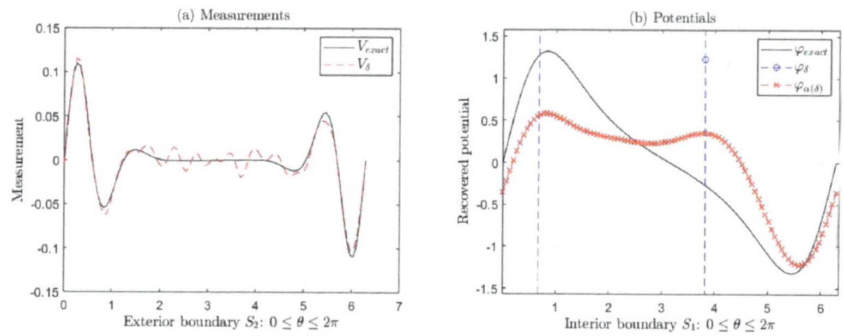

Figure 10. (**a**) Exact measurement V (black line) and with error V_δ (red line). (**b**) Exact potential φ and its approximations $\varphi_{\alpha(\delta)}$ and φ_δ, corresponding to Example 1 for $\beta = 7.9999999$, $m = 8$, and $\delta = 0.1$ (see Table 8). In this case, we take $\alpha(\delta) = 2 \times 10^{-8}$ and $N = 30$.

For $(x, y) \in S_1$. These figures show the following:

(a) The exact measurement V and the measurement with error V_δ.

(b) The exact potential φ and its approximations $\varphi_{\alpha(\delta)}$ (with regularization) and φ_δ (without regularization).

In both cases, as mentioned in the previous paragraph, the errors increase starting at $m = 7$ for the first function and starting at $m = 5$ for the second one, as can be seen in Figures 9b and 10b for $m = 8$, where we can see the ill-posedness of the IP if we do not apply regularization. For example, for the first function, the $RE(\varphi, \varphi_\delta) = 3.3483$ is greater than $RE\left(\varphi, \varphi_{\alpha(\delta)}\right) = 0.0199$ for $\beta = 7.1$, $m = 8$, and $\delta = 0.1$ (see Table 7). For the second one, the $RE(\varphi, \varphi_\delta) = 164.4764$ is greater than $RE\left(\varphi, \varphi_{\alpha(\delta)}\right) = 0.5806$ for $\beta = 7.9999999$, $m = 8$, and $\delta = 0.1$ (see Table 8). In this latter function, the approximate solution $\varphi_{\alpha(\delta)}$ is far from the exact solution φ. In this case, we could apply an iterative method to obtain a better solution, taking $\varphi_{\alpha(\delta)}$ as an initial point.

4.2.4. Case 4: $\beta < m - 1$, for $m = 2,\ldots,12$ and $\delta = 0.1$

In this Section, we consider the case when $\beta < m - 1$ for $m = 2, 3,\ldots,12$ and $\delta = 0.1$. Tables 9 and 10 show the relative errors of the approximations $\varphi_{\alpha(\delta)}$ and φ_δ when $\delta = 0.1$, with the same two exact functions φ considered in the Section 4.1.

In Table 9, we observe that the $RE\left(\varphi, \varphi_{\alpha(\delta)}\right)$ from solutions with regularization $\varphi_{\alpha(\delta)}$ are less than the $RE(V, V_\delta)$. For some values of N, β, and m given in this same table, we can see that $RE\left(\varphi, \varphi_{\alpha(\delta)}\right)$ and $RE(\varphi, \varphi_\delta)$ are of the same order, i.e., the solutions without regularization φ_δ are close to regularized solutions $\varphi_{\alpha(\delta)}$; however, the regularized solutions $\varphi_{\alpha(\delta)}$ are better than the solutions without regularization. The $RE(\varphi, \varphi_\delta)$ increases faster than the $RE(V, V_\delta)$ starting at $m = 3$. Furthermore, we can observe similar results in Table 10, where the $RE\left(\varphi, \varphi_{\alpha(\delta)}\right)$ are of the same order as $RE(V, V_\delta)$ for $m = 2, 3, 4$, with $\beta = 0.5$, $\delta = 0.1$, except for $m = 5, 6, \ldots, 12$. Nevertheless, the $RE(\varphi, \varphi_\delta)$ increases faster than the $RE(V, V_\delta)$ starting at $m = 5$. For the values of $m = 8$, 9, 10, 11, and 12, the $RE\left(\varphi, \varphi_{\alpha(\delta)}\right)$ increases between 40% and 90%, but no more than $RE(\varphi, \varphi_\delta)$. In this case, the TRM does not provide a good approximate solution to the IP. However, as mentioned before, we could use the regularized solution $\varphi_{\alpha(\delta)}$ as an initial point of an iterative method to recover a better solution to the IP. Also, the relative errors of the recovered solutions φ_δ without applying regularization increase suddenly, starting at $m = 7$ and $m = 5$ (see Tables 9 and 10) for the functions $\varphi(x, y) = x^2 - y^2$ and $\varphi(x, y) = e^x \sin(y)$ for $(x, y) \in S_1$, respectively. Here also, as in the previous cases, the parameter of regularization $\alpha(\delta)$ changes depending on V_δ, N, m, and β.

Table 9. Case 4: Numerical results applying TRM to solve the IP related to the fractional Cauchy problem (12) for $\delta = 0.1$ and different values of β and m, where $\varphi(x,y) = x^2 - y^2$, $(x,y) \in S_1$.

N	$\alpha(\delta)$	β	m	$RE(V,V_\delta)$	$RE(\varphi,\varphi_{\alpha(\delta)})$	$RE(\varphi,\varphi_\delta)$
20	1×10^{-4}	0.5	2	0.1037	0.0369	0.0371
25	1×10^{-4}	0.5	2	0.1043	0.0846	0.0850
30	1×10^{-4}	0.5	2	0.1123	0.0947	0.0952
20	1×10^{-7}	0.1	3	0.0953	0.0558	0.3293
20	1×10^{-5}	0.8	3	0.0928	0.0397	0.1176
20	1×10^{-3}	1.6	4	0.1031	0.0692	0.3781
20	1×10^{0}	2.9	4	0.0988	0.0550	0.2616
20	1×10^{-10}	0.5	5	0.1063	0.0493	2.3272
20	1×10^{-2}	3.5	5	0.1002	0.0479	0.6521
20	1×10^{-10}	0.5	6	0.0824	0.0664	1.0640
20	1×10^{-3}	3.4	6	0.1156	0.0541	1.0860
20	1×10^{-16}	0.5	7	0.1093	0.0251	11.3550
20	1×10^{-5}	4.2	7	0.0871	0.0203	4.0183
20	1×10^{-16}	0.5	8	0.0915	0.0280	3.7583
20	1×10^{-6}	3.7	8	0.1097	0.0212	4.3335
20	1×10^{-22}	0.5	9	0.0969	0.0195	21.9486
20	1×10^{-6}	5.2	9	0.0841	0.0345	9.1221
20	1×10^{-22}	0.5	10	0.1022	0.0290	22.9009
20	1×10^{-5}	7.6	10	0.1073	0.0181	7.2447
20	1×10^{-28}	0.5	11	0.0936	0.0300	122.7548
20	1×10^{-8}	6.6	11	0.0886	0.0121	91.7074
20	1×10^{-28}	0.5	12	0.0926	0.0149	30.3739
20	1×10^{-8}	6.3	12	0.1007	0.0329	36.2964

Table 10. Case 4: Numerical results applying TRM to solve IP related to the fractional Cauchy problem (12) for $\delta = 0.1, \beta = 0.5$ and different values of m, where $\varphi(x,y) = e^x \sin(y)$, $(x,y) \in S_1$.

N	$\alpha(\delta)$	β	m	$RE(V,V_\delta)$	$RE(\varphi,\varphi_{\alpha(\delta)})$	$RE(\varphi,\varphi_\delta)$
20	1×10^{-4}	0.5	2	0.1422	0.0274	0.0284
25	1×10^{-4}	0.5	2	0.1537	0.0496	0.0551
30	1×10^{-4}	0.5	2	0.1524	0.0585	0.0627
30	9×10^{-10}	0.5	3	0.1619	0.1437	0.1438
30	5×10^{-8}	0.5	4	0.1960	0.1656	0.2199
30	7×10^{-12}	0.5	5	0.2014	0.1376	13.9836
30	4×10^{-12}	0.5	6	0.2265	0.2261	8.6601
30	5×10^{-17}	0.5	7	0.2419	0.2709	209.3490
30	5×10^{-17}	0.5	8	0.2131	0.4002	205.2009
30	3×10^{-19}	0.5	9	0.2154	0.8873	2.6220×10^3
30	1.4×10^{-19}	0.5	10	0.2010	0.8863	4.1567×10^3
30	6×10^{-24}	0.5	11	0.2148	0.9054	1.3299×10^5
30	3×10^{-24}	0.5	12	0.2021	0.8935	1.6833×10^5

Figures 11 and 12 show the graphs of the exact measurement V and with error V_δ with $\delta = 0.1$, the graphs of the exact potential φ and its approximations $\varphi_{\alpha(\delta)}$ (with regularization) and φ_δ (without regularization), corresponding to the functions $\varphi(x,y) = x^2 - y^2$ and $\varphi(x,y) = e^x \sin(y)$ for $(x,y) \in S_1$, respectively. In both cases, as mentioned in the previous paragraph, the errors increase suddenly, starting at $m = 7$ for the first function and starting at $m = 5$ for the second one, as can be seen in Figures 11b and 12b, where we can see the ill-posedness of the IP if we do not apply regularization. For example, for the first function, the relative error $RE(\varphi, \varphi_\delta) = 36.2964$ is greater than $RE\left(\varphi, \varphi_{\alpha(\delta)}\right) = 0.0329$ for $\beta = 6.3$,

$m = 12$, and y $\delta = 0.1$ (see Table 9). For the second one, the $RE(\varphi, \varphi_\delta) = 205.2009$ is greater than $RE\left(\varphi, \varphi_{\alpha(\delta)}\right) = 0.4002$ for $\beta = 0.5$, $m = 8$, and $\delta = 0.1$ (see Table 10). In this case, we could use the regularized solution $\varphi_{\alpha(\delta)}$ as an initial point of an iterative method to recover a better solution to the IP.

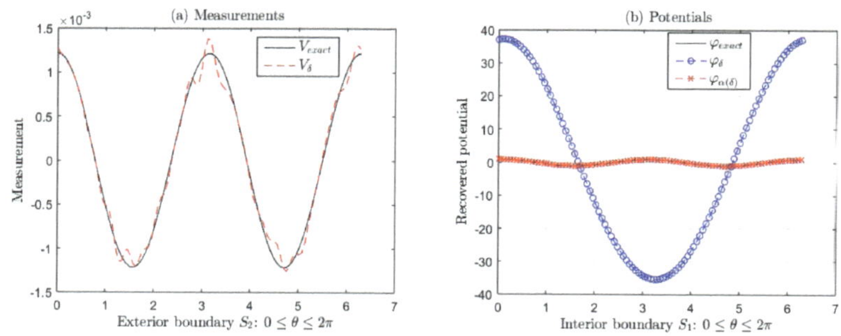

Figure 11. (**a**) Exact measurement V (black line) and with error V_δ (red line). (**b**) Exact potential φ and its approximations $\varphi_{\alpha(\delta)}$ and φ_δ, corresponding to Example 1 for $\beta = 6.3$, $m = 12$, and $\delta = 0.1$ (see Table 9). In this case, we take $\alpha(\delta) = 10^{-8}$ and $N = 20$.

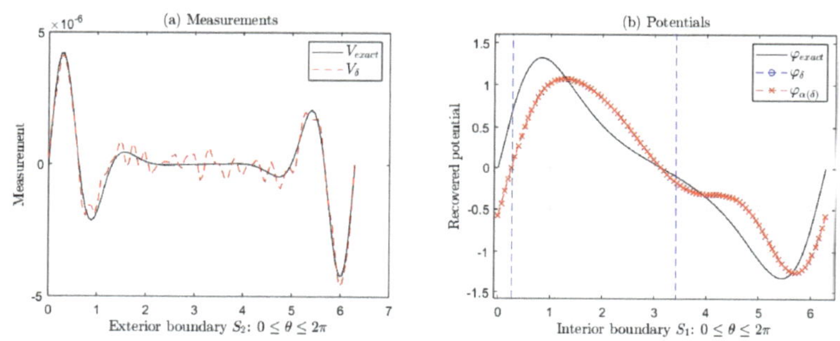

Figure 12. (**a**) Exact measurement V (black line) and with error V_δ (red line). (**b**) Exact potential φ and its approximations $\varphi_{\alpha(\delta)}$ and φ_δ, corresponding to Example 2 for $\beta = 0.5$, $m = 8$, and $\delta = 0.1$ (see Table 10). In this case, we take $\alpha(\delta) = 5 \times 10^{-17}$ and $N = 30$.

4.2.5. Case 5: $\beta < m - 1$, When β Is Next to n or $n - 1$, Where $0 < n \leq m - 1$, for $m = 2, \ldots, 12$ and $\delta = 0.1$

In this section, we consider the case when $\beta < m - 1$, when β is next to n or $n - 1$, where $0 < n \leq m - 1$, for $m = 2, \ldots, 12$ and $\delta = 0.1$. Tables 11 and 12 show the relative errors of the approximations $\varphi_{\alpha(\delta)}$ and φ_δ when $\delta = 0.1$ for the same two exact functions φ considered in Section 4.1.

Table 11. Case 5: Numerical results applying TRM to solve the IP related to the fractional Cauchy problem (12) for $\delta = 0.1$ and different values of β and m, where $\varphi(x, y) = x^2 - y^2$, $(x, y) \in S_1$.

N	$\alpha(\delta)$	β	m	$RE(V,V_\delta)$	$RE(\varphi,\varphi_{\alpha(\delta)})$	$RE(\varphi,\varphi_\delta)$
20	1×10^{-4}	0.1	2	0.0981	0.0294	0.0312
20	1×10^{-4}	0.0001	2	0.0880	0.0665	0.0702
20	1×10^{-4}	0.0000001	2	0.0877	0.0420	0.0436
20	1×10^{-2}	0.9	2	0.0825	0.0252	0.0301
20	1×10^{-2}	0.9999	2	0.0896	0.0355	0.0383
20	1×10^{-2}	0.9999999	2	0.0988	0.0497	0.0524

Table 11. *Cont.*

N	$\alpha(\delta)$	β	m	$RE(V,V_\delta)$	$RE(\varphi,\varphi_{\alpha(\delta)})$	$RE(\varphi,\varphi_\delta)$
20	1×10^{-7}	0.1	3	0.0984	0.0736	0.4041
20	1×10^{-7}	0.0001	3	0.0988	0.0556	0.2586
20	1×10^{-7}	0.0000001	3	0.1063	0.0633	0.5828
20	1×10^{-5}	0.9	3	0.0824	0.0632	0.2671
20	1×10^{-4}	0.9999	3	0.0940	0.0747	0.4997
20	5×10^{-5}	0.9999999	3	0.0843	0.0582	0.2323
20	1×10^{-2}	2.1	4	0.1133	0.0813	0.4297
20	1×10^{0}	2.9	4	0.1097	0.0494	0.2746
20	1×10^{-4}	3.1	5	0.0841	0.0521	0.5721
20	1×10^{-1}	3.9	5	0.0923	0.0284	0.9725
20	1×10^{-1}	4.1	6	0.0982	0.0441	1.3912
20	1×10^{1}	4.9	6	0.1192	0.0261	0.2905
20	1×10^{-2}	5.1	7	0.1027	0.0331	12.2005
20	1×10^{0}	5.9	7	0.0926	0.0324	2.2300
20	1×10^{-3}	5.1	8	0.0928	0.0306	2.3248
20	1×10^{-3}	5.0001	8	0.0936	0.0312	5.2064
20	1×10^{-18}	5.0000001	8	0.1020	0.0174	1.3253
20	1×10^{0}	5.9	8	0.0967	0.0113	3.4116
20	1×10^{0}	5.9999	8	0.0946	0.0260	7.0605
20	1×10^{-15}	5.9999999	8	0.0907	0.0275	1.9174
20	1×10^{-1}	7.1	9	0.0849	0.0118	9.9754
20	1×10^{0}	7.9	9	0.0916	0.0224	16.5564
20	1×10^{-13}	3.1	10	0.0930	0.0114	20.7949
20	1×10^{-11}	3.9	10	0.1054	0.0197	31.4437
20	1×10^{-10}	6.1	11	0.0978	0.0207	79.4649
20	1×10^{-7}	6.9	11	0.1048	0.0109	59.7459
20	1×10^{-15}	4.1	12	0.1035	0.0133	108.9786
20	1×10^{-12}	4.9	12	0.0978	0.0278	131.9629

Table 12. Case 5: Numerical results applying TRM to solve the IP related to the fractional Cauchy problem (12) for $\delta = 0.1$ and different values of β and m, where $\varphi(x,y) = e^x \sin(y)$, $(x,y) \in S_1$.

N	$\alpha(\delta)$	β	m	$RE(V,V_\delta)$	$RE(\varphi,\varphi_{\alpha(\delta)})$	$RE(\varphi,\varphi_\delta)$
30	1×10^{-5}	0.1	2	0.1918	0.0633	0.0669
30	1×10^{-5}	0.0001	2	0.1747	0.0510	0.0551
30	1×10^{-5}	0.0000001	2	0.1932	0.0408	0.0466
30	1×10^{-3}	0.9	2	0.1678	0.0267	0.0338
30	1×10^{-2}	0.9999	2	0.1879	0.0538	0.0996
30	1×10^{-3}	0.9999999	2	0.1757	0.0330	0.0352
30	7×10^{-9}	0.1	3	0.2035	0.1269	0.3288
30	3×10^{-9}	0.0001	3	0.1945	0.0741	0.2384
30	6×10^{-10}	0.0000001	3	0.1863	0.2035	0.2102
30	7×10^{-7}	0.9	3	0.2208	0.1301	0.2493
30	2×10^{-6}	0.9999	3	0.2057	0.0770	0.2885
30	5×10^{-7}	0.9999999	3	0.2013	0.1809	0.2036

Table 12. *Cont.*

N	$\alpha(\delta)$	β	m	$RE(V,V_\delta)$	$RE(\varphi,\varphi_{\alpha(\delta)})$	$RE(\varphi,\varphi_\delta)$
30	1×10^{-5}	2.1	4	0.2001	0.1393	0.1397
30	3×10^{-2}	2.9	4	0.1984	0.1061	0.2367
30	2×10^{-4}	3.1	5	0.2106	0.2788	14.4371
30	6×10^{-3}	3.9	5	0.1827	0.2914	6.9833
30	1×10^{-2}	4.1	6	0.2299	0.3008	4.2311
30	5×10^{-1}	4.9	6	0.1770	0.2723	3.9078
30	3×10^{-3}	5.1	7	0.2095	0.1755	297.5818
30	5×10^{-2}	5.9	7	0.1969	0.3580	51.0331
30	2×10^{-3}	5.1	8	0.2424	0.2924	216.7317
30	3×10^{-3}	5.0001	8	0.2004	0.3714	452.3457
30	1.4×10^{-3}	5.0000001	8	0.1747	0.3650	239.9826
30	2×10^{2}	5.9	8	0.2001	0.3549	263.7365
30	1.3×10^{-1}	5.9999	8	0.1829	0.3781	78.5332
30	5×10^{-1}	5.9999999	8	0.2137	0.2490	311.6103
30	1.5×10^{2}	7.1	9	0.2263	0.8972	4.5033×10^{3}
30	1×10^{4}	7.9	9	0.2228	0.8943	4.3085×10^{3}
30	1×10^{-10}	3.1	10	0.2658	0.8914	7.9703×10^{3}
30	8×10^{-8}	3.9	10	0.2642	0.8948	5.5828×10^{3}
30	1×10^{-4}	6.1	11	0.2378	0.9312	2.8572×10^{5}
30	5×10^{-3}	6.9	11	0.2089	0.8950	5.1113×10^{4}
30	2×10^{-11}	4.1	12	0.2311	0.8945	7.0888×10^{4}
30	4×10^{-9}	4.9	12	0.1994	0.9057	7.9818×10^{4}

In Table 11, we observe that $RE\left(\varphi, \varphi_{\alpha(\delta)}\right) < RE(V, V_\delta)$. For $m = 2$, we can see that $RE\left(\varphi, \varphi_{\alpha(\delta)}\right)$ and $RE(\varphi, \varphi_\delta)$ are of the same order when β is next to 1 or 0 (taking $n = 1$), i.e., the solutions without regularization φ_δ are close to regularized solutions $\varphi_{\alpha(\delta)}$. However, the regularized approximates $\varphi_{\alpha(\delta)}$ are better than the solutions without regularization. The $RE\left(\varphi, \varphi_{\alpha(\delta)}\right)$ increases faster than the $RE(V, V_\delta)$ starting at $m = 3$, as shown in Figure 13b for $\beta = 0.0001$, $m = 3$, and $\delta = 0.1$, where $RE\left(\varphi, \varphi_{\alpha(\delta)}\right) = 0.0556$ and $RE(\varphi, \varphi_\delta) = 0.2586$. These approximations, $\varphi_{\alpha(\delta)}$ and φ_δ, are recovered from measurements with error V_δ, shown in Figure 13a. Also, we can observe similar results in Table 12, where $RE\left(\varphi, \varphi_{\alpha(\delta)}\right)$ and $RE(\varphi, \varphi_\delta)$ are of the same order for $m = 2, 3, 4$, and when β is next to n or $n-1$ (taking $n = 1, 1$, and 3, respectively), for $\delta = 0.1$, nevertheless the $RE\left(\varphi, \varphi_{\alpha(\delta)}\right)$ increases between 17% and 38%, but no more than the $RE(\varphi, \varphi_\delta)$ for $m = 5, 6, 7$, and 8. For the values of $m = 9, 10, 11$, and 12, the $RE\left(\varphi, \varphi_{\alpha(\delta)}\right)$ increases around 90%, but no more than $RE(\varphi, \varphi_\delta)$. In this case, we could use the regularized solution $\varphi_{\alpha(\delta)}$ as an initial point of an iterative method to recover a better solution to the IP. Nevertheless, the relative errors of the recovered solutions φ_δ without applying regularization increase suddenly, starting at $m = 7$ and $m = 5$ (see Tables 11 and 12) for the functions $\varphi(x, y) = x^2 - y^2$ and $\varphi(x, y) = e^x \sin(y)$ for $(x, y) \in S_1$, respectively. Here, the regularization parameters $\alpha(\delta)$ also change depending on V_δ, N, m, and β.

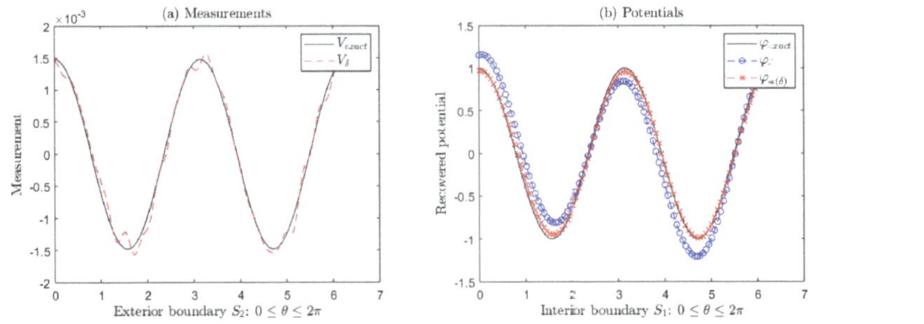

Figure 13. (**a**) Exact measurement V (black line) and with error V_δ (red line). (**b**) Exact potential φ and its approximations $\varphi_{\alpha(\delta)}$ and φ_δ, corresponding to Example 1 for $\beta = 0.0001$, $m = 3$, and $\delta = 0.1$ (see Table 11). In this case, we take $\alpha(\delta) = 10^{-7}$ and $N = 20$.

Figures 13–16 show the graphs of the exact measurement V and with error V_δ with $\delta = 0.1$, the graphs of the exact potential φ and its approximations $\varphi_{\alpha(\delta)}$ (with regularization) and φ_δ (without regularization), corresponding to the functions $\varphi(x,y) = x^2 - y^2$ and $\varphi(x,y) = e^x \sin(y)$ for $(x,y) \in S_1$, respectively. In both cases, as mentioned in the previous paragraph, the errors increase suddenly, starting at $m = 7$ for the first function and starting at $m = 5$ for the second one, as can be seen in Figure 14b, Figure 15b, and Figure 16b, where we can see the ill-posedness of the IP if we do not apply regularization for $m = 12, 8$, and $m = 11$, respectively. For example, for the approximations φ_δ and $\varphi_{\alpha(\delta)}$ shown in Figure 14b of the first function, the $RE(\varphi, \varphi_\delta) = 131.9629$ is much greater than $RE\left(\varphi, \varphi_{\alpha(\delta)}\right) = 0.0278$ for $\beta = 4.9$, $m = 12$, and $\delta = 0.1$ (see Table 11). For the approximations φ_δ and $\varphi_{\alpha(\delta)}$ shown in Figure 15b of the second one, the $RE(\varphi, \varphi_\delta) = 78.5332$ is much greater than $RE\left(\varphi, \varphi_{\alpha(\delta)}\right) = 0.3781$, for $\beta = 5.9999$, $m = 8$, and y $\delta = 0.1$ (see Table 12). Lastly, for the approximations φ_δ and $\varphi_{\alpha(\delta)}$ shown in Figure 16b of the second one, the $RE(\varphi, \varphi_\delta) = 2.8572 \times 10^5$ is greater than $RE\left(\varphi, \varphi_{\alpha(\delta)}\right) = 0.9312$ for $\beta = 6.1$, $m = 11$, and $\delta = 0.1$ (see Table 12). In these last two examples, when the approximate solutions $\varphi_{\alpha(\delta)}$ are not close to the exact solution φ, we could use the regularized solution $\varphi_{\alpha(\delta)}$ as an initial point of an iterative method to recover a better solution to the IP.

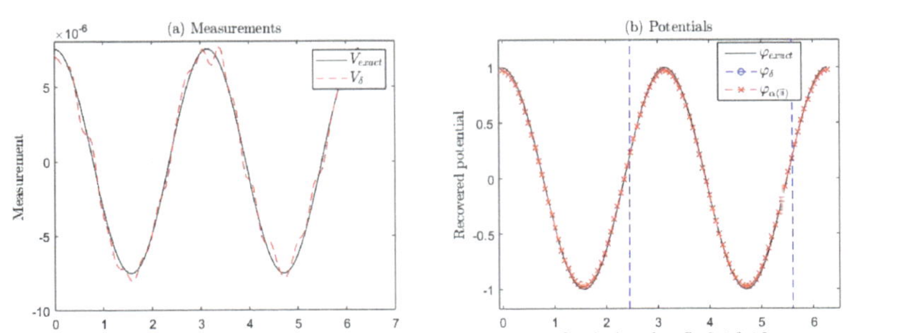

Figure 14. (**a**) Exact measurement V (black line) and with error V_δ (red line). (**b**) Exact potential φ and its approximations $\varphi_{\alpha(\delta)}$ and φ_δ, corresponding to Example 1 for $\beta = 4.9$, $m = 12$, and $\delta = 0.1$ (see Table 11). In this case, we take $\alpha(\delta) = 10^{-12}$ and $N = 20$.

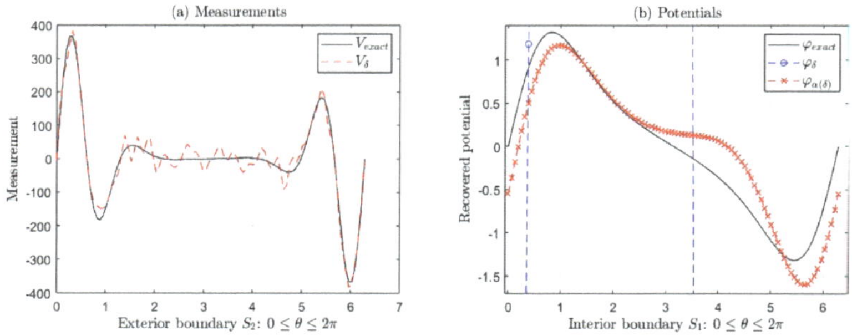

Figure 15. (**a**) Exact measurement V (black line) and with error V_δ (red line). (**b**) Exact potential φ and its approximations $\varphi_{\alpha(\delta)}$ and φ_δ, corresponding to Example 2 for $\beta = 5.9999$, $m = 8$, and $\delta = 0.1$ (see Table 12). In this case, we take $\alpha(\delta) = 1.3 \times 10^{-1}$ and $N = 30$.

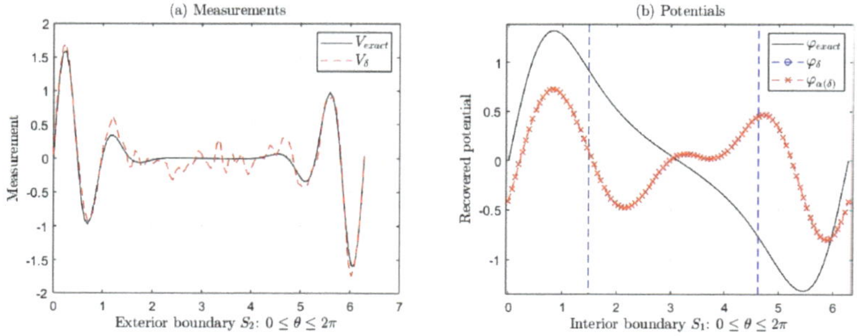

Figure 16. (**a**) Exact measurement V (black line) and with error V_δ (red line). (**b**) Exact potential φ and its approximations $\varphi_{\alpha(\delta)}$ and φ_δ, corresponding to Example 2 for $\beta = 6.1$, $m = 11$, and $\delta = 0.1$ (see Table 12). In this case, we take $\alpha(\delta) = 10^{-4}$ and $N = 30$.

4.2.6. Case 6: $\beta > m$, for $m = 1,2,\ldots,13$ and $\delta = 0.1$

In this case, we consider the case when $\beta > m$, with $m = 1,2,\ldots,13$. for $\delta = 0.1$. Tables 13 and 14 show the relative errors of the approximations $\varphi_{\alpha(\delta)}$ and φ_δ when $\delta = 0.1$ for the same two exact functions φ considered in Section 4.1.

Table 13. Case 6: Numerical results applying TRM to solve IP related to the fractional Cauchy problem (12) for $\delta = 0.1$ and different values of β and m, where $\varphi(x,y) = x^2 - y^2$, $(x,y) \in S_1$.

N	$\alpha(\delta)$	β	m	$RE(V,V_\delta)$	$RE(\varphi,\varphi_{\alpha(\delta)})$	$RE(\varphi,\varphi_\delta)$
20	1×10^{-6}	1.001	1	0.0949	0.0430	0.0432
20	1×10^{14}	1.5	1	0.0860	0.0825	0.0965
20	1×10^{-6}	2.0001	2	0.0911	0.0574	0.0581
20	1×10^{32}	3.1	2	0.0809	0.0783	0.0863
20	1×10^{-1}	3.01	3	0.0894	0.0278	0.1076
20	1×10^{45}	4.5	3	0.1048	0.0639	0.2370
20	1×10^{5}	4.1	4	0.0942	0.0294	0.1592
20	1×10^{95}	7.1	4	0.1051	0.0495	0.2588
20	1×10^{-4}	5.0001	5	0.0792	0.0298	1.3533
20	1×10^{33}	6.1	5	0.0946	0.0165	0.4475
20	1×10^{6}	6.1	6	0.0907	0.0277	0.9987
20	1×10^{48}	7.5	6	0.0867	0.0276	0.6014
20	1×10^{5}	7.1	7	0.0932	0.0109	3.5379
20	1×10^{50}	8.6	7	0.1024	0.0131	3.1745

Table 13. *Cont.*

N	$\alpha(\delta)$	β	m	$RE(V,V_\delta)$	$RE(\varphi,\varphi_{\alpha(\delta)})$	$RE(\varphi,\varphi_\delta)$
20	1×10^{1}	8.001	8	0.1135	0.0277	2.2757
20	1×10^{46}	9.4	8	0.0877	0.0180	3.6123
20	1×10^{-2}	9.0001	9	0.0935	0.0163	22.9725
20	1×10^{53}	10.7	9	0.0979	0.0342	3.1946
20	1×10^{-1}	10.0001	10	0.1046	0.0345	50.3389
20	1×10^{49}	11.5	10	0.0979	0.0179	12.7233
20	1×10^{0}	11.001	11	0.0834	0.0164	12.8627
20	1×10^{53}	12.6	11	0.0915	0.0331	44.9837
20	1×10^{1}	12.0001	12	0.1136	0.0195	121.2423
20	1×10^{47}	13.4	12	0.0822	0.0165	113.4837
20	1×10^{1}	11.5	13	0.0941	0.0155	420.7545
20	1×10^{3}	12.0001	13	0.0950	0.0280	319.2454
20	1×10^{4}	12.5	13	0.1188	0.0187	701.4226
20	1×10^{1}	12.9999	13	0.0925	0.0198	183.4905
20	1×10^{29}	13.5	13	0.0938	0.0219	400.2461
20	1×10^{51}	14.5	13	0.1112	0.0295	319.4336

Table 14. Case 6: Numerical results applying TRM to solve IP related to the fractional Cauchy problem (12) for $\delta = 0.1$ and different values of β and m, where $\varphi(x,y) = e^{x}\sin(y)$, $(x,y) \in S_1$.

N	$\alpha(\delta)$	β	m	$RE(V,V_\delta)$	$RE(\varphi,\varphi_{\alpha(\delta)})$	$RE(\varphi,\varphi_\delta)$
30	3×10^{-5}	1.001	1	0.1942	0.0575	0.1229
30	3×10^{13}	1.5	1	0.1687	0.0687	0.0965
30	6×10^{-6}	2.0001	2	0.1681	0.0606	0.1017
30	1×10^{32}	3.1	2	0.1800	0.0660	0.0763
30	2×10^{-2}	3.01	3	0.1855	0.1200	0.5139
30	5×10^{44}	4.5	3	0.2034	0.1517	1.0728
30	3×10^{3}	4.1	4	0.1930	0.0226	0.3616
30	5×10^{93}	7.1	4	0.2236	0.1517	0.4427
30	3.1×10^{-5}	5.0001	5	0.2254	0.1550	13.8261
30	1.1×10^{32}	6.1	5	0.2046	0.1057	3.5558
30	6.3×10^{-4}	6.0001	6	0.1984	0.1863	8.7178
30	6.4×10^{46}	7.5	6	0.2036	0.1233	4.9854
30	6.5×10^{4}	7.1	7	0.2188	0.3310	253.4635
30	4.4×10^{36}	8.2	7	0.1844	0.2976	36.0590
30	3.4×10^{-4}	8.00001	8	0.1944	0.4695	460.0541
30	1.1×10^{57}	9.8	8	0.2512	0.4601	157.1510
30	3×10^{1}	9.0001	9	0.2299	0.8871	7.2299×10^{3}
30	5.1×10^{57}	10.7	9	0.2143	0.8912	4.4110×10^{3}
30	5×10^{2}	10.0001	10	0.2151	0.8907	2.0204×10^{4}
30	1×10^{53}	11.5	10	0.2191	0.8920	2.4527×10^{3}
30	6×10^{3}	11.0001	11	0.2632	0.8857	3.6429×10^{5}
30	9×10^{50}	12.4	11	0.2290	0.9571	1.8509×10^{5}
30	1.3×10^{5}	12.0001	12	0.2451	0.9564	3.7314×10^{5}
30	2.1×10^{52}	13.4	12	0.2211	0.9192	4.5936×10^{5}
30	5.9×10^{10}	11.5	13	0.2168	0.9964	1.3190×10^{7}
30	5×10^{11}	12.0001	13	0.2607	0.9994	1.0836×10^{5}
30	9×10^{11}	12.5	13	0.2326	0.9904	1.0214×10^{7}
30	1.1×10^{9}	12.9999	13	0.2439	0.9890	1.3107×10^{7}
30	1.3×10^{29}	13.5	13	0.2338	0.9895	5.7405×10^{6}
30	1.7×10^{59}	14.5	13	0.2424	0.9899	1.3997×10^{7}

In Table 13, we observe that $RE\left(\varphi, \varphi_{\alpha(\delta)}\right) < RE(V, V_\delta)$. For $m = 1, 2$, we can see that $RE\left(\varphi, \varphi_{\alpha(\delta)}\right)$ and $RE(\varphi, \varphi_\delta)$ are of the same order when $\delta = 0.1$, i.e., the solutions without regularization φ_δ are close to regularized solutions $\varphi_{\alpha(\delta)}$. However, the regularized approximates $\varphi_{\alpha(\delta)}$ are better than the solutions without regularization. Additionally, the relative errors of the solutions without regularization φ_δ increase faster, starting at $m = 3$. Furthermore, we can observe similar results in Table 14. In this case, the relative errors from solutions with regularization $RE\left(\varphi, \varphi_{\alpha(\delta)}\right)$ are less than the $RE(V, V_\delta)$ for $m = 1, 2, \ldots, 6$, and increase between 29% and 46% for $m = 7, 8$, when $\delta = 0.1$. Nevertheless, the corresponding relative errors of the solutions with regularization increase between 88% and 96% for $m = 9, 10, 11, 12$, but no more than the corresponding $RE(\varphi, \varphi_\delta)$. The relative errors of the recovered solutions φ_δ without regularization increase suddenly, starting at $m \geq 5$. For example, for $\beta = 10.7$ and $m = 9$, the $RE(\varphi, \varphi_\delta) = 4.4110 \times 10^3$. For $m = 13$, the $RE\left(\varphi, \varphi_{\alpha(\delta)}\right)$ increases between 91% and 99%, but no more than the $RE(\varphi, \varphi_\delta)$), for $\beta > m$, as well as for $\beta < m$ and $m - 1 < \beta < m$ with $\delta = 0.1$. Moreover, as in the previous cases, the regularization parameters $\alpha(\delta)$ change depending on the data with error V_δ, the values N, m, and β. Analogous results can be obtained for values $m \geq 14$, as those obtained for $m = 13$, which are not included in this work.

In the following two examples, we have considered non-smooth functions.

Example 3. *We consider the 'exact potential'* $\varphi(x, y) = \sqrt{(x^2 + y^2)} - \pi/2$, *for* $(x, y) \in S_1$, *which in polar coordinates is given by* $\varphi(\theta) = |\theta| - \pi/2$, *for* $\theta \in [-\pi, \pi]$. *Resembling the first example, the 'exact measurement' V and the 'measurement with error'* V_δ *are generated with the first N terms of the Fourier series (22) and (23), respectively, such that* $\| \varphi_N - \varphi \|_{L_2(S_1)} \leq \epsilon_F$, *with* $0 \leq \epsilon_F$. *For* $N = 30$, $\epsilon_F = 0.0056$. *In this case,* $V_0 = \varphi_0 = 0$ *and the Fourier coefficients* φ_k, $1 \leq k \leq N$, *are obtained numerically using the intrinsic function quadl of MATLAB.*

Table 15 shows the numerical results for data without error, applying TRM to solve the IP of the classical Cauchy problem (2), where $N = 20, 25$, and 30. In this table, we can observe similar results to the previous examples where the regularized solutions $\varphi_{\alpha(\delta)}$ are better.

Table 15. Numerical results applying TRM to solve the IP related to the classical Cauchy problem (2), for $\varphi(\theta) = |\theta| - \pi/2, \theta \in [-\pi, \pi]$ and different values of δ and N.

δ	N	$\alpha(\delta)$	$RE(V,V_\delta)$	$RE(\varphi,\varphi_{\alpha(\delta)})$	$RE(\varphi,\varphi_\delta)$
0.1	20	0.1	0.1080	0.1545	1.1839
0.1	25	0.1	0.0953	0.1314	2.1403
0.1	30	0.1	0.1342	0.1630	5.5998
0.01	20	0.01	0.0121	0.0409	0.1102
0.01	25	0.01	0.0114	0.0425	0.1676
0.01	30	0.01	0.0128	0.0370	0.5224
0.001	20	0.001	8.1891×10^{-4}	0.0043	0.0046
0.001	25	0.001	9.1866×10^{-4}	0.0088	0.0139
0.001	30	0.001	0.0014	0.0117	0.0557

Table 16 shows the numerical results for data without error, applying TRM to solve the IP of the fractional Cauchy problem (12) for different values of β, m, and N.

Table 16. Numerical results applying TRM to solve IP related to the fractional Cauchy problem (12) for $\delta = 0.1$ and different values of β and m, where $\varphi(\theta) = |\theta| - \pi/2, \theta \in [-\pi, \pi]$

N	$\alpha(\delta)$	β	m	$RE(V, V_\delta)$	$RE(\varphi, \varphi_{\alpha(\delta)})$	$RE(\varphi, \varphi_\delta)$
20	2×10^{-3}	0.5	1	0.3124	0.1870	0.2188
20	2×10^{-5}	0.1	1	0.1347	0.1118	0.1121
20	1×10^{-2}	0.9	1	0.1180	0.1919	0.2229
20	2×10^{-4}	0.1	2	0.1524	0.0594	0.0974
20	1×10^{-3}	1.1	2	0.1362	0.1985	0.1989
20	1×10^{-3}	1.5	2	0.1047	0.2133	0.2141
20	4.7×10^{0}	1.9	2	0.2123	0.3425	1.3020
10	4.9×10^{-3}	2.1	3	0.0523	0.2633	1.9458
10	5×10^{-3}	2.5	3	0.0527	0.1430	0.3773
10	2.8×10^{-1}	2.9	3	0.0495	0.2678	2.7599
10	2.1×10^{0}	3.5	4	0.0463	0.2386	1.8027
8	1×10^{-1}	4.1	5	0.0397	0.4832	48.7338
8	4.2×10^{-1}	4.5	5	0.0378	0.3875	24.7968
8	2.8×10^{0}	4.9	5	0.0269	0.4535	27.6306
8	1×10^{2}	5.9	6	0.0281	0.2041	23.2792
6	1×10^{1}	6.9	7	0.0207	0.2113	105.7357
6	7.2×10^{1}	7.1	8	0.0257	0.5941	280.7669
4	2.2×10^{1}	9.1	10	0.0416	0.2823	79.7393
4	3.3×10^{0}	10.1	11	0.0265	0.1181	1.4592×10^{3}
4	2.5×10^{2}	11.1	12	0.0306	0.1032	1.0285×10^{3}
4	2.2×10^{2}	12.5	13	0.0348	0.7974	1.0332×10^{4}

Figure 17a,b show the graphs of the exact potential φ and its approximations $\varphi_{\alpha(\delta)}$ (with regularization) and φ_δ (without regularization), corresponding to the function $\varphi(\theta) = |\theta| - \pi/2$, for $\theta \in [-\pi, \pi]$, for different values of β, m, and N, respectively.

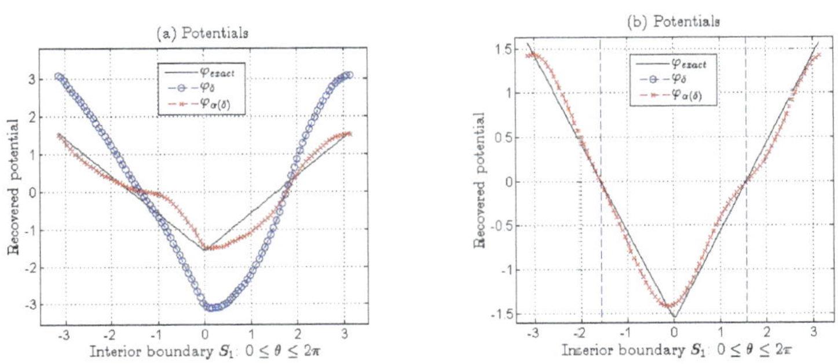

Figure 17. (**a**) Exact potential φ and its approximations $\varphi_{\alpha(\delta)}$ and φ_δ, corresponding to Example 3 for $\beta = 1.9$, $m = 2$, $N = 20$, and $\delta = 0.1$. (**b**) Exact potential φ and its approximations $\varphi_{\alpha(\delta)}$ and φ_δ, corresponding to Example 3 for $\beta = 11.1$, $m = 12$, $N = 4$, and $\delta = 0.1$ (see Table 16).

Example 4. *We consider the 'exact potential' in polar coordinates given by* $\varphi(\theta) = -1$ *if* $\theta \in [-\pi, 0)$ *and 1 if* $\theta \in [0, \pi]$. *Similar to the first example, the 'exact measurement' V and the 'measurement with error'* V_δ *are generated with the first N terms of the Fourier series (22) and (23), respectively, such that* $\| \varphi_N - \varphi \|_{L_2(S_1)} \leq \epsilon_F$, *with* $0 \leq \epsilon_F$. *For* $N = 30$, $\epsilon_F = 0.2913$. *In this case,* $V_0 = \varphi_0 = 0$, *and the Fourier coefficients* φ_k, $1 \leq k \leq N$, *are obtained numerically using the intrinsic function quadl of MATLAB.*

Table 17 shows the numerical results for data without error, applying TRM to solve the IP of the classical Cauchy problem (2), where $N = 20, 25$, and 30. In this table, we can observe similar results to the previous examples where the regularized solutions $\varphi_{\alpha(\delta)}$ are better.

Table 17. Numerical results applying TRM to solve the IP related to the classical Cauchy problem (2), for $\varphi(\theta) = -1$ if $\theta \in [-\pi, 0)$ and 1 if $\theta \in [0, \pi]$, and different values of δ and N.

δ	N	$\alpha(\delta)$	$RE(V,V_\delta)$	$RE(\varphi,\varphi_{\alpha(\delta)})$	$RE(\varphi,\varphi_\delta)$
0.1	20	0.1	0.0666	0.1608	0.7217
0.1	25	0.1	0.0638	0.1702	0.7641
0.1	30	0.1	0.0767	0.1827	2.6840
0.01	20	0.01	0.0080	0.0544	0.0659
0.01	25	0.01	0.0068	0.0761	0.0998
0.01	30	0.01	0.0078	0.0946	0.3552
0.001	20	0.0001	7.6172×10^{-4}	0.0044	0.0048
0.001	25	0.0001	7.2529×10^{-4}	0.0099	0.0118
0.001	30	0.0001	9.2391×10^{-4}	0.0195	0.0353

Table 18 shows the numerical results for data without error, applying TRM to solve the IP of the fractional Cauchy problem (12) for different values of β, m, and N.

Table 18. Numerical results applying TRM to solve IP related to the fractional Cauchy problem (12) for $\delta = 0.1$ and different values of β and m, where $\varphi(\theta) = -1$ if $\theta \in [-\pi, 0)$ and 1 if $\theta \in [0, \pi]$.

N	$\alpha(\delta)$	β	m	$RE(V,V_\delta)$	$RE(\varphi,\varphi_{\alpha(\delta)})$	$RE(\varphi,\varphi_\delta)$
20	4.5×10^{-2}	0.5	1	0.1062	0.2731	1.9413
10	3.9×10^{-4}	0.1	1	0.0645	0.0729	0.1759
10	3.2×10^{-3}	0.5	1	0.0640	0.1624	0.2390
10	4×10^{-3}	0.9	1	0.0609	0.0841	0.0934
10	2.1×10^{-2}	1.2	2	0.0910	0.1530	0.1636
10	1.5×10^{-3}	1.5	2	0.0549	0.2629	0.2699
6	2.1×10^{-2}	2.5	3	0.0352	0.3415	1.3513
6	1.5×10^{0}	3.5	4	0.0523	0.5428	1.1704
6	1×10^{3}	4.2	5	0.0472	0.9410	88.1158
6	3.6×10^{4}	5.2	6	0.0483	0.9354	41.0394
6	1.5×10^{7}	7.2	8	0.0400	0.9343	1.7546×10^{3}
6	4.2×10^{16}	12.8	13	0.0606	0.9825	2.1740×10^{7}

Figure 18a,b show the graphs of the exact potential φ and its approximations $\varphi_{\alpha(\delta)}$ (with regularization) and φ_δ (without regularization), corresponding to the function $\varphi(\theta) = -1$ if $\theta \in [-\pi, 0)$ and 1 if $\theta \in [0, \pi]$, for $\theta \in [-\pi, \pi]$, for different values of β, m, and N, respectively.

Examples 3 and 4 show numerical results analogous to the previous examples for both the classical and fractional cases. However, when we have piecewise constant functions, we truncate the series of approximate solutions $\varphi_{\alpha(\delta)}$ to find a better approximation of the solution to the fractional Cauchy problem.

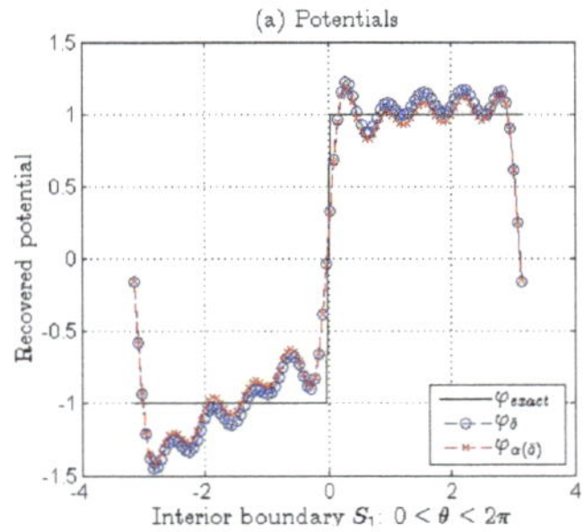

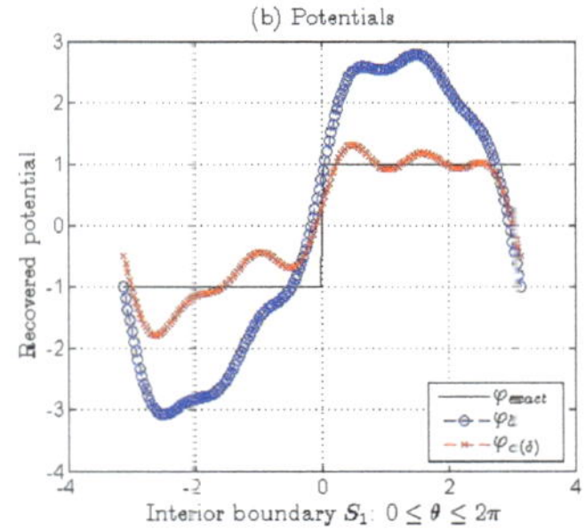

Figure 18. (**a**) Exact potential φ and its approximations $\varphi_{\alpha(\delta)}$ and φ_δ, corresponding to Example 4 for $\beta = 1.2, m = 2, N = 10$, and $\delta = 0.1$. (**b**) Exact potential φ and its approximations $\varphi_{\alpha(\delta)}$ and φ_δ, corresponding to Example 4 for $\beta = 2.5, m = 3, N = 6$, and $\delta = 0.1$ (see Table 18).

4.3. Solution to the IP Related to the Fractional Cauchy Problem Morozov Discrepancy Method and the Criterion of Tikhonov

In this subsection, the numerical results of the approximate solutions of the four examples presented above are calculated by the proposed method, choosing the regularization parameter by Morozov's discrepancy method and by the Tikhonov criterion. The relative errors of the approximations given in Tables 19–26 show that the best approximate solutions to the inverse problem are obtained by the *L-curve criterion*, as shown in Table 13, Table 14, Table 16, and Table 18, taking the same values of N, α, β, and m from these same examples for Tables 19–26, respectively.

Table 19. Numerical results choosing the regularization parameter $\alpha(\delta)$ with the discrepancy principle of Morozov to solve IP related to the fractional Cauchy problem (12) for $\delta = 0.1$ and different values of β and m, where $\varphi(x, y) = x^2 - y^2$, $(x, y) \in S_1$. According to the results of the table and additional experiments, the discrepancy method fails when $\beta > (m + 1)$ and $m > 1$, as well as for some values of $\beta < (m - 1)$.

N	$\alpha(\delta)$	β	m	$RE(V,V_\delta)$	$RE(\varphi,\varphi_{\alpha(\delta)})$	$RE(\varphi,\varphi_\delta)$
20	0.1613	1.001	1	0.0959	1.0470	0.0972
20	7.5089×10^6	1.5	1	0.0983	0.1328	0.1328
20	1.03×10^3	2.0001	2	0.0827	1.0000	0.0418
20	NaN	3.1	2	0.0870	NaN	0.0259
20	0.3137	3.01	3	0.0921	0.0526	0.1290
20	NaN	4.5	3	0.0994	NaN	0.4877
20	17.79	4.1	4	0.1069	0.5115	0.5102
20	NaN	7.1	4	0.0949	NaN	0.1675
20	6.53×10^7	5.0001	5	0.0901	1.0000	1.2477
20	NaN	6.1	5	0.0965	NaN	1.4885
20	4.7570	6.1	6	0.0860	1.9250	1.9237
20	NaN	7.5	6	0.0932	NaN	1.9387
20	0.2678	7.1	7	0.0922	6.1441	6.1514
20	NaN	8.6	7	0.0983	NaN	8.3501
20	0.0909	8.001	8	0.1113	0.0995	1.4459
20	NaN	9.4	8	0.0947	NaN	3.8687
20	0.1659	9.0001	9	0.0913	0.0497	9.9648
20	NaN	10.7	9	0.1018	NaN	21.7190
20	3.7653×10^{-5}	10.0001	10	0.0818	23.2250	36.9474
20	NaN	11.5	10	0.0980	NaN	34.1752
20	3.1198×10^{-5}	11.001	11	0.0828	50.0353	57.7211
20	NaN	12.6	11	0.0827	NaN	105.6237
20	1.0869×10^{-5}	12.0001	12	0.0892	80.4548	94.1772

Table 19. *Cont.*

N	$\alpha(\delta)$	β	m	$RE(V,V_\delta)$	$RE(\varphi,\varphi_{\alpha(\delta)})$	$RE(\varphi,\varphi_\delta)$
20	NaN	13.4	12	0.0852	NaN	13.4034
20	NaN	11.5	13	0.0984	NaN	517.1635
20	5.1280×10^{-6}	12.0001	13	0.1027	780.2431	783.2423
20	4.8277×10^{-5}	12.5	13	0.1031	292.2405	293.0910
20	1.03160×10^{-6}	12.9999	13	0.0971	148.1006	226.6417
20	3.4578×10^{3}	13.5	13	0.1041	597.4911	597.4911
20	NaN	14.5	13	0.0747	NaN	105.6650

Table 20. Numerical results choosing the regularization parameter $\alpha(\delta)$ with the Tikhonov criterion to solve IP related to the fractional Cauchy problem (12) for $\delta = 0.1$ and different values of β and m, where $\varphi(x,y) = x^2 - y^2$, $(x,y) \in S_1$. Similar results have been obtained for other values of α using this criterion.

N	$\alpha(\delta)$	β	m	$RE(V,V_\delta)$	$RE(\varphi,\varphi_{\alpha(\delta)})$	$RE(\varphi,\varphi_\delta)$
20	$\delta^{\frac{3}{2}}$	1.001	1	0.0994	0.9040	0.0769
20	$\delta^{\frac{3}{2}}$	1.5	1	0.0965	0.0938	0.0938
20	$1 + 1 = 2\delta^{\frac{99}{50}}$	2.0001	2	0.1046	0.9022	0.0702
20	δ	3.1	2	0.0984	0.1345	0.1345
20	δ	3.01	3	0.0812	0.0852	0.3750
20	$\delta^{\frac{1}{3}}$	4.5	3	0.0862	0.3665	0.3665
20	$\delta^{\frac{1}{100}}$	4.1	4	0.1050	0.3914	0.3914
20	δ	7.1	4	0.0971	0.8561	0.8561
20	$\delta^{\frac{99}{50}}$	5.0001	5	0.0893	0.5591	1.5529
20	$\delta^{\frac{99}{50}}$	6.1	5	0.0934	1.2559	1.2559
20	$\delta^{\frac{1}{3}}$	6.1	6	0.0929	1.5055	1.5056
20	$\delta^{\frac{1}{2}}$	7.5	6	0.0994	2.0076	2.0076
20	$\delta^{\frac{99}{50}}$	7.1	7	0.0949	2.5857	2.5858
20	$\delta^{\frac{99}{50}}$	8.6	7	0.0913	2.4924	2.4924
20	$\delta^{\frac{1}{100}}$	8.001	8	0.0957	0.0431	5.6199
20	$\delta^{\frac{1}{100}}$	9.4	8	0.0971	13.6930	13.6930
20	$\delta^{\frac{1}{10000}}$	9.0001	9	0.0893	0.3169	24.8451
20	$\delta^{\frac{1}{10000}}$	10.7	9	0.1057	8.4227	8.4227
20	δ	10.0001	10	0.1031	0.0334	25.1034
20	δ	11.5	10	0.0990	28.5194	28.5194
20	$\delta^{\frac{99}{10000}}$	11.001	11	0.0834	0.0225	36.1203
20	$\delta^{\frac{99}{10000}}$	12.6	11	0.0988	66.3421	66.3421
20	$\delta^{\frac{1}{100}}$	12.0001	12	0.1022	0.0279	95.8910
20	$\delta^{\frac{1}{100}}$	13.4	12	0.0926	31.8654	31.8654
20	$\delta^{\frac{1}{10000}}$	11.5	13	0.1099	0.0552	457.5273
20	$\delta^{\frac{1}{10000}}$	12.0001	13	0.1133	0.5851	438.2230
20	$\delta^{\frac{1}{10000}}$	12.5	13	0.1097	4.5644	279.6731
20	$\delta^{\frac{1}{10000}}$	12.9999	13	0.0982	0.0249	358.7193
20	$\delta^{\frac{1}{10000}}$	13.5	13	0.0925	183.5626	183.5626
20	$\delta^{\frac{1}{10000}}$	14.5	13	0.1027	789.7150	789.7150

Table 21. Numerical results applying Morozov discrepancy principle to solve IP related to the fractional Cauchy problem (12) for $\delta = 0.1$ and different values of β and m, where $\varphi(x,y) = e^x \sin(y)$, $(x,y) \in S_1$. According to the results, the discrepancy principle has problems when $\beta > m+1$ and $m > 1$, and for $\beta = 1.001$ and $m = 1$.

N	$\alpha(\delta)$	β	m	$RE(V,V_\delta)$	$RE(\varphi,\varphi_{\alpha(\delta)})$	$RE(\varphi,\varphi_\delta)$
30	NaN	1.001	1	0.1245	NaN	0.0962
30	2.0984×10^6	1.5	1	0.1910	0.1082	0.1082
30	1.03×10^3	2.0001	2	0.0954	1.0000	0.0244
30	NaN	3.1	2	0.1075	NaN	0.0402
30	0.2289	3.01	3	0.1118	0.0657	0.4844
30	NaN	4.5	3	0.1218	NaN	1.4902
30	49.9604	4.1	4	0.1040	0.1739	0.1726
30	NaN	7.1	4	0.1200	NaN	0.3234
30	6.46×10^7	5.0001	5	0.1222	1.0000	1.5949
30	NaN	6.1	5	0.1098	NaN	1.8066
30	0.0725	6.0001	6	0.1133	0.1997	1.9133
30	NaN	7.5	6	0.1058	NaN	1.5889
30	0.1979	7.1	7	0.1126	8.3170	8.3243
30	NaN	8.2	7	0.1207	NaN	4.1606
30	0.0471	8.00001	8	0.1245	0.5043	5.9603
30	NaN	9.8	8	0.1132	NaN	6.3306
30	0.1675	9.0001	9	0.1027	0.0679	15.9717
30	NaN	10.7	9	0.0961	NaN	9.3670
30	0.8716	10.0001	10	0.0954	0.0099	5.1670
30	NaN	11.5	10	0.1075	NaN	8.4426
30	2.7508×10^{-8}	11.0001	11	0.2068	5.0164×10^4	5.0171×10^4
30	NaN	12.4	11	0.1218	NaN	125.4177
30	6.7362×10^{-9}	12.0001	12	0.2635	1.1476×10^5	1.1477×10^5
30	NaN	13.4	12	0.1200	NaN	82.7087
30	1.6116×10^{-10}	11.5	13	0.2236	5.6345×10^6	5.6345×10^6
30	9.6113×10^{-8}	12.0001	13	0.2251	3.6801×10^5	3.6801×10^5
30	3.7943×10^{-9}	12.5	13	0.2458	3.2872×10^6	3.2872×10^6
30	1.2925×10^{-11}	12.9999	13	0.2276	1.0450×10^7	1.0450×10^7
30	NaN	13.5	13	0.2413	NaN	8.8855×10^6
30	NaN	14.5	13	0.2454	NaN	9.6411×10^6

Table 22. Numerical results choosing the regularization parameter $\alpha(\delta)$ with the Tikhonov criterion to solve IP related to the fractional Cauchy problem (12) for $\delta = 0.1$ and different values of β and m, where $\varphi(x,y) = e^x \sin(y)$, $(x,y) \in S_1$. In almost all cases, the L-curve method gives better results.

N	$\alpha(\delta)$	β	m	$RE(V,V_\delta)$	$RE(\varphi,\varphi_{\alpha(\delta)})$	$RE(\varphi,\varphi_\delta)$
30	δ	1.001	1	0.1760	0.9884	0.1546
30	$\delta^{\frac{1}{10000}}$	1.5	1	0.1562	0.1227	0.1227
30	$\delta^{\frac{99}{50}}$	2.0001	2	0.1783	0.9696	0.1104
30	$\delta^{\frac{99}{50}}$	3.1	2	0.1806	0.0813	0.0813
30	$\delta^{\frac{99}{50}}$	3.01	3	0.1774	0.4170	0.5684
30	$\delta^{\frac{1}{10000}}$	4.5	3	0.1939	0.3294	0.3294
30	$\delta^{\frac{1}{10000}}$	4.1	4	0.2089	0.6294	0.6294
30	$\delta^{\frac{1}{10000}}$	7.1	4	0.2157	0.9195	0.9195
30	$\delta^{\frac{99}{50}}$	5.0001	5	0.2132	0.9217	6.2925
30	$\delta^{\frac{1}{10000}}$	6.1	5	0.1985	10.2345	10.2345
30	$\delta^{\frac{99}{50}}$	6.0001	6	0.2047	0.8225	11.8088
30	$\delta^{\frac{1}{10000}}$	7.5	6	0.1893	8.5339	8.5339
30	$\delta^{\frac{1}{10000}}$	7.1	7	0.2272	66.4146	66.7074

Table 22. *Cont.*

N	$\alpha(\delta)$	β	m	$RE(V,V_\delta)$	$RE(\varphi,\varphi_{\alpha(\delta)})$	$RE(\varphi,\varphi_\delta)$
30	$\delta^{\frac{1}{10000}}$	8.2	7	0.2041	65.2733	65.2733
30	$\delta^{\frac{5}{3}}$	8.00001	8	0.1768	1.1965	97.2722
30	δ	9.8	8	0.1955	90.8875	90.8875
30	$\delta^{\frac{5}{3}}$	9.0001	9	0.2093	2.2232	2.6794×10^3
30	$\delta^{\frac{1}{10000}}$	10.7	9	0.2038	3.7798×10^3	3.7798×10^3
30	$\delta^{\frac{1}{10000}}$	10.0001	10	0.2252	4.4215	7.1864×10^3
30	$\delta^{\frac{1}{100}}$	11.5	10	0.2077	4.0842×10^3	4.0842×10^3
30	$\delta^{\frac{1}{100}}$	11.0001	11	0.2059	80.7421	2.1991×10^5
30	$\delta^{\frac{1}{10000}}$	12.4	11	0.2496	3.1294×10^5	3.1294×10^5
30	$\delta^{\frac{1}{2}}$	12.0001	12	0.2130	93.2559	4.0135×10^5
30	$\delta^{\frac{1}{3}}$	13.4	12	0.2228	3.0318×10^5	3.0318×10^5
30	$\delta^{\frac{1}{100}}$	11.5	13	0.2128	785.0281	5.2143×10^6
30	$\delta^{\frac{1}{3}}$	12.0001	13	0.2471	6.1257×10^3	2.1351×10^6
30	$\delta^{\frac{1}{8}}$	12.5	13	0.2025	2.3802×10^5	1.0999×10^7
30	$\delta^{\frac{1}{50}}$	12.9999	13	0.2472	583.7069	2.0067×10^7
30	$\delta^{\frac{1}{100}}$	13.5	13	0.2477	1.8405×10^7	1.8405×10^7
30	$\delta^{\frac{1}{10000}}$	14.5	13	0.2120	1.3707×10^7	1.3707×10^7

Table 23. Numerical results choosing the regularization parameter $\alpha(\delta)$ with the discrepancy principle of Morozov to solve IP related to the fractional Cauchy problem (12) for $\delta = 0.1$ and different values of β and m, where $\varphi(\theta) = |\theta| - \pi/2$, $\theta \in [-\pi, \pi]$. In almost all cases, the L-curve method gives better results.

N	$\alpha(\delta)$	β	m	$RE(V,V_\delta)$	$RE(\varphi,\varphi_{\alpha(\delta)})$	$RE(\varphi,\varphi_\delta)$
20	0.0320	0.5	1	0.2693	0.7013	0.3930
20	6.8073	0.1	1	0.2790	0.9988	0.2787
20	0.0595	0.9	1	0.2171	0.5500	0.2910
20	2.3795	0.1	2	0.2306	0.9972	0.3700
20	0.0297	1.1	2	0.2421	0.4419	0.6114
20	0.0993	1.5	2	0.1942	0.7216	0.7481
20	0.1561	1.9	2	0.2055	1.0241	1.0477
10	0.0079	2.1	3	0.0840	0.8154	1.8753
10	0.0096	2.5	3	0.0958	2.0322	2.6709
10	0.0122	2.9	3	0.0965	2.6035	2.8554
10	0.0231	3.5	4	0.0992	4.7034	4.7820
8	6.6817×10^{-5}	4.1	5	0.0688	59.6092	61.4116
8	1.9688×10^{-4}	4.5	5	0.0577	55.5120	56.1750
8	5.4635×10^{-4}	4.9	5	0.0623	48.1493	48.4213
8	0.0013	5.9	6	0.0620	127.5673	127.6119
6	1.3725×10^{-4}	6.9	7	0.0225	196.1122	196.3843
6	1.2394×10^{-4}	7.1	8	0.0316	342.3864	342.5589
4	4.6244×10^{-4}	9.1	10	0.0577	92.5918	92.7644
4	2.3783×10^{-6}	10.1	11	0.0451	1.7044×10^3	1.7062×10^3
4	2.4299×10^{-5}	11.1	12	0.0491	1.7439×10^3	1.7441×10^3
4	1.0631×10^{-6}	12.5	13	0.0401	1.0360×10^4	1.0361×10^4

Table 24. Numerical results choosing the regularization parameter $\alpha(\delta)$ with the Tikhonov criterion to solve IP related to the fractional Cauchy problem (12) for $\delta = 0.1$ and different values of β and m, where $\varphi(\theta) = |\theta| - \pi/2, \theta \in [-\pi, \pi]$. In almost all cases, the L-curve method gives better results.

N	$\alpha(\delta)$	β	m	$RE(V,V_\delta)$	$RE(\varphi,\varphi_{\alpha(\delta)})$	$RE(\varphi,\varphi_\delta)$
20	$\delta^{\frac{1}{3}}$	0.5	1	0.3117	0.9521	0.5185
20	$\delta^{\frac{1}{3}}$	0.1	1	0.2942	0.9921	0.4911
20	$\delta^{\frac{1}{3}}$	0.9	1	0.2614	0.7377	0.5011
20	$\delta^{\frac{3}{2}}$	0.1	2	0.2944	0.9571	0.4833
20	$\delta^{\frac{3}{2}}$	1.1	2	0.2224	0.5438	0.5251
20	$\delta^{\frac{5}{4}}$	1.5	2	0.2703	0.7070	0.6965
20	δ	1.9	2	0.2060	0.9825	1.3020
10	$\delta^{\frac{99}{50}}$	2.1	3	0.0105	0.1151	1.0675
10	$\delta^{\frac{99}{50}}$	2.5	3	0.0790	1.7172	2.2671
10	$\delta^{\frac{99}{50}}$	2.9	3	0.0955	2.4815	2.6475
10	$\delta^{\frac{1}{3}}$	3.5	4	0.0992	3.7011	5.0290
8	$\delta^{\frac{3}{2}}$	4.1	5	0.0725	2.3038	32.4249
8	$\delta^{\frac{3}{2}}$	4.5	5	0.0822	26.8355	76.0308
8	$\delta^{\frac{1}{5}}$	4.9	5	0.0696	7.6778	60.6948
8	$\delta^{\frac{99}{10000}}$	5.9	6	0.0538	34.4968	43.7691
6	$\delta^{\frac{99}{10000}}$	6.9	7	0.0418	25.5161	284.8013
6	$\delta^{\frac{99}{50}}$	7.1	8	0.0514	414.7308	432.4166
4	$\delta^{\frac{99}{50}}$	9.1	10	0.0432	39.8856	41.6149
4	$\delta^{\frac{1}{100}}$	10.1	11	0.0440	2.1960	1.1303×10^3
4	$\delta^{\frac{1}{100}}$	11.1	12	0.0485	306.1543	1.5216×10^3
4	$\delta^{\frac{1}{10000}}$	12.5	13	0.0387	75.5346	4.6817×10^4

Table 25. Numerical results choosing the regularization parameter $\alpha(\delta)$ with the discrepancy principle of Morozov to solve IP related to the fractional Cauchy problem (12) for $\delta = 0.1$ and different values of β and m, where $\varphi(\theta) = -1$ if $\theta \in [-\pi, 0)$ and 1 if $\theta \in [0, \pi]$. In almost all cases, the L-curve method gives better results. In this Table, $m - 1 < \beta < m$.

N	$\alpha(\delta)$	β	m	$RE(V,V_\delta)$	$RE(\varphi,\varphi_{\alpha(\delta)})$	$RE(\varphi,\varphi_\delta)$
20	0.0084	0.5	1	0.1321	1.1988	1.4773
10	0.1198	0.1	1	0.0763	0.9065	0.3944
10	0.0480	0.5	1	0.0641	0.7212	0.2595
10	0.0292	0.9	1	0.0780	0.1942	0.2717
10	0.0556	1.2	2	0.0905	0.4162	0.4471
10	0.1325	1.5	2	0.0808	0.3404	0.2828
6	0.0026	2.5	3	0.0542	3.2590	3.8754
6	0.0184	3.5	4	0.0555	3.8072	3.8819
6	1.8758×10^{-4}	4.2	5	0.0402	25.9293	27.2371
6	0.0016	5.2	6	0.0417	30.6892	30.8212
6	2.1013×10^{-5}	7.2	8	0.0454	2.3159×10^3	2.3161×10^3
6	1.1443×10^{-9}	12.8	13	0.0415	1.8147×10^7	1.8147×10^7

Table 26. Numerical results choosing the regularization parameter $\alpha(\delta)$ with the Tikhonov criterion to solve IP related to the fractional Cauchy problem (12) for $\delta = 0.1$ and different values of β and m, where $\varphi(\theta) = -1$ if $\theta \in [-\pi, 0)$ and 1 if $\theta \in [0, \pi]$. In almost all cases, the L-curve method gives better results.

N	$\alpha(\delta)$	β	m	$RE(V,V_\delta)$	$RE(\varphi,\varphi_{\alpha(\delta)})$	$RE(\varphi,\varphi_\delta)$
20	$\delta^{\frac{5}{4}}$	0.5	1	0.1447	1.3444	2.7243
10	$\delta^{\frac{3}{2}}$	0.1	1	0.0882	0.8467	0.3194
10	$\delta^{\frac{3}{2}}$	0.5	1	0.0584	0.7281	0.5733
10	$\delta^{\frac{99}{50}}$	0.9	1	0.0602	0.8434	0.3127
10	$\delta^{\frac{99}{50}}$	1.2	2	0.0711	0.5848	0.5872
10	$\delta^{\frac{99}{50}}$	1.5	2	0.0624	0.5957	0.6053
6	$\delta^{\frac{3}{2}}$	2.5	3	0.0334	1.1740	1.6434
6	$\delta^{\frac{1}{10000}}$	3.5	4	0.0602	1.1241	1.2573
6	$\delta^{\frac{1}{50}}$	4.2	5	0.0437	1.0155	88.8735
6	$\delta^{\frac{1}{100}}$	5.2	6	0.0551	23.6302	86.1879
6	$\delta^{\frac{1}{10000}}$	7.2	8	0.0359	387.4002	1.4405×10^3
6	$\delta^{\frac{1}{10000}}$	12.8	13	0.0425	2.0636×10^6	3.2829×10^7

5. Discussion

The numerical tests show that the proposed algorithm usually gives good results. Even if the numerical results are unsatisfactory, they are enough to start an iterative method. In all cases, the regularized method is worth more than the method without regularization. After some numerical tests, we found that the series expansion of the solution to the fractional Cauchy problem can be truncated in $N = 20$, $N = 25$, or $N = 30$.

When $\beta < m$ for $m = 1, 2 \ldots, 7$, the results obtained are similar, i.e., the results obtained with and without regularization almost coincide. One possible explanation can be associated with the smoothing properties of the integral operator to have similar results when $\beta > m$, for $m = 1, 2$. In the other cases, the regularized case is better.

When $m > 7$, the regularized method loses precision. However, the approximate solution obtained can be used as an initial point of a stable iterative method. From the numerical results, we want to emphasize that the solution by the Tikhonov regularization method of the classical Cauchy problem works adequately in all cases.

The Tikhonov regularization parameter was very large in some cases. We do not have an explanation for this situation, but we consider this an interesting topic that must be studied in future works. According to numerical results, in almost all cases, the best approximate solutions to the inverse problem are obtained by the L-curve criterion. According to the results, the discrepancy principle has problems when $\beta > m + 1$ and $m > 1$.

In the classical Cauchy problem, the adjoint operator is associated with a boundary value problem called the adjoint problem. In the fractional Cauchy problem, we calculate the adjoint operator using its definition. One interesting question is whether a boundary value problem is associated with the adjoint operator. If the answer is positive, the following question arises: Can the adjoint operator be used in irregular regions? This is an interesting question whose answer can help us apply numerical methods to find the minimum of the functional since we have to solve boundary value problems. One of the most used methods to find such a minimum is the conjugate gradient method in combination with the finite element method.

6. Conclusions

This work proposes an algorithm to solve the fractional Cauchy problem obtained from the Tikhonov regularization and the circular harmonics. The regularization was obtained using the L-curve method, Morozov's discrepancy principle, and numerical tests by the Tikhonov criterion. The numerical results show that the algorithm is feasible for various parameters. The discrepancy principle presents some problems in finding the regularization parameter for some values of the parameters appearing in the fractional Cauchy problem.

In almost all cases, the L-curve method gives better results than the Tikhonov Criterion and Morozov's discrepancy principle. In all cases, the regularization using the L-curve method gives better results than without regularization. In some cases, despite not being a good approximation, the regularized solution is much better than the solution without regularization. Since the algorithm does not give good results in some cases, it must be improved using an iterative method, which takes the regularized solution as an initial point. This point might be future work.

Author Contributions: Conceptualization, J.J.C.M., J.A.A.V., E.H.M., M.M.M.C., and J.J.O.O.; methodology, J.J.C.M., J.A.A.V., E.H.M., M.M.M.C., and J.J.O.O.; software, J.J.C.M., J.A.A.V. and E.H.M.; validation, J.J.C.M., J.A.A.V., and J.J.O.O.; formal analysis, J.J.C.M., J.A.A.V., E.H.M., M.M.M.C., C.A.H.G., and J.J.O.O.; investigation, J.J.C.M., J.A.A.V., E.H.M., M.M.M.C., C.A.H.G., and J.J.O.O.; resources, J.J.C.M., J.A.A.V., E.H.M., M.M.M.C., C.A.H.G., and J.J.O.O.; data curation, J.J.C.M., J.A.A.V., and J.J.O.O.; writing—original draft preparation, J.J.C.M., J.A.A.V., E.H.M., M.M.M.C., C.A.H.G., and J.J.O.O.; writing—review and editing, J.J.C.M., J.A.A.V., E.H.M., M.M.M.C., and J.J.O.O.; visualization, J.J.C.M., J.A.A.V., E.H.M., M.M.M.C., C.A.H.G., and J.J.O.O.; supervision, J.J.C.M. and J.J.O.O.; project administration, J.J.C.M. and J.J.O.O.; funding acquisition, J.J.C.M., J.A.A.V., E.H.M., M.M.M.C., and J.J.O.O. All authors have read and agreed to the published version of the manuscript.

Funding: This research was funded by VIEP-BUAP withthe support provided, to research project number 151. The National Council for Humanities, Sciences, and Technologies in Mexico (CONAHCYT) also provided partial funding through a PhD scholarship for the second author, with CVU number 1030555.

Data Availability Statement: Data are contained within the article.

Conflicts of Interest: The authors declare that they have no competing interests.

Appendix A. Computation of the Derivatives of J_α

If $J_\alpha(\varphi) = \frac{1}{2} \| K_m^\beta \varphi - V_\delta \|^2_{L_2(S_2)} + \frac{\alpha}{2} \| \varphi \|^2_{L_2(S_1)}$, then

$$
\begin{aligned}
J_\alpha(\varphi+h)-J_\alpha(\varphi) &= \frac{1}{2}\|K_m^\beta(\varphi+h)-V_\delta\|^2_{L_2(S_2)}+\frac{\alpha}{2}\|\varphi+h\|^2_{L_2(S_1)}-\left(\frac{1}{2}\|K_m^\beta\varphi-V_\delta\|^2_{L_2(S_2)}+\frac{\alpha}{2}\|\varphi\|^2_{L_2(S_1)}\right)\\
&= \frac{1}{2}\|K_m^\beta(\varphi+h)-V_\delta\|^2_{L_2(S_2)}-\frac{1}{2}\|K_m^\beta\varphi-V_\delta\|^2_{L_2(S_2)}+\frac{\alpha}{2}\|\varphi+h\|^2_{L_2(S_1)}-\frac{\alpha}{2}\|\varphi\|^2_{L_2(S_1)}\\
&= \frac{1}{2}\langle K_m^\beta\varphi-V_\delta+K_m^\beta h, K_m^\beta\varphi-V_\delta+K_m^\beta h\rangle_{L_2(S_2)}-\frac{1}{2}\langle K_m^\beta\varphi-V_\delta, K_m^\beta\varphi-V_\delta\rangle_{L_2(S_2)}\\
&+\frac{\alpha}{2}\langle\varphi+h,\varphi+h\rangle_{L_2(S_1)}-\frac{\alpha}{2}\langle\varphi,\varphi\rangle_{L_2(S_1)}\\
&= \frac{1}{2}\left[\langle K_m^\beta\varphi-V_\delta, K_m^\beta\varphi-V_\delta\rangle_{L_2(S_2)}+\langle K_m^\beta h, K_m^\beta h\rangle_{L_2(S_2)}+\langle K_m^\beta\varphi-V_\delta, K_m^\beta h\rangle_{L_2(S_2)}\right]\\
&-\frac{1}{2}\langle K_m^\beta\varphi-V_\delta, K_m^\beta\varphi-V_\delta\rangle_{L_2(S_2)}+\frac{1}{2}\langle K_m^\beta h, K_m^\beta\varphi-V_\delta\rangle_{L_2(S_2)}\\
&+\frac{\alpha}{2}\left[\langle\varphi,\varphi\rangle+\langle\varphi,h\rangle_{L_2(S_1)}+\langle h,\varphi\rangle_{L_2(S_1)}+\langle h,h\rangle_{L_2(S_1)}\right]-\frac{\alpha}{2}\langle\varphi,\varphi\rangle_{L_2(S_1)}\\
&= \frac{1}{2}\|K_m^\beta h\|^2_{L_2(S_2)}+\langle (K_m^\beta)^*(K_m^\beta\varphi-V_\delta),h\rangle_{L_2(S_1)}+\langle\alpha\varphi,h\rangle_{L_2(S_1)}+\frac{\alpha}{2}\|h\|^2_{L_2(S_1)}\\
&= \frac{1}{2}\|K_m^\beta h\|^2_{L_2(S_2)}+\langle ((K_m^\beta)^*K_m^\beta+\alpha I)\varphi-(K_m^\beta)^*V_\delta,h\rangle_{L_2(S_1)}+\frac{\alpha}{2}\|h\|^2_{L_2(S_1)}
\end{aligned}
$$

Thus, $DJ_\alpha(\varphi) = \left(\left(K_m^\beta\right)^* K_m^\beta+\alpha I\right)\varphi-\left(K_m^\beta\right)^* V_\delta$. For the second derivative, we have

$$
\begin{aligned}
DJ_\alpha(\varphi+h)-DJ_\alpha(\varphi) &= ((K_m^\beta)^*K_m^\beta+\alpha I)(\varphi+h)-(K_m^\beta)^*V_\delta-[((K_m^\beta)^*K_m^\beta+\alpha I)\varphi-(K_m^\beta)^*V_\delta]\\
&= ((K_m^\beta)^*K_m^\beta+\alpha I)\varphi+((K_m^\beta)^*K_m^\beta+\alpha I)h-(K_m^\beta)^*V_\delta-[((K_m^\beta)^*K_m^\beta+\alpha I)\varphi-(K_m^\beta)^*V_\delta]\\
&= ((K_m^\beta)^*K_m^\beta+\alpha I)h.
\end{aligned}
$$

Hence, $D^2 J_\alpha(\varphi) = \left(K_m^\beta\right)^* K_m^\beta+\alpha I.$

Appendix B. Proof That the Functional J_α $(f) = \frac{1}{2}\| K_m^\beta(f)-V_\delta \|^2_{L_2(S_2)_{L_2(S_2)}}+\frac{\alpha}{2}\| f \|^2_{L_2(S_1)}$ Is Convex

Theorem A1. *The functional*

$$J_1(f) = \frac{\alpha}{2}\| f \|^2_{L_2(S_1)}$$

is convex.

Proof. We want to show that

$$J_1(\lambda f+(1-\lambda)g)\leq\lambda J_1(f)+(1-\lambda)J_1(g),$$

for all $f,g\in L_2(S_1)$, with $f\neq g$, and for all $\lambda\in(0,1)$.

Starting with the left-hand side,

$$
\begin{aligned}
J_1(\lambda f+(1-\lambda)g) &= \frac{\alpha}{2}\| \lambda f+(1-\lambda)g \|^2_{L_2(S_1)}\\
&= \frac{\alpha}{2}\langle\lambda f+(1-\lambda)g,\lambda f+(1-\lambda)g\rangle_{L_2(S_1)}\\
&= \frac{\alpha}{2}\langle\lambda f,\lambda f\rangle_{L_2(S_1)}+\alpha\langle\lambda f,(1-\lambda)g\rangle_{L_2(S_1)}\\
&\quad+\frac{\alpha}{2}\langle(1-\lambda)g,(1-\lambda)g\rangle_{L_2(S_1)}\\
&= \frac{\alpha}{2}\lambda^2\| f \|^2+\alpha\lambda(1-\lambda)\langle f,g\rangle_{L_2(S_1)}\\
&\quad+\frac{\alpha}{2}(1-\lambda)^2\| g \|^2_{L_2(S_1)}
\end{aligned}
$$

By the Cauchy–Schwarz inequality:

$$|\langle f,g\rangle_{L_2(S_1)}|\leq\| f \|_{L_2(S_1)}\| g \|_{L_2(S_1)}, \quad \forall f\neq g$$

we obtain

$$J_1(\lambda f+(1-\lambda)g)\leq \lambda^2\frac{\alpha}{2}\| f \|^2_{L_2(S_1)}+(1-\lambda)^2\frac{\alpha}{2}\| g \|^2_{L_2(S_1)}$$
$$+\alpha\lambda(1-\lambda)\| f \|_{L_2(S_1)}\| g \|_{L_2(S_1)}$$

Applying the inequality: $a\cdot b\leq\frac{a^2+b^2}{2}$, if $a,b\in\mathbb{R}$. Considering that $a = ||f||_{L_2(S_1)}$ and $b = ||g||_{L_2(S_1)}$, then,

$$\alpha\lambda(1-\lambda)||f||_{L_2(S_1)}||g||_{L_2(S_1)}\leq\alpha\lambda(1-\lambda)\frac{1}{2}\left[||f||^2_{L_2(S_1)}+||g||^2_{L_2(S_1)}\right]$$

Then

$$\begin{aligned}
J_1(\lambda f+(1-\lambda)g)&\leq \lambda^2\frac{\alpha}{2}||f||^2_{L_2(S_1)}+(1-\lambda)^2\frac{\alpha}{2}||g||^2_{L_2(S_1)}+\lambda(1-\lambda)\frac{\alpha}{2}\left[||f||^2_{L_2(S_1)}+||g||^2_{L_2(S_1)}\right]\\
&=\lambda^2\frac{\alpha}{2}||f||^2_{L_2(S_1)}+\left(1-2\lambda+\lambda^2\right)\frac{\alpha}{2}||g||^2_{L_2(S_1)}+\left(\lambda-\lambda^2\right)\frac{\alpha}{2}\left[||f||^2_{L_2(S_1)}+||g||^2_{L_2(S_1)}\right]\\
&=\lambda^2\frac{\alpha}{2}||f||^2_{L_2(S_1)}+\frac{\alpha}{2}||g||^2_{L_2(S_1)}-\lambda\alpha||g||^2_{L_2(S_1)}+\lambda^2\frac{\alpha}{2}||g||^2_{L_2(S_1)}+\lambda\frac{\alpha}{2}||f||^2\\
&\quad+\lambda\frac{\alpha}{2}||g||^2_{L_2(S_1)}-\lambda^2\frac{\alpha}{2}||f||^2_{L_2(S_1)}-\lambda^2\frac{\alpha}{2}||g||^2_{L_2(S_1)}\\
&=\lambda\frac{\alpha}{2}||f||^2_{L_2(S_1)}+\frac{\alpha}{2}||g||^2_{L_2(S_1)}-\lambda\frac{\alpha}{2}||g||^2_{L_2(S_1)}\\
&=\lambda\frac{\alpha}{2}||f||^2_{L_2(S_1)}+(1-\lambda)\frac{\alpha}{2}||g||^2_{L_2(S_1)}\\
&=\lambda J_1(f)+(1-\lambda)J_1(f).
\end{aligned}$$

Therefore, $J_1(f) = \frac{\alpha}{2}(||f||^2_{L_2(S_1)}$ is convex. □

Theorem A2. *The functional*

$$J_2(f) = \frac{1}{2}\| K^\beta_m(f)-V_\delta \|^2_{L_2(S_2)}$$

is convex.

Proof. Assuming the operator K^β_m is linear, we have

$$\begin{aligned}
J_2(\lambda f+(1-\lambda)g)&=\frac{1}{2}\| K^\beta_m(\lambda f+(1-\lambda)g)-V_\delta \|^2_{L_2(S_2)}\\
&=\frac{1}{2}\| \lambda\left(K^\beta_m(f)-V_\delta\right)+(1-\lambda)\left(K^\beta_m(g)-V_\delta\right) \|^2_{L_2(S_2)}\\
&=\frac{\lambda^2}{2}\| K^\beta_m(f)-V_\delta \|^2_{L_2(S_2)}+\frac{(1-\lambda)^2}{2}\| K^\beta_m(g)-V_\delta \|^2_{L_2(S_2)}\\
&\quad+\lambda(1-\lambda)\left\langle K^\beta_m(f)-V_\delta,K^\beta_m(g)-V_\delta\right\rangle_{L_2(S_2)},
\end{aligned}$$

where $\lambda\in(0,1)$.

By the Cauchy–Schwarz inequality:

$$\left\langle K^\beta_m(f)-V_\delta,K^\beta_m(g)-V_\delta\right\rangle_{L_2(S_2)}\leq\|K^\beta_m(f)-V_\delta \|_{L_2(S_2)}\| K^\beta_m(g)-V_\delta \|_{L_2(S_2)}$$

and using the inequality $ab \leq \frac{1}{2}(a^2+b^2)$ if $a,b \in \mathbb{R}$. Considering that $a = \|K_m^{\beta}(f) - V_\delta\|_{L_2(S_2)}$ and $b = \|K_m^{\beta}(g) - V_\delta\|_{L_2(S_2)}$, we obtain

$$\lambda(1-\lambda)\left\langle K_m^{\beta}(f) - V_\delta, K_m^{\beta}(g) - V_\delta \right\rangle_{L_2(S_2)} \leq \frac{\lambda(1-\lambda)}{2}\left(\| K_m^{\beta}(f) - V_\delta \|_{L_2(S_2)}^2 + \| K_m^{\beta}(g) - V_\delta \|_{L_2(S_2)}^2\right)$$

Combining these results,

$$\begin{aligned} J_2(\lambda f + (1-\lambda)g) \leq & \frac{\lambda^2}{2} \| K_m^{\beta}(f) - V_\delta \|_{L_2(S_2)}^2 + \frac{(1-\lambda)^2}{2} \| K_m^{\beta}(g) - V_\delta \|_{L_2(S_2)}^2 \\ & + \frac{\lambda(1-\lambda)}{2}\left(\| K_m^{\beta}(f) - V_\delta \|_{L_2(S_2)}^2 + \| K_m^{\beta}(g) - V_\delta \|_{L_2(S_2)}^2\right) \\ & = \frac{\lambda}{2} \| K_m^{\beta}(f) - V_\delta \|_{L_2(S_2)}^2 + \frac{1-\lambda}{2} \| K_m^{\beta}(g) - V_\delta \|_{L_2(S_2)}^2 \\ & = \lambda J_2(f) + (1-\lambda) J_2(g). \end{aligned}$$

Therefore, $J_2(f) = \frac{1}{2}\| K_m^{\beta}(f) - V_\delta \|_{L_2(S_2)}^2$ is convex. □

Thus, from the last two theorems, we have that the functional $J_\alpha(f) = \frac{1}{2} \| K_m^{\beta}(f) - V_\delta \|_{L_2(S_2)}^2 + \frac{\alpha}{2} \| f \|_{L_2(S_1)}^2$ is convex.

References

1. Conde Mones, J.J.; Juárez Valencia, L.H.; Oliveros Oliveros, J.J.; León Velasco, D.A. Stable numerical solution of the Cauchy problem for the Laplace equation in irregular annular regions. *Numer. Methods Partial Differ. Equ.* **2017**, *33*, 1799–1822. [CrossRef]
2. Oliveros, J.; Morín, M.; Conde, J.; Fraguela, A. A regularization strategy for the inverse problem of identification of bioelectrical sources for the case of concentric spheres. *Far East J. Appl. Math.* **2013**, *77*, 1–20.
3. Lee, J.Y.; Yoon, J.R. A numerical method for Cauchy problem using singular value decomposition. *Commun. Korean Math. Soc.* **2001**, *16*, 487–508.
4. Wei, T.; Chen, Y.G. A regularization method for a Cauchy problem of Laplace's equation in an annular domain. *Math. Comput. Simul.* **2012**, *82*, 2129–2144. [CrossRef]
5. Zhou, D.; Wei, T. The method of fundamental solutions for solving a Cauchy problem of Laplace's equation in a multi-connected domain. *Inverse Probl. Sci. Eng.* **2008**, *16*, 389–411. [CrossRef]
6. Chang, J.R.; Yeih, W.; Shieh, M.H. On the modified Tikhonov's regularization method for the Cauchy problem of the Laplace equation. *J. Mar. Sci. Technol.* **2001**, *9*, 113–121. [CrossRef]
7. Gong, X.; Yang, S. A local regularization scheme of Cauchy problem for the Laplace equation on a doubly connected domain. *Bound. Value Probl.* **2023**, *2023*, 30. [CrossRef]
8. Cheng, J.; Hon, Y.C.; Wei, T.; Yamamoto, M. Numerical computation of a Cauchy problem for Laplace's equation. *ZAMM-J. Appl. Math. Mech. Angew. Math. Mech. Appl. Math. Mech.* **2001**, *81*, 665–674. [CrossRef]
9. Borachok, I.; Chapko, R.; Tomas Johansson, B. Numerical solution of a Cauchy problem for Laplace equation in 3-dimensional domains by integral equations. *Inverse Probl. Sci. Eng.* **2016**, *24*, 1550–1568. [CrossRef]
10. Hào, D.N.; Lesnic, D. The Cauchy problem for Laplace's equation via the conjugate gradient method. *IMA J. Appl. Math.* **2000**, *65*, 199–217. [CrossRef]
11. Caubet, F.; Dardé, J.; Godoy, M. On the data completion problem and the inverse obstacle problem with partial Cauchy data for Laplace's equation. *ESAIM Control Optim. Calc. Var.* **2019**, *25*, 30. [CrossRef]
12. Amdouni, S.; Ben Abda, A. The Cauchy problem for Laplace's equation via a modified conjugate gradient method and energy space approaches. *Math. Methods Appl. Sci.* **2023**, *46*, 3560–3582. [CrossRef]
13. León-Velasco, A.; Glowinski, R.; Juárez-Valencia, L.H. On the controllability of diffusion processes on the surface of a torus: A computational approach. *Pac. J. Optim.* **2015**, *11*, 763–790.
14. Conde Mones, J.J.; Estrada Aguayo, E.R.; Oliveros Oliveros, J.J.; Hernández Gracidas, C.A.; Morín Castillo, M.M. Stable identification of sources located on interface of nonhomogeneous media. *Mathematics* **2021**, *9*, 1932. [CrossRef]
15. Berntsson, F.; Lars, E. Numerical solution of a Cauchy problem for the Laplace equation. *Inverse Probl.* **2001**, *17*, 839–853. [CrossRef]
16. Kress, R. Inverse Dirichlet problem and conformal mapping. *Math. Comput. Simul.* **2004**, *66*, 255–265. [CrossRef]

17. Clerc, M.; Kybic, J. Cortical mapping by Laplace-Cauchy transmission using a boundary element method. *Inverse Probl.* **2007**, *23*, 2589–2601. [CrossRef]
18. Denisov, A.M.; Zakharov, E.V.; Kalinin, A.V.; Kalinin, V.V. Numerical solution of an inverse electrocardiography problem for a medium with piecewise constant electrical conductivity. *Comput. Math. Math. Phys.* **2010**, *50*, 1172–1177. [CrossRef]
19. Kalinin, A.; Potyagaylo, D.; Kalinin, V. Solving the inverse problem of electrocardiography on the endocardium using a single layer source. *Front. Physiol.* **2019**, *10*, 58. [CrossRef]
20. Ruiz de Miras, J.; Derchi, C.-C.; Atzori, T.; Mazza, A.; Arcuri, P.; Salvatore, A.; Navarro, J.; Saibene, F.L.; Meloni, M.; Comanducci, A. Spatio-Temporal Fractal Dimension Analysis from Resting State EEG Signals in Parkinson's Disease. *Entropy* **2023**, *25*, 1017. [CrossRef]
21. Sirpal, P.; Sikora, W.A.; Refai, H.H.; Yang, Y. Association between Opioid Dependence and Scale Free Fractal Brain Activity: An EEG Study. *Fractal Fract.* **2023**, *7*, 659. [CrossRef]
22. Perez-Sanchez, A.V.; Valtierra-Rodriguez, M.; Perez-Ramirez, C.A.; De-Santiago-Perez, J.J.; Amezquita-Sanchez, J.P. Epileptic Seizure Prediction Using Wavelet Transform, Fractal Dimension, Support Vector Machine, and EEG Signals. *Fractals* **2022**, *30*, 2250154. [CrossRef]
23. Karaoulanis, D.; Lazopoulou, N.; Lazopoulos, K. On A-Fractional Derivative and Human Neural Network. *Axioms* **2023**, *12*, 136. [CrossRef]
24. Turmetov, B.K.; Nazarova, K.D. On a generalization of the Neumann problem for the Laplace equation. *Math. Nachrichten* **2019**, *293*, 169–177. [CrossRef]
25. Kadirkulov, B.J.; Kirane, M. On solvability of a boundary value problem for the Poisson equation with a nonlocal boundary operator. *Acta Math. Sci.* **2015**, *35B*, 970–980. [CrossRef]
26. Kirsch, A. *An Introduction to the Mathematical Theory of Inverse Problems*, 2nd ed.; Springer: New York, NY, USA, 2011; Volume 120, Applied Mathematical Sciences.

fractal and fractional

Article

On a Preloaded Compliance System of Fractional Order: Numerical Integration

Marius-F. Danca

STAR-UBB Institute, Babes-Bolyai University, 400084 Cluj-Napoca, Romania; m.f.danca@gmail.com

Abstract: In this paper, the use of a class of fractional-order dynamical systems with discontinuous right-hand side defined with Caputo's derivative is considered. The existence of the solutions is analyzed. For this purpose, differential inclusions theory is used to transform, via the Filippov regularization, the discontinuous right-hand side into a set-valued function. Next, via Cellina's Theorem, the obtained set-valued differential inclusion of fractional order can be restarted as a single-valued continuous differential equation of fractional order, to which the existing numerical schemes for fractional differential equations can be applied. In this way, the delicate problem of integrating discontinuous problems of fractional order, as well as integer order, is solved by transforming the discontinuous problem into a continuous one. Also, it is noted that even the numerical methods for fractional-order differential equations can be applied abruptly to the discontinuous problem, without considering the underlying discontinuity, so the results could be incorrect. The technical example of a single-degree-of-freedom preloaded compliance system of fractional order is presented.

Keywords: switch dynamical systems of fractional order; Caputo's derivative; fractional differential equations

Academic Editor: António Lopes

Received: 4 December 2024
Revised: 7 January 2025
Accepted: 21 January 2025
Published: 26 January 2025

Citation: Danca, M.-F. On a Preloaded Compliance System of Fractional Order: Numerical Integration. *Fractal Fract.* **2025**, *9*, 84. https://doi.org/10.3390/fractalfract9020084

1. Introduction

Generally, integer-order (IO) systems with a discontinuous right-hand side are mostly ideal due to the presence of switch-type functions like sgn or Heaviside H. Discontinuity can generally be found in two-dimensional mechanical systems: oscillating systems combined with dry and viscous damping; systems with dry friction; systems with stick and slip modes, forced vibrations, brake processes with locking phases, control synthesis for uncertain systems, and elastoplasticity; and in control theory, game theory, optimization, calculus of variations, physiological, biological systems, electrical (chaotic) circuits, complex networks, power electronics, etc. (see [1], the workshop on the "Nonlinear Dynamics of Mechanical Systems", an attempt to give a standard textbook devoted to nonlinear dynamical systems with discontinuities).

It is important to note that there are two main strategies to approach numerically discontinuous systems of IO: the first one is to ignore the discontinuities ("time stepping" methods) and to rely on a local error estimator such that the error remains acceptably small; the other strategy is to determine a scalar event function defining the discontinuity. In this way, the intersection point serves as the new starting point for continuing the numerical solution ("event-driven" methods).

To integrate discontinuous Ordinary Differential Equations (ODEs) of IO, there are dedicated numerical methods (see [1,2]). However, it is important to note that numerical methods for discontinuous systems could be either inaccurate or even inefficient in the region of the discontinuity. Also, the error analysis often fails if there is not sufficient local smoothness.

On the other hand, in recent years, FO dynamical systems have received increasing attention due to their broad range of applications, such as real-life applications, [3], viscoelastic materials [4], control engineering [5], fractional electrical circuits [6], chaotic FO dynamical systems [7,8], and Lyapunov exponents of FO systems [9]. Fractional differential equations are presented in [10], while for a review of definitions of fractional derivatives, see [11].

FO models can describe complex physics problems more clearly and concisely compared with their IO variant.

Therefore, considering the FO variants of discontinuous is a subject of interest. Most known examples of discontinuous systems are defined as IO. In this paper, we consider their FO variants, a fact that could represent a novel approach. For this purpose, by using the differential inclusions theory, the problem modeling the underlying discontinuous problem is transformed into a continuous one.

The manuscript is organized as follows: Section 2 introduces the class of switch dynamical systems of IO. Section 3 deals with a class of discontinuous dynamical systems of FO, discusses notions and results on differential inclusions used to transform a discontinuous problem into a continuous one and considers the preloaded compliance system of FO. Special attention is paid to the correctness of the numerical integration of underlying IVP. This paper ends with the Conclusion Section.

2. Switch Dynamical Systems of IO

The class of systems analyzed in this paper, with a discontinuous right-hand side (Piece-Wise Continuous (PWC), see Definition 1), usually called "switch" or "relay systems", are modeled by the following discontinuous with respect to the state variable, autonomous Initial Value Problem (IVP):

$$\dot{x} = f(x) := g(x) + As(x), \quad x(0) = x_0, \quad t \in I = [0, T], \tag{1}$$

where $g : \mathbb{R}^n \to \mathbb{R}^n$ is a nonlinear Lipschitz continuous function and $s : \mathbb{R}^n \to \mathbb{R}^n$ is a piecewise function with piecewise constant components $s_i : \mathbb{R} \to \mathbb{R}$, usually sgn or Heaviside $H(x) = (1 + \text{sgn}\,(x))/2$. $T \in \mathbb{R}, T > 0$, and the real matrix $A \in \mathbb{R}^{n\times n}$ is supposed to have at least one nonzero element.

The considered system, briefly described below, which comes from real applications and is called a preloaded compliance system with a single degree of freedom [1], is modeled by the following IVP:

$$m\ddot{y} + c\dot{y} + h(y) = P\cos\omega t, \quad y(0) = y_0, \quad \dot{y}(0) = y_1, \tag{2}$$

where m is the mass of the system, k is the stiffness and c is the viscous damping coefficient (see Figure 1a). The restoring force that has discontinuity at $y = 0$, where its value is $2ky_0$, the segment of the values at the discontinuity point, $[-ky_0, ky_0]$ (presented in [12]), will be rigorously studied in Section 3. The system is excited by the periodic external force $P\cos\omega t$. The preloaded force ky_0 is defined by the function (Figure 1b)

$$h(y) = \begin{cases} k(y + y_0), & y \geq 0 \\ k(y - y_0), & y \leq 0 \end{cases}$$

which outlines the switch character of the system.

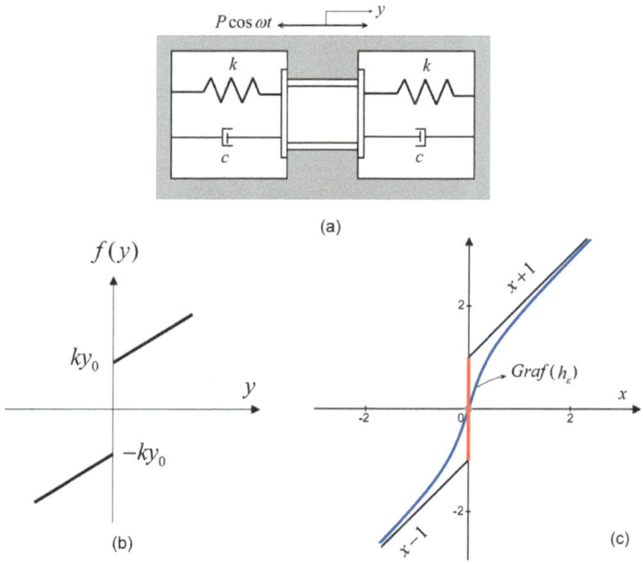

Figure 1. (**a**) A model of the preloaded system (3); (**b**) the characteristic of the restoring force f; (**c**) a graph of the approximate function h_ε.

With the change in the variables $x = y/y_0$ and $\tau = \omega_n t$, after some notations and substitutions (see the details in [12]), the dimensionless form of the system (2) becomes

$$\ddot{x} + 2\gamma\dot{x} + h(x) = P\cos\nu\tau, \tag{3}$$

where $h(x)$, after the mentioned substitutions, has the following form:

$$h(x) = \begin{cases} x+1, & y \geq 0, \\ x-1, & y \leq 0, \end{cases} \tag{4}$$

which can be written as $h(x) = x + \text{sgn}(x)$.

Finally, the standard form of the preloaded compliance system (3) becomes

$$\begin{aligned} \dot{x}_1 &= x_2, \\ \dot{x}_2 &= -2\gamma x_1 - x_1 - \text{sgn}\,(x_1) + P\cos(x_3), \quad x(0) = x_0. \\ \dot{x}_3 &= \nu. \end{aligned} \tag{5}$$

As can be seen, the system is in the form of the general switch system (1) with

$$A = \begin{pmatrix} 0 & 0 & 0 \\ -1 & 0 & 0 \\ 0 & 0 & 0 \end{pmatrix}, \quad g(x) = \begin{pmatrix} x_2 \\ -2\gamma - x_1 + P\cos(x_3) \\ \nu \end{pmatrix}, \quad s(x) = \begin{pmatrix} \text{sgn}(x_1) \\ \text{sgn}(x_2) \\ \text{sgn}(x_3) \end{pmatrix}.$$

The dynamics of the system are presented in [1] and are not considered here.

A few other switch dynamical systems are presented in Table 1. Note that the presence of the term $x_1\text{sgn}(x_1)$ in the function g of the variant of the Shimizu–Morioka system [13] (see Table 1) does not affect the continuity of g.

Table 1. Examples of switch dynamical systems of IO modeled by the IVP (1).

System	Equations	$g(x)$	A
Steam turbine [14]	$\begin{aligned}\dot{x}_1 &= p(x_3 - x_1 + \text{sgn}(x_2)),\\ \dot{x}_2 &= x_1 - x_2,\\ \dot{x}_3 &= -x_2\end{aligned}$	$\begin{pmatrix} px_3 - px_1 \\ x_1 - x_2 \\ -x_2 \end{pmatrix}$	$\begin{pmatrix} 0 & p & 0 \\ 0 & 0 & 0 \\ 0 & 0 & 0 \end{pmatrix}$
Sprott system [15]	$\begin{aligned}\dot{x}_1 &= x_2,\\ \dot{x}_2 &= x_3,\\ \dot{x}_3 &= -x_1 - x_2 - x_3 + p\text{sgn}(x_3)\end{aligned}$	$\begin{pmatrix} x_2 \\ x_3 \\ -x_1 - x_2 - x_3 \end{pmatrix}$	$\begin{pmatrix} 0 & 0 & 0 \\ 0 & 0 & 0 \\ 0 & 0 & p \end{pmatrix}$
Shimizu–Morioka [13] system	$\begin{aligned}\dot{x}_1 &= x_2,\\ \dot{x}_2 &= (1 - x_1)\text{sgn}(x_1) - p_1x_2,\\ \dot{x}_3 &= x_1^2 - p_2x_3\end{aligned}$	$\begin{pmatrix} x_2 \\ -x_1\text{sgn}(x_1) - p_1x_2 \\ x_1^2 - p_2x_3 \end{pmatrix}$	$\begin{pmatrix} 0 & 0 & 0 \\ 1 & 0 & 0 \\ 0 & 0 & 0 \end{pmatrix}$

Due to the broad range of applications of fractional calculus in real-life applications and in diverse and widespread fields of engineering and science [16], considering the fractional approaches of some systems like switch systems represents an open research direction.

3. Switch Dynamical Systems of FO

Considering the FO counterpart of the PWC systems, the counterpart of the autonomous systems modeled by the IVP (1) is as follows:

$$D_*^q x = f(x) := g(x) + A(x)s(x), \quad x(0) = x_0, \quad t \in I = [0, T], \tag{6}$$

where g and A are described in Section 2 and the Caputo derivative is considered, as in usual examples, with a lower limit $t = 0$ as follows:

$$D_*^q x(t) = \frac{1}{\Gamma(n-q)} \int_0^t (t-\tau)^{n-q-1} x^{(n)} \tau d\tau$$

where for $x : [0, T] \to \mathbb{R}$, $q \in (m-1, m)$, $m \in \mathbb{Z}^+$ is a smooth enough function, and $x^{(n)}$ is the classical derivative of the integer order n. Even more restrictive compared to other fractional-order derivatives, Caputo's derivative with a lower limit of 0 has the great advantage of using the initial conditions, like for IO systems, making it fully justified in practical examples.

One of the main restrictions on the numerical implementation of fractional derivatives is the memory-dependent characteristic, i.e., every state variable depends on all previously calculated values. Due to this effect, integration is time-consuming. On the other hand, the variation in real systems where the instantaneous change rate depends on the past state can be better described by the memory effect. In this context, with its entire (full) memory effect, Caputo's derivative describes better the real dynamics of an FO system but to the detriment of the integration time (see e.g., [17]. Compared with Caputo's derivative, the Gründwald–Letnikov derivative, which is the basic extension of the derivative in fractional calculus [18], follows the short memory principle, and its numerical implementation offers a relatively short time integration in numerical problems.

In this paper, the Adams–Bashforth–Moulton (ABM) method is used [19].

As observed in the previous sections, discontinuous ODEs and FDEs might not have classical solutions. Therefore, even though the dedicated numerical methods used to solve ODEs, as well as FDEs, can be used directly to integrate such problems, the numerical results can be wrong, as stated by the following example:

Consider the IVP

$$D_*^q x = f(x) := 2 - 3\text{sgn}(x), \;\; x(0) = x_0. \tag{7}$$

For $t \in [0, \infty)$, there are no solutions in the classical sense that start from x_0. If $x_0 = 0$, the right-hand side of the equation $D_*^q(0) = 0$ is different from the right-hand side $2 - \text{sgn}(0) = 2$. In the case of $x_0 > 0$, there is the solution $x(t) = x_0 - t^q/\Gamma(1+q)$, defined only on the interval $[0, T')$, $T' = (\Gamma(1+q)x_0)^{1/q}$. This solution cannot be extended over a larger interval $[0, T')$. If $x_0 < 0$, there is the solution $x(t) = x_0 + 5t^q/\Gamma(1+q)$, defined only on $[0, T'')$, with $T'' = (\Gamma(1+q)x_0/5)^{1/q}$. Both these solutions tend toward the line $x = 0$, but they cannot extend along this line $x = 0$, whose equation, $x(t) = 0$, does not verify the equation (in Figure 2a, the cases $q = 0.8$ with $x_0 = \pm 1$ are presented). Note that by using a numerical method to solve this FDE, it is possible to obtain some numerical solutions even for $t > T', T''$, but they could be incorrect since we cannot verify the equation for $t \in [0, \infty)$. This aspect will be analyzed below. Therefore, a special approach for solutions to FDEs is required.

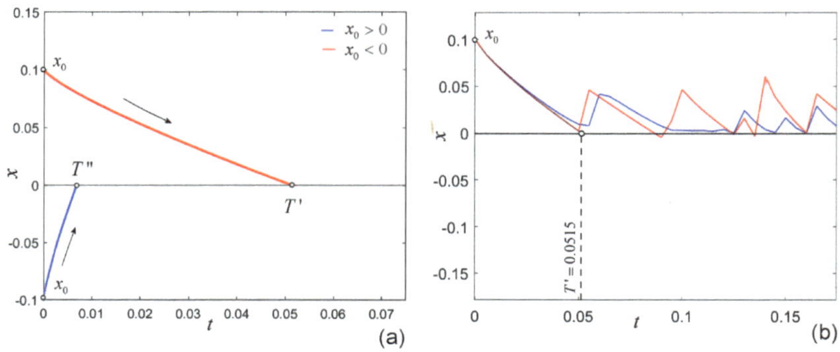

Figure 2. (**a**) Nonclassical solutions of the IVP (7), existing for $t < T', T''$; (**b**) a graph of the exact solution of the discontinuous FO problem (7) with $x_0 >$, on $[0, T']$, with $T' = 0.0505$ (black plot), the solution of the approximated problem (16) (blue plot), and the solution of the discontinuous IVP (7) (red plot) obtained with numerical integration for $t \in [0, T^*]$, $T^* > T'$ (without considering the problem of discontinuity).

3.1. Notions and Utilized Results

Notation 1. *Denote using $\mathcal{M}$ the set of discontinuity points of f.*

The set $\mathcal{M}$ has a zero Lebesgue measure and divides $\mathbb{R}^n$ to several $m > 1$ open disjuncts and connected sub-domains $\mathcal{D}_i \subset \mathbb{R}^n$, $i = 1, 2, ..., m$, $\mathbb{R}^n = \cup_{i=1}^m \overline{\mathcal{D}_i}$ on which f is continuous. Thus, discontinuity points belong to the union of the boundaries of $\mathcal{D}_i$. In Example (7), $\mathcal{M} = \{0\}$, which determines in $\mathbb{R}$ the sub-domains $\mathcal{D}_1 = (-\infty, 0)$ and $\mathcal{M}_2 = (0, \infty)$, where f is continuous.

Definition 1. *$f : \mathbb{R}^n \to \mathbb{R}^n$ is called PWC if it is continuous throughout $\mathbb{R}^n \setminus M$ and, at $\mathcal{M}$, has finite (possible different) limits.*

Definition 2. *A set-valued function $F : \mathbb{R}^n \rightrightarrows \mathbb{R}^n$ is a function that associates with any element $x \in \mathbb{R}^n$, a subset of $\mathbb{R}^n$, $F(x)$ (see the sketch in Figure 3).*

For more details on set-valued functions, see [20–22].

Definition 3. *The graph of a set-valued function $F : X \rightrightarrows Y$, where X, Y are non-empty sets, is defined as follows (see Figure 3):*

$$Graph(F) = \{(x, y) \in X \times Y | y \in F(x)\}.$$

To overcome the difficulty of finding solutions for PWC systems, IVP problems can be restarted by using the Filippov approach [23], as a set-valued Fractional Differential Inclusion (FDI) of

$$D_*^q x(t) \in F(x(t)), \quad \text{for a.e. } t \in I, \tag{8}$$

where F is a set-valued function.

The existence and uniqueness of solutions of DIs of IO are studied in [20,21], while numerical schemes can be found, e.g., in [24–28].

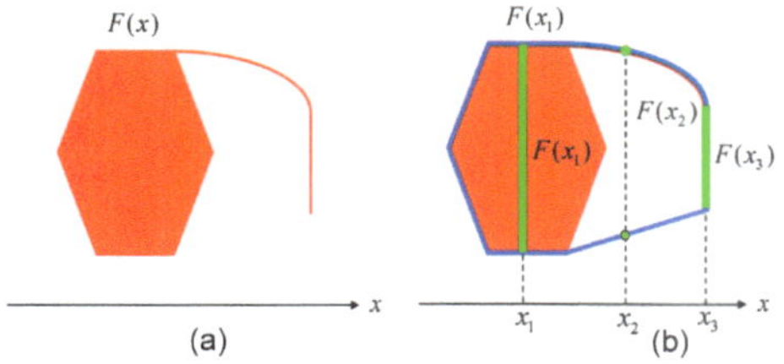

Figure 3. (**a**) The graph of a set-valued function F; (**b**) the values of $F(x)$ for different values of x.

The set-valued function F can be defined in several ways, the most known being Filippov's regularization [23]:

$$F(x) = \bigcap_{\varepsilon>0} \overline{conv(f(\{z \in \mathbb{R}^n : \|z-x\| \le \varepsilon\} \setminus M))}. \tag{9}$$

$F(x)$ defined by (9) is a convex function, its values representing the convex hull of f, and ε represents the radius of the ball centered in x. At the points x where f is continuous, $F(x) = f(x)$, i.e., $F(x)$, consists of a single point $\{f(x)\}$ (see Figure 3), while for $x \in \mathcal{M}$, $F(x)$ is given by (9). It is important to note that for physical meanings, ε should have as many small values as possible (ideally $\varepsilon \to 0$, when the solution of the DI tends to the real motion of the underlying physical system modeled by the discontinuous IVP (6)).

The regularization applied to one of the most used functions to define switch systems, sgn, generates the so-called sigmoid function, denoted usually as Sgn (see Figure 4a,b):

$$\text{Sgn}(x) = \begin{cases} \{-1\}, & x < 0, \\ [-1,1], & x = 0, \\ \{+1\}, & x > 0. \end{cases} \tag{10}$$

Therefore, in the case of example (7), the discontinued problem transforms into the following FDI (Figure 4c,d) for *a.e.* $t \in I$.:

$$\text{Sgn}(x) - 2 \in \begin{cases} \{5\}, & x < 0, \\ [-1,5], & x = 0, \\ \{-1\}, & x > 0. \end{cases} \tag{11}$$

Now, the problem transforms into the FDI, $D_*^q x \in [-1,5]$. When the solutions starting from negative or positive initial conditions reach the axis $x = 0$, the derivative $D_*^q x$ can now take an infinite number of values within the segment of values $[-1,5]$, so that the problem can be solved, even for $x_0 = 0$ (see [20,21]). For $x_0 \neq 0$, the problem reads as a continuous problem: $D_*^q x = 2 - 3\text{sgn}(x) = 5$, for $x_0 \in \mathcal{D}_1 = (-\infty, 0)$ and $D_*^q x = 2 - 3\text{sgn}(x) = -1$ for $x_0 \in \mathcal{D}_2 = (0,\infty)$.

Therefore, Filippov regularization opens the door for the possible existence of solutions to discontinuous systems of IO and even of FO.

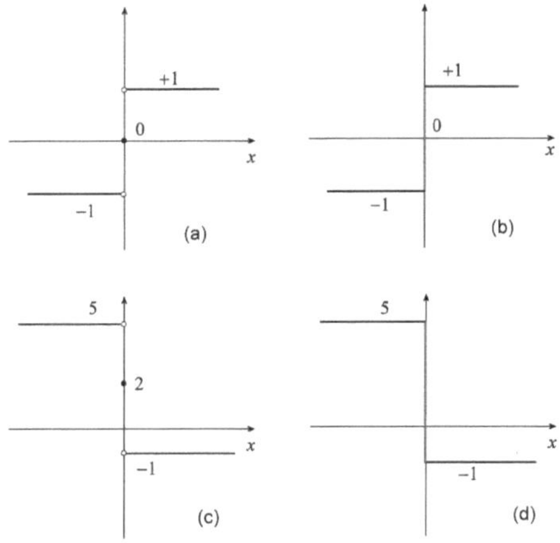

Figure 4. (**a**) A graph of the sgn function; (**b**) a graph of the set-valued function Sgn; (**c**) a graph of the discontinuous right-hand side of (7); (**d**) a graph of the set-valued function of (7).

Definition 4. *The map F is upper-semi-continuous (USC) on $\mathbb{R}^n$ if for each $x \in \mathbb{R}^n$, the set $F(x)$ is a non-empty and closed subset of R^n, and for each open set N of $\mathbb{R}^n$ containing $F(x)$, there is an open neighborhood B of x such that $F(B) \subset N$.*

Remark 1. *Filippov's approach* (9) *ensures enough regularity so that if f is continuous at x, then a solution to the IVP* (8) *satisfies the IVP* (6) *and any classical solution to the IVP* (6) *is a solution to the IVP* (8). *On mild assumptions, DIs of IO, obtained with Filippov regularization, admit generalized (Filippov) solutions [23]. As stipulated in [23], it is justified to call a solution for the IVP* (6) *a solution for the IVP* (8). *For FDIs, existence results are presented in [29].*

Definition 5. *The single-valued function $f : \mathbb{R}^n \to \mathbb{R}^n$ is a selection (approximation) of a set-valued function F, if $f(x) \in F(x)$ for every $x \in \mathbb{R}^n$ (see, e.g., [20,21]).*

Finding numerical solutions for DIs of IO represents a complex task (see e.g., [24–26,30]), while numerical solutions for FDIs are, based on our knowledge, an even more challenging and less studied topic. Therefore, proposing continuous approximation via selection theory represents a successful approach.

Theorem 1. *Cellina's Theorem [20,21]: Let a USC set-valued function $F : \mathbb{R}^n \rightrightarrows \mathbb{R}^n$ with non-empty convex values. Then, for every $\varepsilon > 0$, there is a local Lipschitz selection $f_\varepsilon : \mathbb{R}^n \rightrightarrows \mathbb{R}^n$ such that*

$$Graph(f_\varepsilon) \subset Graph(F) + \varepsilon B$$

showing the sphere of radius ε (see the sketch in Figure 5a).

Generally, a set-valued function admits infinitely many selections, a fact that represents a major advantage in approaching discontinuous systems coming from practical problems.

Remark 2 ([20,21]). *Because of the symmetric interpretation of a set-valued map as a graph, we can say that a set-valued function satisfies a property if and only if its graph satisfies it. For instance, a set-valued function is said to be closed if and only if its graph is closed. For a function f that is locally bounded, its set-valued map obtained with Filippov regularization enjoys the following properties:*

it is USC and has non-empty closed and convex values (Péano functions ([20], Proposition 1, p. 102, [21], Lemma 3, p. 67), which satisfies the existence result of solutions to DIs. For FDIs, similar results can be found in [29].

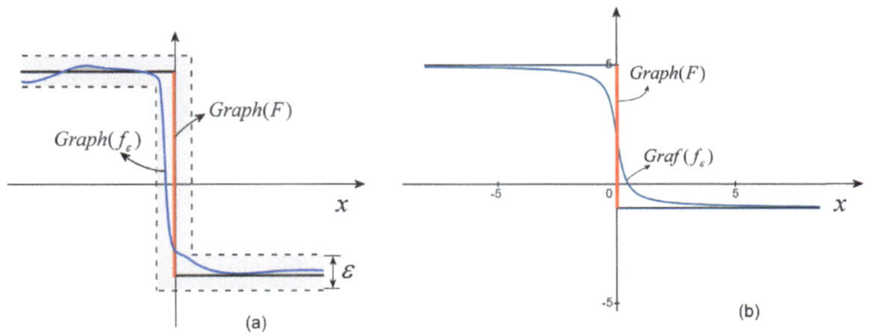

Figure 5. (**a**) A sketch of a continuous approximation f_ε within a ε-neighborhood of the set-valued function F; (**b**) a graph of the continuous ε-approximation for system (7).

The set-valued function defined with Fillipov's regularization verifies Theorem 1 [31].

3.2. Numerical Integration of FO Switch Systems

Consider the FO counterpart of the IVP (1) defined with Caputo's derivative:

$$D_*^q x = f_\varepsilon(x) := g(x) + As(x), \quad x(0) = x_0, \quad t \in I = [0, T]. \tag{12}$$

The steps required to restart the PWC problem into a continuous problem are the following:

(I) Firstly, the set-valued right-hand side, obtained with Filippov's regularization, is

$$F(x) = g(x) + AS(x), \quad x(0) = x_0, \quad t \in I = [0, T], \tag{13}$$

with $S(x) = [\mathrm{Sgn}(x_1), \mathrm{Sgn}(x_2), ..., \mathrm{Sgn}(x_n)]^T$ being the set-valued sigmoid functions given by (10).

(II) Next, the continuous approximation of the component $\mathrm{Sgn}(x)$ (Theorem 1) is

$$S_\varepsilon(x) = [S_\varepsilon(x_1), S_\varepsilon(x_2), ..., S_\varepsilon(x_n)]^T,$$

where $S_\varepsilon(x_i)$, $i = 1, 2, ..., n$, are continuous approximations of the set-valued functions $\mathrm{Sgn}(x_i)$ given by, e.g., (15).

(III) The result regarding the continuous approximation of the IVP (12) is

Theorem 2 ([31]). *The PWC IVP (6) can be restarted as the following continuous IVP of FO*

$$: \mathrm{D}_*^q x = g(x) + A(x)\mathrm{S}_\varepsilon(x), \quad x(0) = x_0, \quad t \in I, \tag{14}$$

Theorem 2 represents the main result according to which it is demonstrated how a PWC problem of IO can be transformed into a continuous one of FO. Thus, by applying the algorithm in Figure 6, the discontinuous preloaded compliance system of FO (12) is transformed into a continuous problem of FO (14), which can be studied both analytically and numerically.

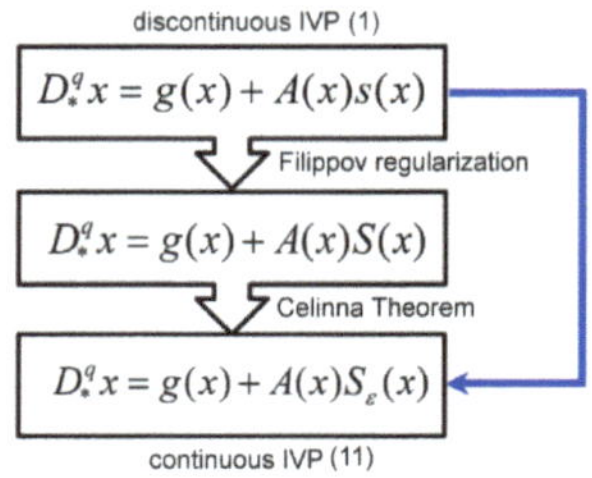

Figure 6. The algorithm of the continuous approximation of the discontinuous IVP (6).

The algorithm described above is presented in Figure 6. It is important to note that in the examples, step II can be omitted and the discontinuous function approximated directly (step III).

One of the used approximations S_ε of the set-valued functions sgn function, which provides enough flexibility, has the following form:

$$S_\varepsilon(x) = \frac{2}{\pi}\arctan\frac{x}{\delta}, \tag{15}$$

where δ is a positive parameter that controls the slope of the approximation in the ε-neighborhood of $x = 0$. A cubic polynomial approximation is presented in [32], where a discussion on the δ influence of the approximation can be found. While approximations like (15) are defined on I, in [32], the polynomial approximation is defined only in a chosen local ε-neighborhood around the discontinuity, a fact that confers better properties to the approximated function.

Once the continuity of the underlying problem is ensured by Theorem 2, the obtained IVP can be numerically integrated via some numerical schemes for FDEs, like the ABM method [19] used in this paper.

Reconsider example (7) by using approximation (15). Then, the discontinuous problem is transformed via Theorem 2 into the following continuous IVP of FO:

$$D_*^q x = 2 - \frac{6}{\pi}\arctan\frac{x}{\delta} \approx 2 - 3S_\varepsilon(x), x(0) = x_0, \quad t \in I, \tag{16}$$

with the approximation of the right-hand side presented in Figure 5b. Now, the problem admits numerical solutions for $t \in I$, which can be determined with numerical schemes for FDEs. In Figure 2b are the exact solution of the discontinuous FO problem on $[0, T']$, $T' = 0.0505$ (black plot), the solution of the approximated problem (16) (blue plot), and the solution of the discontinuous IVP (7) (red plot) obtained with numerical integration for $t \in [0, T^*]$, with $T^* > T'$, without considering the discontinuity $x = 0$. The differences can be seen in Figure 2b.

3.3. Preloaded Compliance System of FO

Consider now the FO of the system modeling the preloaded compliance system (5). Via Theorem 2, by applying the continuous approximation algorithm (see Figure 6) with h_ε approximation presented in Figure 1c), one obtains the single-valued continuous FO system:

$$\begin{aligned} D_*^q x_1 &= x_2, \\ D_*^q x_2 &= -2\gamma x_1 - x_1 - S_\varepsilon(x_1) + P\cos(x_3), \quad x(0) = x_0, \quad t \in I. \\ D_*^q x_3 &= \nu. \end{aligned} \tag{17}$$

Despite the fact that the form (17) is defined with three FDEs (see Figure 7a), required to transform the standard form into an autonomous one, the dynamics of the system are graphically described in the plane (x_1, x_2) (Figure 7b–d).

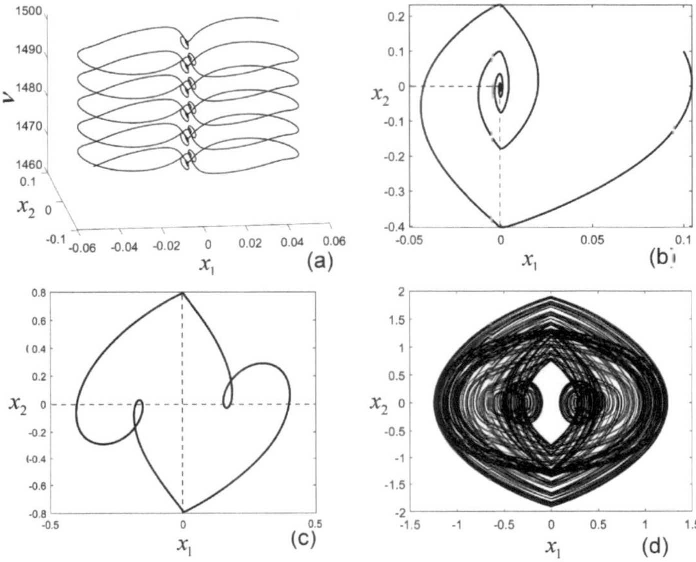

Figure 7. (**a**) The trajectory of the system (17) in space (x_1, x_2, v); (**b**) attractive fixed point-like. The zoomed-in detail reveals the influence of the sgn function; (**c**) periodic-like trajectory; (**d**) chaotic trajectory.

Remark 3. *Compared with the IO variant of the preloaded compliance system* (2), *its obtained FO counterpart* (17) *presents several advantages: (1) The first one is that by applying the proposed algorithm, the obtained problem is a continuous one, thus taking advantage of the existing tools to study continuous systems. (2) The second one is the fact that with this approach, numerical integration is more accessible compared with the IO case, which, as can be seen, e.g., in [1], does not exit a clear unified way to integrate discontinuous systems. Thus, generally, at the discontinuity surface, $\mathcal{M}$, the considered solutions obtained (incorrectly) with some standard methods for continuous systems are just "glued". The third argument in favor of the proposed algorithm, i.e., to transform the IO problem into a FO one, is the fact that, as mentioned in the previous sections, FO variants generally offer a better characterization of the underlying IO system.*

Note: *(1) As known now, FO systems (continuous or discrete) cannot have periodic non-constant solutions [33]. Besides the analytical proof in [33], this property can be explained due to the fact that all the past history has to be taken into account (time-memory of FO systems is continuous-time or discrete). Also, any equilibrium of a general FO nonlinear system described by Caputo's derivative can never be finite-time stable in the sense that the trajectory will never reach the equilibrium in finite time [34].*

(2) Numerical schemes for FDEs (and also for DEs),require at least the continuity of the underlying equations [19]. However, the numerical methods can directly (abruptly) be applied to discontinuous problems too, but this method, which is still poorly understood, can lead to questionable results. Therefore, the correct results for the integration of IVP (6) *should be obtained, e.g., after the continuous approximation of the underlying problem.*

While in the IO variant of the preloaded compliance system (2), there are periodic solutions abruptly changing the stability [12], in (17), this does not happen anymore (see Note 1).

Consider $q = 0.95$ and $\delta = 1e - 3$ and integrate the system (17). In Figure 7b, an attractive fixed point-like element is presented (see Note 1). For $p = 2$, in Figure 7c, a periodic-like trajectory is presented. Even though visually, the orbit looks periodic, it is not (see Note 1). A chaotic trajectory is presented in Figure 7d. The apparent corners, visible in Figure 7b–d, are due to the sgn function. However, due to the continuous approximations, these corners are in fact smooth (see the zoomed-in Figure 7b).

4. Discussion and Conclusions

In this paper, the continuous approximation of discontinuous (switch) IVPs of IO is proposed as a novel technique to approach a class of dynamical systems of FO. Using the theory of differential inclusions and Cellina's Theorem, the discontinuous IVP is restarted as a continuous problem of FO, which can be integrated with some numerical scheme for FDEs. First, it is shown that the Filippov regularization applied to the discontinuous function allows the existence of a sufficiently regular set-valued map, so that the underlying FDI verifies Cellina's Theorem. As a consequence, the discontinuous problem can be restarted as a continuous one. Also, the problem of using numerical integration, without considering the discontinuity, is treated on a concrete example.

In the case of a practical example, a preloaded compliance system is considered.

Funding: This research received no external funding.

Data Availability Statement: The original contributions presented in this study are included in the article. Further inquiries can be directed to the corresponding author.

Conflicts of Interest: The author declares no conflicts of interest.

References

1. Wiercigroch, M.; de Kraker, B. *Applied Nonlinear Dynamics and Chaos of Mechanical Systems with Discontinuities*; Marian Wiercigroch and Bram de Kraker, World Scientific Series on Nonlinear Science; Chua, L.O., Ed.; World Scientific: Singapore, 2000.
2. Acary, V.; Brogliato, B. *Numerical Methods for Nonsmooth Dynamical Systems: Applications in Mechanics and Electronics*; Lecture Notes in Applied and Computational Mechanics; Springer: New York, NY, USA, 2008.
3. Ray S.S.; Atangana, A.; Noutchie, S.C.O.; Kurulay, M.; Bildik, N.; Kilicman, A. Fractional calculus and its applications in applied mathematics and other sciences. *Math. Probl. Eng.* **2014**, *2014* , 849395. [CrossRef]
4. Bagley, R.; Torvik, P. On the fractional calculus model of viscoelastic behavior. *J. Rheol.* **1986**, *30*, 133–155. [CrossRef]
5. Oustaloup, A.; Mathieu, B.; Lanusse, P. The CRONE control of resonant plants: Application to a flexible transmission. *Eur. J. Control* **1995**, *1*, 113–121. [CrossRef]
6. Nakagawa, M.; Sorimachi, K. Basic characteristics of a fractance device. *IEICE Trans. Fundam. Electron. Commun. Comput. Sci.* **1992**, *75*, 1814–1819.
7. Danca, M.-F. Fractional order logistic map: Numerical approach. *Chaos Soliton. Fract.* **2022**, *157*, 111851. [CrossRef]
8. Danca, M.-F. Chaotic hidden attractor in a fractional order system modelling the interaction between dark matter and dark energy. *Commun. Nonlinear Sci.* **2024**, *131*, 107838 [CrossRef]
9. Danca, M.F.; Fečkan, M. Memory Principle of the MATLAB Code for Lyapunov Exponents of Fractional-Order. *Int. J. Bifurcat. Chaos* **2024**, *34*, 2450156. [CrossRef]
10. Podlubny, I. *Fractional Differential Equations*; Academic Press: New York, NY, USA, 1999.
11. Teodoro, G.S.; Machado, J.A.T.; de Oliveira, E.C. A review of definitions of fractional derivatives and other operators. *J. Comput. Phys.* **2019**, *388*, 195–208. [CrossRef]
12. Yoshitake, Y.; Sueoka, A. Forced Self-Excited Vibration with Dry Friction. In *Applied Nonlinear Dynamics and Chaos of Mechanical Systems with Discontinuities*; Marian Wiercigroch and Bram de Kraker, World Scientific Series on Nonlinear Science; Chua, L.O., Ed.; World Scientific: Singapore, 2000.
13. Shimizu, T.; Morioka, N. On the bifurcation of a symmetric limit cycle to an asymmetric one in a simple model. *Phys. Lett. A* **1980**, *76*, 201–204. [CrossRef]

14. Belea, C. *Automatica Neliniara; Teorie, Exemple si Aplicatii*; Editura Tehnica: Bucuresti, Romania, 1983. (In Romanian)
15. Sprott, J.C. A new class of chaotic circuit. *Phys. Lett. A* **2000**, *266*, 19–23. [CrossRef]
16. Tenreiro Machado, J.A.; Silva, M.F.; Barbosa, R.S.; Jesus, I.S.; Reis, C.M.; Marcos, M.G.; Galhano, A.F. Some Applications of Fractional Calculus in Engineering. *Math. Probl. Eng.* **2010**, *2010*, 639801. [CrossRef]
17. Wang, J.-L.; Li, H.-F. Surpassing the fractional derivative: Concept of the memory-dependent derivative. *Comput. Math. Appl.* **2011**, *62*, 1562–1567. [CrossRef]
18. Podlubny, I. *Fractional Differential Equations. An Introduction to Fractional Derivatives, Fractional Differential Equations, to Methods of their Solution and Some of their Applications*; Academic Press: San Diego, CA, USA, 1999.
19. Diethelm, K.; Ford, N.J.; Freed, A.D. A predictor-corrector approach for the numerical solution of fractional differential equations. *Nonlinear Dyn.* **2002**, *29*, 3–22. [CrossRef]
20. Aubin, J.-P.; Cellina, A. *Differential Inclusions Set-Valued Maps and Viability Theory*; Springer: Berlin/Heidelberg, Germany, 1984.
21. Aubin, J.-P.; Frankowska, H. *Set-Valued Analysis*; Birkhuser: Boston, MA, USA, 1990.
22. Deimling, K. *Multivalued Differential Equations*; de Gruyter: Berlin, Germany; New York, NY, USA, 1992.
23. Filippov, A.F. *Differential Equations with Discontinuous Right-Hand Sides*; Kluwer Academic: Dordrecht, The Netherlands, 1988.
24. El-Sayed, A.M.A.; Ibrahim, A.G. Multivalued fractional differential equations of arbitrary orders. *Appl. Math. Comput.* **1995**, *68*, 15–25.
25. Elliot, C.M. On the convergence of a onestep method for the numerical solution of an ordinary differential inclusion. *IMA J. Numer. Anal.* **1985**, *5*, 3–21. [CrossRef]
26. Lempio, F. Modified Euler methods for differential inclusions. In *Set-Valued Analysis and Differential Inclusions*; A Collection of Papers Resulting from a Workshop; Kurzanski, A.B., Veliov, V.M., Eds.; Birkhäuser: Boston, MA, USA, 1993.
27. Lempio, F.; Veliov, V. Discrete approximations of differential inclusions. *Bayreuth. Math. Schr.* **1998**, *54*, 149–232.
28. Dontchev, A.; Lempio, F. Difference Methods for Differential Inclusions: A Survey. *SIAM Rev.* **1992**, *34*, 263–294. [CrossRef]
29. Agarwal, R.; Benchohra, M. A Survey on Existence Results for Boundary Value Problems of Nonlinear Fractional Differential Equations and Inclusions. *Acta Appl. Math.* **2010**, *109*, 973–1033. [CrossRef]
30. Kastner-Maresch, A.; Lempio, F. Difference methods with selection strategies for differential inclusions. *Numer. Funct. Anal. Optim.* **1993**, *14*, 555–572. [CrossRef]
31. Danca, M.-F. On a Class of Discontinuous Dynamical Systems. *Miskolt Math. Notes* **2001**, *2*, 103–116.
32. Danca, M.-F. Continuous approximations of a class of piecewise continuous systems. *Int. J. Bifurcat. Chaos*, **2015**, *25*, 1550146. [CrossRef]
33. Tavazoei, M.S.; Haeri, M. A proof for non existence of periodic solutions in time invariant fractional order systems. *Automatica* **2009**, *45*, 1886–1890. [CrossRef]
34. Shen, J.; Lam, J. Non-existence of finite-time stable equilibria in fractional-order nonlinear systems. *Automatica* **2014**, *50*, 547–551. [CrossRef]

MDPI AG
Grosspeteranlage 5
4052 Basel
Switzerland
Tel.: +41 61 683 77 34

Fractal and Fractional Editorial Office
E-mail: fractalfract@mdpi.com
www.mdpi.com/journal/fractalfract

www.ingramcontent.com/pod-product-compliance
Lightning Source LLC
Chambersburg PA
CBHW041411210926
51850CB00013B/3243